OPTICAL FIBER SENSORS

Advanced Techniques and Applications

Devices, Circuits, and Systems

Series Editor

Krzysztof Iniewski

CMOS Emerging Technologies Research Inc.,
Vancouver, British Columbia, Canada

PUBLISHED TITLES:

Atomic Nanoscale Technology in the Nuclear Industry
Taeho Woo

Biological and Medical Sensor Technologies
Krzysztof Iniewski

Building Sensor Networks: From Design to Applications
Ioanis Nikolaidis and Krzysztof Iniewski

Circuits at the Nanoscale: Communications, Imaging, and Sensing
Krzysztof Iniewski

Design of 3D Integrated Circuits and Systems
Rohit Sharma

Electrical Solitons: Theory, Design, and Applications
David Ricketts and Donhee Ham

Electronics for Radiation Detection
Krzysztof Iniewski

Embedded and Networking Systems: Design, Software, and Implementation
Gul N. Khan and Krzysztof Iniewski

Energy Harvesting with Functional Materials and Microsystems
Madhu Bhaskaran, Sharath Sriram, and Krzysztof Iniewski

Graphene, Carbon Nanotubes, and Nanostuctures: Techniques and Applications
James E. Morris and Krzysztof Iniewski

High-Speed Devices and Circuits with THz Applications
Jung Han Choi

High-Speed Photonics Interconnects
Lukas Chrostowski and Krzysztof Iniewski

High Frequency Communication and Sensing: Traveling-Wave Techniques
Ahmet Tekin and Ahmed Emira

Integrated Microsystems: Electronics, Photonics, and Biotechnology
Krzysztof Iniewski

PUBLISHED TITLES:

Integrated Power Devices and TCAD Simulation
Yue Fu, Zhanming Li, Wai Tung Ng, and Johnny K.O. Sin

Internet Networks: Wired, Wireless, and Optical Technologies
Krzysztof Iniewski

Labs on Chip: Principles, Design, and Technology
Eugenio Iannone

Low Power Emerging Wireless Technologies
Reza Mahmoudi and Krzysztof Iniewski

Medical Imaging: Technology and Applications
Troy Farncombe and Krzysztof Iniewski

Metallic Spintronic Devices
Xiaobin Wang

MEMS: Fundamental Technology and Applications
Vikas Choudhary and Krzysztof Iniewski

Micro- and Nanoelectronics: Emerging Device Challenges and Solutions
Tomasz Brozek

Microfluidics and Nanotechnology: Biosensing to the Single Molecule Limit
Eric Lagally

MIMO Power Line Communications: Narrow and Broadband Standards, EMC, and Advanced Processing
Lars Torsten Berger, Andreas Schwager, Pascal Pagani, and Daniel Schneider

Mobile Point-of-Care Monitors and Diagnostic Device Design
Walter Karlen

Nano-Semiconductors: Devices and Technology
Krzysztof Iniewski

Nanoelectronic Device Applications Handbook
James E. Morris and Krzysztof Iniewski

Nanopatterning and Nanoscale Devices for Biological Applications
Šeila Selimović

Nanoplasmonics: Advanced Device Applications
James W. M. Chon and Krzysztof Iniewski

Nanoscale Semiconductor Memories: Technology and Applications
Santosh K. Kurinec and Krzysztof Iniewski

Novel Advances in Microsystems Technologies and Their Applications
Laurent A. Francis and Krzysztof Iniewski

Optical, Acoustic, Magnetic, and Mechanical Sensor Technologies
Krzysztof Iniewski

Optical Fiber Sensors: Advanced Techniques and Applications
Ginu Rajan

PUBLISHED TITLES:

Organic Solar Cells: Materials, Devices, Interfaces, and Modeling
Qiquan Qiao

Radiation Effects in Semiconductors
Krzysztof Iniewski

Semiconductor Radiation Detection Systems
Krzysztof Iniewski

Smart Grids: Clouds, Communications, Open Source, and Automation
David Bakken

Smart Sensors for Industrial Applications
Krzysztof Iniewski

Soft Errors: From Particles to Circuits
Jean-Luc Autran and Daniela Munteanu

Solid-State Radiation Detectors: Technology and Applications
Salah Awadalla

Technologies for Smart Sensors and Sensor Fusion
Kevin Yallup and Krzysztof Iniewski

Telecommunication Networks
Eugenio Iannone

Testing for Small-Delay Defects in Nanoscale CMOS Integrated Circuits
Sandeep K. Goel and Krishnendu Chakrabarty

VLSI: Circuits for Emerging Applications
Tomasz Wojcicki

Wireless Technologies: Circuits, Systems, and Devices
Krzysztof Iniewski

Wireless Transceiver Circuits: System Perspectives and Design Aspects
Woogeun Rhee

FORTHCOMING TITLES:

Advances in Imaging and Sensing
Shuo Tang, Dileepan Joseph, and Krzysztof Iniewski

Analog Electronics for Radiation Detection
Renato Turchetta

Cell and Material Interface: Advances in Tissue Engineering, Biosensor, Implant, and Imaging Technologies
Nihal Engin Vrana

Circuits and Systems for Security and Privacy
Farhana Sheikh and Leonel Sousa

CMOS: Front-End Electronics for Radiation Sensors
Angelo Rivetti

FORTHCOMING TITLES:

CMOS Time-Mode Circuits and Systems: Fundamentals and Applications
Fei Yuan

Electrostatic Discharge Protection of Semiconductor Devices and Integrated Circuits
Juin J. Liou

Gallium Nitride (GaN): Physics, Devices, and Technology
Farid Medjdoub and Krzysztof Iniewski

Laser-Based Optical Detection of Explosives
Paul M. Pellegrino, Ellen L. Holthoff, and Mikella E. Farrell

Mixed-Signal Circuits
Thomas Noulis and Mani Soma

Magnetic Sensors: Technologies and Applications
Simone Gambini and Kirill Poletkin

MRI: Physics, Image Reconstruction, and Analysis
Angshul Majumdar and Rabab Ward

Multisensor Data Fusion: From Algorithm and Architecture Design to Applications
Hassen Fourati

Nanoelectronics: Devices, Circuits, and Systems
Nikos Konofaos

Nanomaterials: A Guide to Fabrication and Applications
Gordon Harling, Krzysztof Iniewski, and Sivashankar Krishnamoorthy

Optical Imaging Devices: New Technologies and Applications
Dongsoo Kim and Ajit Khosla

Physical Design for 3D Integrated Circuits
Aida Todri-Sanial and Chuan Seng Tan

Power Management Integrated Circuits and Technologies
Mona M. Hella and Patrick Mercier

Radiation Detectors for Medical Imaging
Jan S. Iwanczyk

Radio Frequency Integrated Circuit Design
Sebastian Magierowski

Reconfigurable Logic: Architecture, Tools, and Applications
Pierre-Emmanuel Gaillardon

Terahertz Sensing and Imaging: Technology and Devices
Daryoosh Saeedkia and Wojciech Knap

Tunable RF Components and Circuits: Applications in Mobile Handsets
Jeffrey L. Hilbert

Wireless Medical Systems and Algorithms: Design and Applications
Pietro Salvo and Miguel Hernandez-Silveira

OPTICAL FIBER SENSORS

Advanced Techniques and Applications

EDITED BY **Ginu Rajan**

UNIVERSITY OF NEW SOUTH WALES, SCHOOL OF ELECTRICAL ENGINEERING AND TELECOMMUNICATIONS, UNSW AUSTRALIA

Krzysztof Iniewski MANAGING EDITOR

CMOS EMERGING TECHNOLOGIES RESEARCH INC.
VANCOUVER, BRITISH COLUMBIA, CANADA

CRC Press
Taylor & Francis Group
Boca Raton London New York

CRC Press is an imprint of the
Taylor & Francis Group, an **informa** business

CRC Press
Taylor & Francis Group
6000 Broken Sound Parkway NW, Suite 300
Boca Raton, FL 33487-2742

First issued in paperback 2020

ISBN-13: 978-1-4822-2825-0 (hbk)
ISBN-13: 978-0-367-65605-8 (pbk)

Library of Congress Cataloging-in-Publication Data

Optical fiber sensors (2014)
Optical fiber sensors : advanced techniques and applications / editor, Ginu Rajan.
pages cm -- (Devices, circuits, and systems)
Includes bibliographical references and index.
ISBN 978-1-4822-2825-0 (alk. paper)
1. Optical fiber detectors. I. Rajan, Ginu. II. Title.

TA1815.O6955 2014
681'.25--dc23 2014035111

Visit the Taylor & Francis Web site at
http://www.taylorandfrancis.com

and the CRC Press Web site at
http://www.crcpress.com

Contents

Contents

Preface

In the past several decades since the invention of the laser in 1960 and the development of modern low-loss optical fibers in 1966, optical fiber technology has made a transition from the experimental stage to practical applications. The main focus of the development of optical fiber has always been on telecommunications, but the early 1970s saw some of the first experiments on low-loss optical fibers being used for sensor purposes. The field of optical fiber sensing has continued to progress and has developed enormously since that time. Magnetic, pressure, temperature, acceleration, displacement, fluid level, current, and strain optical fiber sensors were among the first types extensively investigated and explored for sensing and measurement. Compared with other types of sensors, optical fiber sensors exhibit a number of advantages, such as immunity to electromagnetic interference, applicability in high-voltage or explosive environments, a very wide operating temperature range, multiplexing capabilities, and chemical passivity. This book describes the fundamentals of optical fiber sensors, the latest developments in the field, and the practical applications of the optical fiber sensing technology.

This book covers wide aspects of different sensing mechanisms using optical fibers and also demonstrates their use in application areas. Chapters based on new and emerging areas such as photonic crystal fiber sensors, micro-/nanofiber sensing, liquid crystal photonics, acousto-optic effects in fiber and its sensing applications, and fiber laser–based sensing and other well-established areas such as surface plasmon resonance sensors, interferometric fiber sensors, polymer fiber sensors, Bragg gratings in polymer and silica fibers, and distributed fiber sensors are also included in this book. On the application side of optical fiber sensors, humidity sensing applications, smart structure applications, and medical applications are also covered. A future outlook on the fiber sensing research area is also presented to provide the reader with an understanding of its potential.

This book is a collective effort of a number of authors who contributed different chapters in their areas of expertise. With such a wide variety of topics covered in detail, we are hoping that the reader will find this book stimulating to read and will discover the sensing potential of optical fibers and devices. This book is intended to be a comprehensive introduction with a strong practical focus suitable for undergraduate and graduate students as well as a convenient reference for scientists and engineers working in the field.

Editors

Dr. Ginu Rajan is vice-chancellor's research fellow/lecturer at the University of New South Wales, Sydney, Australia. He earned his BSc in physics from the University of Kerala and MSc in applied physics from Mahatma Gandhi University, Kerala, India, in 2000 and 2002, respectively. He worked as a researcher at the Indian Institute of Astrophysics during the period of 2003–2005. He subsequently undertook research in the area of optical fiber sensors, following which he earned a PhD from Dublin Institute of Technology (DIT), Ireland, in 2008.

During 2009–2012, Dr. Rajan worked as a project manager/research associate at the Photonics Research Centre of DIT in collaboration with the Warsaw University of Technology, Poland. He also was a lecturer at DIT during this period. He has published more than 100 journal articles, conference papers, and book chapters. He also holds a patent. Dr. Rajan is currently a reviewer for many scientific journals and also an international reviewer for funding applications of the Portugal Science Foundation and Australian Research Council. His research and teaching interests include optical fiber sensors and their applications in biomedical engineering, fiber Bragg grating (FBG) interrogation systems, photonic crystal fiber sensors, polymer fiber sensors, smart structures, and physics of photonic devices. He can be reached at ginu.rajan@unsw.edu.au or ginurajan@gmail.com.

Dr. Krzysztof (Kris) Iniewski is the R&D manager at Redlen Technologies Inc., a start-up company in Vancouver, Canada. Redlen's revolutionary production process for advanced semiconductor materials enables a new generation of more accurate, all-digital, radiation-based imaging solutions. He is also the president of CMOS Emerging Technologies (www.cmoset.com), an organization of high-tech events covering communications, microsystems, optoelectronics, and sensors.

In his career, Dr. Iniewski held numerous faculty and management positions at the University of Toronto, the University of Alberta, SFU, and PMC-Sierra Inc. He has published more than 100 research papers in international journals and conferences. He holds 18 international patents granted in the United States, Canada, France, Germany, and Japan. He is a frequent invited speaker and has consulted for multiple organizations internationally. He has written and edited several books for IEEE Press, Wiley, CRC Press, McGraw-Hill, Artech House, and Springer. His personal goal is to contribute to healthy living and sustainability through innovative engineering solutions. He can be reached at kris.iniewski@gmail.com.

Contributors

Asrul Izam Azmi
Faculty of Electrical Engineering
Universiti Teknologi Malaysia
Johor, Malaysia

Abolfazl Bahrampour
Department of Physics
Sharif University of Technology
Tehran, Iran

Ali Reza Bahrampour
Department of Physics
Sharif University of Technology
Tehran, Iran

Gilberto Brambilla
Optoelectronics Research Centre
University of Southampton
Southampton, United Kingdom

George Y. Chen
Optoelectronics Research Centre
University of Southampton
Southampton, United Kingdom

Brian Culshaw
Centre for Micro Systems and Photonics
University of Strathclyde
Glasgow, United Kingdom

Gerald Farrell
Photonics Research Centre
Dublin Institute of Technology
Dublin, Ireland

Kun Liu
College of Precision Instrument
and Opto-Electronic Engineering
Tianjin University
Nankai, Tianjin, People's Republic of China

Sunish Mathews
Department of Medical Physics
and Biomedical Engineering
University College London
London, United Kingdom

Vandana Mishra
Chandigarh, Punjab, India

Muhammad Yusof Mohd Noor
Faculty of Electrical Engineering
Universiti Teknologi Malaysia
Johor, Malaysia

Gang-Ding Peng
School of Electrical Engineering
and Telecommunications
University of New South Wales
Sydney, New South Wales, Australia

Kara Peters
Department of Mechanical and Aerospace
Engineering
North Carolina State University
Raleigh, North Carolina

Kent B. Pfeifer
Microsystems-Enabled Detection Department
Sandia National Laboratories
Albuquerque, New Mexico

Ana M.R. Pinto
INESC P&D Brazil
São Paulo, Brazil

Nafiseh Pishbin
Department of Electrical Engineering
Amirkabir University of Technology
Tehran, Iran

Alexandre de Almeida Prado Pohl
Graduate School of Electrical Engineering
and Computer Science
Federal University of Technology
Paraná Curitiba, Brazil

Haifeng Qi
Laser Institute of Shandong Academy of Sciences
Jinan, Shandong, People's Republic of China

Ginu Rajan
School of Electrical Engineering and Telecommunications
University of New South Wales
Sydney, New South Wales, Australia

Manjusha Ramakrishnan
Photonics Research Centre
Dublin Institute of Technology
Dublin, Ireland

Yuliya Semenova
Photonics Research Centre
Dublin Institute of Technology
Dublin, Ireland

Dipankar Sengupta
Department of Information Engineering
University of Padova
Padova, Italy

Balaji Srinivasan
Department of Electrical Engineering
Indian Institute of Technology
Chennai, India

Shiquan Tao
Department of Mathematics, Chemistry and Physics
West Texas A&M University
Canyon, Texas

Steven M. Thornberg (Retired)
Materials Reliability Department
Sandia National Laboratories
Albuquerque, New Mexico

Sara Tofighi
Department of Physics
Sharif University of Technology
Tehran, Iran

Deepa Venkitesh
Department of Electrical Engineering
Indian Institute of Technology
Chennai, India

David J. Webb
Aston Institute of Photonic Technologies
School of Engineering and Applied Science
Aston University
Birmingham, United Kingdom

1 Introduction to Optical Fiber Sensors

Ginu Rajan

CONTENTS

1.1 INTRODUCTION

Sensing has become a key enabling technology in many areas, from entertainment technology to health, transport, and many industrial technologies. In many such advanced applications where miniaturization, sensitivity, and remote measurement are vital, optical fiber–based sensing techniques can provide novel solutions. As a result of this, optical fiber sensing (OFS) technology is developed as a powerful and rich technology that is currently being implemented in a wide variety of applications [1,2].

Research works on OFS are started in the 1960s, but it is the development of modern low-loss optical fibers that has made a transition from the experimental stage to practical applications. Some of the first sensing experiments using low-loss optical fibers [3–6] were demonstrated during the early 1970s. The field of OFS has continued to progress and developed enormously since that time. For instance, distributed fiber-optic sensors have now been installed in bridges and dams to monitor the performance and structural damage of these facilities. Optical fiber sensors are used to monitor the conditions within oil wells and pipelines, railways, wings of airplanes, and wind turbines. Decades of research led to the development of accurate optical fiber measuring instruments, including gyroscopes, hydrophones, temperature probes, and chemical monitors. As the OFS field is complementary to optical communication, it has availed the benefit of advancements on optical fibers and optoelectronic instrumentation. Due to the rapid development of optical fiber communication systems, the cost of optical fiber sensor systems has also significantly reduced due to the commercially viable key components such as light sources and photo detectors.

It is anticipated that research on OFS technology still continues to grow, and the technology will become widespread. Several new types of fiber sensors are progressively emerging, and some of the fiber sensing technologies are in a mature state and has been commercially produced and used in different applications. As light wave technology is becoming the new enabling technology behind many advanced high-end applications, optical fiber sensors can play an integral part of those developments.

The aim of this chapter is to introduce the basic principles of OFS and to provide a brief overview of some of the different types of fiber sensors developed all through the years and also to form the base for the rest of the chapters in this book.

1.2 OPTICAL FIBER SENSORS

For a given process or operation, a system needs to perform a given function that requires qualifying and/or quantifying the condition of one or multiple parameters. Therefore, these parameters are to be measured and processed so that the system can carry out its function. To facilitate the process, a sensor can be used, which is a device that provides these information by measuring any physical/chemical/biological measurand of interest (pressure, temperature, stress, pH, etc.) and converting it into an alternative form of energy (e.g., electrical or optical signal) that can be subsequently processed, transmitted, and correlated to the measurand of interest.

As an optical fiber alternative of traditional sensors, *optical fiber sensors* can be defined as a means through which physical, chemical, and biological measurands interact with light guided through an optical fiber or the light is guided to an interaction region by an optical fiber, to produce a modulated optical signal with information related to the measurement parameter. The basic concept of a fiber-optic sensing system is demonstrated in Figure 1.1, where the fiber (guided light) interacts with an external parameter and carries the modulated light signal from the source to the detector. The input measurand information can be extracted from this modulated optical signal. Depending on the type of fiber sensor and its operating principle, the sensor system can operate either in transmission mode or in reflection mode, which is elaborated in later sections of this chapter.

1.3 MEASUREMENT CAPABILITIES AND ADVANTAGES OF OPTICAL FIBER SENSORS

One of the biggest advantages of fiber sensors is that it has the capability to measure a wide range of parameters based on the end user requirement, once a proper fiber sensor type is used. Some of the measurement capabilities of fiber sensors are given as follows (but not limited to):

- Strain, pressure, force
- Rotation, acceleration
- Electric and magnetic fields

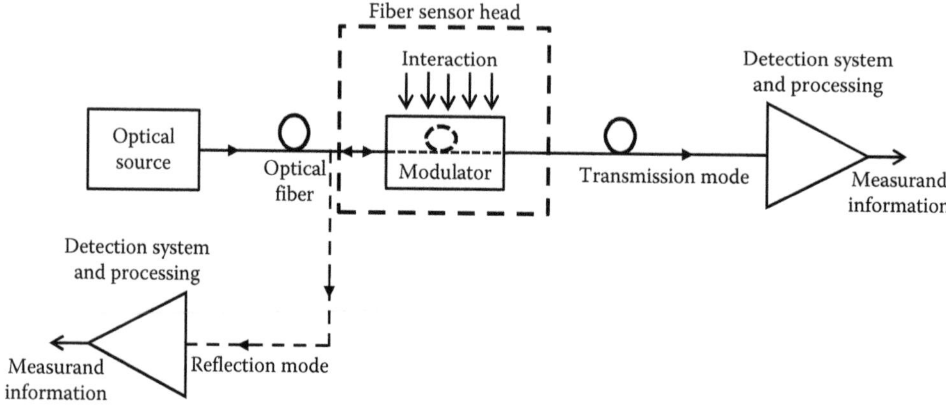

FIGURE 1.1 Basic concept of an optical fiber sensor.

- Acoustics and vibration
- Temperature, humidity
- pH and viscosity
- Biosensing single molecules, chemicals, and biological elements such as DNA, single viruses, and bacteria

Given the advancement in the development of optical fibers and instrumentation, the development and measurement capabilities of optical fiber sensors have also increased and are currently being employed in many structural health monitoring (SHM) and biomedical applications [7–9]. On the advantages of optical fiber sensors, they share all the benefits of optical fibers and are excellent candidates for monitoring environmental/external changes. Some of the advantages of fiber sensors over conventional electronic sensors are listed in the following:

- Lightweight
- Passive/low power
- EMI resistant
- High sensitivity and bandwidth
- Environmental ruggedness
- Complementary to telecom/optoelectronics
- Multiplexing capability
- Multifunctional sensing possibility
- Easy integration into a wide variety of structures
- Robust, more resistant to harsh environments

Though optical fiber sensors have many advantages compared to its electrical counterpart, there are some concerns as well with this technology. Major drawbacks include cost, long-term stability, less efficient transduction mechanism, and complexity in its interrogation systems. Another disadvantage to the fiber sensing system is the unfamiliarity of the end user in dealing with the system. As more research, commercialization, and standardization efforts are ongoing, it is expected that some of the current disadvantages may become extinct in the near future.

1.3.1 Classification of Optical Fiber Sensors

As optical fiber sensors operate by modifying one or more properties of light passing through the fiber, they can be broadly classified as extrinsic or intrinsic. In the extrinsic sensor types, the optical fiber is used as a means to carry light to an external sensing system, while in the intrinsic type, the light does not have to leave the optical fiber to perform the sensing function. The intrinsic fiber sensor types are more attractive and widely researched as this scheme has many advantages compared to extrinsic, such as their in-fiber nature and the flexibility in the design of the fiber sensor head.

Furthermore, depending on the light property that is modified, optical fiber sensors can be mainly classified into four main categories:

- Intensity-modulated sensor
- Phase-modulated (interferometric) sensor
- Polarization-modulated (polarimetric) sensor
- Wavelength-modulated (spectrometric) sensor

Optical fiber sensors can also be further classified based on its spatial positioning as well as based on the measurement capabilities. A summary of the different classifications of OFS is shown in Table 1.1.

As given in Table 1.1, according to the spatial distribution of the measurand, fiber sensors can be classified. For example, a point sensor can be used to sense measurands from discrete points, and

TABLE 1.1
Different Classifications of Optical Fiber Sensors

Classification Based on the Working Principle	Classification Based on the Spatial Positioning	Classification Based on the Measurement Parameters
Intensity-modulated sensors: Detection through light power	*Point sensors*: Discrete points, different channels for each sensor/measurand	*Physical sensors*: Temperature, strain, pressure, etc.
Phase-modulated (interferometric) sensors: Detection using the phase of the light beam	*Distributed*: Measurand determined along a path, surface, or volume	*Chemical sensors*: pH content, gas sensors, spectroscopic study, etc.
Polarimetric sensors: Detection of changes in the state of polarization of the light	*Quasi-distributed*: Variable measured at discrete points along an optical link	*Biosensors*: DNA, blood flow, glucose sensors
Spectrometric sensors: Detection of changes in the wavelength change of the light	*Integrated*: Measurand integrated along an optical link giving a single value output	

with spatial multiplexing capability, multiple channel measurements are possible for these types of sensors. On the other hand, distributed fiber sensors provide spatial and temporal information of the measurand from any point along a single fiber with a certain resolution. Phenomena such as Rayleigh, Raman, and Brillouin scattering in the fiber are the underlying technology behind these types of sensors. In the case of quasi-distributed sensors, measurements are possible from a number of locations on the single fiber, while integrated sensors provide an average measurement of the perturbation over the sensing length of the fiber. Further details of sensors based on spatial classification can be found in Chapter 2.

1.3.1.1 Intensity-Modulated Fiber-Optic Sensors

Intensity-modulated sensors are one of the earliest types and perhaps the simplest type of optical fiber sensors. In the intensity modulation scheme, optical signal is transmitted through optical fibers, and then its intensity is modulated by various means such as fiber bending, reflectance, or changing the medium through which the light is transmitted. The main advantages of intensity-modulated sensors are its ease of fabrication, simple detection system and signal processing requirements, and its low-price performance. A simple example of a reflection-type intensity-modulated fiber-optic sensor (IM-FOS) using a single fiber that can be used to measure distance, pressure, etc., is shown in Figure 1.2.

Compared to the single-fiber type, two-fiber-type reflection sensors are often used as a proximity sensor in a variety of applications. In this case, one fiber is used as the input fiber and a second fiber

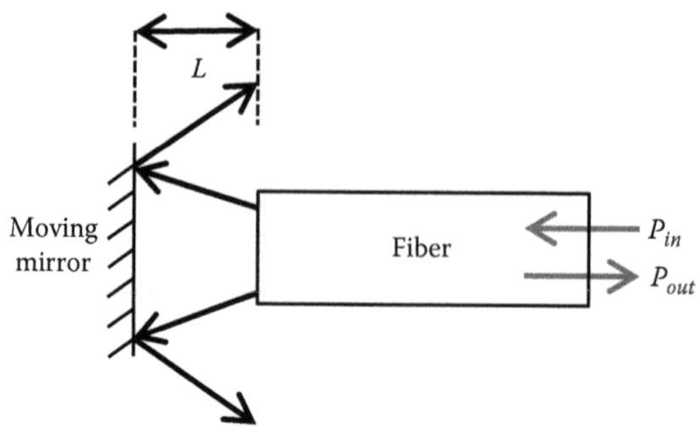

FIGURE 1.2 A simple extrinsic intensity-modulated sensor.

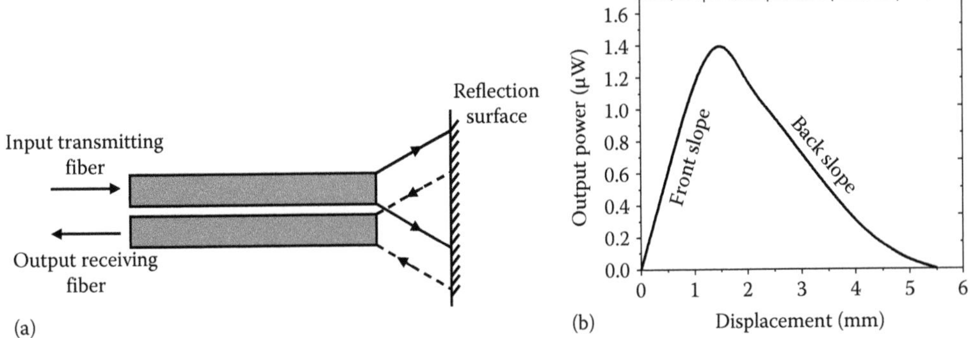

FIGURE 1.3 (a) A two fiber reflection type sensor and (b) its typical characteristic response.

is used to collect the reflected light from a reflecting surface. The schematic of a two-fiber configuration and its typical intensity response is also shown in Figure 1.3. The modulation function of a two-fiber reflection-type sensor is given as [10]

$$M_s = \frac{\Phi_{RS}}{\Phi_E} = \frac{S_R \cos^4\theta}{\pi(2h\text{NA})^2}, \tag{1.1}$$

where

Φ_E are the radiant flux from the emitting fiber
Φ_{RS} is the radiant flux intercepted by the end face of the receiving fiber
S_R is the area of the end face of the receiving fiber
θ is the cable between the axis of the receiving fiber and the line connecting the centers of the receiving fiber and the emitting fiber (virtual image)
NA is the numerical aperture of the fiber

In contrast to other sensing principles, intensity modulation can be obtained by using a simple arrangement; however, optical fiber bending, coupling misalignments, source power fluctuation, etc., can cause signal attenuation and signal intensity instability, which lead to a less reliable sensor system. One solution for this problem is the intensity referencing, where a fraction of the input light (obtained using a fiber splitter) is used for monitoring the input power fluctuation. An example of such a configuration is shown in Figure 1.4 [11], where the power ratio of the modulated signal and reference signal is used.

Other types of IM-FOSs include evanescent wave sensors and microbend and macrobend sensors. Evanescent field–based fiber sensor is one of the most useful and commonly used IM-FOS. When light propagates along an optical fiber, it is not completely confined to the core region but extends into the surrounding cladding region and is normally named as evanescent wave field. Exposing this evanescent field to interact with measurands can change the output power of the optical fiber based on the measurand property. Detailed descriptions on evanescent field sensors and micro- and macrobend fiber sensors can be found in Chapter 2.

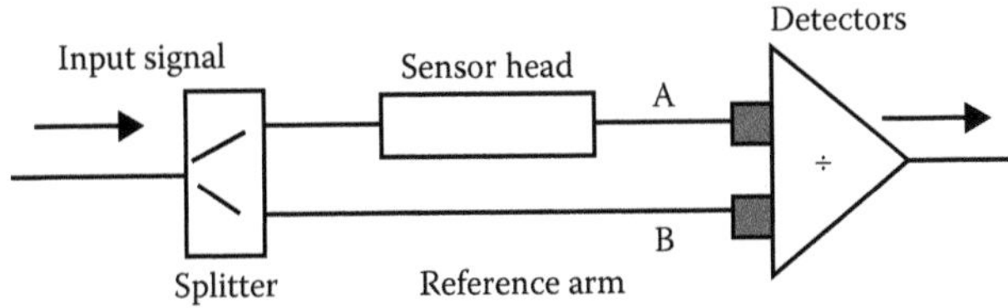

FIGURE 1.4 An example of ratiometric power measurement method for intensity referencing.

1.3.1.2 Phase-Modulated Fiber-Optic Sensors

Phase-modulated sensors, also known as interferometric sensors, are the ones that work based on phase difference of coherent light traveling along two different paths, in the same fiber or different fibers. These sensors are often considered as high-sensitivity sensors due to its capability to respond to small changes in the external measurands. The common interferometric fiber sensors include fiber-optic Mach–Zehnder, Michelson, Fabry–Perot, and Sagnac interferometers. A detailed description of interferometric fiber sensors and its applications can be found in Chapter 3.

1.3.1.3 Spectrometric (Wavelength-Based) Fiber Sensors

Wavelength-modulated fiber sensors exhibit a change in the propagating optical wavelength (spectral modulation) when interacted by an external perturbation. Some common wavelength-based fiber sensors include Bragg grating sensors, fluorescence sensors, and blackbody sensors.

Arguably, one of the most widely used wavelength-based sensors is the Bragg grating sensor. Fiber Bragg gratings (FBGs) are inscribed in an optical fiber by inducing periodic changes in the index of refraction in the core [12]. The periodic index–modulated structure enables the light to be coupled from the forward propagating core mode into a backward propagating core mode generating a reflection response. The reflected Bragg wavelength is sensitive to a range of physical parameters, and thus, an FBG can be used as a sensor in a variety of applications. Theory, fabrication, and applications of FBG are presented in Chapter 9.

Fluorescent-based fiber sensors are widely used for physical and chemical sensing for measurands such as temperature, humidity, and viscosity. Among the different configurations used, the two of the most common ones are the end tip sensor and the blackbody cavity types [13]. In the case of a blackbody sensor, a blackbody cavity is placed at the end of an optical fiber [14], and as the temperature of the cavity rises, it starts to glow and act as a light source. Using suitable detectors and narrow band filters, the profile of the blackbody curve can be determined, which provides the measurand information. Due to the capability to measure temperature to within a few degrees centigrade under intense RF fields, this type of sensors is developed as optical fiber thermometers and has been successfully commercialized.

1.3.1.4 Polarimetric Fiber Sensors

It is quite well known that as the light wave propagates along the optical fiber, its state of polarization changes because of the difference in the phase velocity of the two polarization components in a birefringent fiber. The polarization properties of light propagating through an optical fiber can be affected by stress, strain, pressure, and temperature acting on it, and in a fiber polarimetric sensor, the change in the state of polarization is measured and is used to retrieve the sensing parameter [15]. A symmetric deformation effect or temperature variation in a single-mode fiber influences the propagation constant (β) for every mode because of the changes in the fiber length (L) and the refractive indices of the core and the cladding. Under the influence of longitudinal strain (ε) or temperature (T), for single-mode fiber polarimetric sensors, the change in the phase difference can be written as [16]

$$\frac{\delta(\Delta\Phi)}{\delta X} = \Delta\beta\frac{\partial L}{\partial X} + L\frac{\partial(\Delta\beta)}{\partial X} \tag{1.2}$$

where X stands for temperature, pressure, or strain.

A fiber-optic polarimetric sensor that operates in intensity domain using a polarizer–analyzer arrangement is shown in Figure 1.5. In a typical polarimetric sensor, linearly polarized light is launched at 45° to the principal axes of a birefringent fiber such that both the polarization modes

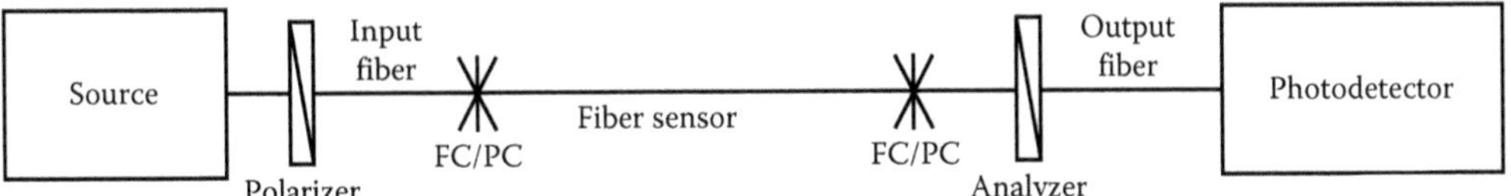

FIGURE 1.5 Schematic of a typical polarimetric fiber sensor.

can be equally excited. The polarization state at the output is converted to intensity by using a polarizer–analyzer oriented at 90° to the input polarization state. For such polarimetric sensors, the change in the output intensity at a wavelength λ due to externally applied perturbation can be described by the formula

$$I_s(\lambda) = \frac{I_0}{2}\left[1+\cos(\Delta\Phi)\right] \tag{1.3}$$

Therefore, a change in polarization can be observed as a change in intensity, and by correlating the output change in intensity to the measurand, a polarimetric fiber sensor can be effectively used as a sensor for a variety of applications. The phase difference between the two orthogonal polarizations can also be extracted using an experimental setup consisting of a tunable laser source and a polarimeter/polarization control system.

For polarimetric fiber sensors, typically high-birefringent optical fibers are used such as panda fiber, bow tie fiber, and elliptical core or polarization maintaining photonic crystal fibers. Due to the high sensitivity of polarimetric fiber sensors to external parameters such as strain and temperature, often cross sensitivity is a major problem to these types of sensors. The capability of polarimetric fiber sensors to measure strain, temperature, and pressure is demonstrated in a variety of applications such as in SHM [16–19]. Fiber polarimetric sensors are also used for current and voltage measurements, where a range of polarization-based effects (optical activity, Faraday effect, and electrooptic effect) are used.

1.4 RECENT DEVELOPMENTS IN OPTICAL FIBER SENSING

Since the first demonstration of the microstructured fiber (MOF) in 1996 (photonic crystal fiber), it has caused enormous attention and excitement in the photonics community. The unique features of the fiber such as the light guidance, dispersion properties, endlessly single-mode nature, higher birefringence, and enhanced nonlinear effects have led MOFs being applied in many fields such as in optical communications, nonlinear optics, and sensing. Unlike conventional fibers, MOFs can be optimized for a large range of applications by tailoring the size, number, and the geometry of the air holes that form the confining microstructure around the core region. This led to a large interest among researchers to exploit the advantages and use MOF-based sensors in a variety of applications. In a short span of time since the first introduction of the MOF, a variety of MOFs are reported ranging from silica to polymer for applications ranging from environmental sciences to medicine, industry, and astronomy [20]. Fascinating physics and applications of MOFs are still of great interest among researchers. Some of the sensing applications of photonic crystal fibers are presented in Chapter 6 (Figure 1.6).

Another rapidly growing fiber sensor types are the optical microfiber–based sensors. Optical microfibers exhibit many desirable characteristics such as large evanescent field, strong optical confinement, bend insensitivity, configurability, compactness, robustness, and feasibility of extremely high-Q resonators [21,22]. The resulting sensors hold numerous advantages over their standard optical fiber counterparts, including high responsivity, fast response, selectiveness, nonintrusiveness, and

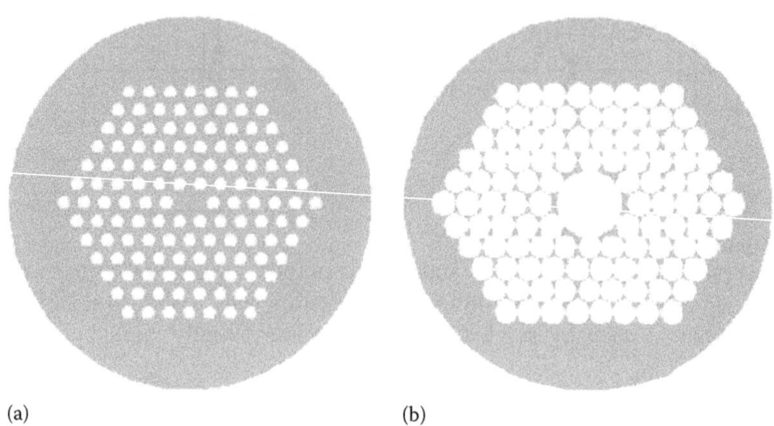

FIGURE 1.6 (a) Solid-core PCF and (b) hollow-core PCF. Colors: white—air, light gray—silica, and dark gray—doped silica (see Chapter 6 for more details).

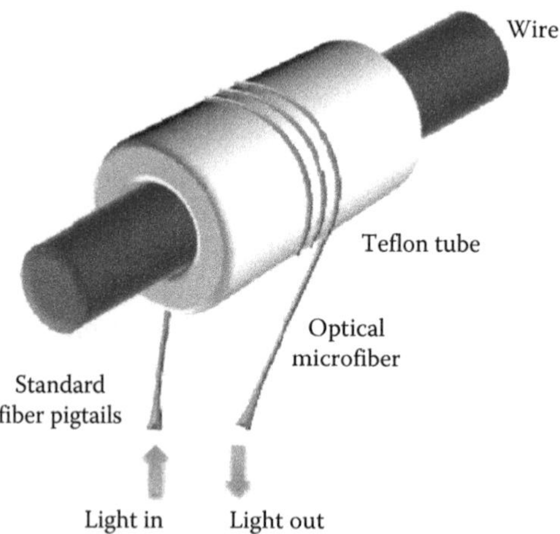

FIGURE 1.7 An optical microfiber temperature sensor (see Chapter 8 for more details).

small size. With sensing areas spanning acceleration, acoustic, bend/curvature, current, displacement, electric/magnetic field, roughness, rotation, and temperature, the future of optical microfiber technology looks exceptionally promising. The latest developments of optical microfiber–based sensors are presented in Chapter 8 (Figure 1.7). Other emerging areas such as slow light in fiber sensing and nanophotonics are discussed in Chapter 18.

1.5 OPTICAL FIBER SENSOR INSTRUMENTATION

Though the sensor head based on optical fibers is relatively cheap, the optical and electronic instrumentation required to interrogate the sensor is often expensive and complicated. Based on the type of fiber sensor used, the instrumentation requirement to interrogate the sensor will vary. The basic characteristics of any fiber-optic sensing instrumentation are its suitability to the type of fiber sensor and sufficient sensitivity and resolution with adequate measurement range.

The simplest instrumentation requirement is for the intensity-based systems, where a simple light source (LED, semiconductor laser, etc.) and an optical receiver with or without an amplifier would be sufficient in most of the applications. However, based on the stability and

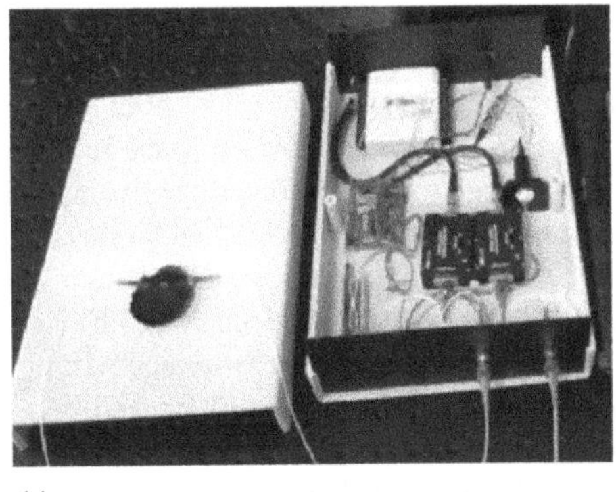

(a)

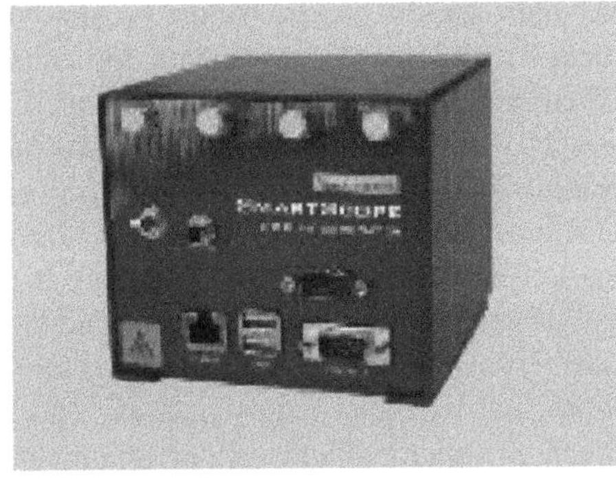

(b)

FIGURE 1.8 (a) An OEM interrogation system for a fiber temperature sensor [23] and (b) a portable FBG interrogator [24].

accuracy required, further modifications would be necessary such as ratiometric measurement (Figure 1.4). For spectrometric fiber sensors, devices that can measure the spectral change are required such as an optical spectrum analyzer or an optical spectrometer. One of the most common types of spectrometric fiber sensor that is widely used is the FBG sensors, and for their interrogation, FBG interrogation systems are widely used. The operating principles of these interrogation units vary, and a range of techniques that can be used to interrogate an FBG are reported (see Chapter 9 for more details). For polarimetric fiber–based sensing systems, polarimeter, polarization controller, etc., are most often used to improve the accuracy and reliability of the system.

In general, optical fiber sensor instrumentation plays a vital part in the overall development of the fiber sensing area, and during these years, we have seen tremendous increase in new developments in the instrumentation area, which is also supported by the growth of the optical fiber network area as well as shares a common interest when it comes to the instrumentation requirement. It should be also noted that the instrumentation requirement for different fiber sensor types is sometimes unique to those sensor types, and as a result, an overall system development is required to realize a complete fiber sensing unit. Some examples of OFS instrumentation are shown in Figure 1.8.

1.6 OPTICAL FIBER SENSORS: INDUSTRY APPLICATIONS AND MARKET

Over the years, optical fiber sensors have made a rapid growth in quality and functionality and have found widespread use in many areas such as in defense, aerospace, automotive, manufacturing, biomedical, and among various other industries. The rapid development on sensor devices and instrumentation over the years has made OFS an attractive alternative to conventional sensors for measuring range of parameters such as pressure, temperature, strain, vibration, acceleration, rotation, magnetic field, electric field, viscosity, and a wide range of other measurands. Improvements in performance and standardization, together with a decrease in price, have helped OFS to make an impact on the industry.

Among the different types of optical fiber sensors, FBGs, fiber gyroscopes, and distributed fiber sensors are the widely used and popular ones. High-speed and reliable interrogation systems for data acquisition, processing, and interpretation are also developed for these devices. They also meet application-specific demands, which are not efficiently satisfied by conventional electronic sensors. Examples of such applications include the use of fiber-optic sensors for pressure and high-temperature monitoring in the oil and gas exploration industry, which is a major market for distributed fiber sensors.

Fiber-optic sensors can contribute to improving safety, by providing an early warning of failure in critical structures and components, for example, in the aeronautical and civil engineering sectors. Therefore, SHM application has emerged as a promising market for optical fiber sensors. As the growth of production of composite parts has increased, the composite/aerospace industry has increasingly focused on damage-/failure-free composite structures and nondestructive techniques for SHM over their lifetime. Compared to conventional nondestructive sensing techniques, optical fiber sensors have achieved wide acceptance due to their attractive properties such as small size and its capability to embed within the host structure. Optical fiber sensors embedded in various structures are very useful for strain and temperature monitoring applications in extreme environmental conditions. For example, issues such as bend loading and icing on wind turbine and helicopter blades can be monitored and avoided by implementing smart composite structures with embedded optical fiber sensors. Such smart structures with embedded optical fiber sensors can enhance the safety of advanced machines, structures, and devices (Figure 1.9).

The medical and biomedical industry also provides many challenging instrumentation requirements for which OFS can provide novel and reliable solutions. One such challenge is the need to monitor within radioactive medical environments and in magnetic resonance imaging (MRI) systems, where electrical sensors cannot operate. Another is the need for in vivo sensing during keyhole surgical procedures and catheter procedures where space is a premium and where a miniature fiber sensor is the largest intrusion that can be tolerated. Recent success with fiber sensors on the

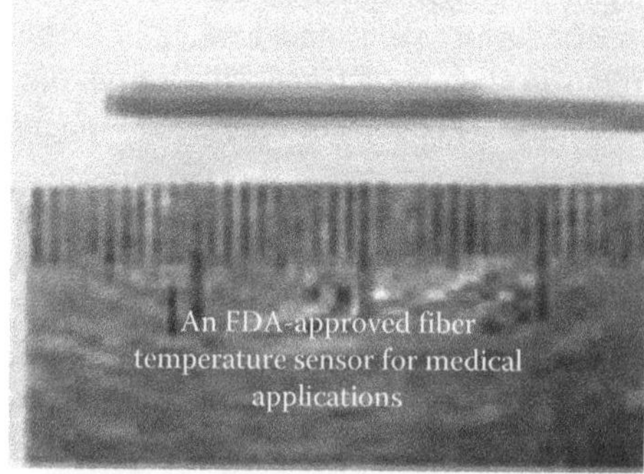

FIGURE 1.9 Some applications of optical fiber sensors in industry.

biomedical front includes commercial use of fiber-based catheters and miniature fiber temperature and pressure transducers for medical applications.

The acceptance of optical fiber sensors among different industries/applications is expected to experience rapid growth during the coming years due to a number of initiatives being taken by the market participants [25]. Another driving force behind the growth of OFS is that a number of research initiatives and efforts to achieve standardization are also ongoing and promoted by various research groups and industry leaders. However, more initiatives are required, such as ensuring the reliability, longevity, and accuracy of fiber sensor products to achieve a wider acceptance of the technology across various industrial sectors. According to the analysis of market research firms (BCC Research), the global fiber-optic sensor market almost reached a billion dollars in 2010 and is expected to reach 2.15 billion dollars by 2015.

1.7 CONCLUSION

This chapter provided a brief summary of the basic and broad principles of different types of optical fiber sensors. Descriptions of fiber sensors based on different classifications such as modulation methods, spatial positioning, and measurement parameters were also discussed. Some of the emerging optical fiber sensors are also introduced, and finally, the fiber-optic instrumentation and market trends are also discussed. With the emergence of a variety of new types of fibers and new configurations, OFS and instrumentation are still growing and expected to further widen the scope of applications in the near future.

REFERENCES

1. Clushaw, B. and Kersey, A., Fiber-optic sensing: A historical perspective, *Journal of Lightwave Technology*, 26(9), 1064–1078, 2008.
2. Bogue, R., Fibre optic sensors: A review of today's applications, *Sensor Review*, 31/4, 304–309, 2011.
3. Bucaro, J. A., Dardy, H. D., and Carome, E. F., Fibre optic hydrophone, *Journal of the Acoustical Society of America*, 52, 1302, 1977.
4. Culshaw, B., Davies, D. E. N., and Kingsley, S. A., Acoustic sensitivity of optical fiber waveguides, *Electronics Letters*, 13(25), 760–761, 1977.
5. Vali, V. and Shorthill, R. W., Fibre ring interferometer, *Applied Optics*, 15(5), 1099–1100, May 1976.
6. Butter, C. D. and Hocker, G. E., Fiber optics strain gauge, *Applied Optics*, 17, 2867, 1978.
7. Mishra, V., Singh, N., Tiwari, U., and Kapur, P., Fiber grating sensors in medicine: Current and emerging applications, *Sensors and Actuators A: Physical*, 167(2), 279–290, 2011.
8. Rajan, G., Ramakrishnan, M., Lesiak, P., Semenova, Y., Boczkowska, A., Woliński, T., and Farrell, G., Composite materials with embedded photonic crystal fiber interferometric sensors, *Sensors and Actuators A: Physical*, 182, 57–67, 2012.
9. Lopez-Higuera, J. M., Cobo, L. R., Incera, A. Q., and Cobo, A., Fiber optic sensors in structural health monitoring, *Journal of Lightwave Technology*, 29(4), 587–608, 2011.
10. Zhao, Z., Lau, W. S., Choi, A. C. K., and Shan, Y. Y., Modulation functions of the reflective optical fiber sensor for specular and diffuse reflection, *Optical Engineering*, 33(9), 2986–2991, 1994.
11. Rajan, G., Semenova, Y., and Farrell, G., An all-fiber temperature sensor based on a macro-bend single-mode fiber loop, *Electronics Letters*, 44, 1123–1124, 2008.
12. Schwab, S. D. and Levy, R. L., In-service characterization of composite matrices with and embedded fluorescence optrode sensor, *Proceedings of SPIE*, 1170, 230, 1989.
13. Grattan, K. T. V., Selli, R. K., and Palmer, A. W., A miniature fluorescence referenced glass absorption thermometer, in *Proceedings of the Fourth International Conference on Optical Fiber Sensors*, Tokyo, Japan, 1986, p. 315.
14. Rao, Y. J., In-fibre Bragg grating sensors, *Measurement Science and Technology*, 8(4), 355–375, 1998.
15. Wolinski, T. R., Polarimetric optical fibers and sensors, *Progress in Optics*, 40, 1–75, 2000.
16. Wolinski, T. R., Lesiak, P., and Domanski, A. W., Polarimetric optical fiber sensors of a new generation for industrial applications, *Bulletin of the Polish Academy of Sciences, Technical Sciences*, 56(2), 125–132, 2008.

17. Murukeshan, V. M., Chan, P. Y., Seng, O. L., and Asundi, A., On-line health monitoring of smart composite structures using fibre polarimetric sensor, *Smart Materials and Structures*, 8, 544–548, 1999.
18. Ma, Ji. and Asundi, A., Structural health monitoring using a fibre optic polarimetric sensor and a fibre optic curvature sensor-static and dynamic test, *Smart Materials and Structures*, 10, 181–188, 2001.
19. Ramakrishnan, M., Rajan, G., Semenova, Y., Boczkowska, A., Domański, A., Wolinski, T., and Farrell, G., Measurement of thermal elongation induced strain of a composite material using a polarization maintaining photonic crystal fibre sensor, *Sensors and Actuators A: Physical*, 190, 44–51, 2013.
20. Pinto, A. M. R. and Lopez-Amo, M., Photonic crystal fibers for sensing applications, *Journal of Sensors*, 2012, 1–21, 2012, Article ID 598178.
21. Chen, G. Y., Ding, M., Newson, T. P., and Brambilla, G., A review of micro fiber and nano fiber based optical sensors, *The Open Optics Journal*, 7(1), 32–57, 2013.
22. Lou, J., Wang, Y., and Tong, L., Microfiber optical sensors: A review, *Sensors*, 14, 5823–5844, 2014.
23. Rajan, G., Semenova, Y., Mathew, J., and Farrell, G., Experimental analysis and demonstration of a low cost temperature sensor for engineering applications, *Sensors and Actuators A: Physical*, 163, 88–95, 2010.
24. http://www.smartfibres.com/FBG-interrogators. Accessed May 2014.
25. Frost & Sullivan Research Service, Fiber optic sensors (technical insights), Frost & Sullivan Research Service, Mountain View, CA, 2010.

2 Optical Fiber Sensing Solutions
From Macro- to Micro-/Nanoscale

Yuliya Semenova and Gerald Farrell

CONTENTS

2.1 INTRODUCTION

Optical fibers have revolutionized communications industry and since the early 1970s have become the communications medium of choice for telephones, Internet, cable television, security cameras, utility, and industrial networks. Enormous information transmission capacity of the optical fibers combined with additional advantages, such as low losses, immunity to electromagnetic interference, small size, lightweight, and electrical insulation, makes them the most effective means for transporting information. Applications of optical fibers extend well beyond communications industry and their number continues to grow. For example, due to fibers' ability to deliver light to confined spaces, optical fibers are used in medicine for viewing internal body parts. Other examples include applications for illumination, in fiber lasers, and in sensing.

In the following sections, we focus on the recent developments in the area of fiber-optic sensing from the perspective of physical scale including an overview of the fundamentals of the phenomena utilized for sensing at different physical scales along with some specific examples of macro- and micro-/nanosensors and their practical applications. This chapter also discusses future challenges and opportunities for fiber-optic technology at different physical scales and some emerging applications of fiber-optic sensors.

2.2 FIBER-OPTIC SENSOR TYPES

A *fiber-optic sensor* converts a physical parameter to an optical output. The key part of every fiber-optic sensor is a *transducer*—device that converts one form of energy associated with the physical parameter into another form of energy. Depending on the transducer type, all fiber-optic sensors fall into two broad categories: *extrinsic* and *intrinsic* sensors (Figure 2.1). In the case of an extrinsic sensor, the optical fiber serves as a delivery mechanism to guide the optical signal to the sensing region outside the fiber, where it is modulated in response to the physical parameter of interest and then collected by the same (or different) optical fiber and guided to a detector for processing. In an intrinsic sensor, the optical fiber acts both as the means for transporting the optical signal to/from the sensing region and as the transducer.

Since the earliest days of fiber-optic communications, the main focus for researchers and engineers has been on minimizing the optical losses and the influence of various external factors on the light propagating through the fiber. Not surprisingly, the earliest developed fiber-optic sensors, such as a fiber-optic proximity probe patented in 1967 (US003327584), were extrinsic. The proximity probe utilized optical fibers as the light-conducting medium to measure minute rotation, vibration, or displacement by analyzing the amount of light reflected from the surface positioned near the fiber probe end. Approximately a decade later, researchers from the University College London suggested a practical application for the phase modulation of light signal resulting from dynamic mechanical stressing of the fiber. Davies and Kingsley [1] recognized that an external perturbation (such as strain) has the potential of modulating the optical carrier within the fiber waveguide without the need to break and join the fibers. Such an approach has the obvious advantage as the need for multiple interfaces between the fiber and transducer is removed. This idea of impressing the external influence on the optical carrier within the fiber marked the beginning of the development of a wide variety of intrinsic fiber-optic sensors.

Another commonly accepted classification of fiber-optic sensors is based on whether they allow measurements of the physical parameter of interest locally, in which case they are called *point sensors*, or provide the value of the physical parameter over a distance and as a function of position along the fiber—*distributed sensors*. Point transducers have a finite length and therefore provide the value of the physical parameter averaged over a given volume of space, corresponding to the length of the point transducer. If multiple-point measurements are required, a number of point sensors need to be used, which in many cases requires installing multiple lead-in/lead-out fibers, arrays of detectors, etc. (Figure 2.2a). The number of point sensors, which defines the spatial resolution of the system, is limited by the cost and therefore becomes insufficient for many practical applications.

Distributed fiber-optic sensing approach (Figure 2.2b) has no real equivalent among other types of sensors and has become an important differentiator of the fiber-optic sensing. Distributed sensors are most often intrinsic and the sensing principles and practical implementation are significantly different from those used for point sensing. Since only one fiber needs to be installed to sense the

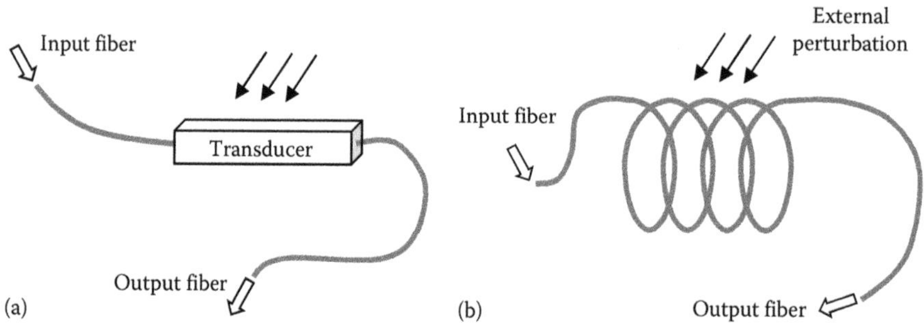

FIGURE 2.1 Fiber-optic sensor types: (a) extrinsic (sensing is performed outside the fiber) and (b) intrinsic (the fiber also acts as a transducer).

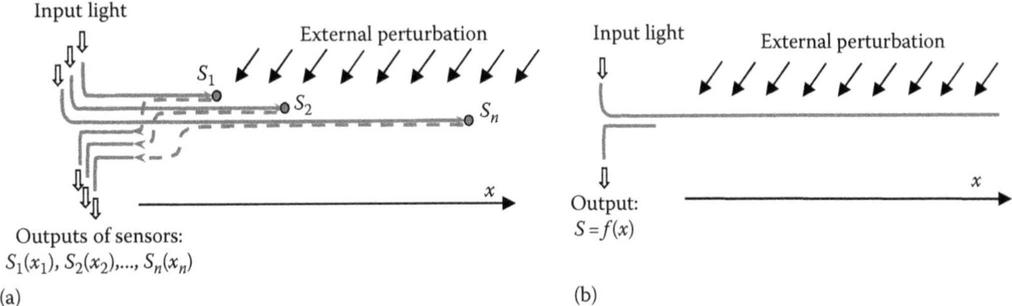

FIGURE 2.2 Schematic illustration: (a) point sensing and (b) distributed sensing.

physical parameter of interest at many points, it is possible to share the optical source and detector and often eliminate other equipment required for multipoint sensing using point sensors. This results in much lower cost per point measurement and also much higher weight and space efficiency of the distributed sensing, making this technique the most powerful monitoring option in many applications, especially for structural monitoring of civil and aerospace structures.

Finally, fiber-optic sensors can be classified based on the light characteristic that is affected by the parameter to be sensed: *intensity*, *wavelength*, *phase*, or *polarization sensors*.

2.3 SENSING ON A VARIETY OF SCALES: THEORY AND PRACTICE

2.3.1 Large-Scale Fiber-Optic Sensing

There are several application areas, such as environmental, safety, and security monitoring, which require large-scale and/or remote sensing approach and where fiber-optic sensors can offer a significant improvement over traditional electronic sensors. The distinctive feature of the fiber-optic technology is its dual functionality, as the fiber can serve as both the sensor and the telemetry channel. This removes the need for additional means for delivery of information from the sensor to the user and thus gives the fiber optics an advantage in comparison with other techniques. Moreover, the telemetry channel based on the optical fiber is immune to electromagnetic interference and harsh environments, allows for avoidance of reflections from natural and artificial obstacles and bad weather interruptions, and exhibits very low transmission losses, thus enabling coverage over long distances. Intrusive data interception is more difficult with the optical fiber when comparing to free-space radio propagation or metal wires, and the fiber's nonelectric nature reduces the risks when being used in combustible environments.

In the past several decades, an increased number of large-scale civil infrastructures have been built all over the world. These large critical structures including buildings, bridges, pipelines, and other civil constructions are characterized by a high cost and usually have a long-term service life during which they are expected to remain safe and even to survive the rare catastrophic events such as earthquakes or fires. In order to ensure the survivability and safety of the civil structures, it is necessary to monitor their key structural and environmental parameters in real time. This technique, referred to as structural health monitoring (SHM), has attracted considerable interest worldwide and became a popular niche application for optical fiber sensors [2].

In general, stress points and damages are randomly distributed throughout the structure and their locations are usually hidden. Also, the structural damages are time dependent. Thus, a cost-effective SHM system should be capable of providing both stress/temperature distribution and real-time local damage assessment simultaneously [3]. In order to satisfy these requirements for large-scale and often remotely located structures, optical fiber sensing networks pose three main challenges [4].

First, in most practical situations, it is desirable to maximize the number of sensors that can be multiplexed on a single network while ensuring good signal quality. The main motivation for fiber-optic sensors multiplexing is the cost, due to the fact that when the optoelectronic unit (optical source and detector), which is much more expensive than the fiber and sensors, is shared between many sensing points, the cost per sensing element decreases [5].

Commercial interrogators for multiple fiber-optic sensors fall into two main categories: time division multiplexing (TDM) and wavelength division multiplexing (WDM) [6]. TDM discriminates between many sensors on a single optical fiber by gauging the time required for a pulse of light to return to the detection system. As the time signature of each sensor changes, TDM instruments translate these time shifts into changes in the parameter of interest. TDM system must be designed to balance the sensor-sampling rate with the distance of the sensor from the light source and detection system. Distance between sensors must typically be greater than a few meters to allow the instrument to clearly distinguish between adjacent sensors.

However, the most popular approach to interrogation of multiple fiber-optic sensors on a single network is WDM. WDM systems discriminate individual sensors by wavelength. Most WDM systems are designed using one of two basic configurations: a broadband source combined with a swept detector [7] or a laser source and a broadband detector [8]. In the second technique, a narrowband tunable laser is swept across the appropriate spectral region, and a reflected signal is observed with a broadband detector only when the laser is precisely tuned to the wavelength range assigned to a particular sensor.

The second challenge for the fiber-optic system is to ensure the continuity of service in the event of point failure(s) on the network. The continued operation of the sensor network after accidental or malicious damage becomes increasingly important when the structure that is being monitored is of high value (e.g., an oil pipeline or a power transmission line), human safety is at risk (bridges, dams chemical storage sites, and nuclear plants), or perimeter security is a concern (airports, banks, etc.) [2].

The third main challenge is to enable the possibility of remote sensing. While the first two requirements are common to all optical fiber networks, the last one is more specific to the situations when there is a need to monitor environmental or structural parameters from a central location positioned tens or hundreds kilometers away from the object or field of interest. Examples of such situations include SHM of gas or oil pipelines, ultralong bridges and tunnels [9,10], riverbanks, and offshore platforms [11,12]. Some other promising applications of remote sensing are as follows: tsunami detection and warning before their arrival to the coast [13]; geodynamical monitoring, such as surveillance of volcanic and tectonic areas to predict the possible evolution toward critical stages or to detect landslides [14]; railway applications, such as train speed measurements, derailment, wheel defects, and rail crack detection [15]; and underwater acoustic applications in sonar and seismic surveying for the oil–gas industry [16].

Advantages of using fiber-optic sensors in these areas are very clear. For example, traditional underwater sonar arrays based on electroceramic sensors require large amount of underwater electronics for multiplexing and data telemetry, which results in a highly complex system [16]. The underwater electronics, associated power, and telemetry cabling can be expensive, heavy, and prone to failure. The problem of reliability is made worse by the need for additional waterproofing. Another example of how cumbersome the present large-scale sensing techniques are is the method of tsunami detection, based on electronic pressure sensors that need to be powered underwater, buoys on the surface of the ocean, and satellites transmitting the data to the surveillance centers [4]. With the optical fiber sensors, there is no need for underwater electronics and the sensors are secure and more reliable and due to their higher speed offer the possibility of real-time measurement. Table 2.1 summarizes the sensors types, measurands, and sensing principles for the optical fiber sensors used in the large-scale applications, and this method of classification is adopted here to organize the following sections.

2.4 POINT SENSORS

Many intensity-based sensors are point sensors, which can measure changes at specific local points within a structure or along the sensing fiber. The simplest examples of such intensity-based devices are macrobend fiber sensors and most of the interferometric fiber-optic sensors.

TABLE 2.1
Fiber-Optic Sensors for Large-Scale Sensing Applications

	Sensor Type	Measurands	Performance Metrics	Modulation Method	Intrinsic/ Extrinsic	Notes
Point sensors	Microbend sensors	Microdisplacement, pressure, temperature, acceleration, flow, local strain, speed	Displacement sensitivity 10^{-10} m/$\sqrt{\text{Hz}}$; temperature sensitivity 10^{-4}/°C; absolute accuracy ±0.05% at T = const; insertion loss 20 dB [17].	Intensity	Intrinsic	Limited multiplexing capacity, temp. compensation, self-referencing required.
	Macrobend sensors	Microdisplacement, high temperature, refractive index (RI), flow	Displacement resolution 50 nm [22,23], refractive index resolution 5×10^{-5} RIU [25], temperature resolution better than 0.5°C [24].	Intensity	Intrinsic	Limited multiplexing capacity, temp. compensation, self-referencing required.
	Interferometric:					
	Fabry–Pérot	Temperature, strain, pressure, displacement, RI, etc.	Dynamic strain resolution picostrain/√Hz (100 HZ–100 kHz) in a remote application (5 km) [30].	Intensity/phase	Intrinsic or extrinsic	High sensitivity but complex interrogation, limited multiplexing capacity, WDM is difficult.
	Mach–Zehnder	Temperature, strain, pressure, displacement, RI, etc.	Resolution ~0.01% gauge length; temperature sensitivities in the order of ~100 pm/°C; RI sensitivities ~250 pm/RIU have been reported [62].	Intensity/phase	Intrinsic or extrinsic	Compactness and high efficiency combined with ease of fabrication, but complex interrogation.
	Michelson	Temperature, strain, pressure, displacement, RI, etc.	Resolution ~0.01% gauge length; temperature sensitivities in the order of ~100 pm/°C; RI sensitivities ~250 pm/RIU have been reported [62].	Intensity/phase	Intrinsic or extrinsic	High accuracy, compactness, good multiplexing capability, complex interrogation.

(Continued)

TABLE 2.1 (*CONTINUED*)
Fiber-Optic Sensors for Large-Scale Sensing Applications

	Sensor Type	Measurands	Performance Metrics	Modulation Method	Intrinsic/ Extrinsic	Notes
Quasi-distributed	Interferometric:					
	Sagnac	Optical gyroscopes, strain, pressure, twist	For gyros bias of 1°/h, noise levels <0.001°/h [26].	Intensity/phase	Intrinsic	Simple and easy to fabricate but suffer from the significant temperature strain cross-sensitivity.
	Fiber Bragg gratings	Temperature, strain, pressure, can be configured to measure displacement, acceleration, etc.	Strain sensitivity (Bragg wavelength shift) typically of the order of 1 pm/με and temperature sensitivity ~10 pm/°C.	Wavelength	Intrinsic	Excellent WDM capability, advantages of wavelength encoding but disadvantages of cross-sensitivity between strain and temperature.
Distributed	Rayleigh scattering	Temperature, strain	Strain resolution 1 με, spatial resolution 10 mm [61].	Intensity (OTDR)	Intrinsic	High strain/temperature resolution and good spatial resolution, but limited length; temperature compensation required
	Raman scattering	Temperature	Typical temperature resolution ~0.1°C, spatial resolution of 1 m over a measurement range up to 8 km [60].	Intensity (OTDR)	Intrinsic	Advantageous for temperature sensing only, since no particular packaging needed to make sensing fiber strain free.
	Brillouin scattering	Strain/temperature	Strain resolution 2 με (stimulated), 30 με (spontaneous), spatial resolution 0.5–5 m (stimulated), 1 m (spontaneous) [61].	Intensity (BOTDR)	Intrinsic	Advantage of relatively stronger signal but synchronization of two laser sources at the opposite ends of the fiber is required.

2.4.1 Fiber Bend Sensors

Effects associated with bending in multimode and single-mode fibers (SMFs) found their application in two early types of intensity-modulated point sensors: *microbend* and *macrobend* fiber-optic sensors. Prefixes *micro* and *macro* in this context refer to the scale of the fiber bends rather than to the scale of the volume within which the parameter of interest is measured. While there are certain similarities between the micro- and macrobending in fibers, there are also significant differences between both concepts.

Microbending occurs in a multimode fiber and is the mechanical perturbation that causes a redistribution of light power among the many modes in the fiber. The more severe the mechanical perturbation or bending, the more light is coupled to radiation modes and is lost, so that the light intensity decreases with mechanical bending [17]. Fields et al. [18] suggested that the microbending effect could be enhanced by squeezing the fiber between a set of corrugated plates called deformer plates or tooth blocks as shown in Figure 2.3. They found that by tuning the spatial frequency of the teeth, the microbending loss could be increased by orders of magnitude. The step-index fiber exhibits a threshold response and the graded-index fiber a resonant response. For a graded-index fiber, it was shown that in order to achieve maximum microbend sensitivity, the spatial period of the deformer must match the expression

$$\Lambda_c = \frac{2\pi a}{(2\Delta)^{1/2}} \tag{2.1}$$

where

a is the core radius

Δ is the refractive index (RI) difference

Typically, the spatial period is within the range of 0.2–2 mm. For example, for a multimode fiber with a core diameter of 62.5 μm and $\Delta \sim 0.014$, the maximum sensitivity period Λ_c is circa 1.2 mm. The amplitude of the fiber bending is less than 5 μm, and the length of the sensor is circa 10 mm.

Although not as sensitive as interferometric sensors, microbenders have good sensitivity to small displacements in the order 10^{-10} m/$\sqrt{\text{Hz}}$ and a very large dynamic range, which is one of the major advantages of these sensors. Microbenders are most useful as contact sensors for the measurement of small displacements in mechanical systems. Over the years, microbend sensors have been configured for the measurement of many different parameters including pressure, temperature, acceleration, flow, local strain, and speed. Microbend sensors arrays have been used in tactile sensing systems and in distributed sensing systems for temperature, strain,

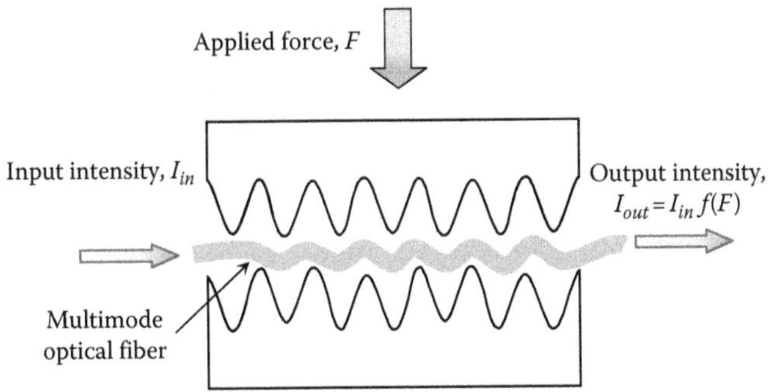

FIGURE 2.3 Schematic of a microbend fiber sensor.

structural monitoring, and water detection. In some of these areas, prototypes have been built and tested, and in a few of the areas, commercial products have been offered [17].

One disadvantage of microbend sensors common with other intensity sensors is sensitivity to optical power level, which may be overcome with using self-referencing methods, for example, a ratiometric scheme [19]. In addition, these sensors suffer from modal sensitivity and high insertion loss (typically ~20 dB). The latter disadvantage limits the number of microbend sensors, which can be multiplexed onto a single fiber to only two or three high-performing units.

Macrobend sensors are another type of intensity-modulated, intrinsic fiber-optic sensors. The operating principle of the macrobend sensors shares certain similarities with microbending; however, there are also significant differences between both concepts. In macrobend sensors, an SMF is usually used and it is bent at relatively large diameters (typically, bend radius is in the order of few centimeters). Power loss in the core of an SMF due to a uniform bend consists of two components: a pure bend loss and a transition loss [20,21]. The pure bend loss results from the loss of guidance at the outer portion of the evanescent field of the fundamental mode. This loss of guidance is due to the phase velocity of the outer part of the evanescent field becoming equal to the speed of light in the cladding (Figure 2.4). The smaller the radius of the bend, the greater the fraction of the evanescent field affected and hence the greater the percentage of light lost at the bend. The transition loss arises from the coupling of light from the fundamental mode to leaky core modes whenever there is a change in curvature of the fiber axis, for example, from a straight to curved or vice versa (Figure 2.4).

Both mechanisms may occur in an SMF simultaneously. It should be noted that not all the light leaving the fundamental mode by the aforementioned methods is permanently lost, as some can reenter by coupling from the leaky core modes and cladding modes to the fundamental mode. Regular telecommunication fiber is usually not enough sensitive to macrobending; therefore, it is usually necessary to use specially bend sensitive SMFs [22]. Macrobend sensors can be used in applications similar to microbend sensors, especially for the measurement of small displacements [22,23], measurement of wide range of temperatures [24], RI [25], airflow, and other measurands. Similarly to the microbend type, the main limitation of the macrobend sensors is their high insertion loss, which limits their multiplexing capability.

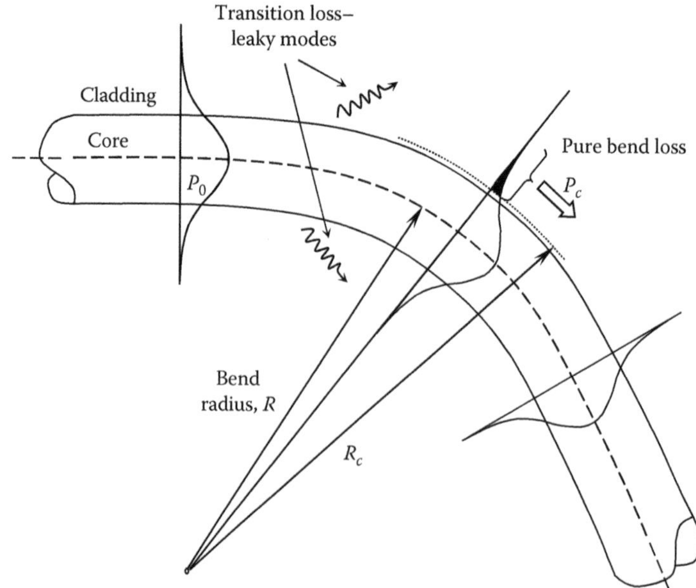

FIGURE 2.4 Schematic illustration of bend loss mechanisms.

2.5 INTERFEROMETRIC FIBER-OPTIC SENSORS

Interferometric fiber-optic sensors are the most sensitive and as a result the most commonly used point sensors. The interferometric technique is based on detecting the optical phase change induced as light propagates along the fiber. Usually, the input light is equally divided between two paths, one of which serves as a reference, while the other one is influenced by the measurand of interest. The two fiber paths recombine at the output of the sensor forming an interference pattern, which is directly related to the optical phase difference (OPD) between the two paths and can be related to the value of the measurand. Relating this phase difference to variations in the fiber environment is relatively straightforward, and typically, phase changes of around 100 rad/m/°C temperature change, 10 μrad/m/με of longitudinal strain, and 10 μrad/m/bar of pressure change are obtained [26]. There are four main types of fiber-optic interferometers: Fabry–Pérot, Mach–Zehnder, Michelson, and Sagnac. In the following sections, we briefly consider the types of interferometric sensors, which are most often used for large-scale sensing applications.

2.5.1 In-Line Fabry–Pérot Interferometers

A Fabry–Pérot interferometer (FPI) is generally composed of two parallel reflectors separated by a certain distance L (cavity length). Mirrors or interfaces between two dielectrics can be used as reflectors. Interference occurs due to the multiple superpositions of both reflected and transmitted beams at the two parallel reflectors. As a result, the reflected and transmitted spectra of such an interferometer are functions of cavity length, index of refraction of the medium, and reflectivity of the mirrors. The phase difference of an FPI is given as

$$\delta_{FPI} = \frac{2\pi}{\lambda} n2L \tag{2.2}$$

where

λ is the wavelength of light
n is the RI of the cavity
L is the cavity length

External perturbation induces a change in the OPD of the interferometer, resulting in a phase variation. For example, applying longitudinal strain to the sensor changes the physical length of the cavity and/or the RI of its material. By measuring the shift of the FPI spectrum, the strain applied to it can be qualitatively determined. The free spectral range (FSR) of the interferometer also depends on the variation of the optical path difference: a shorter OPD results in a larger FSR. Despite the fact that a larger FSR gives a wide dynamic range to the sensor, at the same time, it leads to a degraded resolution due to less sharp spectral peaks.

FPI can be extrinsic or intrinsic, depending on whether the two reflectors are separated by an air gap (or by some material other than fiber) or by an SMF (Figure 2.5).

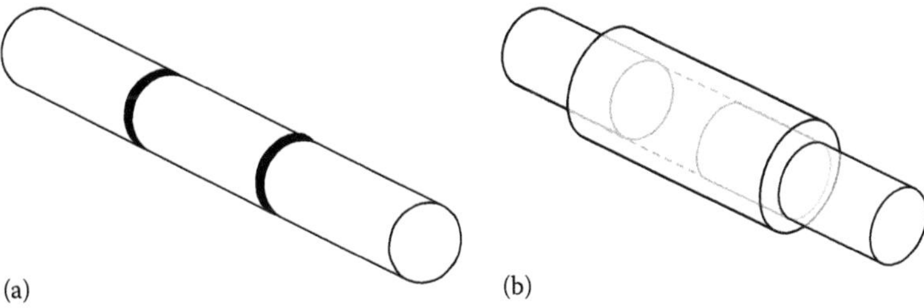

FIGURE 2.5 Basic configurations of fiber FPIs: (a) intrinsic sensor with reflectors formed within the fiber and (b) extrinsic FPI made by forming an air cavity between reflecting fiber surfaces using a supporting structure.

Intrinsic FPIs can be formed by various methods, such as micromachining, fiber Bragg gratings (FBGs), chemical etching, and thin film deposition, and, in comparison with the extrinsic FPIs, suffer from high fabrication cost [27]. Extrinsic FPIs can utilize fiber surfaces as the reflectors or alternatively deposited high-reflectivity mirrors, resulting in a high finesse interference response. The fabrication of extrinsic FPIs is typically simpler and does not involve any high-cost equipment. The main disadvantages of the extrinsic FPIs are low coupling efficiency, the need for careful alignment, and problems with packaging [27]. Overall, FPIs can provide high sensitivity, large dynamic range, and fast response to a number of measurands including temperature, strain, pressure, and displacement. Utilizing a series of in-line point FPI sensors in a fiber network or loop offers the possibility of quasi-distributed measurement. However, such sensors also have some common disadvantages. One of the disadvantages is costly signal interrogation, resulting from generally weak and inherently ambiguous periodic cosine spectral response [28]. The second common disadvantage is the difficulty in using the sensors in wavelength division multiplexed systems.

Since the demonstration of FPIs fabricated by splicing a section of a hollow-core optical fiber between two SMFs for use in monitoring concrete and composites by Liu et al. [29], the FP technology has become probably the most widespread among all the point sensing technologies in health monitoring for strain and temperature analysis. Several FPIs have been developed for remote sensing applications [30–32] with point FPIs measuring dynamic strain located at distances of 50–100 km.

2.5.2 Mach–Zehnder Interferometers

Mach–Zehnder interferometric (MZI) configurations were investigated extensively for early optical fiber sensors, particularly in the context of hydrophone applications [26]. The early fiber MZIs consisted of two independent arms: a signal arm and a reference arm, which was kept isolated from the external perturbation and only the signal arm was exposed to it. Two standard fiber couplers were typically used to split the input light into the two arms and then recombine them at the detector (Figure 2.6). The OPD variation in the sensing arm induced by the external perturbation (temperature, strain, etc.) then could be easily detected by analyzing the changes in the interference signal at the detector.

More recently, a number of in-line MZI configurations have been proposed, where instead of two separate fibers the two separate light paths exist within the same fiber section, for example, a single-mode–multimode–single-mode structure (SMS) [33], a structure created by two airhole-collapsed regions in a section of photonic crystal fiber (PCF) or a PCF spliced with SMFs [34,35], a structure with two in-line long-period gratings (LPGs) [36], a core-mismatched section between two SMFs, a small-core SMF section spliced with standard SMFs [37], and two in-line fiber tapers [38]. In all these configurations, a portion of the core mode of an SMF is uncoupled from the fiber core by a single-mode–multimode interface, single-mode airhole-collapsed region in an LPG, or some other element and then recoupled to the core mode by the second similar interface or element leading to

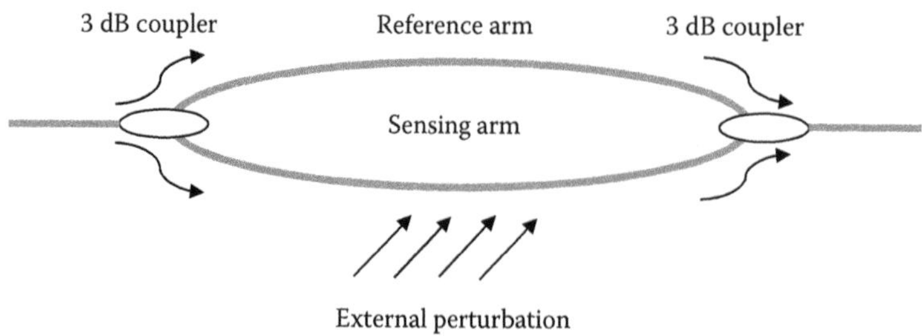

FIGURE 2.6 Basic configuration of a two-path fiber MZI.

the interference between the two beams due to the modal dispersion. Such in-line interferometers offer the advantages of compactness and high efficiency combined with ease of fabrication.

MZIs can be configured as multipoint arrays for interrogation in a TDM scheme as a flexible passive sensor network with remote capability [26].

2.5.3 White-Light Michelson Interferometers

Fiber-based Michelson interferometer (MI) configuration is very similar to the Mach–Zehnder configuration and represents a half of an MZI. MI operation is based on the interference between the beams in two arms with each of the beams reflected from the mirrors placed at the end of each arm. Such a reflection-type configuration has the advantages of compactness and ease of installation but is more vulnerable to reflection-induced instabilities of the source and to stray reflections within the interference path. Employing a short coherence light source in the MI configuration illustrated in Figure 2.7 allows for elimination of the phase ambiguity inherent to interferometric measurement and the problems associated with reflections.

In a white-light interferometer, the fringes of the interferogram are narrowly located in a zero path length difference region [39]. Thus, the phase difference can be determined without ambiguity by measuring the envelope peak of the interferogram. White-light interferometer technology was patented by Belleville and Duplain [40] and commercialized by FISO Technologies (Canada).

Typically, in a white-light MI, of the two arms one is used as a measurement arm and the other as the reference arm. The length of the reference arm is mechanically adjusted to match the signal arm (e.g., using a movable mirror or a piezoelectric [PZT] controller) within a tolerance determined by the source coherence length. Once the two arms of the interferometer are balanced and the OPD in the measurement arm is compensated, the desired measurement can be achieved remotely with high accuracy. Another advantage of the MI configuration is the multiplexing capability with parallel connection of several sensors. Given these advantages, MIs have found extensive applications in civil engineering as extensometers [28] and in highly multiplexed arrays of fiber-optic hydrophones [41].

2.5.4 Interferometric Multiplexing

Being point-type sensors, optical fiber interferometers can only be utilized for sensing of large-scale and/or remote structures or objects if an efficient multiplexing of large sensor arrays can be realized. The need for implementation of such large-scale arrays, particularly for underwater acoustic sensing, has led to the development of many multiplexing approaches and configurations.

In addition to the TDM and WDM approaches mentioned earlier, several other schemes have been implemented, such as frequency division multiplexing (FDM), spatial division multiplexing (SDM), coherence-domain multiplexing (CDM), and various combinations of these techniques. Cranch et al. have designed an array with 96 in-line MI hydrophones based on a dense WDM (DWDM)/TDM architecture [16]. A comprehensive overview of the multiplexing schemes is given by Grattan and Sun in [28].

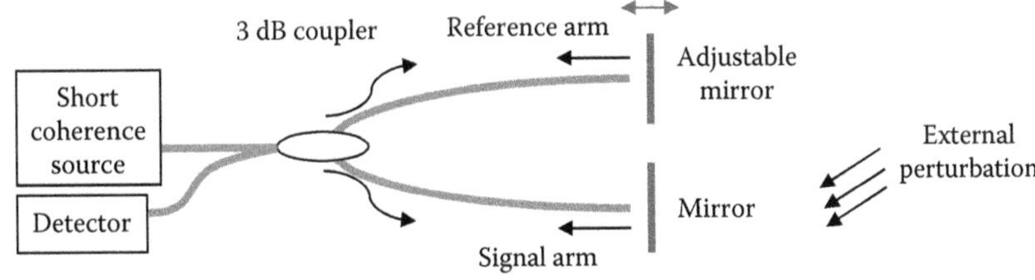

FIGURE 2.7 Basic configuration of a white-light fiber MI.

2.5.5 Sagnac Interferometers

Fiber-optic Sagnac interferometers (SIs) were developed in the 1970s based on the well-known principle of Sagnac interference. In a fiber-optic SI, input light is split into two parts propagating in the opposite directions by a 3 dB fiber coupler and these two counterpropagating beams are combined again at the same coupler as shown in Figure 2.8 [41]. The fabrication of such an interferometer can be simply achieved by connecting the ends of a conventional 3 dB coupler. Highly birefringent fibers or polarization-maintaining fibers (PMFs) are typically utilized as sensing fibers since such fibers maximize the polarization dependence of the signal within the SIs. In order to control the input light polarization, a polarization controller is connected to the sensing fiber.

The signal at the output port of the fiber coupler is a result of interference between the beams polarized along the slow axis and the fast axis. The phase of the interference is given as

$$\delta_{SL} = \frac{2\pi}{\lambda} BL \tag{2.3}$$

where

$B = |n_f - n_s|$ is the birefringent coefficient of the sensing fiber
L is the length of the fiber
n_f and n_s are the effective indices of the fast and slow modes

This class of fiber sensors is one of the most commercially successful to date, mostly for applications as fiber-optic gyroscopes (if the Sagnac loop is rotating, the light propagating in the direction of rotation will arrive to the output slightly ahead of the light propagating in the opposite direction). Such fiber-optic gyroscopes have become a very competitive technology with a bias of 1 degree per hour and better and noise levels of less than 0.001 degrees per hour [26]. Fiber-optic gyroscopes have become essential components in platform stabilizing systems, such as large satellite antennas, subsea navigation, aircraft stabilization and navigation, and many others. Fiber-optic SIs can also be used for measuring various other parameters such as strain, temperature, pressure, and twist, which makes them promising candidates for composite material sensing applications. Embedded in composite materials, fiber-optic-based SIs provide the value of the sensed parameter averaged over the length of the sensor. It should be noted, however, that in such applications, SIs, while offering the advantages of simplicity and ease of fabrication, suffer from the significant temperature–strain cross-sensitivity [41]. Rajan et al. reported that the temperature–strain cross-sensitivity issue can be eliminated by employing polarization-maintaining PCFs [42]. In 1992, Kurmer et al. [43] developed a distributed fiber-optic acoustic sensor based on SI for leak detection in pipelines. Later based on

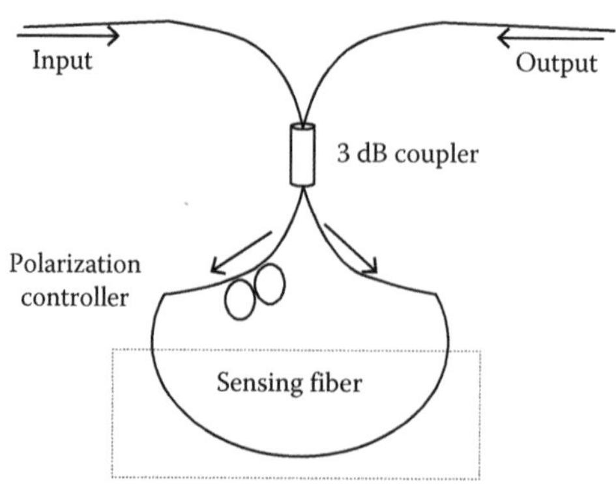

FIGURE 2.8 Basic configuration of a fiber-optic SI.

this technique, a distributed fiber-optic sensor was developed to detect the occurrence of leak incidents. Experimentally, longitudinal position of a disturbance can be determined with an average error of 59 m over a total length of 25 km [44].

2.6 FIBER GRATINGS

Originated from the discovery of photosensitivity of germanium-doped silica, FBG sensing technology has attracted widespread attention and has been the subject of continuous and rapid development since FBGs were used for the first time for sensing purposes in 1989. An elementary FBG comprises a short section of an SMF in which the core RI is modulated periodically using an intense optical interference pattern, typically at UV wavelengths. The change in the core RI is between 10^{-5} and 10^{-3}, and the length of a Bragg grating is usually around 10 mm. A schematic of an FBG is shown in Figure 2.9. The periodic-index-modulated structure enables the light to be coupled from the forward propagating core mode into a backward propagating core mode generating a reflection response. The light reflected by the periodic variations of the RI has a central wavelength λ_G given as

$$\lambda_G = 2n_{eff}\Lambda \tag{2.4}$$

where

n_{eff} is the effective RI of the core
Λ is the periodicity of the RI modulation

The basic principle of operation of any FBG-based sensor based on that any local strain or temperature changes alters the effective index of core refraction and the grating period, followed by the changes in wavelength of the reflected light. The corresponding Bragg wavelength shifts are typically of the order of 1 pm/με and 10 pm/°C. Monitoring of the changes in the reflected wavelength can be achieved by employing edge filters, tunable narrowband filters, tunable lasers, or CCD spectrometers [6,45]. In comparison with other fiber-optic sensors, FBGs have a unique advantage in that they encode the wavelength, which is an absolute parameter and which does not suffer from disturbances in the light path and optical source power fluctuations. This wavelength-encoding nature of the FBGs facilitates WDM of multiple sensors: FBGs sensors could be particularly useful when gratings with different periods are arranged along an optical fiber. Each of the reflected signals will have a unique wavelength and can be easily monitored, thus achieving multiplexing of the outputs of multiple sensors using a single fiber. Thus, strain, temperature, or pressure can be measured at many locations, making FBGs a kind of typical *quasi-distributed sensors*. FBG sensors are preferred in many civil engineering applications and have been successfully employed in many full-scale structures requiring multiple-point sensing distributed over a long range.

Most of the research on FBG sensors has focused on the use of these devices to provide quasi-distributed point sensing of strain or temperature [46]. FBGs have been also used as point sensors

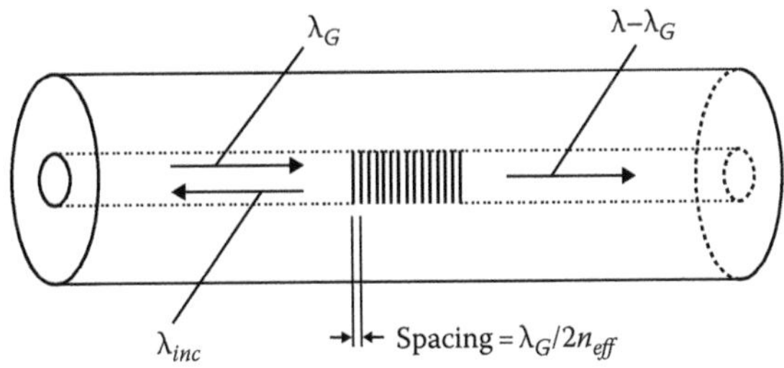

FIGURE 2.9 Schematic of an FBG.

to measure pressure [47,48] and acoustic emissions levels [49–51] and to monitor the curing process of composites [52,53]. One disadvantage of the FBG technology in real engineering applications is associated with their cross-sensitivity effect, which causes simultaneous sensitivity of Bragg wavelength shift to both strain and temperature. Thereby, some methods to discriminate both measurements are usually needed [8,54].

2.7 DISTRIBUTED FIBER SENSORS

Distributed fiber-optic sensors are capable of providing a continuous measurand profile over the length of the optical fiber and thus are most suitable for large structural applications such as bridges, buildings, and pipelines [55–57]. There are three main physical effects utilized for distributed sensing: Rayleigh scattering, Raman scattering, and Brillouin scattering.

Rayleigh scattering effect can be used for both strain and temperature monitoring. It is based on the shifts in the local Rayleigh backscatter pattern that is dependent on the strain and the temperature. Thus, the strain measurements must be compensated for temperature. The main characteristics of this system are high resolution of measured parameters and good spatial resolution (typically a few meters), but the maximal length of sensor is limited to relatively short distances (70 m [58]). This is due to the low-level signal arising as a result of low nonlinear coefficients of the silica, which constitutes the fiber, and thus this gives quite a long system response time, resulting from the necessity to integrate small signals received over many pulses. Thus, this system is suitable for monitoring localized strain changes.

Raman scattering is the result of a nonlinear interaction between the light traveling in silica fiber. When an intense light signal is launched into the fiber, two frequency-shifted components called, respectively, Raman Stokes and Raman anti-Stokes will appear in the backscattered spectrum. The relative intensity of these two components depends only on the local temperature of the fiber. If the light signal is pulsed and the backscattered intensity is recorded as a function of the round-trip time, it becomes possible to obtain a temperature profile along the fiber [59]. The insensitiveness of this parameter to strain is an advantage compared with Rayleigh- and Brillouin-based temperature monitoring, since no particular packaging of the sensor must be made to make sensing fiber strain free. Systems based on Raman scattering are commercialized by SMARTEC (Switzerland), Sensornet, and Sensa (United Kingdom) [60]. Typically, a temperature resolution of the order of 0.1°C and a spatial resolution of 1 m over a measurement range up to 8 km are obtained for multimode fibers. Since the leakage of pipelines, dykes, dams, etc., often changes the thermal properties of surrounding soil, besides the temperature monitoring, the Raman-based systems are used for leakage monitoring in large structures.

Brillouin scattering is the result of the interaction between optical and sound waves in optical fibers. Thermally excited acoustic waves produce a periodic modulation of the RI. Brillouin scattering occurs when light propagating in the fiber is diffracted backward by this moving grating, giving rise to a frequency-shifted component by a phenomenon similar to the Doppler shift. This process is called spontaneous Brillouin scattering. The main challenge in using spontaneous Brillouin scattering for sensing applications resides in the extremely low level of the detected signal. This requires sophisticated signal processing and relatively long integration times [60].

Acoustic waves can also be generated by injecting in the fiber two counterpropagating waves with a frequency difference equal to the Brillouin shift. Through electrostriction, these two waves will give rise to a traveling acoustic wave that reinforces the phonon population. This process is called stimulated Brillouin amplification. Systems based on the stimulated Brillouin amplification have the advantage of working with a relatively stronger signal but face another challenge. To produce a meaningful signal, the two counterpropagating waves must maintain an extremely stable frequency difference. This usually requires the synchronization of two laser sources that must inject the two signals at the opposite ends of the fiber under test [61].

Brillouin scattering sensor systems able to measure strain or temperature variations of fibers with length up to 50 km with spatial resolution down in the meter range [60]. For temperature measurements, the Brillouin sensor is a strong competitor to systems based on Raman scattering, while for strain measurements it has practically no rivals.

DFSs are often categorized based on the interrogation method and the physical effect underpinning the operating principle: (1) optical time domain reflectometry (OTDR) and optical frequency domain reflectometry (OFDR), both based on Rayleigh scattering; (2) Raman OTDR (ROTDR) and Raman OFDR (ROFDR), both based on Raman scattering; and (3) Brillouin OTDR (BOTDR) and Brillouin OFDR (BOFDR), both based on Brillouin scattering [58].

2.8 MICRO- AND NANOSCALE FIBER SENSORS

In the last two decades, advances in microfabrication and a number of innovative manufacturing methods, such as laser drilling, micromilling, and advanced electrical discharge machining, led to the production of a large number of microscale components at economical rates. Some examples include microelectromechanical systems (MEMSs), micromachines, microfluidics, lab-on-a-chip devices and circuits, optical MEMS, miniature flat-panel displays, and solar cells. These components are now integrated into many of everyday consumer products that span a wide range of industries. However, the development of metrology tools for inspecting these micro- and nanoscale features that can function at these scales is not always easy. For example, the main difficulty of contact-based tools is that a tactile sensor probe should normally be at least 10 times smaller than the measured feature to adequately measure the feature's form. Noncontact optical metrology has the ability to rapidly measure parts with high throughput as well as avoiding touching the part, which may lead to surface damage. Conventional noncontact optical methods have many drawbacks including sensitivity to ambient conditions such as lighting, the need to thoroughly clean the part to remove any dirt or containments, and the need for alignment between the part and optics [63].

Many powerful industries such as automotive, biomedical, and defense drive the increasing demand for highly sensitive, selective, nonintrusive, fast-response, compact and robust sensors that can perform in situ micro- and nanoscale measurements of current, displacement, bend, surface, acceleration, force, rotation, acoustic, electric field, and magnetic field in many cases at remote and harsh environments [64]. In the field of bioscience and biomedicine great advances in further research depend on the provision by nanosensors and nanoprobes of critical information for monitoring biomolecular processes within a single living cell. Clinical applications often require sensing in locations that are inaccessible to conventional measurement tools, such as inside the living human brain, next to a developing embryo, or within an artery [65]. Applications related to detection of volatile organic compounds cover several areas such as environmental monitoring, chemical industry, safety at work, or food industry [66].

Conventional macroscale optical fiber sensors described in the previous sections exhibit relatively small dimensions; they are light weight, inert, and highly sensitive to a wide range of physical parameters and therefore could potentially provide attractive sensing solutions in submicron environments. However, the dimensions of the transducer for a conventional fiber sensor are determined by the dimensions of the standard optical fiber, the smallest of which, the fiber diameter, is typically in the order of hundreds microns.

Fabrication of nanoscale sensors requires techniques capable of making reproducible optical fibers with submicron-size-diameter cores or, alternatively, modification of the existing standard fibers by fabricating nanostructures at the end or side of the fibers.

Submicron-size fiber probes are typically manufactured by heating and stretching of regular-sized optical fibers [67]. The result is a biconical taper that provides a smooth lossless connection to other fiberized components. By controlling the pulling rate during the fabrication process, the taper profile can be fine-tuned to suit the application [68,69]. Fabricated using this technique, micro- and nanofibers (MNFs) exhibit remarkable optical and mechanical properties, including large evanescent fields, strong optical confinement, flexibility, configurability, and robustness. Such desirable characteristics that have gathered much attention in recent years make MNFs an excellent platform for optical sensors [64]. Other techniques of standard fibers' modification include fabricating nanostructures at the end or side of the fibers using cladding polishing techniques, femtosecond laser processing [70], metal film deposition, focused ion beam (FIB) milling of nanoapertures in the metallic films deposited on tapered tips of optical fibers [71], and many others. PCFs with their microsize periodicity in the order of a wavelength also offer possibilities for fabrication of a sensor interacting with a local environment on at the microscale [72].

In addition to the challenges associated with fabrication of the micro- and nanosized fiber-optic transducers, specific sensing applications may require additional design and fabrication steps. For example, in biosensing, it is necessary to detect biological species selectively. This means that an ideal biosensor not only has to respond to low concentrations of analytes but also must have the ability to discriminate among different species. To satisfy this requirement, surfaces of fiber probes for biosensing are often functionalized by immobilizing certain recognition molecules to enhance their selectivity. Examples of biological recognition molecules include enzymes, antibodies, and oligonucleotides. Biosensors have wide applications, including biomarker detection for medical diagnostics and pathogen and toxin detection in food and water. Other challenges may include specific mechanism of sample/measurand delivery, additional means for mechanical stability, packaging, and connectivity.

Depending on the fabrication approach and physical mechanism underpinning the operation, micro- and nanoscale fiber sensors can be broadly categorized as *evanescent field sensors, micro- and nanostructured sensors, and surface plasmon resonance (SPR) sensors.* In the following sections, we briefly discuss the key concepts of such fiber sensors and some examples of their applications.

2.9 EVANESCENT FIELD FIBER SENSORS

Evanescent field fiber sensors perform the sensing function along the fiber's length by utilizing the interaction between the evanescent waves and the local environment surrounding the fiber surface. In conventional single- or multimode fibers, light propagates by means of total internal reflection (TIR). However, for light reflecting at angles near the critical angle, a significant part of the electromagnetic field extends into the cladding that surrounds the core (Figure 2.10a). This phenomenon is referred to as evanescent field. The power within this evanescent field

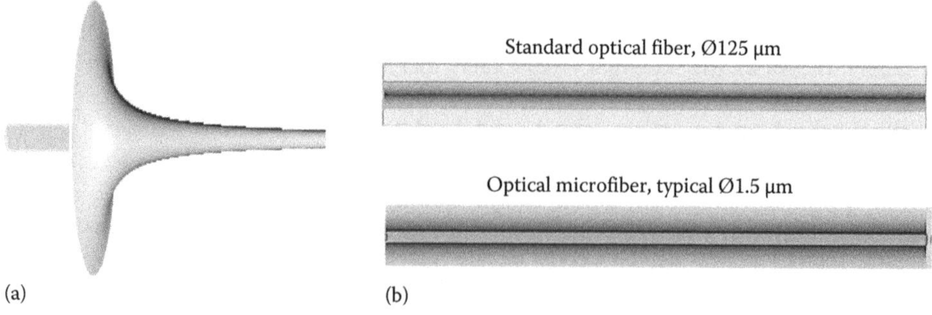

FIGURE 2.10 (a) Schematic illustration of the evanescent field surrounding the fiber core and its sharp decrease with the distance from the fiber surface and (b) comparison of the evanescent fields in the standard fiber (top) and microsized fiber (bottom).

decays exponentially with distance from the core surface. The penetration depth of the evanescent wave, d_p, is the distance at which the strength of the field has decayed to $1/e$ [73]:

$$d_p = \frac{\lambda}{2\pi\sqrt{n_{co}^2 \sin^2\theta - n_{cl}^2}} \tag{2.5}$$

where
- λ is the wavelength of the light source
- θ is the angle of incidence of the light at the core/cladding interface
- n_{co} and n_{cl} are the refractive indices of the core and cladding

In a uniform-diameter fiber, the evanescent field decays to almost zero within the cladding. Thus, light propagating in standard uniform-diameter cladded fibers cannot interact with the fiber's surroundings (Figure 2.10b). While optical fibers were originally intended for light propagation with minimal loss, for sensing, it is critical that the light interacts with the fiber's surroundings. One way to achieve this interaction is to expose the evanescent field of the transmitted light. For example, if the cladding of a fiber is reduced or removed, the evanescent field penetrates into the adjacent to the fiber core medium. The earliest studies of evanescent field sensors involved fiber structures where the cladding was removed uniformly over a distance of the fiber and exposed to the surrounding medium or replaced by the analyte. The measurand is detected through the absorption of the evanescent wave, which creates leakages and modulates the light intensity. Unfortunately in this simple unclad configuration, the penetration depth of the evanescent field is typically far too small for efficient sensing.

To overcome this problem, a number of techniques to increase the interaction with the evanescent field have been proposed to date. Some of them are as follows: the use of special fibers, such as D-fibers or microstructured fibers [74], fiber bending [75,76], altering the light launching angle [77], increasing the wavelength [78], and fiber tapering [79,80]. The latter technique—fiber tapering—has proved to be the most versatile and probably the most popular to date, in both laboratory and industrial settings. Evanescent sensors based on tapered MNFs come in many different forms: tapered fiber tips [71], straight biconical tapers [81], tapers with functionalized surfaces [82], tapers with inscribed FBGs [83], taper-based interferometric structures [38], periodically tapered structures [84], and microfiber coupler structures [85]. Comprehensive reviews of MNF-based sensors, including MNF microresonator structures and their applications, are given in [64,86].

Changes of sensor response arising from the evanescent field are very versatile because they can be detected alone, amplified, or detected in conjunction with various optical methods [87]. The mechanisms used in evanescent sensing depend on the analytes and the application of the sensor. They include changes in output power due to RI changes alone (intensity-based evanescent sensors), evanescent field absorption (absorption-based sensors), fluorescence (fluorescence-based sensors), SPR-based sensors, and whispering-gallery modes (WGMs) (WGM microfiber-resonator-based sensors).

Overall, sensors based on evanescent wave excitation have the advantage of surface-specific detection. The limited penetration depth of the evanescent wave allows spatial separation of surface bound molecules from interfering species present in the bulk solution.

2.10 MICRO- AND NANOSTRUCTURED FIBER SENSORS

Micro- or nanostructured optical fibers cover a wide range of optical fibers, including PCFs and some standard fibers whose structure is modified by fabricating nanofeatures at the end or side of the fibers.

The PCF has a periodic structure whose periodicity is on the order of a wavelength. Depending on whether the effective RI of the cladding is lower or higher than the core RI, PCFs guide light either by the modified TIR (m-TIR) or by the photonic bandgap (PBG) mechanism. Hollow core fibers guide light by the PBG as the effective cladding index of the air–silica cladding is higher than the core RI (air). For the solid silica core fibers, the guidance mechanism is m-TIR owing to the higher core RI of the silica core when compared to the air–silica cladding.

The biggest attraction in PCFs is that by varying the size and location of the cladding holes and/or the core, the fiber transmission spectrum, mode shape, nonlinearity, dispersion, air-filling fraction, and birefringence can be tuned to reach values that are not achievable with conventional fibers. Additionally, the existence of airholes gives the possibility of light propagation in air or alternatively provides the ability to infiltrate liquids or gases into the airholes. This enables a controlled interaction between light and sample leading to new sensing mechanisms [87].

For example, hollow-core PCF has been used for sensing purposes as the cladding channels can be filled with gas or liquid, thus serving as an efficient type of evanescent wave sensor [88]. An important feature of the PCF sensor is that it requires a small sample volume, owing to the long interaction length and the high confinement of the guided mode. However, there are issues to be solved, in that the selective filling of the measurand is somewhat difficult and a preparation time is required.

A humidity sensor proposed in [35] utilizes solid-core PCF in an MI configuration, and its operating principle is based on the excitation and recombination of modes occurring in the region of the PCF in which the voids of the PCF are collapsed. The interference spectrum of the device shifts owing to the adsorption and desorption of water molecules at the air–glass interface within the PCF holes, which is a function of the ambient humidity level.

PCF sensors with tapered or chirped structures have also been widely studied. The presence of gas molecules in the tapered region gives rise to a change in the surface RI, leading to a shift in the interference pattern. An advantage of the proposed device is that there is no need to fill the airholes in the PCFs with the gas measurand, which usually requires lengthy preparation time.

For a compact fiber refractometer based on a PCF half taper probe presented in [89], both the conical taper transition region and the fiber tip and modal propagation (and thus multimode interference) depend on the external RI changes; therefore, the PCF tip can be potentially used as a highly sensitive RI sensor with a high spatial resolution. Other promising PCF-based sensor configurations include chirped FBG or LPG-inscribed PCFs and evanescent field sensing in the cladding holes of an air-suspended solid-core PCF.

Developments in the area of femtosecond laser processing techniques have been used to fabricate microfluidic devices on various structures [70]. A microslit-based RI sensor in [90] takes advantage of the fact that the change in the RI in the microslit in a step-index fiber influences the propagation properties through the fiber.

Benefitting from their small sizes and high-fractional evanescent fields, microfiber sensors have shown special advantages (such as high sensitivity for RI measurement and fast response for temperature and humidity sensing). Also, the tight confinement and surface enhancement of probing light guided along a microfiber are beneficial for achieving high sensitivity with extremely low optical power, which is highly desired for many applications.

2.11 SURFACE PLASMON RESONANCE SENSORS

SPR refers to the excitation of surface plasmon polaritons (SPPs), which are electromagnetic waves coupled with free electron density oscillations on the surface between a metal and a dielectric medium (or air). SPPs or surface plasmon waves (SPWs) propagate along the interface of the metal and dielectric material. For nanoscaled metallic structures such as metallic nanoparticles, or nanorods, a light with the appropriate wavelength can be used to excite the localized oscillation of charges, a process that is referred to as localized SPR (LSPR).

When the energy as well as the momentum of both, the incident light and SPW, match, a resonance occurs that results in a sharp dip in the reflected light intensity. The resonance condition depends on the angle of incidence, wavelength of the light beam, and the dielectric functions of both the metal and the dielectric. In an SPR-based fiber-optic sensor, all the guided rays are launched and hence, the angle of incidence is not varied. In this case, the coupling of evanescent field with surface plasmons strongly depends on wavelength, fiber parameters, probe geometry, and the metal layer properties.

The SPR-based fiber-optic sensors have a large number of applications for quantitative detection of chemical and biological species. These include food quality, medical diagnostics, and environmental monitoring. The sensing principle is based on detecting the change in RI of the medium around the metallic coating. Various fundamental SPR fiber sensor configurations have been proposed to date, and comprehensive overviews of such sensors are given in [91,92].

2.12 FUTURE CHALLENGES AND OPPORTUNITIES

Fiber-optic sensors can be reliable tools for a wide range of applications at a variety of physical scales: from the large-scale and remote sensing in SHM of large civil structures to the micro- and nanosized probes reaching locations normally inaccessible by conventional measurement tools, such as inside the living human organism or even within a single living cell.

Although both ends of the physical scale address very different realities, they both benefit from the intrinsic unique advantages of optical fiber sensors. For SHM applications, the dual functionality of the optical fibers, serving both as the sensors and as the telemetry channels, their long-term reliability in all weathering conditions, and their passive dielectric nature make this technology very attractive. For biomedical applications, miniature size and total immunity to electromagnetic interference are probably the most important advantages over other existing technologies.

At present, the most appropriate sensing solutions for specific applications are often selected from different available technologies offering single-point, long-gauge, quasi-distributed, or distributed sensors. One of the future challenges for the optical fiber sensors will possibly include the development of multiscale integration, multifunctionality, and multiparametric sensing systems.

One of the exciting fields where optical fiber sensors are expected to play a significant role is smart structures, smart materials, and intelligent systems. Smart composite materials with embedded optical fiber sensors often require monitoring of physical parameters, such as stress, strain, or temperature both averaged across the entire structure or composite part and at multiple-point locations (e.g., for an early detection of damage at critical design points). In the near future, sophisticated hybrid sensing systems will be developed by combining different sensing approaches (e.g., distributed BOTDA/R and localized FBGs sensors).

There is a growing need in SHM for realization of systems with multifunctional measurement capabilities. Specifically, fiber-optic sensors can perform static strain measurement on a large scale (thousands of μ strains) at a low speed (hundreds of Hz or less) for potential operational load monitoring and ultrafast strain measurement in a small scale (tens of μ strains) at an ultrafast speed (~kHz) for potential damage detection. This would significantly reduce the complexity of SHM systems and the costs related to SHM.

Simultaneous real-time monitoring of various physical parameters of civil structures and objects and efficient interpretation of the data is yet another future challenge, which will require development of multifunctional sensor systems based on novel or hybrid approaches and common means for sensor interrogation.

Biosensors for medical applications will be an important application field of optical fiber sensors. In this context, likely challenges for the fiber-optic sensors on the micro-/nanoscale will include achieving higher sensitivity to allow for single-molecule detection for biochemical optical sensing, <10 ppb detection limit for gas sensing and $<10^{-6}$ for RI sensing; better selectivity, enabling selective detection of target samples by properly functionalizing the microfiber structure; long-term stability against environmental changes (temperature, displacement, vibrations, airflow, etc.); better robustness that is highly desired for long-term use; and better protection or package for practical applications.

There are several barriers to widespread implementation of optical sensor technology. In some application areas, such as process control, there are conventional sensor technologies, which are well developed and often cost effective. High-quality package, and fabrication facilities, and short testing cycle make conventional sensors more easily to be commercialized. The advantage of fiber-optic sensor, such as immunity to electromagnetic interference and high sensitivity, cannot be sufficient to replace these conventional technologies. Cost is another obstacle to adapting fiber-optic sensors. Electro-optic interface, optical components of high accuracy, or components specially used in fiber-optic sensors other than fiber-optic communication are expensive. Fiber-optic sensors are chosen unless conventional sensors are more expensive or cannot function in some harsh environments. However, future development and penetration of fiber-optic sensors in the civil engineering and medical market sectors and mass production of such sensors should make the technology more accessible and cost efficient for an increasing number of applications.

REFERENCES

1. Davies, D. E. N. and Kingsley, S., Method of phase-modulating signals in optical fibres: Application to optical-telemetry systems, *Electron. Lett.*, 10, 21, 1974.
2. Li, H. and Li, D., Recent applications of fiber optic sensors to health monitoring in civil engineering, *Eng. Struct.*, 26, 1647–1657, 2004.
3. Zhou, Z., He, J., and Ou, J., Integrated optical fiber sensing system by combining large-scale distributed BOTDA/R and localized FBGs, *Int. J. Distrib. Sens. Netw.*, 2012, Article ID 804394, 18 pp., 2012
4. Fernandez-Vallejo, M. and Lopez-Amo, M., Optical fiber networks for remote fiber optic sensors, *Sensors*, 12, 3929–3951, 2012.
5. Dandridge, A. and Kirkendall, C., Passive fiber optic sensor networks, in *Optical Fibre Sensing Technology*, J. M. Lopez-Higuera (ed.). Wiley & Son: Berlin, Germany, 2002, pp. 433–448.
6. Kersey, A. D., Davis, M. A., Patrick, H. J., LeBlanc, M., Koo, K. P., Askins, C. G., Putnam, M. A., and Friebele, E. J., Fiber grating sensors, *J. Lightwave Technol.*, 15(8), 1442–1463, 1997.
7. Askins, C. G., Putnam, M. A., Friebele, E. J., Mille, S. M., Liu, K., and Measures, R. M., A passive wavelength demodulation system for guided wave Bragg grating sensors, *IEEE Photon. Technol. Lett.*, 4, 516–518, 1992.
8. Zhao, Y. and Liao, Y., Discrimination methods and demodulation techniques for fiber Bragg grating sensors, *Opt. Lasers Eng.*, 41, 1–18, 2004.
9. Ko, J. M. and Ni, Y. Q., Technology developments in structural health monitoring of large-scale bridges, *Eng. Struct.*, 27, 1715–1725, 2005.
10. Minakuchi, S., Tsukamoto, H., Banshoya, H., and Takeda, N., Hierarchical fiber-optic based sensing system: Impact damage monitoring of large-scale CFRP structures, *Smart Mater. Struct.*, 20, 085029 (9 pp.), 2011.
11. Merhani, E., Ayoub, A., and Ayoub, A., Evaluation of fiber optic sensors for remote health monitoring of bridge structures, *Mater. Struct./Mater. Construct.*, 42, 183–199, 2009.
12. Rao, Y., Ran, Z., and Chen, R., Long-distance fiber Bragg grating system with a high optical signal-to-noise ratio based on a tunable fiber laser ring configuration, *Opt. Lett.*, 31, 2684–2686, 2006.
13. Hara, T., Imamura, F., and Ito, H., Optical hydraulic pressure sensor using frequency shifted-feedback laser for ocean-bottom-tsunami sensing, in *the 3rd International Workshop on Scientific Use of Submarine Cables and Related Technologies*, Tohoku University, Japan, 2003, pp. 45–50.
14. Ferraro, P., and De Natale, G., On the possible use of optical fiber Bragg gratings as strain sensors for geodynamical monitoring, *Opt. Lasers Eng.*, 37, 115–130, 2002.
15. Da Costa Marques Pimentel, R. M., Barbosa, M. C. B., Costa, N. M. S. et al., Hybrid fiber-optic/electrical measurement system for characterization of railway traffic and its effects on a short span bridge, *IEEE Sens. J.*, 8, 1243–1249, 2008.
16. Cranch, G. A., Nash, P. J., and Kirkendall, C. K., Large-scale arrays of fiber-optic interferometric sensors for underwater acoustic applications, *IEEE Sens. J.*, 3, 19–30, 2003.
17. Berthold, J. W., Historical review of microbend fiber-optic sensors, *J. Lightwave Technol.*, 13, 1193–1199, 1995.
18. Fields, J. N., Asawa, C. K., Ramer, O. G., and Barnaski, M. K., Fiber optic pressure sensor, *J. Acoust. Soc. Am.*, 67, 816–818, 1980.
19. Wang, Q., Rajan, G., Farrell, G., Wang, P., Semenova, Y., and Freir, T., Macrobending fiber loss filter, ratiometric wavelength measurement and application, *Meas. Sci. Technol.*, 18, 3082–3088, 2007.

20. Murakami, Y. and Tsuchiya, H., Bending loss of coated single mode optical fibers, *IEEE. J. Quant. Electron.*, QE-14, 495–501, 1978.
21. Gauthier, R. C. and Ross, R. C., Theoretical and experimental considerations for single-mode fiber-optic bend type sensor, *Appl. Opt.*, 36, 6264–6273, 1997.
22. Wang, P., Brambilla, G., Semenova, Y., Wu, Q., and Farrell, G., A simple ultrasensitive displacement sensor based on a high bend loss singlemode fibre and a ratiometric measurement system, *J. Opt.*, 13, 075402, 2011.
23. Wang, P., Semenova, Y., Wu, Q., and Farrell, G., A bend loss–based singlemode fiber microdisplacement sensor, *Microw. Opt. Technol. Lett.*, 52, 2231–2234, 2010.
24. Rajan, G., Semenova, Y., Mathew, J., and Farrell, G., Experimental analysis and demonstration of a low cost temperature sensor for engineering applications, *Sens. Actuat. A-Phys.*, 163, 88–95, 2010.
25. Wang, P., Semenova, Y., Wu, Q., Farrell, G., Ti, Y., and Zheng, J., Macrobending single-mode fiber-based refractometer, *Appl. Opt.*, 48, 6044–6049, 2009.
26. Culshaw, B. and Kersey, A., Fiber-optic sensing: A historical perspective, *J. Lightwave Technol.*, 26(9), 1064–1078, 2008.
27. Lee, B. H., Kim, Y. H., Park, K. S., Eom, J. B., Kim, M. J., Rho, B. S., and Choi, H. Y., Interferometric fiber optic sensors, *Sensors*, 12, 2467–2486, 2012.
28. Grattan, K. T. V. and Sun, T., Fiber optic sensing technology: An overview, *Sens. Actuat.*, 82, 40–61, 2000.
29. Liu, T., Wu, M., Rao, Y., Jackson, D. A., and Fernando, G. F., A multiplexed optical fibre-based extrinsic Fabry-Perot sensor system for in-situ strain monitoring in composites, *Smart Mater. Struct.*, 7, 550–556, 1998.
30. Chow, J. H., Littler, I. C. M., McClelland, D. E., and Gray, M. B., Long distance, high performance remote strain sensing with a fiber Fabry-Perot by radio-frequency laser modulation, *Proc. SPIE*, 6201, 620121-1, 2006.
31. Chow, J. H., Littler, I. C. M., McClelland, D. E., and Gray, M. B., A 100 km ultra-high performance fiber sensing system, *Proceedings of Conference on Lasers and Electro-Optics, CLEO 2007*, Baltimore, MD, May 6–11, 2007. doi:10.1109/CLEO.2007.4452840.
32. Habel, W. and Hofmann, D., Determination of structural parameters concerning load capacity based on fiber Fabry-Perot interferometers, *Proc. SPIE*, 2361, 176–179, 1994.
33. Wang, Q., Farrell, G., and Yan, W., Investigation on single-mode–multimode–single-mode fiber structure, *J. Lightwave Technol.*, 26(5), 512–519, 2008.
34. Choi, H. Y., Kim, M. J., and Lee, B. H. All-fiber Mach-Zehnder type interferometers formed in photonic crystal fiber, *Opt. Express*, 15, 5711–5720, 2007.
35. Mathew, J., Semenova, Y., Rajan, G., and Farrell, G., Humidity sensor based on a photonic crystal fiber interferometer, *Electron. Lett.*, 46(19), 1341–1343, 2010. doi:10.1049/el.2010.2080.
36. Lim, J. H., Jang, H. S., Lee, K. S., Kim, J. C. and Lee, B. H., Mach-Zehnder interferometer formed in a photonic crystal fiber based on a pair of long-period fiber gratings, *Opt. Lett.*, 29, 346–348, 2004.
37. Wu, Q., Semenova, Y., Mathew, J., Wang, P., and Farrell, G., Humidity sensor based on a single-mode hetero-core fiber structure, *Opt. Lett.*, 36, 10, 1752–1754, 2011.
38. Tian, Z., Yam, S. S.-H., Barnes, J., Bock, W, Greig, P., Fraser, J. M., Loock, H.-P., and Oleschuk, R. D., Refractive index sensing with Mach-Zehnder interferometer based on concatenating two single-mode fiber tapers, *IEEE Photon. Technol. Lett.*, 20, 626–628, 2008.
39. Flourney, P. A., McClure, R. W., and Wyntjes, G., White-light interferometric thickness gauge, *Appl. Opt.*, 11(9), 1907–1915, 1972.
40. Belleville, C. and Duplain, G., Fabry-Perot optical sensing device for measuring a physical parameter, US Patents #5,202,939, 1993; #5,392,117, 1995.
41. Udd, E. and Spillman, W. B., *Fiber Optic Sensors*. John Wiley & Sons, New York, 1991.
42. Rajan, G., Ramakrishnan, M., Lesiak, P., Semenova, Y., Wolinski, T., Boczkowska, A., and Farrell, G. Composite materials with embedded photonic crystal fiber interferometric sensors, *Sens. Actuat. A-Phys.*, 182, 57–67, 2012.
43. Kurmer, J. P., Kingsley, S. A., Laudo, J. S., and Krak S. J., Distributed fiber optic acoustic sensor for leak detection, Distributed and multiplexed fiber optic sensors, *Proc. SPIE*, 1586, 117–128, 1992.
44. Wu, D.-F., Zhang, T.-Z., and Ji, B., Modified Sagnac interferometer for distributed disturbance detection, *Microw. Opt. Technol. Lett.*, 50(6), 1608–1610, 2008.
45. Rao, Y. J., In-fiber Bragg grating sensors, *Meas. Sci. Technol.*, 8, 355–375, 1997.
46. Lee, B., Review of present status of optical fiber sensors, *Opt. Fiber Technol.*, 9, 57–79, 2003.
47. Liu, L., Zhang, H., Zhao, Q., Liu, Y., and Li, F., Temperature-independent FBG pressure sensor with high sensitivity, *Opt. Fiber Technol.*, 13, 78–80, 2007.

48. Hsu, Y. S., Wang, L., Lie, W. F., and Chiang, Y. J., Temperature compensation of optical fiber Bragg grating pressure sensor, *IEEE Photon. Technol. Lett.*, 18, 874–876, 2006.
49. Mohanty, L., Koh, L. M., and Tjin, S. C., Fiber Bragg grating microphone system, *Appl. Phys. Lett.*, 89, 161109, 2006.
50. Lee, J. R. and Tsuda, H., A novel fiber Bragg grating acoustic emission sensor head for mechanical tests, *Scripta Mater.*, 53, 1181–1186, 2005.
51. Tosi, D., Olivero, M., and Perrone, G., Low-cost fiber Bragg grating vibroacoustic sensor for voice and heartbeat detection, *Appl. Opt.*, 47, 5123–5129, 2008.
52. Kuang, K. S. C., Kenny, R., Whelan, M. P., Cantwell, W. J., and Chalker, P. R., Embedded fibre Bragg grating sensors in advanced composite materials, *Compos. Sci. Technol.*, 61, 1379–1387, 2001.
53. Murukeshan, V. M., Chan, P. Y., Ong, L. S., and Seah, L. K., Cure monitoring of smart composites using fiber Bragg grating based embedded sensors, *Sens. Actuat. A*, 79, 153–161, 2000.
54. Jones, J. D. C. and MacPherson, W. N., Discrimination techniques for optical sensors. In *Optical Fibre Sensing Technology*, Lopez-Higuera, J. M. (ed.). Wiley & Son, Berlin, Germany, 2002, pp. 403–420.
55. Adachi, S., Distributed optical fiber sensors and their applications, *SICE Annual Conference*, 2008. IEEE, Yokogawa Electr. Corp., Tokyo, Japan, 2008.
56. Murayama, H., Kazuro, K., Hiroshi, N., Akiyoshi, S., and Kiyoshi, U. Application of fiber-optic distributed sensors to health monitoring for full-scale composite structures, *J. Intell. Mater. Syst. Struct.*, 14, 3–13, 2003.
57. Niklès, M., Fibre optic distributed scattering sensing system: Perspectives and challenges for high performance applications, *Third European Workshop on Optical Fibre Sensor*, International Society for Optics and Photonics, 2007.
58. Lanticq, V., Bourgeois, E., Magnien, P., Dieleman, L., Vinceslas, G., Sang, A., and Delepine-Lesoille, S., Soil-embedded optical fiber sensing cable interrogated by Brillouin optical time-domain reflectometry (B-OTDR) and optical frequency-domain reflectometry (OFDR) for embedded cavity detection and sinkhole warning system, *Meas. Sci. Technol.*, 20, 034018, 2009.
59. Dakin, J. P., Pratt, D. J., Bibby, G. W., and Ross, J. N., Temperature distribution measurement using Raman ratio thermometry, *Proc. SPIE*, 0566, Fiber Optic and Laser Sensors III, 249, 1986.
60. Inaudi, D. and Glisic, B., Distributed fiber optic strain and temperature sensing for structural health monitoring, *IABMAS'06 The Third International Conference on Bridge Maintenance, Safety and Management*, Porto, Portugal, 2006.
61. Glisic, B., Distributed fiber optic sensing technologies and applications—An overview, ACI Special Publication, SP-292, Art. no. 2, 18 pp., 2013.
62. Zhu, T., Wu, D., Liu, M., and Duan, D. W., In-line fiber optic interferometric sensors in single-mode fibers, *Sensors*, 12, 10430–10449, 2012.
63. Woody, S., Pushing the boundaries in nano and microscale metrology, *Qual. Dig.*, 27, 22–25, 2007.
64. Chen, G. Y., Ding, M., Newson, T. P., and Brambilla, G., A review of microfiber and nanofiber based optical sensors, *Open Opt. J.*, 7(Suppl. 1, M3), 32–57, 2013.
65. Zhao, J., Jin, D., Schartner, E. P. et al., Single-nanocrystal sensitivity achieved by enhanced upconversion luminescence, *Nat. Nanotechnol.*, 8, 729–734, 2013.
66. Elosua, C., Bariain, C., and Matias, I. R., Optical fiber sensing applications: Detection and identification of gases and volatile organic compounds. In *Fiber Optic Sensors*, M. Yasin, S. W. Harun and H. Arof (eds.). InTechOpen, 530 pp, 2012.
67. Brambilla, G., Optical fibre nanowires and microwires: A review, *J. Opt.*, 12, 043001, 2010.
68. Stiebeiner, A., Garcia-Fernandez, R., and Rauschenbeutel, A. Design and optimization of broadband tapered optical fibers with a nanofiber waist, *Opt. Express*, 18, 22677–22685, 2010.
69. Birks, T. A. and Li, Y. W. The shape of fiber tapers, *J. Lightwave Technol.*, 10, 432–438, 1992.
70. Brakel, A. V., Grivas, C., Petrovich, M. N., and Richardson, D. J., Micro-channels machined in microstructured optical fibers by femtosecond laser, *Opt. Express*, 15(14), 8731–8736, 2007.
71. Zhang, Y., Dhawan, A., and Vo-Dinh, T., Design and fabrication of fiber-optic nanoprobes for optical sensing, *Nanoscale Res. Lett.*, 6, 18, 2011.
72. Passaro, D., Foroni, M., Poli, F., Cucinotta, A., Selleri, S., Lægsgaard, and J., Bjarklev, A. O., All-silica hollow-core microstructured Bragg fibers for biosensor application, *IEEE Sens. J.*, 8(7), 1280–1286, 2008.
73. Pollock, C. R., *Fundamentals of Optoelectronics*. Irwin, Burr Ridge, IL, 1995.
74. Monro, T. M., Richardson, D. J., and Bennet, P. J., Developing holey fibres for evanescent field devices, *Electron. Lett.*, 35, 1188–1189, 1999.

75. Khijwania, S. K. and Gupta, B. D., Maximum achievable sensitivity of the fiber optic evanescent field absorption sensor based on the U-shaped probe, *Opt. Commun.*, 175(1–3), 135–137, 2000.
76. Hale, Z. M., Payne, F. P., Marks, R. S., Lowe, C. R., and Levine, M. M., The single mode tapered optical fibre loop immunosensor, *Biosens. Bioelectron.*, 11(1–2), 137–148, 1996.
77. Ahmad, M. and Hench, L. L., Effect of taper geometries and launch angle on evanescent wave penetration depth in optical fibers, *Biosens. Bioelectron.*, 20(7), 1312–1319, 2005.
78. Moar, P. N., Huntington, S. T., Katsifolis, J., Cahill, L. W., Roberts, A., and Nugent, K. A., Fabrication, modeling, and direct evanescent field measurement of tapered optical fiber sensors, *J. Appl. Phys.*, 85(7), 3395–3398, 1999.
79. Mignani, A. G., Falciai, R., and Ciaccheri, L., Evanescent wave absorption spectroscopy by means of bi-tapered multimode optical fibers, *Appl. Spectrosc.*, 52(4), 546–551, 1998.
80. Villatoro, J., Luna-Moreno, D., and Monzon-Hernandez, D., Optical fiber hydrogen sensor for concentrations below the lower explosive limit, *Sens. Actuat. B-Chem.*, 110(1), 23–27, 2005.
81. Lacroix, S., Black, R. J., Veilleux, C. et al., Tapered single-mode fibers—External refractive-index dependence, *Appl. Opt.*, 25, 2468–2479, 1986.
82. Diez, A., Andres, M. V., and Cruz, J. L., In-line fiber-optic sensors based on the excitation of surface plasma modes in metal-coated tapered fibers, *Sens. Actuat. B-Chem.*, 73: 95–99, 2001.
83. Liang, W., Huang, Y. Y., Xu, Y. et al., Highly sensitive fiber Bragg grating refractive index sensors, *Appl. Phys. Lett.*, 86, 151122, 2005.
84. Wang, P., Bo, L., Guan, C., Semenova, Y., Wu, Q., Brambilla, G., and Farrell, G., Low-temperature sensitivity periodically tapered photonic crystal-fiber-based refractometer, *Opt. Lett.*, 38(19), 3795–3798, 2013.
85. Bo, L., Wang, P., Semenova, Y., and Farrell, G., High sensitivity fiber refractometer based on an optical microfiber coupler, *Photon. Technol. Lett.*, 25(3), 228–230, 2013.
86. Leung, A., Shankar, M. P., and Mutharasan, R., A review of fiber-optic biosensors, *Sens. Actuat. B*, 125, 688–703, 2007.
87. Frazão, O., Santos, J. L., Araújo, F. M., and Ferreira, L. A., Optical sensing with photonic crystal fibers, *Laser Photon. Rev.*, 2(6), 449–459, 2008.
88. Cox, F. M., Argyros, A., and Large, M. C. J., Liquid-filled hollow core microstructured polymer optical fiber, *Opt. Express*, 14(9), 4135–4140, 2006.
89. Wang, P., Ding, M., Bo, L., Guan, C., Semenova, Y., Sun, W., Yuan, L., Brambilla, G., and Farrell, G., A photonic crystal fiber half taper probe based refractometer, *Opt. Lett.*, 39(7), 2076–2079, 2014.
90. Petrovic, J., Lai, Y., and Bennion, I., Numerical and experimental study of microfluidic devices in step-index optical fibers, *Appl. Opt.*, 47(10), 1410–1416, 2008.
91. Lee, B., Roh, S., and Park, J., Current status of micro- and nano-structured optical fiber sensors, *Opt. Fiber Technol.*, 15, 209–221, 2009.
92. Gupta, B. D. and Verma, R. K., Surface plasmon resonance-based fiber optic sensors: Principle, probe designs, and some applications, *J. Sens.*, (1), 1–12, 2009, doi:10.1155/2009/979761.

3 Interferometric Fiber-Optic Sensors

Sara Tofighi, Abolfazl Bahrampour, Nafiseh Pishbin, and Ali Reza Bahrampour

CONTENTS

3.1 OPTICAL FIBERS AND THEIR CHARACTERISTICS

Fiber-optic lines have revolutionized long-distance phone calls and Internet. Optical fibers are widely used for communications and fiber sensors, thanks to their high speed, large bandwidth, reliability, and low attenuation. They are mainly made of silica or polymers. Depending on the fiber applications, different structures such as conventional fiber, photonic crystal fiber, and quasi-photonic crystal fibers are designed. A brief review of different fiber structures is presented in this section.

3.1.1 Standard Optical Fibers

On the basis of total internal reflection, light is guided by standard optical fibers. A standard optical fiber consists of two different coaxial cylindrical layers. Core is the central region, which is surrounded by a cladding layer. The core's refractive index is greater than the cladding's. Most of the energy of the guided modes propagates in the core and a small fraction of the total energy propagates in the cladding region. The cladding radius compared to the core radius is so large that the surrounding medium has no effect on the light propagating inside the optical fiber. Depending on the application, optical fibers are made from glass, polymers, or crystals. Conventional or communication optical fibers are made of silica (SiO_2) glass; the core and cladding refractive indices are adjusted by employing suitable dopants such as GeO_2, Al_2O_3, and B_2O_3 impurities. The optical fibers are characterized by the core–cladding index difference Δ and normalized frequency v parameters. Δ and v are defined by the following equations $\Delta = (n_1 - n_c)/n_1$ and $v = k_0 a\sqrt{n_1^2 - n_c^2}$, where n_1 and n_c are the core and cladding refractive indices, respectively, $k_0 = 2\pi/\lambda$ is the wave number, a is the core radius, and λ is the light wavelength [1]. The number of modes supported by the optical fiber is determined by v. Optical fibers designed for $v<2.405$ support single-mode operation. The Δ parameter is about 0.003 for a single-mode fiber. The main difference between the single- and multimode fiber is the core size. Single-mode fibers require a core size of $a<5\,\mu m$. An outer radius of $b = 62.5\,\mu m$ is commonly used for both single- and multimode fibers. Optical fibers with low Δ parameters are named weakly guiding fibers. The weakly guiding modes are denoted by $LP_{\mu\nu}$, where μ and ν are integer numbers. The fundamental mode of optical fiber (HP_{11}) corresponds to the LP_{01} weakly guiding mode [2]. The normalized propagation constant β versus the dimensionless frequency is called the dispersion curve.

Obviously, the frequency dependence of group velocity $v_g = (d\beta/d\omega)^{-1}$ leads to pulse broadening because different spectral components of the pulse do not arrive simultaneously at the fiber output. The dispersion parameter $D = d/d\lambda(1/v_g)$ is expressed in units of Ps/(Km · nm). D consists of material dispersion, waveguide dispersion, polarization mode dispersion, and mode dispersion. In single-mode fibers, the mode dispersion parameter vanishes [2]. Fiber loss is another limiting factor that reduces the fiber output power and it depends on the atomic absorption and scattering [3]. Depending on the optical fiber parameters, light wavelength, and coupling conditions; bounded, radiation, and evanescent modes can be excited. However, evanescent modes cannot propagate along the optical fiber and store energy near the excitation source [4].

3.1.2 Photonic Crystal Fibers

Photonic crystal fibers (PCFs) refer to another class of optical fibers that have wavelength-scale morphological microstructures running down their length [5]. PCFs can be divided into different categories

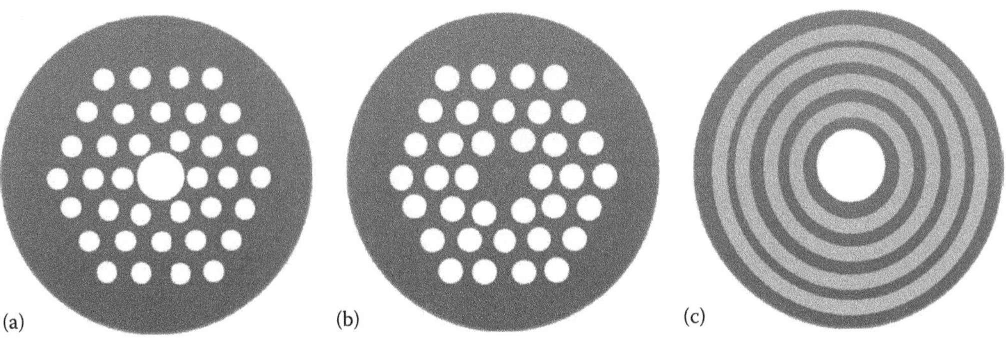

FIGURE 3.1 Schematic cross section of (a) HC-PCF, (b) solid-core PCF (holey fiber), and (c) Bragg fiber.

based on their guidance mechanism or photonic crystal dimensionality in their transverse plane. Based on their guiding mechanisms, they are divided into index-guiding PCFs (IG-PCFs) and photonic bandgap fibers (PBFs). Over the last decade, both types of PCFs have been studied, but particular attention has been given to PBFs due to their lattice-assisted light propagation within a hollow core [6]. This particular feature indeed has a number of advantages such as lower Rayleigh scattering, reduced nonlinearity, novel dispersion characteristics, and potentially lower loss compared to conventional optical fibers [7]. In addition, the hollow-core PCFs (HC-PCFs) also enable enhanced light–material interaction, thus providing a valuable technological platform for ultrasensitive and distributed biochemical sensors [8].

These different types of PCFs are shown in Figure 3.1. As mentioned before, light guidance in HC-PCF (Figure 3.1a) is based on the bandgap effect in photonic crystals [5]. For a given value of propagation constant (β), if the transverse component of the wave vector falls into the bandgap region, no transverse propagation is allowed [9]. So the propagation mode is confined to the hollow defected core of the PCF. It is expected that the attenuation of the hollow PCF is much smaller than a standard optical fiber, but due to the surface capillary wave production during the PCF fabrication, its attenuation is about 1 dB/km [5].

The guidance of light in solid-core PCFs (Figure 3.1b) is due to the modified total internal reflection. The average refractive index of the cladding in a solid-core PCF is lower than the core's refractive index. In fact, here, the average refractive index is not the geometric average, but an effective refractive index (n_{eff}) corresponding to the largest possible value of the photonic crystal propagation constant ($\beta_{max}(\omega)$) at a given frequency. In the effective refractive index method, n_{eff} attributed to the cladding of solid-core PCFs is equal to the modal index of fundamental space-filling model $\left(n_{eff} = \beta_{max}(\omega)c/\omega\right)$ [10,11]. Light at a given frequency and a β greater than β_{max} cannot propagate in the photonic crystal structure; hence, modified internal reflection is a specific case of bandgap guidance. Some authors also use the term microstructured optical fiber (MOP) for referring to PCFs where guidance results from bandgap effect [5].

Bragg fibers are a special case of PCFs. A Bragg fiber is a concentric arrangement of dielectric layers wrapped around a core, which may or may not be hollow. The periodicity of the photonic crystal cladding in Bragg fiber is 1D. Light propagation in the Bragg fiber (Figure 3.1c) is based on Bragg diffraction effect. Bragg fibers with full circular symmetry and without birefringence can be single-mode and single-polarization fibers.

3.1.3 Quasi-Photonic Crystal Fibers

The geometry of the cladding (microstructured) plays an important role in the bandgap regime, but all HC-PCF fibers are based on periodic structures and thus are limited to a few geometries such as a triangular, square, honeycomb, or Kagome periodic array lattice. However, recently, few works showed that it is possible to explore new functionalities and increase the degree of freedom

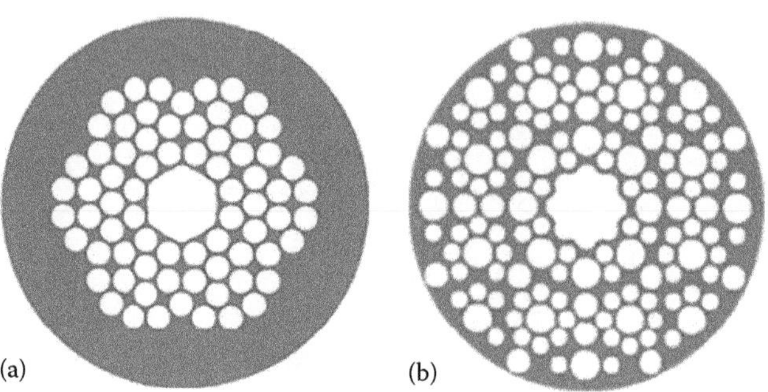

FIGURE 3.2 (a) 12-fold PQ fiber and (b) modified 8-fold.

of HC-PCFs by studying new kinds of fibers based on photonic quasi-crystals (PQs). We define PQs as aperiodic photonic crystal (1D, 2D, or 3D) that lack translational symmetry and instead are extremely rich in rotational symmetry [12]. PQ structures have presented intriguing achievements in optics and crystallography. Optical properties such as complete photonic bandgap, laser micro-cavity, and guided resonance in different aperiodic structures have been investigated and reported [13–16]. An opulent rotational symmetry provides a bandgap in the lower refractive index region, which can lead to HC fibers with minimum air–glass refractive index contrast [13]. Recently, PQs have been investigated principally for index-guiding mechanism [17] and Sun et al. [18] for the first time proposed an HC PQ fiber based on a 12-fold symmetric structure as shown in Figure 3.2a. They demonstrated that 12-fold fibers exhibit a double photonic bandgap. Different structures such as a 12-fold [19] and a modified 8-fold symmetric structure (Figure 3.2b) have been proposed and simulated where a double bandgap with $\lambda/\Lambda < 1$ (λ is the wavelength and Λ is a defined pitch) has been demonstrated. Other simulations on PQ fibers confirm an extremely lower loss (nominally) [12].

3.1.4 Polarization-Maintained Optical Fibers

Birefringent optical fibers are fibers with anisotropic cross sections that have two distinct principle axes with different refractive indices, called the fast and slow axis. These are named so because there exist two different light velocities that depend on the polarization alignment of the incident light with one of the principle axes. For a light beam whose polarization is aligned with one of the principle axes of the birefringent fiber, the light's polarization is kept constant during propagation along that optical fiber. The birefringence parameter of the fiber is defined by the difference between the two refractive indices corresponding to the two principle axes $B = n_s - n_f$, where n_s and n_f are the refractive indices of the slow and fast axis, respectively. The fiber beat length $L_B = \lambda/B$ is defined as the fiber length over which the phase difference between the fast and slow waves becomes 2π radians. To preserve the polarization direction, perturbation periods introduced in the drawing process as well as the physical bend and twists must be greater than the beat length. This kind of optical fibers are called polarization-maintained fibers (PMF). PMFs are employed in different interferometric optical fiber sensors (IOFS) such as optical fiber gyroscopes and intruder detection systems. PCFs with asymmetric microstructured in either cladding or the core region could exhibit strong birefringent compared to the conventional PMFs. Figure 3.3 shows the cross section of high-birefringence photonic crystal fiber (HiBi-PCF) and polarization-maintaining photonic crystal fiber (PM-PCF).

3.1.5 Slab Optical Waveguides

Optical fibers are suitable for long-length sensors such as oil pipeline leak detection systems and intruder sensors. In many applications such as integrated circuits and miniaturized sensors, the transmission length is less than a few millimeters, so slab optical waveguides are used for these

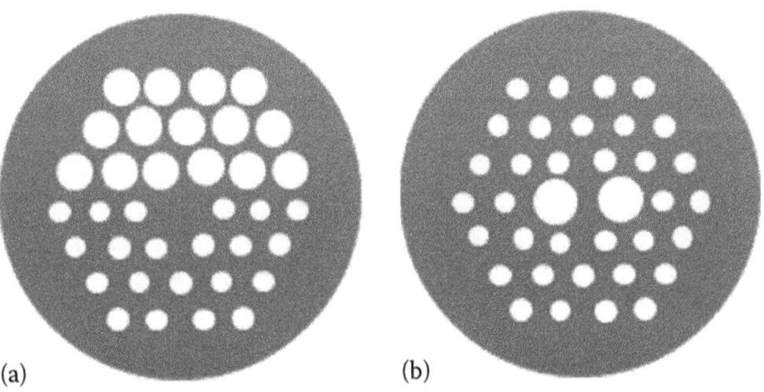

FIGURE 3.3 Schematic cross section of (a) HiBi-PCF and (b) PM-PCF.

short-length applications. A dielectric waveguide consists of a dielectric with refractive index n_1, which is deposited on a lower refractive index substrate; the refractive index of the surrounding medium is also smaller than n_1. The modes propagating in the slab waveguides are transverse electric (TE) and transverse magnetic (TM) modes. The mathematical model analysis of slab waveguides can be found in any standard text book [20]. The narrow dielectric strip waveguide modes are also classified into bounded, radiation, and evanescent modes [4,20]. The slab optical waveguides can also be employed in quantum computing unitary gates.

3.2 ELECTROMAGNETIC INTERFERENCE

Interference is a wave effect that can be observed in any type of wave whether it be electromagnetic, acoustic, elastic, or matter waves. Generally, field or first-order interference is just called interference, while higher-order interferences can be employed for measurement of the electromagnetic wave characteristics such as statistical parameters [21]. The interference behavior is formulated by the correlation function.

3.2.1 Field Interference and First-Order Correlation Function

Most of the well-known interferometers such as Fabry–Pérot, Sagnac, Michelson, and Young double-slit interferometers are operating on the basis of field interference. The intensity corresponding to the field $E(x_1,t_1,x_2,t_2) = \alpha E(x_1,t_1) + \beta E(x_2,t_2)$ results in interference fringes. As an example, in Young's double-slit experiment, $E(x_1,t_1)$ and $E(x_2,t_2)$ are the fields on the screen corresponding to the slits at positions (x_1,t_1) and (x_2,t_2), respectively. The intensity at the observation point (x,t) is given by

$$I(x,t) = |\alpha|^2 I(x_1,t_1) + |\beta|^2 I(x_2,t_2) + 2Re\left(\alpha\beta^* E(x_1,t_1)E^*(x_2,t_2)\right) \quad (3.1)$$

The third term on the right-hand side of Equation 3.1 corresponds to the field interference. In conventional interferometer, light can propagate from (x_1,t_1) and (x_2,t_2) points, through an arbitrary medium to the observation point (x,t). In optical fiber interferometers (OFIs), the transformation media may be one or more optical fibers.

The first-order correlation function $G^{(1)}(x,t;x',t')$ is defined by

$$G^{(1)}(x,t;x',t') = \left\langle E^{(-)}(x,t)E^{(+)}(x',t')\right\rangle \quad (3.2)$$

where

$E^{(+)}(x)$ and $E^{(-)}(x)$ are the positive and negative frequency parts of the electric field

$\langle\ldots\rangle$ stands for the ensemble average

For a stationary field, the statistical description is invariant under time-variable displacements $\left(G^{(1)}(t,t')=G^{(1)}(t-t'=\tau)\right)$. The random classical fields are usually stationary and have the ergodic property; therefore, the ensemble average has the same value as the time-averaged correlation function $\mathcal{T}^{(1)}(t-t')$ [22–24]:

$$G^{(1)}(x,x',\tau)=\mathcal{T}^{(1)}(x,x',\tau)=\lim_{T\to\infty}\frac{1}{T}\int_0^T E^{(-)}(x,t_1+\tau)E^{(+)}(x',t_1)dt_1 \tag{3.3}$$

No fringes will be observed if the correlation function $G^{(1)}(x,x')$ vanishes and it can be considered that the fields at x and x' are incoherent. On the other hand, the highest degree of coherence is associated with a field that exhibits the strongest possible interference fringe. The strength of interference is defined by the visibility factor as follows:

$$V=\frac{I_{max}-I_{min}}{I_{max}+I_{min}}=\frac{2\sqrt{G^{(1)}(x,x)G^{(1)}(x',x')}}{G^{(1)}(x,x)+G^{(1)}(x',x')} \tag{3.4}$$

If the fields incident on the two pinholes have equal intensity, the visibility varies between $V=0$ for incoherent light and $V=1$ for first-order coherent light. Generally, the first-order normalized correlation function is

$$g^{(1)}(x,x')=\frac{G^{(1)}(x,x')}{\sqrt{G^{(1)}(x,x)G^{(1)}(x',x')}} \tag{3.5}$$

The necessary condition for coherence is $\left|g^{(1)}(x,x')\right|=1$ that is equivalent to the factorization property of correlation function $\left(G^{(1)}(x,x')=F(x)F(x')\right)$ and normalized correlation function $\left(g^{(1)}(x,x')=f(x)f(x')\right)$ [22,23].

3.2.2 Second- and Higher-Order Interferences

In second-order interferometry, the intensities at two different positions are measured, multiplied, and averaged. In Hanbury Brown and Twiss (HB-T) interferometer, the intensities at two different positions r_i $(i=1,2)$ are detected individually [25]. The detectors' outputs, which are in the low-frequency range, are transmitted to a central correlating device where they are multiplied and the product is averaged:

$$\left|E^{(+)}(r_i,t)\right|^2=\left|E_k\right|^2+\left|E_{k'}\right|^2+E_kE_{k'}^*e^{i(k-k')\cdot r_i}+E_k^*E_{k'}e^{-i(k-k')\cdot r_i} \tag{3.6}$$

where E_k, $E_{k'}$, k, and k' are the amplitudes and wave vectors of the incoming fields, respectively.

The advantage of HB-T method was to detect the signals first, then filter the high-frequency components so that the signals are of relatively low frequency and can transmit undisturbed over large distances. This method deals with signal intensities and is quite different from the field interferometric method that works with the average of the product of two random fields. Hence, the inputs of HB-T experiment have no rapid oscillations. The average of the product of the two intensities from (3.6) is

$$\left\langle\left|E^{(+)}(r_1,t)\right|^2\left|E^{(+)}(r_2,t)\right|^2\right\rangle=\left\langle\left(\left|E_k\right|^2+\left|E_{k'}\right|^2\right)^2\right\rangle+2\left\langle\left|E_k\right|^2\left|E_{k'}\right|^2\right\rangle\cos[(k-k')\cdot(r_1-r_2)] \tag{3.7}$$

Clearly, the third term on the right-hand side of Equation 3.7 represents an interference effect. The second-order correlation function $G^{(2)}(r_1, t_1; r_2, t_2)$ is defined as

$$G^{(2)}(r_1,t_1;r_2,t_2) = \left\langle E^{(-)}(r_1,t_1) E^{(-)}(r_2,t_2) E^{(+)}(r_2,t_2) E^{(+)}(r_1,t_1) \right\rangle \tag{3.8}$$

$G^{(2)}(r_1, t_1; r_2, t_2)$ is defined by ensemble averages rather than the time averages. That is, it is assumed that the incoming wave is a stochastic electromagnetic radiation and has the ergodic property. As shown in Equation 3.8, $G^{(2)}(r_1, t_1; r_2, t_2)$ is a measure of the strength of intensity interference. The normalized second-order correlation function is defined by

$$g^{(2)}(x_1,x_2) = \frac{G^{(2)}(x_1,x_2)}{\sqrt{G^{(1)}(x_1,x_1) G^{(1)}(x_2,x_2)}} \tag{3.9}$$

The second-order coherent light is defined by $\left| g^{(2)}(x_1,x_2) \right| = 1$, which is equivalent to the factorization property of the second-order correlation function. The first- and second-order coherency can be found in radiation generated by natural sources. In general, man-made sources such as lasers and radio transmitters can have much higher regularity than is ever possible for natural sources. In order to better understand the concept of coherence, higher-order correlation functions are defined by [23]

$$G^{(n)}(x_1,\ldots,x_n;x_n,\ldots,x_1) = \left\langle E^{(-)}(x_1)\ldots E^{(-)}(x_n) E^{(+)}(x_n)\ldots E^{(+)}(x_1) \right\rangle \tag{3.10}$$

The nth-order normalized correlation function is

$$g^{(n)}(x_1,\ldots,x_n) = \frac{G^{(n)}(x_1,\ldots,x_n;x_n,\ldots,x_1)}{\prod_{j=1}^{n} \left\{ G^{(1)}(x_j,x_j) \right\}^{1/2}} \tag{3.11}$$

The Mth order coherence field is defined by $\left| g^{(n)}(x_1,\ldots,x_{2n}) \right| = 1$ for all $n \le M$ and all combinations of arguments of x_i. Full coherence requires coherency for all orders of correlation function.

3.2.3 Quantum Theory of Correlation Functions

The most popular quantum interference arrangement is the Hong–Ou–Mandel interferometer [26,27]. As shown in Figure 3.4, in Hong–Ou–Mandel interferometer, a pair of photons enters a beam splitter (BS) that can be a lossless optical fiber coupler (OFC) or a conventional BS. In general, there are four possible outputs for the BS:

(a) Both are reflected.
(b) Both are transmitted.
(c and d) One is transmitted, while the other is reflected.

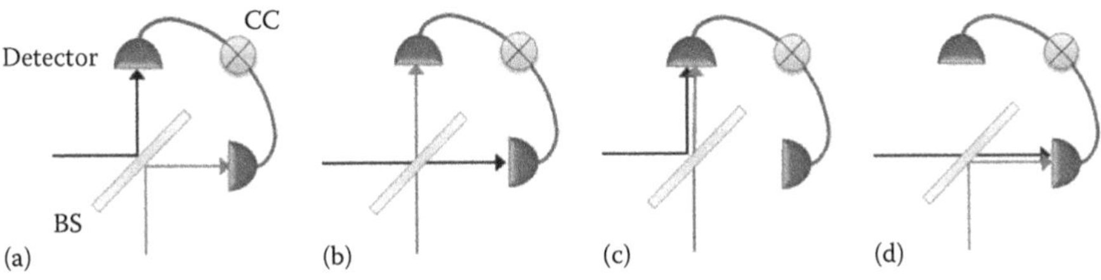

FIGURE 3.4 4 Four possible outputs of BS in Hong–Ou–Mandel interferometer: (a) both photons are reflected, (b) both photons are transmitted, (c and d) one photon is transmitted while the other is reflected. CC and BS stand for coincident count and beam splitter, respectively.

Here, (c) and (d) are degenerate states. Due to energy conservation in a symmetric BS, there is an overall π-phase difference in the photons of the first two states (a) and (b) that leads to destructive interference. Hence, the first two cases completely cancel each other out. If the input state is $|1,1\rangle$, then the output state will be $|\Psi\rangle_{out} = 1/\sqrt{2}\left(|2,0\rangle - |0,2\rangle\right)$.

Due to the destructive interference described earlier, the $|1,1\rangle$ term disappears at the output of the BS $\left(|\Psi\rangle_{out}\right)$. The Bosonic property of the photons dictates that the photons will have the tendency to go together to either side of the BS.

Generally, for a quantum system which is characterized by a matrix density (ρ), and an expectation value of an arbitrary observable (O), which is given by $O = Tr\{\rho O\}$. The average counting rate of an ideal photodetector which operates based on single-photon absorption is proportional to the expectation value of the observable $E^{(-)}(x)E^{(+)}(x)$ ($Tr\{\rho E^{(-)}(x)E^{(+)}(x)\}$), where $E^{(+)}(x)$ and $E^{(-)}(x)$ are the positive and negative frequency part of the electric field operator $Tr\{\rho E^{(-)}(x)E^{(+)}(x)\}$. Similarly, the n-photon absorption rate is obtained by nth-order quantum correlation function as

$$G^{(n)}\left(x_1,\ldots,x_n;x_n,\ldots,x_1\right) = Tr\left\{\rho E^{(-)}\left(x_1\right)\ldots E^{(-)}\left(x_n\right)E^{(+)}\left(x_n\right)\ldots E^{(+)}\left(x_1\right)\right\} \tag{3.12}$$

If there is an upper bound (M) on the number of photons present in the field, then the function $G^{(n)}$ vanishes for all orders higher than that fixed M.

3.3 MULTIPORT OPTICAL FIBER

Linear multiport OFCs are a generalization of the BS. The BS has an essential role in many experiments of classical and quantum optics. The simplest BS can be considered as a four-port device. An optical fiber BS (OFBS) or OFC is a pair of coupled optical fibers with their cores brought alongside one another. The evanescent field of each fiber core excites the mode of the other fiber and these modes are then coupled. OFBSs are fabricated by placing a pair of single-mode fibers side by side, twisting together and fusing while elongating the contact region. In order to better understand the operation of multiport OFCs, a simple BS will be studied first in the following section.

3.3.1 Beam Splitter

A BS can be described by two input and two output modes of the radiation field. The operation of a lossless BS can be described by a unitary scattering matrix. In a lossless BS, the output fields ($\vec{E}_{out}$) are related to the input fields ($\vec{E}_{in}$) through a 2×2 unitary scattering matrix ($\mathbb{U}$):

$$\begin{pmatrix} E_{1out} \\ E_{2out} \end{pmatrix} = \begin{pmatrix} \sin\Omega & e^{i\varphi}\cos\Omega \\ \cos\Omega & -e^{i\varphi}\sin\Omega \end{pmatrix} \begin{pmatrix} E_{1in} \\ E_{2in} \end{pmatrix} \tag{3.13}$$

φ is the phase difference between the input fields, which can be generated by placing an external phase shifter (PS) placed before the BS in one of the input ports. The parameter Ω is related to the reflectivity and transmittance of the BSs via $R = \cos^2\Omega$ and $T = \sin^2\Omega$, respectively [28,29].

The transfer matrix of an optical fiber BS $\mathbb{U}$ can be described by a coupling constant κ and a coupling length L. The coupling coefficient κ is determined by the wavelength of light and is an exponentially decaying function of interfiber distance:

$$\mathbb{U} = \begin{pmatrix} cos\kappa L & isin\kappa L \\ isin\kappa L & cos\kappa L \end{pmatrix} \tag{3.14}$$

For $\kappa L = \pi/4$, a symmetric 2×2 coupler (or 3 dB coupler) is obtained [30,31].

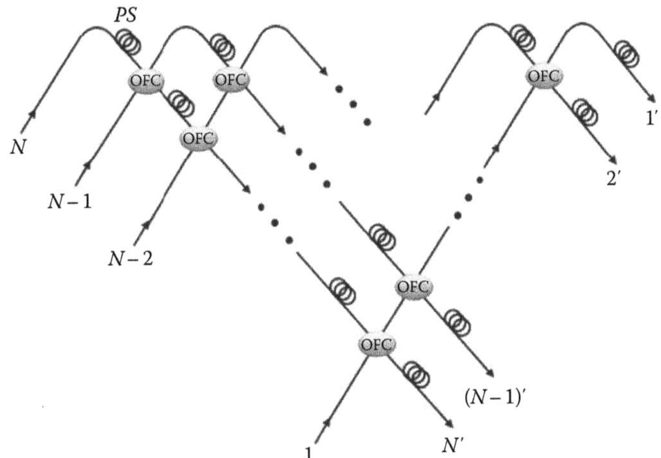

FIGURE 3.5 A schematic diagram of multiport OFBS built of 2 × 2 OFCs and PSs.

3.3.2 Multiport Optical Fiber Beam Splitter

A generalized BS can be built from commercial optical components. As shown in Figure 3.5, the implementation of a generalized N port OFBS can be achieved by a suitable configuration of 2×2 OFBSs and PSs. For tritters, PSs $\varphi_1 = \tan(1/3)$ and $\varphi_2 = \varphi_1/2$ are needed, while for quarter, a PS of $\varphi = \pi/2$ is required.

A multiport OFBS consists of N coupled optical fibers. N optical fibers are placed side by side, twisting and fusing them together while elongating the contact region. A generalized OFBS is a $2N$-port optical fiber network. Due to the energy conservation law, any $2N$-port optical fiber network can be described by an $N\times N$ unitary matrix. The matrix elements of a generalized OFBS are wavelength dependent and can be determined versus the distances between the fiber cores. The matrix element $\mathbb{U}_{ij}$ is the ratio of the electric field at port j to the electric field at port i, while there is no input light at the other ports. In many cases, the forward and backward modes can be considered as the input and output of the multiport interferometer, respectively.

3.3.3 Unitary Transformation as a Multiport Optical Fiber Beam Splitter

As mentioned in the previous section, due to the energy conservation law, any lossless multiport OFBS can be described by a unitary matrix. Moreover, it was shown that for any unitary matrix, there exists an experimental setup consisting of PSs and simple 2×2 BSs [29,32]. The multiport BS can be viewed as a black box transforming N external inputs into N external outputs. The internal inputs (outputs) are those that are internally connected to the output (inputs) of another BS inside the system. The transformation matrix of a multiport BS can be obtained by employing the scattering matrices of the 2×2 BSs and PSs. After eliminating the internal inputs and outputs of 2×2 BSs, the external input–output matrix of multiport BS, which consists of four submatrices, is obtained as follows:

$$\begin{pmatrix} \vec{E}_{out}^{(ext)} \\ \vec{E}_{out}^{(int)} \end{pmatrix} = \begin{pmatrix} S_{ee} & S_{ei} \\ S_{ie} & S_{ii} \end{pmatrix} \begin{pmatrix} \vec{E}_{in}^{(ext)} \\ \vec{E}_{in}^{(int)} \end{pmatrix} \tag{3.15}$$

For example, the submatrix S_{ei} describes the transformation from the internal inputs of the BS to the external outputs. The connection between BSs is described by a connection matrix $\mathcal{P}$. The matrix elements $\mathcal{P}_{ij}$ are the phase shifts internally accumulated by fields evolving between the BSs:

$$\vec{E}_{out}^{(int)} = \mathcal{P}\,\vec{E}_{in}^{(int)} \tag{3.16}$$

The input–output relation of the whole system can be obtained by solving Equations 3.15 and 3.16 simultaneously:

$$\vec{E}_{out}^{(ext)} = \left(S_{ee} + S_{ei}\left(\mathcal{P} - S_{ii}\right)^{-1} S_{ie}\right)\vec{E}_{in}^{(ext)} \tag{3.17}$$

The unitary matrix of a complex multiport BS involving many BSs and PSs is simply obtained by

$$\mathbb{U} = S_{ee} + S_{ei}\left(\mathcal{P} - S_{ii}\right)^{-1} S_{ie} \tag{3.18}$$

Symmetric multiports are a special class of unitary multiports. All the matrix elements are of the same modulus. A field at any of the inputs is a coherent superposition of all output modes with equal modulus of the amplitude. If intensity I enters one input of a symmetric $N \times N$ multiport, the intensity at any output is I/N.

The generic form of the transfer matrix of a symmetric $N \times N$ multiport [32] can be expressed versus the roots of unity $z_N = \exp(i(2\pi/N))$ as

$$\mathbb{U}_{mk}^{(N)} = \frac{1}{\sqrt{N}} z_N^{(m-1)(k-1)} \tag{3.19}$$

As an example, the transfer matrix of 3×3 symmetric multiport (which is called a tritter) is as follows:

$$\mathbb{U}^{(3)} = \frac{1}{\sqrt{3}}\begin{pmatrix} 1 & 1 & 1 \\ 1 & \alpha & \alpha^2 \\ 1 & \alpha^2 & \alpha \end{pmatrix} \tag{3.20}$$

where $\alpha \equiv e^{i2\pi/3}$. Symmetric multiports that can be transformed into one another by simple renumbering of inputs and outputs or by including PSs at the inputs and outputs can be considered as an equivalent class. Each of the 2×2 BSs and 3×3 symmetric multiports has only one equivalence class [29].

3.3.4 Matrix Theory of Multiport Optical Fibers

In the previous sections, it was assumed that light was linearly polarized in all parts of the multiport and there was no element in the optical fiber network to change the polarization. In this section, the Jones vectors and matrices [24,33] are employed to introduce a method for analyzing complex interferometric fiber systems, including polarization and back reflection effects [34].

The polarization state of quasi-plane wave-front is described by a Jones vector in a complex 2D vector space $E = \left(a\hat{i} + b\hat{j}e^{i\delta}\right)E_0 e^{i\omega t - ikz}$, where $|a|^2 + |b|^2 = 1$ and δ is the phase difference between the x and y components of the electric field. Any common phase in a and b can be taken out to be absorbed by the phase term $e^{i\omega t - ikz}$.

Every linear optical element in each branch of the interferometer is represented by a 2×2 matrix. Systems with bidirectional propagation may be faithfully simulated with scattering matrices, that is, bidirectional Jones matrices can be generalized to an arbitrary OFI. A two-port device is described by a 4×4 scattering matrix, while a four-port device such as fiber-optic coupler is presented by 8×8 scattering matrices. In general, a bidirectional vectorial 2N-port OFI can be described by a 4N×4N scattering matrix. The forward and backward electric field vectors are denoted by $E_f = \left(E_f^{(1)}, E_f^{(2)}, \ldots, E_f^{(2N)}\right)^T$ and $E_b = \left(E_b^{(1)}, E_b^{(2)}, \ldots, E_b^{(2N)}\right)^T$, respectively, where $E_f^{(i)} = \left(E_{fx}^{(i)}, E_{fy}^{(i)}\right)$ and $E_b^{(i)} = \left(E_{bx}^{(i)}, E_{by}^{(i)}\right)$. The forward and backward fields are related through the scattering matrix $S(E_b = SE_f)$ where the matrix elements S^{ij} are 2×2 block matrices given by the backward field at port i when the forward fields at other ports except port j are zero $E_b^{(i)} = S^{ij} E_f^{(j)}$. The 2×2 block matrices S^{ij} are quasi-Jones matrices, and their elements S_{kl}^{ij} $(k,l = x, y)$ are the ratio of $E_{bl}^{(i)}/E_{fk}^{(j)}$ when all other field components $E_{fm}^{(q)}$ $(q \neq j, m \neq k)$ are zeros.

3.4 MULTIPORT OPTICAL FIBER INTERFEROMETERS

Interferometry is based on superimposing two or more light beams to measure the phase difference between them. These beams have the same frequency. In classical experiments and interferometric sensors, all the light beams are generated by a given light source, even though from the quantum point of view the interference of the beams from different sources is of great importance. Typically, an incident light beam in an interferometer is split into two or more parts and then recombined together to create the interference pattern. To consider the interference fringes, at least two optical paths are necessary for an interferometry experiment. These paths can be in a single- or multimode optical fiber. In OFIs utilizing multimode optical fibers such as the Sagnac interferometer, each mode defines an individual optical path. In Sagnac interferometer, the optical paths are defined by the clockwise (CW) and counterclockwise (CCW) optical fiber modes. The optical path can be defined by separate single-mode optical fibers such as in Mach–Zehnder OFI. The maximum and minimum points of the fringes correspond to even and odd numbers of half-wavelength optical path differences (OPDs), respectively. There are many interferometer configurations that have been realized with the optical fiber. Some configurations such as Fabry–Pérot, fiber Bragg gratings (FBGs), Sagnac birefringence OFI, Mach–Zehnder, Michelson, and Moiré interferometer are presented in this section.

3.4.1 Fabry–Pérot Interferometer

The simplest OFI is the Fabry–Pérot interferometer (FPI). It consists of two parallel reflectors with reflection coefficient $R_1(\omega)$ and $R_2(\omega)$ separated by a cavity length L. These reflectors can be mirrors, interface of two dielectrics, or FBGs. The cavity may be an optical fiber or any other medium. The FP reflectance R_{FP} and transmittance T_{FP} versus the mirror's reflection coefficient are obtained [35] by

$$R_{FP} = \frac{R_1 + R_2 + 2\sqrt{R_1 R_2}\cos\phi}{1 + R_1 R_2 + 2\sqrt{R_1 R_2}\cos\phi} \tag{3.21}$$

$$T_{FP} = \frac{T_1 T_2}{1 + R_1 R_2 + 2\sqrt{R_1 R_2}\cos\phi} \tag{3.22}$$

where

$\phi = 4\pi nL/\lambda$ is the round-trip propagation phase shift in the interferometer
n is the refractive index between the reflectors
λ is the free-space optical wavelength

The FP transmittance has its maximum at the resonance frequencies corresponding to the round-trip propagation phase $\phi_m = (2m + 1)\pi$, where m is an integer number. The detuning phase is defined by $\Delta = \phi - \phi_m$. For high reflectance mirrors, the transmission coefficient near the resonance frequencies can be written as follows:

$$T_{FP} = \frac{T^2}{(1-R)^2 + R\Delta^2} \tag{3.23}$$

where $R = R_1 = R_2$, $T = 1-R$, and $\delta = \pm(1-R)/\sqrt{R}$ is the phase corresponding to the FP bandwidth. The FP finesse can be written as

$$F = \frac{\pi\sqrt{R}}{1-R} \tag{3.24}$$

In an interferometer with lossless mirrors $R = R_1 = R_2 = 0.99$, the finesse is equal to $F = 312.6$, which is considered very high. For the mirrors' reflectance $R = R_1 = R_2 \ll 1$, the FP reflectance and transmittance can be approximated by

$$R_{FP} \cong 2R(1+\cos\phi) \tag{3.25}$$

$$T_{FP} \cong 1 - 2R(1+\cos\phi) \tag{3.26}$$

The concept of finesses is not suitable for $R \ll 1$ FPs. The finesse equals one for $R = 0.172$ and is undefined for $R < 0.172$ [35].

Due to the energy conservation law, the reflection and transmission coefficients of FP follow $R + T = 1$ equation. This equation made it possible to measure the FPI resonance frequency by employing the FP transmission or reflection coefficient.

If in the optical fiber FPI the mirrors are separated by a single-mode optical fiber, then the optical fiber FPI is called *intrinsic fiber FP interferometer* (IFFPI). However, in the extrinsic fiber FP interferometer (EFFPI), the two mirrors are separated by an air gap or some material other than fiber. Light from emitter to the FP and from FP to the detector is generally transmitted by single-mode fibers. Three schematic configurations of IFFPI are shown in Figure 3.6. As shown in Figure 3.6a, one end of the fiber is polished as a mirror. For higher-quality factor ($Q = \omega/\Delta\omega$), the polished end is coated with suitable dielectric layers. The second mirror of IEFPI is an internal mirror that can be made by splicing polished fibers or polished coated fibers.

The mirrors of an IFFPI presented in Figure 3.6b are internal fiber mirrors, while those used in the IFFPI shown in Figure 3.6c are FBG reflectors. Depending on the application of IFFPI, one of the presented configurations can be used.

Four different configurations for EFFPI are also shown in Figure 3.6. Figure 3.6d shows an EFFPI with air gap cavity bounded by the ends of a polished fiber and a diaphragm mirror. The cavity length is of the order of several microns and can be increased by convex mirror diaphragm. To increase the EFFPI quality factor as shown in Figure 3.6e, a thin film of transparent solid material is also coated on the end of the optical fiber. The EFFPIs of Figure 3.6d and e are operating based on the reflection coefficient measurement. The air gap cavity between two polished fiber surfaces where the fibers are aligned in a hollow tube is another EFFPI configuration (Figure 3.6f). The structure of the in-line fiber etalon is shown in Figure 3.6g. This structure is an HC fiber spliced between two single-mode fibers. The diffraction loss imposes a limit of the order of a few hundreds of microns on the EFFPI's practical length [35].

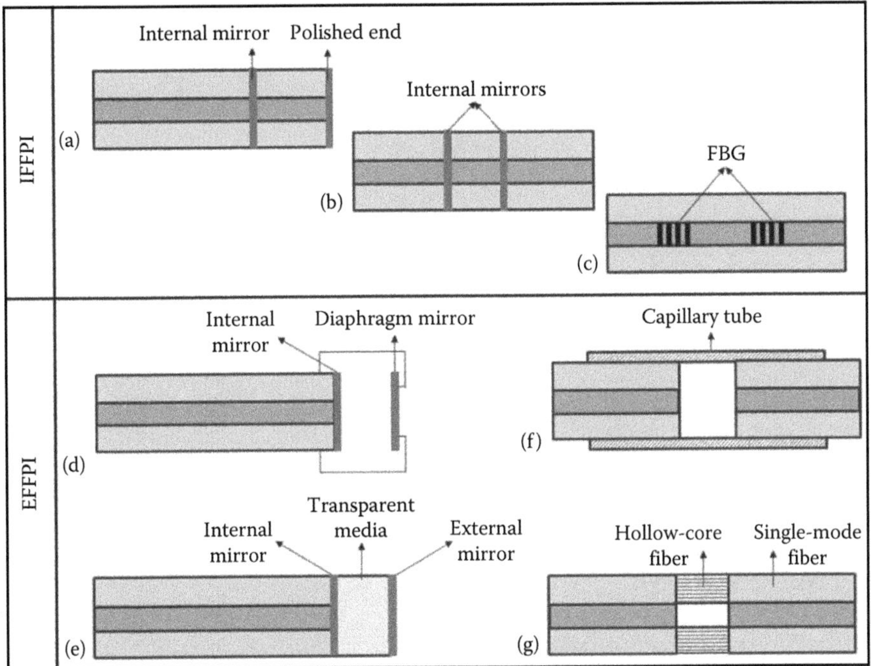

FIGURE 3.6 Different configurations of (a–c) IFFPI and (d–g) EFFPI.

3.4.2 Fiber Bragg Gratings

The forward and backward modes of optical fibers propagate without coupling to each other in the absence of any perturbation. Mode coupling can be controlled by changing the refractive index along the optical fiber. The coupled mode theory (CMT) is a standard subject in related text books that explain mode propagation in slightly nonuniform media [36]. In a periodic structure, where the refractive index of the optical fiber core varies periodically from a lower index n_0 to a higher index n, the scattering from different periods can produce constructive interference for some frequencies and destructive interference for other frequencies in the forward and backward modes. This well-known effect is called Bragg diffraction. Depending on the period length (Λ), periodic structures are classified as long-period grating (LPG) or FBG. The periods of the LPG and FBG are of the order of microns and nanometers, respectively.

The LPG operation is based on the coupling of fundamental core modes to higher-order copropagating cladding modes. The coupling wavelength is obtained by the phase matching condition or linear momentum conservation law $\lambda = (\beta_1 - \beta_2)\Lambda$, where β_1 and β_2 are propagation constants of the core and cladding modes, respectively [37].

FBGs are employed as a frequency-selective mirror, multilayer mirrors, or a polarization-selective rotator. Backward waves at selected frequencies have constructive interference, while forward waves have destructive interference. The backward constructive interference occurs in a narrow range of wavelength around the Bragg condition $\lambda_B = 2n_{eff}\Lambda$, where n_{eff} is the effective refractive index of the core.

The strong and weak grating limits are defined as $\Delta nL \gg \lambda_B$ and $\Delta nL \ll \lambda_B$, respectively, where $\Delta n = n - n_0$ is an index modulation and L is the FBG length. For the strong and weak FBGs, the reflection bandwidth is proportional to Δn and $1/L$, respectively [38]. The FBG bandwidth is typically below one nanometer. In the polarization rotator, a mode with a given polarization is coupled to another mode with a different polarization. The conservation laws of energy and momentum can be employed to obtain the governing equations of FBG interferometer. The period of FBG depends

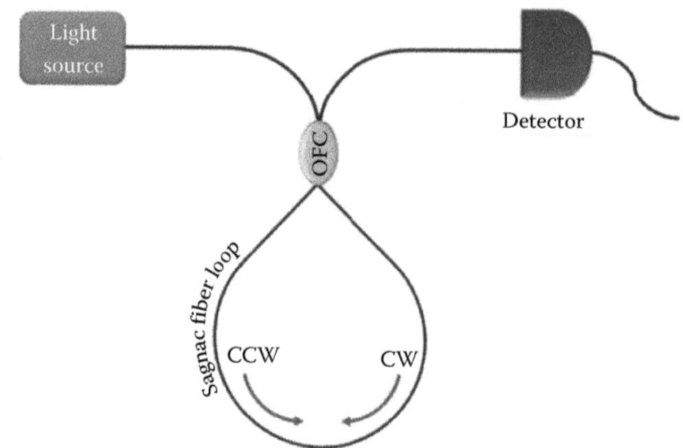

FIGURE 3.7 A schematic of Sagnac interferometer.

on its application and can vary systematically or randomly along the optical fiber core. In a chirped FBG, the period varies monotonically along the optical fiber and has many applications in sensors and optical fiber networks [39].

3.4.3 Sagnac Interferometer

A schematic of Sagnac OFI is presented in Figure 3.7. A single-mode stabilized semiconductor laser or erbium-doped optical fiber laser is employed as a light source for the interferometer. The laser beam is well collimated with uniform phase and splits into two parts with equal intensity by a 3 dB OFC. The two parts travel around a single-mode optical fiber coil (Sagnac coil) in opposite directions. The output of the Sagnac coil is guided toward a single detector. The CW and CCW modes are in phase in a nonrotating fiber Sagnac interferometer, while in a rotating one due to the rotation velocity, the optical path of one of the modes is shorter than the other one. The interference spectrum depends on the angular frequency of the interferometer [40]. Theoretical analysis is based on the Doppler shifts of the CW and CCW modes. It is assumed that the rotational axis is oriented along the optical fiber's coil axis. The phase difference between the CW and CCW mode $\Delta\phi$ versus the free-space laser wavelength λ, the coil area A, the number of coil turns N, and the coil angular frequency Ω is $\Delta\phi = 8\pi NA\Omega/(\lambda c)$ [41,42]. The sensitivity $\left(S = \Delta\varphi/\Omega = 8\pi NA/\left(\lambda c\right)\right)$ increases by increasing the coil radius, total fiber length, and laser frequency. It is obvious that the total fiber length is restricted by the optical fiber attenuation and the coil radius is limited by the packaging size.

Since the Sagnac interferometer's output is independent of source noise, a superfluorescent optical source can be employed instead of a narrow linewidth laser source. Minimum polarization fading is another advantage of the Sagnac interferometer. The Sagnac interferometer fiber is insensitive to low frequencies.

The Sagnac interferometer has been used for rotation sensing primarily. An optical gyroscope based on the Sagnac interferometer is commercially available. Fiber Sagnac interferometers are also employed for detecting current, acoustic wave, strain, and temperature [43–47]. Even a single-photon Sagnac interferometer with a net visibility of up to $99.2 \pm 0.04\%$ for gyro and quantum application was reported [48].

3.4.4 Fiber Ring Resonator Interferometer

As shown in Figure 3.8a, a ring resonator is a fiber ring which is coupled to the input–output fibers through an optical fiber directional coupler. At frequencies where the incoming light and countered fields are in-phase at the input of the ring resonator, constructive interference causes field

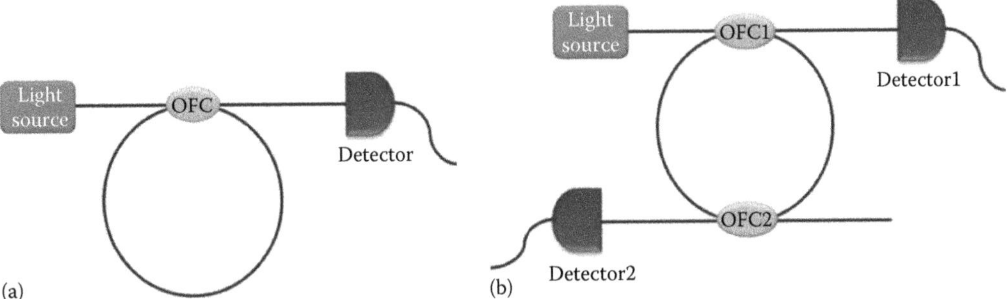

FIGURE 3.8 A schematic configuration of (a) fiber ring resonator and (b) double-coupler fiber ring resonator.

enhancement proportional to the fiber ring quality factor. In lossless optical fibers, the absolute value of transmission coefficient of the ring resonator is unity at all frequencies, while its phase abruptly changes around the resonance frequency. In other words, a lossless ring resonator is an all-pass filter, while for a lossy optical fiber at the resonance frequency, the field in the ring resonator increases and the output decreases. Commonly, in ring resonators, the bandwidth is inversely proportional to the fiber ring quality factor [30].

One of the advantages of fiber ring resonator interferometer in comparison with other feedback fiber devices—such as FP interferometer and FBG—is that a multiple input/output design can easily be achieved. For instance, Figure 3.8b shows the structure of a double-coupler ring resonator, in which the input is coupled through coupler 1 and the output can come from coupler 1 or 2. The output of OFC2 has its maximum value at the resonance frequency. Therefore, the lossless double-coupler fiber ring resonator has the advantage of not depending on phase measurement compared to fiber ring resonator. The resonance frequency depends on the fiber length and its refractive index ($nL = m\lambda; m \in \mathbb{N}$). The length of ring resonators can be from several meters to a few micrometers. Microrings are usually used in photonic integrated circuits.

By exploiting nonlinearity (such as Kerr effect or saturable absorption) in the fiber ring resonator, various types of instabilities (such as bistability, monostability, periodic pulse generation, optical turbulence, and chaos) can occur in fiber ring resonator, given suitable conditions [49,50]. This wide range of dynamic behavior makes fiber ring resonators a multipurpose element (having applications in clock pulse generation, timing control, and all-optical memory) for utilization in optical fiber communication networks.

3.4.5 Mach–Zehnder Optical Fiber Interferometer

The Mach–Zehnder optical fiber interferometer (MZOFI) is so flexible that it can be employed for many diverse applications. An isolated laser diode is employed as the light source of long coherence length. In the two-legged MZOFI, the light is split into two similar parts by a symmetric OFBS and coupled to the two legs of MZOFI. Typically, the difference of the optical path lengths can be detected by a homodyne demodulator.

The N-path MZOFI can be easily constructed using $N \times N$ couplers and single-mode optical fibers. Figure 3.9a shows an N-path Mach–Zehnder interferometer. Each $2N$-port coupler in the presence of parasite reflection and polarization is described by a $4N \times 4N$ unitary matrix. For linear polarized (LP) fields and in the absence of polarization changing devices, the coupler can be characterized by an $N \times N$ matrix. The tritters that are commercially available are described by a 3×3 unitary matrix [51]. For LP modes propagating in single-mode fibers, a scalar analysis is sufficient. As an example, a three-path Mach–Zehnder interferometer is described by the product of two coupler matrices $\mathbb{U}$ and an optical path matrix $\mathcal{P} = diag\left(e^{i\varphi_1}, e^{i\varphi_2}, e^{i\varphi_3}\right)$ via $M = \mathbb{U}\mathcal{P}\mathbb{U}$, where $\varphi_i (i = 1,2,3)$ is the phase of the ith path. When only one of the input fields is nonzero $\left(E_{in} = \left(E_{in}, 0, 0\right)\right)$, the output

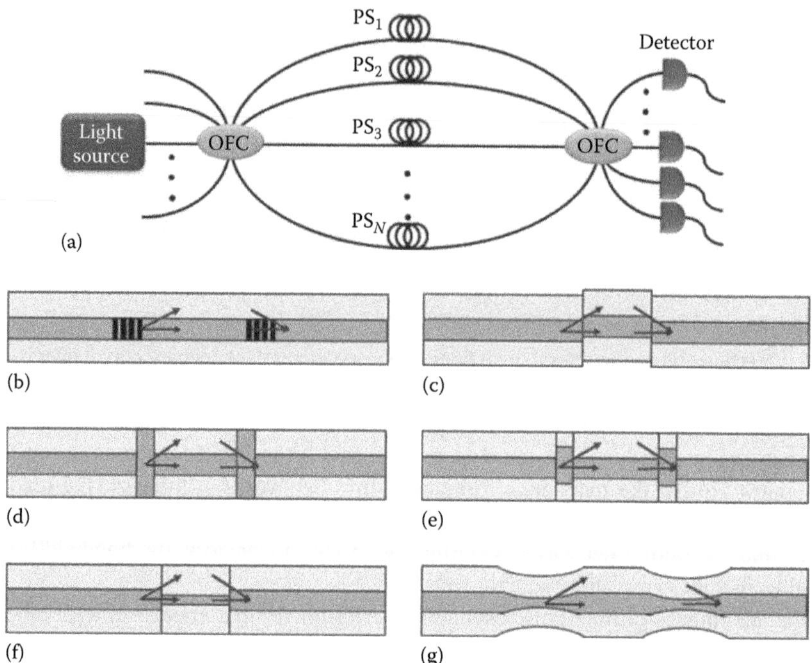

FIGURE 3.9 A schematic of (a) N-path MZOFI and (b–g) various configurations of in-line MZOFI.

fields are obtained by the Mach–Zehnder transformation matrix ($E_{out} = ME_{in}$). The output intensities I_n (n = 1,2,3) versus the optical path phase differences $\varphi_{ij} = \varphi_i - \varphi_j$ (i,j = 1,2,3) and input intensity I_0 are given by

$$I_n = \frac{I_0}{9}\left[3 + 2\cos(\varphi_{12} + \theta_n) + 2\cos(\varphi_{23} + \theta_n) + 2\cos(\varphi_{31} + \theta_n)\right] \quad n = 1,2,3 \tag{3.27}$$

where $(\theta_1, \theta_2, \theta_3) = (0, -2\pi/3, 2\pi/3)$. In the presence of loss or gain in optical fibers, the optical path matrix is $\mathcal{P} = diag(a_1, a_2e^{i\varphi_{12}}, a_3e^{i\varphi_{13}})$, where $a_n(n = 1, 2, 3)$ is the transmission coefficient of the nth optical fiber branch. In this case, the output intensities are obtained by

$$I_n = \frac{I_0}{9}\left[a_1^2 + a_2^2 + a_3^2 + 2a_1a_2\cos(\varphi_{12} + \theta_n) + 2a_1a_3\cos(\varphi_{13} + \theta_n) + 2a_2a_3\cos(\varphi_{23} + \theta_n)\right] \tag{3.28}$$

In the interference pattern of an N-path OFI, N–2 side lobes are observed between the main peaks. This is similar to the interference pattern of an N-slit illuminated by a monochromatic plane wave. Since the slopes and main peaks are steeper in an N-path interferometer, its sensitivity is higher than a conventional MZOFI. In addition, due to the phase differences φ_{ij} being sensitive to the environmental parameters such as temperature and strain, MZOFI is a suitable choice for environmental parameter measurement.

Due to the higher sensitivity of the cladding modes to the changes in the surrounding environment, conventional MZOFI are replaced by in-line waveguide MZIs in which the core and cladding modes are employed as the two arms of the interferometer [52,53]. As shown in Figure 3.9b–g, several configurations for the in-line MZOFI have been proposed and investigated, both theoretically and experimentally. In-line waveguide MZIs use different methods to couple core modes to the cladding and then recouple them to the core. The recoupled cladding mode creates an interference with the uncoupled core mode, which makes this structure very compact and efficient.

As shown in Figure 3.9b, a part of the mode guided in the single-mode fiber is coupled to the cladding modes of the fiber via an LPG and then recoupled to the core mode by another LPG. Due to modal dispersion, the core and cladding modes have different optical path lengths. The recombination of core and cladding mode in the core produces an interference pattern [53–59]. The LPG-Mach–Zehnder interferometer can be employed for multiparameter measurement. As shown in Figure 3.9c, the core–cladding mode splitter can be made by a very small lateral offset of the fiber cores. Due to the fiber offset, a part of the core mode is coupled to the cladding modes and then recoupled to the core mode by a second core–cladding mode splitter. The coupling coefficient of offset fibers is nearly wavelength independent; hence, the offset method can be employed at any wavelength and is cost-effective compared to the in-line MZI composed of a pair of LPGs. The in-line MZOFI can be simply made by a commercial fusion splicing. The cladding mode and insertion loss can be controlled by the amount of lateral offset. This lateral offset can be adjusted so that only one cladding mode is dominant.

As shown in Figure 3.9d, another proposed method for in-line MZOFI manufacturing is collapsing air holes of a PCF, which does not need any alignment or cleaving processes. The PCF mode is expanded at the collapsed region and part of its energy is coupled to the cladding modes. In this structure, the coupling to several cladding modes is observed and controlling the number of cladding modes is not so simple [60]. The insertion loss of this structure is high in comparison to the offset method, although by combining the LPG and collapsing methods, this insertion loss can be reduced [61].

Fibers with different core sizes can be used for beam splitting [62,63]. One method is splicing a short piece of multimode fiber between two single-mode fibers as shown in Figure 3.9e. The light exiting the single-mode fiber is spread at the multimode region and then coupled into the core and cladding of the next single-mode fiber [62].

A small core fiber can be inserted between two conventional single-mode fibers to make an in-line MZI. As shown in Figure 3.9f at small core fiber region, part of the light is guided as a cladding mode [63]. An in-line MZOFI can be obtained by tapering a single-mode fiber at two points along the fiber, as shown in Figure 3.9g [64,65]. At the tapering points, the core mode diameter increases and part of it couples to cladding modes. This structure is very simple but the tapering regions are mechanically weak. Many other configurations for in-line MZOFI such as using double cladding fiber [66], microcavities [67], and a twin-core fiber [68] have been investigated.

3.4.6 Michelson Optical Fiber Interferometer

As shown in Figure 3.10a, the fabrication method and operation principle of conventional Michelson OFI (MOFI) is very similar to the MZOFI. The main difference is the mirrors at the end of the interferometer legs, which cause the MOFI to become a folded MZOFI. In a conventional MOFI, the high coherent light is split into two optical paths by a 2×2 OFC. The reflected light by mirror M_1 and M_2 are recombined by the OFC to produce interference patterns at the detector. The compact in-line configuration of MOFI is also possible, which is depicted in Figure 3.10b. Part of the core mode is coupled to the cladding modes by core–cladding mode splitter. Both the core and cladding modes are reflected by a common reflector at the end of the fiber [69–72]. The LPG can be used as a core–cladding BS in the in-line MI structures. For some applications, to prevent the environmental effects on LPG behavior, metal-coated LPGs are employed [73]. To reduce the temperature effect on the measurements, fused silica PCF can be employed [74].

As shown in Figure 3.10c, the conventional MOFI can be generalized to an N-path MOFI. The generalized BS is characterized by an $N\times N$ matrix (U) in a scalar model. The optical path matrix and the mirror reflection matrix are $\mathcal{P} = diag(e^{-i\varphi_1},\ldots,e^{-i\varphi_N})$ and $M = diag(-1,-1,\ldots,-1)$, respectively. The output fields are related to the input fields by the scattering matrix ($S = U\mathcal{P}M\mathcal{P}U$) via $E_{out} = SE_{in}$, where $E_{in} = \left(E_f^{(1)},E_f^{(2)},\ldots,E_f^{(N)}\right)$ and $E_{out} = \left(E_b^{(1)},E_b^{(2)},\ldots,E_b^{(N)}\right)$ are the input and output

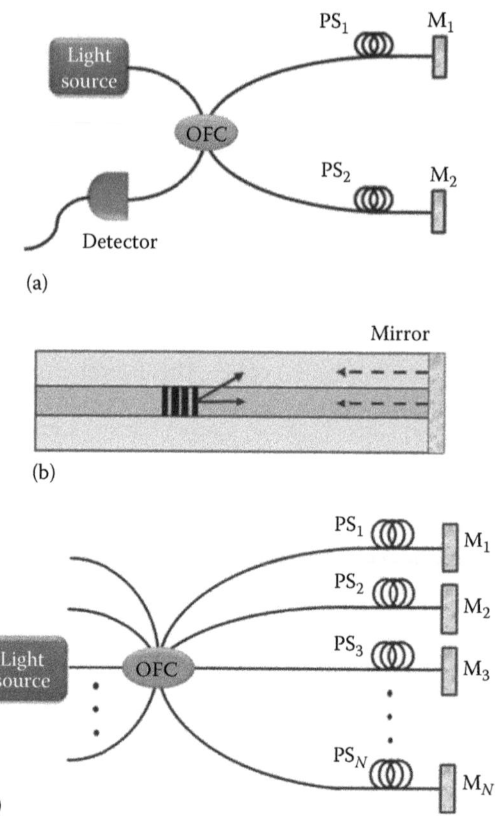

FIGURE 3.10 A schematic configuration of (a) basic MOFI, (b) compact in-line MOFI, and (c) *N*-path MOFI.

fields and $E_f^{(i)}$ and $E_b^{(i)}$ are the forward and backward fields in the *i*th optical fiber. It is assumed that only one of the input fields is nonzero. The input field vector is $E_{in} = (\varepsilon_1, 0, 0, \ldots, 0)$ and the output intensities at the *j*th port I_j are as follows:

$$I_j = \left| \varepsilon_1 \sum_{m=1}^{N} U_{jm} U_{m1} e^{-2i\varphi_m} \right|^2 \tag{3.29}$$

The sensitivity of the multipath MOFI is also greater than the conventional two-path ones.

3.4.7 Modal Optical Fiber Interferometer

Different modes of multimode fibers have different velocities and the modal interferometers are established on the basis of this effect (dispersion). Typically, LP_{01} and LP_{11} modes or HE_{11} and HE_{21} modes of step index optical fibers can be employed to design the modal interferometers. Moreover, the two eigen polarizations of PMF can be employed for modal interferometry [75]. The unique properties of holey fibers can be employed to design modal OFI. The PCF interferometers have the advantage of detecting, sensing, or spectroscopic analyzing of gas and liquids [76]. The holey and hollow fibers have their own advantages. In holey fibers that are also called index-guiding fibers or solid-core PCFs, the desired gas or liquid interacts with cladding evanescent fields that have a few percentage of the total light power, while in HC-PCF, the gas or liquid interacts with the central part of the fundamental mode, which is more than 90% of the total light power. The bandwidth of silicon-core single-mode PCF is more than one thousand nanometers, which is greater than those of an air-core PCF fiber. A nanolayer of rare metal coating on the surface of the core and voids causes plasmon–light interaction in PCF and extremely enhances the interferometer's sensitivity [77]. Depending on the modal fiber interferometer, the birefringent PCF or PANDA fiber can be employed [78].

3.4.8 Moiré Optical Fiber Interferometer

Overlapping of two or more gratings at different angles (θ) creates fringe patterns, which are the basis of Moiré interferometry. The desired fringe pattern can be achieved by choosing a suitable arrangement of optical fibers. N polarization-maintained optical fibers are employed for generation of interference grid pattern. The polarization angle of the jth fiber relative to the x-axis is denoted by θ_j ($j = 1,2,\ldots,N$) and its center coordinate in the $z = 0$ plane is (a_j,b_j). The field in the $z = D$ plane at the point (x, y) is given by

$$E(x,y) = \sum_{j=1}^{N} \vec{E}_j(a_j,b_j)e^{-i\left[\frac{k}{D}\left(xa_j+b_jy\right)+\varphi_j\right]} + c.c. \tag{3.30}$$

where φ_j is the phase of the field of the jth fiber at $z = 0$ plane. The field intensity at the (x,y) point in the observation plane versus light intensity corresponding to the ith fiber (I_i ($i = 1,2,\ldots,N$)) is

$$I = \sum_{i=1}^{N} I_i + \sum_{i\neq j} \sqrt{I_iI_j}\cos(\theta_i-\theta_j)\cos\left\{\frac{k}{D}\left(a_i-a_j\right)x+\left(b_i-b_j\right)y-\varphi_{ij}\right\} \tag{3.31}$$

where

φ_{ij} is the phase difference between the ith and jth optical fiber
k is the incident light wave number [79]

The desired fringe pattern can be obtained by choosing suitable fiber coordinates and polarizations. As an example, consider a system of three fibers centered at $P(0,0)$, $P(2a,0)$, and $P(0,2a)$, where a is the radius of the PMF. The fibers' arrangement and interference patterns are shown in Figure 3.11a. The horizontal and vertical patterns correspond to the interference of fibers 1 and 3 and fibers 1 and 2, respectively. The oblique lines in Figure 3.11a are due to the interference of fibers 2 and 3. By employing perpendicular polarizations for fiber 2 and 3, the interference between them cancels and oblique lines are eliminated. Setting an angle of 45 degrees between the polarization of fiber 1 and the x-axis ensures the occurrence of interference between both fibers 2 and 3 with fiber 1. To obtain a desired interferometric pattern, an arbitrary configuration of PMF is chosen with variable positions ((a_i, b_j) and θ_j). In order for the generated intensity distribution to be the closest to the desired distribution, in the sense of the metric of the L^2 (R^2) space, the parameters a_i, b_j and θ_j must be chosen wisely.

3.4.9 White Light Optical Fiber Interferometer

In coherent light interferometry, the coherent length of the narrowband sources such as lasers is much greater than the optical path length difference in the interferometers. The fringes have a periodic structure, so the interferometric measurement suffers from an integer multiple of 2π phase ambiguity. Hence, coherent interferometry does not produce absolute data unless extra complexity is added to the interferometer itself. Using wideband light sources, the phase ambiguity is eliminated. Such an interferometry is called low coherency or white light interferometry (WLI).

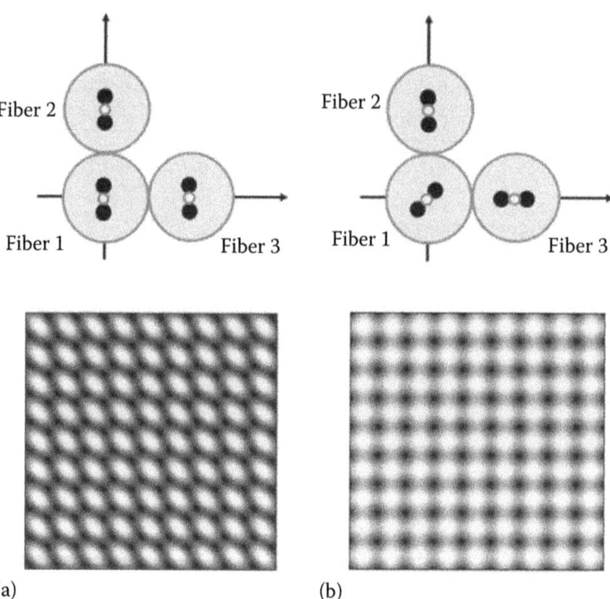

FIGURE 3.11 (a-b) Two specific arrangements of (top) fiber-optic Moiré interference system and (bottom) their corresponding interference patterns.

In WLI, corresponding to each wavelength, a separate fringe system is produced. The fringes of different wavelengths will no longer coincide as moving away from the center of the pattern. The electric field at any point of observation is the sum of electric fields of these individual patterns. The overall pattern is a sequence of colors whose saturation decreases rapidly. The WLI is adjusted such that the OPD is zero at the center of the field of view, so the electric field of different wavelengths exhibits a maximum at the center point. So by measuring the fringe peak or envelop of the interferogram, the phase difference can be obtained without any ambiguity [80].

The light sources such as tungsten lamps, fluorescent lamps, super-luminescent diodes (SLDs), light-emitting diode (LEDs), laser diodes near threshold, and optically pumped erbium-doped fibers can be used in WLIs. The spectral width of SLD and LED is between 20 and 100 nm and coherence length is less than 20 μm at the operating 1.3 μm wavelength.

Generally, WLI operates on the basis of balancing the two arms of the interferometer by compensating for the OPD in the reference arm. The length of the reference arm can be controlled by different methods such as moving mirrors or piezoelectric (PZT) devices.

The intensity of the interference fringe drops from a maximum to a minimum value by increasing the OPD between the two paths of WLI. Measuring the position of the central fringe is of utmost importance in WLI. The distance between the central fringe and its adjacent side fringe is so small that in the presence of noise there are some ambiguities in determining the central fringe position. These ambiguities can be removed by employing a combinational source of two or three multimode laser diodes with different wavelengths [81,82].

WLFIs are designed on different topologies of single- or multimode fiber interferometers. Single- and multimode fibers have their own advantages and disadvantages. White light single-mode fiber interferometer provides stable and large signal-to-noise ratio, whereas interferometers based on the multimode fibers employ cheaper optical components [81,82]. There are several WLFIs corresponding to each of the standard OFIs or their combinations [83,84]. Figure 3.12a shows a WLFI based on the MOFI working in the spatial domain. LED light splits by a lossless 2 × 2 OFC and couples to the arms of the MOFI. The reflected beam recombines on the avalanche photodiode (APD) of the MOFI. The mirror M_2 is adjusted to a maximum output corresponding to the position of the central fringe.

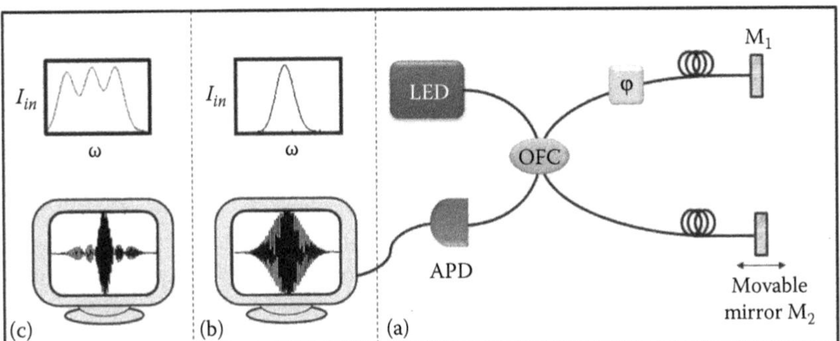

FIGURE 3.12 (a) A schematic configuration of WLI Michelson interferometer, (b) the spectrum of a normal LED (up) and the corresponding fringe pattern of the WLFI (down), and (c) the spectrum of three peak LED (up) and the corresponding fringe pattern of the WLFI (down).

Figure 3.12b shows the fringe pattern of the WLFI based on the MOFI, for OPDs less than coherent length of the source. The position of the highest amplitude corresponds to exactly zero OPD. In Figure 3.12c, the results of three peak LEDs are compared with those of a normal LED to see how the precision increases when the multiwavelength wideband light source is employed in comparison to the single-wavelength wideband interferometer.

3.4.10 Composite Optical Fiber Interferometer

Some first-order or field interferometers were already discussed in Sections 3.4.1 through 3.4.9. In all of the mentioned interferometers, the OPD between two or among many paths can be measured on the basis of field interference. Evidently, combination of different topologies can produce a new topology with advantageous properties. For instance, Michelson–Sagnac interferometer has new properties for applications in quantum optomechanics [85]. The reentrant topologies have feedback loops that have considerable effect on the stability of the interferometer. For instance, the double-loop interferometer such as that shown in Figure 3.13 has phase and polarization stability [86]. By employing the proper loop's parameters, the fiber interferometer operates in the stable regime.

The higher-order interferometers are not in the scope of this chapter. However, Hanbury HB-T interferometer that is a well-known configuration for second-order interferometry was introduced briefly in Section 3.2.2. As an example, the intensity interferometry can be used in astronomy to measure the angle between the light sources such as stellar.

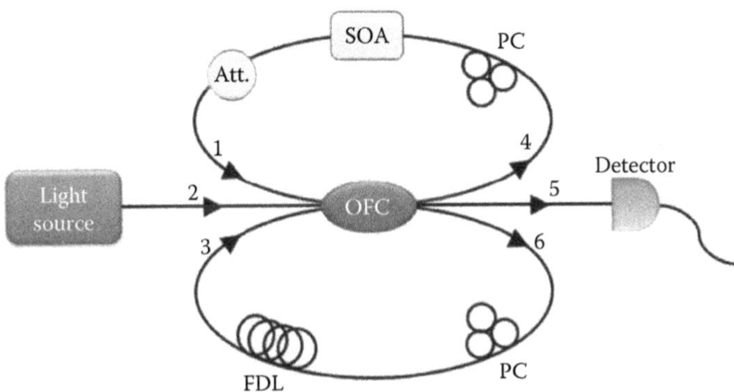

FIGURE 3.13 A schematic configuration of double-loop OFI. FDL, PC, Att., and SOA stand for fiber delay line, polarization controller, attenuator, and signal optical amplifier, respectively.

3.5 SIGNAL RECOVERING AND NOISE SOURCE IN OPTICAL FIBER INTERFEROMETRY

Different configurations of interferometric fiber sensors can be classified into balanced and unbalanced paths design. The balanced-paths interferometer can be made by making the reference and sensing arm similar to each other by employing some devices such as PZTs. The balanced scheme is very popular because it usually does not require high-coherence sources. The imbalanced-paths configuration is very sensitive to laser noise and requires a very long coherence length laser. In the unbalanced fiber-optic interferometric sensors, the phase difference is measured by signal recovery method.

The accuracy of OPD measurement by different optical fiber interferometry is limited by the noise sources. In this section, the signal recovery methods and sources of noise in OFIs are briefly reviewed.

3.5.1 Signal Recovery Method

As mentioned in Section 3.4, the OPDs are converted to the optical phase differences in different types of optical interferometers. The phase difference variation has sufficiently low-frequency components. The phase difference can be measured both in time and spatial domain. When the phase differences are converted to light intensity on the spatial observation plane, then the photodiode matrix or charge-coupled devices (CCDs) can be employed to measure the interferometric fringe pattern. Generally, the outputs are proportional to the cosine of the phase difference ($\cos\phi$). In the absence of $\sin\phi$, there is a π radian ambiguity in the phase recovery. If the phase difference is greater than 2π radians in addition to the sine and cosine values, the history of phase variation must be tracked to obtain the precise phase difference and corresponding OPD. Methods based on the production of new frequencies are called heterodyne detection; other methods are named homodyne. The three main recovering methods described in the following sections are phase-generated carrier homodyne detection, fringe-rate measurement, and homodyne detection methods.

3.5.2 Phase-Generated Carrier Homodyne Detection

The semiconductor laser source of the interferometer is driven by a biased sinusoidal current. The laser frequency and its output power are modulated by the laser current variation. Corresponding to given OPDs, the phase differences at the output of the interferometer changes by varying the laser output wavelength. Therefore, the current frequency is modulated to the photo diode output.

The current frequency and its harmonics can be observed in the side bands of the demodulated outputs. Two copies of sidebands are chosen by band-pass filters. Choosing filters with proper characteristics guaranties copies with the same amplitude. The filter outputs are employed as the inputs of an electronic mixer. The two outputs of the mixer are proportional to the cosine and sine of the phase difference of interest. Sometimes, the method is also called pseudoheterodyne detection (PHD) [87]. In synthetic heterodyne detection [88], instead of laser frequency modulation, the phase shift modulation is produced by wrapping one arm of the interferometer around a PZT device, which is driven by a sinusoidal voltage.

3.5.3 Fringe-Rate Method

In order to measure phase shifts greater than 2π radians, the fringe-counting and fringe-rate demodulation methods are introduced [89,90]. These methods are based on the transition of interferometric output across some central value. The fringe-counting methods are based on the digital counting of the detector output in a suitable period of time. The frequency is the ratio of the counting number to the counting time. Due to the Heisenberg's inequality principle, the resolution of frequency is proportional to the inverse of the counting time. The phase difference can be obtained by integrating the instantaneous frequency.

The fringe-rate method is based on a frequency to voltage converter (FVC). The detector output is employed as an input of FVC. The FVC output is integrated to obtain the phase difference. The minimum detectable signal is of the order of π radians and can be improved by axillary circuits.

3.5.4 Homodyne Method

As mentioned in previous sections, the range of phase difference measurement by the fringe-counting and fringe-rate methods have a lower limit of π radians, while that of synthetic heterodyne method has an upper limit of π radians. The combination of both methods has no limitation on the phase difference measurement. In principle, the phase difference of different paths is precisely measurable, since the outputs of the OFIs are related to the input fields through a unitary matrix and unitary matrices are invertible. A number of homodyne techniques are used to bridge the gap between the measurement range of synthetic heterodyne detection method and that of fringe-counting and fringe-rate methods.

OFIs usually have two outputs. In the homodyne method, an external 2×2 OFC is employed to create the interference patterns from the OFI's outputs. Due to the unitarity of the scattering matrix, it is easy to show that the two outputs of OFC are 180° out of phase from one another. Hence, when all energy is presented in one of the OFC's outputs, the other one is completely dark and vice versa. Therefore, no orthogonal components can be found in the OFC's output. The orthogonal component can be produced by the heterodyne method. In order for the orthogonal component to directly appear in the outputs of the external coupler, a generalized OFC can be employed. For example, a 3×3 coupler is used as the external coupler of interferometer to create outputs with orthogonal components without employing the heterodyne detection technique [91].

3.5.5 Noise Sources in OFIs

The precision and the minimum detectable phase differences of OFIs are determined by their output signal-to-noise ratio. The sources of noise can be originating from the light source, optical fibers, electronic circuits, and the surrounding environment. In practice, the output noise also depends on the OFI topology. In other words, any random process taking place in each part of the OFI acts as a noise source in it.

Laser is the source of light in narrowband OFIs, while LED or any other wideband light generator is used as a light source for WLOFIs. Laser sources operate on stimulated emission, while wideband light generators are based on spontaneous emission of atoms and molecules. The spontaneous and stimulated emission of atoms and molecules are quantum effects, which are stochastic processes and can produce phase and amplitude noise in the light source output [92–94]. On the other hand, the interaction between laser cavity and its surrounding heat bath modes through the mirror coupling is another noise source for the laser output [95]. The cavity as a narrowband filter reduces the laser noise significantly [96,97]. The presence of phase and amplitude noise causes the bandwidth of laser light to increase. By employing proper cavity in a single-mode laser, the bandwidth can be reduced to several kilohertz corresponding to several tens of kilometers for coherence length. Cross-saturation and mode competition effects are new noise sources in multimode lasers that can make multimode laser more suitable for WLFI than the narrowband OFI.

Rayleigh scattering, Mie scattering, core–cladding interface scattering, absorption, amplification, Brillouin scattering, and Raman scattering are the main sources of noise in OFI arms and connectors. The first five cases are linear and the two latter ones are nonlinear noise sources. Some fraction of light that is deviated by scattering is trapped in the guided region and can propagate in both forward and backward directions along the optical fiber axis. This fraction of scattered light contributes to phase and amplitude noise simultaneously. Other parts that are scattered out of the optical fiber affect the amplitude noise only. The nonlinear effects such as Brillouin and Raman scattering have two different components: Stokes and anti-Stokes frequencies. The Brillouin and Raman shifts are due to the light interaction with acoustic (±25 GHz) and optical phonons (13 THz), respectively [1]. Beating between Stokes, anti-Stokes with laser light cannot be observed at the output of any realistic detector and will be filtered intrinsically by the detector as a low pass filter. The Brillouin and Raman nonlinear losses are considered as a source of amplitude noise. The Raman, Brillouin, and Rayleigh scattering are symmetrically distributed with respect to forward and backward propagation directions, while those of Mie and core–cladding interference scattering are mainly in the forward propagation direction. In multimode fiber interferometers, the mode coupling is another source of noise.

APD, PIN diode, CCD, and photomultiplier (PhM) are used as an electronic detector in OFIs. Thermal noise, dark current noise, shot noise, background noise, and flicker noise are common noises in all optical detectors. The electron–hole generation and recombination are stochastic processes and are noise sources in semiconductor detectors. The avalanche effect is also a random process and is a noise source in APDs. The same effect can be found on the anodes of PhMs and create noise in the detector. The Johnson noise, shot noise, burst noise, and flicker noise of different electronic elements are the final intrinsic noise of OFIs.

The optical fiber parameters can be affected by the environmental physical variations such as mechanical vibration, acoustic agitation, pressure, tension, and thermal variations. These effects are channels for transferring environmental fluctuations to the OFI output as noise.

In conclusion, the output signal is affected by all noise sources. The output signal can be denoised by signal processing methods. Depending on the signal's behavior, an appropriate denoising method must be used. For instance, the *windowed Fourier method* can be employed to denoise the music-like signals of a fiber intruder detector that operates based on the birefringent fiber interferometer [75]. Wavelet analysis methods such as Fourier regularized deconvolution (FoRD) and Fourier-wavelet regularized deconvolution (ForWaRD) are employed to denoise the transient signals [98].

3.6 INTERFEROMETRIC OPTICAL FIBER SENSORS (IOFSs)

The optical path length of each arm of the OFI—which is the product of fiber refractive index and geometrical length of the arm—can be perturbed by one or more environmental physical parameters. The optical path length variation can be measured precisely by the optical fiber interferometry. This is the basic concept of IOFS operation. By measuring the phase difference, one can sense the changes in the environmental physical parameter.

3.6.1 Principle of Operation of IOFSs

The IOFSs can be classified into resonance and nonresonance types. The resonance types such as optical fiber FPIs, ring resonators, and FBGs are operating on the basis of constructive and destructive interference. In the resonance-based IOFSs, the resonance frequency changes as a function of the environmental physical parameters. The resonance frequency or bandwidth measurement can be employed to sense the environmental parameter changes. The output amplitude or phase change at a given frequency can provide some information about IOFSs' environment.

The nonresonance IOFSs are operating on the basis of comparing the OPD of the different arms of the OFI. One or several arms are kept isolated from external variations and only the sensing arm is exposed to the environmental variations. The OPD can be easily detected by analyzing the variation in the interference signal. The phase of resonators is highly sensitive to the optical length variation in the vicinity of resonance frequency. Hence, employing resonators in the sensing arm of the nonresonance OFI improves the sensitivity of the sensor versus the environmental variations.

In some applications such as chemical and biological sensors, cooperation of plasmons in the sensing arm causes an enhancement in the sensitivity of the IOFSs.

3.6.2 Principle of Plasmon

When an incoming electric field is coupled to the collective oscillations of free electrons on a metal surface (surface plasmon resonances), the surface electromagnetic waves that are called surface plasmon polaritons (SPPs) are excited and propagate along the metal–dielectric interface. SPP modes can be excited at the flat and curved metal–dielectric interfaces such as metal nanostructures. The SPPs are able to concentrate electromagnetic waves beyond the diffraction limit and enhance the field strength by several orders of magnitude [99,100].

The simplest geometry for sustaining SPPs is the flat interface between a dielectric and a conductor. The relative permittivity (or dielectric constant) of dielectric medium is denoted by ε_d. The dielectric constant of a conductor depends strongly on the incident light's frequency and is called the dielectric function $\varepsilon_m(\omega)$. Since TE waves (S-polarized) are purely transverse waves, they cannot be coupled to the longitudinal electron oscillations in metal and only TM (P-polarized) waves can excite the SPP waves at the metal–dielectric interface. By applying the boundary conditions at the interface, the required condition for exciting SPP modes can be obtained directly. For $\varepsilon_d>0$, the confinement of electromagnetic wave at the conductor–dielectric interface requires that $Re\{\varepsilon_m(\omega)\}<0$ [100]. This condition is satisfied at frequencies below the bulk plasmon frequency (ω_p). In lossless metals with $Re\{\varepsilon_m(\omega)\}<0$ because of the pure imaginary wave vector, there are no propagating waves $\left(k_x^2+k_z^2=\varepsilon_m(\omega)k_0^2<0\right)$. Typically, ε_m is much greater than one $\left(|\varepsilon_m|\gg 1\right)$, which means the penetration depth in metals $|k_z|^{-1}\cong\lambda_0/\left(2\pi\sqrt{|\varepsilon_m|}\right)$ is rather small and can be on the order of nanometer scale [99].

By introducing a TM-polarized electric field as $E(z>0)=\left(E_x^0,0,E_z^d\right)\cdot e^{i(k_{sp}x-\omega t)}$ $exp\left(-z\sqrt{k_{sp}^2-\varepsilon_d k_0^2}\right)$ and $E(z<0)=\left(E_x^0,0,E_z^m\right)\cdot e^{i(k_{sp}x-\omega t)}exp\left(-z\sqrt{k_{sp}^2-\varepsilon_m k_0^2}\right)$ into Helmholtz wave equation and employing the boundary conditions at the metal–dielectric interface, one can derive the SPP dispersion relation as follows:

$$k_{sp}(\omega)=\frac{\omega}{c}\sqrt{\frac{\varepsilon_d\varepsilon_m(\omega)}{\varepsilon_d+\varepsilon_m(\omega)}} \tag{3.32}$$

The necessary conditions for an SPP excitation are $Re\{\varepsilon_d\varepsilon_m(\omega)\}<0$ and $\varepsilon_d+Re\{\varepsilon_m(\omega)\}<0$. In the long-wavelength range of the visible spectrum and in the infrared region, these conditions are satisfied for most metals and dielectrics.

The other key point is the SPP wavelength $\lambda_{sp}=2\pi/Re(k_{sp})$, which is always smaller than the incident light wavelength in the dielectric. The damping of SPPs corresponding to ohmic loss of electron oscillations is a very important factor that restricts both the lower limit of SPPs' wavelength and its propagation length [99,101]. The amplitude of normal and tangential electric field components in the dielectric and metal versus ε_d and ε_m is obtained [99] by

$$E_z^d=i\sqrt{\frac{-\varepsilon_m}{\varepsilon_d}}E_x^0 \tag{3.33}$$

$$E_z^m=-i\sqrt{\frac{-\varepsilon_d}{\varepsilon_m}}E_x^0 \tag{3.34}$$

Typically, $|\varepsilon_m| \gg \varepsilon_d$; therefore, the normal component in the dielectric and the tangential component in the metal are dominant, which is the nature of SPPs. It is easy to show that the penetration depth of SPPs in the dielectric (d_d) and metal (d_m) are related to each other by $\varepsilon_d d_d = |\varepsilon_m| d_m$. The decay lengths of SPPs in metal and dielectric are given by [99,100]

$$d_d = \frac{\lambda}{2\pi}\sqrt{-\frac{\varepsilon_d + Re\{\varepsilon_m(\omega)\}}{\varepsilon_d^2}} \tag{3.35}$$

$$d_m = \frac{\lambda}{2\pi}\sqrt{-\frac{\varepsilon_d + Re\{\varepsilon_m(\omega)\}}{Re\{\varepsilon_m(\omega)\}^2}} \tag{3.36}$$

The penetration depth in metal is generally much shorter than in the dielectric. The decay length in metals is of the order of a few tens of nanometers and weakly depends on the light's wavelength. While in the dielectric, it strongly depends on the wavelength and is of the order of a few hundreds of nanometers for long wavelengths. For long wavelengths, the decay lengths are approximated by the following relations [99]:

$$d_d \approx \left(\frac{\lambda}{2\pi}\right)^2 \frac{\omega_p}{c\varepsilon_d} \tag{3.37}$$

$$d_m \approx \frac{c}{\omega_p} \tag{3.38}$$

where

$\omega_p = ne^2/\varepsilon_0 m_e$ is the plasma frequency
n is the electron density
e and m_e are charge and mass of the electron, respectively

Since the propagation constant of SPPs ($k_{sp}(\omega)$) will never match the dispersion relation of light travelling in the dielectric medium $\left(\beta = k_0/\sqrt{\varepsilon_d}\right)$, SPPs on the flat dielectric–metal interface cannot be excited directly. The presence of a second dielectric layer having a higher dielectric constant or a grating can be used to match the wave vectors.

In multilayer systems, many different SPP modes can be excited. The most famous multilayer structures are a thin metal film abutted with dielectric layers or a thin dielectric layer surrounded by metal layers, often called the insulator–metal–insulator (IMI) and metal–insulator–metal (MIM) configurations, respectively. For the E_z component in a symmetric IMI, there are even and odd-symmetric modes. The even-symmetric modes have considerably smaller attenuation than the odd-symmetric SPP modes. The even- and odd-symmetric SPP modes are called the long-range (LR-SPP) and short-range (SR-SPP) SPP modes, respectively [99,100].

On the surface of metal-coated fiber, SPPs are excited and enhance the field intensity on the metal surface. Inside the holes of a PCF, SPPs can be excited on the metal coating [102]. SPP propagation in a metal-coated fiber is analyzed by solving Maxwell equations in the cylindrical coordinates [103].

In addition, the conduction electrons of metallic nanostructures coupled to the electromagnetic fields excite the nonpropagating or localized surface plasmons. Localized SPP modes arise from monochromatic light scattered by subwavelength conductive nanoparticles. The curved surface of the nanoparticle exerts an effective restoring force on the driven electrons, so that a resonance can occur. This resonance leads to the field enhancement inside and in the near-field zone around the particle. This effect is called localized plasmon resonance for metal nanoparticles. This plasmon resonance frequency falls into the visible region. Plasmon resonances can be excited by direct light illumination without any phase matching arrangement, which is necessary in SPPs. Due to the small size of the nanoparticles (d) relative to the incident light wavelength ($d \ll \lambda$), the propagation effects are negligible and electrostatic approximation can be employed. The harmonic time dependence can be added to the solution obtained from the electrostatic problem. Adding metal nanoparticles in the vicinity of the fiber core causes an enhancement of the evanescent field in the fiber core surrounding.

3.6.3 Plasmon Cooperated IOFS

The intensity enhancement property of SPPs in cooperation with the resonance IOFS made it suitable for nonlinear spectroscopy. Nonlinear effects such as Raman spontaneous and stimulated emission can be observed in high-intensity pumps. The presence of metal nanoparticles around an FP or fiber ring resonator causes an increase in the pump intensity and generates Raman frequencies. The Raman frequencies are coupled to the resonator to detect the desired material in its environment. Single-mode and multimode optical fibers can also be employed to excite the SPPs and sense the refractive changes of their environment.

3.6.4 Fabrication Methods of IOFS

Depending on the application of IOFS, OFIs are used in discrete, quasi-distributed, and distributed fiber sensors. In some applications such as intruder detectors, which are based on the distributed or quasi-distributed detection, OFIs can be manufactured by commercial optical fiber and other components, which are designed for optical fiber communication applications. In most applications such as optical fiber gyroscope or temperature and strain measurements, there is no need for any special optical fiber components. The jacketing and cladding of the optical fiber sensor must be suitably designed for its specific application. As an example, the optical fiber cable structure for strain sensing must be resistive to temperature change and prevent fast heat transfer to the optical fiber.

In some applications such as chemical and biosensors that are operating on the basis of refractive index change measurement or different spectroscopy methods such as SPP spectroscopy, the cladding of the fiber sensor must be removed. The molecules in the surrounding environment can be illuminated by evanescent waves around the fiber core for environmental spectroscopy and refractive index measurement. There are different methods for removing fiber cladding; the most famous ones are chemical etching by hydrogen fluoride (HF) acid and fiber tapering by CO_2 laser or hydrogen torch.

The basic elements of IOFS are Fabry–Pérot, Mach–Zehnder, Michelson, and Sagnac OFIs. The miniaturization of IOFS for microscale and in-line applications is the current trend of OFIs. In-line structures have two optical paths in one physical line. Owing to their several advantages such as easy alignment, high coupling efficiency, and high stability, they have been widely investigated. The optical fiber FPI is an in-line OFI intrinsically. In the EFFPI, the cavity is formed out of the optical fiber by external mirrors and supporting structure. Besides the high finesse interference signal, the EFFPIs have the advantage of a relatively simple fabrication method with low-cost equipment. However, they have some disadvantages such as low coupling efficiency, precise alignment, and packaging problems [104]. PCFs are employed to overcome the problem of low coupling efficiency in EFFPIs. By introducing a fiber laser or by employing a CO_2 laser or electric discharge of fiber splicer, the air holes of the PCF will collapse and a lens will easily form on the PCF end. The introduced lens acts as a reflector and collimator simultaneously [105,106].

In IFFPIs, the mirrors are inside the optical fiber itself. IFFPIs are fabricated using high-cost equipment [53]. Multicavity FPI sensors have unique and interesting characteristics [107–109]. A double cavity FPI sensor is made of a small piece of holey optical fiber (HOP) inserted between a piece of a multimode fiber and a single-mode fiber.

3.7 APPLICATIONS

Besides being used as a transmission line, optical fiber can be employed as a fiber sensor as well. During the last half century, optical fiber sensors and their applications have developed nearly in all branches of science, medicine, and technology. Gyroscope and acoustic fiber sensors are some of the initial research achievements for military and aerospace applications [110,111]. Optical fiber sensors can be divided into the lumped, quasi-distributed, and distributed classes. Lumped sensors are employed to measure the physical parameters at a given position. Distributed fiber sensors are used to continuously measure a given parameter along the fiber length, while quasi-distributed sensors are suitable for measuring physical parameters at finite number of positions along the optical fiber. Interferometric optical fiber sensors are used in all of the three configurations mentioned earlier. In quasi-distributed optical fiber sensors, multiplexing techniques such as

1. Time division multiplexing (TDM) [112–114]
2. Frequency division multiplexing (FDM) [115]
3. Wavelength division multiplexing (WDM) [116]
4. Code division multiplexing (CDM) [117,118]
5. Space division multiplexing (SDM) [119]
6. Hybrid multiplexing schemes [120,121]

are employed to transmit sensors' output data on an optical fiber transmission line [35]. The multiplexing schemes are chosen on the basis of the number of sensors, dynamic range, cross-talk, sensor's frequency response, and cost-effectiveness. TDM techniques are categorized into the transmissive and reflective types. Figure 3.14a shows a transmissive TDM configuration, which is usually employed in hydrophone array. Figure 3.14b and c are reflective types that are operating on the basis of Michelson and FBG interferometers, respectively [122,123].

In TDM schemes, quasi-continuous lasers are employed. Depending on the number of sensors, a laser with an appropriate pulse width is chosen that is typically of the order of 1 ms or less.

Figure 3.15 shows a schematic of an FDM hydrophone array. In FDM technique, each interferometer in the array has a different path length, so a different phase shift is induced. As shown in Figure 3.15, the laser diode is driven by a ramp current to generate a ramp frequency modulated laser. Fringes are tracked by a separate phase-locked loop to extract the signal from different interferometers. In this scheme, the interferometers are connected in series.

In WDM systems, many lasers with different wavelengths are launched into the sensor system based on one of the multiplexing schemes mentioned previously. Each laser may set up one sensor array. If there are N number of wavelengths and n sensors in a single array, then the total number of sensors will be $n \times N$. Figure 3.16 shows a diagram of N-channel WDM based on Michelson interferometer using FBG pairs as reflection elements. To increase the number of sensors, it is always possible to employ various multiplexing techniques simultaneously.

3.7.1 Temperature Measurement–Based Sensors

Temperature variation sensors can be employed in many industries and medicine. In ultralong fiber sensors such as oil pipe leak detection systems, generally Brillouin or Raman scattering methods are employed for temperature measurement. All types of OFIs can be used as a temperature variation sensor. In MZOFI and MOFI, the reference arms are kept at a given temperature and the sensing

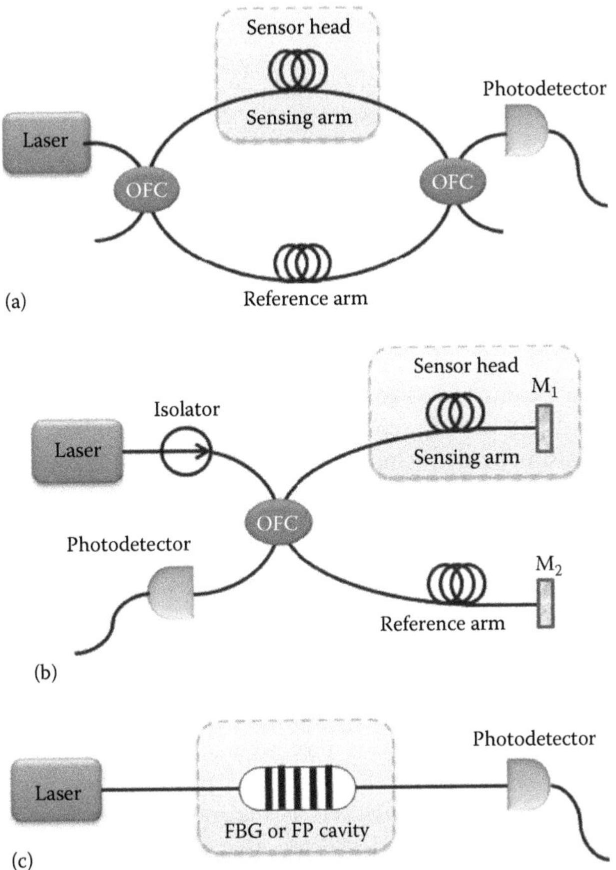

FIGURE 3.14 (a) Transmissive TDM configuration, (b) reflective TDM based on Michelson interferometer, and (c) reflective TDM based on FBG interferometer.

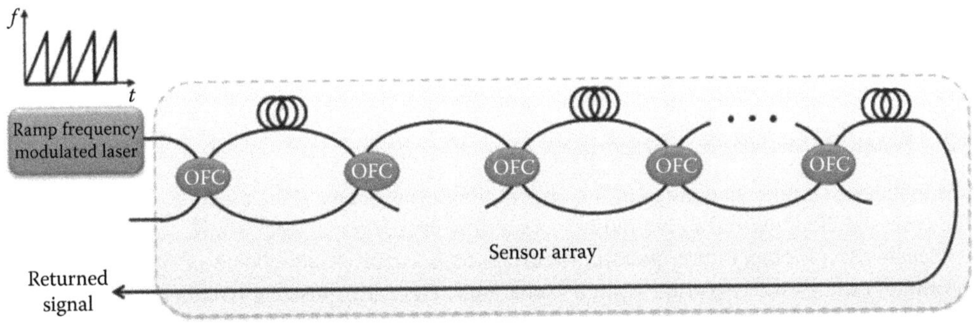

FIGURE 3.15 A schematic of an FDM hydrophone array.

arm is taken into the desired position, while in optical fiber FPI and FBG interferometers by changing the temperature, resonance frequency and Bragg reflection wavelength varies, respectively.

FBG sensors can be employed as a multiparameter measurement device. By enclosing the FBG sensor in a protection sleeve, a strain-free probe can be obtained. The outer diameter of the sleeve is of the order of 0.5 mm that is thick enough to support large transverse stress without affecting the FBG. The sleeve's inner diameter is a few times larger than the FBG diameter, so under the maximum transverse stress threshold, the FBG will have no contact with the sleeve wall.

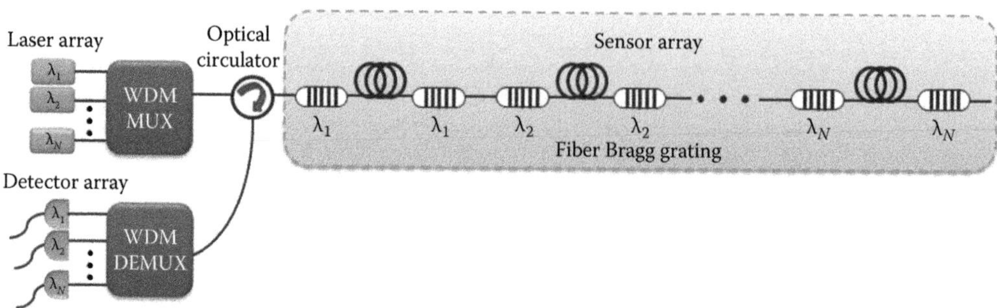

FIGURE 3.16 A schematic of an *N*-channel WDM based on Michelson interferometer using FBG pairs.

FBG temperature sensors can be used to realize quasi-distributed sensor systems with a single fiber link. The quasi-distributed FBG sensors are suitable for use in the petroleum industry and can be employed to measure the well temperature distribution for oil recovery optimization [124]. The production efficiency of oil recovery can be improved by injecting steam into the well where the trapped oil is heated by the steam and flows to the reservoir from the borehole, the temperature profile of this heat flow is used to optimize oil recovery.

In medicine, the heart's efficiency can be evaluated by FBG sensors. The evaluation is based on the flow-directed thermodilution method. In this method, a cold solution is injected by a catheter to the right atrium of the patient's heart to measure the blood temperature in the pulmonary artery. According to the blood temperature and pulse rate, it can be determined how much blood the heart pumps. Conventional thermistor or thermocouple catheters can be replaced with catheters with FBG sensors [125]. It is found that the FBG is more accurate than the thermistor and thermocouple catheters.

In-line MZI and MI can also be employed as integrated temperature sensors; however, to decrease the number of cladding modes, LPGs are used as BSs. The thermooptic coefficient of Germanium-doped core had a stronger thermo-optic coefficient than the Boron codoped core [55]. When using LPGs, the operation wavelength is limited due to the phase matching phenomenon, so to reach maximum performance, identical LPGs should be used [126]. Since the in-line MI has only one LPG, it will have better performance relative to the in-line MZI. In the Sagnac interferometer for temperature sensing applications, the fiber of Sagnac loop is doped such that a large thermal expansion coefficient is obtained. This large thermal expansion coefficient causes high-birefringence variation [127].

3.7.2 Strain Measurement–Based Sensors

Strain induces a significant change in the optical fiber dimensions, which lead to a strong variation in the core and cladding refractive indices. As a drawback, optical fiber strain sensors suffer from cross sensitivity to temperature and strain variation. An appropriate choice of fiber shield and its composition ensures the two effects can be separated. Thus, it is possible to counterbalance the different effects and produce temperature-insensitive fiber sensors or strain-insensitive sensors. The temperature-insensitive fibers can be employed for strain sensors and vice versa. The strain optical fiber sensors have many applications such as in hydrophones, geophones, seismographs, and intruder detector systems. An acoustic optical fiber sensor that is called a fiber hydrophone has high sensitivity and a large dynamic range relative to the conventional PZT ceramic sensors. Optical fiber hydrophones are immune to electromagnetic interference. In some applications such as seismic exploration for oil reserves, a large number of hydrophones are required [128,129]. Only fiber hydrophones allow multiplexing a large number of sensor outputs on a single optical fiber transmission line.

OFI strain sensors are categorized into transmission and reflection types. MZI can be used to design a transmission-type sensor array and has been widely used in the early optical fiber

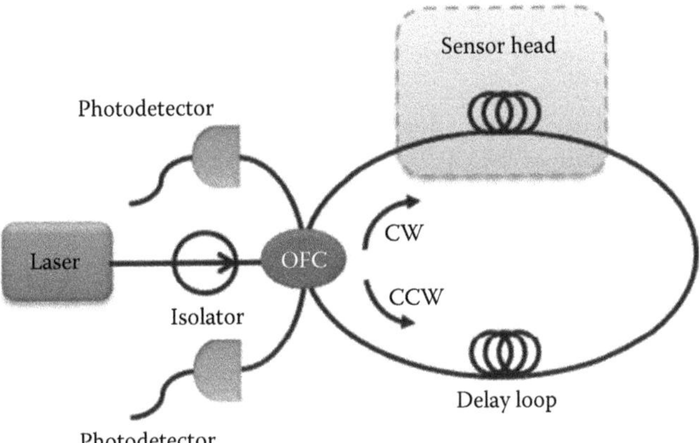

FIGURE 3.17 A schematic of Sagnac fiber hydrophone.

hydrophone systems. The MI configuration in turn is appropriate for building a reflective fiber hydrophone array. The circulating light passes through the sensing arm twice; thus, an MI sensor is twice as sensitive as an MZI.

The optical fiber Fabry–Pérot hydrophone is the simplest transmission-type strain sensor. Its advantages are compactness and high sensitivity, while its disadvantage is a small dynamic range.

A schematic diagram of Sagnac fiber hydrophone is shown in Figure 3.17. Light enters the middle port of a 3×3 fiber coupler and couples to the CCW and CW modes of the Sagnac interferometer. The CW mode will see the delay loop first and then the fiber sensing head, while the CCW mode sees the fiber sensing head first and the delay loop later. Due to the asymmetric configuration, the phase difference of the two modes is converted into intensity modulation at the output of the 3×3 fiber coupler.

FBGs and LPGs can be employed as strain sensors and are used in transmission array of hydrophones. FBG sensors have distinguishing advantages over other implementations of optical fiber sensors, such as low cost, direct measurement, and unique wavelength multiplexing capability.

For medical applications such as ultrasound surgery, ultrasound sensors are required to monitor the output power from diagnostic ultrasound equipment. Due to its unique multiplexing capability, the FBG ultrasound sensors can be employed to simultaneously measure the ultrasound fields at several points. The length of the FBG must be greater than the ultrasound wavelength. Hence, the maximum measurable frequency range is limited.

In structural engineering, fiber-optic sensors are very attractive for quality control during the construction and afterwards to monitor the health of the building. The ability to measure strain and temperature simultaneously by multiplexed interferometric optical fiber sensors makes them appropriate for structure health monitoring. Many applications of IOFS in important structures such as aircraft, marine vehicles, dams, and bridges have been demonstrated [130–132]. IOFS are also used in the electric power industry for remote load monitoring of power transmission lines, for instance.

Distributed optical fiber intrusion sensors are designed for monitoring long perimeter with the aim of online detection and locating intruder crossing point. The principle of fiber-optic intrusion detection can be explained easily as follows. The pressure of the intruder above the buried sensor induces a change in a parameter of the light propagating along the fiber. The induced change depends on the type of implemented optical fiber. As an example, for a single-mode optical fiber, the stress causes the phase of propagating light to shift [133]. While for a polarization-maintaining optical fiber, the stress leads to mode coupling between the HE_x^{11} and HE_y^{11} modes [75]. Based on which beam parameter (amplitude, frequency, polarization, and phase) is modulated under the action of strain, the optical fiber intruder sensors are classified into different categories. The optical time domain reflectometer (OTDR), which is an amplitude fiber-optic sensor, operates based on Rayleigh scattering. Optical pulses are injected into the single-mode fiber and the Rayleigh backscattered light is monitored by photodetector to determine the fiber loss and intruder crossing point. Nevertheless, the OTDR measurements for this purpose are usually high in cost, because the backscattered signal is very weak and a high-power laser source is required to improve the sensitivity and signal-to-noise ratio of OTDR.

In Brillouin OTDR (BOTDR), which is based on Brillouin scattering, the environmental parameters affect the frequency of scattered light. By evaluating the spectral content of the output beam, one can determine the applied force such as intruder strain.

In the polarization modulation–based optical fiber multiintruder sensor, a birefringent optical fiber and an x-polarized ramp frequency laser are employed. The presence of an intruder above the buried PMF fiber causes mode coupling to occur between x- and y-polarized modes. The intruder crossing point is mapped onto the beating frequency at the output photodetector [75].

Phase changes resulting from the pressure of the intruder can be sensed by a phase-sensitive OTDR (φ-OTDR). φ-OTDR requires a highly coherent stabilized laser source and a stabilized sensing fiber. The induced phase changes are detected by subtracting φ-OTDR trace (plot of returned optical power vs. time) from an earlier stored one. The intruder crossing point location can be determined by the time at which the phase is perturbed.

Fiber-optic intrusion sensors that apply interferometric methods to sense the strain-induced phase shifts are usually more sensitive and capable of detecting intruders in real time. A wide range of interferometric configurations are proposed to be employed as an interferometric optical fiber intrusion sensor such as Sagnac, Sagnac/Mach–Zehnder [134], Sagnac/Michelson [135], and Sagnac/Sagnac interferometer [136].

3.7.3 Refractive Index Measurement–Based Sensors

The optical path length of light propagating in an optical fiber is affected by the refractive index of its environment. The changes in the environment's refractive index can be sensed by an OFI. Side-polished fiber and D-fiber have been demonstrated for refractive index measurement [137,138].

In refractive index sensors based on FBG operation, the central wavelength of FBG varies with refractive index change [139]. The LPG is more sensitive to the environment's refractive index change compared to FBG [139] so any variation of the surrounding refractive index modifies the transmission spectrum. In environmental pollution control, optical fiber sensors are employed for sensing gases such as methane, hydrocarbons, nitrogen oxide, sulfur oxide, and ammonia [140]. In principle, any chemical agent (such as hexanol, methylcyclohexane, hexadecane, ethanol, $CaCl_2$, and NaCl) can be detected by the LPG method when its loss peak is lying in the refractive index range of 1.3–1.45 [141,142]. The LPG sensitivity to temperature change must be compensated though.

The SPP resonance (SPPR) is employed to detect the local refractive index changes. SPPR technique is quite fast and sensitive, so it is suitable for real-time monitoring. Refractive index measurement–based OFIs have found widespread applications in biological and chemical sensors. The passive conductive nature of IOFS makes them safe in hazardous environments and chemical industries.

Biosensors are classified to the catalytic-based biosensors and bioaffinity-based optrodes. Optrode is a combination of the words *optical* and *electrode* and refers to fiber-optic sensors that can measure the concentration of a chemical compound or a group of compounds.

The catalytic-based biosensor recognizes the analyte when a catalyzed chemical reaction causes the analyte to convert from a nondetectable form to a detectable form. Isolated enzymes, microorganisms, subcellular organelle, or a tissue slice can be used as the biocatalyst. Most enzymes do not have intrinsic optical properties that can be directly detected; therefore, in fiber-optic enzymatic biosensors, an outside agent is employed that interacts with the analyte to indicate the change of analyte concentration in the sample. Bioaffinity-based sensors are based on affinity interaction when a range of components are selected from complex mixture of biomolecules. The immunoassay and nucleic acid optical fiber biosensors are two examples of bioaffinity sensors.

Immunoassay optical biosensor operation is based on the optical signals generated by antibody–antigen binding. Immunoassay optical fiber–based biosensors have three modes of operation: direct, competitive, and sandwich. Nucleic acid optical biosensors are implemented by using the affinity of single-stranded DNA to form double-stranded DNA with complementary sequences. This type of biosensor can be employed to detect chemically induced DNA damage or to sense microorganisms through the hybridization of species-specific sequences of DNA. Fiber-optic biosensors have also been used for monitoring drug delivery [35].

The sensitivity of the biosensor could be improved by integrating a fiber with long interaction length with surface plasmon resonances. An SPPR-enhanced biosensor is made of single-mode optical fiber in which half the core along the fiber is polished away and a thin-film layer of gold is deposited on the polished surface. Despite traditional SPPR technique in which light intensity is measured, SPR-enhanced biosensor measures the phase difference to reach a very high sensitivity. Interferometric optical fibers based on localized SPR also enhance the sensitivity of biosensors and have a faster response time [143,144].

3.7.4 Quantum Mechanical Applications

The ultrahigh sensitivity of laser interferometers makes them a promising candidate for detecting faint signals such as gravitational waves (GWs). GWs are distortions in the fabric of space–time, propagating at the speed of light. Gravitational radiation emitted by an accelerating mass can be decomposed into a series of multipoles. Due to the mass conservation as well as linear and angular momentum conservation, the strongest form of GWs would be expected for quadruple radiation [145,146]. The two perpendicular arms in Michelson interferometer make it compatible with the quadruple nature of GW and thus suitable for detecting GWs. When a GW with a polarization along the interferometer's arms propagates normal to the plane of the interferometer, one of the interferometer's arms is slightly stretched and the perpendicular one is compressed [147]. The quadruple strain field produced by GWs ($h(r,t)$) changes the space–time metric elements and causes the effective speed of light travelling in the horizontal and vertical arms of MI to become $c/\sqrt{1 \pm h(r,t)} \cong (1 \mp (h(r,t)/2))c$ [146]. Thus, it is acceptable to think that the effective refractive index of vacuum is changed in the presence of GW. A conservative estimate of GW amplitude radiated from astrophysical sources (at frequency of about 1 kHz) on earth is of the order of 10^{-21} [148]. Hence, for a long-baseline GW detector such as LIGO and Virgo, a highly sensitive detector is required to measure the path difference between the arms on the order of attometer (10^{-18} m). The sensitivity of ground-based GW detectors is limited by many sources of noise such as seismic noise, acoustic noise, thermal noise, gravity-gradient noise, scattered light, residual gas, and electromagnetic fluctuations of the vacuum [147]. Among all of these noise sources, shot noise and radiation pressure noise that arise from the quantum fluctuations in a number of circulating photons (Δn) determine the ultimate limit to the sensitivity of these gigantic detectors [149,150].

As it is well known in quantum optics, there exists a number-phase Heisenberg uncertainty principle (HUP) as

$$\Delta n \Delta \varphi \geq 1 \tag{3.39}$$

where Δn is the uncertainty or fluctuation in the photon number, which depends on the quantum state of light. For a coherent state of light such as laser used in LIGO, the shot noise limit (SNL) is equal to $\Delta\varphi_{SNL} = 1/\sqrt{\bar{n}} = \sqrt{I_0/I}$, where $\bar{n}$ is the average number of photons, $I_0 = \hbar\omega/(\epsilon_0 V)$ is intensity of a single photon, and ϵ_0 and V are the vacuum permittivity and mode volume for electromagnetic field, respectively [150]. In the long-baseline GW detector, the optimum power of laser is of the order of several kilowatts.

Caves demonstrated that by exploiting a nonclassical state of light known as squeezed state, one can improve the sensitivity of the interferometer beyond the SNL [151]. It was also shown that by choosing an optimum quantum state of light for the dark input of the interferometer, quantum noise and optimum laser power decrease one order of magnitude relative to the conventional interferometer [152].

Since the fluctuation in number of photons cannot exceed the mean number of photons in the laser, the Heisenberg limit (HL) is obtained by inserting $\Delta n = \bar{n}$ into the inequality (3.39), which is proportional to the inverse of the laser intensity $\Delta\varphi_{HL} = I_0/I$.

Moreover, the idea of using other interferometry configurations such as Sagnac interferometer was proposed for GW detection [153,154]. In comparison to Fabry–Pérot–Michelson interferometer, Sagnac interferometer operates more robustly with high-power laser. The Sagnac interferometer measures the phase difference between two counterpropagating beams that have travelled in both arms of the interferometer. Since the two beams follow the same path, the interferometer only detects the high-frequency optical path changes (compared to inverse of storage time). In order to detect GWs in the frequency range 10 Hz–1 kHz, the storage time between 1 and 100 ms is required, which is equivalent to the optical path length of several thousand of kilometers. To achieve a desired storage time, a delay line can be employed.

In this regard, optical fiber GW interferometric detectors open a new frontier for achieving small, cheap, and simple GW observatory in amateur laboratories [155–157].

According to this brief discussion, it becomes evident that reaching the HL (minimum phase uncertainty) is the ultimate goal for detecting GWs as well as in other branches of science such as quantum metrology and quantum lithography. As the squeezing parameter of the light entering the dark port of the interferometer approaches infinity, the phase fluctuations approach zero. This infinite squeezing is very hard to achieve in the laboratory, so the expected limit is somewhere between SNL and HL [158]. The experimental situation is a lot closer to the SNL than to the HL.

To increase the photon intensity fluctuations and to reach the HL, there is another approach in which path-entangled photon number states are exploited [159]. Figure 3.18 depicts a schematic diagram of a folded two-mode optical interferometer consisting of a source, PS, and detector. Let us consider the $1/\sqrt{2}\left(|N,0\rangle + |0,N\rangle\right)$ state, where the first number is the number of photons in the upper arm and the second one is the number of photons in the lower arm of the interferometer. This state is called the NOON state. The Hamiltonian of the device is $H = \hbar\omega\left(a_1^{+}a_1 + a_2^{+}a_2\right)$, where a_1 and a_2 are the annihilation operators of the upper and lower fiber modes, respectively. By applying the time evolution operator on the input state, one can obtain the output state. The phase shift operator is defined as $U(\varphi) \equiv exp(i\varphi\hat{n})$, where $\hat{n}$ is the photon number operator. The phase shift for the coherent state is independent of the number of photons, while a Fock state $|N\rangle$ has a phase shift N times greater than that of the coherent state. The NOON state is affected by passing through the PS as $\left(1/\sqrt{2}\right)\left(|N,0\rangle + |0,N\rangle\right) \xrightarrow{PS} \left(1/\sqrt{2}\right)\left(|N,0\rangle + e^{iN\varphi}|N,0\rangle\right)$, which is the desired quantum state to improve the sensitivity of the interferometer. The output signal of N-photon detecting analyzer $M(\phi) = I_A \cos N\phi$ oscillates N times faster than the case in which the coherent state is an input of the interferometer. This effect is commonly known as superresolution. The phase distortion for

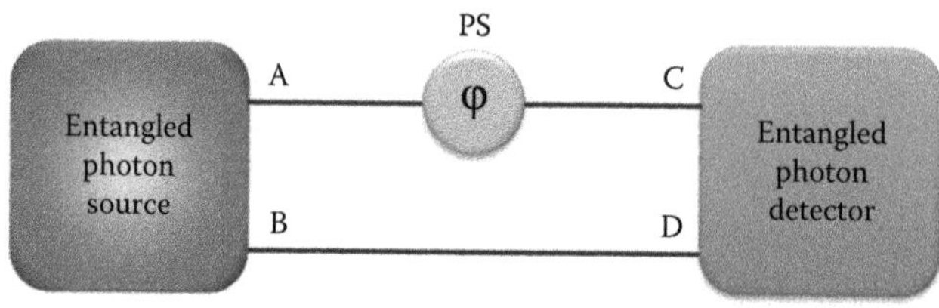

FIGURE 3.18 Schematic diagram of a folded two-mode optical interferometer. PS stands for phase shifter.

the NOON sate is $\Delta\varphi_{\mathrm{NOON}} = 1/N$, which is precisely the HL. Indeed, the NOON state owes its minimum phase uncertainty to the largest uncertainty in its photon number. Here, the photon number uncertainty is the uncertainty in the number of photons distributed over the two-photon path, A and B. In NOON state, there exists a complete uncertainty about the mode (A or B) in which one can find all N photons. In other words, since the photons are nonlocal and the photon number is completely uncertain $\Delta n = N$, the phase uncertainty reaches the HL.

Hence, the production of high-photon NOON state is of prime importance in quantum metrology. One of the significant applications of ultralow-loss fiber interferometer in quantum mechanics is quantum state engineering. OFIs are very essential and practical for quantum metrology and sensing.

When a high-power UV laser pumps the birefringent crystal, unentangled twin-number state $|N/2\rangle_1 |N/2\rangle_2$ (N even) is produced in the spontaneous parametric down conversion process (SPDC). The number states of the form $|N/2\rangle_1 |0\rangle_2$ can be obtained by a photon number measurement on the second mode of a twin-number state. For this purpose, one should check to see if there exist $N/2$ photons in the second mode and then allow what's in the first mode to enter the interferometer. This process is called heralded number state production [150].

The one-photon and two-photon NOON states are obtained if $|1,0\rangle$ and $|1,1\rangle$ are incident on a 50/50 BS, respectively. The production of one-photon NOON state is due to the fact that the single photon cannot be split into two while the two-photon NOON state is illustrated by the Hong–Ou–Mandel interferometer. The 50/50 BS transforms the twin-number state $|N/2\rangle|N/2\rangle$ into $C_N|N,0\rangle + C_{N-2}|N-2,2\rangle + \cdots + C_{N-2}|N-2,0\rangle + C_N|N,0\rangle$, where C_N are probability amplitude weight factors. The two end terms are the desired NOON components and have the highest probability. By exploiting the non-NOON components, supersensitivity can be achieved; but reaching the HL actually requires the pure NOON state [160]. Hence, the non-NOON components must be eliminated. One approach to remove these middle terms is to use a quantum filter [161]. Another method is to employ the magic BS as shown in Figure 3.19 [162]. The idea is based on the cross-Kerr coupling of two MZIs. The single photon in the upper MZI-1 controls the phase shift in the lower MZI-2. It is assumed that the presence of single photon in the upper MZI-1 causes a π radian phase shift in the lower MZI-2. If the Kerr phase shift is set to zero and $|N\rangle_A |0\rangle_B$ is incident on the input AB, the output CD is $|0\rangle_C |N\rangle_D$, where C and D are the bright and dark ports, respectively. If the Kerr phase shift is π radians, the output will be $|N\rangle_C |0\rangle_D$. The presence of $|1,0\rangle$ at the input of upper MZI-1 causes the single photon of MZI-1 to be in an equal superposition of upper and lower paths. Therefore, the single photon in the MZI-1 induces a superposition of zero and π phase shift in the MZI-2, which leads to the NOON state production at the output of MZI-2. Since the Kerr nonlinearity coefficient of optical material is so low [163], some methods (such as exploiting optical cavity [164] and electromagnetically induced transparency (EIT) [165]) are required to enhance Kerr effect. However, the

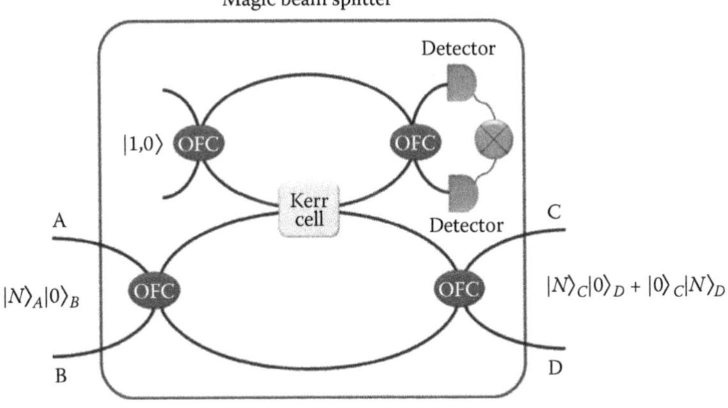

FIGURE 3.19 Schematic diagram of magic BS for generating NOON state.

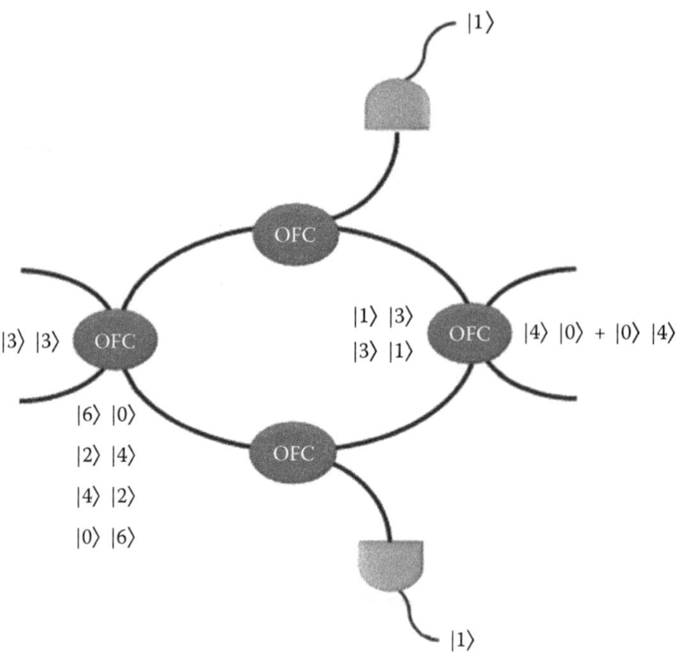

FIGURE 3.20 Schematic diagram of all-linear optical device based on heralded process for generating four-photon N00N state.

implementation of these methods is not so easy and N00N state generation based on the all-linear optical devices and detection [166] is preferable.

Figure 3.20 shows a linear optical system for four-photon N00N state generation. The system is based on MZI, where the upper and lower arms are connected to two detectors through additional OFCs. A $|3\rangle|3\rangle$ twin-number state is incident on the input of MZI. After the first 50/50 OFC, one will have a coherent superposition of $|6,0\rangle$, $|4,2\rangle$, $|2,4\rangle$, and $|0,6\rangle$ states. The upper and lower detectors are checked for detection of one and only one photon simultaneously. This is a heralded process. If both detectors click *one*, a total of two photons are missing from the interferometer that means $|6,0\rangle$ and $|0,6\rangle$ states have collapsed by quantum measurement. So after the intermediate detection, only $|3,1\rangle$ and $|1,3\rangle$ states are left. Due to the Hong–Ou–Mandel effect, it is easy to show that after the second 50/50 OFC, the four-photon N00N state is obtained. Under the perfect condition in which the coupling coefficients of OFCs are optimum, the N00N state only emerges about 1 time in 10. This method can be concatenated or stacked to produce an arbitrary number of photons (N) [167]. In such schemes, the probability of N00N generation worsens exponentially as N increases. The N00N state can be employed to the miniaturized sensors, interferometric Lidar systems, and quantum computing circuits.

REFERENCES

1. G. Agrawal, *Nonlinear Fiber Optics*, 2nd edn. San Diego, CA: Academic Press, 2007.
2. K. Okamoto, *Fundamentals of Optical Waveguides*, 2nd edn. San Diego, CA: Academic Press, 2006.
3. B. D. Gupta, *Fiber Optic Sensors Principles and Applications*. New Delhi, India: New India Publishing Agency, 2006.
4. A. Snyder and J. Love, *Optical Waveguide Theory*. Berlin, Germany: Springer, 1983.
5. F. Poli, A. Cucinotta, and S. Selleri, *Photonic Crystal Fibers: Properties and Applications*. Dordrecht, the Netherlands: Springer, 2007.
6. A. Cusano, D. Paladino, and A. Iadicicco, Microstructured fiber Bragg gratings, *J. Lightwave Technol.*, 27(11), 1663–1697, 2009.

7. L. Vincetti, F. Poli, and S. Selleri, Confinement loss and nonlinearity analysis of air-guiding modified honeycomb photonic crystal fibers, *IEEE Photon. Technol. Lett.*, 18(3), 508–510, 2006.
8. P. Ghenuche, H. Rigneault, and J. Wenger, Hollow-core photonic crystal fiber probe for remote fluorescence sensing with single molecule sensitivity, *Opt. Express*, 20(27), 28379–28387, 2012.
9. J. D. Joannopoulos, S. G. Johnson, J. N. Winn, and R. D. Meade, *Photonic Crystals: Molding the Flow of Light*. Singapore: Princeton University Press, 2008.
10. K. Porsezian and V. C. Kuriakose, *Optical Solitons Theoretical and Experimental Challenges*. Berlin, Germany: Springer, 2002.
11. S. K. Varshney, M. P. Singh, and R. K. Sinha, Propagation characteristics of photonic crystal fibers, *J. Opt. Commun.*, 24, 856, 2003.
12. A. Bahrampour, New hollow core fiber design and porphyrin thin film deposition method towards enhanced optical fiber sensors, Thesis, University Naples "federico II", Naples, Italy, 2013.
13. M. E. Zoorob, M. Charlton, G. J. Parker, J. J. Baumberg, and M. Netti, Complete photonic bandgaps in 12-fold symmetric quasicrystals, *Nature*, 404, 740–743, 2000.
14. M. Florescu, S. Torquato, and P. J. Steinhardt, Complete band gaps in two-dimensional photonic quasicrystals, *Phys. Rev.*, 80, 155112, 2009.
15. A. Ricciardi, I. Gallina, S. Campopiano, G. Castaldi, M. Pisco, V. Galdi, and A. Cusano, Guided resonances in photonic quasicrystals, *Opt. Express*, 17, 6335–6346, 2009.
16. A. Ricciardi, M. Pisco, A. Cutolo, and A. Cusano, Evidence of guided resonances in photonic quasicrystal slabs, *Phys. Rev. B*, 84, 085135, 2011.
17. S. Kim, C. Kee, and J. Lee, Novel optical properties of six-fold symmetric photonic quasicrystal fibers, *Opt. Express*, 15, 13221–13226, 2007.
18. X. Sun and D. Hu, Air guiding with photonic quasi-crystal fiber, *IEEE Photon. Technol. Lett.*, 22(9), 607–609, 2011.
19. A. Bahrampour, A. Iadicicco, A. R. Bahrampour, S. Campopiano, A. Cutolo, A. Cusano, Design and analysis of photonic quasi-crystal hollow core fibers, *Proc. SPIE 8794, Fifth European Workshop on Optical Fibre Sensors*, 87942H, 2013.
20. M. Adams, *An Introduction to Optical Waveguides*. New York: John Wiley & Sons Ltd, 1981.
21. L. W. E. Mandel, *Optical Coherence and Quantum Optics*. New York: Cambridge University Press, 1995.
22. C. Gerry and P. Knight, *Introductory Quantum Optics*. New York: Cambridge University Press, 2005.
23. R. J. Glauber, *Quantum Theory of Optical Coherence: Selected Papers and Lectures*. Weinheim, Germany: Wiley-VCH Verlag GmbH & Co. KGaA, 2007.
24. M. Born and E. Wolf, *Principles of Optics*, 7th edn. London, U.K.: Cambridge University, 2003.
25. R. H. Brown and R. Q. Twiss, A test of a new type of stellar interferometer on Sirius, *Nature*, 178, 1046–1048, 1956.
26. C. K. Hong, Z. Y. Ou, and L. Mandel, Measurement of subpicosecond time intervals between two photons by interference, *Phys. Rev. Lett.*, 59(18), 2044–2046, 1987.
27. Z. Y. J. Ou, *Multi-Photon Quantum Interference*. New York: Springer, 2007.
28. B. Yurke, S. L. McCall, and J. R. Klauder, SU(2) and SU(1,1) interferometers, *Phys. Rev. A*, 33(6), 4033–4053, 1986.
29. M. H. A. Reck, Quantum interferometry with multiports: Entangled photons in optical fibers, Thesis, The University of Innsbruck, Innsbruck, Austria, 1996.
30. G. P. Agrawal, *Applications of Nonlinear Fiber Optics*, 2nd edn. Boston, MA: Academic Press, Elsevier, 2008.
31. R. Hui and M. O'Sullivan, *Fiber Optic Measurement Techniques*. Burlington, VT: Elsevier Academic Press, 2009.
32. P. Törmä, I. Jex, and S. Stenholm, Beam splitter realizations of totally symmetric mode couplers, *J. Mod. Opt.*, 43(2), 245–251, 1996.
33. D. Goldstein, *Polarized Light*. New York: CRC Press, 2003.
34. R. P. Dahlgren, Matrix operators for complex interferometer analysis, in *21st International Conference on Optical Fiber Sensors*, Ottawa, Ontario, Canada, 2011.
35. S. Yin, P. B. Ruffin, and F. Yu, *Fiber Optic Sensors*, 2nd edn., Optical Science and Engineering Series. Boca Raton, FL: CRC Press, 2008.
36. H. Huang, *Coupled Mode Theory as Applied to Microwave and Optical Transmission*. Utrecht, the Netherlands: VNU Science Press, 1984.
37. R. Kashyap, *Fiber Bragg Gratings*. San Diego, CA: Academic Press, 1999.
38. W. J. Bock, I. Gannot, and S. Tanev, *Optical Waveguide Sensing and Imaging*. Dordrecht, the Netherlands: Springer, 2006.

39. B. P. Pal, *Guided Wave Optical Components and Devices: Basics, Technology, and Applications.* Burlington, VT: Elsevier Academic Press, 2006.
40. G. Sagnac, L'ether lumineux demontre par l'effet du vent relatif d'ether dans un interferometre en rotation uniforme. The demonstration of the luminiferous aether by an interferometer in uniform rotation, *C. R. Acad. Sci.*, 157, 708–710, 1913.
41. W. Burns, *Optical Fiber Rotation Sensing*, 1st edn. San Diego, CA: Academic Press, 1993.
42. V. Vali and R. Shorthill, Fiber ring interferometer, *Appl. Opt.*, 15(5), 1099–1100, 1976.
43. K. Bohnert, P. Gabus, J. Nehring, and B. H., Temperature and vibration insensitive fiber optic current sensor, *J. Lightwave Technol.*, 20(2), 267–276, 2002.
44. W. Lin, C. Chang, C. Wu, and M. Chen, The configuration analysis of fiber optic interferometer of hydrophones, in *Proceeding of OCEANS '04. MTTS/IEEE TECHNO-OCEAN '04*, 2004.
45. A. Starodumov, L. Zenteno, and E. De La Rosa, Fiber Sagnac interferometer temperature sensor, *Appl. Phys. Lett.*, 70(1), 19–21, 1997.
46. X. Dong and H. Tam, Temperature-insensitive strain sensor with polarization maintaining photonic crystal fiber based Sagnac interferometer, *Appl. Phys. Lett.*, 90(15), 151113, 2007.
47. H. Fu, C. Wu, M. Tse, L. Zhang, K. Cheng, H. Tam, B. Guan, and C. Lu, High pressure sensor based on photonic crystal fiber for downhole application, *Appl. Opt.*, 49(14), 2639–2643, 2010.
48. G. Bertocchi, O. Alibart, D. B. Ostrowsky, S. Tanzilli, and P. Baldi, Single-photon Sagnac interferometer, *J. Phys. B: At. Mol. Opt. Phys.*, 39, 1011, 2006.
49. S. Tofighi and A. R. Bahrampour, Analysis of transient response and instability in fiber ring resonators containing an erbium-doped fiber amplifier and quantum dot-doped fiber saturable absorber, *J. Opt. Soc. Am. B*, 30(12), 3215–3224, 2013.
50. N. Dou and N. Li, Optical bistability in fiber ring resonator containing an EDFA, *Opt. Commun.*, 281, 2238–2242, 2008.
51. G. Weihs, M. Reck, H. Weinfurter, and A. Zeilinger, All-fiber three-path Mach–Zehnder interferometer, *Opt. Lett.*, 21(4), 302–304, 1996.
52. E. Dianov, S. Vasiliev, A. Kurkov, O. Medvedkov, and V. Protopopov, Infiber Mach-Zehnder interferometer based on a pair of long-period gratings, in *Proceedings of 22nd European Conference on Optical Communication—ECOC'96*, Oslo, Norway, 1996.
53. B. H. Lee, Y. H. Kim, K. S. Park, J. B. Eom, M. J. Kim, B. S. Rho and H. Y. Choi, Interferometric fiber optic sensors, *Sensors*, 12, 2467–2486, 2012.
54. J. Lim, H. Jang, K. Lee, J. Kim, and B. Lee, Mach-Zehnder interferometer formed in a photonic crystal fiber based on a pair of long-period fiber gratings, *Opt. Lett.*, 29, 346–348, 2004.
55. Y. Kim, U. Paek, and B. Lee, Measurement of refractive-index variation with temperature by use of long-period fiber gratings, *Opt. Lett.*, 27, 1297–1299, 2002.
56. T. Allsop, R. Reeves, D. Webb, and I. Bennion, A high sensitivity refractometer based upon a long period grating Mach-Zehnder interferometer, *Rev. Sci. Instrum.*, 73, 1702–1705, 2002.
57. Y. Kim, M. Kim, M. Park, J. Jang, K. Kim, and B. Lee, Hydrogen sensor based on a palladium-coated long-period fiber grating pair, *J. Opt. Soc. Korea*, 12, 221–225, 2008.
58. M. Kim, Y. Kim, and B. Lee, Simultaneous measurement of temperature and strain based on double cladding fiber interferometer assisted by fiber grating pair, *IEEE Photon. Technol. Lett.*, 20, 1290–1292, 2008.
59. J. Ding, A. Zhang, L. Shao, J. Yan, and S. He, Fiber-taper seeded long-period grating pair as a highly sensitive refractive-index sensor, *IEEE Photon. Technol. Lett.*, 17, 1247–1249, 2005.
60. H. Choi, M. Kim, and B. Lee, All-fiber Mach-Zehnder type interferometers formed in photonic crystal fiber, *Opt. Express*, 15, 5711–5720, 2007.
61. H. Choi, K. Park, and B. Lee, Photonic crystal fiber interferometer composed of a long period fiber grating and one point collapsing of air holes, *Opt. Lett.*, 33, 812–814, 2008.
62. L. Ngyuen, D. Hwang, S. Moon, D. Moon, and Y. Chung, High temperature fiber sensor with high sensitivity based on core diameter mismatch, *Opt. Express*, 16, 11369–11375, 2008.
63. J. Zhu, A. Zhang, T. Xia, S. He, and W. Xue, Fiber-optic high-temperature sensor based on thin-core fiber modal interferometer, *IEEE Sens. J.*, 10, 1415–1418, 2010.
64. Z. Tian, S. Yam, J. Barnes, W. Bock, P. Greig, J. Fraser, H. Loock, and R. Oleschuk, Refractive index sensing with Mach-Zehnder interferometer based on concatenating two single-mode fiber tapers, *IEEE Photon. Technol. Lett.*, 20, 626–628, 2008.
65. P. Lu, L. Men, K. Sooley, and Q. Chen, Tapered fiber Mach-Zehnder interferometer for simultaneous measurement of refractive index and temperature, *Appl. Phys. Lett.*, 94, 131110, 2009.
66. F. Pang, H. Liu, H. Guo, Y. Liu, X. Zeng, N. Chen, Z. Chen, and T. Wang, In-fiber Mach-Zehnder interferometer based on double cladding fibers for refractive index sensor, *IEEE Sens. J.*, 11, 2395–2400, 2011.

67. L. Jiang, J. Yang, S. Wang, B. Li, and M. Wang, Fiber Mach-Zehnder interferometer based on microcavities for high-temperature sensing with high sensitivity, *Opt. Lett.*, 36, 3753–3755, 2011.
68. O. Frazao, S. Silva, J. Viegas, J. Baptista, J. Santos, J. Kobelke, and K. Schuster, All fiber Mach-Zehnder interferometer based on suspended twin-core fiber, *IEEE Photon. Technol. Lett.*, 22, 1300–1302, 2010.
69. L. Yuan, L. Zhou, and J. Wu, Fiber optic temperature sensor with duplex Michelson interferometric technique, *Sens. Actuat. A*, 86, 2–7, 2000.
70. R. Kashyap and B. Nayar, An all single-mode fiber Michelson interferometer sensor, *J. Lightwave Technol.*, LT-1, 619–624, 1983.
71. K. O'Mahoney, R. O'Byrne, S. Sergeryev, L. Zhang, and I. Bennion, Short-scan fiber interferometer for high-resolution Bragg grating array interrogation, *IEEE Sens. J.*, 9, 1277–1281, 2009.
72. Y. Zhao and F. Ansari, Intrinsic single-mode fiber-optic pressure sensor, *IEEE Photon. Technol. Lett.*, 13, 1212–1214, 2001.
73. D. Kim, Y. Zhang, K. Cooper, and A. Wang, In-fiber reflection mode interferometer based on a long-period grating for external refractive-index measurement, *Appl. Opt.*, 44, 5368–5373, 2005.
74. K. Park, H. Choi, S. Park, U. Paek, and B. Lee, Temperature robust refractive index sensor based on a photonic crystal fiber interferometer, *IEEE Sens. J.*, 10, 1147–1148, 2010.
75. A. Bahrampour, M. Bathaee, S. Tofighi, A. Bahrampour, F. Farman, and M. Vali, Polarization maintaining optical fiber multi-intruder sensor, *Opt. Laser Technol.*, 44(7), 2026–2031, 2012.
76. J. Villatoro, M. Kreuzer, R. Jha, V. Minkovich, V. Finazzi, G. Badenes, and V. Pruneri, Photonic crystal fiber interferometer for chemical vapor detection with high sensitivity, *Opt. Express*, 17(3), 1447–1453, 2009.
77. A. Hassani and M. Skorobogatiy, Design of the microstructured optical fiber-based surface plasmon resonance sensors with enhanced microfluidics, *Opt. Express*, 14(24), 11616–11621, 2006.
78. J. Villatoro, V. Minkovich, and D. Hernández, Compact modal interferometer built with tapered microstructured optical fiber, *IEEE Photon. Technol. Lett.*, 18(11), 1258–1260, 2006.
79. L. Yuan, Y. Liu, and W. Sun, Fiber optic Moiré interferometric profilometry, *Proc. SPIE* 5633, *Advanced Materials and Devices for Sensing and Imaging II*, 55, 2005.
80. P. Flourney, R. McClure, and G. Wyntjes, White-light interferometric thickness gauge, *Appl. Opt.*, 11(9), 1907–1915, 1972.
81. G. Song, X. Wang, and Z. Fang, White-light interferometer with high sensitivity and resolution using multi-mode fibers, *Optik*, 112(6), 245–249, 2001.
82. L. Manojlovi, A simple white-light fiber-optic interferometric sensing system for absolute position measurement, *Opt. Lasers Eng.*, 48(4), 486–490, 2010.
83. L. Yuan, Multiplexed, white-light interferometric fiber-optic sensor matrix with a long-cavity, Fabry-Perot resonator, *Appl. Opt.*, 41(22), 4460–4466, 2002.
84. J. Mercado, A. Khomenko, and A. Weidner, Precision and sensitivity optimization for white-light interferometric fiber-optic sensors, *J. Lightwave Technol.*, 19(1), 70–74, 2001.
85. A. Xuereb, R. Schnabel, and K. Hammerer, Dissipative opto mechanics in Michelson Sagnac interferometer, *Phys. Rev. Lett.*, 107, 213604, 2011.
86. W. Ya-Ping, W. Chong-Qing, and Y. Ping, Polarization stability of a double-loop interferometer based on a planar 3 × 3 coupler, *Chin. Phys. Lett.*, 29(4), 044212, 2012.
87. D. Jackson, A. Kersey, M. Corke, and J. Jones, Pseudo heterodyne detection scheme for optical interferometers, *Electron. Lett.*, 18(25), 1081–1083, 1982.
88. C. Strauss, Synthetic-array heterodyne detection: A single-element detector acts as an array, *Opt. Lett.*, 19(20), 1609–1611, 1994.
89. F. Barone, E. Calloni, R. De Rosa, L. Di Fiore, F. Fusco, L. Milano, and G. Russo, Fringe-counting technique used to lock a suspended interferometer, *Appl. Opt.*, 33(7), 1194–1197, 1994.
90. C. M. Crooker and S. L. Garrett, Fringe rate demodulator for fiber optic interferometric sensors, in *Proceeding of Fiber Optic and Laser Sensors*, San Diego, CA, 1987.
91. M. Choma, C. Yang, and J. Izatt, Instantaneous quadrature low-coherence interferometry with 3 × 3 fiber-optic couplers, *Opt. Lett.*, 28(22), 2162–2164, 2003.
92. D. Linde, Characterization of the noise in continuously operating mode-locked lasers, *Appl. Phys. B*, 39(4), 201–217, 1986.
93. T. Clark, T. F. Carruthers, P. J. Matthews, and I. N. Duling III, Phase noise measurements of ultrastable 10 GHz harmonically modelocked fiber laser, *Electron. Lett.*, 35, 720–721, 1999.
94. H. Tsuchida, Correlation between amplitude and phase noise in a modelocked Cr:LiSAF laser, *Opt. Lett.*, 23, 1686–1688, 1998.
95. M. Scully and M. Zubairy, *Quantum Optics*. Cambridge, U.K.: Cambridge University Press, 2001.

96. S. Sanders, N. Park, J. Dawson, and K. Vahala, Reduction of the intensity noise from an erbium-doped fiber laser to the standard quantum limit by intracavity spectral filtering, *Appl. Phys. Lett.*, 61(16), 1889–1891, 1992.
97. J. Cliche, Y. Painchaud, C. Latrasse, M. Picard, I. Alexandre, and M. Têtu, Ultra-narrow Bragg grating for active semiconductor laser linewidth reduction through electrical feedback, in *Proceeding in Bragg Gratings, Photosensitivity, and Poling in Glass Waveguides*, Quebec City, Canada, 2007.
98. A. R. Bahrampour and A. A. Askari, Fourier-wavelet regularized deconvolution (ForWaRD) for lidar systems based on TEA CO_2 laser, *Opt. Commun.*, 257(1), 97–111, 2006.
99. S. I. Bozhevolnyi, *Plasmonic Nanoguides and Circuits*. Singapore: Pan Stanford Publishing, 2009.
100. S. A. Maier, *Plasmonics: Fundamentals and Applications*. New York: Springer, 2007.
101. L. Novotny and B. Hecht, *Principles of Nano-Optics*. New York: Cambridge University Press, 2006.
102. X. Yu, Y. Zhang, Sh. Pan, P. Shum, M. Yan, Y. Leviatan and Ch. Li, A selectively coated photonic crystal fiber based surface plasmon resonance sensor, *J. Opt.*, 12(1), 015005-4, 2010.
103. S. Ramo, J. Whinnery, and T. V. Duzer, *Fields and Waves in Communication Electronics*, 3rd edn. New York: John Wiley & Sons, 1994.
104. J. Sirkis, D. Brennan, M. Putman, T. Berkoff, A. Kersey, and E. Friebele, In-line fiber etalon for strain measurement, *Opt. Lett.*, 18, 1973–1975, 1973.
105. G. Mudhana, K. Park, and B. Lee, Dispersion measurement of liquids with a fiber optic probe based on a bi-functional lensed photonic crystal fiber, *Opt. Commun.*, 284, 2854–2858, 2011.
106. G. Mudhana, K. Park, S. Ryu, and B. Lee, Fiber-optic probe based on a bifunctional lensed photonic crystal fiber for refractive index measurements of liquids, *IEEE Sens. J.*, 11, 1178–1183, 2011.
107. H. Choi, K. Park, S. Park, U. Paek, B. Lee, and C. E.S., Miniature fiber-optic high temperature sensor based on a hybrid structured Fabry-Perot interferometer, *Opt. Lett.*, 33, 2455–2457, 2008.
108. H. Choi, G. Mudhana, K. Park, U. Paek, and B. Lee, Cross-talk free and ultra-compact fiber optic sensor for simultaneous measurement of temperature and refractive index, *Opt. Express*, 18, 141–149, 2009.
109. K. Park, Y. Kim, J. Eom, S. Park, M. S. Park, J. Jang, and B. Lee, Compact and multiplexible hydrogen gas sensor assisted by self-referencing technique, *Opt. Express*, 19, 18190–18198, 2011.
110. L. Schenato, L. Palmieri, G. Gruca, D. Iannuzzi, G. Marcato, A. Pasuto, A. Galtarossa, Fiber optic sensors for precursory acoustic signals detection in rockfall events, *JEOS:RP*, 7, 12048, 2012.
111. J. P. Dakin and M. Volanthen, Distributed and multiplexed fibre grating sensors, including discussion of problem areas, *IEICE Trans. Electron.*, E83-C(3), 391–399, 2000.
112. A. D. Kersey, A. Dandridge, and A. B. Tveten, Time-division multiplexing of interferometric fiber sensors using passive phase-generated carrier interrogation, *Opt. Lett.*, 12, 775–777, 1987.
113. S. C. Huang, W. W. Lin, and M. H. Chen, Time-division multiplexing of polarization-insensitive fiber-optic Michelson interferometric sensors, *Opt. Lett.*, 20, 1244–1246, 1995.
114. A. D. Kersey, A. Dandridge, and K. L. Dorsey, Transmissive serial interferometric fiber sensor array, *J. Lightwave Technol.*, 7(5), 846–854, 1989.
115. A. Dandridge and A. D. Kersey, Multiplexed interferometric fiber sensor arrays, *Proc. SPIE*, 1586, 176–183, 1991.
116. A. D. Kersey, Array topologies for implementing serial fiber Bragg grating interferometer arrays. U.S. Patent 5,987,197, Nov. 1999.
117. A. D. Kersey, A. Dandridge, and M. A. Davis, Low-cross-talk code-division multiplexed interferometric array, *Electron. Lett.*, 28(4), 351–352, 1992.
118. C. P. Jacobson and H. K. Whitesel, Code division multiplexing of optical fiber sensors for shipboard applications, *Proc. Ship Control Systems Symp.*, 1, 381–390, 1997.
119. R. Ryf, S. Randel, A. H. Gnauck, C. Bolle, R-J Essiambre, P. Winzer, D. W. Peckham, A. McCurdy, and R. Lingle, Space-division multiplexing over 10 km of three-mode fiber using coherent 6 × 6 MIMO processing, *Optical Fiber Communication Conference*, Los Angeles, CA, 2011.
120. P. J. Nash and G. A. Cranch, Multichannel optical hydrophone array with time and wavelength division multiplexing, *Proc. SPIE*, 3746, 304–307, 1999.
121. Y. J. Rao, A. B. L. Ribeiro, D. A. Jackson, L. Zhang, and I. Bennion, In-fiber grating sensing network with a combined SDM, TDM, and WDM topology, in *Proceedings—Lasers and Electro-Optics Society Annual Meeting*, 1996, p. 244.
122. R. Ashoori, Y. M. Gebremichael, S. Xiao, J. Kemp, K. T. V. Grattan, and A. W. Palmer, Time domain multiplexing for a Bragg grating strain measurement sensor network, in *Proceedings of the 13th International Conference on Optical Fiber Sensors (OFS-13)*, Kyongju, South Korea, 1999, Vol. 3746, pp. 308–311.
123. R. S. Weis, A. D. Kersey, and T. A. Berkoff, A four-element fiber grating sensor array with phase-sensitive detection, *IEEE Photon. Technol. Lett.*, 6(12), 1469–1472, 1994.

124. A. D. Kersey, Optical fibre sensors for downwell monitoring applications in the oil and gas industry, in *Proceedings of OFS'13*, Kyongju, South Korea, 1999, pp. 326–331.
125. Y. J. Rao, D. A. Jackson, D. J. Webb, L. Zhang, and I. Bennion, In-fibre Bragg grating flow-directed thermodilution catheter for cardiac monitoring, in *Optical Fiber Sensors*, Williamsburg, VA, 1997.
126. B. Lee and U. M. Paek, Multislit interpretation of cascaded fiber gratings, *J. Lightwave Technol.*, 20, 1750–1761, 2002.
127. D. Moon, B. Kim, A. Lin, G. Sun, T. Han, W. Han, and Y. Chung, The temperature sensitivity of Sagnac loop interferometer based on polarization maintaining side-hole fiber, *Opt. Express*, 15, 7962–7967, 2007.
128. M. H. Houston, B. N. P. Paulsson, and L. C. Knauer, Fiber optic sensor systems for reservoir fluids management, in *Proceedings of the Annual Offshore Technology Conference*, Houston, TX, 2000.
129. E. B. Wooding, K. R. Peal, and J. A. Collins, The ORB ocean-bottom seismic data logger, *Sea Technol.*, 39(8), 85–89, 1998.
130. U. Udd, *Fibre-Optic Smart Structures*. New York: Wiley, 1995.
131. A. Mendez, T. E. Morse, and E. Mendez, Applications of embedded optical fibre sensors in reinforced concrete buildings and structures, *Proc. SPIE*, 1170, 60–69, 1989.
132. R. M. Measures, Smart structures in nerves of glass, *Prog. Aerosp. Sci.*, 26, 289, 1989.
133. G. Righini, A. Tajani, and A. Cutolo, *An Introduction to Optoelectronics Sensors*. Singapore: World Scientific Publishing, 2009.
134. J. Wang, J. Xu, Q. Liang, and J. Li, A novel Mach-Zehnder and Sagnac interferometer for distributed optic fiber sensing, in *Photonics and Optoelectronics Meetings (POEM) 2011: Optoelectronic Sensing and Imaging*, 2011.
135. S. Spammer, P. Swart, and A. Chtcherbokov, Merged Sagnac–Michelson interferometer for distributed disturbance detection, *J. Lightwave Technol.*, 15(6), 972–976, 1997.
136. X. Fang, A variable-loop Sagnac interferometer for distributed impact sensing, *J. Lightwave Technol.*, 14, 2250–2254, 1996.
137. G. Meltz, S. J. Hewlett, and J. D. Love, Fibre grating evanescent-wave sensors, *Proc. SPIE*, 2836, 342–350, 1996.
138. W. Ecke, K. Usbeck, V. Hagemann, R. Mueller, and C. R. Willsch, Chemical Bragg grating sensor network basing on side-parallel optical fiber, *Proc. SPIE*, 3555, 457–466, 1998.
139. B. H. Lee, Y. Liu, S. B. Lee, and S. S. Choi, Displacement of the resonant peaks of a long-period fiber grating induced by a change of ambient refractive index, *Opt. Lett.*, 22, 1769, 1997.
140. O. S. Wolfoeis, *Fibre Optic Chemical Sensors*, Vols. I and II. Boca Raton, FL: CRC Press, 1991.
141. Z. Zhang and J. S. Sirkis, Temperature-compensated long period grating chemical sensor, in *Proceedings of the 12th International Conference on Optical Fibre Sensors*, Williamsburg, VA, 1997.
142. R. Falciai, A. G. Mignani, and A. Vannini, Optical fibre long-period gratings for the refractometry of aqueous solutions, *Proc. SPIE*, 3555, 447–450, 1998.
143. L. M. Lechuga, A. Calle, and F. Prieto, Optical sensors based on evanescent field sensing. Part I. Surface plasmon resonance sensors, *Quím Anal.*, 19, 54–60, 2000.
144. L. Chau, Y. Lin, S. Cheng, and T. Lin, Fiber-optic chemical and biochemical probes based on localized surface plasmon resonance, *Sens. Actuat. B-Chem.*, 113(1), 100–105, 2006.
145. H. C. Ohanian and R. Ruffini, *Gravitation and Spacetime*. New York: Cambridge University Press, 2013.
146. B. F. Schutz, *A First Course in General Relativity*. Cambridge, U.K.: Cambridge University Press, 1985.
147. J. Hough and S. Rowan, Laser interferometry for the detection of gravitational waves observatories, *J. Opt. A: Pure Appl. Opt.*, 7, S257–S264, 2005.
148. S. A. Hughes, S. Marka, P. L. Bender, and C. J. Hogan, New physics and astronomy with the new gravitational-wave observatories, in *Proceedings of the 2001 Snowmass Meeting*, LIGO Report No. LIGO-P010029-00-D, ITP Report No. NSF-ITP-01-160.
149. C. M. Cave, Quantum-mechanical radiation-pressure fluctuations in an interferometer, *Phys. Rev. Lett.*, 45, 75–79, 1980.
150. J. P. Dowling, Quantum optical metrology—The lowdown on high-N00N states, *Contemp. Phys.*, 49(2), 125–143, 2008.
151. C. M. Cave, Quantum mechanical noise in an interferometer, *Phys. Rev. D*, 23, 1693, 1981.
152. S. Tofighi, A. R. Bahrampour, and F. Shojai, Optimum quantum state of light for gravitational-wave interferometry, *Opt. Commun.*, 283, 1012–1016, 2010.
153. P. Beyersdorf, The polarization Sagnac interferometer for gravitational wave detection. Dissertation, Stanford University, Stanford, CA, 2001.
154. P. T. Beyersdorf, M. M. Fejer, and R. L. Byer, Polarization Sagnac interferometer with postmodulation for gravitational-wave detection, *Opt. Lett.*, 24(16), 1112–1114, 1999.

155. R. T. Cahill, Optical-fiber gravitational wave detector: Dynamical 3-space turbulence detected, *Prog. Phys.*, 4, 63–68, 2007.
156. R. Cahill and F. Stokes, Correlated detection of sub-mHz gravitational waves by two optical-fiber interferometers, *Prog. Phys.*, 2, 103–110, 2008.
157. V. K. Sacharov, Linear relativistic fiber-optic interferometer, *Laser Phys.*, 11(9), 1014–1018, 2001.
158. T. Corbitt and N. Mavalvala, Quantum noise in gravitational-wave interferometers, *J. Opt. B*, 6, S675–S683, 2004.
159. P. Kok, W. Munro, K. Nemoto, T. Ralph, J. Dowling, and G. Milburn, Linear optical quantum computing with photonic qubits, *Rev. Mod. Phys.*, 79, 135–174, 2007.
160. G. Durkin and J. Dowling, Local and global distinguishability in quantum interferometry, *Phys. Rev. Lett.*, 99, 070801, 2007.
161. H. Cable and J. P. Dowling, Efficient generation of large number-path entanglement using only linear optics and feed-forward, *Phys. Rev. Lett.*, 99, 163604, 2007.
162. H. Lee, P. Kok, and J. Dowling, A quantum Rosetta stone for interferometry, *J. Mod. Opt.*, 49, 2325–2338, 2002.
163. P. Kok, S. Braunstein, and J. Dowling, Quantum lithography, entanglement and Heisenberg-limited parameter estimation, *J. Opt. B*, 6, S811–S815, 2004.
164. Q. Turchette, C. Hood, W. Lange, H. Mabuchi, and H. Kimble, Measurement of conditional phase-shifts for quantum logic, *Phys. Rev. Lett.*, 75, 4710–4713, 1995.
165. M. Lukin, Trapping and manipulating photon states in atomic ensembles, *Rev. Mod. Phys.*, 75, 457–472, 2003.
166. H. Lee, P. Kok, N. Cerf, and J. Dowling, Linear optics and projective measurements alone suffice to create large-photon-number path entanglement, *Phys. Rev. A*, 65, 030101, 2002.
167. P. Kok, H. Lee, and J. Dowling, Creation of large photon number path entanglement conditioned on photodetection, *Phys. Rev. A*, 65, 052104, 2002.

4 Polymer Optical Fiber Sensors

Kara Peters

CONTENTS

4.1 INTRODUCTION

The majority of optical fiber sensors are based on silica optical fibers due to their wide use in telecommunication applications and the wide availability of components, instrumentation, and optical fiber specifications. More recently, polymer optical fibers (POFs) have experienced a surge in applications for short haul telecommunication systems. Many researchers have realized their unique properties for sensing of strain, temperature, humidity, etc. Simultaneously, researchers are also developing POFs with new properties including single-mode fibers, microstructured fibers, and fibers for propagation of terahertz (THz) wavelengths. These unique properties have been used to expand the envelope of sensing applications beyond those previously realized with silica optical fiber sensors.

The goal of this chapter is to highlight the unique benefits of POF-based sensors, as well as challenges to their application. The unique benefits of POFs include the low stiffness and high ductility of polymer materials, as well as their low weight. These permit POF sensors to be applied in applications where high levels of strain are expected and applications where precision measurements must be made at very low force levels. While POF properties vary across the wide variety of polymer optical materials, the inherent response of POFs to strain and temperature differ from silica fibers. Generally, their sensitivity to strain and temperature is higher than that of silica, while their temperature sensitivity is sometimes of the opposite sign. Due to their low stiffness, POFs can also support bending more easily than equivalent silica fibers of the same diameter. Most polymers are also biocompatible, extending the use of optical fiber sensors into biological sensing, without the need for specialized coatings.

It is also critical to understand the limits of POFs including their low thermal threshold and generally lower fabrication quality as compared to silica fibers. In addition, the viscoelastic nature of polymer material means that the full loading history must be known to predict their response for some applications. Finally, coupling POFs with small diameter cores can be difficult due to cross-sectional variances and difficulties in preparing the end-faces to be coupled. For applications requiring substantial lengths of optical fiber, the higher inherent attenuation of polymeric materials, particularly at near-infrared wavelengths, can also be an issue.

This chapter presents a summary of the properties of POFs that are critical to their sensing performance, including their optical transmission, strain and temperature sensitivity, and their response to humidity. Practical issues including coupling to POFs are also discussed. Afterwards, examples of sensors using a variety of sensing mechanisms are reviewed. These sensing mechanisms are the same as used in silica fiber–based sensors; however, the focus in this chapter is on sensing examples that take advantage of the unique POF properties. Due to space limitations, not all POF sensor examples can be presented; therefore, the examples discussed in this chapter are biased towards more recent examples. Finally, comments are made on the outlook of POF sensors and suggestions for future research directions to enable new sensing capabilities are presented.

4.2 PROPERTIES OF POFs

There are several critical advantages of polymer over silica optical fibers, including their large elastic and plastic strain deformation capabilities, negative thermal sensitivities, high numerical aperture, and lower stiffness. These properties will be described in the following sections. Typical POF materials include polymethyl methacrylate (PMMA), polystyrene (PS), and amorphous fluorinated polymer (CYTOP) [1]. Due to the challenges in fabricating POFs, commercially available POFs are typically multimode at their operating wavelengths. As a result, these fibers are low cost and are easier to cut and connect, as compared to single-mode silica optical fibers. Multimode POFs are commercially available with different cross-sectional index distributions, including both step index and gradient index configurations. At the same time, a large variety of POF diameters are available, all larger than conventional single-mode silica optical fibers.

Recent research and development has also produced POFs that are single mode in the visible wavelength range. The fabrication of these fibers is challenging due to their high numerical aperture and small core size that must achieved [2]. Single-mode POFs were first developed by Koike [3] and recent research has continued to decrease the intrinsic attenuation of these fibers [4]. However, doped, single-mode POFs are still experimental, with one commercial manufacturer, Paradigm Optics.

The second class of specialty POFs is microstructured POFs (MPOFs). MPOFs are characterized by the presence of airholes distributed in the fiber cross section. The air fraction decreases the material perpendicular to the propagation direction and therefore decreases the attenuation of the lightwave as it propagates through the optical fiber [1]. Additionally, lightwave guidance in the MPOF can be controlled by the hole arrangement, providing single-mode guidance over a wide range of wavelengths and even endlessly single-mode optical fibers [5].

4.2.1 Optical Properties

The attenuation spectrum of polymer optical materials is considerably different than that of silica, changing the wavelength windows in which POF sensors typically operate. Figure 4.1 plots the intrinsic attenuation loss of common optical polymers as a function of wavelength. Three wavelength windows are commonly applied for multimode POF sensors: the visible wavelength range (400–700 nm) where the intrinsic attenuation is low; near 850 nm where common telecom, low-cost components are available; and the near-infrared range above 1300 nm where the specialty

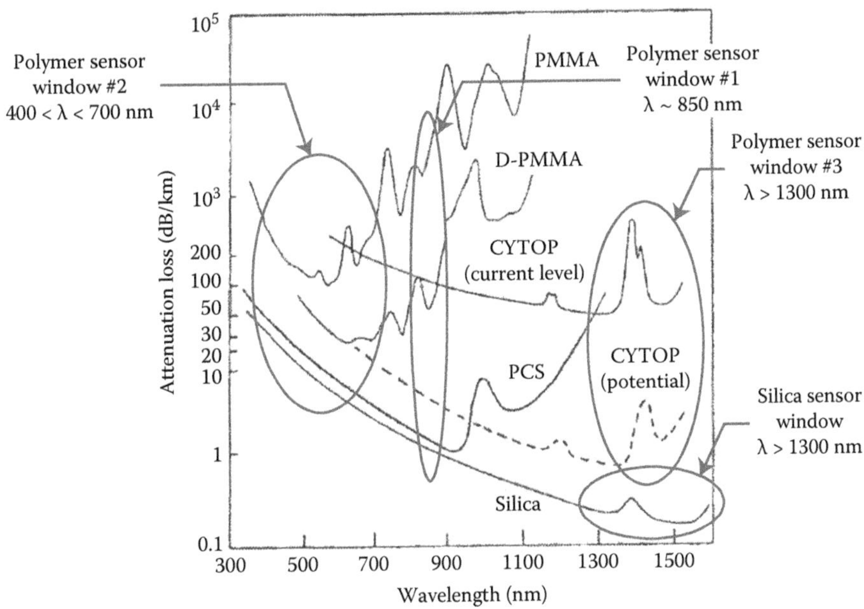

FIGURE 4.1 Attenuation loss of common optical polymers as a function of wavelength. (Adapted from Zubia, J. and Arrue, J., *Opt. Fiber Technol.*, 7, 101, 2001.)

amorphous fluorinated polymer (CTOP) has low attenuation. At all wavelengths, the intrinsic attenuation of polymer materials are significantly larger than that of silica; therefore, the design of long-distance POF sensors must consider signal losses. Dispersion losses are also important, particularly for multimode POFs. Graded-index profile POFs (GIPOFs) significantly improve the resolution of sensors based on signals transmitted over long lengths of the fiber [1].

4.2.2 Mechanical Properties

The mechanical properties of POFs vary significantly and are highly influenced by the fiber drawing process and dopant concentration (if used to increase the core index of refraction) [7,8]. One critical factor in the drawing process is the annealing process to remove internal residual stresses and anisotropy in the polymer [8]. As a result, it is critical to calibrate the mechanical and thermal behavior of a specific POF prior to its use as a strain or temperature sensor. Beyond the fabrication process, the mechanical properties of POFs are also dependent upon loading conditions and rates and are further affected by environmental conditions such as high temperature or humidity [9,10].

Figure 4.2 shows typical true stress versus strain curves measured for a PMMA optical fiber at different applied strain rates. These curves are typical of viscoelastic materials. POF properties typically fall in the ranges of initial elastic modulus of 1–5 GPa, yield strain of 1%–6%, and ultimate strains of 6%–100%, as compared to 1%–5% for silica optical fibers [1,8,12]. We observe that the initial stiffness of the material and the yield strain are both a function of the applied strain rate. In addition, we observe that the material behavior is nonlinear beyond the yield strain and therefore the loading history is also critical to predict hysteresis and cyclic behavior of the material [13]. When POFs are strained beyond their yield limit, plastic deformation occurs in the fiber, resulting in a residual deformation even when the fiber is unloaded. For sensing applications, this residual deformation can be considered a *shape memory effect* that can store the maximum strain information to be extracted later [12]. PMMA also has a lower density (1195 kg m^{-3}) than silica (2200 kg m^{-3}), reducing the weight of distributed optical fiber sensors [2].

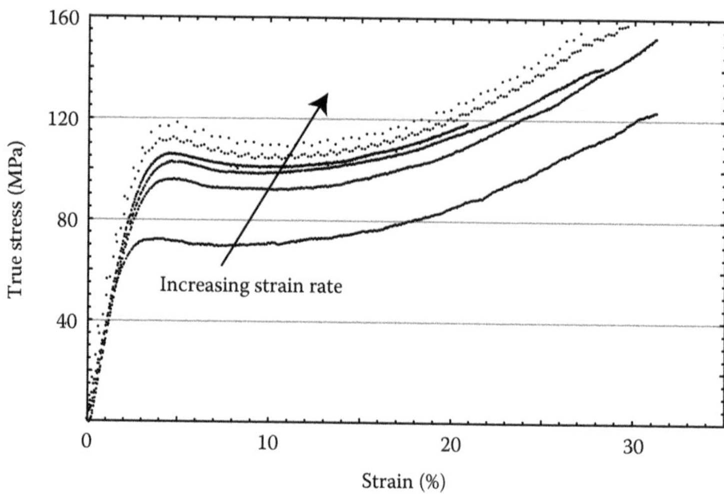

FIGURE 4.2 Measured, true stress–strain curves for single-mode PMMA-doped core POF at strain rates of 0.01, 0.30, 0.60, 0.90, 1.22, and 3.05 min^{-1}. (From Kiesel, S. et al., *Meas. Sci. Technol.*, 18, 3144, 2007.)

A second method to reveal the dependence of the material behavior on applied loading rates is through dynamic mechanical analysis (DMA) in which cyclic loads are applied to the materials at different frequencies. From DMA, Young's modulus at different frequencies can be determined. Figure 4.3 plots Young's modulus starting at 7 Hz loading frequency for a solid PMMA POF, a microstructured PMMA POF, and a silica optical fiber for comparison [14]. The particular measurement system had a mechanical resonance at 1 kHz affecting results near this frequency. All measurements were made in the low-deformation regime. As compared to the silica fiber, we observe the lower Young's modulus at low frequencies of the polymer materials and the start of a frequency-dependent modulus at lower frequencies in the polymer materials. Stefani et al. [14] also demonstrated an excellent fit of measured material properties for PMMA and TOPAS optical fibers to Maxwell's model for viscoelastic materials.

Finally, we define the strain sensitivity of an optical fiber to be the phase change in a lightwave propagating through a unit length of the fiber per unit axial strain. Here, we consider the linear

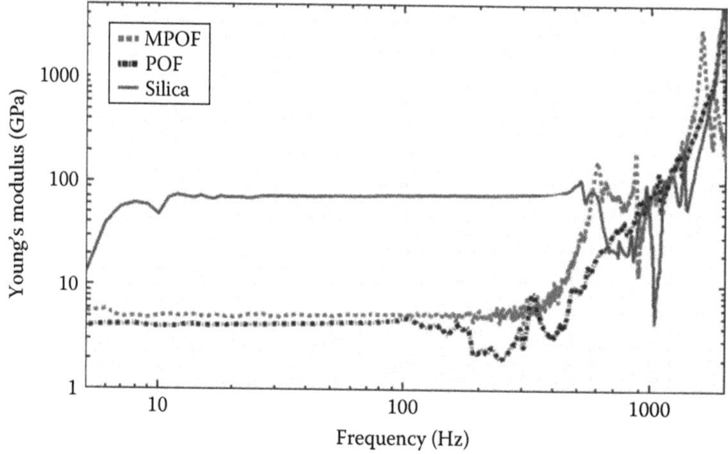

FIGURE 4.3 Dynamic Young's modulus of PMMA MPOF, step index POF, and silica SMF28. (From Stefani, A. et al., *IEEE Sens. J.*, 12, 3047, 2012.)

region of the stress–strain curve. Again, this sensitivity varies depending upon the POF material and fiber drawing conditions. As an example, applying the Pockels constants for bulk PMMA, the strain sensitivity of a PMMA optical fiber is theoretically predicted to be 132.6×10^5 rad m^{-1} [15]. This sensitivity is approximately 15% larger than that of a silica optical fiber.

4.2.3 Thermal Properties

At the same time, we define the temperature sensitivity to be the phase shift per unit change in temperature per unit length of the optical fiber. For bulk PMMA, the thermal sensitivity is −154.3 rad m^{-1} K^{-1} [15]. Again, this is significantly larger than that of silica (by approximately 57%) but is also of the opposite sign. This negative temperature sensitivity (which is not universal for all polymers) presents new possibilities for strain and temperature compensation in optical fiber sensors.

As for silica optical fiber or electrical resistance sensor systems, the inherent temperature response of a POF must be known for temperature compensation of strain measurements. For example, Huang et al. [16] reports a strain–temperature cross talk in a multimode POF of 33 $\mu\varepsilon$ $°C^{-1}$. Recent work has also highlighted the coupling between the response of the POF to temperature and humidity [17,18]. As most temperature measurements are performed in nonmoisture-controlled environments, and polymeric materials typically have high sensitivities to relative humidity (RH), it is difficult to separate these two effects. To better understand the response of PMMA POFs to strictly temperature variations, Zhang and Tao [17] performed thermal measurements in a controlled RH environment. The typically presented coefficient of thermal expansion (COT) of POF is approximately -1×10^{-4} $°C^{-1}$, which is an order of magnitude (and of opposite sign) than that of silica, 1×10^{-5} $°C^{-1}$. In contrast, Zhang and Tao [17] found that when RH is accurately controlled as the temperature is varied, the COT is nonlinear with temperature and ranged from 5% of the predicted value at 20°C to 20% of the predicted value at 60°C. They explained the discrepancies between these measured COTs and those predicted from bulk PMMA parameters to be due to the orthotropic alignment of the polymer chains created during drawing of the optical fibers. These aligned chains significantly decrease the thermal expansion of the POF in the axial direction as compared to the isotropic bulk PMMA.

4.2.4 Moisture Properties

As discussed in the previous section, POFs have an inherent sensitivity to humidity. In contrast, humidity sensors based on silica fibers require specialized coatings (often polymeric) that absorb moisture and induce strain on the silica fiber, which can then be measured. Zhang et al. [19] demonstrated that an increase in RH caused an increase in the index of refraction and swelling of PMMA, up to a maximum water content in the PMMA of 2 wt%. Etching the solid PMMA optical fibers to reduce the diameter decreased the response time of the POF to RH. Figure 4.4 plots the wavelength change of a fiber Bragg grating (FBG) sensor written in the PMMA POF as a function of temperature at two separate RH levels. This response to humidity must be accounted for as it can affect the temperature measurements as described previously or other measurements in biological environments where large amounts of moisture are present. The application of a POF sensor would require a thorough understanding of the environment if the temperature and moisture conditions were expected to vary.

4.2.5 Chemical Infiltration

The intrinsic ability of POFs to absorb moisture can also make them sensitive to chemical infiltration. For example, Hamouda et al. [20] measured the effect of vinyl ester and epoxy resins on the integrity and signal transmission of perfluorinated POFs embedded in these resin systems for sensing applications. The more aggressive of the two resin systems, vinyl ester, penetrated into the POF during curing of the resin, causing a significant increase in backscattering level in the POF and eventual signal transmission loss. Figure 4.5 shows the visible change in the POF cross section

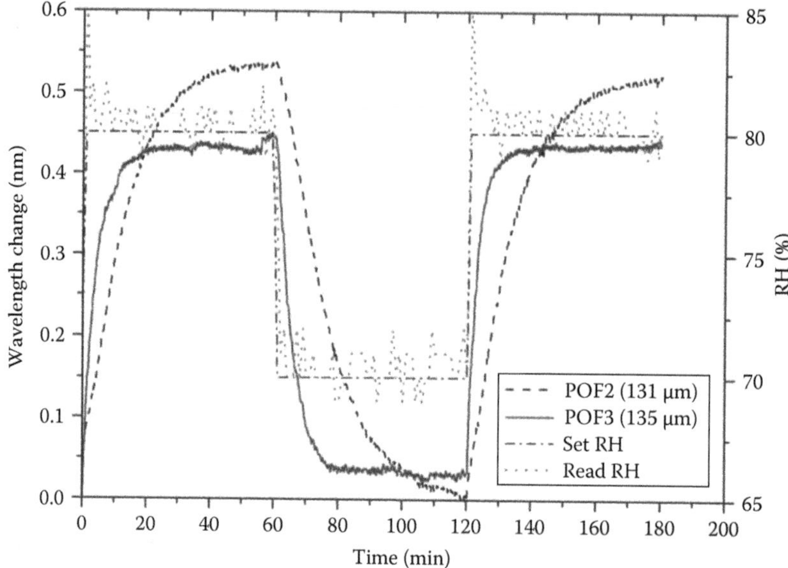

FIGURE 4.4 Full-cycle responses of two POF FBG sensors. RH varied from 80% to 70% and back to 80%, while the temperature was kept constant at 25°C. (From Zhang, W. et al., *J. Lightwave Technol.*, 30, 1090, 2012.)

before and after exposure to the vinyl ester resin. In contrast, the epoxy resin did not penetrate the POF during cure, creating no increase in backscattering level.

4.2.6 Connecting POFs

As mentioned in the introduction, one of the benefits to multimode POFs is their large diameter and ease of splicing [1]. The most common method to prepare multimode POFs for connecting is by cutting the optical fiber with a blade. The end-face surface quality can be improved by gripping the optical fiber in a precision holder prior to cutting. In addition, heating the blade prior to cutting can also improve the smoothness of the cut fiber section by making the polymer more ductile prior

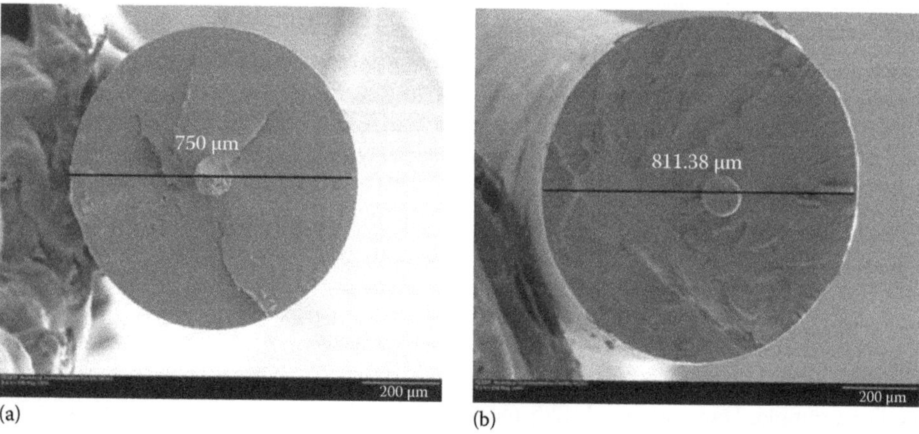

FIGURE 4.5 Scanning electron microscope image of (a) original graded-index perfluorinated polymer optical fiber (GI-PF-POF) and (b) the same fiber type after embedment in vinyl ester resin. (From Hamouda, T. et al., *Smart Mater. Struct.*, 21, 055023, 2012.)

to cutting. After cleaving, POFs are typically butt-coupled together in a ferrule, as fusion splicing is not an option due to their low melting temperature. Again, the large diameters of these fibers reduce the losses associated with butt-coupling.

In contrast, due to their extremely small core sizes, single-mode POFs are difficult to couple. At the same time, the high attenuation properties of single-mode POFs often requires that the POF sensor be connected to silica optical fibers to serve as leads to and from the instrumentation [21]. Connecting MPOFs for sensing applications presents similar challenges. Again, it is important to produce a smooth cleaved surface so as to introduce minimal distortion to the cross-sectional microstructure. In addition, the cleaved cross section must be clean to prevent microparticles from blocking the cross-sectional holes and the temperature must be well controlled to prevent thermal collapse of the holes [22]. The fabrication of single-mode POFs is not as well controlled as their silica counterparts; therefore, the placement of the core can vary within the cross section, requiring significant alignment of the optical fibers prior to coupling.

For silica optical fibers, cleaving is performed by first notching the outer surface to produce a microcrack in the fiber cross section and then bending the optical fiber to propagate the crack across the cross section. Due to the viscoelastic nature of polymers and their low stiffness, applying this cleaving method to POFs produces a combination of cutting and tearing, which reduces the quality of the cleaved surface [23]. The ductile behavior of the polymer produces plastic deformation at the fracture surface that warps the cross section. When the polymer is in a brittle state, the material anisotropy can create turning of the crack front and chipping, creating debris that is dragged over the fracture surface. Therefore, preparing POFs for coupling must be performed through alternate methods.

The most popular technique to prepare single-mode or MPOFs for coupling is hot-knife cleaving of the fiber [24–26]. Prior to application, optimization of the blade and fiber temperatures should be performed to minimize distortion of the final POF cross section. In contrast, chilling of the POF to low temperatures causes chipping of the optical fiber and produces poor quality surfaces [23,24]. Alternate methods include focused-ion-beam machining and UV laser cleaving [22,24], although these are considerably more resource intensive and not suitable for field applications.

Once cleaved, challenges with fiber alignment again make coupling of single mode or MPOFs challenging. One solution is that of Abang and Webb [27] who demonstrated a demountable ferrule connector/angled physical contact (FC/APC) connection between a single-mode silica fiber and a 50 mm core MPOF with minimal coupling losses. The MPOF was first etched with acetone to reduce the cladding diameter to fit into a standard FC/APC ferrule. The protruding end of the MPOF was cleaved with a heated blade and fixed into the ferrule with a UV cure adhesive. Finally, the authors polished the cleaved MPOF.

4.3 SENSOR EXAMPLES

This section will present several examples of optical fiber–based sensors that highlight the advantages of POFs discussed previously. The sensors examples are grouped by the particular sensing mechanisms that they use to convert the physical parameters into properties of propagating lightwaves. For brevity, this is not an extensive list of all POF-based sensors, but rather a selection from each type to highlight the potential of POF sensors for the reader.

4.3.1 Optical Loss

We will start with sensors based on the measurement of optical power losses, as these are typically the simplest and lowest cost POF-based sensor systems. The cost of such sensors is low because commercially available multimode POFs and inexpensive light sources can be used and a simple photodetector is required to convert the optical fiber power transmitted through the optical fiber into a voltage output. In the following examples, power losses as introduced into the system through

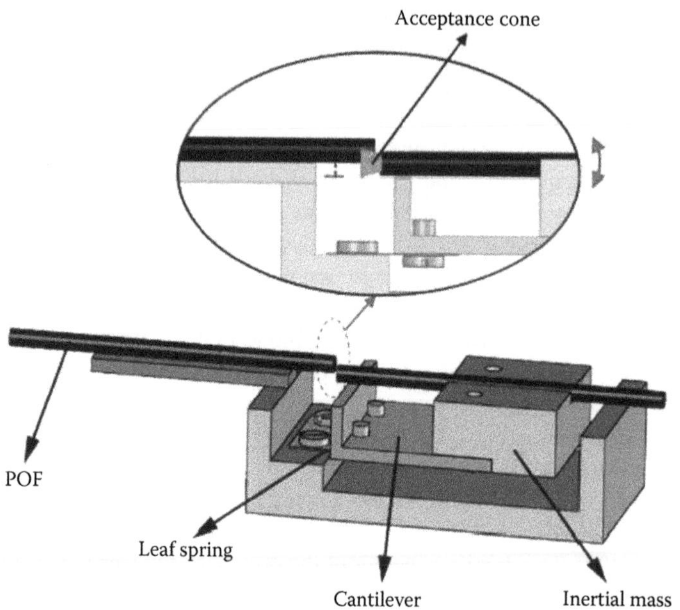

FIGURE 4.6 Schematic of POF-based accelerometer. The inset shows a magnification of the fiber gap region. (From Antunes, PFC et al., *IEEE Sens. J.*, 13, 1716, 2013.)

either changing the amount of input optical power coupled into the POF or through creating losses along the length of the optical fiber (e.g., through critical bending of the POF).

In the first case, a repeatable coupling loss is introduced through a misalignment between the input POF and the sensing POF. The misalignment is then tuned to be a function of the particular parameter to be measured. For example, Antunes et al. [28] implemented a POF accelerometer based on the transfer of lightwaves between two multimode POFs. One POF was fixed to the inertial frame and the other moved with the object whose acceleration is to be measured, as shown in Figure 4.6. Acceleration of the object moved the POF mounted on the cantilever beam in a direction parallel to the cross section, creating a coupling loss into the sensor fiber. Advantages to this accelerometer are its low cost, ease of fabrication, and small size. Disadvantages are the low resolution inherent in the power coupling measurements. In a similar concept, Mohanty and Kuang [29] designed a respiration rate sensor based on the relative displacement, and therefore coupling efficiency, of two POFS. Kulkarni et al. [30] developed a pressure sensor based on coupling loss in which the sensor POF was mounted to silicone rubber (PDMS) block. As the block was compressed due to pressure, the optical properties changed, specifically the scattering coefficient. By measuring the total power coupled into a POF on the other side of the block, the applied pressure was determined.

Some of the earliest examples of POF sensors were based on the second case of inducing transmission losses in the POF through localized bending or damage to the POF. Takeda [31] embedded multimode, PMMA POFs into carbon fiber–epoxy laminates for the detection of cracking. Light was coupled into the POF from a light-emitting diode (LED) and the output power was measured with a photodetector. The POF output power was shown to be highly sensitive to the local crack density in the laminate. Due to its high strain limit, the POF survived beyond the failure of the laminate.

Kuang et al. [32] modified a multimode POF to be sensitive to bending losses by removing a portion of the 980 μm diameter cross section. The resulting cross section was not symmetric; therefore, the bending direction could also be identified. The POF was adhered to a metallic specimen and acted as a strain gauge through bending of the POF. Two configurations were applied: a test in which the POF was along the axis of a beam during bending and a second test in which the POF was mounted in a curved path such that tensile and compressive loading caused a change in curvature of the POF. Kuang et al. [33] later applied this sensor to detect cracks in concrete beams loaded in bending. The presence of cracks produced localized necking of the POF and therefore large decreases in optical transmission. Other researchers have increased the sensitivity of the transmission power loss to axial strain by chemical tapering of the POF [34] or removing a curved section (groove) of the cross section [35]. The low cost and ease of implementation of these bending sensors also make them ideal for medical textile applications. For example, Bilro et al. [36] and Stupar et al. [37] incorporated side-polished and notched POFs into a cloth knee support brace for gait and joint curvature monitoring, respectively.

Remouch et al. [38] took advantage of the large thermooptic coefficient of polymers to design a temperature sensor based on a bent POF. The optical loss through the bend changed as a function of applied temperature, with a theoretical thermal sensitivity one order of magnitude greater than an equivalent silica optical fiber–based sensor.

An additional configuration to create bending losses in the POF is the geometry of U-shaped POF sensors. In such configurations, the radius of the curved portion is well controlled such that the bending losses are repeatable in the POF. By changing the index of refraction of a fluid external to the POF, the fraction of light coupled into the surrounding fluid is changed. Such low-cost sensors can therefore be calibrated to serve a liquid level sensor, for example, for fuel level monitoring [39]. The advantage here is that the optical sensor does not create a spark hazard near the fuel. Researchers have also demonstrated a U-shaped POF sensor for salinity measurements in water and TNT concentrations in alcohol, based on the index of refraction change with concentration [40–42].

4.3.2 Interferometry-Based Sensors

One sensor configuration that takes advantage of the low cost and ease of use of multimode POFs is the time of flight interferometer. This is an incoherent interferometer configuration and therefore does not require phase measurements. For some structural applications, it is sufficient to measure the integrated strain along the POF. The time-of-flight measurement system provides sufficient displacement resolution for a full-scale structure, yet at orders of magnitude lower cost than optical time domain reflectometry (OTDR) and scattering systems described in the following section (however, these systems provide distributed strain information). For example, Gomez et al. [43] and Durana et al. [44] applied time-of-flight measurements to monitor the global displacement of a vibrating aircraft wing flap. A diagram of the voltage-controlled oscillator (VCO)-driven interrogator is shown in Figure 4.7, along with a photograph of the aircraft flap with the surface mounted sensor POF. The phase of the optical sensor signal could also be compared to the original electrical signal used to modulate the laser source if a reference fiber is not required [45]. The interrogator shown in Figure 4.7 is entirely constructed of low-cost telecommunication components. The system is also portable and durable (since no moving parts are required) and has relatively low power requirements. The measurement displacement range is determined by the oscillator modulation frequency and can be quite large compared to other interrogation methods.

Coherent interferometry sensors using POF fibers have also been demonstrated. As coherent interferometry requires control of the propagating modes, this has been applied in solid-core single-mode POFs and is significantly more expensive than the time-of-flight measurements, however, with the benefit of higher resolution sensing. Coherent interferometry in single-mode POFs enables high-precision, large-deformation optical fibers for a variety of applications. Kiesel et al. [46] demonstrated coherent interferometry in a single-mode PMMA POF in-fiber Mach–Zehnder interferometer up to 15.8% elongation of the POF. This strain range is well beyond that previously measured with silica optical fiber sensors. As a result of the POF viscoelastic behavior, the measured phase-displacement sensitivity was not constant over the usable strain range and varied with strain

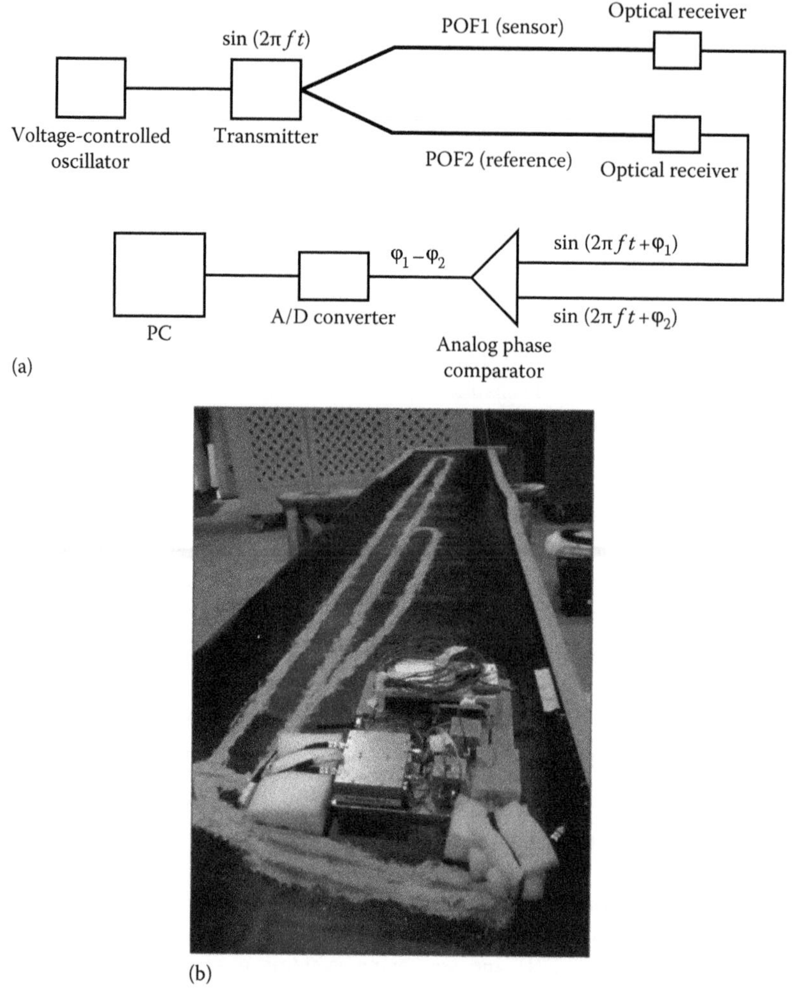

FIGURE 4.7 (a) Schematic of VCO interrogator used for time-of-flight measurements. (b) Photograph of upper side of aircraft flap showing POF adhered to surface and prototype instrumentation. (From Gomez, J. et al., *Appl. Opt.*, 48, 1436, 2009.)

rate. It is therefore important to both calibrate and understand the source of nonlinearities in the phase response of the POF.

Silva-López et al. [15] first measured the sensitivity of doped single-mode POFs to strain and temperature. The POFs were designed to be single mode at 850 nm, with an acrylic cladding and doped PMMA core. The authors operated the fibers with a visible red light source, outside of the single-mode region; however, they only observed the fundamental mode propagating through the fiber. Using a Mach–Zehnder interferometer arrangement and loading the optical fiber on a translation stage, they measured a phase sensitivity to displacement of 1.31×10^7 rad m^{-1}. This phase sensitivity is in good agreement with the properties of bulk PMMA. The measurements of Silva-López et al. [15] were made in the strain range of 0–0.04%, which is a limited portion of the strain range over which single-mode POF sensors can be applied.

To extend this strain range, Kiesel et al. [11] derived a formulation for the phase-displacement response of an optical fiber in uniaxial tension considering large strain magnitudes, including the finite deformation of the optical fiber and potential nonlinear photoelastic effects. Kiesel et al. [11]

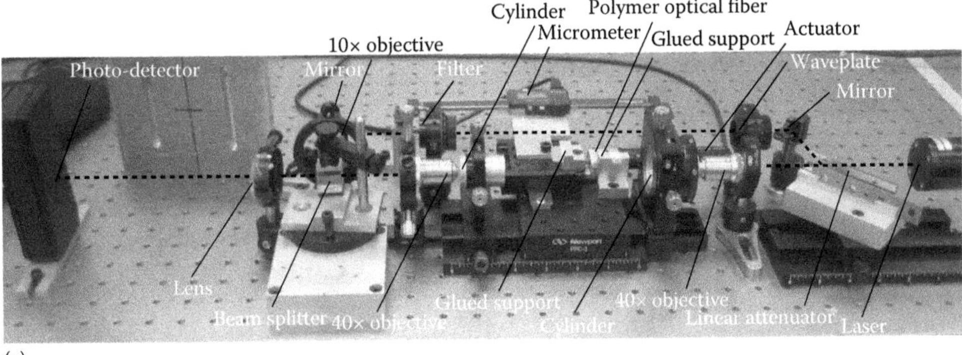

(a)

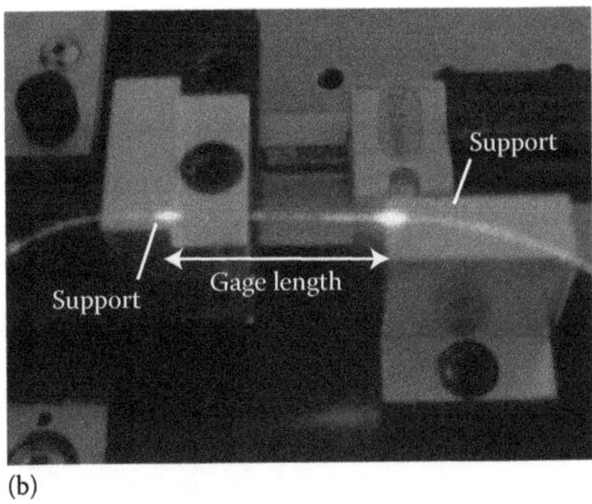

(b)

FIGURE 4.8 (a) Photograph of Mach–Zehnder interferometer for the measurement of POF phase sensitivities. Path of the reference beam is shown as a dashed line. (b) POF sample during loading (red light attenuation is visible). (From Kiesel, S. et al., *Meas. Sci. Technol.*, 20, 034016, 2009.)

then performed a series of tests on single-mode PMMA POFs to calibrate the mechanical nonlinearities and demonstrated their importance at as little as 1% axial strain in the POF. Figure 4.8 shows the Mach–Zehnder interferometer arrangement used for the calibration and an example single-POF segment under tension. Finally, Kiesel et al. [47] measured the phase-displacement response of the POF specimens over the maximum strain range. It was demonstrated that the contribution of the photoelastic nonlinearity is of the same order of magnitude as the finite deformation for the PMMA optical fiber and therefore cannot be neglected in predicting the sensor response to strain.

Abdi et al. [21] applied the in-fiber POF Mach–Zehnder interferometer to a tensile specimen and compared the sensor performance to an independent strain measurement from an extensometer. For this experiment, the reference arm was a silica optical fiber and the phase response was measured using a 3×3 coupler interrogator. The 3×3 coupler arrangement was modified to compensate for the power imbalance in the measurement and reference arms (due to the different attenuation properties of silica and PMMA) and to permit the extraction of the changing intensity in the sensor arm that would be expected near the elastic strain limit of the POF. This interrogator is shown in Figure 4.9a. The POF was first loaded in a single cycle up to 10% elongation and confirmed the extensometer measurements. A photograph of the experiment is shown in Figure 4.9b. The phase shift–strain response was extremely repeatable between specimens, and no hysteresis was observed at cycles up to 4% elongation. The nonlinearity was greater than that predicted in the formulation of Kiesel et al. [11,47], presumably due to the behavior of the adhesive bonding of the POF to the tensile specimen.

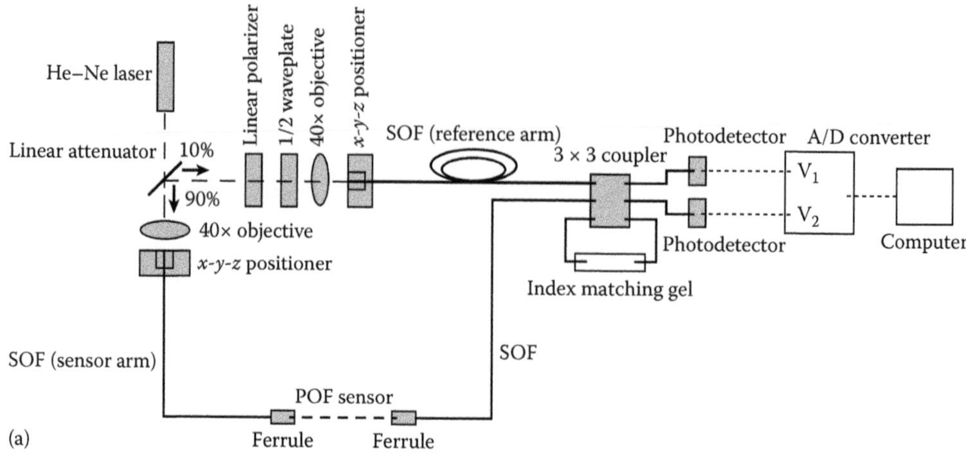

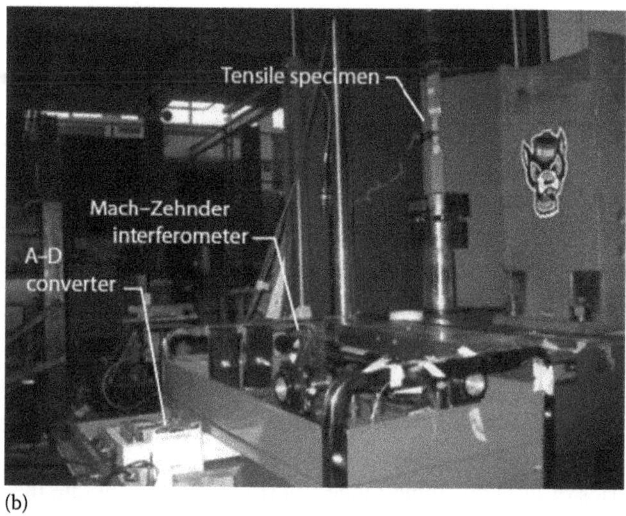

(b)

FIGURE 4.9 (a) Schematic of the Mach–Zehnder interferometer for the measurement of the relative phase shift in the POF sensor. (b) Experimental components for tensile specimen testing. (From Abdi, O. et al., *Meas. Sci. Technol.*, 22, 075207, 2011.)

4.3.3 OTDR, OFDR, and Scattering

Several researchers have also recently demonstrated applications of distributed sensing exploiting the unique properties of POF sensors. In particular, plastic deformation in a POF significantly changes the scattering and loss properties along the optical fiber, which can be easily measured. As silica fibers demonstrate a brittle behavior, this behavior is not observed in silica based sensors. It is important to note that the backscattering signal in POFs is not highly sensitive to small strains but has a high sensitivity to large strains, well beyond those that can be supported by silica optical fibers. Therefore, measurements of these properties can provide strain information throughout large structures in applications for which large strains are important.

The nonlinear stress–strain behavior of POFs at large strain values (typically beyond the yield point shown in Figure 4.2) is encoded along the POF and can be measured through the scattering or other loss profiles along the POF, for example, through OTDR [48]. Liehr et al. [48] embedded large diameter, multimode PMMA POFs into geotextiles for the monitoring of geotechnical and masonry structures.

The geotextiles were embedded in a railway embankment for monitoring of soil displacements. The large core diameter allowed easy connection to and handling of the sensors at the construction site, while the use of the standard POF itself as the sensor permitted monitoring of a large area at low cost. Additionally, the high ultimate strain of the PMMA allowed the POF to elongate with the large soil deformations. The OTDR measurements were limited by the attenuation and dispersion characteristics of the POF. Replacing the POF with a low loss, graded-index, perfluorinated POF (PF-GIPOF) significantly improved both the measurement resolution and maximum fiber length, up to 500 m, as a result of the reduced dispersion and attenuation in the fiber [48].

The resolution and speed of such measurements can also be enhanced by applying incoherent optical frequency domain resolution (OFDR) rather than OTDR [49]. The authors demonstrated OFDR measurements along a POF at 2 kHz data acquisition rates with high spatial resolution on the order of a few micrometers. The POF signal was sensitive to large strain magnitudes; therefore, this technique was applied to the measurement of seismic induced strains (up to 125%).

Such distributed sensing has also been applied to medical sensing, for example, Witt et al. [50] integrated a POF textile material into a belt for respiration monitoring of a patient. The elongation of the POF during breathing was measured using OTDR, from which data on the respiration rate was extracted. This particular application was enabled by the choice of POF for the sensing element, as the maximum strain measured was around 3%. Furthermore, as the sensor signal was entirely optical, respiration monitoring could be performed in environments with electromagnetic noise, for example, during MRI imaging of the patient. To accentuate the POF transmission losses due to strain, Liehr et al. [51] also micromachined defects in a POF using femtosecond laser pulses prior to measuring the scattering profile along the POF using OTDR.

Similar effects can be measured when observing Brillouin scattering coefficients along the length of POFs. Hayashi et al. [12] measured the Brillouin gain spectrum in a perfluorinated, graded-index POF up to 20% strain. The measured Brillouin gain spectrum frequency shift was nonmonotonic with applied strain; however, combining this measurement with the measurement of the Stokes power loss (which increased rapidly starting at 10% axial strain) provided a unique measurement of strain. In particular, the POF could function as a recorder of the highest strain value reached and its location, through the plastic deformation induced *memory effect*.

4.3.4 Fiber Bragg Gratings

FBG sensors are one of the most widely applied silica optical fiber sensors as they can provide local sensing information and can be multiplexed in large numbers along a single optical fiber. In addition, the fact that the sensing information is wavelength encoded means that it is invariant to fluctuations in laser power and coupling losses. Based on this success, numerous researchers have developed techniques to inscribe FBGs in POFs with the motivation to exploit the large tuning range of POF FBGs as compared to those in silica optical fibers. This topic is covered in detail in Chapter 10; however, this section presents a summary of these sensors for completeness.

Xiong et al. [52] first demonstrated the writing of FBGs in POFs through photosensitivity. The FBGs were fabricated by UV exposure of PMMA, single-mode optical fibers with dye-doped cores to increase the photosensitivity of the polymer. Liu et al. [53] later wrote FBGs operating in near IR wavelengths with a transmission loss of >28 dB and a bandwidth of <0.5 nm. By applying tensile strain to the FBG, the authors shifted the Bragg wavelength by a total of 52 nm [54], which is an order of magnitude larger than that previously achieved with FBGs in silica optical fibers. The large wavelength shift was partially due to the increased strain sensitivity of the POF and partially due to the large failure strain of the POF. The POF FBG also demonstrated a 22% increase in sensitivity as compared to a silica optical fiber POF of the same Bragg wavelength. Applying thermal loading, the authors also tuned the FBG over 18 nm with a temperature change of 50°C [55]. No hysteresis was observed during the cyclic thermal loading.

One demonstration that highlights the unique properties of FBGs in POFs is that of high-sensitivity pressure measurements to detect small pressure changes. Rajan et al. [56] first inscribed an FBG into a single-mode POF and then etched the cladding of the POF to significantly reduce the cladding diameter. Combining the low elastic modulus of the polymer and small cladding diameter produced an FBG with high sensitivity to small axial loads on the POF. The authors then attached the POF to a vinyl diaphragm to transfer the surrounding pressure to an axial force on the POF and demonstrated a complete sensor with the extremely high pressure sensitivity of 1.32 pm Pa^{-1} over the range of 0.1–5.0 kPa.

Due to the difficulties inherent in fabrication of single-mode POFs and FBG inscription in these fibers, researchers have also pursued the development of FBG inscription in multimode POFs. Luo et al. [57] wrote FBGs in a POF, doped with benzil dimethyl ketal (BDK) for increased photosensitivity. More than 10 peaks were observed in the reflected FBG spectrum, due to the numerous modes propagating through the POF. An example reflected spectrum is shown in Figure 4.10, before and after loading of the FBG. The use of the multimode POF permitted much easier coupling to and from the POF and significantly reduced the cost of the sensor; however, the output spectrum of the FBG was not easily controlled as it depended upon the coupling conditions through the amount of power coupled into each mode. This could present significant issues for repeatability in applications of multimode POFs. The reflected peak wavelength shift was similar for most modes during mechanical loading of the POF; however, this was not the case for unloading of the POF. It was proposed that yielding in the POF changed the local mode coupling behavior at the site of the FBG. The authors also measured the temperature sensitivity of the multimode POF, demonstrating a negative shift in peak wavelength of the FBG with increasing temperature [58].

Challenges still are present in the application of POF FBG sensors. Some of these challenges come from the inherent properties of the POF, including the low maximum temperature threshold and viscoelastic strain response. In addition, thermal erasing of POF FBGs can occur when the grating is exposed to thermal loads for extended periods of time. The process of thermal erasing is not fully understood for FBGs in POFs. The physical mechanisms for grating fabrication are varied and different in POFs than in silica optical fibers. In POFs, the dominant mechanism is typically

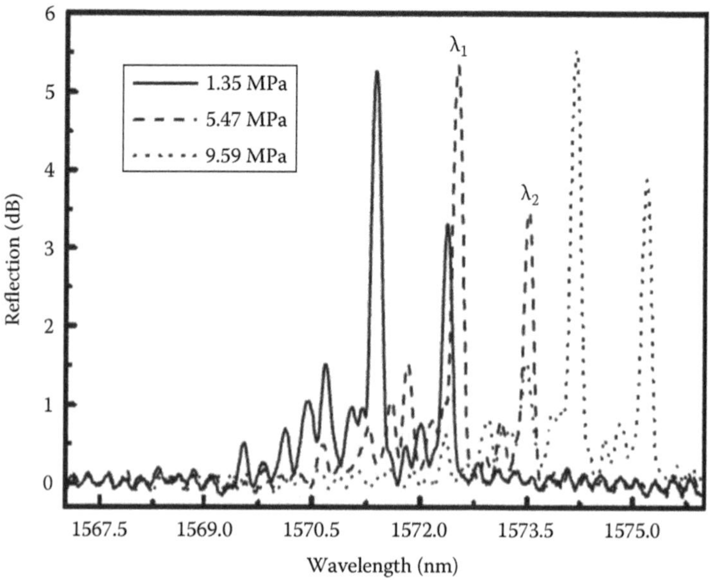

FIGURE 4.10 Shift in reflection spectra of multimode POF FBG for different applied axial stresses. (From Luo, Y. et al., *Opt. Fiber Technol.*, 17, 201, 2011.)

through photopolymerization, whereas in silica optical fibers, it is through trapping of the ultraviolet excited charge carriers [55]. Similar to the writing process for FBGs in silica fibers, the grating depth increases with exposure time until a threshold is reached at which point damage to the polymer fiber occurs and transmission losses are introduced into the fiber [59]. Temporary thermal erasing has also been observed during the fabrication of POF FBGs. Liu et al. [59] observed that when FBGs were written with lower power UV exposures, the FBG peak depth increased, reached a maximum, remained constant, then began to erase during the exposure time. Once the FBG was completely erased, the UV exposure was stopped and the FBG reappeared over a period of 8 h and then was permanent and stable. The authors speculate that the heating of the fiber during the UV exposure temporarily changed the index of refraction, counteracting the change due to photosensitivity.

4.3.5 Long-Period Gratings

An alternative method to produce grating structures in optical fibers is through the inscription of long-period gratings (LPGs). LPGs rely on the coupling between the core mode in the optical fiber and cladding modes, with the overlap region between the two modes occurring primarily in the cladding. In addition, LPG periods are typically orders of magnitude greater than those of comparable FBGs. As a result, LPGs are generally easier to inscribe and can be written directly into the cladding from the external surface of the optical fiber. Such sensors are highly sensitive to index of refraction changes external to the optical fiber.

Li et al. [60] wrote LPGs in low-mode POFs through photoetching of the PMMA POF cladding. Although the POF was low-mode at the wavelength of the grating resonance (1570 nm), interference between multiple modes was not observed. Such an LPG could also be applied as a strain/temperature/humidity sensor; however, Li et al. [60] did not measure the sensitivity of the grating resonance to these parameters.

More recently, Yan et al. [61] demonstrated measurements in the THz wavelength range exploiting an LPG in a POF (although the authors refer to the sensors as THz FBGs, the coupling mechanisms is that of an LPG). The LPGs were written through CO_2 laser inscription of the cladding of a step index PS POF. This particular POF material demonstrates extremely low attenuation in the 0.2–0.5 THz regime. Figure 4.11a plots the transmission spectrum of this sensor. In this early demonstration of the sensor performance, the thickness of a surrounding paper, transparent to THz waves, was measured through the induced wavelength shift in the transmission loss resonance, as seen in Figure 4.11b. The authors obtained a shift to thickness sensitivity of −0.067 GHz μm^{-1}.

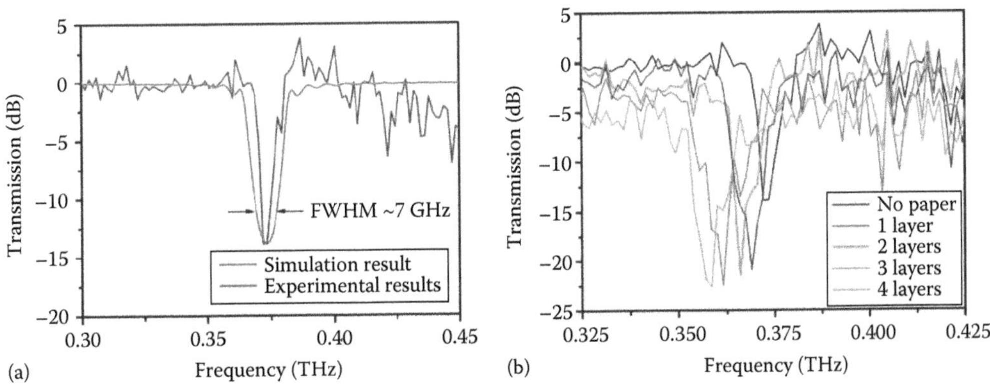

FIGURE 4.11 (a) Comparison between simulated and experimental transmission spectra of terahertz FBG (TFBG). (b) Experimental TFBG spectra for a different number of paper layers placed in direct contact with TFBG. (From Yan, G. et al., *Opt. Lett.*, 38, 2200, 2013.)

4.3.6 POF-Based Biosensing

One of the unique features of POFs, as compared to silica optical fibers, is their biocompatibility. This inherent material property has motivated many researchers to apply POF sensors for chemical detection in medical and other fields [62]. Some of these sensors have targeted low-cost disposable sensing applications using multimode commercial POFs, such as the pH sensor based on a solgel hybrid material drop on the end of the multimode POF [63]. Additional sensor designs have been based on a specialty film applied to the POF that is activated by certain chemicals or environmental conditions [62]. Example parameters measured include radiation exposure [64], temperature [65], and ammonia [66]. In contrast, Wang and Wang [67] coated a gradient-index (GRIN) lens on the end of a POF with a fluorescent pH sensing layer. Groups of POFs have also been bundled together to measure field distributions, for example, to measure radiation dose distributions on a patient in proton beam therapy dosimetry [68].

Higher-cost POF chemical sensors with high sensitivities have also been successfully demonstrated. Grassini et al. [69] etched the cladding of a multimode POF, then plasma sputtered a nanostructured Ag sensing film on the outer surface of the POF. The sensing film selectively reacts with different gas components. The sensing film condition was measured through the lightwave transmittance through the POF, a function of the interaction quality between the evanescent field of the lightwave propagating through the POF and the surrounding medium. The authors then applied this sensor to the detection of hydrogen fluoride (HF) gas [70]. Figure 4.12 shows field emission scanning electron microscopy (FESEM) images of the nanostructured sensing film before and after exposure to the HF gas. The gas exposure clearly degraded the quality of the sensing film, as was confirmed by the transmittance data.

A final sensing configuration that has been successfully employed in POF chemical sensors is surface plasmon resonance (SPR) [71]. Such sensors require etching the POF to increase the overlap of the propagating lightwave and the surrounding medium. Through the application of specialty coatings, resonance conditions can be arranged between a surface plasmon wave in the coating and the propagating lightwave through the fiber. This resonance condition is dependent upon the surrounding medium index of refraction and can be applied to measure small changes in this index of refraction, for example, in the detection of antibodies [72].

4.3.7 Microstructured POF Sensors

We have seen several examples in the previous sections on the use of multimode and single-mode POFs to yield unique sensing information. However, while the single-mode POF option permits high-resolution sensing, challenges in their fabrication and coupling to these POFs mean that power losses in such sensors can be extremely high. As a result, only short sensing lengths of single-mode POFs can be applied [21]. A promising solution is the use of MPOFs, first developed

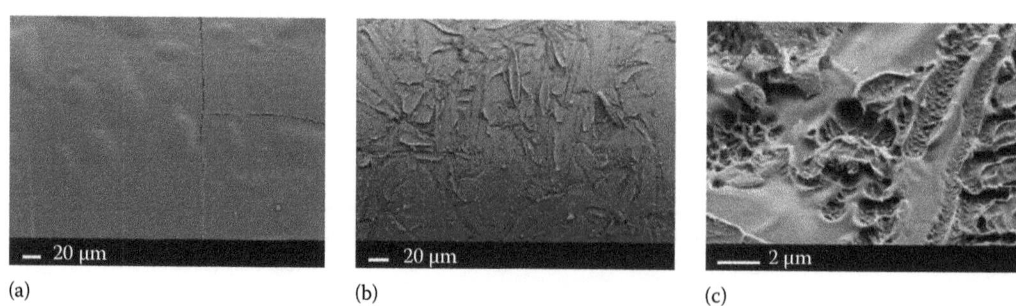

FIGURE 4.12 FESEM images of the coated fiber before (a) and after (b, c) exposure to the HF vapors. (From Grassini, S. et al., *J. Instrum.*, 7, 12006, 2012.)

by researchers at the University of Sydney [2]. In this section, we will describe new sensors that have been enabled by the unique properties of MPOFs.

MPOFs have a large air fraction in the cross section of the fiber that significantly reduces the attenuation of a lightwave propagating through the fiber. Additionally, the confinement effect on the lightwave, created by the holes in the microstructure, enables single-mode propagation over a wide wavelength range. These advantages therefore make MPOFs ideal for the inscription of Bragg gratings to serve as highly tunable FBG sensors without the power losses inherent in solid-core single-mode POFs. Bragg grating inscription is more difficult in MPOFs due to the multiple interfaces encountered when exposing the MPOF to side illumination; however, researchers have successfully written FBG sensors in the 850 nm Bragg wavelength range in both PMMA and TOPAS MPOFs [73–75]. The 850 nm window is a region where polymers typically have low attenuation properties and telecommunication sources and detectors are commercially available. Stefani et al. [76] applied an MPOF FBG into an accelerometer, demonstrating a sensitivity four times that of an equivalent silica optical fiber–based accelerometer. In addition, the ultimate strain of the polymer material is much higher than that of silica; therefore, the maximum acceleration potential is also much higher. Yuan et al. [77] also took advantage of the high strain capabilities of POFs and wrote a dual-wavelength FBG with a single setup, by writing a single grating and then tensioning the POF prior to writing a second FBG. They achieved a 1 nm separation between the two wavelengths.

In contrast to PMMA, TOPAS has a very low moisture absorption rate and a significantly high glass transition temperature (Tg), around 135°C. Markos et al. [78] inscribed FBGs into a TOPAS MPOF with a Bragg wavelength of 853 nm. The authors were able to perform strain sensing over a large strain range at elevated temperatures, up to 110°C. The low moisture absorption of TOPAS did not create apparent thermal strain in the sensor. Figure 4.13a shows a microscope image of the MPOF cross section, as well as the reflected spectrum of the FBG sensor. The FBG was exposed to 110°C for seven hours, after which the MPOF was returned to room temperature, as seen in Figure 4.13b. The thermally induced blue shift in the Bragg wavelength was not recovered at room temperature, indicating that permanent changes had occurred to the MPOF during the extended thermal loading. This result again emphasizes that the loading and thermal history of a POF sensor must be taken into account when analyzing data from the sensor.

LPGs have also been demonstrated in MPOFs, written through mechanical stamping of a heated MPOF [79] and UV inscription [80]. Sáez-Rodríguez et al. [80] applied the LPG for humidity sensing, using the high diffusion coefficient of PMMA (6.7×10^{-9} cm^2 s^{-1} at room temperature) producing a 250 nm shift in one of the cladding mode resonances at maximum moisture absorption, orders of magnitude larger than can be obtained with silica optical fiber–based sensors.

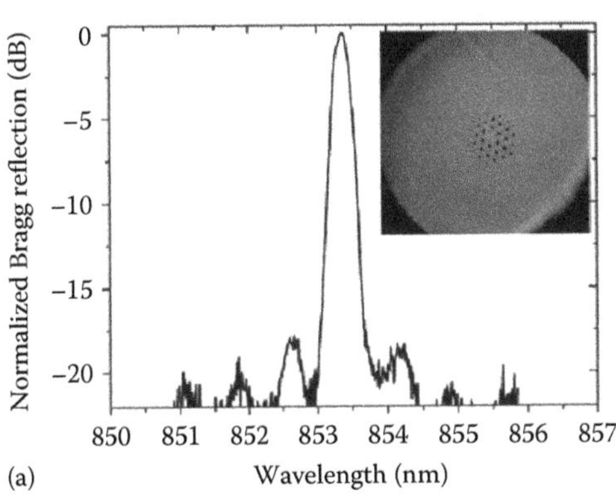

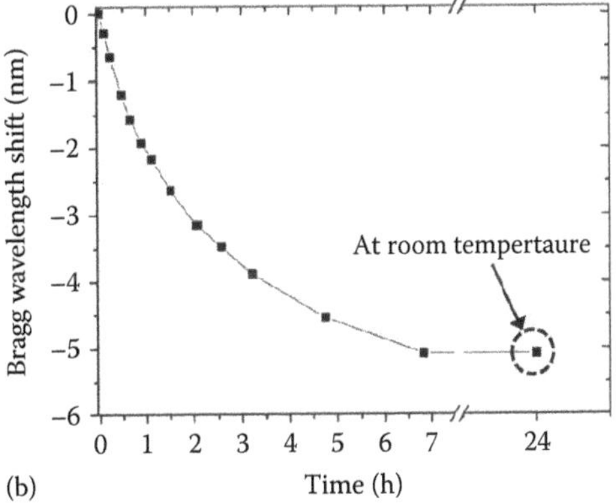

FIGURE 4.13 (a) Reflection spectrum of high-Tg TOPAS FBG at room temperature. Inset shows microscope image of the end facet of the MPOF. (b) Variation of Bragg wavelength with time. The MPOF was heated for 7 h at 110°C and then left at room temperature. (From Markos, C. et al., *Opt. Express*, 21, 4758, 2013.)

Even without inscribed grating structures, the unique cross sections of MPOFs can be designed for force or temperature measurements. For example, Szczurowski et al. [81] measured the polarimetric sensitivity to hydrostatic pressure, axial strain, and temperature of a birefringent dual-core MPOF. In this example, high levels of birefringence can be designed into the cross-sectional microstructure, while the polymer material permits high-sensitivity sensing to low load levels.

A final, unique advantage to MPOFs as sensor elements is that the individual holes have extremely small diameters and can be filled with fluids. Since the holes extend the length of the MPOF, the interaction surface area of a propagating lightwave with a fluid trapped in one or more of these holes is very long. For biosensing applications, this means that extremely small volume small samples can be analyzed with a high sensitivity. The interior surfaces of the holes can be coated with coatings that selectively react to different chemicals. For example, Peng et al. [82] created an ammonia gas probe by coating the holes of a PMMA MPOF with a fluorescent film that was sensitive to ammonia. Similarly, Li et al. [83] coated the hole walls of a MPOF with a rhodamine-doped titanium dioxide film for high-sensitivity detection of hydrogen peroxide. Emiliyanov et al. [84] performed antibody detection in TOPAS MPOFs by coating the walls with a fluorescent coating and again exploiting the low moisture property of TOPAS, important since the MPOF are filled with fluids. Further, the authors demonstrated serial biosensing by coating two different sections along the MPOF with coatings sensitive to two different antibodies. Applying a different measurement strategy, Markos et al. [85] coated select walls of a double-core MPOF with a coating to which antibodies could bind. As the layer thickness increased with increasing concentrations of antibodies, the coupling between lightwaves in the two cores was changed. Therefore, measurement of the coupling efficiency served to calibrate the antibody concentration.

4.4 FUTURE OUTLOOK

The examples of POF sensors presented in this chapter highlight the unique features of POFs for measurement applications. In some applications, it is the low cost and ease of connection of multimode fibers that are beneficial; in other applications, it is their high ultimate strain, low stiffness, and low weight or unique moisture absorption properties. The advantages of the POFs themselves are as varied as the different polymer materials from which they have been fabricated. There are many more examples of POF sensors that can be found through a search of the literature and the reader is recommended to do so.

Recent advances in new POFs such as microstructured fibers, TOPAS fibers, and POFs designed with low attenuation for lightwave propagation in the THz wavelength ranges hold exciting new opportunities for sensor designs. The use of the THz wavelength range will permit high bandwidth information and multiwavelength sensing with even wider differences in individual wavelengths. The recent surge in MPOF manufacturing and development also releases some of the initial issues with attenuation in single-mode POFs and the high-precision sensors designed based on these fibers. We have also seen the new possibilities in biochemical sensing of very low volumes through the cross-sectional holes. Designing new MPOF cross sections also opens the gates for entirely new coupling behaviors, interactions, and even multiaxis sensing. Again, the biocompatibility of POFs will allow researchers to take these sensors into a large variety of applications both in the laboratory and for patient care. We have already seen examples where extremely small quantities of fluid can be tested for the detection of antibodies, which will continue to grow as a field.

In conclusion, researchers are understanding better and better the properties of POFs in sensing environments and how to exploit these properties in new ways. With the close collaboration of POF manufacturers, designers, and sensor research groups, the author expects that we will see a rapidly growing number of such sensors, both in research laboratories and in the commercial market, in the near future.

REFERENCES

1. Ziemann O, Krauser J, Zamzow PE, and Daum W. 2008. *POF Handbook* (Berlin, Germany: Springer), Chapter 8.
2. Large MCJ, Poladian L, Barton GW, and van Eijkelenborg MA. 2007. *Microstructure Polymer Optical Fibers* (Berlin, Germany: Springer).
3. Koike Y. 1992. High bandwidth and low loss polymer optical fiber. In *Proceedings of the First International Conference on Plastic Optical Fibres and Applications*, Paris, France, pp. 15–19.
4. Kuzyk MG. 2007. *Polymer Fiber Optics: Materials, Physics and Applications* (Boca Raton, FL: CRC/ Taylor & Francis).
5. van Eijekelenborg MA et al. 2001. Microstructured polymer optical fibre. *Optics Express* 9, 319–327.
6. Zubia J and Arrue J. 2001. Plastic optical fibers: An introduction to their technological processes and applications. *Optical Fiber Technology* 7, 101–140.
7. Bosc D and Toinen C. 1993. Tensile mechanical properties and reduced internal stresses of polymer optical fiber. *Polymer Composites* 14, 410–413.
8. Jiang C, Kuzyk MG, Ding JL, Johns WE, and Welker DJ. 2002. Fabrication and mechanical behavior of dye-doped polymer optical fiber. *Journal of Applied Physics* 92, 4–12.
9. Ziemann O, Krauser J, Zamzow PE, and Daum W. 2008. *POF Handbook* (Berlin, Germany: Springer), Chapter 9.
10. Ziemann O, Daum W, Bräher A, Schlick J, and Frank W. 2000. Results of a German 6,000 h accelerated aging test of PMMA POF and consequences for the practical use of POF. In *Proceedings of POF 2000*, Boston, MA, pp. 133–137.
11. Kiesel S, Peters K, Hassan T, and Kowalsky M. 2007. Behaviour of intrinsic polymer optical fibre sensor for large-strain applications. *Measurement Science and Technology* 18, 3144–3154.
12. Hayashi N, Mizuno Y, and Nakamura K. 2012. Brillouin gain spectrum dependence on large strain in perfluorinated graded-index polymer optical fiber. *Optics Express* 20, 21101–21106.
13. Abang A and Webb DJ. 2013. Influence of mounting on the hysteresis of polymer fiber Bragg grating strain sensors. *Optics Letters* 38, 1376–1378.
14. Stefani A, Andresen S, Yuan W, and Bang O. 2012. Dynamic characterization of polymer optical fibers. *IEEE Sensors Journal* 12, 3047–3053.
15. Silva-López M, Fender A, MacPherson WN, Barton JS, Jones JDC, Zhao D, Dobb H, Webb DJ, Zhang L, and Bennion I. 2005. Strain and temperature sensitivity of a single-mode polymer optical fiber. *Optics Letters* 30, 3129–3131.
16. Huang J, Lan X, Wang H, Yuan L, Wei T, Gao Z, and Xiao H. 2012. Polymer optical fiber for large strain measurement based on multimode interference. *Optics Letters* 37, 4308–4310.
17. Zhang ZF and Tao XM. 2013. Intrinsic temperature sensitivity of fiber Bragg gratings in PMMA-based optical fibers. *IEEE Photonics Technology Letters* 25, 310–312.
18. Zhang C, Zhang W, Webb DJ, and Peng GD. 2010. Optical fibre temperature and humidity sensor. *Electronics Letters* 46, 643–644.
19. Zhang W, Webb DJ, and Peng GD. 2012. Investigation into time response of polymer fiber Bragg grating based humidity sensors. *Journal of Lightwave Technology* 30, 1090–1096.
20. Hamouda T, Peters K, and Seyam A-FM. 2012. Effect of resin type on the signal integrity of an embedded perfluorinated polymer optical fiber. *Smart Materials and Structures* 21, 055023.
21. Abdi O, Peters K, Kowalsky M, and Hassan T. 2011. Validation of a single-mode polymer optical fiber sensor and interrogator for large strain measurements. *Measurement Science and Technology* 22, 075207.
22. Canning J, Buckley E, Groothoff N, Luther-Davies B, and Zagari J. 2002. UV laser cleaving of air-polymer structured fibre. *Optics Communications* 202, 139–143.
23. Law SH, van Eijkelenborg MA, Barton GW, Yan C, Lwin R, Gan J, and Large MCJ. 2006. Cleaved end-face quality of microstructured polymer optical fibres. *Optics Communications* 265, 513–520.
24. Abdi O, Wong KC, Hassan T, Peters KJ, and Kowalsky MJ. 2009. Cleaving of single mode polymer optical fiber for strain sensor applications. *Optics Communications* 282, 856–861.
25. Law SH, Harvey JD, Kruhlak RJ, Song M, Wu E, Barton GW, van Eijkelenborg MA, and Large MCJ. 2006. Cleaving of microstructured polymer optical fibers. *Optics Communications* 258, 193–202.
26. Stefani A, Nielsen K, Rasmussen HK, and Bang O. 2012. Cleaving of TOPAS and PMMA microstructured polymer optical fibers: Core-shift and statistical quality optimization. *Optics Communications* 285, 1825–1833.
27. Abang A and Webb DJ. 2012. Demountable connection for polymer optical fiber grating sensors. *Optical Engineering* 51, 080503.

28. Antunes PFC, Varum H, and André PS. 2013. Intensity-encoded polymer optical fiber accelerometer. *IEEE Sensors Journal* 13, 1716–1720.
29. Mohanty L and Kuang KSC. 2010. A breathing rate sensor with plastic optical fiber. *Applied Physics Letters* 97, 073703.
30. Kulkarni A, Kim H, Choi J, and Kim T. 2010. A novel approach to use of elastomer for monitoring of pressure using plastic optical fiber. *Review of Scientific Instruments* 81, 145108.
31. Takeda N. 2002. Characterization of microscopic damage in composite laminates and real-time monitoring by embedded optical fiber sensors. *International Journal of Fatigue* 24, 281–289.
32. Kuang KSC, Cantwell WJ, and Scully PJ. 2002. An evaluation of a novel plastic optical fibre sensor for axial strain and bend measurements. *Measurement Science and Technology* 13, 1523–1534.
33. Kuang KSC, Akmaluddin, Cantwell WJ, and Thomas C. 2003. Crack detection and vertical deflection monitoring in concrete beams using plastic optical fibre sensors. *Measurement Science and Technology* 14, 205–216.
34. Wong YM, Scully PJ, Bartlett RJ, Kuang KSC, and Cantwell WJ. 2003. Plastic optical fibre sensors for environmental monitoring: Biofouling and strain applications. *Strain* 39, 115–119.
35. Chen Y, Xie WF, Ke YL, and Chen LW. 2008. Power loss characteristics of a sensing element based on a grooved polymer optical fiber under elongation. *Measurement Science and Technology* 19, 105203.
36. Bilro L, Oliveira JG, Pinto JL, and Nogueira RN. 2011. A reliable low-cost wireless and wearable gait monitoring system based on a plastic optical fibre sensor. *Measurement Science and Technology* 22, 045801.
37. Stupar DZ, Bajić JS, Manojlović LM, Slankamenac MP, Joža AV, and Živanov MB. 2012. Movements monitoring based on fiber-optic curvature sensor. *IEEE Sensors Journal* 12, 3424–3431.
38. Remouche M, Mokdad R, Chakari A, and Meyrueis P. 2007. Intrinsic integrated optical temperature sensor based on waveguide bend loss. *Optics & Laser Technology* 39, 1454–1460.
39. Montero DS, Lallana PC, and Vázquez C. 2012. A polymer optical fiber fuel level sensor: Application to paramotoring and powered paragliding. *Sensors* 12, 6186–6197.
40. Wang J and Chen B. 2012. Experimental research of optical fiber sensor for salinity measurement. *Sensors and Actuators A* 184, 53–56.
41. Chu F. 2012. Experimental study of plastic optical fiber TNT sensor based on evanescence absorption. *Optical Engineering* 51, 054403.
42. Chu F and Yang J. 2012. Coil-shaped plastic optical fiber sensor heads for fluorescence quenching based TNT sensing. *Sensors and Actuators A* 175, 43–46.
43. Gomez J, Zubia J, Aranguren G, Arrue J, Poisel H, and Saez I. 2009. Comparing polymer optical fiber, fiber Bragg grating, and traditional strain gauge for aircraft structural health monitoring. *Applied Optics* 48, 1436–1443.
44. Durana G, Kirchhof M, Luber M, Sáez de Ocáriz I, Poisel H, Zubia J, and Vázquez C. 2009. Use of a novel fiber optical strain sensor for monitoring the vertical deflection of an aircraft flap. *IEEE Sensors Journal* 9, 1219–1225.
45. Jiang G, van Vickle P, Peters K, and Knight V. 2007. Oscillator interrogated time-of-flight fiber interferometer for global strain measurements. *Sensors and Actuators A* 135, 443–450.
46. Kiesel S, Peters K, Hassan T, and Kowalsky M. 2008. Large deformation in-fiber polymer optical fiber sensor. *IEEE Photonics Technology Letters* 20, 416–418.
47. Kiesel S, Peters K, Hassan T, and Kowalsky M. 2009. Calibration of a single-mode polymer optical fiber large-strain sensor. *Measurement Science and Technology* 20, 034016.
48. Liehr S, Lenke P, Wendt M, Drebber K, Seeger M, Thiele E, Metschies, H, Gebreselassie B, and Münich JC. 2009. Polymer optical fiber sensors for distributed strain measurement and application in structural health monitoring. *IEEE Sensors Journal* 9, 1330–1338.
49. Liehr S and Krebber K. 2012. Application of quasi-distributed and dynamic length and power change measurement using optical frequency domain reflectometry. *IEEE Sensors Journal* 12, 237–245.
50. Witt J, Narbonneau F, Schukar M, Krebber K, de Jonckheere J, Jeanne M, Kinet D et al. 2012. Medical textiles with embedded fiber optic sensors for monitoring of respiratory movement. *IEEE Sensors Journal* 12, 246–254.
51. Liehr S, Burgmeier J, Krebber K, and Schade W. 2013. Femtosecond laser structuring of polymer optical fibers for backscatter sensing. *Journal of Lightwave Technology* 31, 1418–1425.
52. Peng GD, Xiong Z, and Chu PL. 1999. Photosensitivity and gratings in dye-doped polymer optical fibers. *Optical Fiber Technology* 5, 242–251.
53. Liu HY, Peng GD, and Chu PL. 2002. Polymer fiber Bragg gratings with 28-dB transmission rejection. *IEEE Photonics Technology Letters* 14, 935–937.

54. Liu HY, Liu HB, and Peng GD. 2005. Tensile strain characterization of polymer optical fibre Bragg gratings. *Optics Communications* 251, 37–43.
55. Liu HY, Peng GD, and Chu PL. 2001. Thermal tuning of polymer optical fiber Bragg gratings. *IEEE Photonics Technology Letters* 13, 824–826.
56. Rajan G, Liu B, Luo Y, Ambikairajah E, and Peng GD. 2013. High sensitivity force and pressure measurements using etched singlemode polymer fiber Bragg gratings. *IEEE Sensors Journal* 13, 1794–1800.
57. Luo Y, Yan B, Li M, Zhang X, Wu W, Zhang Q, and Peng GD. 2011. Analysis of multimode POF gratings in stress and strain sensing applications. *Optical Fiber Technology* 17, 201–209.
58. Luo Y, Wu W, Wang T, Cheng X, Zhang Q, Peng GD, and Zhu B. 2012. Analysis of multimode BDK doped POF gratings for temperature sensing. *Optics Communications* 285, 4353–4358.
59. Liu HB, Liu HY, Peng GD, and Chu PL. 2004. Novel growth behaviors of fiber Bragg gratings in polymer optical fiber under UV irradiation with low power. *IEEE Photonics Technology Letters* 16, 159–161.
60. Li Z, Tam HY, Xu L, and Zhang Q. 2005. Fabrication of long-period gratings in poly(methyl methacrylate-co-methyl vinyl ketone-co-benzyl methacrylate)-core polymer optical fiber by use of a mercury lamp. *Optics Letters* 30, 1117–1119.
61. Yan G, Markov A, Chinifooroshan T, Tripathi SM, Bock WJ, and Skorobogatiy M. 2013. Resonant THz sensor for paper quality monitoring using THz fiber Bragg gratings. *Optics Letters* 38, 2200–2202.
62. Kulkarni A, Kim H, Amin R, Park SH, Hong BH, and Kim T. 2012. A novel method for large area grapheme transfer on the polymer optical fiber. *Journal of Nanoscience and Nanotechnology* 12, 3918–3921.
63. Rovati L, Fabbri P, Ferrari L, and Pilati F. 2011. Construction and evaluation of a disposable pH sensor based on a large core plastic optical fiber. *Review of Scientific Instruments* 82, 023106.
64. Yoo WJ, Heo JY, Jang KW, Seo JK, Moon JS, Park JY, Park BG, Cho S, and Lee B. 2011. Measurements of spectral responses for developing fiber-optic pH sensor. *Optical Review* 18, 139–143.
65. Yoo WJ, Jang KW, Seo JK, Heo JY, Moon JS, Jun JH, Park JY, and Lee B. 2011. Development of optical fiber-based respiration sensor for noninvasive respiratory monitoring. *Optical Review* 18, 132–138.
66. Jala AH, Yu J, and Nnanna AGA. 2012. Fabrication and calibration of Oxazine-based optic fiber sensor for detection of ammonia in water. *Applied Optics* 41, 3768–3775.
67. Wang J and Wang L. 2012. An optical fiber sensor for remote pH sensing and imaging. *Applied Spectroscopy* 66, 300–303.
68. Jang KW, Yoo WJ, Moon J, Han KT, Park BG, Shin D, Park SY, and Lee B. 2012. Multi-dimensional fiber-optic radiation sensor for ocular proton therapy dosimetry. *Nuclear Instruments and Methods in Physics Research A* 695, 322–325.
69. Grassini S, Angelini E, Parvis M, and Faraldi F. 2012. Surface modification plasma treatments of PMMA optical fibres for sensing applications. *Surface and Interface Analysis* 44, 1068–1071.
70. Grassini S, Ishtaiwi M, Parvis M, Benussi L, Bianco S, Colafranceschi S, and Piccolo D. 2012. Gas monitoring in RPC by means of non-invasive plasma coated POF sensors. *Journal of Instrumentation* 7, 12006.
71. Cennamo N, Massarotti D, Conte L, and Zeni L. 2011. Low cost sensors based on SPR in a plastic optical fiber for biosensor implementation. *Sensors* 11, 11752–11760.
72. Cennamo N, Varriale A, Pennacchio A, Staiano M, Massarotti D, Zeni L, and D'Auria S. 2013. An innovative plastic optical fiber-based biosensor for new bio/applications. The case of celiac disease. *Sensors and Actuators B* 176, 1008–1014.
73. Johnson IP, Kalli K, and Webb DJ. 2010. 827 nm Bragg grating sensor in multimode microstructured polymer optical fibre. *Electronics Letters* 46, 1595.
74. Stefani A, Yuan W, Markos C, and Bang O. 2011. Narrow bandwidth 850-nm fiber Bragg gratings in few-mode polymer optical fibers. *IEEE Photonics Technology Letters* 23, 660–662.
75. Yuan W, Khan L, Webb DJ, Kalli K, Rasmussen HK, Stefani A, and Bang O. 2011. Humidity insensitive TOPAS polymer fiber Bragg grating sensor. *Optics Express* 19, 19731–19739.
76. Stefani A, Andresen S, Yuan W, Herholdt-Rasmussen N, and Bang O. 2012. High sensitivity polymer optical fiber-Bragg-grating-based accelerometer. *IEEE Photonics Technology Letters* 24, 763–765.
77. Yuan W, Stefani A, and Bang O. 2012. Tunable polymer fiber Bragg grating (FBG) inscription: Fabrication of dual-FBG temperature compensated polymer optical fiber strain sensors. *IEEE Photonics Technology Letters* 24, 401–403.
78. Markos C, Stefani A, Nielsen K, Rasmussen HK, Yuan W, and Bang O. 2013. High-Tg TOPAS microstructured polymer optical fiber for fiber Bragg grating strain sensing at 110 degrees. *Optics Express* 21, 4758–4765.
79. Durana G, Gómez J, Aldabaldetreku G, Zubia J, Montero A, and Sáez de Ocáriz. 2012. Assessment of an LPG mPOF for strain sensing. *IEEE Sensors Journal* 12, 2668–2673.

80. Sáez-Rodríguez D, Cruz JL, Johnson I, Webb DJ, Large MCJ, and Argyros A. 2010. Water diffusion into UV inscripted long period grating in microstructured polymer fiber. *IEEE Sensors Journal* 10, 1169–1173.
81. Szczurowski MK, Martynkien T, Statkiewicz-Barabach G, Urbanczyk W, and Webb DJ. 2010. Measurements of polarimetric sensitivity to hydrostatic pressure, strain and temperature in birefringent dual-core microstructured polymer fiber. *Optics Express* 18, 12077–12087.
82. Peng L, Yang X, Yuan L, Wang L, Zhao E, Tian F, and Liu Y. 2011. Gaseous ammonia fluorescence probe based on cellulose acetate modified microstructured optical fiber. *Optics Communications* 284, 4810–4814.
83. Li D and Wang G. 2010. Fluorescence hydrogen peroxide probe based on a microstructured polymer optical fiber modified with a titanium dioxide film. *Applied Spectroscopy* 64, 514–519.
84. Emiliyanov G, Høiby PE, Pedersen LH, and Bang O. 2013. Selective serial multi-antibody biosensing with TOPAS microstructured polymer optical fiber. *Sensors* 13, 3241–3251.
85. Markos C, Yuan W, Vlachos K, Town GE, and Bang O. 2011. Label-free biosensing with high sensitivity in dual-core microstructured polymer optical fibers. *Optics Express* 19, 7790–7798.

5 Surface Plasmon Resonance Fiber-Optic Sensors

Kent B. Pfeifer and Steven M. Thornberg

CONTENTS

5.1 INTRODUCTION

The first report of the effect of surface plasmon resonance (SPR) was made by Wood in 1902 when he noticed the absence of light in narrow spectral bands from a diffraction grating [1]. He also reported in 1935 that these *anomalies*, as he called them, had never yet been properly modeled [2]. It was subsequently left to others to derive a model for the origin of the observed effect [3,4]. The significant advance in SPR research that transformed it from an *anomaly* to a valuable analytical technique was the recognition by Kretschmann and Raether that a thin metallic film supported by a dielectric substrate could have surface plasmon waves excited on the ambient exposed metallic surface when optical excitation was from the substrate–metal interface. Thus, the probe light beam would be incident on one side of the metal film, and chemistry could be performed on the other making a useful sensor system [5].

The following discussion is based on our *IEEE Sensors Journal* paper from 2010 as well as a book chapter from 2014 and follows a similar development [6–8].

SPR spectroscopy has been applied to a number of analytical problems due to its high sensitivity to variations in the electronic nature of a surface. In particular, reports of sensing of chemicals and of biological samples using functionalized surfaces are widespread in the literature [10–17]. The majority of examples employ the open beam optical arrangement known as a Kretschmann configuration that has a single angle of incidence on a metallic layer and a single angle of exit to an optical spectrometer [18]. The SPR is supported in a thin metallic layer, and the chemistry of interest takes place on the side of the metal layer opposite the incident and reflected light. In practice, this thin metallic layer is usually deposited on the diagonal facet of a right-angle prism allowing the thin film access to the chemical system while having structural support for the film provided by the prism.

5.2 PLANAR SPR THEORY OVERVIEW

There are a number of excellent references that detail the mathematical development of the conditions for SPR in a planar geometry such as the Kretschmann configuration and that development will not be repeated in detail here except to summarize the results and to clarify the origin of the important equations of SPR [19–22]. If we consider the geometry shown in Figure 5.1, a wave equation can be solved at the boundary ($x = 0$) for surface waves propagating in the z-direction and decaying exponentially as a function of the x-direction. We will demonstrate that a decaying exponential solution can be derived where the sign convention for the x-direction of positive implies above the material interface and negative implies below the material interface as indicated in Figure 5.1. This assumes that there is no net charge in the system and no external currents. This wave equation (also known as a Helmholtz equation) is formulated for the electric field $\vec{E}$ and the magnetic field $\vec{H}$ as follows:

$$\nabla^2\vec{E} - \varepsilon\mu\frac{\partial^2\vec{E}}{\partial t^2} = 0$$
$$\nabla^2\vec{H} - \varepsilon\mu\frac{\partial^2\vec{H}}{\partial t^2} = 0 \tag{5.1}$$

where ε and μ are the electric permittivity and magnetic permeability of the materials and t is the time.

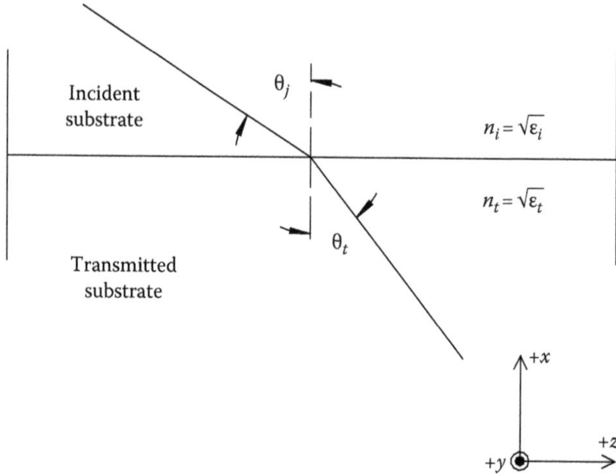

FIGURE 5.1 Diagram illustrating a two-material interface that can support SPR with materials of dielectric constants ε_i, and ε_t. The subscripts i and t represent incident and transmitted light, respectively.

The general solution to these equations for a surface plasmon wave traveling in the z-direction and confined to the surface of the metal dielectric interface, assuming that they are separable functions, is traveling waves of the following form (β_z is the propagation constant, $\vec{r}$ is the location of the wave, and ω is the natural frequency):

$$\vec{E}(\vec{r}) = \vec{E}(x)e^{j(\beta_z z - \omega t)}$$
$$\vec{H}(\vec{r}) = \vec{H}(x)e^{j(\beta_z z - \omega t)} \tag{5.2}$$

Equation 5.2, substituted into Equation 5.1, leads to the following forms of the wave equation for transverse electric (TE) and transverse magnetic (TM) modes, respectively.

$$\frac{1}{\vec{E}(x)}\frac{\partial^2 \vec{E}(x)}{\partial x^2} + \frac{1}{e^{j\beta_z z}}\frac{\partial^2 e^{j\beta_z z}}{\partial z^2} + \varepsilon\mu\omega^2 = 0$$
$$\frac{1}{\vec{H}(x)}\frac{\partial^2 \vec{H}(x)}{\partial x^2} + \frac{1}{e^{j\beta_z z}}\frac{\partial^2 e^{j\beta_z z}}{\partial z^2} + \varepsilon\mu\omega^2 = 0 \tag{5.3}$$

As is indicated earlier, Equation 5.3 is a separable partial differential equation that we can solve for the separation constants in the usual way as $\beta_x^2 - \beta_z^2 = 0$ [23,24]. Since we are only concerned with the bounded modes, we solve Equation 5.3 in the y-dimension as Equation 5.4 for the TE mode and similarly for the TM mode (Equation 5.5) where β_x is the separation constant in the partial differential equation:

$$\frac{\partial^2 E_y(x)}{\partial x^2} + \left(\varepsilon\mu\omega^2 - \beta_x^2\right)E_y(x) = 0 \tag{5.4}$$

$$\frac{\partial^2 H_y(x)}{\partial x^2} + \left(\varepsilon\mu\omega^2 - \beta_x^2\right)H_y(x) = 0 \tag{5.5}$$

Equations 5.4 and 5.5 are now ordinary differential equations, the solutions of which, when combined with Equation 5.2, yield

$$E_y(x,y,z,t) = E_{y0}e^{\pm j\sqrt{\left(\varepsilon\mu\omega^2 - \beta_x^2\right)}\,x}\,e^{j(\beta_z z - \omega t)}$$
$$H_y(x,y,z,t) = H_{y0}e^{\pm j\sqrt{\left(\varepsilon\mu\omega^2 - \beta_x^2\right)}\,x}\,e^{j(\beta_z z - \omega t)} \tag{5.6}$$

The derivation of Equation 5.1 from Maxwell's equation implies that the wave equation is a consequence of Maxwell's formulas; however, solutions to the wave equation are not necessarily solutions to Maxwell's equations. Thus, the components of the H vector for the TE case and the E vector for the TM case must be found by applying Maxwell's equations to the solutions in Equation 5.6 [25]. In Equation 5.6, the sign in front of the x-dependent exponential is positive for negative x values and is negative for positive x in order to keep the function bounded in x.

Thus, the following field components are found for the TE case:

$$\vec{\nabla}\times\vec{E}_y(x,y,z,t)=\begin{pmatrix}-\dfrac{\partial E_y(x,z,t)}{\partial z}\\ 0\\ \dfrac{\partial E_y(x,z,t)}{\partial x}\end{pmatrix}=-\mu\begin{pmatrix}\dfrac{\partial H_x}{\partial t}\\ 0\\ \dfrac{\partial H_z}{\partial t}\end{pmatrix}$$

$$H_x=\int\frac{j\beta_z}{\mu}E_y(x,z,t)\,dt=\frac{-\beta_z}{\omega\mu}E_y(x,z,t) \tag{5.7}$$

$$H_z=\mp j\int\frac{\sqrt{(\varepsilon\mu\omega^2-\beta_x^2)}}{\mu}E_y(x,z,t)\,dt=\frac{\pm\sqrt{(\varepsilon\mu\omega^2-\beta_x^2)}}{\omega\mu}E_y(x,z,t)$$

Similarly, for the TM case, we get the following:

$$\vec{\nabla}\times\vec{H}_y(x,y,z,t)=\begin{pmatrix}-\dfrac{\partial H_y(x,z,t)}{\partial z}\\ 0\\ \dfrac{\partial H_y(x,z,t)}{\partial x}\end{pmatrix}=\varepsilon\begin{pmatrix}\dfrac{\partial E_x}{\partial t}\\ 0\\ \dfrac{\partial E_z}{\partial t}\end{pmatrix}$$

$$E_x=-\int\frac{j\beta_z}{\varepsilon}H_y(x,z,t)\,dt=\frac{\beta_z}{\omega\varepsilon}H_y(x,z,t) \tag{5.8}$$

$$E_z=\pm j\int\frac{\sqrt{(\omega^2\varepsilon\mu-\beta_x^2)}}{\varepsilon}H_y(x,z,t)\,dt=\frac{\mp\sqrt{(\omega^2\varepsilon\mu-\beta_x^2)}}{\omega\varepsilon}H_y(x,z,t)$$

By applying the continuous tangential component boundary conditions $H_{zi}=H_{zt}$, $E_{zi}=E_{zt}$, $H_{yi}=H_{yt}$, and $E_{yi}=E_{yt}$ at $x=0$, where i implies the incident medium and t implies the transmitted medium, we can write the following set of simultaneous equations:

$$\begin{aligned}&\frac{\sqrt{(\varepsilon_i\mu\omega^2-\beta_x^2)}}{\omega\mu}E_{yi}(x)-\frac{\sqrt{(\varepsilon_t\mu\omega^2-\beta_x^2)}}{\omega\mu}E_{yt}(x)=0\\ &E_{yi}(x)-E_{yt}(x)=0\end{aligned} \tag{5.9}$$

$$\begin{aligned}&\frac{\sqrt{(\varepsilon_i\mu\omega^2-\beta_x^2)}}{\omega\varepsilon_i}H_{yi}(x)-\frac{\sqrt{(\varepsilon_t\mu\omega^2-\beta_x^2)}}{\omega\varepsilon_t}H_{yt}(x)=0\\ &H_{yi}(x)-H_{yt}(x)=0\end{aligned} \tag{5.10}$$

In order to solve for the eigenvalue β_x, we must set the determinate of Equations 5.9 and 5.10 equal to zero. The solution of Equation 5.9 leads to the conclusion that $\varepsilon_i=\varepsilon_t$. This is a nonsensical result and implies that there are no bounded modes for the TE case, which has been predicted and demonstrated experimentally [26].

The interesting result comes from Equation 5.10, which leads to a propagation constant for the surface plasmon waves at the surface (Equation 5.11). Since we have a nonzero propagation constant for the x-direction, we conclude that *SPR can only be excited by light polarized in the TM mode*:

$$\frac{\varepsilon_i \mu_0 \omega^2 - \beta_x^2}{\varepsilon_i^2} = \frac{\varepsilon_t \mu_0 \omega^2 - \beta_x^2}{\varepsilon_t^2}$$

$$\beta_x = \omega\sqrt{\frac{\mu_0 \varepsilon_i \varepsilon_t}{\varepsilon_i + \varepsilon_t}} \tag{5.11}$$

Thus, we have shown that there exists the potential for a bound electromagnetic mode that is confined to the surface according to Equation 5.6. It has a propagation constant that is dependent on the material properties of the two substrates according to Equation 5.11 and can only be excited by TM polarized light.

Equation 5.11 demonstrates that the surface plasmon wave will have a propagation constant that is complex since the transmitting material is a metal. Examination of Equation 5.6 illustrates the relationship between the propagation constant and the material properties of the system. If the term $\sqrt{\varepsilon_i \mu \omega^2 - \beta_x^2}$ is complex, then the wave is an exponentially decreasing function of x implying a confined mode at the interface between the two materials. This occurs when $|\varepsilon_t| > \varepsilon_i$. This is the surface plasmon mode. Similarly, if $\sqrt{\varepsilon_i \mu \omega^2 - \beta_x^2}$ is real, then the wave is an evanescent mode into the metal, which is a nonpropagating mode [27].

In order to use SPR as a sensor, a three-layer medium is necessary where the incident medium (dielectric) is labeled subscript 3, the metallic medium is labeled subscript 1, and surrounding media outside the metal are labeled subscript 2. The reflectivity of the three-layer structure can be found by calculating the film scattering matrix (S) as follows [28]:

$$S = \frac{1}{\tau_{31}}\begin{pmatrix} 1 & \rho_{31} \\ \rho_{31} & 1 \end{pmatrix}\begin{pmatrix} e^{-j\gamma_1} & 0 \\ 0 & e^{j\gamma_1} \end{pmatrix}\frac{1}{\tau_{12}}\begin{pmatrix} 1 & \rho_{12} \\ \rho_{12} & 1 \end{pmatrix} = \frac{1}{\tau_{31}\tau_{12}}\begin{pmatrix} e^{-j\gamma_1} + \rho_{12}\rho_{31}e^{j\beta_1} & \rho_{12}e^{-j\gamma_1} + \rho_{31}e^{j\gamma_1} \\ \rho_{12}e^{j\gamma_1} + \rho_{31}e^{-j\beta_1} & e^{j\gamma_1} + \rho_{12}\rho_{31}e^{-j\gamma_1} \end{pmatrix} \tag{5.12}$$

In Equation 5.12, ρ_{ij} and τ_{ij} are the TM Fresnel amplitude reflection coefficient and amplitude transmission coefficients, respectively, for the transition from layer i to layer j. For completeness, these equations are reproduced here where ε_j is the permittivity of the layer material, θ_i and θ_j are the incident and transmitted angles, respectively, d is the thickness of the metal film, and λ is the wavelength of the incident light [29–31]:

$$\begin{aligned} \tau_{ij} &= \frac{2\sqrt{\varepsilon_i}\cos\theta_i}{\sqrt{\varepsilon_j}\cos\theta_i + \sqrt{\varepsilon_i}\cos\theta_j} \\ \rho_{ij} &= \frac{\sqrt{\varepsilon_j}\cos\theta_i - \sqrt{\varepsilon_i}\cos\theta_j}{\sqrt{\varepsilon_j}\cos\theta_i + \sqrt{\varepsilon_i}\cos\theta_j} \end{aligned} \tag{5.13}$$

$$\gamma_j = \frac{2\pi}{\lambda}\sqrt{\varepsilon_j}\, d\cos\theta_j \tag{5.14}$$

The reflectance (R) is then found as the following:

$$R_{31} = \left|\frac{S_{12}}{S_{22}}\right|^2 = \left|\frac{\rho_{31} + \rho_{12}e^{-j2\gamma_1}}{1 + \rho_{12}\rho_{31}e^{-j2\gamma_1}}\right|^2 \tag{5.15}$$

Finally, the resonance condition is found by recognizing that the component of the propagation constant vector in the incident medium that is parallel to the surface must be equal to the surface plasmon propagation constant found in Equation 5.11. Thus, the resonance condition for a planar surface plasmon geometry is the following [32]:

$$\frac{\omega n_i}{c}\sin\theta_i = \omega\sqrt{\frac{\mu_0\varepsilon_1\varepsilon_2}{\varepsilon_1+\varepsilon_2}}$$
$$\sin\theta_i = \frac{c}{n_i}\sqrt{\frac{\mu_0\varepsilon_1\varepsilon_2}{\varepsilon_1+\varepsilon_2}} \tag{5.16}$$

In the aforementioned development, we have derived the basic notions of SPR in a planar system. Namely, that the angle of resonance is related to the material properties of the system but is independent of the frequency of excitation except for the variation of permittivity due to dispersion (Equation 5.16). Thus, any shift in the resonance angle is a function of the dispersion of the light in the medium only and not the excitation frequency. Second, the propagation constant of surface plasmons is found as in Equation 5.11 and is only derived from TM polarized electromagnetic energy implying that TE mode light will not lead to SPR. This has been extensively confirmed using the model developed previously for a planar system. Finally, the reflectivity for a three-layer system can be found from Equation 5.15.

5.3 SPR SENSORS

Equations 5.15 and 5.16 illustrate the relationship between the material properties of the system, the frequency of the excitation light, the thickness of the optical films, and the angle of resonance. For a system to become a practical sensor, there has to be a perturbation term added that represents a small change in one of the fundamental parameters of the resonance condition. For an SPR sensor, a perturbation in the dielectric constant or material thickness can be used effectively as a sensor parameter to modulate the angle of resonance. Alternatively, for a fixed angle of resonance, the amplitude of the reflected light will change if the angle of resonance is modulated. According to Equation 5.15, a variation in either of these parameters will result in a change in the reflectivity at a given fixed incident angle. Thus, by fixing the excitation angle and only measuring the reflectivity of the system, sensing can be accomplished. This is the basis of the Kretschmann geometry that is used in most laboratory-scale SPR instruments.

Figure 5.2 is a diagram of the Kretschmann coupling condition showing incident TM light exciting surface plasmons at the interface between the prism and a thin metal layer on the facet of the prism.

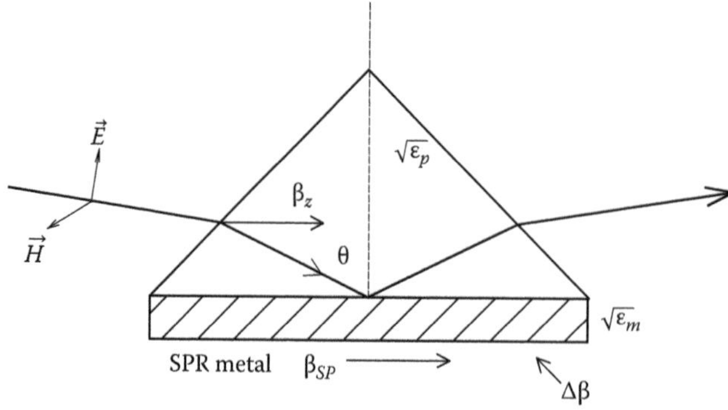

FIGURE 5.2 Diagram illustrating TM mode light incident on a prism coated with an SPR metal film (hashed layer). SPR occurs at the resonance angle (θ_r). The angle of resonance is perturbed by the term $\Delta\beta$, which represents a variation in the fundamental parameters of the resonance condition (dielectric constant, film thickness, etc.). In the aforementioned illustration, ε_p is the dielectric constant of the prism, ε_m is the dielectric constant of the metal film, β_z is the component of the propagation vector in the z-direction (parallel to the interface), and β_{SPR} is the propagation constant of the surface plasmon wave at the interface.

If the incident angle is fixed, then any perturbation due to chemical exposure at the outside surface of the metal film will appear as a variation in the surface plasmon propagation constant ($\Delta\beta$). Since the reflectivity of the system is coupled to the SPR propagation constant through the dielectric of the metal film and the thickness of the film, a chemical variation in the surface will result in a change in the amplitude of the reflected light exiting the prism on the right-hand side of Figure 5.2 according to Equation 5.15. It should be noted that the quiescent thickness of the metal film is critical to the signal-to-noise of the sensor system. If the metal film is too thin, then the width of the resonance is too wide and the sensitivity of the reflectivity is reduced as a function of ($\Delta\beta$). Likewise, metal films that are too thick reduce or eliminate the interaction between the surface plasmons and the exterior surface of the film reducing the sensitivity.

Chemical sensitivity can be produced in the configuration of Figure 5.2 using various means [33]. The first method could be to add a second film on top of the SPR metal (commonly Au) that is sensitive to a target species such as a polymer film that adsorbs an organic vapor or, as will be demonstrated later in this paper, a dielectric film such as SiO_2 that can be used to adsorb moisture. This approach relies on the chemistry of the added layer to scavenge the target species from the surrounding atmosphere and then diffuse into the film to the metal layer for interaction between the composite and the surface plasmon fields. Choice of this approach requires that the added sensing layer be thin enough that the chemical perturbation is observed throughout the entire thickness of the added layer. In addition, for reversible sensing, the chemical interaction between the added film and the analyte of interest must be *weak* in order for partitioning between the atmosphere and the film to be dependent on the partial pressure of the analyte in the environment [34,35].

A second approach that will not be demonstrated here but is important in the study of biological systems is to functionalize the metal layer with antibodies that are specific to various bacteria. As the film is exposed to the bacteria, they bind to the surface, changing the effective propagation constant of the surface plasmon field and hence the amplitude of the reflectivity as before [8–16]. Appropriate choice of antibodies produces an extremely selective but generally irreversible sensing system.

A third approach is to choose an SPR metal that is sensitive to a particular chemical or class of chemicals. For example, Pd will form a reversible hydride (PdH_x) [50]. Since the hydride has a dielectric constant that is different from the pure metal, the surface plasmon propagation constant will again change resulting in a modulation of the reflected amplitude of the light. This case will be studied in more detail later in this paper. Other nonnoble metals such as Cu and Al have been demonstrated as oxidation sensors using SPR [38]. Finally, the nonreversible reaction with Ag and H_2S will be demonstrated using SPR.

The Kretschmann coupling condition described in Equations 5.15 and 5.16 describes mathematically the relationship between the coupling of incident light energy and the SPR field but does not provide an intuitive explanation of the resonance condition. However, if we consider a *ray-optics* description of the geometry of Figure 5.2, we can describe the resonance condition as the angle, at a given wavelength, when the z-component (axes of Figure 5.2 are the same as for Figure 5.1) of the propagation vector of the light incident on the SPR metal is equivalent to the propagation constant calculated in Equation 5.11. An illustration of this condition is given in Figure 5.3, which is a plot of the electromagnetic solution to the SPR problem under the Kretschmann condition (Equation 5.15) compared to the *ray-optics* solution described earlier. The gray-scale contour map is the reflectivity as a function of incident angle and wavelength. The dashed white line is the *ray-optics* solution where the z-component of the propagation constant is set equal to the SPR propagation constant. The solutions show good agreement over the space, implying the general accuracy of the argument.

Figure 5.4 is a plot of the modeled response for a three-layer SPR experiment in the Kretschmann geometry for two different substrate indices of refraction and two different thicknesses of an Au film with a complex index of refraction as noted. These conditions result in a narrow resonance peak that is very sensitive to the thickness of the Au layer. Thus, for SPR measurements in a laboratory setting where the geometry of the optical system can be conveniently constructed, the Kretschmann configuration is optimum due to the narrow resonances measured.

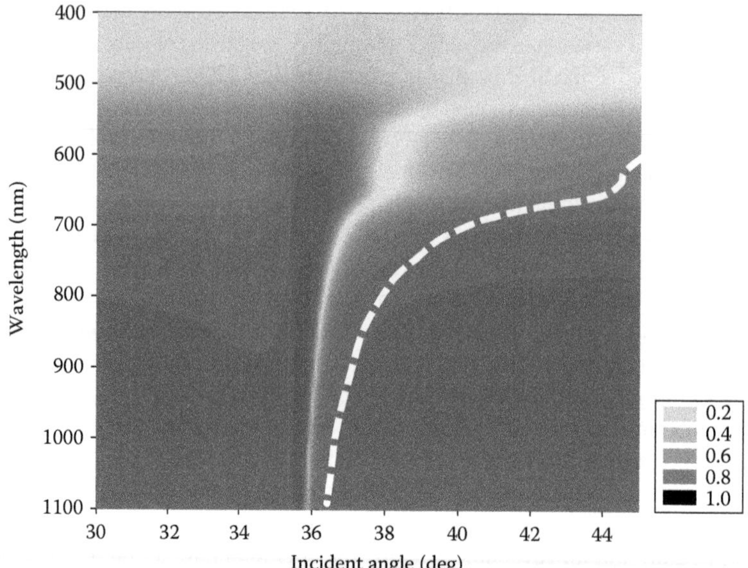

FIGURE 5.3 Plot of the electromagnetic reflectivity solution for an SPR (Equation 5.15) as a gray-scale contour map compared to the *ray-optics* solution (dashed line). This indicates that the resonance condition occurs when the z-component of the propagation vector of the incident light is equivalent to the propagation constant of the SPR field that was calculated in Equation 5.11. Calculation is done for ~50 nm Au film on a quartz prism at a wavelength of 633 nm (He–Ne laser).

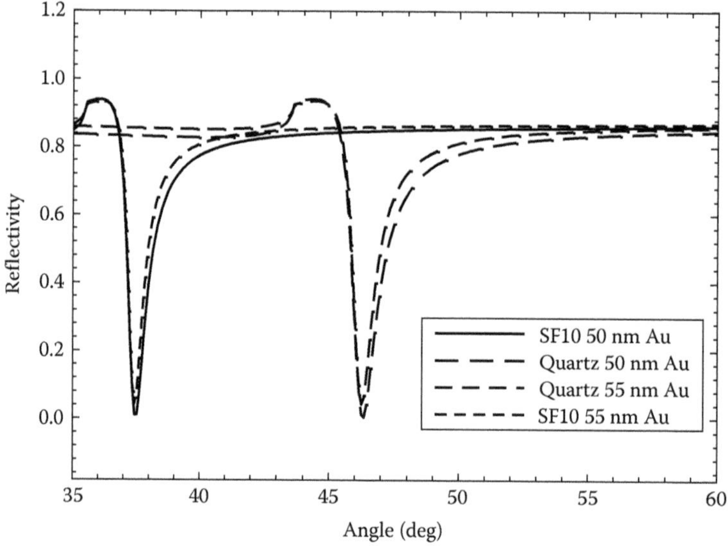

FIGURE 5.4 Plot of the normalized reflectivity from a Kretschmann configured SPR experiment with two different thicknesses of Au ($n_{Au} = 0.1726 + j3.4218$) and two different dielectric substrates ($n_{SF10} = 1.723$ and $n_{quartz} = 1.515$). Note that the position of the resonance is strongly influenced by the thickness of the Au film but is not strongly dependent on the index of refraction of the dielectric material.

5.4 OPTICAL FIBER RATIONALE

For monitoring of sealed systems in field environments where internal volume and external access are limited by the general function of the system, the Kretschmann geometry is often difficult or impossible to implement. For such systems, fiber-optic-based SPR systems have been applied. A fiber-optic-based SPR sensor is normally made from a single fiber that has its cladding etched down to the optical core of the fiber over a known section of length. The exposed core is then coated with an SPR supporting metal, and the sensor is used in a single-pass configuration [36–39]. A fundamental difference between an optical fiber and the common Kretschmann configuration is that in a multimode optical fiber, the allowed incident angles are a continuum from the critical angle of the fiber as set by the numerical aperture (NA) to $\pi/2$ from normal. As has been demonstrated both experimentally and theoretically, and this has the effect of significantly broadening the SPR peak [36,37].

A single-fiber version of the fiber-optic SPR sensor that employs a retroreflecting metal film on the end of a multimode fiber similar in geometry (Figure 5.5) to the intensity-based *micromirror* sensors documented in the literature has been demonstrated [40–43]. While the geometry is similar to previous work, the previous work only sampled the reflectance of the reflecting mirror on the fiber. In our system, light is injected into the fiber that then travels by way of the optical coupler to the SPR-coated fiber where it interacts with the sensing film. The light is then retroreflected back through the fiber to the coupler where half returns to the light source and is lost and the other half is directed via the second leg of the coupler to the monochromator where it is analyzed. The reflecting end serves the purpose of returning the light back through the fiber to the monochromator but does not participate in the sensing since the film is optically thick (100 nm) and each ray only interacts with the end film once as compared to multiple times for the axial coatings. In addition, the end is made from an inert noble metal such as Au or overcoated with a sealant. The retroreflecting coating was chosen to be optically thick to minimize any signal due to a chemical change in reflectivity of the retroreflector. The calculated skin depth for bulk Au at 550 nm is less than 10 nm indicating that a 100 nm film is at least a factor of 10 greater. In addition, Butler and Ricco [41] reported minimal reflectivity changes due to chemical reactions on the surface of optically thick Au films deposited on optical fibers.

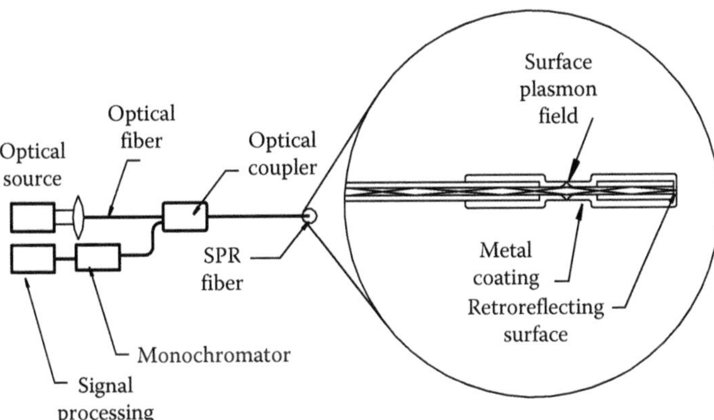

FIGURE 5.5 Schematic diagram of the SPR sensor showing the optical system. Broad-band light is injected into the fiber, which then is sent to the SPR end via a three-port coupler. The light is passed through the SPR section of the fiber and reflected back by a retroreflecting end through the coupler and is detected by the monochromator.

Since the fiber is single ended, it becomes practical to use this sensor to monitor the atmospheres of complex subsystems for contamination and aging effects. For example, deterioration of some important compounds found in packaging can lead to production of sulfur compounds that can corrode connectors and lead to electronic failures. In addition, batteries, transformers, and thermal degradation of packaging can evolve H_2 leading to metallic embrittlement, water formation, and explosive environments. Further, leakage of seals can allow moisture to infiltrate and condense on critical components contributing to corrosion and other failure mechanisms.

5.5 FIBER-OPTIC SPR THEORY

A numerical model was used to determine the approximate thickness for the SPR films required. These calculations were accomplished using a 600 μm core fused quartz fiber and assuming that the index of refraction was constant over the spectral range of 250–1000 nm at a value of $n_q \cong 1.46$ and an NA of 0.22. Comparison to the fiber data sheet confirms the validity of this constant assumption to within 3% over this wavelength range. These quantities allow the calculation of the critical angle in the fiber, which establishes the lower limit of integration in the model as follows [44]:

$$\theta_c = \frac{\pi}{2} - \sin^{-1}\left(\frac{NA}{n_q}\right) \tag{5.17}$$

For a single pass system with no retroreflector, the reflectance (R) as a function of wavelength (λ) of the SPR surface is given, where $p(\theta)$ is the incident power on the film at the angle θ, $r^N(\theta)$ is the reflectance for each individual ray at the angle θ, N is the number of reflections from the surface as determined by the incident angle and the length of the SPR supporting metal (L), and λ is the wavelength of the light:

$$R(\lambda) = \frac{\int_{\theta_c}^{\pi/2} p(\theta) r^N(\theta)\, d\theta}{\int_{\theta_c}^{\pi/2} p(\theta)\, d\theta} \tag{5.18}$$

$$N = \frac{L}{d_2 \tan(\theta)}$$

A complete description of the method is outlined in detail in Xu et al. [36]. In Equation 5.18, the argument of the numerator integral is the power that remains after multiple reflections of any single TM ray at any single angle θ. Recognizing that this a continuum of rays, integration was performed on the guided rays from the critical angle θ_c below which the light will be lost to fiber cladding leakage to an angle of incidence of $\pi/2$ (grazing incidence with the core–cladding interface). A Lambertian model was employed for our light source to model the tungsten-halogen lamp used to illuminate our sample fibers. A Lambertian source has the property of having a uniform radiance that is independent of the angle into which the radiation is directed [45,46]. Thus, the optical power can be expressed as the following function [36]:

$$p(\theta) = p_o n_q^2 \sin\theta\cos\theta \tag{5.19}$$

In Equation 5.19, p_o is the nominal power from the light source. Since the power function appears in both the numerator and the denominator of Equation 5.18, the absolute magnitude of the power is normalized, leaving only the ratio of the reflected light to the incident light or the reflectance of the fiber-film system.

As shown previously, a fundamental concept of SPR is that SPR can only be varied for the TM polarized light as a function of angle; thus, the transmission through a coated section of fiber must be expressed as the following:

$$\begin{aligned} T(\lambda) &= \frac{\Phi_{TE} + R(\theta)\Phi_{TM}}{\Phi_{total}} \\ &= \frac{1}{2} + \frac{R(\theta)}{2} \end{aligned} \tag{5.20}$$

In Equation 5.20, Φ is the transmitted optical power in each of the TM and TE modes. Light is assumed to be evenly distributed between the two modes, resulting in half the transmission coming from the TE mode that is undisturbed by SPR, and the rest coming from the total reflection from the coating divided by two since only half of the original incident light can participate in SPR.

A double-pass optical configuration was employed as shown in Figure 5.5, which requires a metallic film to be deposited on the end facet of the fiber to retroreflect the light back toward the detector. Thus, an additional correction to $T(\lambda)$ must be made to compensate for the spectral influence of the retroreflector on the incident light. Thus, again assuming that half of the light is TE mode and does not contribute to SPR but is modified by the retroreflector according to Fresnel's equations and the other half of the light is modified by both the SPR and the retroreflecting surface, Equations 5.18 and 5.20 can be modified as follows [47]:

$$T(\lambda) = \frac{1}{2}\frac{\int_{\theta_c}^{\pi/2} p(\theta) R_{TE}(\theta)\, d\theta}{\int_{\theta_c}^{\pi/2} p(\theta)\, d\theta} + \frac{1}{2}\frac{\int_{\theta_c}^{\pi/2} p(\theta) R_{TM}(\theta)\, d\theta}{\int_{\theta_c}^{\pi/2} p(\theta)\, d\theta} \tag{5.21}$$

where R_{TE} is defined as [48]

$$R_{TE}(\theta) = \left(\frac{n_2\cos\theta_i - n_{rr}\cos\theta_t}{n_2\cos\theta_i + n_{rr}\cos\theta_t}\right)^2 \times \left(\frac{n_2\cos\theta - n_1\cos\theta'}{n_2\cos\theta + n_1\cos\theta'}\right)^{4N} \tag{5.22}$$

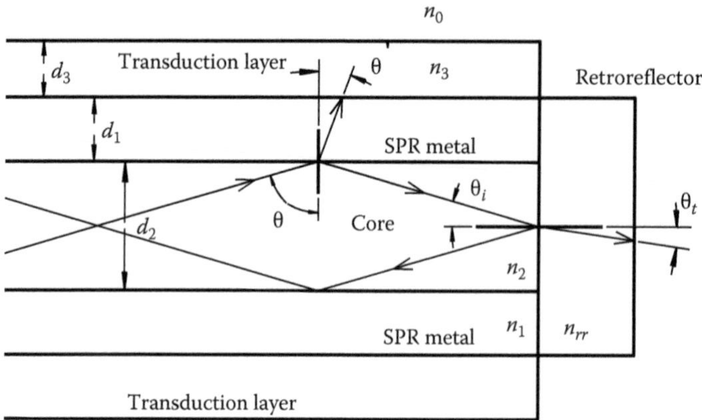

FIGURE 5.6 Diagram of the geometry used to derive equations where the subscript 0 represents the outside atmosphere, 1 is the SPR supporting metal layer, 2 is the core, and 3 is an optional second transduction layer. (From Pfeifer, K.B. et al., *IEEE Sens. J.*, 10(8), 1360, 2010. With permission.)

and R_{TM} is defined as

$$R_{TM}(\theta) = \left(\frac{n_{rr} \cos\theta_i - n_2 \cos\theta_t}{n_{rr} \cos\theta_i + n_2 \cos\theta_t} \right)^2 r^{2N}(\theta) \tag{5.23}$$

The angles in Equations 5.22 and 5.23 are defined in Figure 5.6 and Equation 5.24. In Equation 5.22, the left-hand term is the Fresnel equation for light polarized such that the electric field is perpendicular to the plane of incidence or TE as defined earlier for the single reflection off of the retroreflector. The right-hand term is the same Fresnel equation configured for the multiple reflections off of the fiber core to SPR film interface again in TE mode. Note this term is raised to the 4*N* rather than 2*N* power because the length of the SPR supporting metal is twice as long due to the double-pass nature of the fiber architecture and there are *N* individual reflections in each direction. The right-hand term in Equation 5.23 is the SPR term derived by Xu et al. (Equation 5.25), and the left-hand term is the Fresnel equation for reflection off of the retroreflector for TM mode light as a function of incident angle [36]. This term is raised to the 2*N* rather than *N* power as in Equation 5.18 again because of the double-pass nature of the fiber architecture. The relationships between the various angles and the integration angle are as follows:

$$\begin{aligned} \theta_i &= \frac{\pi}{2} - \theta \\ \theta' &= \sin^{-1}\left(\frac{n_2}{n_1} \sin\theta \right) \\ \theta_t &= \sin^{-1}\left(\frac{n_2}{n_{rr}} \cos\theta \right) \end{aligned} \tag{5.24}$$

The reflectance function due to SPR ($r(\theta)$) is given by the following [36]:

$$r(\theta) = \left| \frac{r_{21} + r_{130} e^{2ik_1 d_1}}{1 + r_{21} r_{130} e^{2ik_1 d_1}} \right|^2 \tag{5.25}$$

where

$$\begin{aligned}
r_{130} &= \frac{z_{10} - iz_{43} \tan(k_3 d_3)}{n_{10} - in_{43} \tan(k_3 d_3)} \\
r_{12} &= \frac{z_{21}}{n_{21}} \\
z_{l,m} &= k_l \varepsilon_m - k_m \varepsilon_l \\
n_{l,m} &= k_l \varepsilon_m + k_m \varepsilon_l \\
k_l &= \left[\varepsilon_l \left(\frac{2\pi}{\lambda} \right)^2 - \left(\frac{2\pi}{\lambda} n_2 \sin\theta \right)^2 \right]^{1/2}, \quad l = 1,2,3 \\
\varepsilon_4 &= \frac{\varepsilon_0 \varepsilon_1}{\varepsilon_3}, \quad k_4 = \frac{k_0 k_1}{k_3}
\end{aligned} \tag{5.26}$$

and the subscripts represent the layers as defined in Figure 5.6.

5.6 MODELING RESULTS

A Mathcad* program was constructed to solve for the transmission through the optical system and vary the thickness parameters of the fibers based on the formalism developed previously. Results were applied to choosing various thicknesses of metal films in order to locate the SPR minimum and estimate the response to various exposures. The results from the models, in general, do not exactly describe the measured behavior of the films tested; however, the general trends are described and employed to direct the sensor design.

Figure 5.7 is a plot of the calculated response of Pd to exposure to H_2 as given by the model and literature values of refractive index [49,50]. The values of refractive index for PdH_x were the values reported by von Rottkay et al. [48] and are for thin films of Pd exposed to an atmosphere of 10^5 Pa of H_2. This concentration is significantly higher than any measurements that were made on our fiber structures; however, they represent an upward bound on the refractive index values as a function of hydride state. These simulations indicated that significant variations between pure Pd and PdH_x should be expected in the 200–400 nm range, which enables the detection of H_2. Calculations were done with 600 μm fiber with an NA of 0.22 and various thicknesses of PdH_x. These types of calculations were employed to direct the choice of metal thicknesses for the experimental studies but do not exactly model the position of the minimum response. In the example of Figure 5.7, various thicknesses of films were simulated in order to locate thicknesses that had substantial resonance peaks in the visible spectrum and could be easily fabricated. Similar studies were performed on Au/SiO_2 and Ag films.

5.7 EXPERIMENTAL

5.7.1 Preparation of the Fiber

Following the ray-optics model described by Xu et al. [36], the metal thickness was calculated for monitoring of H_2 using deposited Pd films, H_2S using Ag films, and moisture using a cover layer of SiO_2 on an Au film. In addition, the core diameter of the optical fiber was chosen using this theoretical model.

* PTC, 140 Kendrick St., Needham, MA 02494, United States.

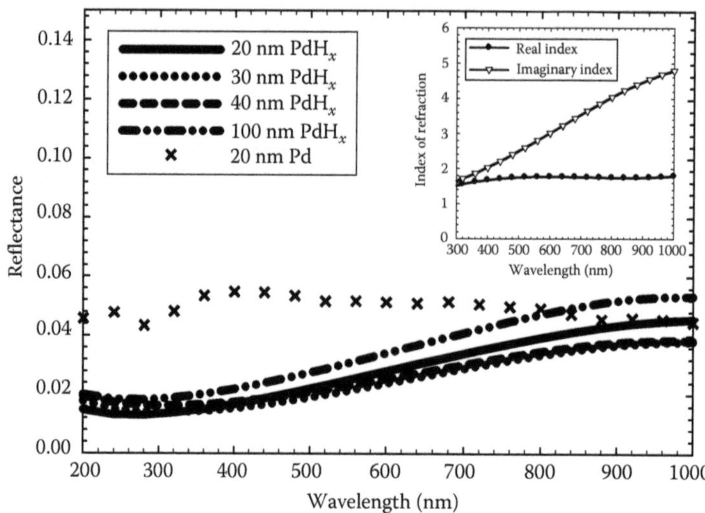

FIGURE 5.7 Plot of calculated reflectance for several thicknesses of Pd film exposed to H_2 to form PdH_x as calculated and compared to a measured unexposed 12.5 mm length, 20 nm thick Pd film (·) in the SPR geometry. Data indicate that these are good candidates for fabrication and testing due to their broad absorption resonance being predicted in the 250–400 nm range are films with thicknesses of 20–40 nm. The real and imaginary parts of the index of refraction were obtained from the literature for PdH_x and are plotted in the inset graph [47,48]. (From Pfeifer, K.B. et al., *IEEE Sens. J.*, 10(8), 1360, 2010. With permission.)

Figure 5.8 is a simulation for an Au film on a fiber of various core diameters from 100 to 1000 μm with the real and complex index of refraction components of Au plotted as a function of wavelength in the inset. The trend is for higher throughput and deeper SPR minima as the core diameter becomes larger. However, in practice, 600 μm couplers are the largest conveniently available, and therefore, fused quartz fiber of this diameter was chosen (Ocean Optics Fiber-600-UV*) for the experiment.

Sensing is accomplished by first etching off the cladding material and then coating a cylindrical section of the fiber with an SPR supporting metal film as illustrated in Figure 5.5. This geometry has the distinct advantage of doubling the sensitivity per unit length of the SPR film since the light travels through the sensitive section twice. In principle, multiple regions could be etched to the core and coated with different metals to vary the location of the SPR peak and allow sensing of several compounds using the same probe fiber.

Based on the results from the numerical simulations, a series of 600 μm core fibers were fabricated and coated with SPR supporting metals. The general process was to first cleave sections of optical fiber approximately 20 cm in length and then pyrolyze the last ~3 cm of polyimide coating to allow the etching chemical access to the fused silica. The pyrolysis was accomplished by inserting the optical fiber in an alumina cylinder with a coil of nichrome wire wrapped around the cylinder and then using a power supply to heat the nichrome to incandescence. The fiber was then cleaned and etched by dipping the last 1.25 cm into a solution of 5 g NH_4HF_2 dissolved in 50 mL deionized H_2O. The initial outside diameter of the fiber is 660 μm, and the nominal core diameter is 600 μm; thus, the fibers were etched until they measured an outside diameter on the order of 590 nm to guarantee that the cladding had been completely removed and the core material alone was exposed. The measurement was accomplished using a digital micrometer. The etching of the cladding required time intervals on the order of 30 min, but the etching was stopped based on the micrometer measurement rather than the elapsed time. In addition, scanning electron microscope measurements of the surface revealed that variations in diameter of <250 nm were observed over the etched lengths.

* Ocean Optics, 830 Douglas Ave. Dunedin, FL 34698, United States.

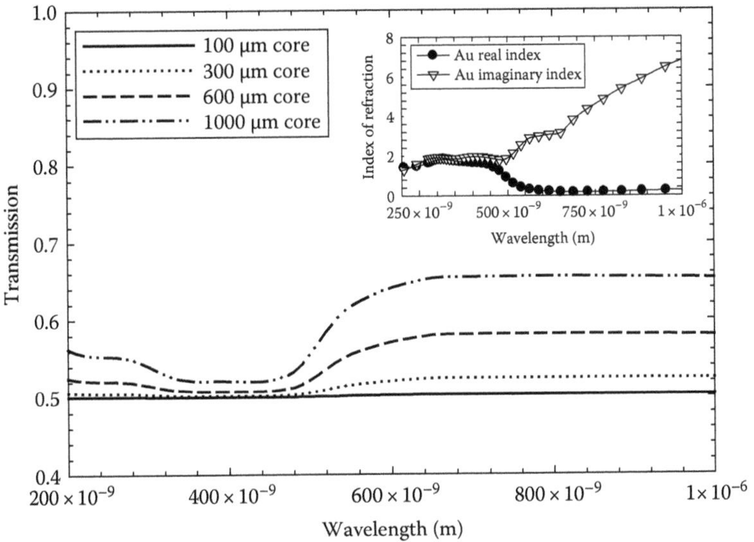

FIGURE 5.8 Plot of optical transmission for a 20 nm Au-coated fiber as a function of wavelength for several different core diameters using the theory of Xu et al. [36]. The plot illustrates that for SPR, the larger diameter gives a higher signal-to-noise ratio SPR peak allowing better wavelength discrimination. (From Pfeifer, K.B. et al., *IEEE Sens. J.*, 10(8), 1360, 2010. With permission.)

Next, the fibers were rinsed in deionized water and then placed in a vacuum-compatible rotating fixture. The fibers were then placed in an electron beam deposition system and pumped down to a base pressure of below 100 μPa. Next, they were coated with the respective SPR metal at rates of approximately 1 Å/s for Pd and Ag and a rate of 4 Å/s for Au. The fibers were rotated at 3 RPM with a substrate temperature of ~24°C until the quartz crystal thickness monitor indicated that the desired thickness was achieved. Finally, the fibers were masked from the source except for the cleaved fiber end, and the retroreflecting coating was applied. In all cases, the retroreflecting coating is a 100 nm Au film deposited using the same parameters as the SPR Au films. The thicknesses were verified using a stylus profilometer and found to be within ±20% of the expected values.

The responses of the fibers were then measured using the system depicted in Figure 5.5, and the data were analyzed. The system in Figure 5.5 consists of a tungsten-halogen light source (Ocean Optics LS-1) connected to a three-port optical coupler (Ocean Optics 600 μm bifurcated fiber assembly). The spectrometer is an Ocean Optics USB-2000 system interfaced to a PC for data collection. An example of the raw data is shown in Figure 5.9, which illustrates the change in the spectral response as a function of exposure. The film in Figure 5.9 is a 30 nm Pd film deposited on ~1.25 cm of exposed fiber core. The plot shows the spectrum of an unexposed film in vacuum (200 Pa air) and the spectrum of a film hydrated in a partial pressure of ~27 Pa of H_2 in N_2. Since this is a large concentration, the effect is visible in the raw spectral data. However, in most cases, the effect is small and must be extracted using numerical methods written in MATLAB®.*

5.7.2 MATLAB® Algorithms

The MATLAB algorithm employed takes the spectrum from the unexposed fiber and averages several scans. This becomes the reference or background spectrum. All the scans are then normalized such that the maximum value is fixed at a value of unity. By doing this, effects such as light source aging, connector loss, and other drift mechanisms in the optical system are eliminated. This is key to the

* The MathWorks, Inc. 3 Apple Hill Drive, Natick, MA 01760-2098, United States.

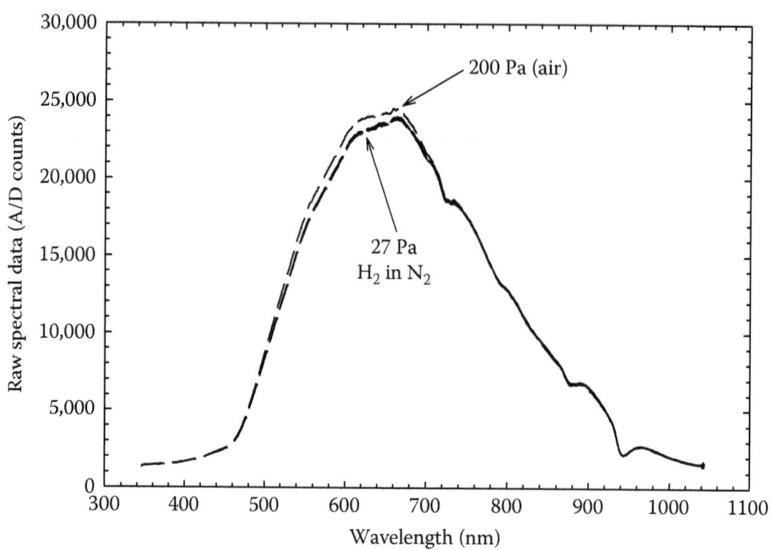

FIGURE 5.9 Plot of the raw spectral response from a 30 nm Pd-coated fiber under conditions of near vacuum pumped from an air background and exposed to ~27 Pa of H_2. (From Pfeifer, K.B. et al., *IEEE Sens. J.*, 10(8), 1360, 2010. With permission.)

application of this technique for long-term unpowered applications where it is neither practical nor desirable to continuously measure the spectrum. The rest of the data is then normalized to the beginning value and plotted as a function of time to look for trends in the spectral response of the fiber.

5.7.3 Gas Sample Generation

Vapor generation in the experiment was accomplished by constructing a crossed tubing arrangement with the fiber inserted into one leg and sealed using a solid Teflon compression ferrule. The ferrule was drilled to accommodate the fiber using a wire drill that was slightly larger than the fiber diameter. By placing the fiber in the ferrule and compressing the ferrule into the fitting, a gas tight seal was achieved. The other legs of the cross arrangement were connected to (1) a venturi pump for evacuation of the volume, (2) a thermocouple gauge for measurement of the pressures, and (3) a length of tubing connected to a sample gas source via a sealed bellows valve.

The measurements were made by first evacuating the volume and then backfilling the volume with standard gas mixtures while monitoring the partial pressure of the volume with the thermocouple gauge. Our standards were 1.00% H_2 in N_2 and 1.09 ppm H_2S in N_2. Moisture samples were generated in the laboratory using a Thunder Scientific Model 3900 two-pressure, two-temperature, low-humidity generator* rather than a gas bottle, and the output was connected in a flow mode past the fiber end.

5.8 RESULTS

5.8.1 Pd/H_2

Typical data for the Pd fiber are plotted in Figure 5.10, which illustrates the response of a 40 nm Pd film with a 100 nm Au retroreflecting film exposed to various partial pressures of H_2 in N_2. The data at each wavelength have been normalized and then compared to its initial value. The data illustrate that at partial pressures below 27 Pa, the response of the sensor is linear, and the sign of the response is wavelength dependent. A linear response to H_2 at low concentrations has been reported previously

* Thunder Scientific Corporation, 623 Wyoming Blvd., SE Albuquerque, New Mexico 87123-3198, United States.

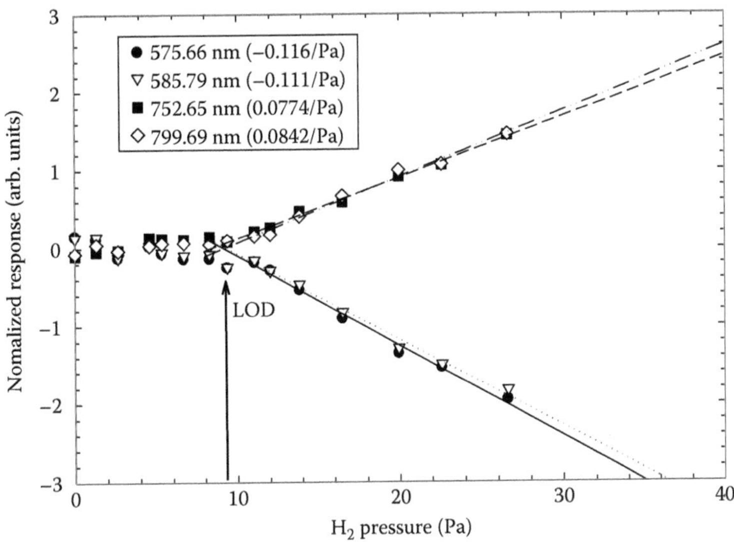

FIGURE 5.10 Plot of the normalized response of a 40 nm Pd film with a 100 nm Au retroreflector as a function of H_2 partial pressure at several wavelengths indicating that the sign of the response is wavelength dependent. Limit of detection is on the order of 9 Pa of H_2. (From Pfeifer, K.B. et al., *IEEE Sens. J.*, 10(8), 1360, 2010. With permission.)

for an SPR experiment in a Kretschmann configuration with a 633 nm light source [51]. It has been demonstrated that the linear trend continues over a sampling of wavelengths, but the absolute magnitude changes as a function of wavelength as well as the sign of the response, making it suitable for multivariate spectral analysis. As illustrated in the graph, changes are no longer observable as a function of concentration below about 9 Pa corresponding to a signal-to-noise ratio of unity in the system. The predominant noise sources in the system are from the spectrometer. The spectrometer was set to average 20 scans with an integration interval of 100 ms.

Palladium hydride (PdH_υ) forms for low partial pressures of H_2 in the α-phase. Transformation from the α-phase to the β-phase of PdH_υ occurs on exposure to higher concentrations of H_2 (2000–2700 Pa) where $\upsilon \sim 0.03$ and is undesirable since it causes a 3.5% increase in the lattice spacing and will result in a mechanical failure of the Pd film on the fiber [52]. Our application is concerned with trace concentrations of H_2 well below the β-phase; hence, concentrations below 27 Pa of H_2 partial pressure were tested. PdH_υ formation is linear with partial pressure in this range at a constant temperature [53].

Figure 5.11 is a plot of the measured response of a 40 nm Pd film exposed to H_2 plotted as a function of wavelength and as a function of file number. Each file was acquired on a 30 s interval implying that the entire scan occurred over approximately 74 min. Results demonstrate that the measurement is strongly a function of wavelength with the maximum response wavelength in the same general wavelength region as was calculated using the model in Figure 5.7.

In addition, the response is reversible and indicates the feasibility of using this technique for the measurement of H_2 in a sealed environment.

5.8.2 Ag/H_2S

Similar measurements have been conducted using Ag films exposed to 0.04 Pa H_2S in N_2 for extended intervals of 7 days that demonstrate a nonreversible SPR response at wavelengths starting around 450 nm. This is about 100 nm shorter wavelength than the Pd examples. These experiments were performed on the same 600 μm fiber coated with similar thicknesses of Ag. Figure 5.12 is

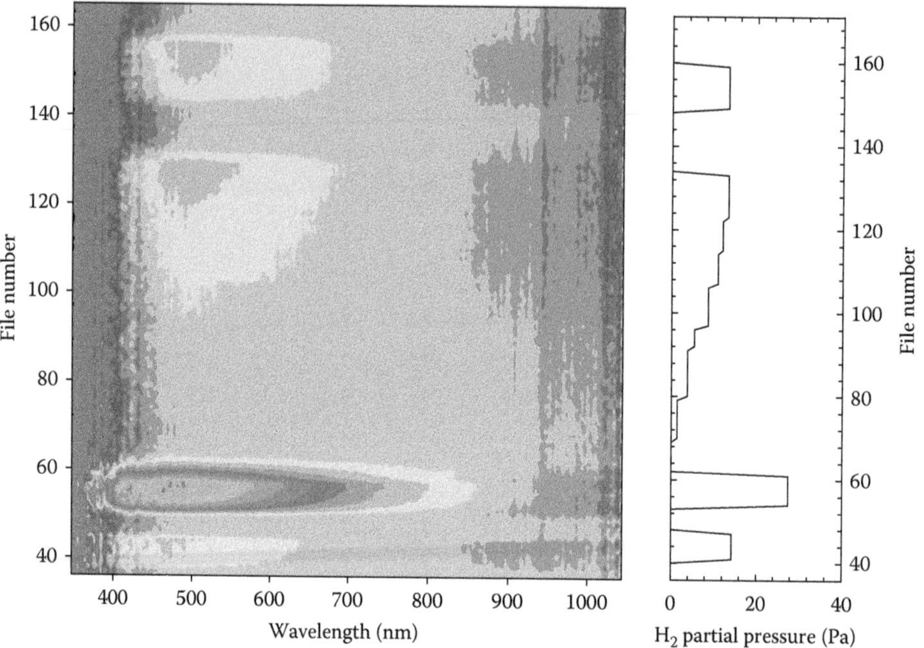

FIGURE 5.11 Plot of sensor response as a function of wavelength and file number plotted along with the H_2 partial pressure. Data illustrate that the sensor is reversible and has a strong dependence on wavelength. Each separate file was acquired on a 30 s interval. (From Pfeifer, K.B. et al., *IEEE Sens. J.*, 10(8), 1360, 2010. With permission.)

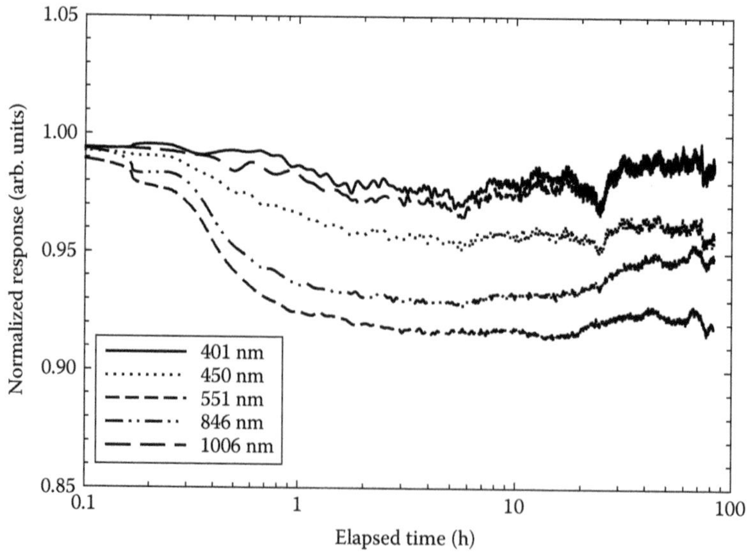

FIGURE 5.12 Plot of the normalized response for 401, 450, 551, 846, and 1006 nm for a 40 nm Ag SPR film on the fiber exposed to 0.04 Pa of H_2S in N_2. Data illustrate that response is complete after approximately 10 h of exposure and is wavelength dependent. (From Pfeifer, K.B. et al., *IEEE Sens. J.*, 10(8), 1360, 2010. With permission.)

a plot of the response of five separate wavelengths as a function of time on exposure to 0.04 Pa H_2S in N_2. Data illustrate that the response is nominally complete after approximately 10 h and is strongly wavelength dependent as expected. Data at 401 and 1006 nm illustrate very little response, while the response in the 450–550 nm region of the spectrum is significantly larger. In the figure, H_2S is introduced after approximately 0.3 h after beginning the data collection. We believe that the majority of this signal is due to SPR interaction with the coatings on the walls of the fiber and not due to interaction with the sulfur bonding to the retroreflecting Au coating on the end of the fiber. This is because the 100 nm thick Au coating is optically thick at visible wavelengths and the light would not interact with the exposed surface of the Au film [41].

5.8.3 SiO_2/H_2O

An additional system consisting of 20 nm of Au deposited on a fiber with a 20 nm cover layer of SiO_2 was used to monitor moisture changes in atmospheres with dew/frost points of 10°C to −70°C in N_2. The experimental results at around 550–600 nm show the most dramatic change in this moisture range and indicate that good discrimination between the spectra at −70°C and −10°C allows SPR measurements of moisture ingress in sealed systems using this SPR geometry. Example data illustrating the increase in dew point from −10°C to +10°C are shown in Figure 5.13.

Figure 5.13 is the normalized response at five different wavelengths from 415 to 701 nm and illustrates that the films are sensitive to moisture at all the wavelengths plotted, but the sensitivity is greater in the range of 455–557 nm.

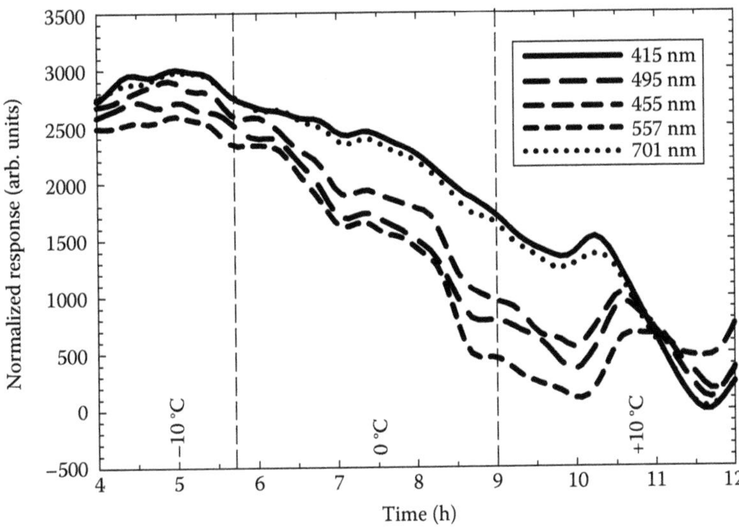

FIGURE 5.13 Plot of response of a 20 nm Au film overcoated with a 20 nm SiO_2 film for moisture monitoring. Data illustrate the normalized response of five different wavelengths as a function of dew/frost point from −10°C to +10°C at atmospheric pressure. Data illustrate that wavelengths over a wide band respond to changes in moisture level but wavelengths around between 455 and 557 nm respond with greater sensitivity than wavelengths away from the center of the band (415 and 701 nm). Each step of the moisture generator was programmed to remain constant for 4.5 h. As is observed in the 0°C zone, the sensor response was slower than this and therefore was not stable when the generator changed to a dew point of +10°C resulting in the constant downward drift of the response. (From Pfeifer, K.B. et al., *IEEE Sens. J.*, 10(8), 1360, 2010. With permission.)

5.9 MULTICOATING SPR FIBER

The ability to measure several contaminants of interest and to discriminate between them by monitoring a series of key wavelengths implies that a single fiber with multiple regions etched and coated with different metals could be used to produce a multicomponent sensor (Figure 5.14). Such a sensor has been demonstrated that was coated with a 12.5 mm length 20 nm Pd film followed by a 12.5 mm length 20 nm Ag film followed by a 12.5 mm length 20 nm Au/20 nm SiO_2 film stack for detection of H_2, sulfur compounds, and moisture, respectively (Figure 5.14). Data from these tests were first acquired using the unexposed fiber to detect 27 Pa H_2 (Figure 5.15a) and then exposed to 0.04 Pa of H_2S (Figure 5.15b). The Ag section of the fiber was then monitored until the nonreversible reaction was complete. The fiber was then reexposed to 27 Pa H_2 with the permanent H_2S/Ag response normalized into the data to determine if the sulfur atmosphere poisoned the Pd film (Figure 5.15c). The data indicate that the resonance peak that appears on exposure to H_2 is essentially unchanged compared to Figure 5.15a in shape and magnitude after the exposure of the Ag film to H_2S.

Multiple measurements of the response of the Pd fiber as a function of H_2 partial pressure were conducted both prior to H_2S exposure and then again post-H_2S exposure. These measurements consisted of stepping the concentration exposed to the fiber to various partial pressures between 0 and 27 Pa of H_2 in N_2 and then back to a vacuum baseline. The response of the fiber after exposure to the H_2S at four discrete wavelengths (547.7, 574.2, 586.4, and 751.3 nm) agrees to within 30% with the pre-H_2S exposures. In addition, the response time prior to H_2S exposure for a 27 Pa to 0 Pa H_2 concentration change is on the order of 2.5 min. The response time post exposure to H_2S is very similar, indicating that the response time is also essentially unchanged.

It should be noted that Figure 5.15a and b is normalized to the unexposed spectrum. Figure 5.15c is renormalized after the H_2S exposure and thus includes the spectral features introduced on the exposure of the Ag film to H_2S. Since the two metals have nonoverlapping SPRs, the responses to the two analytes can be separately resolved. The resulting response indicated that low background levels of sulfur exposure for several days do not appreciably degrade the response of the Pd section to H_2. In all cases, a horizontal line at 1 would represent the reflectivity of the unexposed, normalized optical spectrum.

Figure 5.15d is a plot of the response of the SiO_2/Au section of the fiber to drying from a frost point of 0°C to −70°C. Again, the response is with respect to normalization of the response at 0°C. These data illustrate that the moisture sensing section of the fiber is not poisoned by the H_2S and that monitoring of multiple components in the system is feasible by judicious choice of wavelengths in the SPR spectrum. The wavelength of maximum SPR response is similar in both the Pd and SiO_2/Au cases indicating that multiple wavelengths would need to be monitored simultaneously and analyzed with multivariate spectral analysis to separate H_2O and H_2. Binary or ternary mixtures have not yet been tested using this approach.

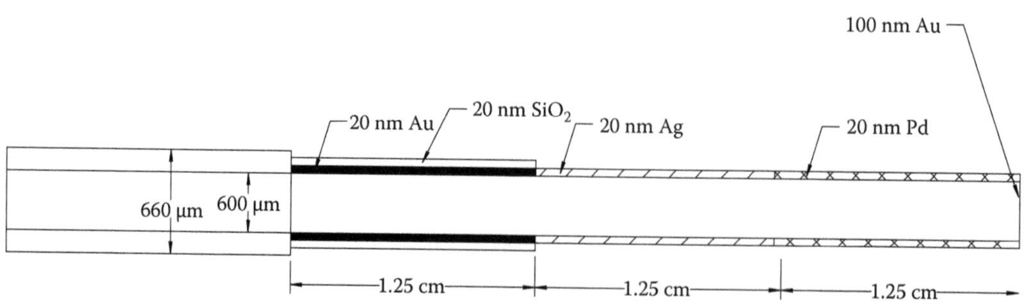

FIGURE 5.14 Diagram of three-section SPR sensor showing Au, Ag, and Pd regions on a 600 μm optical fiber core. (From Pfeifer, K.B. et al., *IEEE Sens. J.*, 10(8), 1360, 2010. With permission.)

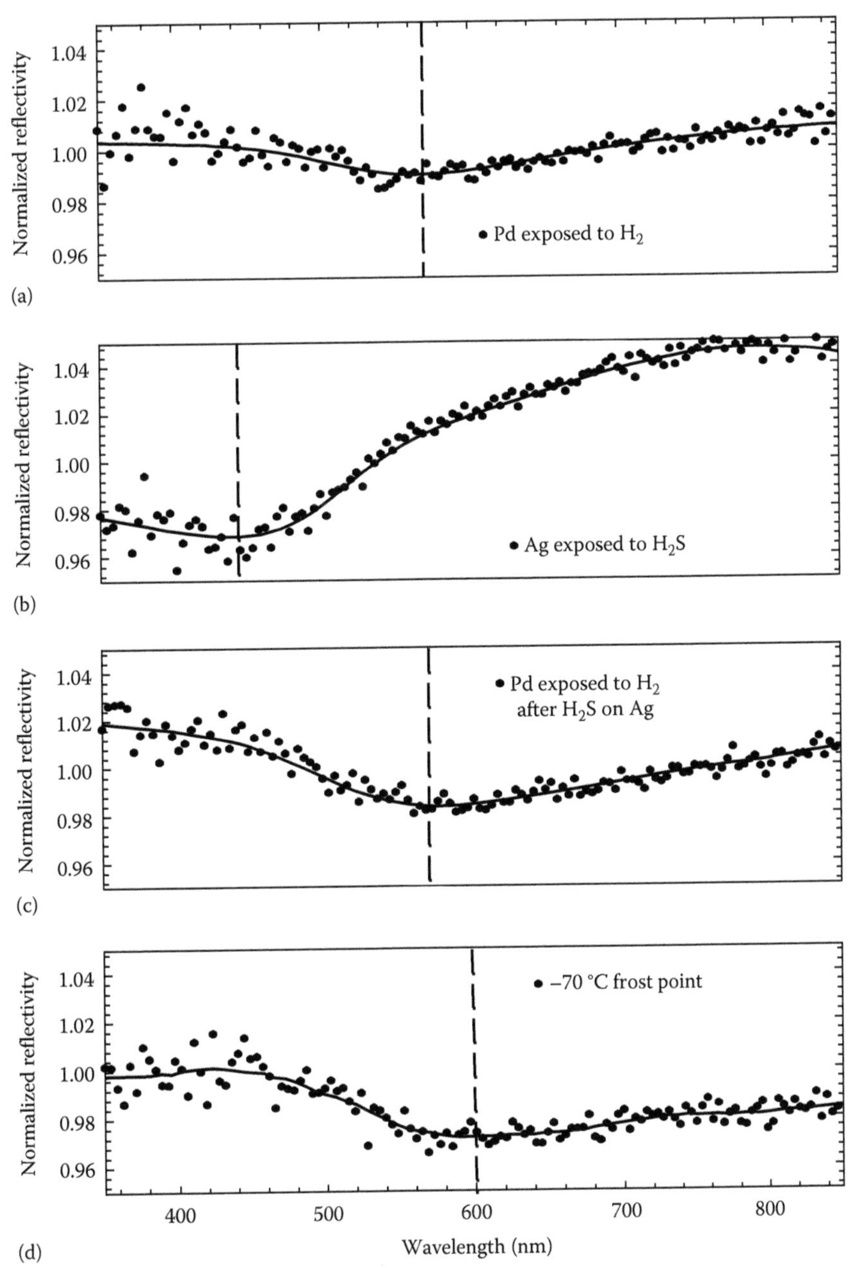

FIGURE 5.15 Plots showing the response to (a) 27 Pa H_2 in Pd, (b) 0.4 Pa H_2S on Ag, and (c) 27 Pa H_2 in Pd and (d) the response of the fiber when dried from 0°C dew point to –70°C frost point on a single SPR fiber. The time progression is from top to bottom, and the data are renormalized after the H_2S on Ag exposure since that change is not reversible. Thus, (c) and (d) are normalized to (b). The SPR wavelengths are separated with the minimum for Ag at approximately 450 nm, the minimum for Pd at about 550 nm, and the minimum for Au at about 600 nm. The dots are the raw data from the spectrometer, and the solid lines are running averages of the data. The vertical dashed lines locate the minimum of the averaged data. (From Pfeifer, K.B. et al., *IEEE Sens. J.*, 10(8), 1360, 2010. With permission.)

5.10 CONCLUSIONS

We have developed the background mathematics that predicts SPR in a planar geometry. This development was then used to illustrate the application of sensing using a planar SPR–type sensor including the origin of the resonance condition. These models were then expanded to a single-ended cylindrical geometry that can be made into a practical single-ended fiber-optic probe. Mathematical models to simulate the response of a fiber-optic-based SPR sensor using a single fiber with a retroreflective coating have then been demonstrated. Several SPR supporting metals were then modeled and used to design sensors for testing including H_2 sensing using Pd films, H_2S sensing using Ag films, and H_2O sensing using SiO_2 films on an Au film. Data illustrate that different metals allow variation in the spectral response to various analytes allowing the manufacture of a three-region SPR sensor that is sensitive to all three analytes using the single fiber and experimental system. It has been demonstrated that the H_2 response is unchanged after an extended interval of exposure to H_2S and that moisture can be detected at low frost points (–70°C) after repeated exposure to H_2 and prolonged exposure to H_2S. Thus, a simplified fiber-optic SPR geometry using a single-ended fiber has been demonstrated and shown to have multiple responses to several chemical species that can be present due to aging effects in sealed systems.

Sandia National Laboratories is a multiprogram laboratory managed and operated by Sandia Corporation, a wholly owned subsidiary of Lockheed Martin Corporation, for the US Department of Energy's National Nuclear Security Administration under contract DE-AC04-94AL85000.

REFERENCES

1. R. W. Wood, On a remarkable case of uneven distribution of light in a diffraction grating spectrum, *Philos. Mag.*, 4(19–24), 396–402, 1902.
2. R. W. Wood, Anomalous diffraction gratings, *Phys. Rev.*, 48, 928–937, 1935.
3. U. Fano, The theory of anomalous diffraction grating and of quasi-stationary waves on metallic surfaces (Sommerfeld's waves), *J. Opt. Soc. Am.*, 31, 213–222, 1941.
4. R. H. Ritchie, Plasma losses by fast electrons in thin films, *Phys. Rev.*, 196(5), 874–881, 1957.
5. E. Kretschmann and H. Raether, Radiative decay of non-radiative surface plasmons excited by light, *Z. Naturforsch.*, 23a, 2135, 1968.
6. K. B. Pfeifer and S. M. Thornberg, Surface plasmon sensing of gas phase contaminants using a single-ended multiregion optical fiber, *IEEE Sens. J.*, 10(8), 1360, 2010.
7. K. B. Pfeifer and S. M. Thornberg, Multi-region surface plasmon resonance fiber-optic sensors, in *Technologies for Smart Sensors and Sensor Fusion,* K. Yallup and K. Iniewski, eds. CRC Press, Boca Raton, FL, 2014, p. 89.
8. K. B. Pfeifer, Multi-region surface plasmon resonance fiber-optic sensor for monitoring high-consequence systems, *CMOS Emerging Technologies 2013*, Whistler, British Columbia, Canada, July 2013.
9. S. Ekgasit, C. Thammacharoen, F. Yu, and W. Knoll, Influence of the metal film thickness on the sensitivity of surface plasmon resonance biosensors, *Appl. Spectrosc.*, 59(5), 661–667, 2005.
10. I. Stemmler, A. Brecht, and G. Gauglitz, Compact surface plasmon resonance-transducers with spectral readout for biosensing applications, *Sens. Actuat. B*, 54, 98–105, 1999.
11. P. T. Leung, D. Pollard-Knight, G. P. Malan, and M. F. Finlan, Modeling of particle-enhanced sensitivity of the surface-plasmon-resonance biosensor, *Sens. Actuat. B*, 22, 175–180, 1994.
12. C. M. Pettit and D. Roy, Surface plasmon resonance as a time–resolved probe of structural changes in molecular films: Consideration of correlating resonance shifts with adsorbate layer parameters, *Analyst*, 132, 524–535, 2007.
13. A. Ikehata, K. Ohara, and Y. Ozaki, Direct determination of the experimentally observed penetration depth of the evanescent field via near-infrared absorption enhanced by the off-resonance of surface plasmons, *Appl. Spectrosc.*, 62(5), 512–516, 2008.
14. S. A. Love, B. J. Marquis, and C. L. Haynes, Recent advances in nanomaterials plasmonics: Fundamental studies and applications, *Appl. Spectrosc.*, 62(12), 346A–362A, 2008.
15. W. Yuan, H. P. Ho, C. L. Wong, S. K. Kong, and C. Lin, Surface plasmon resonance biosensor incorporated in a Michelson interferometer with enhanced sensitivity, *IEEE Sens. J.*, 7, 70–73, 2007.

16. S. Machaelis, J. Wegener, and R. Robelek, Label-free monitoring of cell-based assays: Combining impedance analysis with SPR for multiparametric cell profiling, *Biosens. Bioelectron.*, 49, 63–70, 2013.
17. X. Sun, L. Wu, J. Ji, D. Jiang, Y. Zhang, Z. Li, G. Zhang, and H. Zhang, Longitudinal surface plasmon resonance assay enhanced by magnetosomes for simultaneous detection pefloxacin and microcystin-LR in seafoods, *Biosens. Bioelectron.*, 47, 318–323, 2013.
18. E. Kretschmann, Decay of non-radiative surface plasmons into light on rough silver films. Comparison of experimental and theoretical results, *Opt. Commun.*, 6(2), 185–187, 1972.
19. H. Raether, *Surface Plasmons on Smooth and Rough Surfaces and on Gratings*. Springer-Verlag, Berlin, Germany, 1988, pp. 118–120.
20. J. Homola, Electromagnetic theory of surface plasmons, in *Surface Plasmon Resonance Based Sensors*, J. Homola, ed. Springer-Verlag, Berlin, Germany, 2006, pp. 1–10.
21. M. Yamamoto, Surface plasmon resonance (SPR) theory: Tutorial, *Rev. Polarography*, 48(3), 209–237, 2002.
22. A. K. Sharma, R. Jha, and B. D. Gupta, Fiber-optic sensors based on surface plasmon resonance: A comprehensive review, *IEEE Sens. J.*, 7(8), 1118–1129, 2007.
23. J. D. Jackson, *Classical Electrodynamics*, 2nd edn. John Wiley & Sons, New York, 1975, pp. 68–71.
24. S. Ramo, J. R. Whinnery, and T. Van Duzer, *Fields and Waves in Communications Electronics*, 3rd edn. John Wiley & Sons, New York, 1994, pp. 385–387.
25. J. R. Reitz, F. J. Milford, and R. W. Christy, *Foundations of Electromagnetic Theory*. Addison-Wesley Publishing Company, Reading, MA, 1980, p. 341.
26. A. D. Boardman, Hydrodynamic theory of plasmon-polaritions on plane surfaces, in *Electromagnetic Surface Modes*, A. D. Boardman, ed. John Wiley & Sons, Chichester, U.K., 1982, p. 17.
27. M. J. Adams, *An Introduction to Optical Waveguides*. John Wiley & Sons, Chichester, U.K., 1981, pp. 64–67.
28. M. V. Klein and T. E. Furtak, *Optics*, 2nd edn. John Wiley & Sons, New York, 1986, pp. 295–300.
29. M. V. Klein and T. E. Furtak, *Optics*, 2nd edn. John Wiley & Sons, New York, 1986, pp. 76–80.
30. E. Hecht and A. Zajac, *Optics*. Addison-Wesley, Menlo Park, CA, 1979, pp. 72–75.
31. C. A. Balanis, *Advanced Engineering Electromagnetics*. John Wiley & Sons, New York, 1989, p. 191.
32. H. Raether, *Surface Plasmons on Smooth and Rough Surfaces and on Gratings*. Springer-Verlag, Berlin, Germany, 1988, p. 11.
33. J. Homola and M. Piliarik, Surface plasmon resonance (SPR) sensors, in *Surface Plasmon Resonance Based Sensors*, J. Homola, ed. Springer-Verlag, Berlin, Germany, 2006, pp. 46–67.
34. A. J. Ricco, G. C. Frye, and S. J. Martin, Determination of BET surface areas of porous thin films using surface acoustic wave devices, *Langmuir*, 5(1), 273–276, 1989.
35. K. B. Pfeifer, Heat of adsorption and thin film surface area studies of a silica sol-gel film exposed to HCl and H_2O, *Langmuir*, 11(12), 4793–4796, 1995.
36. X. Bévenot, A. Trouillet, C. Veillas, H. Gagnaire, and M. Clément, Surface plasmon resonance hydrogen sensor using an optical fibre, *Meas. Sci. Technol.*, 13, 118–124, 2002.
37. Y. Xu, N. B. Jones, J. C. Fothergill, and C. D. Hanning, Analytical estimates of the characteristics of surface plasmon resonance fiber-optic sensors, *J. Mod. Opt.*, 47(6), 1099–1110, 2000.
38. A. K. Sharma and B. D. Gupta, Absorption-based fiber optic surface plasmon resonance sensor: A theoretical evaluation, *Sens. Actuat. B*, 100, 423–431, 2004.
39. M. Mitsushio, K. Miyashita, and M. Higo, Sensor properties and surface characterization of the metal-deposited SPR optical fiber sensor with Au, Ag, Cu, and Al, *Sens. Actuat. A*, 125, 296–303, 2006.
40. M. A. Butler and A. J. Ricco, Reflectivity changes of optically-thin nickel films exposed to oxygen, *Sens. Actuat.*, 19, 249–257, 1989.
41. M. A. Butler and A. J. Ricco, Chemisorption-induced reflectivity changes in optically thin silver films, *Appl. Phys. Lett.*, 53(16), 1471–1473, 1988.
42. M. A. Butler, A. J. Ricco, and R. J. Baughman, Hg adsorption on optically thin Au films, *J. Appl. Phys.*, 67(9), 4320–4326, 1990.
43. K. B. Pfeifer, R. L. Jarecki, and T. J. Dalton, Fiber-optic polymer residue monitor, in *Proceedings of the SPIE*, Boston, MA, 1998, Vol. 3539, pp. 36–44.
44. B. E. A. Saleh and M. C. Teich, *Fundamentals of Photonics*. John Wiley & Sons, Inc., New York, 1991, p. 18.
45. W. L. Wolfe, *Introduction to Radiometry*. SPIE Optical Engineering Press, Bellingham, WA, 1998, p. 17.
46. J. M. Palmer and B. G. Grant, *Art of Radiometry*. SPIE Press, Bellingham, WA, 2010, p. 27.
47. K. B. Pfeifer, S. M. Thornberg, M. I. White, and A. N. Rumpf, Surface plasmon sensing of gas phase contaminants using optical fiber, Albuquerque, NM: Sandia National Laboratories Report SAND2009-6096, 2009, p. 8.

48. E. Hecht and A. Zajac, *Optics*. Addison Wesley, Reading, MA, 1979, p. 74.
49. M. A. Ordal, L. L. Long, R. J. Bell, S. E. Bell, R. R. Bell, R. W. Alexander, and C. A. Ward, Optical properties of the metals Al, Co, Cu, Au, Fe, Pb, Ni, Pd, Pt, Ag, Ti, and W in the infrared and far infrared, *Appl. Optics*, 22(7), 1983, 1099–1119.
50. K. V. Rottkay, M. Rubin, and P. A. Duine, Refractive index changes of Pd-coated magnesium lanthanide switchable mirrors upon hydrogen insertion, *J. Appl. Phys.*, 85(1), 1999, 408–413.
51. B. Chadwick, J. Tann, M. Brungs, and M. Gal, A hydrogen sensor based on the optical generation of surface plasmons in a palladium alloy, *Sens. Actuat. B*, 17, 215–220, 1994.
52. M. A. Butler, Optical fiber hydrogen sensor, *Appl. Phys. Lett.*, 45(10), 1007–1009, 1984.
53. R. R. J. Maier, B. J. S. Jones, J. S. Barton, S. McCulloch, T. Allsop, J. D. C. Jones, and I Bennion, Fibre optics in palladium-based, hydrogen sensing, *J. Opt. A: Pure Appl. Opt.*, 9, S45–S59, 2007.

6 Photonic-Crystal Fibers for Sensing Applications

Ana M.R. Pinto

CONTENTS

6.1 INTRODUCTION

Optical fiber (OF) sensing systems emerged as a consequence of two huge scientific developments in the 1960s: the laser in 1960 and the low-loss OF in 1966. In the 1970s, the first experiments on low-loss OFs were performed, not for telecommunications but for sensing purposes. The main drive for OF's first application does not surprise us nowadays. OFs are efficient solutions for sensing due to their high sensitivity, small size, robustness, flexibility, ability for remote monitoring, and multiplexing. Another property of great importance for sensing applications is its aptitude for operational work in the presence of unfavorable environmental conditions such as noise, strong electromagnetic fields and high voltages, nuclear radiation, explosive or chemically corrosive media, and high temperatures. In these characteristics lies the recipe for the success of OF sensing systems: in undertaking difficult measurement situations where the use of conventional electrical sensors is not adequate.

The appearance of photonic crystal fibers (PCFs) was a breakthrough in fiber-optic technology, particularly in the sensing domain. Its characteristic structure enables a considerable enhancement in fiber design flexibility, providing a new breach for new and enhanced sensing solutions relative to the situation where the choice was limited to the standard OF technology. Despite its youth in the sensing field, PCFs have awakened the interest of many scientific groups due to their promising characteristics. The prime attraction of PCFs is related to the fact that by changing the size and location of the cladding holes and/or the core, properties like the fiber transmission spectrum, mode shape, nonlinearity, dispersion, air filling fraction, and birefringence can be tuned to reach values that are not achievable with conventional OFs. Additionally, the existence of air holes offers the opportunity to propagate light in air. As will be seen all along this chapter, PCF's diversity of features makes feasible a large number of new and improved applications in the sensing domain. In this chapter, PCF characteristics will be presented, concerning geometry and light guidance mechanisms, in order to understand its based sensing systems. An overview of physical, gas, and biochemical PCF-based sensors will be also made in this chapter.

6.2 PHOTONIC CRYSTAL FIBERS

Even though standard OFs present an excellent performance in fiber-optic telecommunications, their intrinsic properties have imposed restrictions in the evolution of this technology. The first evident restriction is the material selection for the core and cladding, in order to have matching thermal, chemical, and optical properties. Other limitations are related to its geometry and refractive index profile, which do not allow freely engineering fiber-optic characteristics such as inherent losses, dispersion, nonlinearity, and birefringence for progress in applications such as high-power lasers or fiber sensors. These limitations and restrictions had been refined during 30 years of exhaustive research [1,2], which made the scientific community desire for fibers that had more versatility.

In 1996, the first PCF was fabricated, rewarding scientists and engineers with the step forward in fiber-optic technology they had been so anxiously waiting. PCF geometry is characterized by a periodic arrangement of air holes running along the entire length of the fiber centered on a solid or hollow core. The major difference between common OFs and PCFs relies on the fact that the waveguide properties of PCFs are not due to spatially varying glass composition, as in conventional fibers, but from an arrangement of very tiny and closely spaced air holes that go through the whole fiber length. PCF's air holes provide the opportunity to create dramatic morphologic changes, which leads to enhanced characteristics (transmission, birefringence, nonlinearities, dispersion, etc.) as well as to the ability of fabricating a single material OF. Even more, PCFs also offer the possibility of light guiding in a hollow core, opening new perspectives in fields such as nonlinear fiber optics, fiber lasers, supercontinuum generation, particle guidance, and optical sensing [3,4].

6.2.1 Historical Overview

Since the 1980s, optical physicists recognized that the ability to structure materials to the scale of the optical wavelength would allow the development of new fibers. In 1991, Philip Russell first proposed to develop a new kind of fiber, which he initially called *holey fiber*, and it is now known as PCF [5]. The main idea was that there are photonic bandgaps (PBGs) in which there is a suppression of all optical vibrations within the range of wavelengths spanned by it, leading to a range of wavelengths for which no propagation states exist. So, if properly designed, a PCF cladding would be able to present the bandgap characteristic, allowing guidance in a hollow core [2,6]. The first successful PCF was materialized in November of 1995, even though it did not had the properties aimed (PBG guidance). Despite Philip Russell's primordial idea, this first PCF, presented to the world in 1996, had a solid core surrounded by a hexagonal array of small air holes running all through the fiber length and guided light through total internal reflection (TIR). This first exceptional structure analysis showed a means to provide enhanced interaction between light and gas (in the cladding air holes) in order to be used as a gas sensor or to study nonlinear effects [7]. Rapidly, the field of PCFs became extremely popular and numerous research groups all around the world started making research in this area. With PCFs, almost everything seemed feasible: having single-mode guidance at all wavelengths [8], not having a technological limit on the core size for monomode guidance [9], extremely high values of birefringence [10], guiding light in a hollow core [11], broad optical transmission bands [12], enhancement of gas-based nonlinear effects [13], and even controlling the dispersion with unprecedented freedom [14,15]. The discovery of these possibilities brought novel prospects for new and old fields of application in the fiber-optic technology.

6.2.2 Geometry

Conventional OF's geometry often entails a doped silica solid core surrounded by a pure silica solid cladding (Figure 6.1a), ensuring that the core refractive index is higher than the one of the cladding. PCF's geometry is characterized by a microstructured air hole cladding running along the entire length of the fiber, which surrounds a core that can be solid or hollow. As so, PCFs can be divided into two families based on their geometry: solid-core (SC) PCFs and hollow-core (HC) PCFs. SC PCFs present a solid core surrounded by a periodic array of microscopic air holes, running along its entire length (Figure 6.1b). HC PCFs present an air hole as the core, which is surrounded by a microstructured air hole cladding (Figure 6.1c).

During modeling, as well as during the manufacturing process, there are different physical parameters that can be controlled depending on the kind of fiber to be produced: for a standard OF, the only parameter to take into account is the diameter of the core -ρ-, while for a PCF, there are three physical parameters to be controlled—the core diameter -ρ- (which for SC PCF is defined as the diameter of the ring formed by the innermost air holes), the diameter of the air holes in the cladding -*d*-, and the pitch -Λ- (distance between the center of two consecutive air holes). These three physical parameters in combination with the choice of the material refractive index and the type of lattice make the fabrication of PCFs very flexible, opening up the possibility to manage its properties, leading to a freedom of design not possible with common fibers.

6.2.3 Guidance Mechanisms

Different geometric characteristics combined with different core/cladding materials in the fabrication will imply different structural designs such as to enable different guidance mechanisms:

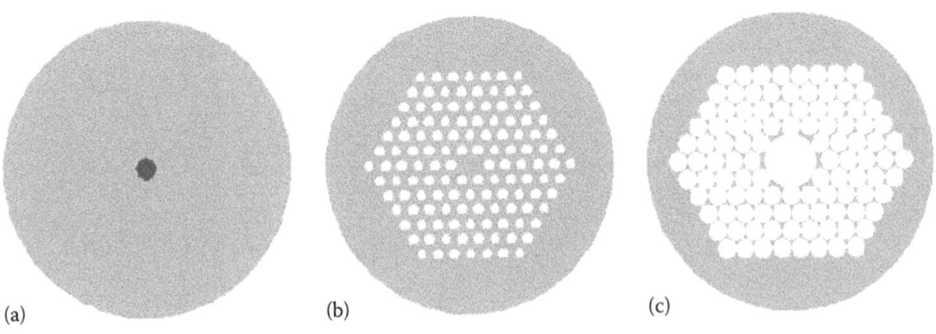

FIGURE 6.1 Illustration of the cross-sectional geometry of (a) conventional OF, (b) SC PCF, and (c) HC PCF. Colors: white, air; light gray, silica; and dark gray, doped silica.

TIR and/or PBG effect. Taking into consideration the guidance mechanism in PCFs, these can be divided into mainly four types of PCFs [4]:

1. Index-guiding PCF—guides in a solid core through TIR
2. Photonic bandgap PCF (PBPCF)—guides through PBG effect in a hollow core
3. All-solid PBPCF—guides by PBG in a solid core
4. Hybrid PCF—guides through PBG and TIR simultaneously

6.2.3.1 Index-Guiding PCF

Index-guiding PCFs present a cross section with an array of air holes surrounding a solid core. In Figure 6.2, an illustration of this type of PCF cross section, as well as its refractive index profile, is presented. Since the SC and the cladding have the same material, its cross-sectional configuration leads to a lowering of the cladding's effective refractive index, allowing the guidance mechanism to be the TIR. The cladding mode index is reduced due to the air holes in it, rather than by using different materials, as in common OFs. The core–cladding refractive index difference in index-guiding PCFs will be then much higher than in conventional OFs (typically 1%–2%).

For a common OF, the V-parameter depends on the core radius and the wavelength of operation. In index-guiding PCFs, this parameter depends in addition on the nature of the air hole arrangement [8,16]:

$$V_{PCF}(\lambda) = \frac{2\pi}{\lambda}\Lambda\sqrt{n_{co}^2 - n_{cl(PCF)}^2} \tag{6.1}$$

While for conventional OFs, the *V-number* rises without limit as the wavelength is decreased or as the core size is increased (resulting inevitably in high-order mode guidance), $V_{PCF}(\lambda)$ presents a sharp contrast in behavior. In the short wavelength limit ($\lambda \sim d$), $V_{PCF}(\lambda)$ is finite and independent of λ and Λ, being its limiting value dependent on the relative size of the air holes. In the long wavelength limit, where λ is far greater than the characteristic distance scales of the PCF (d and Λ), the optical field in the cladding region can only be very weakly modulated by the air holes [17]. Because of this unusual interaction of the guided mode with the photonic crystal cladding, monomode waveguiding is possible over most of the transmission window, startling a feature that appears to be unique to this type of structure—*endlessly single-mode (ESM) guidance*. The ESM behavior is characterized by certain d/Λ ratios [8,17]. Therefore, it is always possible to tailor d/Λ in order to obtain $V < V_{PCF}$, thus ensuring single-mode guidance in the PCF.

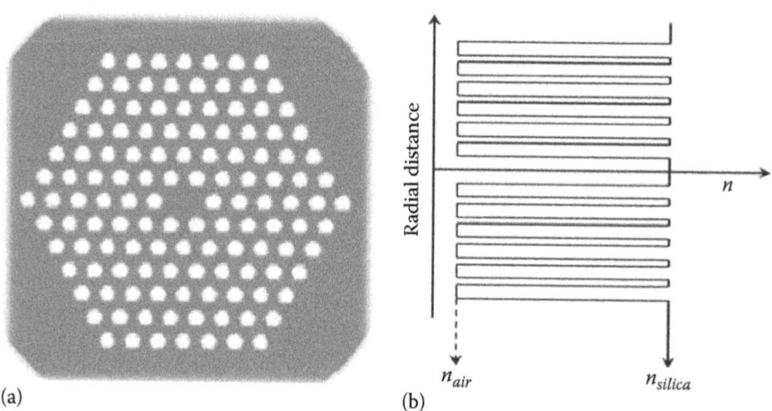

FIGURE 6.2 Illustration of (a) SC PCF cross section and (b) respective refractive index profile. Colors: white, air, and light gray, silica.

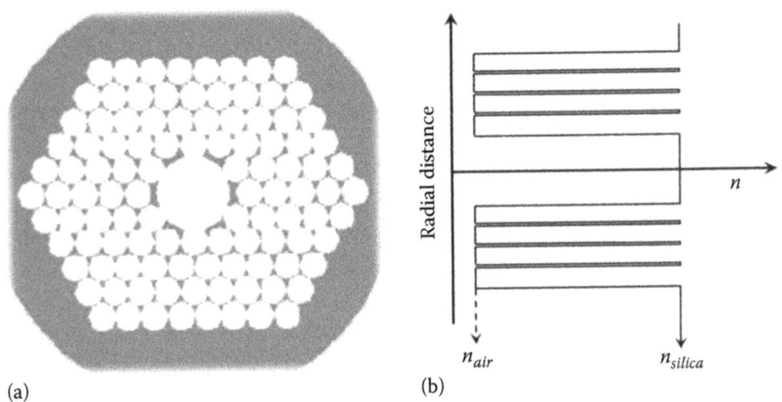

FIGURE 6.3 Illustration of (a) HC PCF cross section and (b) respective refractive index profile. Colors: white, air, and light gray, silica.

6.2.3.2 Photonic Bandgap PCF

PBPCF cross section presents a periodic array of air holes surrounding a hollow core. This cross-sectional structure leads to a negative core–cladding refractive index difference (as can be seen in Figure 6.3), as so they cannot operate via TIR.

Nevertheless, an appropriately designed photonic crystal cladding can prevent the light to escape from the hollow core. Periodically distributed air holes can form a 2D photonic crystal structure with lattice constant similar to the wavelength of light. In 2D crystal structures, PBGs exist that prevent the propagation of light within a certain range of frequencies. If the periodicity of the structure is broken with a defect, a special region with different optical properties can be created. The defect region can support modes with frequencies falling inside the PBG, but since around this defect lies the PBG, the light within the defect will remain confined in the vicinity of the defect. The modes falling outside the defect will be refracted, whereas the modes falling inside the defect region will be strongly confined to the defect and guided throughout the entire length of the fiber [18]. This effect is illustrated in Figure 6.4. Suppose an all-silica HC PCF is designed to work in the dark gray visible region of the electromagnetic spectrum.

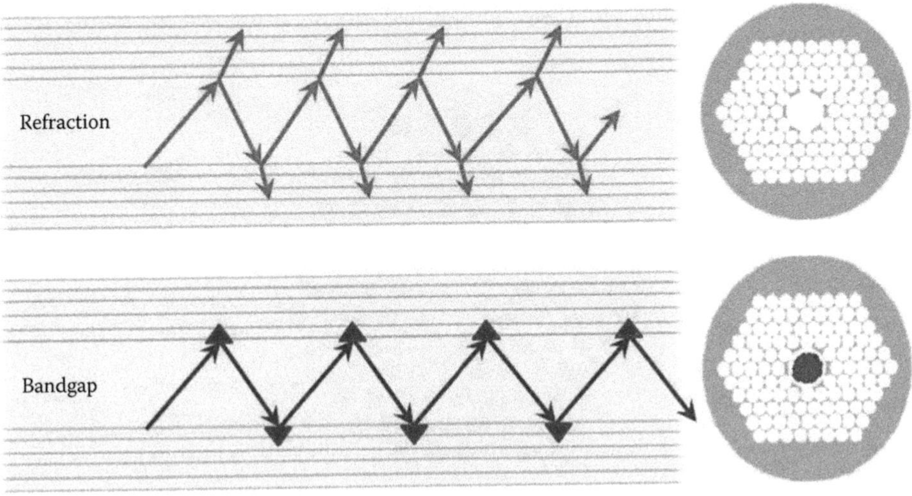

FIGURE 6.4 Refraction and bandgap effect in an HC PCF. Frequencies falling outside the range of guided frequencies are reflected (light gray lines); and frequencies in the range of accepted frequencies are tunneled inside the HC PCF by the bandgap effect (dark gray lines).

When the PCF is illuminated by frequencies falling outside this range (light gray in Figure 6.4), all light will be refracted and no light will be guided by the fiber. On the other hand, if the PCF is illuminated by a broadband light source, only the blue light (represented by the dark gray in Figure 6.4) will be guided throughout the fiber appearing at its end, while all other frequency components of light will be refracted.

This kind of guidance mechanism allows light guidance in air, not possible with standard OFs. Moreover, it presents noteworthy advantages like less interaction between guided modes and fiber material, allowing transmission powers not possible with conventional OFs; extremely small Fresnel reflections, due to the low refractive index discontinuity between the outside environment and the fiber mode; and even more, the ability to fill up the hollow core with gases or liquids, which enables a well-controlled interaction between light and sample leading to new sensing applications that could not ever be considered with standard OFs. These advantages open up fascinating applications based in HC PCFs: high power transmission [19], gas-based nonlinear optics [20], optical tweezers propulsion and particle guidance in liquids [21,22], photochemical reactions [23,24], gas detection [25,26], and selective sensing of antibodies [27].

6.2.3.3 All-Solid Photonic Bandgap PCF

Although the PBG can allow guidance in air, this effect can also be observed in a solid core. PBG guidance can be accomplished in an SC PCF by an arrangement of high-index solid inclusions around a low-index core. The first all-solid PBG fiber was presented in 2004 [28], where the bandgap formation was attributed to antiresonances of the high-index strands in the cladding. An illustration of an all-solid PBG can be found in Figure 6.5.

The transmission properties of solid PBPCFs are determined by the PBGs of the cladding material, which can be described by the antiresonant reflecting optical waveguide (ARROW) model [29].

The spectral transmission through the low-index core takes the form of a number of high- and low-loss regions as the various higher-order modes of the cladding are cut off [30]. All-solid PBPCFs main features include low attenuation, straightforward fabrication, unusual dispersion characteristics, and guidance in a range of fixed frequency windows [4]. The unusual spectral properties of these fibers lead to interesting applications, such as high-power fiber lasers [31].

6.2.3.4 Hybrid PCF

A hybrid PCF is a PCF that presents two types of guidance mechanisms simultaneously: TIR and PBG. Despite the fact that the first PCFs developed guided light only through one of the two different guiding mechanisms, other waveguides had already presented the ability for hybrid

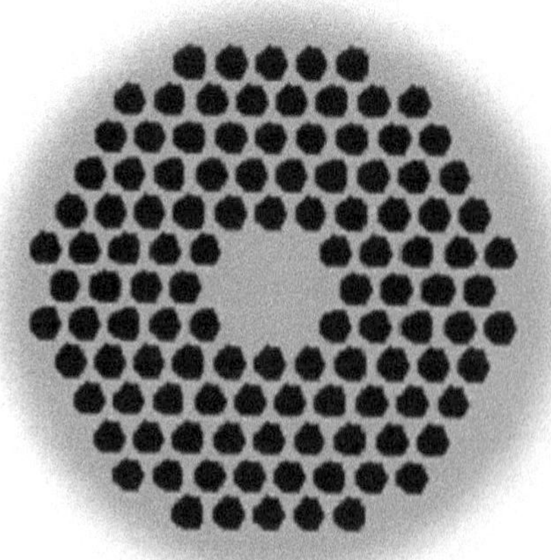

FIGURE 6.5 Illustration of an all-solid PBG cross section. Colors: light gray, low refractive index material, and dark gray, high refractive index material.

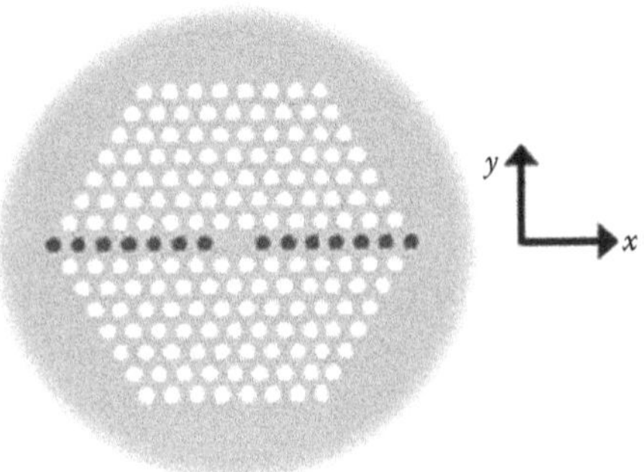

FIGURE 6.6 Illustration of the cross section of a hybrid PCF. Colors: light gray, low refractive index material; white, air; and dark gray, high refractive index material.

guiding [32]. The first hybrid PCF appeared in 2006 [33], composed by air holes arranged in a hexagonal pattern, where a single row of air holes was replaced by a high refractive index material, both surrounding a silica core, as can be seen in Figure 6.6.

Taking Figure 6.6 as reference, guidance of light in a hybrid PCF can be explained as follows:

- *Along the y-axis*: The core has a higher refractive index than the cladding effective refractive index, as so light can be guided by TIR.
- *Along the x-axis*: The cladding rods have a higher index than the core. Consequently, TIR is not possible. Light confinement only occurs at restricted frequency bands, which coincide with the PBGs, which occur at the antiresonant conditions of the high-index inserts.

The hybrid PCF flexibility of the design allows adjusted birefringence, which can be applied for strain and temperature measurement [34], as well as broadband single-polarization guidance for interferometric sensors and fiber lasers [35].

6.3 PHYSICAL SENSING

Physical sensors measure physical parameters such as pressure, temperature, strain/displacement, and electric field. The measurement, monitoring, and control of these parameters are of vast interest for several applications, such as structural health monitoring of structures like buildings, piles, bridges, pipelines, tunnels, dams, and high-voltage areas [36]. OF sensors are perfect for these purposes, since they provide *in situ*, real-time continuous measurement and analysis of key structural and environmental parameters under operating conditions. In this section, several physical sensors will be presented. Unfortunately, due to the limited length of this chapter, it is not possible to make an overview of all PCF-based sensors. For more information on sensors not covered in this chapter, I encourage the reader to examine Ref. [37].

6.3.1 Pressure Sensors

Pressure monitoring is required for several purposes such as control of structural health of civil structures even in extremely harsh environments as turbine engines, compressors, oil and gas exploitations, power plants, and material processing systems. Using periodically tapered long-period gratings (LPGs) written in an ESM PCF, measurements of hydrostatic pressure up to 180 bar showing a pressure sensitivity of 11.2 pm/bar were carried out [38]. By using a fiber Bragg grating (FBG) in a two-hole PCF, a sensitivity of 17.6–26 pm/(N/cm) was shown [39]. By embedding a highly birefringent (Hi-Bi) PCF, with an FBG inscribed, in a reinforced composite material, a sensitivity of 15.3 pm/MPa was achieved [40]. A four-hole suspended-core fiber (SCF) was inscribed with FBGs demonstrating sensitivities going from 2.19 pm/(N/cm) up to 12.23 pm/(N/cm) depending on where the load was applied [41].

Several authors reported polarimetric measurements of pressure based in the commercial Hi-Bi PCF: an intensity measurement of pressure with a sensitivity of 2.34×10^{-6} MPa^{-1} was demonstrated [42]; and a wavelength measurement of pressure variation was shown to provide a sensitivity of 3.38 nm/MPa [43], which later was applied to tsunami detection in the ocean bottom [44]. A photograph of the cross section of a Hi-Bi PCF and a representation of a polarimetric setup can be seen in Figure 6.7. Other polarimetric sensors were developed based in homemade Hi-Bi PCFs: A specially designed fiber showed a pressure of −10 rad/(MPa·m) [45]; by using two fibers with a small number of cladding holes with different diameters, in order to induce birefringence, a sensitivity up to 23 rad/MPa·m was obtained [46]; and using two different germanium (Ge)-doped-core Hi-Bi PCFs to measure pressure, a sensitivity that exceeded 43 rad/MPa·m was demonstrated [47]. A Hi-Bi PCF with larger air holes on one axis presented a sensitivity of ~2.17 nm/(N/cm) [48].

A pressure sensor based on Sagnac loop filter using Hi-Bi PCF operating at 850 nm was demonstrated using a frequency and phase detection technique, showing an improvement in pressure sensitivity of about three times when compared with sensors operating at 1550 nm [49]. Also, using a Hi-Bi PCF–based Sagnac interferometer between two polymer foils, a pressure sensitivity of 1.764 nm/MPa was obtained [50]. Using different types of SC PCFs in a Sagnac configuration, individually and spliced together, it was shown that the combinations of these fibers, spliced together make good sensors, thereby offering the best sensitivity for pressure sensing [51]. Using a commercial Hi-Bi PCF coiled inside a Sagnac loop mirror, a pressure sensor with a sensitivity of 4.21 and 3.24 nm/MPa at 1320 and 1550 nm, respectively, was obtained [52]. Two Sagnac interferometers were developed based on the same large-mode-area (LMA) PCF: In a PCF coil, a sensitivity of 90.4 nm/mm was obtained [53], and by direct measurement of the transmission spectrum shift, a sensitivity of 0.519 nm N^{-1}/mm was achieved [54]. Through the use of a four-hole SCF, a Fabry–Perot (FP) interferometer was made with a sensitivity of 4.68×10^{-5} nm/psi, and a Sagnac interferometer was obtained with 0.032 nm/psi of sensitivity [55]. An interferometric FP cavity based on HC ring PCF exhibited a sensitivity of 0.82 nm/MPa to pressure variations [56].

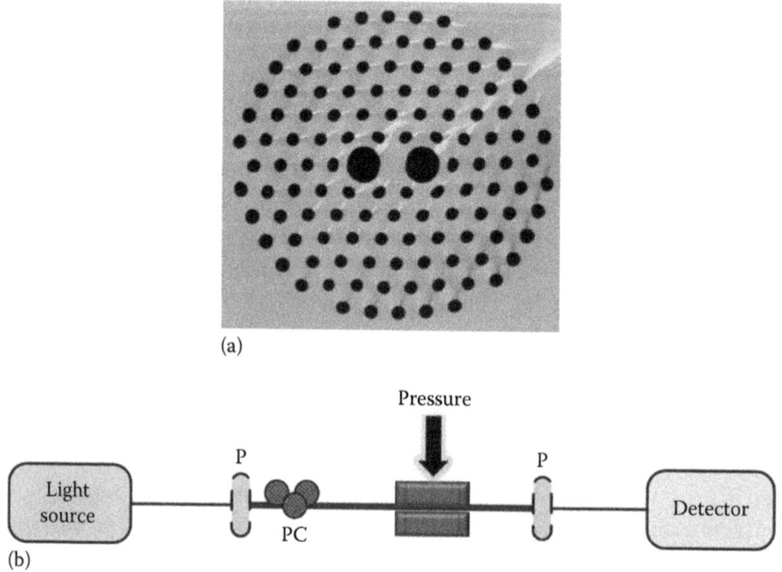

FIGURE 6.7 (a) Microscope photograph of a Hi-Bi PCF cross section (Courtesy of the Max Planck Institute for the Science of Light, Erlangen, Germany) and (b) schematic illustration of a polarimetric pressure sensing setup: P, polarizer; PC, polarization controller; hairline, SMF; and thick line, PCF.

6.3.2 Temperature Sensors

Fiber-optic temperature sensors have proven to be very useful in areas such as glass productions, furnaces of all sorts, semiconductor industry, high-temperature processing, chemical industries, power generation operations, as well as civil, aerospace, and defense industries. This kind of fiber-optic sensor is one of the most required in the commercial market due to its vast number of applications. Using modal interference, several sensors were reported: Using an LMA PCF spliced to a single-mode fiber (SMF), a three-beam path modal interferometer with a sensitivity of 8.17 pm/°C was reported [57]; by tapering a silica-core PCF, a sensitivity of 12 pm/°C was obtained for measurements up to 1000°C [58]; temperature sensing was accomplished using a core offset of a nonlinear PCF in between a multimode fiber (MMF) and an SMF (sensitivity ~73 pm/°C) [59]; and a sensitivity of 78 pm/°C was exhibited by a modal interferometer using a very short PCF stub with a shaped Ge-doped core [60].

Sagnac interferometers were built making use of filled PCFs: Using a two-hole birefringent PCF filled with metal indium, it was possible to achieve a sensitivity of 6.3 nm/K [61]; by filling a Hi-Bi PCF with alcohol, a sensitivity as high as 6.6 nm/°C was reached [62]; and by using a selectively filled Hi-Bi PCF, a sensitivity of 2.58 nm/°C was obtained [63]. Hybrid FP structures were also developed: Two FP cavities were formed by fusion splicing a PCF with a short piece of HC fiber and an SMF in series [64]; by splicing an SMF to a short piece of SCF and interrogating it with a dual-wavelength Raman fiber laser, unambiguous temperature recovery (sensitivity ~0.84 deg/°C) was possible [65]; and by inserting the FP interferometer in a laser cavity as a mirror for simultaneous sensing and lasing, a sensitivity of ~6 pm/°C was achieved [66].

Mach–Zehnder interferometers (MZIs) were reported: Using LPGs in a PCF, a sensitivity of 42.4 pm/°C was achieved [67]; through an all-solid PBPCF (doped fiber), enhancement of temperature sensitivity (71.5 pm/°C) was demonstrated [68]; and by an in-line fully liquid-filled PCF, a sensitivity of 1.83 nm/°C was obtained [69]. The inscription of LPGs in an SC PCF showed a sensitivity of 10.9 pm/°C [70]. A polarimetric interrogation of a Hi-Bi PCF showed a sensitivity of 0.136 rad/°C at 1310 nm [71].

By filling a fiber with quantum dots (QDs), two different temperature sensors were developed: by depositing QD nanocoatings using the layer-by-layer technique in the inner holes of an all-silica LMA ESM PCF with a resultant sensitivity of 0.1636 nm/°C [72] or by inserting luminescent CdSe/ZnS nanocrystals in a hollow core of a PCF obtaining a sensitivity of 70 pm/°C [73]. By filling the cladding air holes of an SC PCF with Fe_3O_4 nanoparticle fluid, a sensitivity of 0.045–0.06 dB/°C was accomplished [74]. A 10 cm SC PCF filled with ethanol was reported with a sensitivity of 0.315 dB/°C [75]. When selectively infiltrating one of the air holes of an SC PCF with a 1.46 refractive index liquid, the sensitivity was proved to be ~54.3 nm/°C [76]. By selective liquid filling an index-guiding PCF, a temperature sensitivity of 5.5 nm/°C was experimentally confirmed [77].

Distributed Brillouin temperature sensing was also accomplished using PCFs: through the use of a germanium-doped-core birefringent PCF with variations of sensitivity of 0.96–1.25 MHz/°C [78] and by using the birefringent effect of a transient Brillouin grating in a polarization-maintaining PCF leading to a sensitivity of 23.5 MHz/°C [79].

6.3.3 Strain/Displacement Sensors

Monitoring strain/displacement-induced changes is important for several application areas such as experimental mechanics, aeronautics, metallurgy, and health monitoring of complex structures, among others. A polarimetric strain sensor based in an all-silica Hi-Bi PCF was demonstrated with a sensitivity of 1.3 pm/με [80]. An in-reflection strain sensor based in a Hi-Bi PCF demonstrated a sensitivity of ~7.96 dB/mε [81]. Using a short piece of HC PCF in reflection, sensitivities of 0.22, 0.685, and 0.727 μW/μm were obtained for cavities with 5, 2.5, and 1.25 cm, respectively [82] (Figure 6.8).

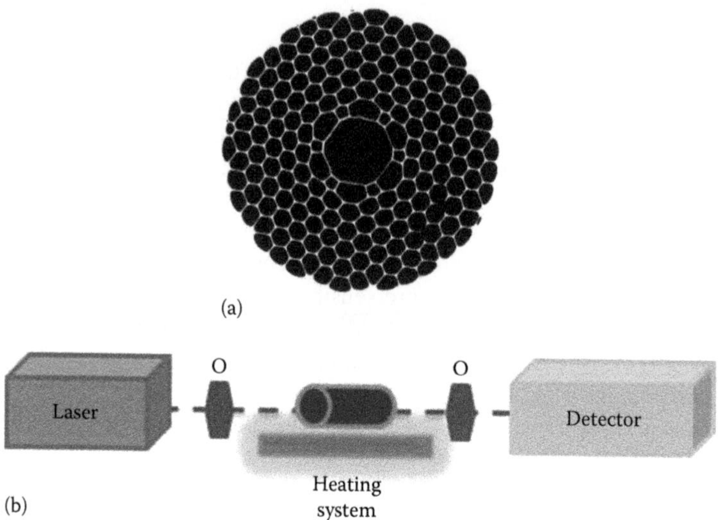

FIGURE 6.8 (a) Microscope photograph of an HC PCF cross section (Courtesy of the Max Planck Institute for the Science of Light, Erlangen, Germany) and (b) illustration of an experimental setup for temperature sensing by filling the PCF core: O, objectives.

A modal interferometer was constructed through tapering an SC silica PCF for strain sensing [83]. A ~2.8 pm/με sensitive wavelength encoded strain sensor was reported, in which the sensing head worked through the modal interference obtained by splicing a piece of PCF to an SMF [84]. Another modal interferometer was obtained through an SMF–PCF–SMF structure with a core offset at one of the joints presenting a sensitivity of 0.0024 dB/μm [85].

Strain sensors developed through Hi-Bi PCF Sagnac interferometers were reported showing temperature insensitivity, using wavelength-based measurement (~1.11 pm/με) [86] and intensity-based measurement (~0.0027 dB/με) [87]. Using a Hi-Bi PCF in a Sagnac interferometer, a displacement sensor was reported with a sensitivity of 0.28286 nm/mm [88]. Through the use of a three-hole SCF in a Sagnac configuration, a displacement sensor was developed with high precision (~0.45 μm) [89]. Figure 6.9 illustrates a Sagnac loop mirror setup for strain/displacement measurement as well as a photograph of the cross section of an SCF.

Miniature in-line FP interferometers were also accomplished for strain sensing: by splicing a small length of HC PBG fiber between two SMFs (sensitivity of 1.55 pm/με) [90] or even by

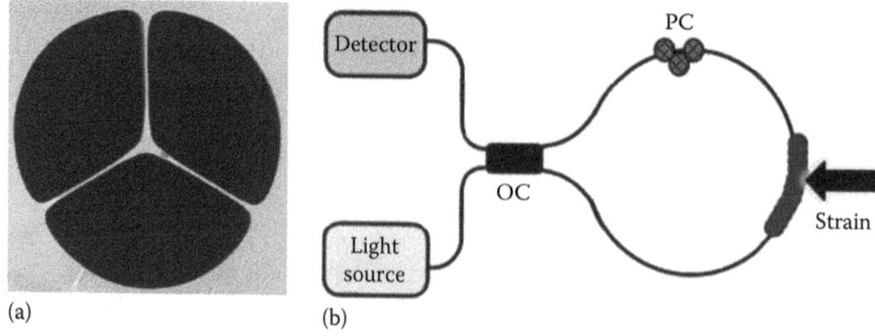

FIGURE 6.9 (a) Microscope photograph of an SCF (or Mercedes fiber) cross section (Courtesy of the Max Planck Institute for the Science of Light, Erlangen, Germany) and (b) schematic illustration of a Sagnac loop mirror for displacement sensing: PC, polarization controller; OC, optical coupler; hairline, SMF; and thick line, PCF.

multiplexing several FP interferometers based in HC-PBPCFs between two SMFs to obtain a strain sensing system [91].

An MZI was fabricated by splicing a short length of PCF between two SMFs with collapsed air holes over a short region in the two splicing points, which showed a high strain sensitivity of ~0.21 μs^{-1}/mε and a minimum detectable strain of ~3.6 με [92]. A strain sensor based in an MZI using a twin-core PCF was developed, achieving a sensitivity of 0.31 pm/με [93].

6.3.4 Electric- and Magnetic-Field Sensors

Electric- and magnetic-field sensing is of extreme importance in the electric power industry. Due to the metallic content of conventional sensors, they can potentially perturb the measured parameter. Fiber-optic sensors have proven themselves in these harsh environments due to their high sensitivity, wide bandwidth, high operation temperature, immunity to electromagnetic interference, lightweight, and long life. Most importantly, they can provide true dielectric isolation between the sensor and the interrogation system in the presence of very high electromagnetic fields. A polarimetric sensing scheme with selectively liquid crystal (LC) infiltrated in a Hi-Bi PCF was demonstrated for electric-field sensing with a sensitivity of ~2 dB/kV$_{rms}$/mm [94]. An LC infiltrated PCF probe was also demonstrated, showing higher sensitivity to the electric-field component aligned along the Hi-Bi PCF axis [95]. An intensity measurement–based electric-field sensor was reported by infiltrating an LMA PCF with LC demonstrating a sensitivity of ~10.1 dB/kV$_{rms}$/mm in transmission and ~4.55 dB/kV$_{rms}$/mm in reflection [96]. The development of a spun elliptically birefringent PCF [97] showed the advantages of this PCF over conventional spun stress birefringence fibers. Figure 6.10 presents an illustration of an LC-infiltrated PCF probe for electric-field measurement.

The development of a magneto-optic Faraday effect in a miniature coil wound from a six-hole spun PCF [98] provided information about the ability of this PCF to efficiently accumulate Faraday phase shift in a magnetic field. By using a Hi-Bi PCF injected with a small amount of Fe_3O_4 nanofluid, a sensitivity of 242 pm/mT was shown [99]. A magnetic-field sensor based on the integration of a Hi-Bi PCF and a composite material made of Terfenol particles and an epoxy resin was demonstrated with a sensitivity of 0.006 nm/mT over a range from 0 to 300 mT with a resolution of ±1 mT [100]. A magnetic-field sensor with ~33 pm/Oe of sensitivity was obtained using an HC PCF [101]. A Ge-doped PCF filled with magnetic fluid presented a maximum resolution of 4.98 Oe [102]. By filling the air holes of the cladding layer of an index-guiding PCF, a resolution up to 0.09 Oe was achieved [103]. A magnetic field sensor based on the magnetic fluid and MZI achieved a sensitivity of 2.367 pm/Oe [104]. Through the combination of magnetic fluid and tunable PBG, a sensitivity of 1.56 nm/Oe was reported [105].

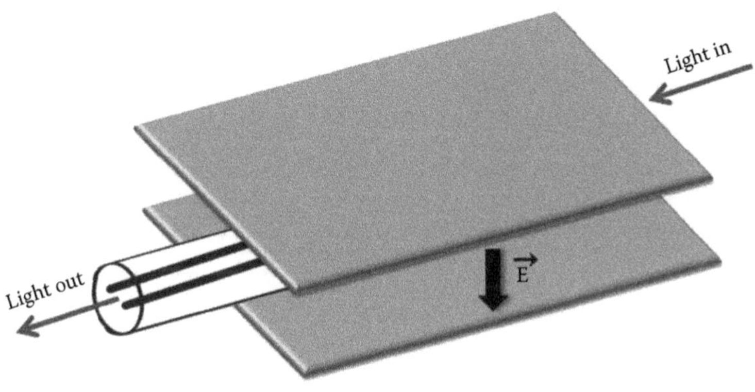

FIGURE 6.10 Illustration of an LC infiltrated in a Hi-Bi PCF probe for electric-field sensing (black lines represent the LC filling the big holes of the fiber).

6.4 GAS SENSING

Industries that work in chemical processing, glass melting, metal casting, paper, and energy produce different amounts and types of gaseous emissions. In the global environmental awareness context, it is ever more important to monitor and control gas emission. There are even other industries (chemical, biochemical, and military) for which gas diffusion is an important parameter to analyze. Therefore, it is of great importance to develop gas sensing techniques that are selective, quantitative, fast acting, and not susceptible to external poisoning.

6.4.1 Acetylene Sensors

Acetylene is highly flammable (normally used in welding) and unstable in its pure form, hence the importance of its detection. An SC hexagonal lattice PCF was used for evanescent-wave absorption in acetylene detection [106], as well as an SC random-hole PCF using the *in situ* bubble formation technique [107]. Positive results were also reported using a three-hole SCF [108] and, even more, by using an SC PCF with FBGs inscribed, presenting a sensitivity of ~0.017–0.022 dB/% to acetylene concentration [109]. By measuring the attenuation of the guided light due to the absorption by the gas sample, gas diffusion inside of an HC PCF was successfully monitored [110]. The use of saturation absorption spectroscopy inside of a large-core HC PCF (20 μm core) showed narrower transition and cleaner signals than when using smaller-core HC PCFs [24]. A cell containing an acetylene volume lower than 5 μL, made through filling an HC PCF with gas and then splicing it to SMFs, was used to measure its concentration through correlation absorption spectroscopy [111]. Acetylene absorption lines were detected using a C-type fiber and a Ge-doped ring PCF [112]. Through the measurement of the gas diffusion coefficient, acetylene and air were detected in an SC PCF [113].

6.4.2 Methane Sensors

Detecting methane is of special relevance in many industrial and safety applications. Using absorption spectroscopy, a methane sensor working at 1670 nm was developed inside of an HC PCF with a minimum detection of 10 ppmv [114]. Using similar PCFs as gas cells, a sensitivity of 49 ppmv-m at 1300 nm was reported [115], as well as two other different wavelength bands in the near-infrared region [116]. And through periodic side opening microchannels in an HC PCF (see illustration of Figure 6.11), a fast response methane sensor was demonstrated with a sensitivity of ~647 ppm [117].

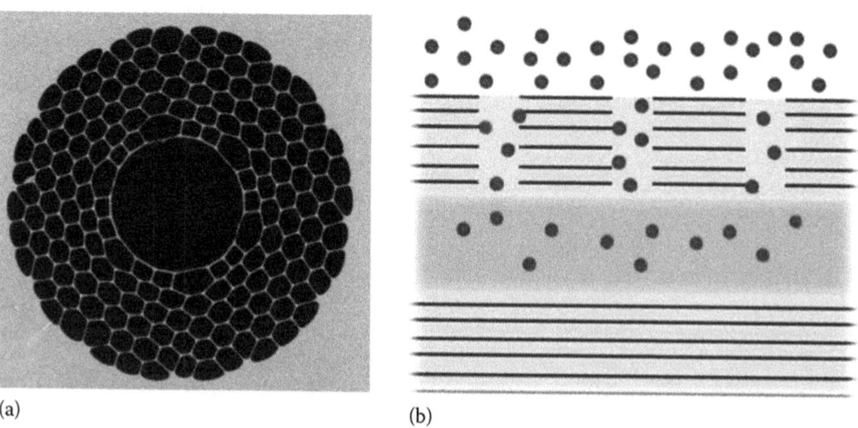

FIGURE 6.11 (a) Microscope photograph of an HC PCF cross section (Courtesy of the Max Planck Institute for the Science of Light, Erlangen, Germany) and (b) schematic drawing of side opening microchannels in an HC PCF allowing gases or liquids to diffuse in its core; while light remains trapped in the core. Colors: dark gray, gas particles, and light gray, light.

A polarization-maintaining PCF-based Sagnac loop filter was developed for methane concentration detection with a sensitivity of 410 ppm [118].

6.4.3 Other Gases

The measurement of other gases has also been done via PCFs. Using a tapered SC PCF with collapsed air holes coated with thin layers, a hydrogen sensor was reported [119]. The detection of volatile organic compounds was demonstrated, without the need of any permeable material, through an in-reflection PCF interferometer [120]. The characterization of acetylene, hydrogen cyanide, methane, and ammonia was shown using a PBPCF spliced to an SMF in one end and filled with gas through the other [121]. A gas cell was developed based in an HC PCF, achieving minimum detectable concentration values of 300 ppm for carbon dioxide and 5 ppm for acetylene [122]. Saturated absorption in overtone transitions of acetylene and hydrogen cyanide molecules confined in a PBPCF was observed [123]. Through spontaneous Raman scattering, the detection of methane, ethane, and propane was accomplished in a PBPCF [124], as well as oxygen and nitrogen [125]. The implementation of a PBPCF-based gas sensor was demonstrated for the detection of methane and acetylene gas [126]. A sensor based in two pieces of PBPCF with drilled lateral microchannels for gas access was reported [127], showing the ability to quantitatively measure gas mixtures. An oxygen gas sensor was also reported using an array of microholes in an SC PCF, with a sensitivity of 10.8 (defined as I_0/I_{100}) and a quick response time (~50 ms) [128]. An HC PCF was used for the detection of the concentration of volatile organic compounds through the Rayleigh scattering effect, showing a sensitivity of 0.022 dB/ppm to ethanol [129]. Through the combination of Raman spectroscopy and an HC PCF, a highly sensitive detection of various gases (ambient nitrogen, oxygen, and carbon dioxide) and vapors (toluene, acetone, and 1,1,1-trichloroethane) was reported [130].

6.5 BIOCHEMICAL SENSING

OF-based sensors are small in size and flexible with remote sensing ability, making them suitable for *in vivo* experiments. Even more, due to the fact that these waveguides are electrically passive, they do not represent a risk for patients during medical exams, allowing real-time and multiparameter measurement. A wide range of optical sensors have been developed for selective biomolecule detection. Most of them have reliability issues as they employ very fragile antibodies as sensing elements [131].

PCFs offer a number of exceptional advantages for biochemical sensing applications: PCFs have the unique ability to accommodate biological and chemical samples in gaseous or liquid forms in the immediate vicinity of the fiber core (air holes) and/or even inside the core, allowing simultaneously for fluidic channel role and light guidance with high light/sample overlap; the air holes can be functionalized with biorecognition layers that can bind and progressively accumulate target biomolecules, thus enhancing sensor sensitivity and specificity; and more importantly, the volume needed for sample analysis using PCFs is of the order of hundreds of nanoliters to tens of microliters, while conventional optics measurement techniques need volumes of order of 1–10 mL [132,133], as illustrated in Figure 6.12.

6.5.1 Refractive Index Sensors

Refractive index is a fundamental material property. Industries like food and beverage, medicine, and biology need the accurate measurement of the refractive index as a part of its quality control and development. A tapered LMA PCF refractometer was developed with a resolution of 1×10^{-5} RIU [134] and another one with 1600 nm/RIU of sensitivity using only 2 cm of PCF [135]. Another approach taken was an in-reflection PCF interferometer obtained by splicing an SMF to an

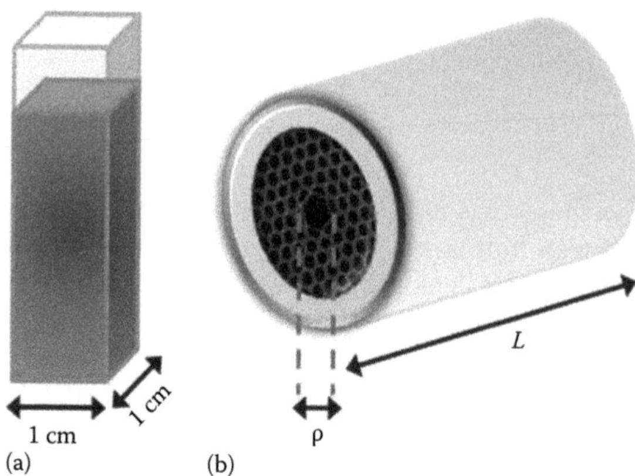

FIGURE 6.12 Illustration of (a) 1 cm cuvette, normally using volumes of 1–5 mL, and (b) kagome PCF (Courtesy of the Max Planck Institute for the Science of Light, Erlangen, Germany) that can be used as a nanoreactor, typically using volumes from nL to μL. ρ, PCF core size (typically tens of μm), and *L*, length of PCF (typically a few cm). Figures not to scale.

LMA PCF [136]. An ultraresponsive PCF-based refractive index sensor was reported with a sensitivity of 30100 nm/RIU, by inserting the fluid in one of the air holes adjacent to the core [137].

Modal interferometers were developed for refractive index sensing: based on an LMA spliced between two SMFs presenting a maximum resolution of ~2.9×10^{-4} RIU [138] and another based in two large-core air-clad PCFs spliced in between SMFs in series, presenting a resolution of ~3.4×10^{-5} RIU [139]. An MZI refractometer was also developed using an LMA PCF in a cavity ringdown loop, presenting a resolution of 7.8×10^{-5} RIU [140]. A refractometer reaching a sensitivity of 320 nm/RIU was reported based in PCF interferometry, proving aptitude for biosensing [141]. Taking advantage of four-wave mixing in a PCF, a high sensitivity of ~8.8×10^3 nm/RIU was obtained [142].

A refractometer was accomplished by observing the wavelength shift, due to refractive index variations, falling inside the PBG of a PBPCF (resolution of 2×10^{-6} RIU) [143]. A refractive index sensor based in LPGs inscribed in an ESM PCF (in Figure 6.13) was reported presenting a wavelength sensitivity of 440 nm/RIU and an intensity sensitivity of 2.2 pm/°C [144], presenting suitable characteristics for applications such as label-free biosensing [145]. By inscribing LPGs in air- and water-filled SC PCFs, a sensitivity as high as ~10^{-7} RIU was shown [146]. A highly sensitive approach (1500 nm/RIU) was reported based in LPGs inscribed in an LMA PCF [147]. Figure 6.13 presents an illustration of a PCF with LPGs inscribed and its principle of work. Through a short piece of LMA PCF spliced between MMFs and surface plasmon resonance, a sensitivity of 731%/RIU to refractive index was accomplished [148].

6.5.2 Humidity and pH Sensors

Humidity measurements are required for meteorological services, chemical and food processing industry, civil engineering, air-conditioning, horticulture, and electronic processing. A humidity sensor was developed through the use of an LMA PCF–based interferometer operating in reflection, with a sensitivity of 5.6–24 pm/%RH [149]. Using an air-guided PCF spliced to an SMF on one side and open in the other, and using direct absorption spectroscopy, humidity detection with a resolution of 0.2%RH was obtained [150]. An LMA PCF–based interferometer was proposed for humidity measurement showing sensitivities between 60.3 and 188.3 pm/%RH [151]. Through polyvinyl alcohol–coated PCF, different sensors were obtained: by collapsing the holes of the PCF in both ends (sensitivity of 0.6 nm/%RH) [152] and by splicing a short length of the PCF between two SMFs (sensitivity of 40.9 pm/%RH) [153]. A high sensitive humidity sensor based on agarose-coated PCF

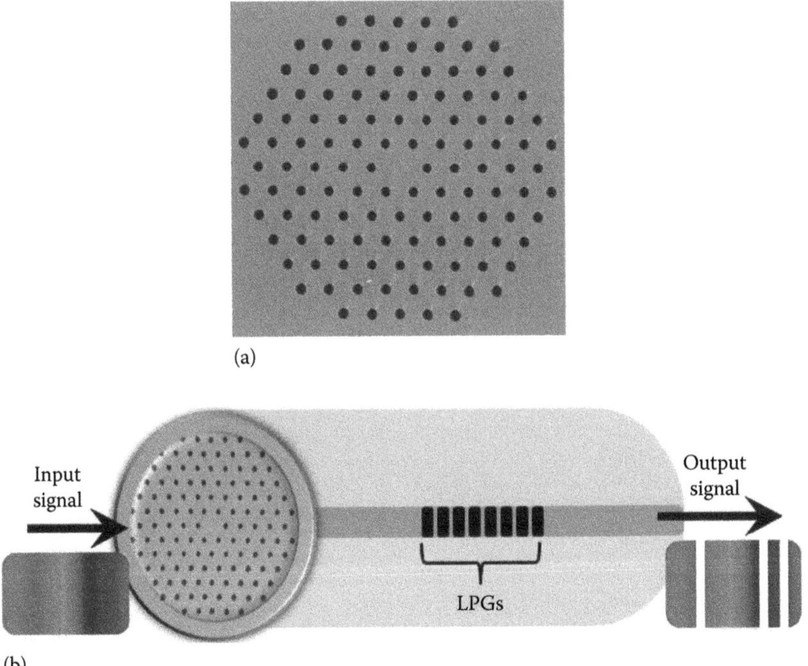

FIGURE 6.13 (a) Microscope photograph of an LMA ESM PCF (Courtesy of the Max Planck Institute for the Science of Light, Erlangen, Germany) and (b) schematic drawing of LPGs inscribed in a PCF and its principle of work.

showed a wavelength shift of 56 nm for a humidity change of 58%RH [154]. And by infiltrating agarose in a PCF, an interferometer was obtained that presented a change in its reflected power of 12 dB for a humidity change of 84%RH [155].

Application fields such as medicine, environmental sciences, agriculture, food science, or biotechnology need to screen pH. The sensing process is carried out in the air holes of the PCF, which greatly enhances the specific surface area for sensing, allowing microanalysis. Using a pH-sensitive fluorescence dye-doped cellulose acetate thin-film modified polymer PCF, a pH sensor was demonstrated [156], showing the possibility to tailor the pH response range through doping the sensing film with a surfactant.

6.5.3 Protein Sensors

The detection of specific protein entails the immobilization of antibodies for selective binding of antigens and/or fluorescent labeling of such proteins. Protein concentration was measured labeling it with QDs and using fluorescence in a soft-glass three-hole SCF, with a detection limit of 1 nM and extremely small sample volume (of the order of 10 picoliter) [157]. Through the immobilization of antibodies within the holes of a three-hole glass SCF, detection of proteins was demonstrated [158]. The recognition of proteins throughout binding with antibodies was made through fluorescence labeling, opening up the possibility for the measurement of multiple biomolecules via immobilization of multiple antibodies.

6.5.4 Rhodamine Sensors

Rhodamine is often used as a tracer dye, being extensively used in biotechnology applications such as fluorescence microscopy and flow cytometry. A double-clad PCF was used to detect fluorescence

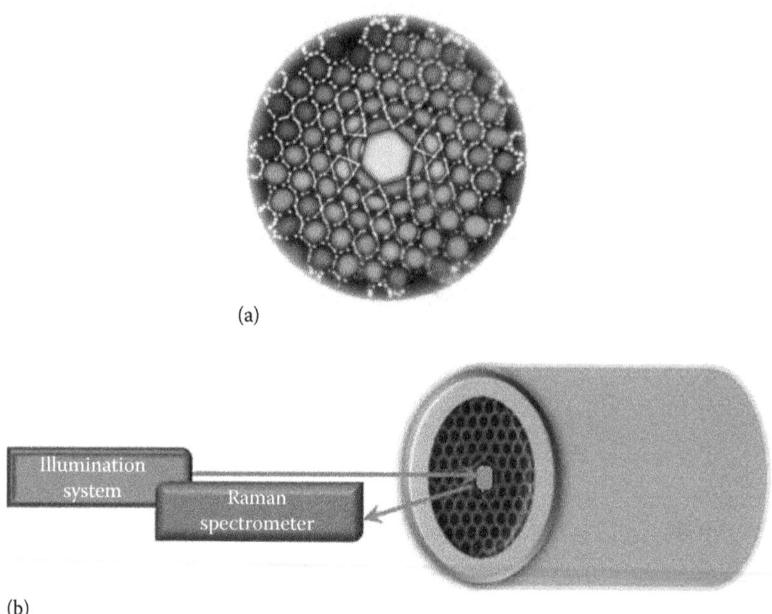

FIGURE 6.14 (a) Kagome PCF illuminated by a broadband light source (Courtesy of the Max Planck Institute for the Science of Light, Erlangen, Germany) and (b) schematic illustration of SERS detection of rhodamine inserted in an HC PCF.

of a rhodamine 6G dye sample, showing enhanced detection efficiency [159]. Using a highly sensitive gold-coated side-polished D-shaped PCF, the fluorescence emission of rhodamine B was found to be enhanced through surface plasmon resonance technology [160]. Rhodamine B detection using a four-hole SCF with gold nanoparticles was presented, showing a large interaction volume between the excitation light and the nanoparticles [161]. By filling HC PCFs with aqueous solutions and using surface-enhanced Raman scattering (SERS), rhodamine 6G was detected with the lowest detectable concentration of 10^{-10} M [162]. Figure 6.14 illustrates a common setup for SERS detection. When comparing SC PCFs and three-hole SCFs, it was observed that the SCFs are more adequate for the purpose, demonstrating a sensitivity of 10^{-10} M to rhodamine 6G in an aqueous solution with only ~7.3 μL of volume [163].

6.5.5 DNA Sensors

DNA analysis techniques are often performed by immobilizing a single strand of DNA on a functionalized glass chip and by checking the hybridization of this strand to its complementary, through the measurement of the fluorescence signal produced by the labeled sample. By using evanescent-wave detection of fluorophore-labeled biomolecule positioned in the air holes of an HC PCF, DNA has been detected [164]. Also, through a 16 mm long piece of functionalized HC PCF incorporated into an optic-fluidic coupler chip, a specific single-stranded DNA string was captured, by immobilizing a sensing layer on the microstructured internal surfaces of the fiber [165]. With an FBG inscribed in an SC PCF, the detection of selected single-stranded DNA was reported, through hybridization to a biofilm in the air holes of this fiber and the measurement of its interaction with the fiber modes [166]. By immobilizing a layer of biomolecules on the sides of the air holes of an LMA PCF with LPGs inscribed, the thickness of double-stranded DNA was measured [167]. By functionalizing a three-hole SCF, selective detection of DNA through hybridization of immobilized peptide nucleic acid probes, a sensor was demonstrated [168]. In Figure 6.15, a diagram of a process for DNA detection using PCFs is presented.

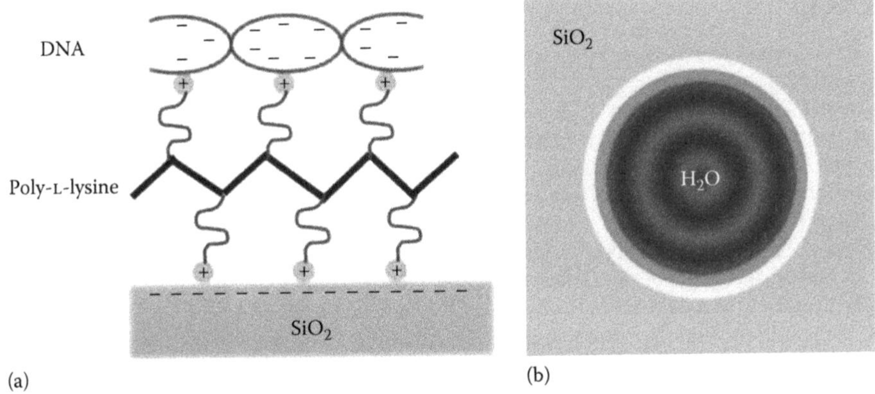

FIGURE 6.15 Illustration of the process used for detecting DNA inside of the PCF holes: (a) negatively charged PCF silica surface (SiO_2) with the molecular structure of poly-L-lysine (+ charges) immobilized onto it; negatively charged DNA (– charges) is immobilized on the poly-L-lysine; and (b) a hole of a PCF (as seen from the top of its cross section) with the inside coated by the monolayers described in (a). Figures not to scale.

6.5.6 Other Biochemical Sensors

Water molecules can be detected through their vibrations related to Raman resonance. This detection was verified on a water-filled HC PCF, suggesting phase-matched coherent anti-Stokes Raman scattering in the inner fiber walls [169]. Different synthesis stages of ZnO nanoparticles, placed inside an HC PCF, were observed using very low pump powers and Raman spectroscopy, using concentrations lower than 1% [170]. By filling an HC PCF with volumes as small as ~0.1 μL, the detection of thiocyanate anions (sensitivity of 1.7×10^{-7} M), water, and ethanol was demonstrated by measuring the trace volume using SERS [171]. Using an SC PCF with silver nanoparticle clusters, the detection of 4-mercaptobenzoic acid was reported [172]. Inserting methylene blue in the cladding holes of an SC PCF, its absorption spectrum was measured [173] and its catalytic reactions monitored in the hollow core of a kagome PCF [174], both using evanescent-field sensing. Also using evanescent-field sensing in a filled three-hole SCF, aqueous $NiCl_2$ solution absorption was measured [175]. Cobalt chloride concentration variations were measured by filling the cladding holes of an SC PCF with it, through absorption spectroscopy, with a sensitivity of 1.6 M^{-1} [176]. A salinity sensor was developed using a polyimide-coated Hi-Bi PCF Sagnac interferometer based on coating-induced radial swelling, achieving 45 times more sensitivity than a polyimide-coated FBG [177]. Glucose detection in an HC PCF was possible through Raman spectroscopy [178]. A PCF surface plasmon resonance biosensor was developed to monitor the binding kinetics of the IgG complexes [179].

6.6 CONCLUSIONS

The PCF sensing field is growing by the day, and as so, it was not possible to mention every work done, or every detail of each sensor. Nevertheless, it can be observed that the diversity of unusual features of PCFs leads to new and improved sensors. Physical sensing is one of the most developed areas of application, with a vast list of published works. PCFs also show a huge potential for gas and biochemical detection, due to their noteworthy characteristics. The amount of interest shown by the scientific community in applying these fibers to these areas shows that PCFs are a technology with an outstanding potential for sensing applications.

ACKNOWLEDGMENTS

The author thanks the Russell Division of Max Planck Institute for the Science of Light (Erlangen, Germany) for the photographs of the PCF's cross sections.

REFERENCES

1. K. T. V. Grattan and T. Sun, Fiber optic sensor technology: An overview, *Sensors and Actuators A: Physical* **82**, 40–61 (2000).
2. P. Russell, Photonic crystal fibers, *Science* **299**, 358–362 (2003).
3. P. R. Laurent Bigot and P. Roy, Fibres à cristal photonique: 10 ans d'existence et un vaste champ d'applications, *Images de la physique*, 71–80 (2007).
4. S. A. Cerqueira, Recent progress and novel applications of photonic crystal fibers, *Reports on Progress in Physics* **73**, 024401 (2010).
5. P. Russell, Photonic crystal fibers: A historical account, *IEEE LEOS Newsletter* **21**, 11–15 (2007).
6. R. D. Meade, A. M. Rappe, K. D. Brommer, and J. D. Joannopoulos, Nature of the photonic band gap: Some insights from a field analysis, *Journal of the Optical Society of America B* **10**, 328–332 (1993).
7. J. C. Knight, T. A. Birks, P. S. Russell, and D. M. Atkin, All-silica single-mode optical fiber with photonic crystal cladding, *Optics Letters* **21**, 1547–1549 (1996).
8. T. A. Birks, J. C. Knight, and P. S. Russell, Endlessly single-mode photonic crystal fiber, *Optics Letters* **22**, 961–963 (1997).
9. J. C. Knight, T. A. Birks, R. F. Cregan, P. S. Russell, and J. P. de Sandro, Large mode area photonic crystal fibre, *Electronics Letters* **34**, 1347–1348 (1998).
10. J. C. K. A. Ortigosa-Blanch, W. J. Wadsworth, J. Arriaga, B. J. Mangan, T. A. Birks, and P. St. J. Russell, Highly birefringent photonic crystal fibers, *Optics Letters* **25**, 1325–1327 (2000).
11. B. J. M. R. F. Cregan, J. C. Knight, T. A. Birks, P. St. J. Russell, P. J. Roberts, and D. C. Allan, Single-mode photonic band gap guidance of light in air, *Science* **285**, 1537–1539 (1999).
12. F. B. F. Couny and P. S. Light, Large-pitch kagome-structured hollow-core photonic crystal fiber, *Optics Letters* **31**, 3574–3576 (2006).
13. F. Benabid, J. C. Knight, G. Antonopoulos, and P. S. J. Russell, Stimulated Raman scattering in hydrogen-filled hollow-core photonic crystal fiber, *Science* **298**, 399–402 (2002).
14. T. A. B. D. Mogilevtsev and P. St. J. Russell, Group-velocity dispersion in photonic crystal fibers, *Optics Letters* **23**, 1662–1664 (1998).
15. J. C. Knight, J. Arriaga, T. A. Birks, A. Ortigosa-Blanch, W. J. Wadsworth, and P. S. Russell, Anomalous dispersion in photonic crystal fiber, *IEEE Photonics Technology Letters* **12**, 807–809 (2000).
16. N. A. Mortensen, J. R. Folkenberg, M. D. Nielsen, and K. P. Hansen, Modal cutoff and the V parameter in photonic crystal fibers, *Optics Letters* **28**, 1879–1881 (2003).
17. J. C. Knight, T. A. Birks, P. S. J. Russell, and J. P. de Sandro, Properties of photonic crystal fiber and the effective index model, *Journal of the Optical Society of America A: Optics Image Science and Vision* **15**, 748–752 (1998).
18. P. S. J. Russell, Photonic-crystal fibers, *Journal of Lightwave Technology* **24**, 4729–4749 (2006).
19. R. F. Cregan, B. J. Mangan, J. C. Knight, T. A. Birks, P. S. Russell, P. J. Roberts, and D. C. Allan, Single-mode photonic band gap guidance of light in air, *Science* **285**, 1537–1539 (1999).
20. D. G.Ouzounov, F. R. Ahmad, D. Müller, N. Venkataraman, M. T. Gallagher, M. G. Thomas, J. Silcox, K. W. Koch, and A. L. Gaeta, Generation of megawatt optical solitons in hollow-core photonic band-gap fibers, *Science* **301**, 1702–1704 (2003).
21. F. Benabid, J. Knight, and P. Russell, Particle levitation and guidance in hollow-core photonic crystal fiber, *Optics Express* **10**, 1195 (2002).
22. T. G. Euser, M. K. Garbos, J. S. Y. Chen, and P. St. J. Russell, Precise balancing of viscous and radiation forces on a particle in liquid-filled photonic bandgap fiber, *Optics Letters* **34**, 3674–3676 (2009).
23. J. S. Y. Chen, T. G. Euser, N. J. Farrer, P. J. Sadler, M. Scharrer, and P. St. J. Russell, Photochemistry in photonic crystal fiber nanoreactors, *Chemistry—A European Journal* **16**, 5607–5612 (2010).
24. A. Cubillas, M. Schmidt, M. Scharrer, T. G. Euser, B. Etzold, N. Taccardi, P. Wasserscheid, and P. St. J. Russell, Monitoring of catalytic reactions in photonic crystal fiber, *European Conference on Lasers and Electro-Optics (CLEO/Europe)*, Munich, Germany, May 22, 2011.
25. F. Magalhaes, J. P. Carvalho, L. A. Ferreira, F. M. Araujo, and J. L. Santos, Methane detection system based on wavelength modulation spectroscopy and hollow-core fibres, *IEEE Sensors*, 1277–1280 (2008).
26. R. Thapa, K. Knabe, M. Faheem, A. Naweed, O. L. Weaver, and K. L. Corwin, Saturated absorption spectroscopy of acetylene gas inside large-core photonic bandgap fiber, *Optics Letters* **31**, 2489–2491 (2006).
27. A. Duval, M. Lhoutellier, J. B. Jensen, P. E. Hoiby, V. Missier, L. H. Pedersen, T. P. Hansen, A. Bjarklev, and O. Bang, Photonic crystal fiber based antibody detection, *Proceedings of IEEE Sensors*, vol. 3, 1222–1225 (2004).

28. F. Luan, A. K. George, T. D. Hedley, G. J. Pearce, D. M. Bird, J. C. Knight, and P. S. J. Russell, All-solid photonic bandgap fiber, *Optics Letters* **29**, 2369–2371 (2004).
29. N. Litchinitser, S. Dunn, P. Steinvurzel, B. Eggleton, T. White, R. McPhedran, and C. de Sterke, Application of an ARROW model for designing tunable photonic devices, *Optics Express* **12**, 1540–1550 (2004).
30. J. C. Knight, F. Luan, G. J. Pearce, A. Wang, T. A. Birks, and D. M. Bird, Solid photonic bandgap fibres and applications, *Japanese Journal of Applied Physics* **45**, 6059–6063 (2006).
31. E. M. Dianov, M. E. Likhachev, and S. Fevrier, Solid-core photonic bandgap fibers for high-power fiber lasers, *IEEE Journal of Selected Topics in Quantum Electronics* **15**, 20–29 (2009).
32. S. G. Johnson, S. Fan, P. R. Villeneuve, J. D. Joannopoulos, and L. A. Kolodziejski, Guided modes in photonic crystal slabs, *Physical Review B* **60**, 5751–5758 (1999).
33. S. A. Cerqueira Jr., F. Luan, C. M. B. Cordeiro, A. K. George, and J. C. Knight, Hybrid photonic crystal fiber, *Optics Express* **14**, 926–931 (2006).
34. M. Pang, L. M. Xiao, W. Jin, and S. A. Cerqueira, Birefringence of hybrid PCF and its sensitivity to strain and temperature, *Journal of Lightwave Technology* **30**, 1422–1432 (2012).
35. S. A. Cerqueira, D. G. Lona, I. de Oliveira, H. E. Hernandez-Figueroa, and H. L. Fragnito, Broadband single-polarization guidance in hybrid photonic crystal fibers, *Optics Letters* **36**, 133–135 (2011).
36. B. Culshaw, Fiber optics in sensing and measurement, *IEEE Journal of Selected Topics in Quantum Electronics* **6**, 1014–1021 (2000).
37. A. M. R. Pinto and M. Lopez-Amo, Photonic crystal fibers for sensing applications, *Journal of Sensors* **2012**, 21 (2012).
38. W. J. Bock, J. Chen, P. Mikulic, T. Eftimov, and M. Korwin-Pawlowski, Pressure sensing using periodically tapered long-period gratings written in photonic crystal fibres, *Measurement Science & Technology* **18**, 3098–3102 (2007).
39. C. Jewart, K. P. Chen, B. McMillen, M. M. Bails, S. P. Levitan, J. Canning, and I. V. Avdeev, Sensitivity enhancement of fiber Bragg gratings to transverse stress by using microstructural fibers, *Optics Letters* **31**, 2260–2262 (2006).
40. T. Geernaert, G. Luyckx, E. Voet, T. Nasilowski, K. Chah, M. Becker, H. Bartelt et al., Transversal load sensing with fiber Bragg gratings in microstructured optical fibers, *IEEE Photonics Technology Letters* **21**, 6–8 (2009).
41. C. M. Jewart, T. Chen, E. Lindner, J. Fiebrandt, M. Rothhardt, K. Schuster, J. Kobelke, H. Bartelt, and K. P. Chen, Suspended-core fiber Bragg grating sensor for directional-dependent transverse stress monitoring, *Optics Letters* **36**, 2360–2362 (2011).
42. H. K. Gahir and D. Khanna, Design and development of a temperature-compensated fiber optic polarimetric pressure sensor based on photonic crystal fiber at 1550 nm, *Applied Optics* **46**, 1184–1189 (2007).
43. F. C. Favero, S. M. M. Quintero, C. Martelli, A. M. B. Braga, V. V. Silva, I. C. S. Carvalho, R. W. A. Llerena, and L. C. G. Valente, Hydrostatic pressure sensing with high birefringence photonic crystal fibers, *Sensors* **10**, 9698–9711 (2010).
44. Y. S. Shinde and H. K. Gahir, Dynamic pressure sensing study using photonic crystal fiber: Application to tsunami sensing, *IEEE Photonics Technology Letters* **20**, 279–281 (2008).
45. T. Nasilowski, T. Martynkien, G. Statkiewicz, M. Szpulak, J. Olszewski, G. Golojuch, W. Urbanczyk et al., Temperature and pressure sensitivities of the highly birefringent photonic crystal fiber with core asymmetry, *Applied Physics B: Lasers and Optics* **81**, 325–331 (2005).
46. T. Martynkien, M. Szpulak, G. Statkiewicz, G. Golojuch, J. Olszewski, W. Urbanczyk, J. Wojcik et al., Measurements of sensitivity to hydrostatic pressure and temperature in highly birefringent photonic crystal fibers, *Optical and Quantum Electronics* **39**, 481–489 (2007).
47. T. Martynkien, G. Statkiewicz-Barabach, J. Olszewski, J. Wojcik, P. Mergo, T. Geernaert, C. Sonnenfeld et al., Highly birefringent microstructured fibers with enhanced sensitivity to hydrostatic pressure, *Optics Express* **18**, 15113–15121 (2010).
48. H. M. Kim, T. H. Kim, B. Kim, and Y. Chung, Enhanced transverse load sensitivity by using a highly birefringent photonic crystal fiber with larger air holes on one axis, *Applied Optics* **49**, 3841–3845 (2010).
49. L. H. Cho, C. Wu, C. Lu, and H. Y. Tam, A highly sensitive and low-cost Sagnac loop based pressure sensor, *IEEE Sensors Journal* **13**, 3073–3078 (2013).
50. Y. H. Yang, J. Li, W. Q. Duan, X. Zhang, W. Jin, and M. W. Yang, An embedded pressure sensor based on polarization maintaining photonic crystal fiber, *Measurement Science and Technology* **24**, 1–5 (2013).
51. M. Karimi, M. Fabian, L. R. Jaroszewicz, K. Schuster, P. Mergo, T. Sun, and K. T. V. Grattan, Lateral force sensing system based on different photonic crystal fibres, *Sensors and Actuators A: Physical* **205**, 86–91 (2014).

52. H. Y. Fu, C. Wu, M. L. V. Tse, L. Zhang, K. C. D. Cheng, H. Y. Tam, B. O. Guan, and C. Lu, High pressure sensor based on photonic crystal fiber for downhole application, *Applied Optics* **49**, 2639–2643 (2010).
53. C. F. Fan, C. L. Chiang, and C. P. Yu, Birefringent photonic crystal fiber coils and their application to transverse displacement sensing, *Optics Express* **19**, 19948–19954 (2011).
54. P. Zu, C. C. Chan, Y. X. Jin, Y. F. Zhang, and X. Y. Dong, Fabrication of a temperature-insensitive transverse mechanical load sensor by using a photonic crystal fiber-based Sagnac loop, *Measurement Science and Technology* **22**, 025204 (2011).
55. S. H. Aref, M. I. Zibaii, M. Kheiri, H. Porbeyram, H. Latifi, F. M. Araujo, L. A. Ferreira et al., Pressure and temperature characterization of two interferometric configurations based on suspended-core fibers, *Optics Communication* **285**, 269–273 (2012).
56. M. S. Ferreira, J. Bierlich, H. Lehmann, K. Schuster, J. Kobelke, J. L. Santos, and O. Frazao, Fabry-Perot cavity based on hollow-core ring photonic crystal fiber for pressure sensing, *IEEE Photonics Technology Letters* **24**, 2122–2124 (2012).
57. S. S. Li, Z. D. Huang, X. S. Song, S. Y. Zhang, Q. Zhong, F. Xu, and Y. Q. Lu, Photonic crystal fibre based high temperature sensor with three-beam path interference, *Electronics Letters* **46**, 1394–1396 (2010).
58. D. Monzon-Hernandez, V. P. Minkovich, and J. Villatoro, High-temperature sensing with tapers made of microstructured optical fiber, *IEEE Photonics Technology Letters* **18**, 511–513 (2006).
59. S. M. Nalawade and H. V. Thakur, Photonic crystal fiber strain-independent temperature sensing based on modal interferometer, *IEEE Photonics Technology Letters* **23**, 1600–1602 (2011).
60. F. C. Favero, R. Spittel, F. Just, J. Kobelke, M. Rothhardt, and H. Bartelt, A miniature temperature high germanium doped PCF interferometer sensor, *Optics Express* **21**, 30266–30274 (2013).
61. B. H. Kim, S. H. Lee, A. X. Lin, C. L. Lee, J. Lee, and W. T. Han, Large temperature sensitivity of Sagnac loop interferometer based on the birefringent holey fiber filled with metal indium, *Optics Express* **17**, 1789–1794 (2009).
62. W. W. Qian, C. L. Zhao, S. L. He, X. Y. Dong, S. Q. Zhang, Z. X. Zhang, S. Z. Jin, J. T. Guo, and H. F. Wei, High-sensitivity temperature sensor based on an alcohol-filled photonic crystal fiber loop mirror, *Optics Letters* **36**, 1548–1550 (2011).
63. Y. Cui, P. P. Shum, D. J. J. Hu, G. H. Wang, G. Humbert, and X. Q. Dinh, Temperature sensor by using selectively filled photonic crystal fiber Sagnac interferometer, *IEEE Photonics Journal* **4**, 1801–1808 (2012).
64. H. Y. Choi, K. S. Park, S. J. Park, U. C. Paek, B. H. Lee, and E. S. Choi, Miniature fiber-optic high temperature sensor based on a hybrid structured Fabry-Perot interferometer, *Optics Letters* **33**, 2455–2457 (2008).
65. A. M. R. Pinto, O. Frazao, J. L. Santos, M. Lopez-Amo, J. Kobelke, and K. Schuster, Interrogation of a suspended-core Fabry-Perot temperature sensor through a dual wavelength Raman fiber laser, *Journal of Lightwave Technology* **28**, 3149–3155 (2010).
66. A. M. R. Pinto, M. Lopez-Amo, J. Kobelke, and K. Schuster, Temperature fiber laser sensor based on a hybrid cavity and a random mirror, *IEEE Journal of Lightwave Technology* **30**, 1168–1172 (2011).
67. J. Ju, W. Jin, and H. L. Ho, Compact in-fiber interferometer formed by long-period gratings in photonic crystal fiber, *IEEE Photonics Technology Letters* **20**, 1899–1901 (2008).
68. Y. F. Geng, X. J. Li, X. L. Tan, Y. L. Deng, and Y. Q. Yu, Sensitivity-enhanced high-temperature sensing using all-solid photonic bandgap fiber modal interference, *Applied Optics* **50**, 468–472 (2011).
69. G. Youfu, L. Xuejin, T. Xiaoling, D. Yuanlong, and H. Xueming, Compact and ultrasensitive temperature sensor with a fully liquid-filled photonic crystal fiber Mach-Zehnder interferometer, *IEEE Sensors Journal* **14**, 167–170 (2014).
70. Y. N. Zhu, P. Shum, H. W. Bay, M. Yan, Y. X, J. J. Hu, J. Z. Hao, and C. Lu, Strain-insensitive and high-temperature long-period gratings inscribed in photonic crystal fiber, *Optics Letters* **30**, 367–369 (2005).
71. J. Ju, Z. Wang, W. Jin, and M. S. Demokan, Temperature sensitivity of a two-mode photonic crystal fiber interferometric sensor, *IEEE Photonics Technology Letters* **18**, 2168–2170 (2006).
72. B. Larrion, M. Hernandez, F. J. Arregui, J. Goicoechea, J. Bravo, and I. R. Matias, Photonic crystal fiber temperature sensor based on quantum dot nanocoatings, *Journal of Sensors* **2009**, 932471 (2009).
73. A. Bozolan, R. M. Gerosa, C. J. S. de Matos, and M. A. Romero, Temperature sensing using colloidal-core photonic crystal fiber, *IEEE Sensors Journal* **12**, 195–200 (2012).
74. Y. P. Miao, B. Liu, K. L. Zhang, Y. Liu, and H. Zhang, Temperature tunability of photonic crystal fiber filled with Fe_3O_4 nanoparticle fluid, *Applied Physics Letters* **98**, 021103–021105 (2011).
75. Y. Q. Yu, X. J. Li, X. M. Hong, Y. L. Deng, K. Y. Song, Y. F. Geng, H. F. Wei, and W. J. Tong, Some features of the photonic crystal fiber temperature sensor with liquid ethanol filling, *Optics Express* **18**, 15383–15388 (2010).

76. W. Ying, Y. Minwei, D. N. Wang, and C. R. Liao, Selectively infiltrated photonic crystal fiber with ultra-high temperature sensitivity, *IEEE Photonics Technology Letters* **23**, 1520–1522 (2011).
77. Y. Peng, J. Hou, Y. Zhang, Z. H. Huang, R. Xiao, and Q. S. Lu, Temperature sensing using the bandgap-like effect in a selectively liquid-filled photonic crystal fiber, *Optics Letters* **38**, 263–265 (2013).
78. L. F. Zou, X. Y. Bao, and L. A. Chen, Distributed Brillouin temperature sensing in photonic crystal fiber, *Smart Materials and Structures* **14**, S8–S11 (2005).
79. Y. K. Dong, X. Y. Bao, and L. Chen, Distributed temperature sensing based on birefringence effect on transient Brillouin grating in a polarization-maintaining photonic crystal fiber, *Optics Letters* **34**, 2590–2592 (2009).
80. Y. G. Han, Temperature-insensitive strain measurement using a birefringent interferometer based on a polarization-maitaining photonic crystal fiber, *Applied Physics B: Lasers and Optics* **95**, 383–387 (2009).
81. S. Rota-Rodrigo, A. M. R. Pinto, M. Bravo, and M. Lopez-Amo, An in-reflection strain sensing head based on a Hi-Bi photonic crystal fiber, *Sensors (Basel)* **13**, 8095–8102 (2013).
82. A. Pinto, J. Baptista, J. Santos, M. Lopez-Amo, and O. Frazão, Micro-displacement sensor based on a hollow-core photonic crystal fiber, *Sensors (Basel)* **12**, 17497–17503 (2012).
83. J. Villatoro, V. P. Minkovich, and D. Monzon-Hernandez, Compact modal interferometer built with tapered microstructured optical fiber, *IEEE Photonics Technology Letters* **18**, 1258–1260 (2006).
84. J. Villatoro, V. Finazzi, V. P. Minkovich, V. Pruneri, and G. Badenes, Temperature-insensitive photonic crystal fiber interferometer for absolute strain sensing, *Applied Physics Letters* **91**, 091109 (2007).
85. B. Dong and E. J. Hao, Temperature-insensitive and intensity-modulated embedded photonic-crystal-fiber modal-interferometer-based microdisplacement sensor, *Journal of the Optical Society of America B* **28**, 2332–2336 (2011).
86. O. Frazao, J. M. Baptista, and J. L. Santos, Temperature-independent strain sensor based on a Hi-Bi photonic crystal fiber loop mirror, *IEEE Sensors Journal* **7**, 1453–1455 (2007).
87. W. W. Qian, C. L. Zhao, X. Y. Dong, and W. Jin, Intensity measurement based temperature-independent strain sensor using a highly birefringent photonic crystal fiber loop mirror, *Optics Communication* **283**, 5250–5254 (2010).
88. H. Zhang, B. Liu, Z. Wang, J. H. Luo, S. X. Wang, C. H. Jia, and X. R. Ma, Temperature-insensitive displacement sensor based on high-birefringence photonic crystal fiber loop mirror, *Optics Applied* **40**, 209–217 (2010).
89. M. Bravo, A. M. R. Pinto, M. Lopez-Amo, J. Kobelke, and K. Schuster, High precision micro-displacement fiber sensor through a suspended-core Sagnac interferometer, *Optics Letters* **37**, 202–204 (2012).
90. Q. Shi, F. Y. Lv, Z. Wang, L. Jin, J. J. Hu, Z. Y. Liu, G. Y. Kai, and X. Y. Dong, Environmentally stable Fabry-Perot-type strain sensor based on hollow-core photonic bandgap fiber, *IEEE Photonics Technology Letters* **20**, 237–239 (2008).
91. Q. Shi, Z. Wang, L. Jin, Y. Li, H. Zhang, F. Y. Lu, G. Y. Kai, and X. Y. Dong, A hollow-core photonic crystal fiber cavity based multiplexed Fabry-Perot interferometric strain sensor system, *IEEE Photonics Technology Letters* **20**, 1329–1331 (2008).
92. W. J. Zhou, W. C. Wong, C. C. Chan, L. Y. Shao, and X. Y. Dong, Highly sensitive fiber loop ringdown strain sensor using photonic crystal fiber interferometer, *Applied Optics* **50**, 3087–3092 (2011).
93. K. Karim Qureshi, Z. Liu, H.-Y. Tam, and M. Fahad Zia, A strain sensor based on in-line fiber Mach–Zehnder interferometer in twin-core photonic crystal fiber, *Optics Communication* **309**, 68–70 (2013).
94. S. Mathews, G. Farrell, and Y. Semenova, All-fiber polarimetric electric field sensing using liquid crystal infiltrated photonic crystal fibers, *Sensors and Actuators A: Physical* **167**, 54–59 (2011).
95. S. Mathews, G. Farrell, and Y. Semenova, Directional electric field sensitivity of a liquid crystal infiltrated photonic crystal fiber, *IEEE Photonics Technology Letters* **23**, 408–410 (2011).
96. S. Mathews, G. Farrell, and Y. Semenova, Liquid crystal infiltrated photonic crystal fibers for electric field intensity measurements, *Applied Optics* **50**, 2628–2635 (2011).
97. A. Michie, J. Canning, I. Bassett, J. Haywood, K. Digweed, B. Ashton, M. Stevenson, J. Digweed, A. Lau, and D. Scandurra, Spun elliptically birefringent photonic crystal fibre for current sensing, *Measurement Science and Technology* **18**, 3070–3074 (2007).
98. Y. K. Chamorovskiy, N. I. Starostin, M. V. Ryabko, A. I. Sazonov, S. K. Morshnev, V. P. Gubin, I. L. Vorob'ev, and S. A. Nikitov, Miniature microstructured fiber coil with high magneto-optical sensitivity, *Optics Communications* **282**, 4618–4621 (2009).
99. H. V. Thakur, S. M. Nalawade, S. Gupta, R. Kitture, and S. N. Kale, Photonic crystal fiber injected with Fe_3O_4 nanofluid for magnetic field detection, *Applied Physics Letters* **99**, 161101 (2011).
100. S. M. M. Quintero, C. Martelli, A. M. B. Braga, L. C. G. Valente, and C. C. Kato, Magnetic field measurements based on terfenol coated photonic crystal fibers, *Sensors* **11**, 11103–11111 (2011).

101. Y. Zhao, R. Q. Lv, Y. Ying, and Q. Wang, Hollow-core photonic crystal fiber FabryPerot sensor for magnetic field measurement based on magnetic fluid, *Optics and Laser Technology* **44**, 899–902 (2012).
102. Y. Zhao, Y. Zhang, D. Wu, and Q. Wang, Magnetic field and temperature measurements with a magnetic fluid-filled photonic crystal fiber Bragg grating, *Instrumentation Science & Technology* **41**, 463–472 (2013).
103. R. Gao, Y. Jiang, and S. Abdelaziz, All-fiber magnetic field sensors based on magnetic fluid-filled photonic crystal fibers, *Optics Letters* **38**, 1539–1541 (2013).
104. P. Zu, C. C. Chan, W. S. Lew, L. M. Hu, Y. X. Jin, H. F. Liew, L. H. Chen, W. C. Wong, and X. Y. Dong, Temperature-insensitive magnetic field sensor based on nanoparticle magnetic fluid and photonic crystal fiber, *IEEE Photonics Journal* **4**, 491–498 (2012).
105. P. Zu, C. C. Chan, T. X. Gong, Y. X. Jin, W. C. Wong, and X. Y. Dong, Magneto-optical fiber sensor based on bandgap effect of photonic crystal fiber infiltrated with magnetic fluid, *Applied Physics Letters* **101**, 241118 (2012).
106. Y. L. Hoo, W. Jin, H. L. Ho, D. N. Wang, and R. S. Windeler, Evanescent-wave gas sensing using microstructure fiber, *Optics Engineeering* **41**, 8–9 (2002).
107. G. Pickrell, W. Peng, and A. Wang, Random-hole optical fiber evanescent-wave gas sensing, *Optics Letters* **29**, 1476–1478 (2004).
108. A. S. Webb, F. Poletti, D. J. Richardson, and J. K. Sahu, Suspended-core holey fiber for evanescent-field sensing, *Optics Engineering* **46**, 010503 (2007).
109. G. F. Yan, A. P. Zhang, G. Y. Ma, B. H. Wang, B. Kim, J. Im, S. L. He, and Y. Chung, Fiber-optic acetylene gas sensor based on microstructured optical fiber Bragg gratings, *IEEE Photonics Technology Letters* **23**, 1588–1590 (2011).
110. Y. L. Hoo, W. Jin, H. L. Ho, J. Ju, and D. N. Wang, Gas diffusion measurement using hollow-core photonic bandgap fiber, *Sensors and Actuators B: Chemical* **105**, 183–186 (2005).
111. E. Austin, A. van Brakel, M. N. Petrovich, and D. J. Richardson, Fibre optical sensor for C_2H_2 gas using gas-filled photonic bandgap fibre reference cell, *Sensors and Actuators B: Chemical* **139**, 30–34 (2009).
112. S. H. Kassani, J. Park, Y. Jung, J. Kobelke, and K. Oh, Fast response in-line gas sensor using C-type fiber and Ge-doped ring defect photonic crystal fiber, *Optics Express* **21**, 14074–14083 (2013).
113. Y. L. Hoo, W. Jin, H. L. Ho, and D. N. Wang, Measurement of gas diffusion coefficient using photonic crystal fiber, *IEEE Photonics Technology Letters* **15**, 1434–1436 (2003).
114. A. M. Cubillas, M. Silva-Lopez, J. M. Lazaro, O. M. Conde, M. N. Petrovich, and J. M. Lopez-Higuera, Methane detection at 1670-nm band using a hollow-core photonic bandgap fiber and a multiline algorithm, *Optics Express* **15**, 17570–17576 (2007).
115. A. M. Cubillas, J. M. Lazaro, M. Silva-Lopez, O. M. Conde, M. N. Petrovich, and J. M. Lopez-Higuera, Methane sensing at 1300 nm band with hollow-core photonic bandgap fibre as gas cell, *Electronics Letters* **44**, 403–405 (2008).
116. A. Cubillas, J. Lazaro, O. Conde, M. Petrovich, and J. Lopez-Higuera, Gas sensor based on photonic crystal fibres in the $2\nu3$ and $\nu2 + 2\nu3$ vibrational bands of methane, *Sensors* **9**, 6261–6272 (2009).
117. Y. L. Hoo, S. J. Liu, H. L. Ho, and W. Jin, Fast response microstructured optical fiber methane sensor with multiple side-openings, *IEEE Photonics Technology Letters* **22**, 296–298 (2010).
118. L. Duan, F. Songnian, T. Ming, P. Shum, and L. Deming, Comb filter-based fiber-optic methane sensor system with mitigation of cross gas sensitivity, *Journal of Lightwave Technology* **30**, 3103–3109 (2012).
119. V. P. Minkovich, D. Monzon-Hernandez, J. Villatoro, and G. Badenes, Microstructured optical fiber coated with thin films for gas and chemical sensing, *Optics Express* **14**, 8413–8418 (2006).
120. J. Villatoro, M. P. Kreuzer, R. Jha, V. P. Minkovich, V. Finazzi, G. Badenes, and V. Pruneri, Photonic crystal fiber interferometer for chemical vapor detection with high sensitivity, *Optics Express* **17**, 1447–1453 (2009).
121. T. Ritari, J. Tuominen, H. Ludvigsen, J. C. Petersen, T. Sorensen, T. P. Hansen, and H. R. Simonsen, Gas sensing using air-guiding photonic bandgap fibers, *Optics Express* **12**, 4080–4087 (2004).
122. H. Ding, X. L. Li, J. H. Cui, S. F. Dong, and L. Yang, An all-fiber gas sensing system using hollow-core photonic bandgap fiber as gas cell, *Instrumentation Science & Technology* **39**, 78–87 (2011).
123. J. Henningsen, J. Hald, and J. C. Petersen, Saturated absorption in acetylene and hydrogen cyanide in hollow-core photonic bandgap fibers, *Optics Express* **13**, 10475–10482 (2005).
124. M. P. Buric, K. P. Chen, J. Falk, and S. D. Woodruff, Enhanced spontaneous Raman scattering and gas composition analysis using a photonic crystal fiber, *Applied Optics* **47**, 4255–4261 (2008).
125. M. P. Buric, K. P. Chen, J. Falk, and S. D. Woodruff, Improved sensitivity gas detection by spontaneous Raman scattering, *Applied Optics* **48**, 4424–4429 (2009).

126. J. P. Parry, B. C. Griffiths, N. Gayraud, E. D. McNaghten, A. M. Parkes, W. N. MacPherson, and D. P. Hand, Towards practical gas sensing with micro-structured fibres, *Measurement Science and Technology* **20**, 075301 (2009).
127. H. Lehmann, H. Bartelt, R. Willsch, R. Amezcua-Correa, and J. C. Knight, In-line gas sensor based on a photonic bandgap fiber with laser-drilled lateral microchannels, *IEEE Sensors Journal* **11**, 2926–2931 (2011).
128. X. H. Yang, L. R. Peng, L. B. Yuan, P. P. Teng, F. J. Tian, L. Li, and S. Z. Luo, Oxygen gas optrode based on microstructured polymer optical fiber segment, *Optics Communications* **284**, 3462–3466 (2011).
129. L. Niu, C.-L. Zhao, J. Kang, S. Jin, J. Guo, and H. Wei, A chemical vapor sensor based on Rayleigh scattering effect in simplified hollow-core photonic crystal fibers, *Optics Communications* **313**, 243–247 (2014).
130. X. Yang, A. S. P. Chang, B. Chen, C. Gu, and T. C. Bond, High sensitivity gas sensing by Raman spectroscopy in photonic crystal fiber, *Sensors and Actuators B: Chemical* **176**, 64–68 (2013).
131. O. S. Wolfbeis, Fibre-optic sensors in biomedical sciences, *Pure and Applied Chemistry* **59**, 683–672 (1987).
132. M. Skorobogatiy, Microstructured and photonic bandgap fibers for applications in the resonant bio- and chemical sensors, *Journal of Sensors* **2009**, 524237 (2009).
133. T. M. Monro, S. Warren-Smith, E. P. Schartner, A. Francois, S. Heng, H. Ebendorff-Heidepriem, and S. Afshar, Sensing with suspended-core optical fibers, *Optical Fiber Technology* **16**, 343–356 (2010).
134. V. P. Minkovich, J. Villatoro, D. Monzon-Hernandez, S. Calixto, A. B. Sotsky, and L. I. Sotskaya, Holey fiber tapers with resonance transmission for high-resolution refractive index sensing, *Optics Express* **13**, 7609–7614 (2005).
135. C. Li, S.-J. Qiu, Y. Chen, F. Xu, and Y.-Q. Lu, Ultra-sensitive refractive index sensor with slightly tapered photonic crystal fiber, *IEEE Photonics Technology Letters* **24**, 1771–1774 (2012).
136. R. Jha, J. Villatoro, and G. Badenes, Ultrastable in reflection photonic crystal fiber modal interferometer for accurate refractive index sensing, *Applied Physics Letters* **93**, 191106 (2008).
137. D. K. C. Wu, B. T. Kuhlmey, and B. J. Eggleton, Ultrasensitive photonic crystal fiber refractive index sensor, *Optics Letters* **34**, 322–324 (2009).
138. R. Jha, J. Villatoro, G. Badenes, and V. Pruneri, Refractometry based on a photonic crystal fiber interferometer, *Optics Letters* **34**, 617–619 (2009).
139. S. Silva, J. L. Santos, F. X. Malcata, J. Kobelke, K. Schuster, and O. Frazao, Optical refractometer based on large-core air-clad photonic crystal fibers, *Optics Letters* **36**, 852–854 (2011).
140. W. C. Wong, W. Zhou, C. C. Chan, X. Dong, and K. C. Leong, Cavity ringdown refractive index sensor using photonic crystal fiber interferometer, *Sensors and Actuators B: Chemical* **161**, 108–113 (2012).
141. D. J. J. Hu, J. L. Lim, M. K. Park, L. T. H. Kao, Y. Wang, H. Wei, and W. Tong, Photonic crystal fiber-based interferometric biosensor for streptavidin and biotin detection, *IEEE Journal of Selected Topics in Quantum Electronics* **18**, 1293–1297 (2012).
142. M. H. Frosz, A. Stefani, and O. Bang, Highly sensitive and simple method for refractive index sensing of liquids in microstructured optical fibers using four-wave mixing, *Optics Express* **19**, 10471–10484 (2011).
143. J. Sun and C. C. Chan, Photonic bandgap fiber for refractive index measurement, *Sensors and Actuators B: Chemical* **128**, 46–50 (2007).
144. Y. N. Zhu, Z. H. He, and H. Du, Detection of external refractive index change with high sensitivity using long-period gratings in photonic crystal fiber, *Sensors and Actuators B: Chemical* **131**, 265–269 (2008).
145. Z. He, F. Tian, Y. Zhu, N. Lavlinskaia, and H. Du, Long-period gratings in photonic crystal fiber as an optofluidic label-free biosensor, *Biosensors and Bioelectronics* **26**, 4774–4778 (2011).
146. Z. H. He, Y. N. Zhu, and H. Du, Long-period gratings inscribed in air- and water-filled photonic crystal fiber for refractometric sensing of aqueous solution, *Applied Physics Letters* **92**, 044105 (2008).
147. L. Rindorf and O. Bang, Highly sensitive refractometer with a photonic-crystal-fiber long-period grating, *Optics Letters* **33**, 563–565 (2008).
148. T. Zhi Qiang, C. Chi Chiu, W. Wei Chang, and C. Li Han, Fiber optic refractometer based on cladding excitation of localized surface plasmon resonance, *IEEE Photonics Technology Letters* **25**, 556–559 (2013).
149. J. Mathew, Y. Semenova, G. Rajan, and G. Farrell, Humidity sensor based on photonic crystal fibre interferometer, *Electronics Letters* **46**, 1341–1343 (2010).
150. M. Y. Mohd Noor, N. Khalili, I. Skinner, and G. D. Peng, Optical humidity sensor based on air guided photonic crystal fiber, *Photonics Sensors* **2**, 277–282 (2012).

151. M. Y. M. Noor, N. M. Kassim, A. S. M. Supaat, M. H. Ibrahim, A. I. Azmi, A. S. Abdullah, and G. D. Peng, Temperature-insensitive photonic crystal fiber interferometer for relative humidity sensing without hygroscopic coating, *Measurement Science and Technology* **24**, 105205 (2013).
152. W. C. Wong, C. C. Chan, L. H. Chen, T. Li, K. X. Lee, and K. C. Leong, Polyvinyl alcohol coated photonic crystal optical fiber sensor for humidity measurement, *Sensors and Actuators B: Chemical* **174**, 563–569 (2012).
153. L. Tao, D. Xinyong, C. Chi Chiu, N. Kai, Z. Shuqin, and P. P. Shum, Humidity sensor with a PVA-coated photonic crystal fiber interferometer, *IEEE Sensors Journal* **13**, 2214–2216 (2013).
154. J. Mathew, Y. Semenova, and G. Farrell, Experimental demonstration of a high-sensitivity humidity sensor based on an Agarose-coated transmission-type photonic crystal fiber interferometer, *Applied Optics* **52**, 3884–3890 (2013).
155. J. Mathew, Y. Semenova, and G. Farrell, Relative humidity sensor based on an agarose-infiltrated photonic crystal fiber interferometer, *IEEE Journal of Selected Topics in Quantum Electronics* **18**, 1553–1559 (2012).
156. X. H. Yang and L. L. Wang, Fluorescence pH probe based on microstructured polymer optical fiber, *Optics Express* **15**, 16478–16483 (2007).
157. Y. L. Ruan, E. P. Schartner, H. Ebendorff-Heidepriem, P. Hoffmann, and T. M. Monro, Detection of quantum-dot labeled proteins using soft glass microstructured optical fibers, *Optics Express* **15**, 17819–17826 (2007).
158. Y. Ruan, T. C. Foo, S. Warren-Smith, P. Hoffmann, R. C. Moore, H. Ebendorff-Heidepriem, and T. M. Monro, Antibody immobilization within glass microstructured fibers: a route to sensitive and selective biosensors, *Optics Express* **16**, 18514–18523 (2008).
159. M. T. Myaing, J. Y. Ye, T. B. Norris, T. Thomas, J. R. Baker, W. J. Wadsworth, G. Bouwmans, J. C. Knight, and P. S. J. Russell, Enhanced two-photon biosensing with double-clad photonic crystal fibers, *Optics Letters* **28**, 1224–1226 (2003).
160. X. Yu, D. Yong, H. Zhang, H. Li, Y. Zhang, C. C. Chan, H.-P. Ho, H. Liu, and D. Liu, Plasmonic enhanced fluorescence spectroscopy using side-polished microstructured optical fiber, *Sensors and Actuators B: Chemical* **160**, 196–201 (2011).
161. H. Yan, J. Liu, C. X. Yang, G. F. Jin, C. Gu, and L. Hou, Novel index-guided photonic crystal fiber surface-enhanced Raman scattering probe, *Optics Express* **16**, 8300–8305 (2008).
162. X. Yang, C. Shi, D. Wheeler, R. Newhouse, B. Chen, J. Z. Zhang, and C. Gu, High-sensitivity molecular sensing using hollow-core photonic crystal fiber and surface-enhanced Raman scattering, *Journal of the Optical Society of America A: Optics Image Science and Vision* **27**, 977–984 (2010).
163. M. K. K. Oo, Y. Han, J. Kanka, S. Sukhishvili, and H. Du, Structure fits the purpose: Photonic crystal fibers for evanescent-field surface-enhanced Raman spectroscopy, *Optics Letters* **35**, 466–468 (2010).
164. J. B. Jensen, L. H. Pedersen, P. E. Hoiby, L. B. Nielsen, T. P. Hansen, J. R. Folkenberg, J. Riishede et al., Photonic crystal fiber based evanescent-wave sensor for detection of biomolecules in aqueous solutions, *Optics Letters* **29**, 1974–1976 (2004).
165. L. Rindorf, P. E. Hoiby, J. B. Jensen, L. H. Pedersen, O. Bang, and O. Geschke, Towards biochips using microstructured optical fiber sensors, *Analytical and Bioanalytical Chemistry* **385**, 1370–1375 (2006).
166. N. Burani and J. Laegsgaard, Perturbative modeling of Bragg-grating-based biosensors in photonic-crystal fibers, *Journal of Optical Society of America B* **22**, 2487–2493 (2005).
167. L. Rindorf, J. B. Jensen, M. Dufva, L. H. Pedersen, P. E. Hoiby, and O. Bang, Photonic crystal fiber long-period gratings for biochemical sensing, *Optics Express* **14**, 8224–8231 (2006).
168. E. Coscelli, M. Sozzi, F. Poli, D. Passaro, A. Cucinotta, S. Selleri, R. Corradini, and R. Marchelli, Toward a highly specific DNA biosensor: PNA-modified suspended-core photonic crystal fibers, *IEEE Journal of Selected Topics in Quantum Electronics* **16**, 967–972 (2010).
169. S. O. Konorov, A. B. Fedotov, A. M. Zheltikov, and R. B. Miles, Phase-matched four-wave mixing and sensing of water molecules by coherent anti-Stokes Raman scattering in large-core-area hollow photonic-crystal fibers, *Journal of Optical Society of America B* **22**, 2049–2053 (2005).
170. J. Irizar, J. Dinglasan, J. B. Goh, A. Khetani, H. Anis, D. Anderson, C. Goh, and A. S. Helmy, Raman spectroscopy of nanoparticles using hollow-core photonic crystal fibers, *IEEE Journal of Selected Topics in Quantum Electronics* **14**, 1214–1222 (2008).
171. Y. Han, M. K. K. Oo, Y. N. Zhu, L. M. Xiao, M. S. Demohan, W. Jin, and H. Du, Index-guiding liquid-core photonic crystal fiber for solution measurement using normal and surface-enhanced Raman scattering, *Optics Engineering* **47**, 040502 (2008).
172. Z. Xie, Y. Lu, H. Wei, J. Yan, P. Wang, and H. Ming, Broad spectral photonic crystal fiber surface enhanced Raman scattering probe, *Applied Physics B: Lasers and Optics* **95**, 751–755 (2009).

173. C. M. B. Cordeiro, M. A. R. Franco, G. Chesini, E. C. S. Barretto, R. Lwin, C. H. B. Cruz, and M. C. J. Large, Microstructured-core optical fibre for evanescent sensing applications, *Optics Express* **14**, 13056–13066 (2006).
174. A. M. Cubillas, M. Schmidt, M. Scharrer, T. G. Euser, B. J. M. Etzold, N. Taccardi, P. Wasserscheid, and P. S. J. Russell, Ultra-low concentration monitoring of catalytic reactions in photonic crystal fiber, *Chemistry—A European Journal* **18**, 1586–1590 (2012).
175. T. G. Euser, J. S. Y. Chen, M. Scharrer, P. S. J. Russell, N. J. Farrer, and P. J. Sadler, Quantitative broadband chemical sensing in air-suspended solid-core fibers, *Journal of Applied Physics* **103**, 103108 (2008).
176. X. Yu, Y. Sun, G. B. Ren, P. Shum, N. Q. Ngo, and Y. C. Kwok, Evanescent field absorption sensor using a pure-silica defected-core photonic crystal fiber, *IEEE Photonics Technology Letters* **20**, 336–338 (2008).
177. C. Wu, B. O. Guan, C. Lu, and H. Y. Tam, Salinity sensor based on polyimide-coated photonic crystal fiber, *Optics Express* **19**, 20003–20008 (2011).
178. X. Yang, A. Zhang, D. Wheeler, T. Bond, C. Gu, and Y. Li, Direct molecule-specific glucose detection by Raman spectroscopy based on photonic crystal fiber, *Analytical and Bioanalytical Chemistry* **402**, 687–691 (2012).
179. W. C. Wong, C. C. Chan, J. L. Boo, Z. Y. Teo, Z. Q. Tou, H. B. Yang, C. M. Li, and K. C. Leong, Photonic crystal fiber surface plasmon resonance biosensor based on protein G immobilization, *IEEE Journal of Selected Topics in Quantum Electronics* **19**, 4602107 (2013).

7 Liquid Crystal Optical Fibers for Sensing Applications

Sunish Mathews and Yuliya Semenova

CONTENTS

7.1 INTRODUCTION

Liquid crystal (LC), due to their unique crystalline phase characterized by the partial order of their constituent molecules along with physical fluidity, can be easily incorporated into desired configurations for a variety of device applications [1,2]. The large anisotropy of LCs dependent on physical parameters, such as temperature and electric and magnetic fields, makes LC materials suited for implementing optical sensors for such parameters. Optical sensing research has developed several successful fiber sensor types, for example, fiber Bragg grating sensors for strain sensing have been widely applied in structural health monitoring. Fiber-optic sensing [3] offers many advantages such as dielectric isolation, immunity from electromagnetic interference, chemical passivity, multiplexing capabilities, wide bandwidth, wide operating temperature range, environmental ruggedness, and safety in explosive conditions.

Electric field sensing is a common need in the electrical power generation and distribution industries and other areas such as industrial automation. Optical fiber–based electric field sensing is attractive, as it can inherit the many advantages of optical fiber sensing, but the natural electromagnetic immunity of optical fibers is a barrier to the development of this type of sensing. As a result, fiber-optic-based electric field sensing is a research area that has received less attention compared to other sensing areas due to the practical issues involved in creating sensitivity to an electric field, which so far have resulted in solutions with notable disadvantages, for example, the difficulty in realizing a compact sensor. Conventional electric field sensors use antennas, conductive electrodes, or metal connections and because of their metallic content are often likely to perturb the unknown field. Furthermore, many conventional sensors involve a conductive path from the sensor to the interrogation system that can be problematic in high field strength or high-voltage environments due to the risk of electrical breakdown. Fiber-optic sensors for voltage and electric field measurement are widely used both in electromagnetic compatibility measurements and in the electric power industry [4]. Unlike their conventional counterparts, fiber-optic-based electric field sensing techniques minimally disturb the electric field, and apart from the sensor head, the connecting fibers are inherently immune to electromagnetic interference [5]. They have several economic and performance advantages over conventional electrode-based sensors, and most importantly, they can provide true dielectric isolation between the sensor and the interrogation system in the presence of very high electric fields or voltages. A wide variety of fiber-optic-based electric field sensing schemes have been proposed and reported to date. These are mostly based on electro-optical crystals employed either in bulk optics or free-space-type configurations or in an integrated waveguide-type configuration [6–10]. However, such schemes have a number of disadvantages such as high losses due to the presence of bulk optics, high coupling losses, costly integration with interconnecting optical fiber, and limited mechanical reliability and are difficult to produce in quantity. Small size, simple design, and an all-fiber configuration with high measurement accuracy are major requirements for fiber-based electric field sensors.

Photonic-crystal fibers (PCFs) [11] with their unique periodic transverse microstructure geometry have been instrumental in extending the functionality of optical fibers, by both improving well-established properties and introducing new features. The efficient light–sample interaction within the microstructure of the PCF enables various applications in sensing, spectroscopy, and nonlinear optics. The infiltration of LCs into the holes of the PCF provides the means of achieving active control on the PCF propagation properties [12–15]. With LCs within the holes of the PCF, the structures are more commonly referred to as liquid-crystal photonic-crystal fibers (LCPCFs). LCPCFs provide a platform to address the requirements of an all-fiber device configuration for the development of optical fiber sensors for applications in electric field sensing.

7.2 TUNABILITY OF LC-FILLED PCFs

7.2.1 LC Tunability

LCs are materials exhibiting properties that are intermediate between solids and liquids. LCs have attractive anisotropic properties that arise due to the typical elongated shape of their constituent organic molecules that impart them with a permanent dipole moment. LC molecules display a tendency to orient in a particular direction with their axes aligned in parallel to each other. The refractive indices of LC materials are highly dependent on ambient conditions such as temperature and externally applied electric or magnetic fields. LCs show anisotropy of their electrical, magnetic, thermal, and optical properties [16,17]. LC materials are characterized by their nematic–isotropic phase transition temperature or clearing point below which they exhibit a liquid crystalline phase. Below the isotropic temperature, the LC material can exhibit a number of distinct liquid crystalline phases. The LC materials used in the studies of this chapter are nematic LC materials [2].

To be used for applications in optical sensing and particularly for electric field sensing, it is important to understand the electrical tunability of LC materials.

The dielectric anisotropy of LCs is given as

$$\Delta\varepsilon = \varepsilon_{\parallel} - \varepsilon_{\perp} \tag{7.1}$$

where $\varepsilon_{\parallel}$ and $\varepsilon_{\perp}$ are the dielectric permittivities in the directions parallel and perpendicular to the director of the LC, respectively, as shown in Figure 7.1. The dielectric anisotropy of an LC is

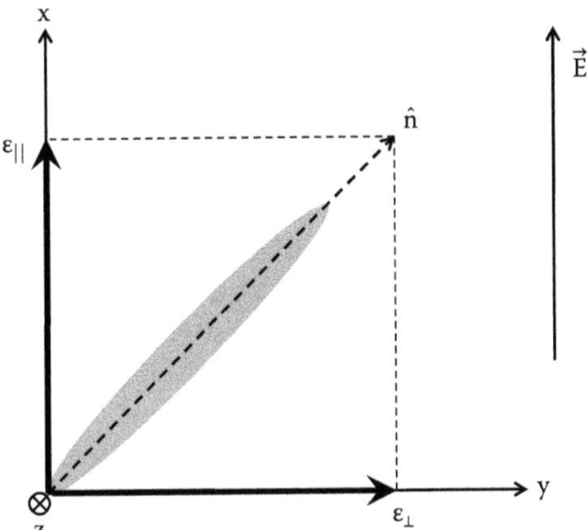

FIGURE 7.1 Parallel and perpendicular components of dielectric constant of the LC, with respect to an applied electric field direction.

determined by two factors, which are the anisotropy of polarizability for the elongated molecules of the LC and the dipole orientation effect [18].

To achieve electrical tunability with LCs, it is important to understand the behavior of the LCs under the influence of an externally applied electric field, particularly the reorientation of LC molecules under the influence of an electric field.

The application of an electric field $\vec{E}$ to an LC produces a dipole moment per unit volume, often referred to as the polarization $\vec{P}$. The polarization depends linearly on the electric field at low field intensities, but in general, the vectors $\vec{P}$ and $\vec{E}$ can have different directions. $\vec{P}$ and $\vec{E}$ are related by a tensor, $\overleftrightarrow{\chi_e}$, called the electric susceptibility, as follows [18]:

$$\vec{P} = \varepsilon_0 \overrightarrow{\chi_e}\vec{E} \quad \text{or} \quad \begin{pmatrix} P_x \\ P_y \\ P_z \end{pmatrix} = \varepsilon_0 \begin{pmatrix} \chi_{e\perp} & 0 & 0 \\ 0 & \chi_{e\perp} & 0 \\ 0 & 0 & \chi_{e\parallel} \end{pmatrix} \begin{pmatrix} E_x \\ E_y \\ E_z \end{pmatrix} \tag{7.2}$$

The induced polarization depends on the orientation of the LC director with respect to the applied field direction. For an arbitrary angle of orientation of the LC director with respect to the applied field, the applied field can be decomposed into components parallel and perpendicular to the LC director.

The induced polarization is given as

$$\vec{P} = \varepsilon_0\chi_{\parallel}\left(\vec{E}\cdot\vec{n}\right)\vec{n} + \varepsilon_0\chi_{\perp}\left[\vec{E} - \left(\vec{E}\cdot\vec{n}\right)\vec{n}\right] = \varepsilon_0\left[\chi_{\parallel}\vec{E} + \Delta\chi\left(\vec{E}\cdot\vec{n}\right)\vec{n}\right] \tag{7.3}$$

Since the dielectric constants $\varepsilon_{\parallel}$ and $\varepsilon_{\perp}$ are related to the susceptibilities as $\varepsilon_{\parallel} = 1 + \chi_{\parallel}$ and $\varepsilon_{\perp} = 1 + \chi_{\perp}$, $\Delta\chi = \chi_{\parallel} - \chi_{\perp} = \varepsilon_{\parallel} - \varepsilon_{\perp} = \Delta\varepsilon$. The electric energy of the LC per unit volume is approximately given by

$$F_{\text{electric}} = -\frac{1}{2}\vec{P}\cdot\vec{E} = -\frac{1}{2}\varepsilon_0\left[\chi_{\perp}\vec{E} + \Delta\chi\left(\vec{E}\cdot\vec{n}\right)\vec{n}\right] = -\frac{1}{2}\varepsilon_0\chi_{\perp}E^2 - \frac{1}{2}\varepsilon_0\Delta\varepsilon\left(\vec{E}\cdot\vec{n}\right)^2 \tag{7.4}$$

If the LC has a positive dielectric anisotropy ($\Delta\varepsilon > 0$), the electrical energy is minimized when the LC director is parallel to the applied field. Conversely, if the dielectric anisotropy is negative

($\Delta\varepsilon < 0$), then the electric energy is low when the LC molecules align perpendicular to the applied electric field.

Rotational viscosity is an important parameter to be taken into account with reorientation of LC molecules on the application of an electric field. The dynamic response of LCs to externally applied electric fields strongly depends on the rotational viscosity of LC. On the application of an external perturbation as in the case with electric fields, the elastic constants of the LC material determine the restoring torque that arises as the system is perturbed from its equilibrium state. When an electric field is applied to reorient the molecules to control the effective birefringence in an electro-optical device, it is the balance between the electric and elastic torque that determines the static deformation of the LC director. Any deformation of the LC director can be divided into a combination of the three possible deformation modes, namely, splay, twist, and bend [19].

The elastic free energy of an LC given by the Oseen–Frank theory incorporating the three deformation modes is given as [20]

$$F = \frac{1}{2}K_{11}\left(\nabla\cdot\vec{n}\right)^2 + \frac{1}{2}K_{22}\left(\vec{n}\cdot\nabla\times\vec{n}\right)^2 + \frac{1}{2}K_{33}\left(\vec{n}\times\nabla\times\vec{n}\right)^2 \quad (7.5)$$

where

K_{11}, K_{22}, and K_{33} are the splay, twist, and bend elastic constants
$\vec{n}$ is the LC director

The elastic constants are temperature-dependent parameters, and they play a larger role in highly confined LC systems. The free energy density is minimized when the director is spatially uniform as often happens in the case with LCs in confined geometries [21].

The response of the LCs to external fields is also dependent in most cases on the interactions between the LC molecules and the forces at the surface boundaries. Reorientation of the LC does not necessarily occur as the applied field rises from zero; rather, there can be a threshold phenomenon such that reorientation only occurs above a critical or threshold field. The dielectric anisotropy and the elastic constant are the parameters that determine the threshold electric field for the LC. For an LC layer of thickness d, with the field applied perpendicularly to the plane of the layer, the threshold field for LC reorientation is given as [22]

$$\text{Splay} \quad E_{th} \approx \left(\frac{\pi}{d}\right)\sqrt{\frac{K_{11}}{\varepsilon_0\Delta\varepsilon}} \quad (7.6)$$

$$\text{Twist} \quad E_{th} \approx \left(\frac{\pi}{d}\right)\sqrt{\frac{K_{22}}{\varepsilon_0\Delta\varepsilon}} \quad (7.7)$$

and

$$\text{Bend} \quad E_{th} \approx \left(\frac{\pi}{d}\right)\sqrt{\frac{K_{33}}{\varepsilon_0\Delta\varepsilon}} \quad (7.8)$$

This reorientation of the director under the influence of an electric field is known as the Freedericksz transition, and the threshold field required for the reorientation is referred to as threshold field E_{th}. The Freedericksz transition threshold for a planar aligning LC is a crucial parameter for LC-based electric field sensors.

7.2.2 Tuning and Sensing Mechanism of LCPCFs

7.2.2.1 Effect of LC Infiltration on PCF Propagation

PCFs guide light either by the modified total internal reflection mechanism (m-TIR) or the photonic bandgap (PBG) mechanism depending on whether the effecting refractive index of the cladding is lower or higher than the core refractive index [23]. For example, commercially available hollow-core fiber HC-1550-02 (NKT Photonics) is an air-guiding fiber that guides light by the PBG mechanism as the effective cladding index of the PCF (air–silica cladding) is higher than the core refractive index (air). For the solid silica core LMA-10 PCF, the guidance mechanism is m-TIR owing to the higher core refractive index of the silica core when compared to the air–silica cladding. Infiltration of materials into the holes of the PCF may lead to the switching of the guidance mechanism of the PCF between PBG and m-TIR mechanisms (Figure 7.2).

With LCs within the holes of the PCF structures, their propagation properties are defined by the refractive indices of the infiltrated LC and the 2D cross-sectional geometry. The LCPCF structure is unique in that it combines the technology of fiber optics along with the variable anisotropy of the LCs. On infiltration with LC, the LCPCF propagation properties will be easily influenced by external electric field and thus can be used for the fabrication of all-fiber-based optical sensors for electric field sensing.

7.2.2.2 Electrical Tuning of Photonic Bandgaps of LCPCF

Most commercially available PCFs are drawn from silica. LCs usually have refractive indices (both ordinary n_o and extraordinary n_e) higher than that of silica (~1.458). On infiltration of LC materials in solid silica core PCFs, the effective refractive index of the cladding becomes higher than that of the silica core. Under these conditions, the propagation properties of the LCPCF are governed by the PBG mechanism. The LCPCF forms a 2D photonic-crystal structure along its cross section with high-index cladding inclusions, and the core, acting as the defect in the PBG structure, guides light through the fiber.

The most important difference between an index-guiding PCF and PBG-based PCF is that in an index-guiding PCF, there is a continuum of guided modes, whereas in PBG fibers, there is a discrete set of guided modes separated by PBGs. As a result, the PBG transmission spectrum is characterized by high (maxima) and low (minima) transmission regimes. The theoretical background for guidance of light through an LCPCF can be analytically explained using the antiresonant reflecting optical waveguide (ARROW) model [24]. The LCPCF structure can be considered as a layered structure consisting of a low-index silica core surrounded by an array of high- and low-index cladding layers. The guiding properties of PBG waveguides are primarily governed by antiresonant reflection from multiple cladding layers. The wavelengths corresponding to the minima of the transmission coefficients are referred to as resonant wavelengths while the wavelengths corresponding to transmission maxima of the spectrum are called antiresonant wavelengths. The transmission maxima originate

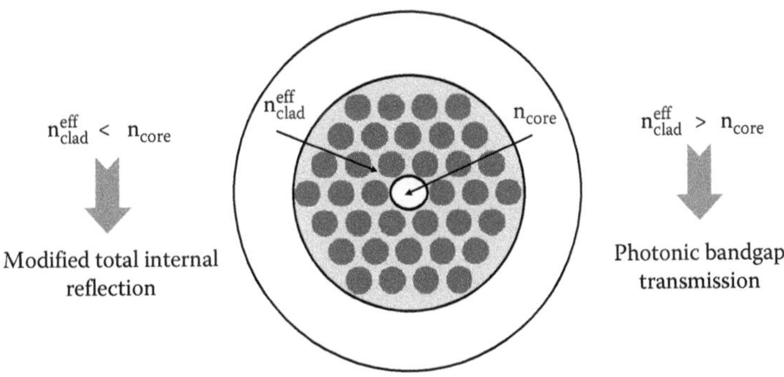

FIGURE 7.2 Schematic of the PCF cross section with conditions for the guidance mechanisms.

from the antiresonant nature of the individual cladding layers with respect to the transverse propagation constants. Each layer can be considered a Fabry–Perot (FP)-like resonator. The narrowband resonance of this FP resonator corresponds to transmission minima for the light propagating in the core or as the resonant wavelengths of the low-index core waveguide. The wideband antiresonances of the FP resonator, which are wavelengths experiencing low leakage as a result of destructive interference in the FP, correspond to a high transmission coefficient for the low-index core waveguide.

The positions of the maxima and minima can be changed by changing the refractive index of the infiltrated material within the holes. As LC materials have variable refractive indices dependent on external electric fields, this facilitates electrical tuning of the PBG and the transmission properties on infiltration within the holes of the PCF.

7.3 ELECTRIC FIELD SENSING USING LCPCFs

An electric field sensor should in principle be able to detect the parameters that define an unknown electric field such as its amplitude, frequency in the case of varying electric fields, and its directional components. LCPCFs allow to fabricate miniature all-fiber electric field (e-field) probe that can detect and measure these parameters of the electric field. In this section, the fabrication of miniature electric field sensing probe for electric field intensity measurement is explained along with the calibration of the sensor in both transmission and reflection modes as in-line- and endpoint-type e-field sensors, respectively. The effect of applied electric field frequency on the propagation properties of LCPCF is explained in this section, and the ability of the miniature sensor to measure both electric field amplitude and frequency is demonstrated. The directional electric field sensitivity of the LCPCF is studied and explained in the section. The capability of the sensing probe to simultaneously measure the transverse directional e-field components and intensity is also demonstrated and presented.

7.3.1 All-Fiber Probe for E-Field Parameters

The underlying principle of LCPCF for electric field sensing is the electric field–dependent PBG transmission of cladding infiltrated solid silica core PCFs. When the nematic liquid crystal (NLC)-infiltrated cladding region of the PCF is subjected to an electric field, the NLC molecules undergo reorientation that changes the effective refractive index of the cladding and under certain conditions allows for tuning of PBG transmission. For LC mixtures that create planar alignment within the holes of the PCF, the propagation through the PCF core is governed by the ordinary refractive index of the infiltrated NLC, in the absence of external field. On the application of an electric field above the threshold, referred to as the Freedericksz transition threshold [22], the LC molecules undergo reorientation within the micro-holes. In the case of an LC layer of thickness d, the threshold electric field E_{th} for LC reorientation is given as

$$E_{th} = \left(\frac{\pi}{d}\right)\sqrt{\frac{K_{11}}{\varepsilon_0\left(\varepsilon_{\parallel} - \varepsilon_{\perp}\right)}} \tag{7.9}$$

where

K_{11} is the splay elastic constant for the nematic LC
$\varepsilon_{\parallel}$, $\varepsilon_{\perp}$ are the dielectric permittivities

As a simple approximation and assuming planar anchoring conditions, the hole diameter of the PCF can be treated as the effective thickness of the LC layer with a planar geometry.

Below the threshold field, the propagation properties of the structure are governed by the ordinary refractive index of the LC given the long-range orientation order of the LC molecules, aligned along the axis of the fiber. Above the threshold, the LC molecules reorient themselves along the direction of the applied field, and the propagation is governed by the effective refractive index, which is partially set by the extraordinary refractive index of the NLC. As the LC molecules begin to reorient under the action of the electric field, there is a loss in the uniformity of the long-range LC molecular alignment. With an increase in the applied electric field intensity, the formation of reverse tilted domains takes place that causes a loss in the LC long-range orientation order. As a result, the PBG condition for transmission in the operating wavelength range is disturbed as light is coupled into the cladding region. In effect as a result, there is a gradual decrease in the optical power transmitted through the core of the LC-infiltrated PCF with an increase in the electric field intensity.

LC materials are characterized by electric permittivity [25], which is frequency dependent in the range from THz to frequencies below 1 kHz [26–28]. As a result, infiltration of LCs into a PCF leads to the propagation properties of the LCPCF being highly dependent on the frequency of the applied electric field. Given the wide range of frequencies that can arise with electric fields, it is important to understand how the frequency of the electric field influences such a sensor.

7.3.1.1 Miniature Electric Field Sensing Probe

The sensor probe fabricated for the demonstration of all-fiber electric field sensing was a <1 cm long MDA-05-2782 LC-infiltrated section of LMA-8 PCF [29], as shown in Figure 7.3. To ensure effective core-to-core light coupling from the single mode fiber (SMF) input fiber to the PCF, a section of the PCF is spliced with an SMF-28 fiber using a standard fusion splicer. The splicing conditions are optimized in order to ensure minimal air-hole collapse at the splice joint [30] and also to provide sufficient mechanical strength for manual fiber handling.

This is achieved by optimizing the fusion current, fusion time, and the number of re-arcs applied during the splicing. The coupling loss estimated after splicing was less than 2 dB for LMA-8 with SMF-28 fiber. The spectral response of the PCF in the wavelength range from 1500 to 1600 nm showed no interference pattern formation after splicing, which suggests that the splicing had not altered the endlessly single-mode operation of LMA-8.

The open end of the PCF was infiltrated with MDA-05-2782 at room temperature by dipping the cleaved end into the LC material. An infiltration length of ~0.5 cm was obtained. The electric field was applied to the infiltrated PCF using a combination of a high-voltage power supply modulated by a standard waveform generator. This provides a positive polarity voltage waveform that varies in time in a sinusoidal fashion from zero volts up to a peak value, V_{peak}, with an average value of $V_{peak}/2$. A sinusoidal waveform was selected since the most common artificial electric fields are sinusoidal. The maximum value of V_{peak} used in the experiment was 1200 V. The frequencies used were 1 kHz and 50 Hz, chosen to reflect a commonly used frequency for LC characterization and the typical frequency of mains power, respectively. A common practice in defining electric field intensity is to define it as volts-RMS per mm. Given the nature of the waveform, the relationship between the V_{peak} value and the RMS value is given by $V_{rms} = (3/8)^{1/2}\ V_{peak}$. Therefore, given

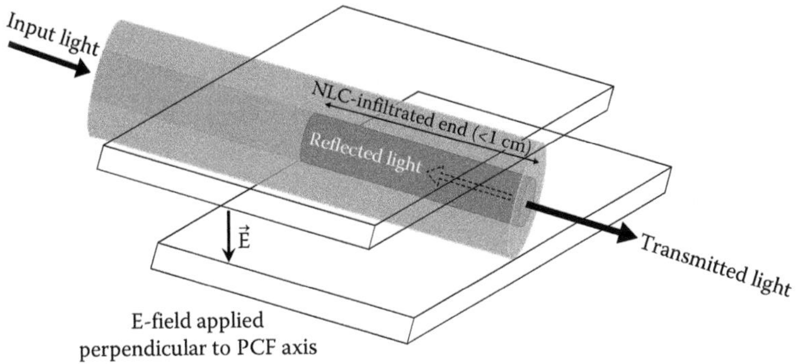

FIGURE 7.3 Nematic LC-infiltrated PCF probe for electric field sensing with electrodes.

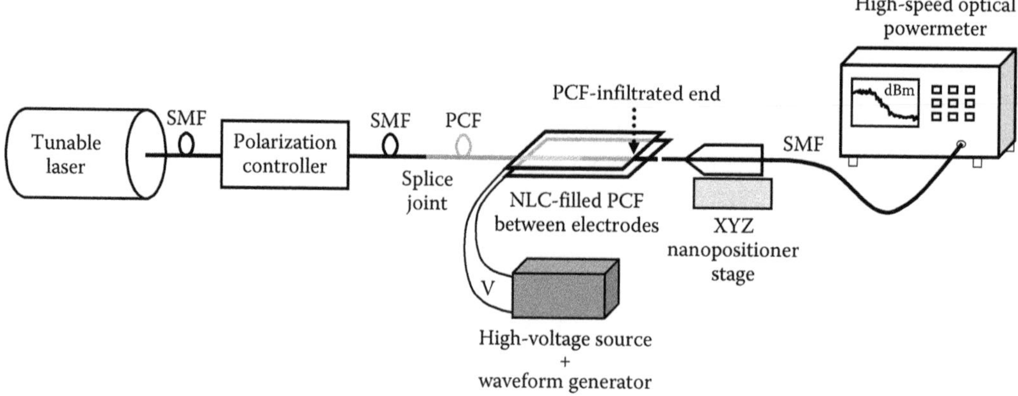

FIGURE 7.4 Schematic of the experimental setup to study the transmission response and temperature dependence of the NLC-infiltrated PCF.

the maximum value of V_{peak} utilized, in effect, the sensing device was subjected to electric field intensities in the range from 0 to 6.0 kV_{rms}/mm given that the distance between the electrodes was ~125 μm. The experimental setup to study the influence of electric field on the transmission properties of the MDA-05-2782-infiltrated LMA-8 is shown in Figure 7.4. Linearly polarized light from a tunable laser source operating at 1550 nm was coupled into the SMF fiber and passed through the infiltrated PCF ends. A polarization controller employed as a linear polarizer at the input was used to minimize polarization-induced instabilities in the output transmittance. In the absence of an applied electric field, LCPCFs exhibit small birefringence as a result of imperfections in the planar (along the fiber axis) alignment of the LC molecules. At voltages above the threshold of reorientation, the LCPCFs become highly birefringent. The output from the infiltrated end of the LCPCF is coupled to another SMF fiber using a butt-coupling technique involving an XYZ nanopositioner stage, with the SMF output coupled to a high-speed optical detector.

The output from the infiltrated end of the PCF is coupled to another SMF fiber by butt coupling using an XYZ nanopositioner stage and is connected to a high-speed optical power meter to record the transmittance as the applied electric field is varied. To study the infiltrated PCF sample in reflection mode, the experimental setup as shown in Figure 7.5 was employed. The light

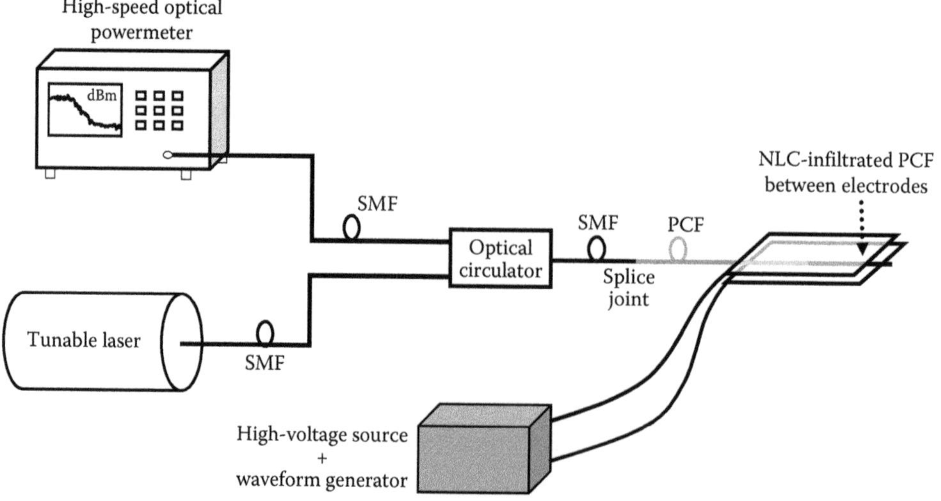

FIGURE 7.5 Schematic of the experimental setup to study reflected power response of the infiltrated PCF.

from the tunable laser source is passed through a fiber circulator and is coupled to the infiltrated PCF. In this case, the infiltrated end of the PCF is left open, and the reflected light from the air–PCF interface is coupled using the circulator to a high-speed optical powermeter for detection and measurement.

7.3.1.2 Measurement of Electric Field Intensity

7.3.1.2.1 Transmission Mode: In-Line-Type Sensor

In the experiments, a polarization controller employed as a linear polarizer was utilized to maximize the extinction ratio of the transmitted power at a high electric field intensities when the infiltrated PCF becomes highly birefringent. The measurement of the transmission response was performed with the state of input polarization set in order to maximize the transmission extinction ratio and to improve the sensitivity of the device to an electric field. The transmission response of the device with increasing electric field intensity is shown in Figure 7.6.

Between the field intensity of zero and that of circa 2.35 kV_{rms}/mm, the transmission response remains unchanged. Above this threshold field intensity, the NLC molecules begin to reorientate, which results in a gradual decrease in transmission through the infiltrated PCF with the increasing electric field until an electric field intensity of ~4.89 kV_{rms}/mm is reached. The transmission response in the range of electric field intensities from 2.35 to 4.89 kV_{rms}/mm is close to linear. The linear part of the transmission response for the infiltrated PCF in this electric field range is shown in Figure 7.7.

A linear fit performed for this electric field range shows that the slope of the response is ~10.1 dB-kV_{rms}/mm, confirming that MDA-05-2782-infiltrated LMA-8 can be used for electric field intensity measurements in this electric field range. Assuming an accuracy of 0.01 dB for the optical power measurement system, the estimated resolution of the device over the usable e-field intensity measurement range is ~1.0 V_{rms}/mm.

Given that one of the most likely applications of the sensor could be in a 50 Hz/60 Hz AC high-voltage transmission environment, the device was also tested in the transmission mode for its sensitivity to a sinusoidally varying electric field at 50 Hz. The transmission response is as shown in Figure 7.8. At 50 Hz, a higher threshold electric field was observed (~2.93 kV_{rms}/mm). Due to the limitations posed by the high-voltage power supply in delivering peak voltages above 1200 V, it was not possible to extract the full linear range of the device with an applied 50 Hz AC

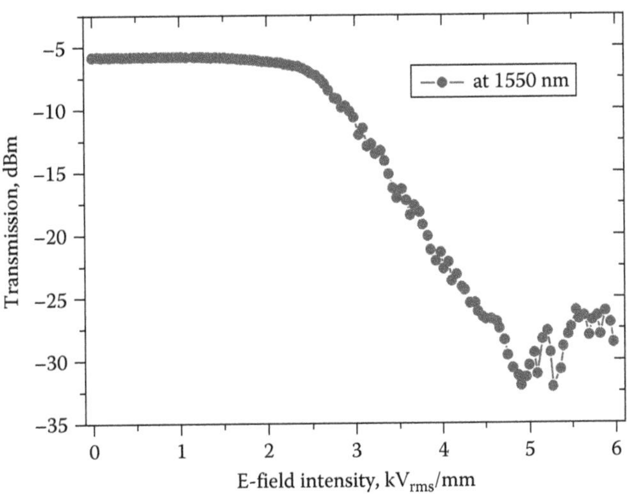

FIGURE 7.6 Transmission response of MDA-05-2782-infiltrated LMA-8 with a changing electric field intensity (1 kHz) at 1550 nm measured at room temperature.

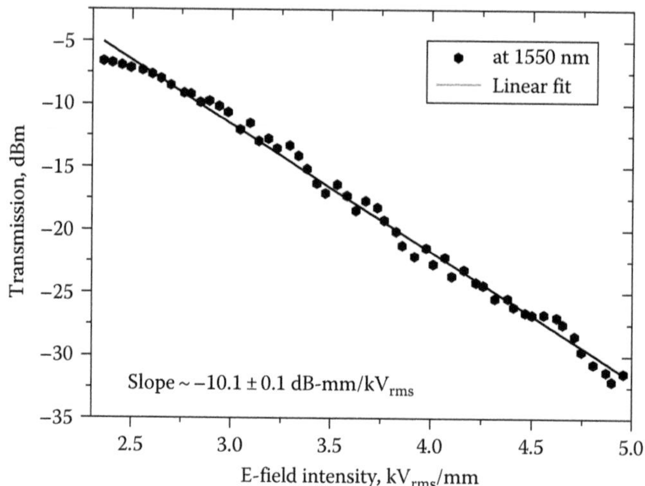

FIGURE 7.7 Linear part of the transmission response of MDA-05-2782-infiltrated LMA-8 with electric field intensity (1 kHz) at 1550 nm shown with a linear fit.

voltage. However, the transmission response in Figure 7.8 confirms that the sensor can be used for 50 Hz/60 Hz AC voltage and electric field sensing [29].

The performance and sensitivity of the device can be further enhanced by employing the sensor in a ratiometric power measurement scheme. The threshold effect of the infiltrated LC puts a limit on the lower measurable electric field intensity.

7.3.1.2.2 Reflection Mode: Endpoint-Type Sensor

The reflected power response at 1550 nm with an increasing electric field (1 kHz) is as shown in Figure 7.9. The reflected power from the infiltrated PCF is relatively low (~20 dB), due to the low reflectance from the air–silica interface (~4%). The response with a varying electric field is found to be similar to the transmission response for the infiltrated PCF.

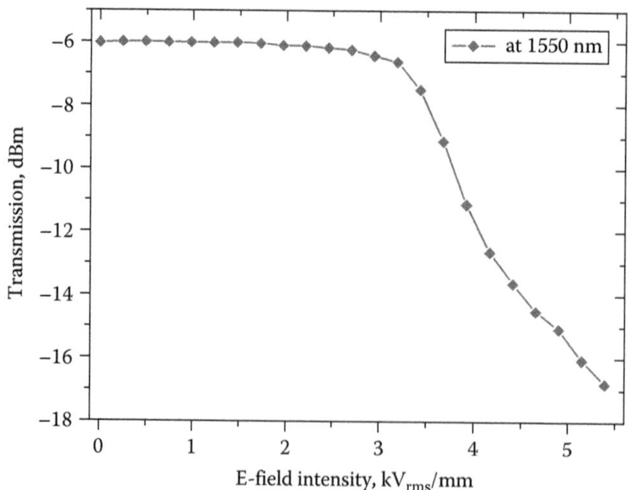

FIGURE 7.8 The transmission response of MDA-05-2782-infiltrated LMA-8 with electric field intensity (50 Hz) at 1550 nm measured at room temperature.

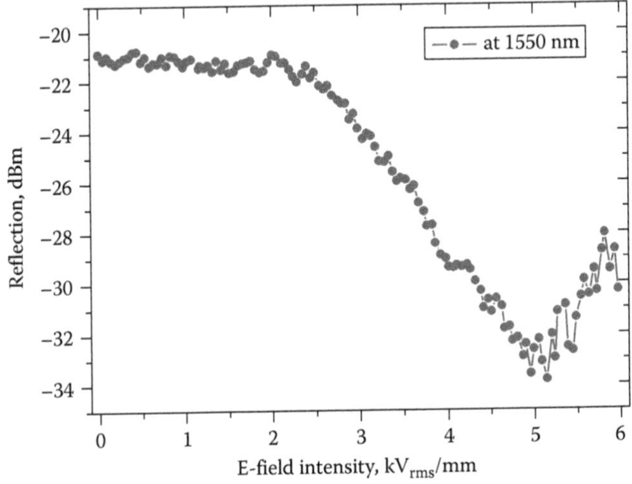

FIGURE 7.9 Reflected power response of MDA-05-2782 infiltrated LMA-8 with a changing electric field intensity (1 kHz) at 1550 nm measured at room temperature.

Due to the expected low measured optical power, arising from the loss upon reflection from the silica–air interface of the cleaved PCF end, there was greater uncertainty in the measured values due to the system noise. It should be noted that the use of the polarization controller in the reflection mode resulted in no change in the reflected power due to the fact that the degree of maintaining of the initial polarization state in the reflection mode configuration is significantly lower than that in the transmission mode. Above the threshold field, the reflected power decreased with an increasing electric field, and the response was again found to be linear with a changing electric field. The linear sensing range is from ~2.35 to ~5.14 kV_{rms}/mm. The linearity of the reflected power with an increasing electric field is depicted in Figure 7.10, shown with the linear fit. The linear fit for the plot provides an estimate of the sensitivity of the device in reflection mode as ~4.55 dB-mm/kV_{rms}. Assuming an accuracy of 0.01 dB for the optical detector in the measurement system, the estimated resolution of the device operated in the reflection mode is ~2.2 V_{rms}/mm. The reduced sensitivity of the device in reflection mode is due to the fact that unpolarized light was used and that the state of

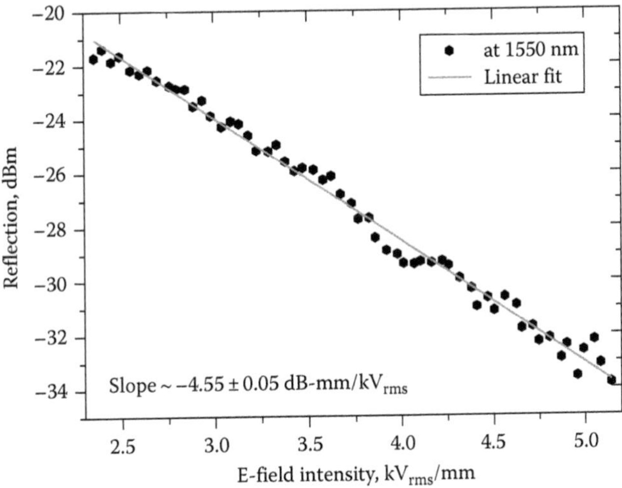

FIGURE 7.10 Linear part of the reflected power response of MDA-05-2782-infiltrated LMA-8 with electric field intensity (1 kHz) at 1550 nm shown with a linear fit.

TABLE 7.1
Comparison of E-Field Sensing Parameters of MDA-05-2782-Infiltrated LMA-8 in Transmitted and Reflected Modes

Sensor Configuration	Sensitivity (dB-kV$_{rms}$/mm)	Measurable E-Field Intensity Range (kV$_{rms}$/mm)	Estimated Resolution (V$_{rms}$/mm)
Transmitted mode (in-line type)	~10.1	2.35–4.95	~1
Reflected mode (endpoint type)	~4.55	2.35–5.14	~2.2

polarization changes after reflection from the open-ended PCF. The increased level of system noise in the reflection mode has also contributed to the reduced sensitivity of the device. In the reflective configuration, the infiltrated PCF can be used as an endpoint e-field sensor, which is an added advantage for the fabrication of sensors for e-field measurement probes in confined spaces.

Table 7.1 summarizes the e-field sensing parameters estimated for MDA-05-2782 infiltrated LMA 8 PCF in both transmitted and reflection modes. It is possible to increase reflected power using a reflective coating at the open end of the PCF, improving the sensitivity of the device in reflection mode.

7.3.1.3 Applied E-Field Frequency Response of LCPCFs

7.3.1.3.1 Background on E-Field Frequency Dependence of LCPCF

On the application of an external electric field, the NLC molecules undergo reorientation that changes the effective refractive index of the holey cladding region and allows for the tuning of PBG transmission. Since the response times of NLC mixtures (switch ON and switch OFF) are usually limited to the order of ~ms, the time response of the transmission through the LCPCF will strongly depend on the frequency of the applied electric field. The application of an AC electric field with a frequency in the order of 1 kHz will produce an amplitude-modulated time-varying transmission response strongly dependent on the applied electric field frequency.

The NLC molecules within the PCF micro-holes reorient under the influence of the electric field above the Freedericksz transition threshold. Above the threshold field, the molecules tend to reorient increasingly toward the direction of the applied field that results in a change in the effective refractive index of the cladding. The maximum field-induced change in the effective refractive index of the cladding in this case results when the field is acting perpendicular to the fiber axis.

On the application of the AC field above the threshold electric field strength and as the amplitude of the electric field signal undergoes a sudden change, the NLC molecules reorientate. The reaction time within which the NLC molecules reorientate when the electric field is turned on is given as [16]

$$\tau_{ON} = \frac{\gamma_1}{\left[(\varepsilon_{\parallel} - \varepsilon_{\perp})\cdot(E^2 - E_{th}^2)\right]} \tag{7.10}$$

where

γ_1 is the rotation viscosity of the NLC

$\varepsilon_{\parallel}$ and $\varepsilon_{\perp}$ are the dielectric permittivities of the NLC that are dependent on the applied electric frequency

E_{th} is the threshold electric field

The relaxation time within which the NLC director decays under the influence of the LC molecular forces, when the field is switched off, is given as [16]

$$\tau_{OFF} = \frac{\gamma_1 d^2}{\left(K_{11}\pi^2\right)} \tag{7.11}$$

where K_{11} is the splay elastic constant of the NLC. The NLC molecules reorientate dynamically under the influence of the alternating field and produce a time-varying transmission response for the LCPCF that is dependent on the reaction and relaxation times of the infiltrated LC mixture.

7.3.1.3.2 Analysis of the Time-Varying Transmission Response of LCPCF

For these experiments, the commonly available solid-core PCF, LMA-10, was used. The LC used for infiltration of the PCF was the low-molecular-weight NLC 5CB (Merck). In order to study the time-varying transmission response of the infiltrated PCF at 1550 nm, a square wave electric field signal with an intensity of ~1.8 kV_{rms}/mm was applied, and the time response was recorded with the high-speed powermeter. Figure 7.11 shows the input square wave signal (at 5 Hz) along with the infiltrated PCF transmittance.

As can be seen, the transmittance response is similar for both the low and high half periods of the applied square wave signal. The NLC molecules undergo a reorientation when there is a change in the electric field strength, and this happens twice during each cycle of the input signal.

With the reference to Figure 7.11, at the beginning of each half cycle and as the electric field amplitude undergoes a change, the NLC molecules reorientate with a time constant as given by Equation 7.10. Initially, the LC molecule reorientation causes degradation of the periodic structure quality in the holey region of PCF, and the decrease in the transmission response takes place. Once the LC molecules assume the new orientation (in this case along the direction of the electric field), the periodic structure quality improves, and the transmission returns to its high value. This switch ON response is comparatively fast and depends on the applied electric field strength. The switch ON time in this case is estimated as τ_{ON} ~ 4 ms. At high electric field intensities above the threshold electric field, the switch ON time of the NLC reduces to values in the order of 1 ms in accordance with Equation 5.1. Given the millisecond order switch ON time of the infiltrated NLC, the LC molecules in the holes of the PCF will reorientate dynamically in phase with applied electric field at frequencies below 1 kHz.

As a result, an LC-infiltrated PCF can be used for detection and measurement of the frequency and amplitude of an applied electric field in the frequency range from 50 Hz to 1 kHz. On the application of a sinusoidally varying electric field, the transmission response of the infiltrated PCF is observed to be periodical with a frequency that is twice that of the frequency of the input signal. This is in accordance with the explanation given earlier. Figure 7.12 shows the transmission

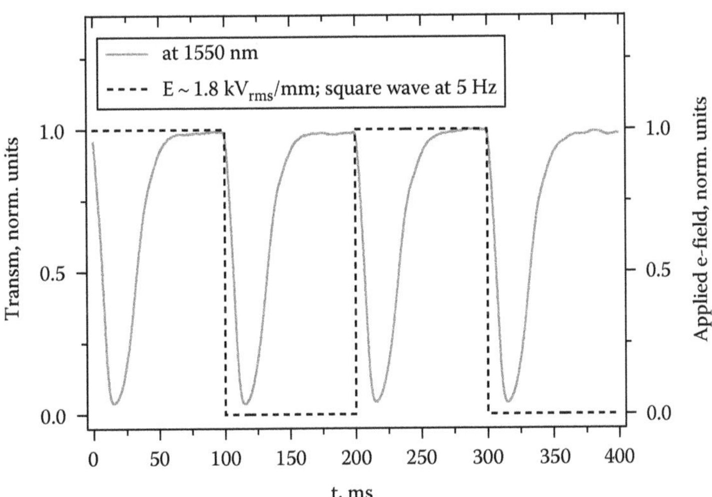

FIGURE 7.11 LCPCF transmission response to a 5 Hz square wave signal at 1.8 kVrms/mm, shown with the input waveform.

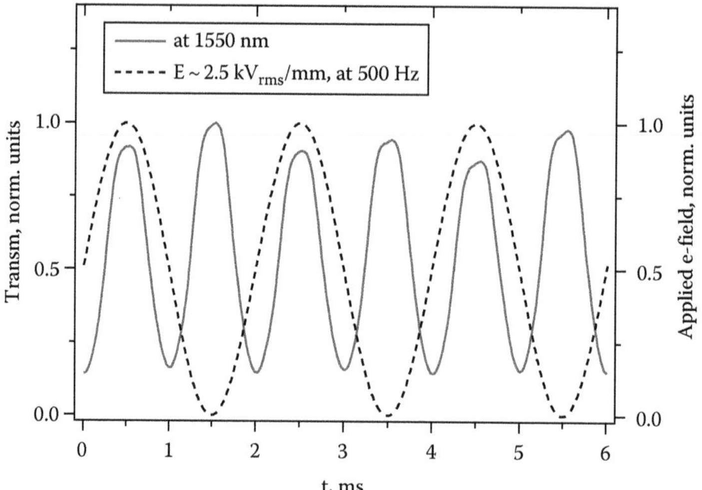

FIGURE 7.12 LCPCF transmission time response to a sinusoidal electric field waveform with an amplitude of ~2.5 kV_{rms}/mm and a frequency of 500 Hz.

response of the LCPCF at 1550 nm when a sinusoidally varying electric field with the amplitude of ~2.5 kV_{rms}/mm is applied at a frequency of 500 Hz. As can be observed, the transmission response is periodically varying with a frequency of 1 kHz.

7.3.1.3.3 Combined Influence of Electric Field Frequency and Intensity

The transmission response of the LCPCF at different frequencies of the applied electric field from 50 Hz to 1 kHz with the electric field intensity varying from 0 to 5.0 kV_{rms}/mm is shown in Figure 7.13. It should be mentioned that all the measurements for this section using the high-speed powermeter are averaged over time, with a sampling rate of ~1000 measurements per reading. For all frequencies, the LCPCF transmittance varies with increasing electric field intensity above the threshold electric field. The threshold electric field for molecular orientation decreases as the frequency of the input signal is increased from 50 Hz to 1 kHz. This behavior for the LCPCF is

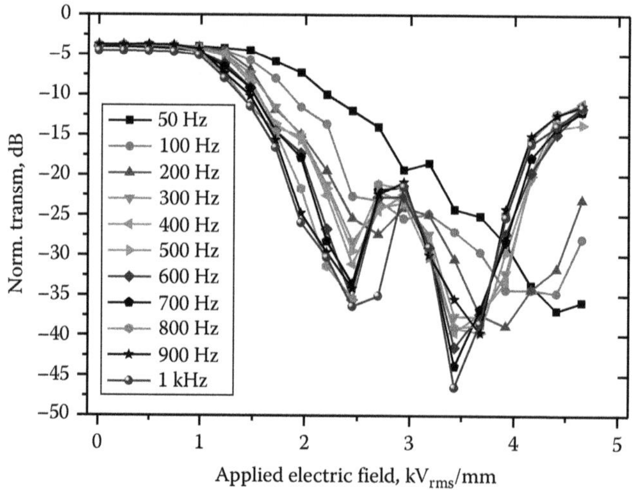

FIGURE 7.13 Frequency dependence of the transmission response of LCPCF with varying electric field intensity at 1550 nm.

found to be similar to that obtained by Scolari et al. [31]. The transmittance response above the threshold field is observed to be close to linear with an increasing electric field, but the region of linearity is found to increase as the frequency is lowered. A notable feature in the transmittance response with increasing electric field is the appearance of transmission peaks in the electric field range from 2.5 to 3.5 kV_{rms}/mm at higher frequencies.

This is because on the application of an electric field above the threshold and with the reorientation of the LC, anisotropy is introduced. As a result, the infiltrated section of the PCF under the action of the electric field behaves as a variable retarder introducing a phase delay between both orthogonal components of the modes propagating through the photonic LC fiber.

This phase retardation is expressed as [32]

$$\delta = \frac{2\pi l \Delta n}{\lambda} \tag{7.12}$$

where

Δn is the effective birefringence of the LC
l is the length of the infiltrated section of the PCF

The effective birefringence of the infiltrated section of the PCF is set by the average orientational state attained by the NLC molecules at each electric field intensity. In this case, the transmission of the linearly polarized input light from the tunable laser source by the LCPCF with an increasing electric field shows oscillatory behavior as the transmittance is related to δ as [33]

$$T = \sin^2\left(\frac{\delta}{2}\right) \tag{7.13}$$

At 50 Hz, the transmission response is observed to be linear for the entire electric field range above the threshold electric field from 1.5 to 4.5 kV_{rms}/mm. At frequencies above 100 Hz due to lower thresholds for molecular reorientation, transmission responses show several minima and maxima at electric field intensities above 2.5 kV_{rms}/mm. At very high field intensities (e.g., as in the case of fields of the order of 3.5 kV_{rms}/mm), the transmission increases as a consequence of the improved periodicity of the holey region infiltrated with the LC. As the LC molecules regain their long-range orientation order in the direction of the applied field, the larger LC–silica refractive index contrast provides better mode confinement, and the guided light throughput increases.

The dynamic properties of LCPCF in responding to a periodically varying electric field offer the possibility to measure an unknown repetition frequency for an applied electric field. In order to demonstrate this, the transmission response of the LCPCF was recorded at an electric field intensity of ~2.5 kV_{rms}/mm for frequencies from 50 Hz to 1 kHz at increments of 50 Hz. The time responses of the infiltrated PCF transmission for an applied electric field with frequencies of 250 Hz, 500 Hz, 750 Hz, and 1 kHz are shown in Figure 7.14. In order to estimate the frequency (f) and other parameters such as the amplitude (A) of the applied electric field from the captured time response of the LCPCF, it is necessary to use a fitted function. A good candidate function is a $\sin^2$ function as follows [34]:

$$T(t) = A\left\{\sin^2\left(2\pi \cdot f \cdot t\right)\right\} \tag{7.14}$$

where

A is the amplitude
f is the frequency of the applied waveform

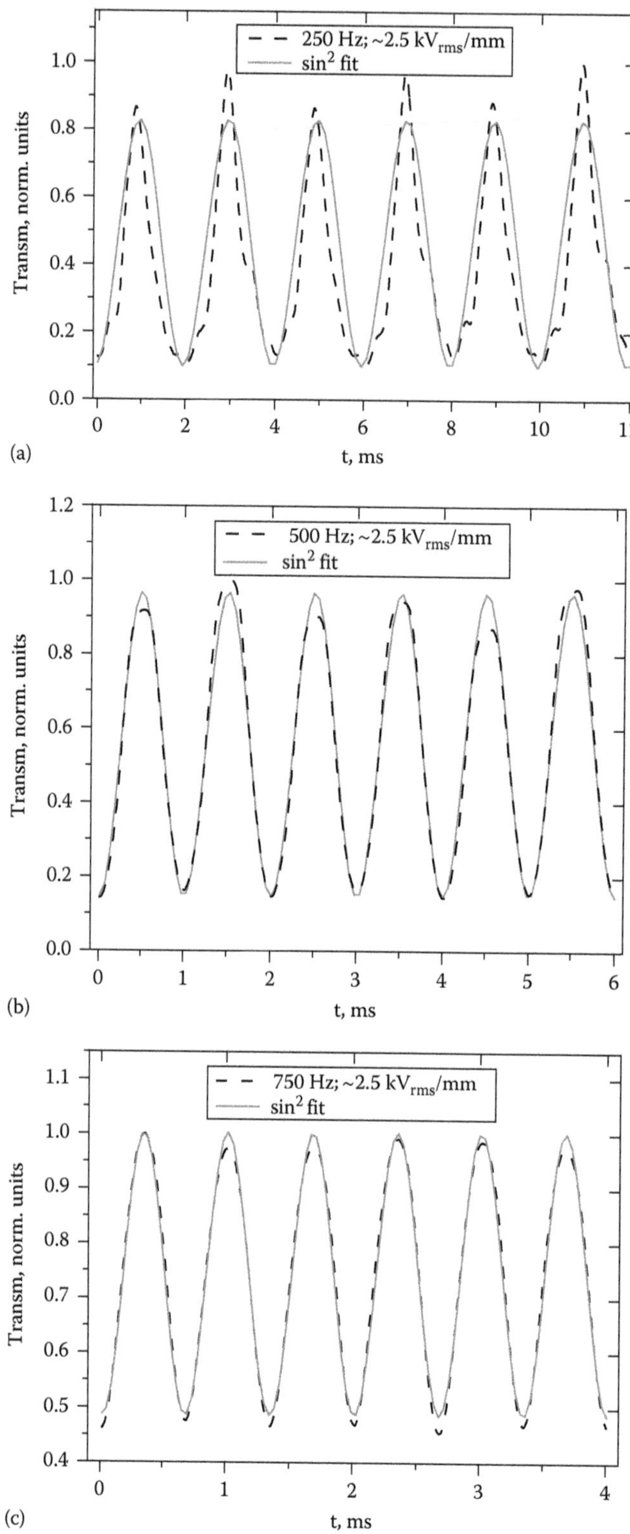

FIGURE 7.14 LCPCF transmission time response at frequencies (a) 250 Hz, (b) 500 Hz, (c) 750 Hz, shown along with a sin^2 fitting.

(Continued)

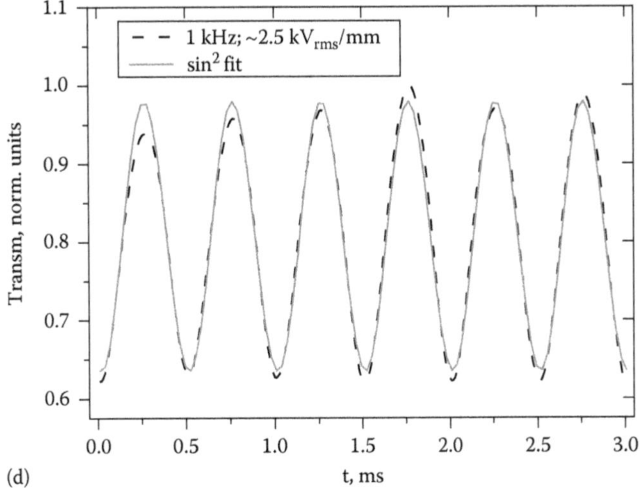

FIGURE 7.14 (CONTINUED) LCPCF transmission time response at frequencies (d) 1 kHz, shown along with a sin^2 fitting.

For each case in Figure 7.14, a sinusoidal fitting based on Equation 5.5 is also shown. To demonstrate the ability to measure an unknown frequency, the repetition frequency was estimated using this sinusoidal fitting. The estimated frequency in each case, obtained as a result of fitting, when compared to the known applied frequency was found to be within a ~±0.5% error margin. Figure 7.15 shows the plot of applied frequency with the percentage error in the measurement of frequency using the LCPCF device in the frequency range from 50 Hz to 1 kHz. This suggests that Equation 5.5 is a good approximation for the time response of the LCPCF transmission in this frequency range and can be used to extract the frequency of an externally applied sinusoidally varying electric field with fixed amplitude. It should be mentioned that slight changes in the frequency (~10 Hz) of the external electric field can lead to a change in the shape of the transmission response waveform, but the use of the fitting function can compensate for this and will still allow the frequency to be estimated correctly.

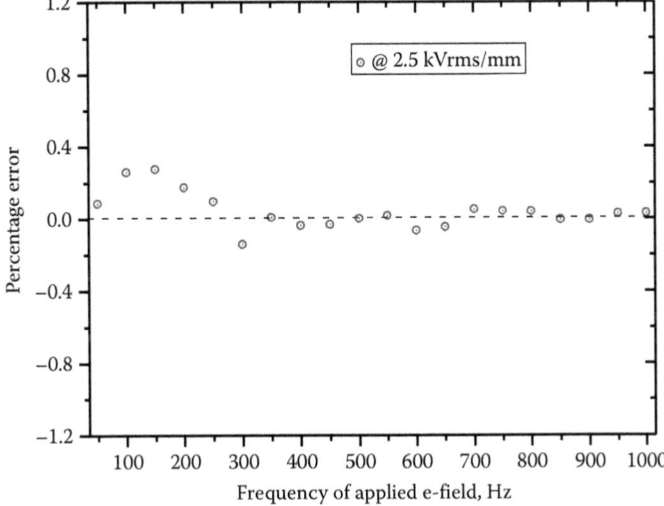

FIGURE 7.15 Applied electric field frequency plotted along with percentage error in the measurement of frequency.

7.3.2 Polarimetric Electric Field Sensing Scheme

In a polarimetric measurement scheme, an optical sensing device, under the influence of the parameter to be sensed, is located between an optical polarizer and a polarization analyzer. The sensing device splits the incident light wave into two orthogonal linearly polarized waves. These two waves propagate with different phase velocities in the polarizing device, and their phases are shifted by a quantity φ. The magnitude of this phase will depend on the parameter being sensed, and thus by measuring the phase shift, it is possible to quantify the unknown physical parameter, assuming a suitable calibration has taken place. Polarimetric optical fiber sensors based on highly birefringent fibers have been extensively investigated for sensing of hydrostatic pressure, strain, vibration, temperature, acoustic waves, etc. [35,36]. Most of these sensors are based on interferometric schemes employing high birefringent (HB) fibers. A polarization-maintaining photonic-crystal fiber (PMPCF) with its elliptical core is an HB fiber and has been employed for various sensing applications for pressure, stress, and vibrations [37–39].

The use of an LC-infiltrated circular-core PCF allows for the measurement of electric field intensity in a range of fields as determined by the threshold electric field and electric field–induced PBG of the LCPCF structure, as described in the previous section. In this study, a nematic LC-infiltrated elliptical-core PMPCF with an optimized infiltration length is used. The optimization of the length of the infiltrated region of the LCPCF subjected to an electric field makes it possible to obtain a linear transmission response for a given electric field range. A lower-threshold electric field and improvement in overall measurable electric field range are obtained with the use of splay-aligning nematic LCs. With the use of an elliptical-core PCF, the sensor head is also shown to have a useful directional sensitivity to the applied electric field. This property of the LC-filled PMPCF is made use for the demonstration of an all-fiber-based directional electric field sensor.

7.3.2.1 Polarimetric Electric Field Sensing Principle

A cross-sectional view of the PMPCF used for the experiments is shown in Figure 7.16. The birefringent axis of the PMPCF that is along a direction orthogonal to the axis passing through the center of the two large holes is also shown. On infiltration of the two large holes of the PMPCF with

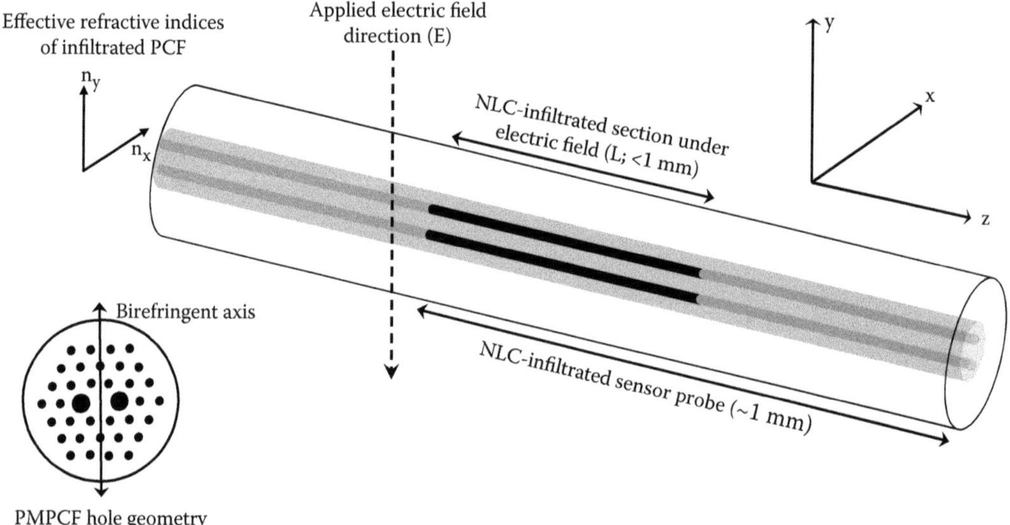

FIGURE 7.16 Selectively infiltrated PMPCF probe showing the axes orientation and applied field direction. The hole geometry of the PMPCF used for infiltration is shown in the bottom left.

an NLC mixture, the birefringent axis undergoes a 90° rotation [40]. The refractive index of the two holes is now set by the effective refractive index of the infiltrated NLC mixture.

On the application of the electric field, the LC molecules undergo reorientation and align themselves along the field direction. The reorientation of the NLC molecules changes the effective refractive index within the large holes that change the phase birefringence of the fiber. The overlap of the propagating core mode of the PMPCF with the infiltrated LC in the two large holes allows for the fiber to exhibit a large variable birefringence on the application of an external electric field. Under these conditions, the infiltrated PCF behaves as a variable phase retarder, with the retardance increasing with an increase in the length of the infiltrated section subjected to the applied electric field.

In a fashion similar to a polarization-maintaining (PM) fiber, the PMPCF with its elliptical core supports two eigenmodes, with dominant electric field components in x and y directions (Figure 7.16). The eigenmodes are characterized by their effective refractive indices n_x and n_y. On infiltration and on the application of the electric field, the fiber birefringence changes to

$$\Delta N = \left(n_{y0} - n_{x0}\right) - \left(n_{yE} - n_{xE}\right) \tag{7.15}$$

where

n_{y0} and n_{x0} are the effective refractive indices in the absence of an electric field
n_{yE} and n_{xE} are the effective indices with the applied electric field

The phase retardance for the light propagating through the infiltrated PCF is given as

$$\phi = \phi_0 + \phi_E \quad \text{where } \phi_E = \kappa_0 \Delta N L \tag{7.16}$$

where

ϕ_0 is the inherent retardance of the infiltrated PCF
ϕ_E is the phase retardance induced by the electric field
$\kappa_0 = 2\pi/\lambda$ is the wave number and λ is the free-space wavelength
L is the length of the infiltrated section under the influence of the applied electric field

Direct measurement of the effective indices of the modes of the infiltrated PMPCF is difficult, and thus in order to characterize the retardance induced by the applied electric field, it is convenient to express the field-induced phase retardance using the characterization term E_π as follows:

$$\phi_E = \pi \frac{\Delta E}{E_\pi} \tag{7.17}$$

where ΔE is the change in electric field intensity. The sensor characterization term E_π can be measured by converting the phase shift into a change in the light intensity transmitted by the fiber using a linear polarizer. If the input and output polarizers are parallel and are at 45° with respect to the infiltrated PCF optical axis (n_y), the output intensity is given as

$$I = \frac{I_0}{2}\left\{1 + \sin\left(\phi_0 + \pi \frac{\Delta E}{E_\pi}\right)\right\} \tag{7.18}$$

The transmitted intensity varies sinusoidally with the electric field intensity with a period set by the electric field–induced phase retardance term that depends on the length of the infiltrated section under the influence of the external electric field. The sensitivity of such a PMPCF sensor to an electric field increases with a decrease in the value of E_π. Since E_π is inversely proportional to the length of infiltration, the sensitivity of the device will increase as the length of the infiltrated section subjected to the applied electric field increases. The optimization of the length of the infiltrated region of the LCPCF subjected to an electric field makes it possible to obtain a linear transmission response for a given electric field range.

7.3.2.2 Selective Infiltration and Infiltration Length Control

The PMPCF used for the experiments is the commercially available PM-1550-01. The presence of two large holes around the core of the PCF introduces a nonaxisymmetric distribution of effective refractive index around the core. A noncircular core combined with a large air–glass refractive index step in the PCF creates a strong form birefringence [41]. It has five rings of air-holes around its solid core. The small holes are of diameter ~2.2 μm, and the two large holes defining the birefringent axis of the fiber have a diameter of 4.5 μm. The fiber has noncircular core, with a major axis 5.4 μm long and a minor axis 4.3 μm long. The intrinsic birefringence of the PMPCF is ~4.0×10^{-4} [42].

For the experiments, the two large holes of the PMPCF were infiltrated with NLC mixtures. The nematic LC mixtures used for infiltration were MDA-05-2782 and MLC-7012 (Merck). The ordinary and extraordinary refractive indices of MDA-05-2782 are ~1.49 and ~1.61, respectively, measured at 589.3 nm, and the isotropic temperature of the material is ~106°C (Merck datasheet). The MLC-7012 material has an ordinary refractive index of ~1.464 and an extraordinary refractive index of ~1.53 measured at 589.3 nm (Merck datasheet). It has an isotropic temperature of 91°C. In order to examine and ascertain the alignment of the nematic LC mixtures in the holes of PM-1550-01 PCF, the NLC mixtures were infiltrated into a silica capillary with ~5 μm inner diameter, which is of the order of the diameter of the large holes of PM-1550-01. The NLC-infiltrated capillaries were then observed under a polarizing microscope. The alignment of MDA-05-2782 was found to be planar, whereas MLC-7012 was found to have a splayed alignment.

A section (~30 mm) of the PMPCF is initially spliced to a PM fiber pigtail using a standard fusion splicer. The fusion current, fusion time, and number of arc discharges are optimized to achieve minimal air-hole collapse of the PMPCF at the splice joint and also to ensure minimal degradation of the PM fiber structure at the splice joint. The splice loss for the PMPCF to PM fiber joint is estimated to be ~6 dB. The sensor head is a <1 mm NLC-infiltrated section of the PMPCF (Figure 7.16). Selective infiltration of the two large holes was performed by collapsing the smaller holes around the core using a standard fusion splicer. In order to achieve this, the cleaved end of the PMPCF is kept between the electrodes of the fusion splicer, and controlled arc discharges are applied. The offset distance from the center of the electrode axis, fusion current, and fusion time are optimized in order to collapse all the smaller holes around the core leaving only the two large holes open [43]. Subsequently, the PCF was infiltrated with a nematic LC mixture by dipping the cleaved end of the PCF into a drop of the NLC mixture at room temperature. The infiltrated PCF is observed under a polarizing microscope to ascertain the quality of infiltration, and care is taken to ensure that both the holes of the PMPCF are evenly infiltrated.

Uneven infiltration results in high insertion loss and sinusoidal spectral interference patterns. After infiltration, the collapsed end is cleaved off resulting in the total length of infiltration within the PCF after cleaving in the order of ~1 mm. This is done to minimize the insertion loss of the sensor, which increases with an increase in the length of infiltration. For MDA-05-2782-filled PMPCF, an infiltration length of ~0.8 mm was observed using the polarizing microscope by cleaving after infiltration, and for the MLC-7012-filled sample, the infiltration length achieved using the same method was ~1.6 mm. The disparity between the lengths achieved and the 1 mm desired length is a result of the limited accuracy with which the cleave length can be set. Furthermore, due to the limitations of the translation stage of the polarizing microscope, the measurement accuracy for the total infiltration length is ±0.1 mm. The infiltration of the LC mixture into the PCF introduces an additional insertion loss of ~6 dB at 1550 nm for both the NLC samples. To test the feasibility of utilizing the selectively infiltrated PMPCF sensor head for polarimetric electric field sensing, the experimental setup shown in Figure 7.17 was employed.

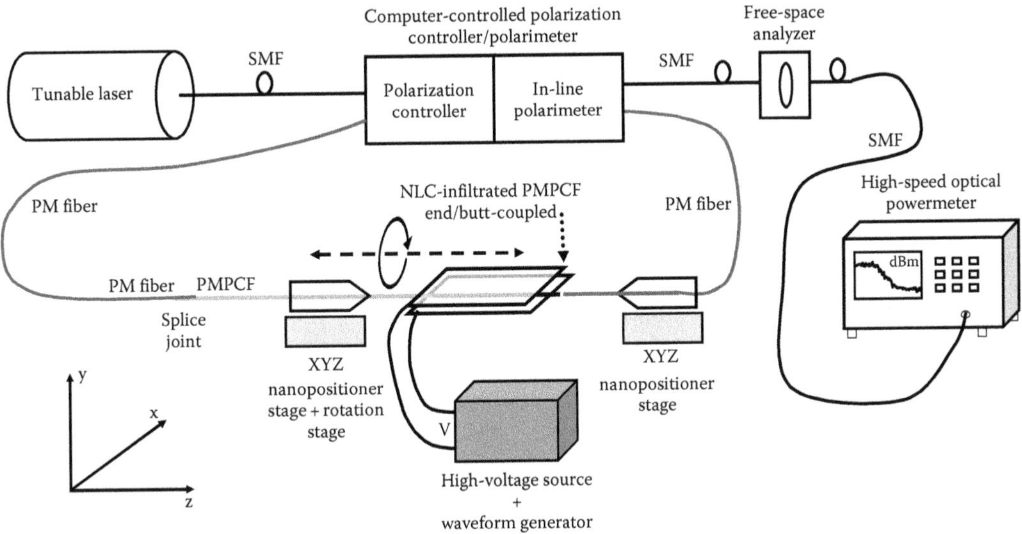

FIGURE 7.17 Schematic of the experimental setup for polarimetric e-field sensing.

Light at 1550 nm from a tunable laser source was linearly polarized using a polarization controller (DPC5500, Thorlabs) and coupled into the input PM fiber with the selectively infiltrated PMPCF spliced to its other end.

The PMPCF is clamped on to a precision bare fiber rotator, and the rotator is mounted on an XYZ nanopositioner stage (1.0 μm translation accuracy). The infiltrated end of the PCF is positioned between two electrodes with a spacing of ~150 μm between them. The fiber axis (n_y) of the PCF is arranged to be parallel with the electric field direction as in Figure 7.16. This was done by using the fiber rotator and by viewing the end facet of the PMPCF under a high-resolution digital microscope. The transmission of the infiltrated PCF in the presence of an external electric field is found to be dependent on the orientation of the fiber optical axis with respect to the direction of the electric field. With the axis of the fiber aligned along the electric field direction, the electric field–induced birefringence change is maximized.

The light transmitted by the PMPCF after passing through the infiltrated section was butt coupled to an output PM fiber pigtail and then coupled to an in-line polarimeter (IPM5300 Thorlabs). The butt coupling at the infiltrated end was done using another XYZ nanopositioner stage. The output from the polarimeter is passed through a free-space analyzer and coupled to the optical detector to record the value of transmittance.

The electric field was applied to the infiltrated PCF using a combination of a high-voltage power supply modulated by a standard waveform generator operating at 1 kHz. This provides a positive polarity voltage waveform that varies in time sinusoidally from 0 V up to a peak value, V_{peak}, with an average value of $V_{peak}/2$. The maximum value of V_{peak} used in the experiment is 1000 V. The relationship between the V_{peak} value and the RMS value is given by the following: $V_{rms} = (3/8)^{1/2} V_{peak}$. Since the distance between the electrodes in the experimental arrangement was ~150 μm, in effect, the sensing device is subjected to electric field intensity in the range from 0 to 4.1 kV_{rms}/mm.

In order to study the influence of the length of the infiltrated section within the electric field on the output transmittance, the infiltrated section was translated between the electrodes, using the input nanopositioner stage, in fixed length steps. The polarized transmittance was recorded in the range from 0 to 4.1 kV_{rms}/mm. It should be mentioned that although a sinusoidally varying AC field was applied to the sensor head, the polarized transmission data provided by the high-speed optical power meter are averaged over time. It is also possible to alter the infiltration length by altering the quantity of LC material introduced into the PMPCF, but achieving the equivalent fine increments in the infiltrated length for the purpose of this study would involve microliter control of the infiltrated

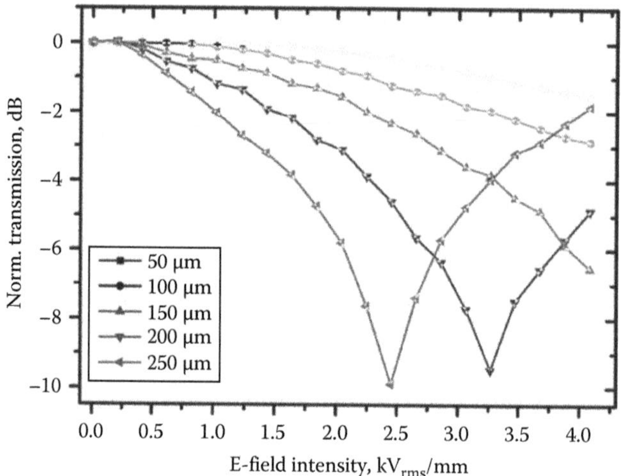

FIGURE 7.18 Transmission response of MLC-7012-filled PM-1550-01 at 1550 nm with varying electric field intensity obtained for different lengths of infiltrated section between the electrodes.

LC quantity that is difficult to realize experimentally. It is for this reason that translation of the infiltrated section with respect to the electrodes is used as a more practical alternative.

Nematic LC mixtures with a splayed alignment do not show threshold effects [44]. Figure 7.18 shows the polarized transmittance at 1550 nm for MLC-7012-infiltrated PMPCF recorded for different lengths of the infiltrated section within the electrodes.

The initial adjustment for the length of the infiltrated section was carried out as explained in the section earlier. The infiltration length was incremented in steps of 50 μm from the reference point for this sample. Unlike the MDA-05-2782, the reorientation of the LC molecules within the holes of the PCF takes place at very low values of the electric field. Since the birefringence of MLC-7012 is lower than that of MDA-05-2782, longer lengths of infiltrated section had to be subjected to the electric field to attain a phase retardance of the same order of magnitude. A near linear response was obtained in the case with 150 μm length of the infiltrated section within the electrodes for the entire applied electric field range. Figure 7.19 shows the normalized

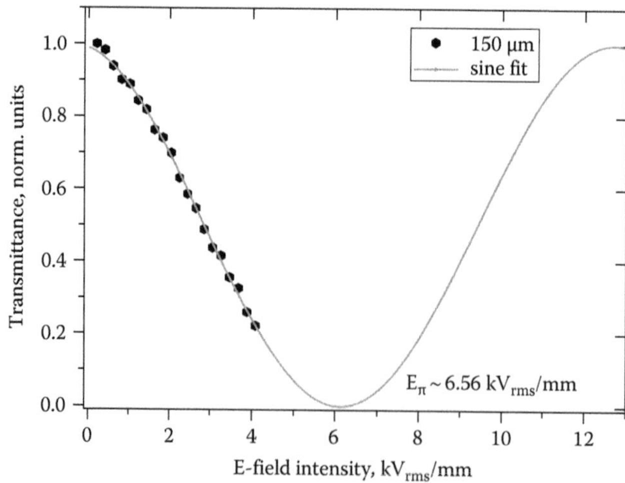

FIGURE 7.19 Transmission response with 150 μm length of infiltrated section within the electrodes and its sine fit.

transmission response obtained at this length of the infiltrated section within the electrodes. A fit based on Equation 7.18 for an electric field range from 0 to 4.0 kV_{rms}/mm is also shown. The value of E_π for the MLC-7012-filled sample estimated from the fit was ~6.56 kV_{rms}/mm. The MLC-7012-infiltrated PMPCF has a useful electric field measurement range from 0 to 4.0 kV_{rms}/mm, and by careful adjustment of the length of the infiltrated section within the electrodes, the sensor can be fabricated so it has a linear electric field response using a measurement scheme based on transmission intensity.

In this case, a linear electric field range can be assumed for the 150 μm sample from 1.0 to 4.1 kV_{rms}/mm. On performing a linear fit, the sensitivity in this case is estimated as ~2.0 dB-kV_{rms}/mm. Assuming a resolution of 0.01 dB for the optical power measurement system, the estimated resolution of the device over this e-field intensity range is ~5×10^{-3} kV_{rms}/mm (~5×10^{3} V_{rms}/m). The MDA-05-2782-infiltrated PMPCF sensor has a lower E_π value when compared to an MLC-7012-infiltrated PMPCF sensor and therefore displays a significantly higher sensitivity to electric field intensity and thus a higher measurement resolution than the MLC-7012-infiltrated PMPCF sensor. However, the advantage of the MLC-7012-infiltrated PMPCF sensor is that it can operate from e-field intensities close to zero, whereas the MDA-05-2782-infiltrated PMPCF sensor only operates above a threshold electric field value.

With an appropriate choice of an LC mixture for infiltration and by controlling the length of infiltration, these structures can be customized for the measurement of electric field intensity in a specified electric field range [45]. Given the high birefringence of the NLC mixtures available and the hole size of the PMPCF used, it is estimated that control on the infiltration length with an accuracy of ~1 μm is desirable, in order to achieve precise control of the electric field–induced phase retardance.

It should be mentioned that precise control on the length of infiltration (~μm accuracy) is difficult to achieve with the standard procedure used by various authors for infiltrating PCFs (also used in our experiments). Acceptable control on the infiltration length within the PCF can be achieved by injecting a known quantity of LC mixture into the PCF using a controlled syringe pump arrangement with ~1 μL volume delivery and/or by fiber cleavers with high precision control of the cleave length to obtain a required infiltration length by cleaving.

7.3.2.3 Electric Field Directional Sensitivity of LCPCFs

Directionally sensitive electric field sensors allow for the measurement of electric field components and can be used for electric field mapping. In [46], a technique employing a GAAs crystal integrated with an optical fiber for the detection of electric field components was demonstrated. Electric field mapping using a sensor based on an electro-optic crystal has also been demonstrated in [7,47], where the symmetry properties of the electro-optical crystal were utilized. These approaches involve the integration of fibers and electro-optical crystals and therefore have a number of disadvantages such as high coupling losses, high cost, and limited mechanical reliability and are difficult to produce on a large scale.

In the previous section, the capability of an LC-filled PMPCF probe to measure the electric field intensity was demonstrated. Due to the elliptical core of the PMPCF combined with an electric field–induced phase birefringence on LC infiltration, the LCPCF probe is sensitive to the direction of the applied electric field. The electric field directional sensitive sensor probe consists of a 1 mm long selectively infiltrated section of a PMPCF. Figure 7.20 shows the infiltrated PCF orientation with respect to the electric field direction. The angle between the electric field direction and the fiber polarization axis (n_y) is θ. Figure 7.21 shows the schematic of the large diameter holes of the PCF along with the average orientation of the LC molecules (NLC director) on the application of electric field. Within each hole, as the molecules orient along the field direction, the NLC director component in the direction of the field is given by $n_E = n_e \sin \varphi$, as shown in Figure 7.22, where φ is the angle between the fiber propagation direction (z) and the NLC director and n_e is the extraordinary refractive index of the NLC.

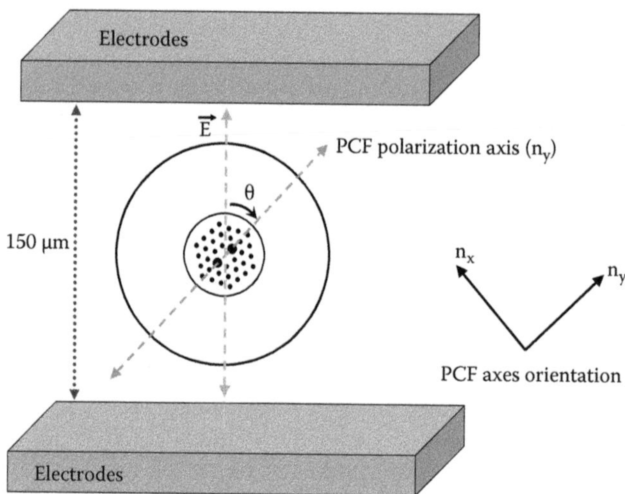

FIGURE 7.20 Infiltrated PCF polarization axis orientation and electric field direction.

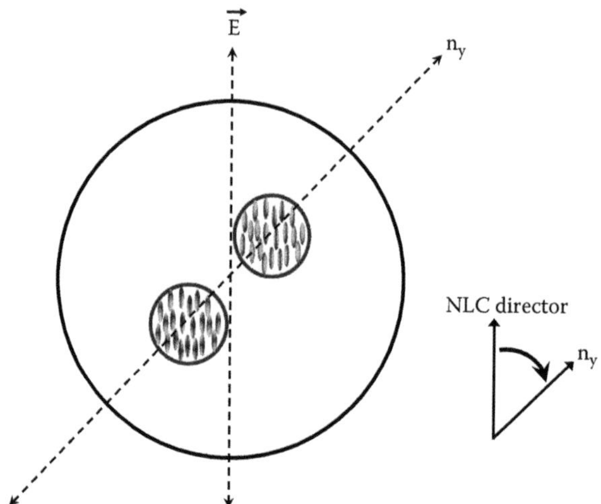

FIGURE 7.21 Large holes of PMPCF with NLC molecular alignment on application of electric field.

As the NLC molecules reorient along the electric field direction (increase in φ), the component n_E increases in magnitude. Linearly polarized light with a polarization direction at 45° with respect to the PCF axis (n_y) in this case will undergo increasing retardance with an increase in electric field intensity for a fixed length of the infiltrated section within the electrodes. For a fixed electric field intensity, on rotation (increase in angle θ), the component of n_E along n_y given by $n_y = n_E \cos\theta$ decreases as θ increases from 0° to 90°, so in this case, the phase retardance experienced by the light decreases. Using the same explanation as earlier, an increase in θ from 90° to 180° will produce the same retardance as θ going from 90° to 0°. As a result, the polarized transmission response of the infiltrated PMPCF with the polarization axis of the PMPCF rotated from 90° to 180°, with respect to the electric field direction, will be the same as when rotation is from 90° to 0°.

The PMPCF probe selectively infiltrated with MLC-7012 was chosen for this study. This sample was chosen as the MLC-7012 showed a splayed alignment. As mentioned and demonstrated in the previous section, splay-aligned LC mixtures within PCF holes have a low electric field threshold for LC molecular reorientation and provide a larger useful electric field measurable range.

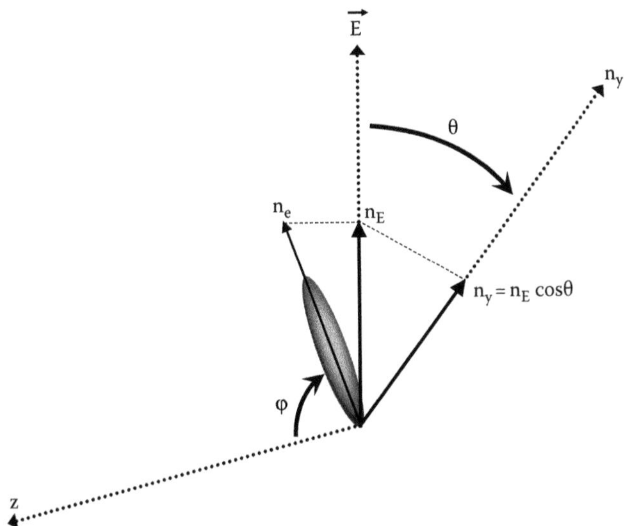

FIGURE 7.22 Average molecular orientation and its projection on the electric field direction and PCF polarization axis.

The experimental setup used for the study is the same as shown in Figure 7.17. The electric field was applied in the form of a sinusoidally varying positive polarity waveform at 1 kHz frequency, using a combination of a high-voltage (1000 V) power supply modulated by a standard waveform generator. By careful adjustment with the XYZ translation stage and by observing the polarized transmission, the infiltration length under the electrodes can be optimized for a particular phase change. The length of the infiltrated section within the electrodes is initially adjusted to have a phase change of ~π/2 for the entire applied electric field range from 0 to 4.0 kV_{rms}/mm. This ensured that a monotonically decreasing transmittance response was obtained for an increasing electric field strength in the range from 0 to 4.0 kV_{rms}/mm.

The sensitivity of the infiltrated PMPCF to the direction of an electric field is studied in a polarimetric scheme, with the length optimized infiltrated section of the PMPCF being the sensing element [48]. In order to change the orientation of the fiber with respect to the electric field direction, the infiltrated fiber section was mounted on a precision fiber rotator and was rotated within the fixed electrodes to study the effect of change in fiber orientation with respect to applied electric field direction.

Figure 7.23 shows the polarized transmission response of the selectively infiltrated fiber at 1550 nm for different orientation angles of the PCF polarization axis with respect to the electric field direction. As can be observed from the plots, the polarized transmittance through the fiber decreases monotonically with the electric field intensity for each orientation angle. For each case of the orientation of the PCF polarization axis, the $n_e \cos \varphi$ component increases with electric field intensity, resulting in higher phase retardance of the propagating light so that a decrease in the polarized transmission occurs. From Figure 7.22, it can also be observed that as the electric field intensity increased, there is an increase in the difference between the transmitted power at $\theta = 0°$ and at $\theta = 90°$. It was also observed that the slope of the transmission response with electric field intensity reduced with an increase in θ from 0° to 90° as the fiber was rotated, in the electric field range from 1.0 to 4.0 kV_{rms}/mm. A sinusoidal fitting based on Equation 7.18 on the polarized transmission response at $\theta = 0°$ and at $\theta = 90°$ yielded an estimate of the E_π values as ~7.02 and ~4.82 kV_{rms}/mm, respectively. The sensor has a higher sensitivity to electric fields oriented parallel to the PCF polarization axis. The sensitivity decreases as the angle θ is increased and is found to be lowest when the polarization axis is orientated orthogonally to the electric field direction. Figure 7.24 shows the angular dependence of the transmitted power versus orientation of the PCF polarization axis at a fixed electric field intensity of 3.67 kV_{rms}/mm. A linear fit performed on the

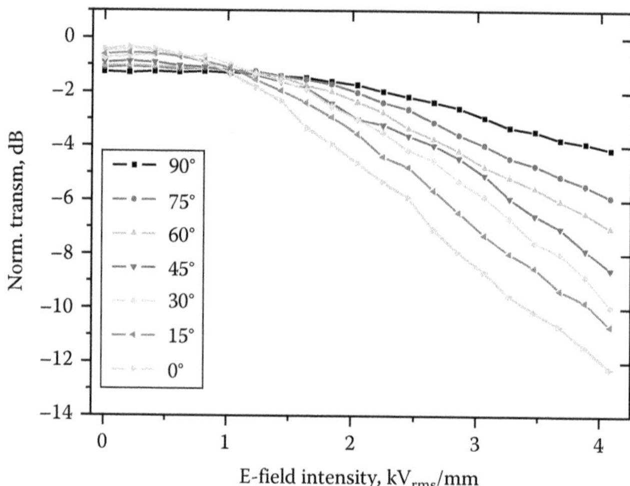

FIGURE 7.23 Polarized transmission response at 1550 nm for different orientations of the PCF polarization axis with respect to electric field direction with increasing electric field intensity.

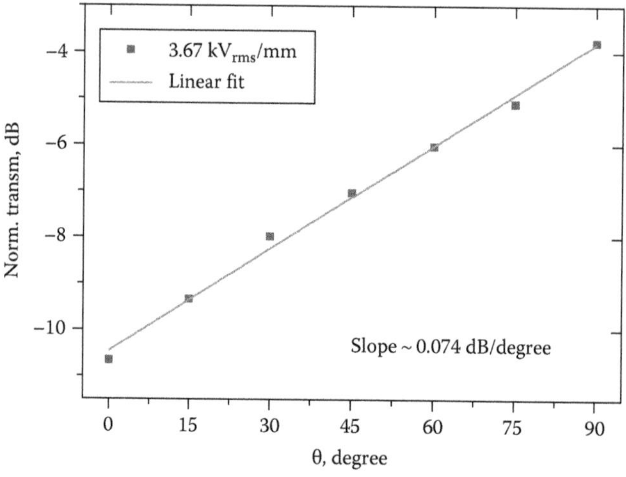

FIGURE 7.24 Angular dependence of transmission on the orientation of the PCF polarization axis with respect to the direction of electric field at an electric field intensity of 3.67 kVrms/mm, shown here with the linear fit.

data gives an estimate of the angular sensitivity of the PMPCF orientation with respect to the electric field as approximately −0.07 dB/degree at a fixed electric field intensity of 3.67 kV$_{rms}$/mm. It should be noted that while the data in Figure 7.24 are plotted with a linear intensity scale, the relationship is nonlinear as expected from Equation 7.18.

A dB intensity scale was used in order to provide an estimate of the angular sensitivity of the device. As explained above (see Figure 7.22), at a fixed value of the electric field intensity, the $n_E \cos \theta$ component decreases in magnitude as θ is increased from 0° to 90° by rotation of the PCF. Correspondingly, the phase retardance suffered by the light decreases, thereby increasing the polarized transmittance. For a fixed electric field intensity, the sensor allows for the determination of the direction of the externally applied electric field from the polarized transmission response. It should be noted that, the response of the sensor is similar for θ going from 0° to 90° and from 180° to 90°; this gives rise to a directional ambiguity that can be resolved with the use of two such similar sensors, placed orthogonally to each other in the electric field.

7.4 CONCLUSIONS AND FUTURE OUTLOOK

In this chapter, the use of LC-infiltrated PCF as an all-fiber probe for sensing electric field and its various parameters are studied experimentally and presented. A brief review of electric field sensors based on various methods is presented. An all-fiber electric field sensing probe based on the electrical tunability of the nematic LC-infiltrated PCF is shown to have a linear transmission and reflection responses with a varying electric field intensity. The temperature dependence of the electric field sensor is evaluated and presented. The LCPCF-based sensor can provide a resolution of ~1 V_{rms}/mm and with the simple all-fiber design of the sensor ensures a compact form factor and allows for easy integration and coupling with fiber optics. The capability of the sensing device to operate as an in-line sensor in the transmitted mode and also as an endpoint sensor in the reflected mode is demonstrated. As an inherent electric field sensing device, the sensor can also be used for high-voltage sensing in the range from 500 V to 15 kV, with the voltage being applied using a fixed electrode configuration. A disadvantage of the use of planar aligning LCs is the presence of a threshold that limits the measurable electric field range.

The dependence of the transmission response of a nematic LC-infiltrated PCF on the frequency of an externally applied electric field was experimentally studied in the frequency range from 50 Hz to 1 kHz. The LCPCF transmission response in time on the application of a sinusoidally varying electric field is found to vary periodically. The time-varying transmission response can be approximated using an appropriate scaled $\sin^2$ function. The fitted function can be used to retrieve the input electric field waveform parameters such as frequency and amplitude. The selective infiltration and infiltration length optimization for the LC-filled PMPCF are studied for all-fiber electric field sensing applications. A polarimetric scheme was employed wherein the phase retardance of the selectively infiltrated PMPCF was controlled by carefully adjusting the length of the infiltrated section subjected to the electric field. By appropriate adjustment of the infiltration length within the electrodes, the sensor can be optimized to provide a linear transmittance response to an external electric field in a polarimetric electric field sensing scheme.

The directional sensitivity of a splay-aligning nematic LC-infiltrated PMPCF in an electric field for sensing of electric field components is also demonstrated. The LCPCF-based sensor is capable of simultaneous detection and measurement of the direction and magnitude of the applied electric field making it a true all-fiber directional electric field sensor. The sensor probe has higher sensitivity to electric field component aligned along the PMPCF polarization axis. The sensitivity was lowest when the polarization axis is oriented orthogonal to the electric field direction. A linear relationship was established between the angle of orientation of the polarization axis with respect to the electric field direction and the transmitted power at fixed electric field intensity. The sensor allows for the determination of the direction of an electric field at fixed electric field intensities, thereby allowing for the measurement of the components of an externally applied electric field.

LC normally has a very low sensitivity to magnetic fields. The doping of ferromagnetic particles enhances the magnetic field sensitivity of LCs. These new types of LCs are referred to as ferronematic LCs. Infiltration of ferromagnetic particle doped nematic LC into PCF will lead to a significant increase in the sensitivity of the PCF propagation properties to magnetic fields, which has scope for future applications in the area of magnetic field sensing and current sensing. With appropriate device configurations, such structures can be used for the fabrication of all-fiber-based magnetic field and current sensors. This area also presents the opportunity to synthesize new LC materials with different doping materials and optimized doping concentrations to enhance the sensitivity of the host LC materials and in turn the LCPCF to external magnetic fields on infiltration. The measurement in real time of current and voltage level is required in a wide range of electrical and electronic applications ranging from fault diagnostics to metering of power consumption to overload protection. In this regard, fiber-optic sensors with the capability to simultaneously measure electric voltage and current offer various advantages when compared to the conventional methods. With the use of ferroelectric nanoparticle doped LC, a common platform for all-optical sensing of current and voltage sensing can be realized.

REFERENCES

1. D. K. Yang and S. T. Wu, *Fundamentals of Liquid Crystal Devices* (John Wiley & Sons, Chichester, U.K., 2002).
2. P. J. Collings, *Liquid Crystals: Nature Delicate Phase of Matter*, 2nd edn. (Princeton University Press, Princeton, NJ, 2002).
3. E. Udd (ed.), *Fiber Optic Sensors: An Introduction for Engineers and Scientists* (John Wiley & Sons, New York, 2006).
4. B. Culshaw, Fiber optics in sensing and measurements, *IEEE J. Sel. Top. Quant. Electron.*, 6, 1014–1021, 2000.
5. V. M. N. Passaro, F. Dell'Olio, and F. De Leonardis, Electromagnetic field photonic sensors, *Prog. Quant. Electron.*, 30, 45–73, 2006.
6. C. G. Martinez, J. S. Aguilar, and R. O. Valiente, An all-fiber and integrated optics electric field sensing scheme using matched optical delays and coherence modulation of light, *Meas. Sci. Technol.*, 15, 3223–3229, 2007.
7. H. Togo, N. Kukutsu, N. Shimizu, and T. Nagatsuma, Sensitivity-stabilized fiber-mounted electrooptic probe for electric field mapping, *J. Lightwave Technol.* 26, 2700–2705, 2008.
8. M. Bernier, G. Gaborit, L. Du Villaret, A. Paupet, and J. L. Lasserre, Electric field and temperature measurement using ultra wide bandwidth pigtailed electro-optic probes, *Appl. Opt.*, 47, 2470–2476, 2008.
9. C. Li and T. Yoshino, Optical voltage sensor based on electrooptic crystal multiplier, *J. Lightwave Technol.*, 20, 843–849, 2002.
10. A. Michie, I. Bassett, and J. Haywood, Electric field and voltage sensing using thermally poled silica fibre with a simple low coherence interferometer, *Meas. Sci. Technol.*, 17, 1229–1233, 2006.
11. P. Russell, Photonic crystal fibers, *Science*, 299, 358–362, 2003.
12. T. R. Wolinski, S. Ertman, P. Lesiak, A. W. Domanski, A. Czapla, R. Dabrowski, E. N. Kruszelnicki, and J. Wojcik, Photonic liquid crystal fibers—A new challenge for fiber optics and liquid crystal photonics, *Opto-Electron. Rev.*, 14, 329–334, 2006.
13. T. R. Wolinski, D. Budaszewski, M. Chychlowski, A. Czapla, S. Ertman, P. Lesiak, K. Rutkowska, M. Sierakowski, M. Tefelska, and A. W. Domanski, Photonic liquid crystal fibers: Towards highly tunable photonic devices, *IEEE International Conference on Photonics (ICP)*, Langkawi, Malaysia, 2010.
14. A. Lorennz, R. Schuhmann, and H. S. Kitzerow, Infiltrated liquid crystal fiber: Experiments and liquid crystal scattering model, *Opt. Express*, 18, 3519–3530, 2010.
15. T. T. Alkeskjold, D. Noordegraaf, L. Laegagaard, J. Weirich, L. Wei, G. Tartarini, P. Bassi, S. Gauza, S. T. Wu, and A. Bjarklev, Integrating liquid crystal based optical devices in photonic crystal fibers, *Opt. Quant. Electron.*, 39, 1009–1019, 2007.
16. D. Demus, J. Gooby, G. W. Gray, H. W. Spiess, and V. Vill, *Physical Properties of Liquid Crystals* (Wiley-VCH, Weinheim, Germany, 1999).
17. S. Singh and D. A. Dunmar, *Liquid Crystals: Fundamentals* (World Scientific, Singapore, 2002).
18. P. G. de Gennes, Short range order effects in isotropic phase of nematic and cholesterics, *Mol. Cryst. Liq. Cryst.*, 12, 193–214, 1971.
19. J. Nehring and A. Saupe, Elastic theory of uniaxial liquid crystals, *J. Chem. Phys.*, 54, 337, 1971.
20. J. L. Ericksen, Liquid crystals with variable degree of orientation, *Arch. Rat. Mech. Anal.*, 113, 97, 1991.
21. C. G. Crawford and S. Zumer, *Liquid Crystals in Complex Geometries* (Taylor & Francis Ltd, London, U.K., 1996).
22. V. Frederiks, and V. Zolina, Forces causing the orientation of an anisotropic liquid, *Trans. Faraday Soc.*, 29, 919, 1933.
23. F. Poli, A. Cucinotta, and S. Selleri, *Photonic Crystal Fibers* (Springer, New York, 2007).
24. M. A. Duguay, Y. Kokobun, T. L. Koch, and L. Pfeiffer, Antiresonant reflecting optical waveguides in SiO_2-Si multilayer structures, *Appl. Phys. Lett.*, 49(1), 13–15, 1986.
25. S. Kumar, *Liquid Crystals: Experimental Study of Physical Properties and Phase Transitions* (Cambridge University Press, Cambridge, U.K., 2001).
26. N. Vieweg, C. Jansen, M. K. Scheller, N. Krumbhloz, R. Wilk, M. Mikulics, and M. Koch, Molecular properties of liquid crystals in the terahertz frequency range, *Opt. Express*, 18, 6097–6107, 2010.
27. V. V. Meriakri, E. E. Chigray, C. L. Pan, R. P. Pan, and M. P. Parkhomenko, Dielectric properties of liquid crystals in terahertz frequency range, *IEEE International Conference on Infrared, Millimetre and Terahertz Waves*, Busan, Korea, September 2009.

28. M. Yazdanpanahi, S. Bulja, D. Mirshekar-Syahkal, R. James, S. E. Day, and F. A. Fernandez, Measurement of dielectric constants of nematic liquid crystals at mm-wave frequencies using patch resonator, *IEEE Trans. Instrum. Meas.*, 59, 3079–3085, 2010.
29. S. Mathews, G. Farrell, and Y. Semenova, Liquid crystal infiltrated photonic crystal fiber for electric field intensity measurements, *Appl. Opt.*, 50(17), 2628–2635, 2011
30. L. Xiao, M. S. Demokan, W. Jin, Y. Wang, and C. Zhao, Fusion splicing photonic crystal fibers and conventional single-mode fibers: Microhole collapse effect, *J. Lightwave Technol.*, 25, 3563–3574, 2007.
31. L. Scolari, S. Guaza, H. Xianyu, L. Zhai, L. Eskildsen, T. T. Alkeskjold, S. T. Wu, and A. Bjarklev, Frequency tunability of solid core photonic crystal fibers filled with nano-particle doped liquid crystals, *Opt. Express*, 17, 3754–3764, 2009.
32. B. E. A. Saleh and M. C. Teich, *Fundamentals of Photonics* (John Wiley & Sons, New York, 1991).
33. S. Wu, U. Efron, and L. Hess, Birefringence measurements of liquid crystals, *Appl. Opt.*, 23, 3911–3915, 1984.
34. S. Mathews, G. Farrell, and Y. Semenova, Experimental study on the frequency dependence of liquid crystal infiltrated photonic crystal fibers, *IEEE Sens. J.*, 12(5), 1018–1024, 2012.
35. T. R. Wolinski, Polarimetric optical fibers and sensors, *Prog. Optics*, XL, 1–75, 2000.
36. F. T. S. Yu and S. Yin, *Fiber Optic Sensors* (Marcel Dekker Inc., New York, 2002).
37. H. K. Gahir and D. Khanna, Design and development of temperature compensated fiber optic polarimetric pressure sensor based on photonic crystal fiber at 1550 nm, *Appl. Opt.*, 46, 1184–1189, 2007.
38. X. Dong, H. Y. Tam, and P. Shum, Temperature-insensitive strain sensor with polarization maintaining photonic crystal fiber based Sagnac interferometer, *Appl. Phys. Lett.*, 90, 151113, 2007.
39. H. Y. Fu, S. K. Khijwania, H. Y. Tam, P. K. A. Wai, and C. Lu, Polarization maintaining photonic crystal fiber based all-optical polarimetric torsion sensor, *Appl. Opt.*, 49, 5954–5958, 2010.
40. S. Ertman, A. Czapla, K. Nowecka, P. Lesiak, A. W. Domanski, T. R. Wolinski, and R. Dabrwoski, Tunable highly-birefringent photonic liquid crystal fibers, *IEEE Instrumentation and Measurement Technology Conference*, 2007, pp. 1–6.
41. PM-1550-01 specifications sheets from NKT photonics. http://nktphotonics.com/files/files/PM-1550-01.pdf.
42. M. Szpulak, T. Martynkien, and W. Urbanczyk, Effects of hydrostatic pressure on the phase and group modal birefringence in microstructured holey fibers, *Appl. Opt.*, 43, 4739–4744, 2004.
43. L. Xiao, W. Jin, M. S. Demokan, H. L. Ho, H. Y. Hoo, and C. Zhao, Fabrication of selective injection microstructured optical fibers with a conventional fusion splicer, *Opt. Express*, 13, 9014–9022, 2005.
44. L. Scolari, T. T. Alkeskjold, J. Riishede, A. Bjarklev, D. S. Hermann, A. Anawati, M. D. Nielsen, and P. Bassi, Continuously tunable devices based on electrical control of dual frequency liquid-crystal filled photonic bandgap fibers, *Opt. Express*, 13, 7483–7496, 2005.
45. S. Mathews, G. Farrell, and Y. Semenova, All-fiber polarimetric electric field sensing using liquid crystal infiltrated photonic crystal fiber, *Sensors Actuators A: Phys.*, 167(1), 54–59, 2011.
46. K. M. Bohnert and J. Nehring, Fiber optic sensing of electric field components, *Appl. Opt.*, 27, 4814–4818, 1998.
47. K. Yang, P. B. Katehi, and J. F. Whitaker, Electric field mapping system using an optical fiber based electro-optic probe, *IEEE Microw. Wireless Compon. Lett.*, 11, 164–166, 2001.
48. S. Mathews, G. Farrell, and Y. Semenova, Directional electric field sensitivity of a liquid crystal infiltrated photonic crystal fiber, *IEEE Photon. Technol. Lett.*, 23(7), 408–410, 2011.

8 Optical Microfiber Physical Sensors

George Y. Chen and Gilberto Brambilla

CONTENTS

8.1 INTRODUCTION

Optical microfibers and nanofibers (MNFs) are a new lineage of optical fiber that confines light due to a refractive index (RI) difference between the solid core and the external medium cladding. The fraction of light that propagates in the cladding depends on the ratio of the fiber diameter to the wavelength of light. It has now been a decade since the first experiments on low-loss MNFs were conducted by Tong et al. [1]. Interest in this early work quickly mushroomed out to laboratories worldwide, generating a thriving research community that has sustained a strong focus on this technology through the 2000s up to the present time. The motivation behind this ongoing research is to establish micro- and nanoscale optical fibers as a new sensor technology that can create devices that are extremely small size and ultralightweight and have the potential to be manufactured at low cost. This chapter provides an introduction to MNFs in terms of the fabrication techniques and optical and mechanical properties and an overview of some MNF-based sensors.

To describe MNFs in more detail, they are the uniform waists of biconical fiber tapers, with diameters comparable to the wavelength of light. MNFs are usually manufactured by heating and stretching [2] regular-sized optical fibers whose diameters are often in excess of 100 μm. The result is a biconical taper that provides a smooth, lossless connection to other fiberized components. By controlling the pull rate during the fabrication process, the taper diameter profile can be fine-tuned to suit the application [3,4]. Optical materials other than silica have been used to manufacture MNFs, including phosphate [5], tellurite [5], lead silicate [6], bismuthate [6], chalcogenide glasses [7], and a variety of polymers [8–11]. The remarkable optical and mechanical properties exhibited by MNFs make them an excellent platform for optical sensors. In the next section, the waveguiding properties of MNFs will be discussed, followed by an introduction to resonator-/non-resonator-type MNF-based sensors.

8.2 FABRICATION

Applications of MNFs have previously been limited because of the difficulties in fabricating low-loss submicron structures. With the development of computer-controllable tapering rigs, it is now possible to manufacture precise and subwavelength tapers. The most conventional way to reduce the diameter of the fiber is to taper it. Optical fiber tapers are made by stretching a heated fiber, forming a structure comprising a narrow stretched filament (i.e., the taper waist), each end of which is linked to an unstretched fiber by a conical section (i.e., the taper transition), as shown in Figure 8.1. In the down-taper transition, the mode confinement changes from a core–cladding to a cladding–surrounding interface. That is, the original core disappears and the original cladding becomes the new core, with the external medium being the new cladding. If the transition is adiabatic [12,13], the optical loss of the transition regions is negligible.

In the following sections, the fabrication techniques are either classified as top-down or bottom-up. Bottom-up techniques grow MNFs from a seed of a few nanometers. Top-down techniques manufacture MNFs by reducing the size of macroscopic samples and thus can provide much longer MNFs.

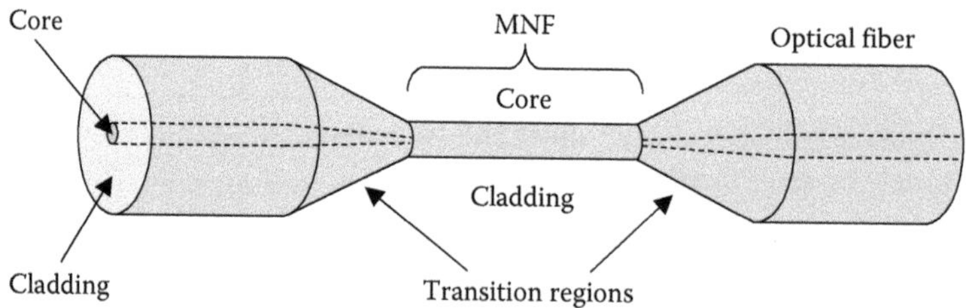

FIGURE 8.1 Schematic diagram of an MNF.

8.2.1 Flame-Brushing Technique

The flame-brushing technique was initially developed for manufacturing fiber tapers and fused couplers. The basic idea involves a moving small flame under the optical fiber that is being stretched. Both the burner and the optical fiber ends are fixed onto stages and controlled by a computer. By fine-tuning the pull rate and the flame movement, the taper shape can be tailored to a high degree of accuracy. Moreover, it has been known to provide the longest MNFs (e.g., 110 mm) [14] with the lowest measured loss (e.g., 0.015 dB/mm) [15]. Finally, the flame-brushing technique enables both ends of the MNF to be pigtailed to a standard optical fiber. This feature is extremely important for practical applications where connectivity and system integration are highly essential.

8.2.2 Modified Flame-Brushing Technique

The modified flame-brushing technique replaces the flame by a ceramic microheater [6] or a sapphire capillary tube heated by a CO_2 laser beam [16]. In the center of the microheater is a heating element whose temperature can be set by tuning the current level. For the sapphire tube/CO_2 laser approach, the temperature is controlled by varying the focus of the laser beam onto the sapphire tube. The modified flame-brushing technique provides more flexibility than the original technique due to the adjustable temperature range. This technique can be applied to manufacture MNFs from a wide range of low softening-temperature glasses. For silica optical fibers, this technique can fabricate MNFs with much lower OH content compared with those manufactured by the flame-brushing technique, which generates water vapor as a combustion by-product [6].

8.2.3 Self-Modulated Taper Drawing

This technique begins with the tapering of a standard optical fiber to a diameter of several micrometers using the conventional flame-brushing technique. Then the taper waist is broken into two halves with one end wrapped around a hot sapphire rod and the other end bent and further drawn to a submicron diameter. The sapphire tip is heated with a flame at a distance from the fiber, in order to maintain a steady temperature distribution. Initially, to draw a thick wire would require a relatively large force. Coincidentally, the center of bending occurs at the thicker part of the taper, which will produce a tensile force to stretch the fiber. As the taper length increases and the waist diameter decreases, the bend loosens and the center of bending moves toward the thinner end of the taper. This results in smaller forces for drawing thinner wires, which helps to keep the taper from breaking under unpredictable drawing conditions. Although this approach involves a complex fabrication procedure and a relatively high loss (at least one order of magnitude higher than the other techniques), the so-called self-modulated taper-drawing technique is capable of providing extremely small-diameter MNFs [17].

8.2.4 Acid-Etching Technique

Etching techniques are mostly used to create or modify MNF tips rather than uniform waists. To initiate the etching process, an acid droplet–filled dish is raised by a translation stage to immerse a standard optical fiber. The droplet shape, position, immersion depth, and time allowed the length and diameter of the waist region to be tailored for the intended application [18].

8.2.5 Direct Drawing from Bulk

For optical materials that are unavailable in fiber form, it is possible to manufacture MNFs straight from the bulk material [5]. In this approach, a small hot sapphire rod is brought into contact with the bulk glass for a localized softening. Then the sapphire rod is promptly moved away, drawing a glass strand with micrometer/submicron diameter. This technique is extremely flexible and does not require expensive equipment, but the MNF uniformity and diameter are difficult to control.

8.2.6 Self-Assembly from Silica Nanoparticles

In a rather different approach, uniform MNFs can be created by evaporative self-assembly [19]. The process initiates from a colloidal silica dispersion of nanoparticles with a small amount of NH_4^+ ions to prevent aggregation in the solution. MNF formation occurs as a consequence of rising stresses during the evaporation of the immobilized drop. These stresses arise due to the van der Waals forces that bind nanoparticles together into a closely packed structure. During the course of drying, the constriction of the so-called coffee-stain effect leads to inward-directional stress around the ring. This packing-in effect counteracts the stresses. As the self-assembly propagates following the evaporation front, the rapid buildup of radial stresses leads to fractures. Although stresses are temporary relieved, they build up again and the fractured planes act as seeds for further fracturing. The maintenance of this cycle forms the basis for the fabrication of MNFs.

8.2.7 Electrospinning from Glass-Forming Melt

To date, the direct production of submicron-diameter glass fibers has relied on mechanical drawing techniques. Electrospinning [20] is a popular technique that has been known for decades for its ability to produce submicron-diameter fibers from a wide range of polymer solutions. Part of the attraction is its simplicity to implement. By applying a voltage to the solution and charging its surface, a liquid jet is ejected from the surface to form continuous glass fibers of nanometer diameter [21].

To summarize the various fabrication methods, one should consider using the modified flame-brushing technique if the application requires long length and low loss. Due to the tunable processing temperature, more fabricators are choosing the modified version over the original. If the emphasis is on producing the smallest possible diameter MNF, the self-modulated taper drawing is the best choice. If the material happens to be in a nonfiber form, one can choose between direct drawing from the bulk, self-assembly from nanoparticles (silica only), or electrospinning from a glass-forming melt. For a nonprecise MNF geometry, the first method is the simplest and quickest to carry out. The second method is still in the early stages of development, so the control over the MNF geometry is unlikely to be optimized. The third method is well established, and therefore, it may serve as the ideal candidate.

8.3 OPTICAL AND MECHANICAL PROPERTIES

In this section, the unique optical and mechanical properties of MNFs are introduced, including mode propagation, evanescent field, optical confinement, propagation loss, bend loss, mechanical strength, dispersion, and nonlinearity.

8.3.1 Mode Propagation

In MNFs, light is guided by the cladding–surrounding interface rather than by the core–cladding interface for two reasons. First, the original core after tapering is more than one order of magnitude smaller than the wavelength of light. Second, the RI difference at the cladding–surrounding (e.g., air) interface can be up to 100 times larger than the RI difference at the core–cladding interface. Therefore, the core has a marginal influence on the guiding properties and can be neglected. As light propagates, the effective index decreases monotonically along the down-taper transition, and the mode becomes guided by the cladding–surrounding interface. For the rest of this chapter, the cladding will be considered as the core of the MNF.

Since the weakly guiding approximation is not valid as a result of the large RI difference between the cladding and its surrounding medium, the exact eigenvalue equation is used for the propagation constant (β) of various hybrid and transverse modes [22].

For HE_{vm} and EH_{vm} modes,

$$\left[\frac{J_v'(U)}{UJ_v(U)}+\frac{K_v'(W)}{WK_v(W)}\right]\cdot\left[\frac{J_v'(U)}{UJ_v(U)}+\frac{n_{sur}^2K_v'(W)}{n_{MNF}^2WK_v(W)}\right]=\left(\frac{v\beta}{kn_{MNF}}\right)^2\cdot\left(\frac{V}{UW}\right)^4 \tag{8.1}$$

where the β solutions are denoted with two indices v and m. v originates from the detailed calculations of the fields and describes the azimuthal dependence. m denotes the mth root of the eigenvalue equation. J_v is the vth-order Bessel function of the first kind and K_v is the vth-order modified Bessel function of the second kind. n_{MNF} and n_{sur} are the RIs of the MNF and its surrounding medium, respectively.

For TE_{0m} modes,

$$\frac{J_1(U)}{UJ_0(U)}+\frac{K_1(W)}{WK_0(W)}=0 \tag{8.2}$$

For TM_{0m} modes,

$$\frac{n_{MNF}^2J_1(U)}{UJ_0(U)}+\frac{n_{sur}^2K_1(W)}{WK_0(W)}=0 \tag{8.3}$$

where

$$U=r\cdot\sqrt{k_0^2n_{MNF}^2-\beta^2}$$

$$W=r\cdot\sqrt{\beta^2-k_0^2n_{sur}^2}$$

$$V=\sqrt{U^2+W^2}=k_0\cdot r\cdot\sqrt{n_{MNF}^2-n_{sur}^2},\ k_0=\frac{2\pi}{\lambda}$$

k_0 is the propagation constant of light in free space. r is used to denote the core radius and corresponds to the core–cladding interface from the center in standard optical fibers and likewise the cladding–surrounding interface from the center in MNFs.

8.3.2 Optical Confinement

MNFs can confine light to the diffraction limit for lengths that are only limited by loss. As shown in Figure 8.2, when r starts to decrease, the V number decreases and light becomes more tightly confined until the mode field diameter (ω) reaches minima (A). For smaller V, the core–cladding interface does not confine light anymore and the spot size expands into the cladding, causing ω to reach maxima (B). Further reducing V, light is then guided only by the cladding–surrounding interface. ω decreases until it reaches minima around $V \approx 2$ (C), before increasing again. The region below $V < 2$ is typical of MNFs, where ω can be much greater than r and a large fraction of the power resides in the evanescent field. For $V < 0.6$, ω can continue to expand until it becomes orders of magnitude larger than r.

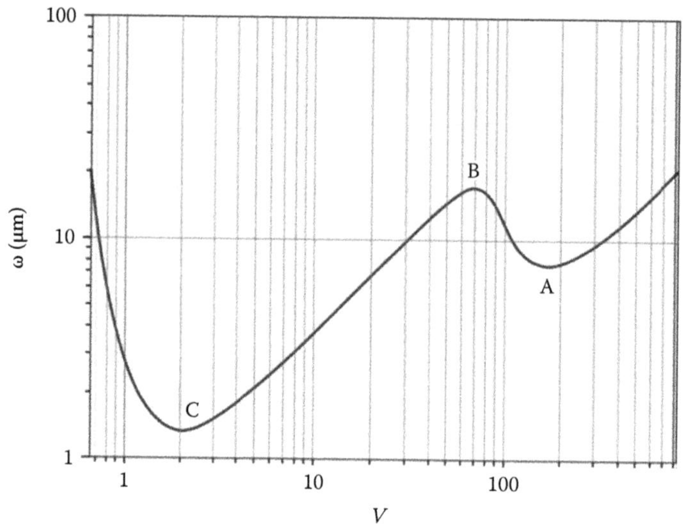

FIGURE 8.2 Dependence of the mode field diameter on the V number of a micro-/nanofiber. (From Chen, G.Y. et al., *Open Opt. J.*, 7, 32, 2014.)

8.3.3 Evanescent Field

When $V = 1$, a large portion of the optical power resides in the evanescent field outside the MNF. The extension of the evanescent field and the fraction of power (η_{EF}) propagating in it depend on the λ/r ratio and can be obtained from the component of the Poynting vector along the direction of the beam propagation S_z [22]:

$$\eta_{EF} = \frac{\int_{out} S_z dA}{\int_{in} S_z dA + \int_{out} S_z dA} = \frac{\int_r^{\infty} S_z dA}{\int_0^r S_z dA + \int_r^{\infty} S_z dA} \quad (8.4)$$

where $\int_{in} S_z dA$ and $\int_{out} S_z dA$ are over the MNF cross section inside and outside the MNF, respectively. Figure 8.3 shows the dependence of η_{EF} on the λ/r ratio for silica MNFs with different surrounding RIs. When the surrounding medium is air, η_{EF} reaches 0.5 at $\lambda/r = 4$, meaning that half of the power is propagating outside the MNF when its radius is a quarter of the wavelength of light. η_{EF} increases with the increasing surrounding RI for the same λ/r ratio. Therefore, to enhance the evanescent field, low-loss polymers can be used to embed (i.e., submerse) the MNF. To improve the η_{EF} even further, hollow MNFs have also been proposed [24]. A large evanescent field is particularly important in resonators, where a significant fraction of the propagating power needs to interact with the surrounding medium.

8.3.4 Propagation Loss

The greatest contributions to loss come from surface roughness, diameter nonuniformity, and impurities associated with the MNF and its surrounding medium [25]. The loss increases for decreasing MNF diameter, regardless of the fabrication technique. This can be explained by the stronger interaction between the field intensity of the guided light and the surface of thinner MNFs. A theory of nonadiabatic intermodal transitions was developed to investigate what is the smallest MNF that

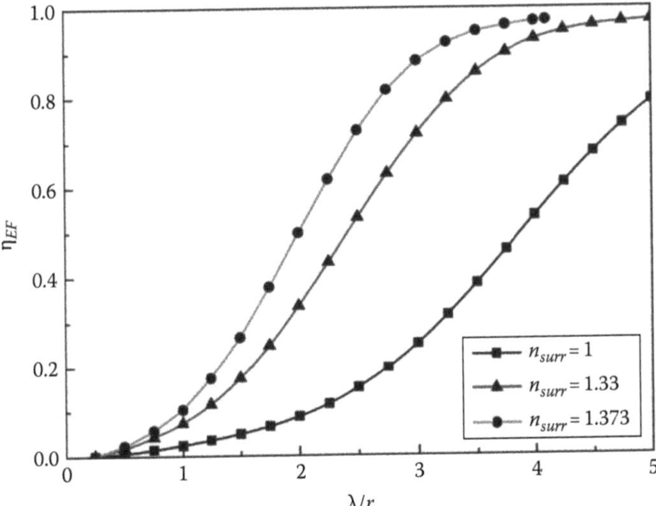

FIGURE 8.3 Dependence of the power fraction propagating in the evanescent field on the wavelength-to-radius ratio in different RI surroundings. (From Chen, G.Y. et al., *Open Opt. J.*, 7, 32, 2014.)

can still transmit light [26,27]. The guided mode was found to vanish at a threshold value of what is approximately one order of magnitude smaller than the wavelength of light. The propagation loss is also time dependent [28], and a high temperature treatment is necessary to restore the loss to its near-initial condition [14]. MNFs with smaller diameters were found to degrade faster in air, and this was attributed to the formation of cracks at the surface as a consequence of water absorption [14]. A method of reducing the time-related degradation was demonstrated by embedding MNFs in low-RI materials such as Teflon ($n = 1.315$) [29].

8.3.5 Bend Loss

MNFs have excellent mechanical properties that allow them to be bent and manipulated without inducing physical damage. Due to the large RI contrast between silica and air, bend radii of the order of micrometers can be readily achieved with relatively low bend-induced loss (e.g., <1 dB for 90° turn with a bend radius of 5 μm in an air-clad 530 nm diameter silica MNF [30]). This gives rise to highly compact devices with complex geometry.

8.3.6 Mechanical Strength

Although MNFs have very small dimensions, they possess an extraordinarily high strength due to the smaller flaw size of surface imperfections. The tensile ultimate strength of MNFs fabricated by the modified flame-brushing technique was found to be much higher than those made by the self-modulated taper-drawing technique [17].

8.3.7 Dispersion

MNFs have a remarkable potential for tailoring the dispersion properties, to a greater extent via the waveguide contribution rather than the material contribution. By controlling the MNF diameter, it is possible to move the zero dispersion to a shorter wavelength [31]. Furthermore, since a considerable fraction of the mode can propagate outside the MNF's physical boundary, its dispersion properties can also be modified by changing the surrounding medium (e.g., using transparent liquids).

8.3.8 Nonlinearity

The optical nonlinearity (γ) of an MNF is at its maximum when ω reaches its minimum in the high confinement region ($V \approx 2$ in Figure 8.2). While standard optical fibers such as the telecom single-mode fiber (SMF-28) have $\gamma \approx 10^{-3}$ W^{-1} m^{-1}, γ in silica MNFs are about 70 times larger due to the inverse relationship between the effective γ and ω. By producing MNFs from highly nonlinear materials (e.g., lead silicate, bismuth silicate, and chalcogenide glasses), γ can be up to five orders of magnitude larger than that of SMF-28 [6]. This makes them superior hosts for observing nonlinear effects.

8.4 PHYSICAL SENSORS

This section introduces a selection of the MNF-based sensors reported to date [23,32] for the measurement of physical measurands such as acceleration, acoustic, bend/curvature, current, displacement/strain, electric field, force/pressure, magnetic field, RI, rotation, roughness, and temperature.

8.4.1 Classification

Non-resonator-type (i.e., no closed-loop interference) MNF-based sensors come in many different forms, as illustrated by Figure 8.4. The straight MNF shown in Figure 8.4a is a common configuration that exploits the strong evanescent field of the guided modes to interact with its surrounding medium. The measured effect is usually a direct change in power or an indirect change in power from a change in phase of the transmitted light. The straight MNF is either connected to standard optical fiber pigtails or evanescently coupled to single-ended fiber tapers, as shown in Figure 8.4b. To enhance the sensitivity to certain measurands, the surface of the MNF can be functionalized to respond to specific chemical or biological species, as shown in Figure 8.4c. Fiber Bragg gratings (FBGs) (illustrated in Figure 8.4d) or long-period gratings (LPGs) can be inscribed on the straight MNF that undergoes a spectral shift in response to a change in the ambient conditions.

Figure 8.4e shows that two strands of MNF can be manipulated to form a Mach–Zehnder interferometer (MZI) that features high sensitivity to the phase difference between the sensing and reference arms. Alternatively, abrupt tapered sections can be exploited for modal interferometry in the configuration of either MZI (Figure 8.4f) or Michelson interferometer (MI) (Figure 8.4g). The MNF tip (shown in Figure 8.4h) is widely established as a sensor head for probing and manipulating atoms and molecules. The subwavelength cross section of the tip facilitates minimally invasive analysis of extremely small areas. Lastly, MNF in an uncoupled helical arrangement (illustrated in Figure 8.4i) can be employed for detecting rotations in the state of polarization induced by variations in the local magnetic field.

Resonator-type MNF-based sensors comprise all sensors that exploit resonant structures. MNFs (mostly optical microfibers) have been used to manufacture homogeneous resonant sensors in the arrangements of loop, knot, and coil. Coiling an MNF onto itself allows the guided modes in adjacent turns to evanescently overlap and couple, thereby creating compact resonators with predicted Q-factors as high as ~10^9. These resonators have many advantages including small size, low fabrication cost, and robustness that have attracted much interest to develop them as optical sensors. A variety of configurations have been reported in the literature, and they are described in the following paragraph.

The optical microfiber loop resonator (MLR) (shown in Figure 8.5a) is the simplest form of the homogeneous (all-MNF) microresonator. Fabrication begins with manufacturing the MNF and bending it into a self-coupling loop. Translation and/or rotation stages can be used to adjust the loop to the desired size. The resulting MLRs show strong self-coupling due to the close proximity of the waveguide with itself at the coupling region, with Q-factors of the order of 10^5. The geometrical shape of the MLR is maintained by electrostatic and van der Waals interactions at the point of

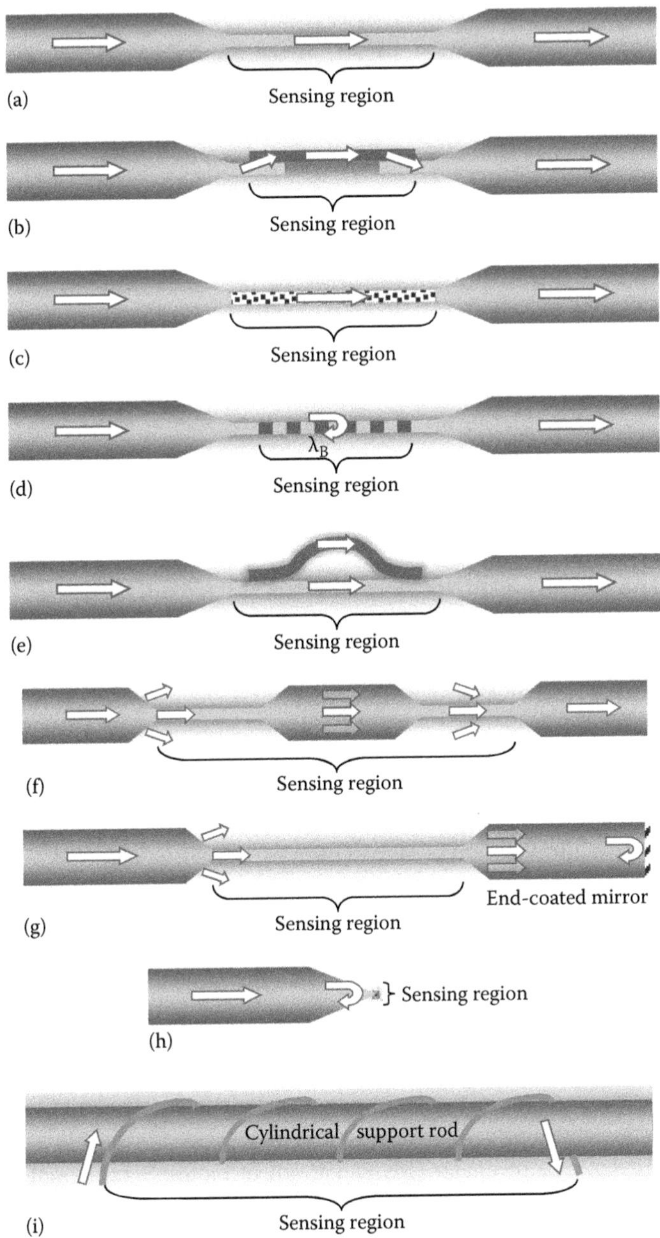

FIGURE 8.4 Schematic diagrams of common non-resonator-type MNF arrangements: (a) tapered core and cladding, (b) evanescently coupled MNF, (c) surface-coated MNF, (d) MNF with Bragg grating, (e) MNF-based Mach–Zehnder interferometer, (f) abrupt-MNF-based Mach–Zehnder interferometer, (g) abrupt-MNF based Michelson interferometer, (h) MNF tip, and (i) uncoupled MNF coil.

coupling, meaning it suffers from a limited stability dependence on the environment. Embedding the MLR in a polymer has been the preferred solution to provide long-term stability, though it can considerably modify the transmission spectrum. Fusing the loop contact points with a CO_2 laser has been proposed. Although this technique has generally negative effects relating to the Q-factor, resonators with $Q > 10^5$ have nonetheless been demonstrated. The Q-factor is defined as 2π times the ratio of the stored energy to the energy dissipated per oscillation cycle, or equivalently, the ratio of the resonant wavelength to the full width at half maximum (FWHM) linewidth of the

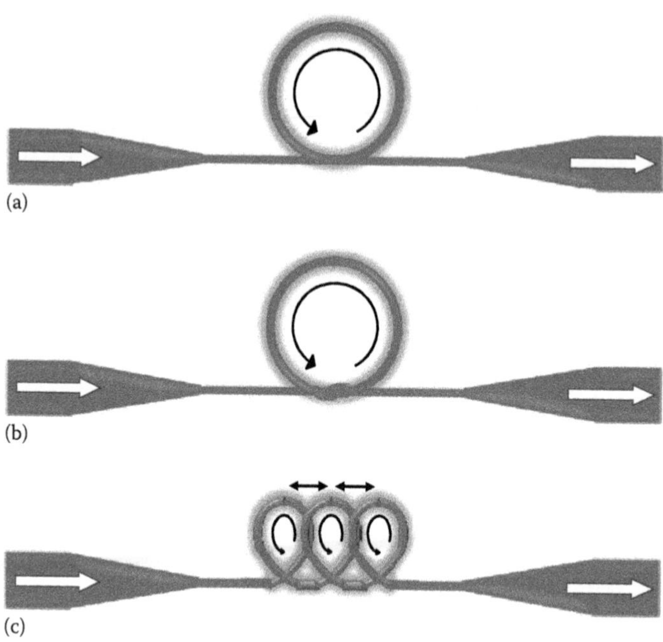

FIGURE 8.5 Schematic diagrams of the optical microfiber (a) loop resonator, (b) knot resonator, and (c) coil resonator. The evanescent field couples power between different sections of the same fiber. The directions of light propagation and coupling are shown as arrows.

resonance shape. Similarly, CO_2 lasers have also been deployed to splice together different silica and soft glass MNFs. An alternative approach to increase the MLR long-term stability relies on the use of a copper support rod to preserve its geometry. Critical coupling (i.e., coupling coefficient optimized based on propagation loss to attain maximum interference visibility) has been demonstrated by tuning the resonator with thermal effects induced by current flowing in the conducting rod, achieving a Q-factor up to 4000 and an extinction ratio of 30 dB.

The optical microfiber knot resonator (MKR) (shown in Figure 8.5b) can be fabricated by forming a simple knot to couple adjacent MNF sections. This design requires less alignment precision than the MLR and benefits from improved stability. Q-factors of up to 57,000 and finesses of 22 have been achieved. The Q-factor and spectral properties can be tuned by changing the knot radius and tightness.

The optical microfiber coil resonator (MCR) (shown in Figure 8.5c) is a 3D resonator consisting of self-coupled adjacent loops in a helix arrangement. The theoretical Q-factor is of the order of 10^9, competing with those achieved using whispering-gallery-mode (WGM) resonators. MCRs were first experimentally demonstrated in 2007 and subsequently implemented in various applications. The highest Q-factor experimentally achieved in an MCR ($Q = 470{,}000$) is still a few orders of magnitude smaller than the theoretical maximum. Fabrication involves coiling an MNF around a support rod of mm/cm scale diameter using a translation stage that controls the pitch between the turns of the coil and a rotation stage to adjust the rotation angle of the winding. Postfabrication, the resonator can be packaged in the polymer to improve its lifetime by preventing the ingress of dust and moisture. The multiturn MCR transmission can be analyzed during fabrication to identify the eigenmodes present.

A large number of resonant sensors utilize a single MNF to excite or collect light from high-Q resonators such as microrings, microspheres, microtoroids, microcapillaries, or bottle resonators, which are classified as heterogeneous. By matching the propagation constants of the mode in the MNF and the microresonator, coupling efficiencies in excess of 90% have been demonstrated.

Evanescent sensing in these types of high-Q resonators has been used to monitor chemical and biological elements positioned in proximity of the resonator surface.

8.4.2 Acceleration

Accelerometers are well established in the commercial landscape. They are employed in earthquake monitoring, guidance systems, inertial navigation, platform stabilization, vibration monitoring in portable electronics, machinery, vehicles, and vessels. Such optical sensors are highly desirable since they are unaffected by electromagnetic interference (EMI) from static electricity, strong magnetic fields, and surface potentials. MNF-based devices can operate in very confined spaces and this gives them an advantage in many applications.

Conventional flexural disk (FD) accelerometers are well known for their relatively high detection bandwidth and high sensitivity that is proportional to the disk radius. However, for applications in need of compact solutions, such sensor designs cannot perform adequately. The minimum bend radius of MNFs allows very small device dimensions to be feasible without the problems of bend loss or depolarization. In addition, the low volume-to-length ratio of MNFs reduces loading effects during the flexing motion of the disk and thus enables a higher sensitivity (e.g., unit of radians/g, where $g = 9.81$ m/s^2).

An MNF-based FD accelerometer was first suggested and demonstrated by Chen et al. [33]. Fixed wavelength light from a tunable laser source (TLS) was used to excite the MI and the phase/intensity modulation was detected by a photoreceiver. With an FD, axial acceleration induces extensive and compressive strain in the MNF depending on the direction of motion, as shown in Figure 8.6. The changes in the optical path difference (OPD) then translate to a relative phase shift and subsequently an intensity modulation. The influence of design parameters was studied and an efficient surface winding technique for the MNF was proposed [34].

A sensitivity of 4 rad/g and a detection bandwidth in excess of 1 kHz have been demonstrated for a 60 mm length MNF embedded on a 25 mm diameter FD composed of pyrolytic graphite.

To increase the sensitivity, a longer length of MNF must be used. The limiting factor is the maximum length of MNF that can be fabricated. A disk of larger radius and lower thickness with a material of lower Young's modulus and higher density would also be beneficial. The phase (unit of radians) and acceleration detection limits (e.g., unit of g) can be improved with a balanced detection scheme to eliminate common-mode intensity noise from the laser source. In addition, the interferometer can be temperature compensated by attaching the reference arm to the underside of the FD, so that any thermal expansion would equally affect both optical path lengths (OPLs). To increase the fundamental frequency/detection bandwidth would require a disk of smaller radius and greater thickness, with ideally a material of higher Young's modulus, lower density, and higher Poisson's ratio. Hence, there is an optimization issue with the disk radius, thickness, Young's modulus, and density to deliver a good balance between sensitivity and detection bandwidth. The response time can be reduced by minimizing the length of fibers between the sensor head and the detection system.

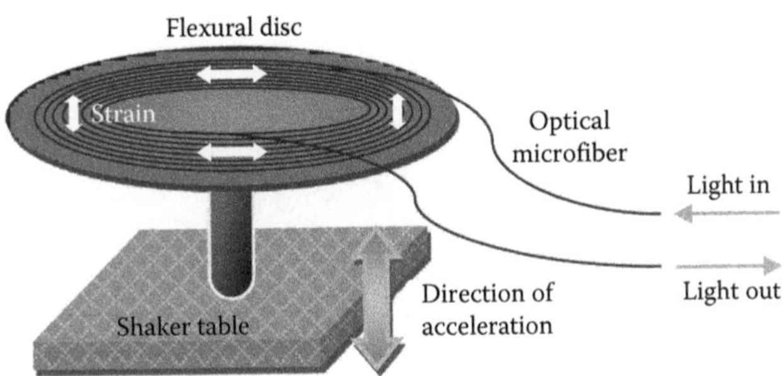

FIGURE 8.6 Schematic diagram of the MNF-based acceleration sensor.

The ultimate compactness of the sensor head is limited by the minimum bend radius of the MNF, followed by the minimum diameter of the central support.

8.4.3 Acoustic

The commercial use of acoustic sensing spans several decades, and it is one of the most successful applications in the field of fiber-optic sensors. Rapid advances in this area have been driven by the increasing demand from industries such as defense. The acoustic signatures of tanks, aircrafts, helicopters, and submarines can be monitored for high-precision battlefield awareness and surveillance. MNF-based devices offer less intrusion and easier deployment, making them an attractive choice.

Chen et al. [35] proposed and demonstrated a compact microphone (shown in Figure 8.7) comprising a 2 μm diameter, 35 mm length MNF coiled around a 3 mm diameter air-backed mandrel (ABM). Fixed wavelength light from a TLS was used to excite the polarimetric interferometer and the phase/intensity modulation was detected by a balanced detector. Acoustic waves induce local pressure variations that change the mandrel diameter and thus the OPD of the eigenmodes propagating along the MNF. This translates to a relative phase shift and subsequently an intensity modulation.

The device shown in Figure 8.7 exhibited an average acoustic sensitivity of −137 and −142 dB re. rad/m · μPa between 40–500 Hz and 1.5–4 kHz, respectively. The average detection limits measured with no noise averaging between 40–500 Hz and 1.5–4 kHz are 38.9 and 44.3 dB_{SPL}, respectively.

To increase the sensitivity, a longer length of MNF must be used. A mandrel material of lower Young's modulus and lower Poisson's ratio would also be preferable. Moreover, the wall thickness of the mandrel and the polymer packaging can be reduced to lower the stiffness while still maintaining clearance of the central support. The detection limit can be improved by applying a running average on the measurement data. To increase the fundamental frequency/detection bandwidth, the ABM must be assembled using a mandrel material of higher Young's modulus, lower density, and shorter length. Hence, there is an optimization issue with increasing Young's modulus to extend the detection bandwidth and decreasing it to improve the sensitivity. The response time can be reduced by minimizing the length of the fibers between the sensor head and the detection system. The ultimate compactness of the sensor head is limited by the diameter of the mandrel, which has to be larger than the diameter of the support pin. It is then followed by the minimum bend radius of the MNF.

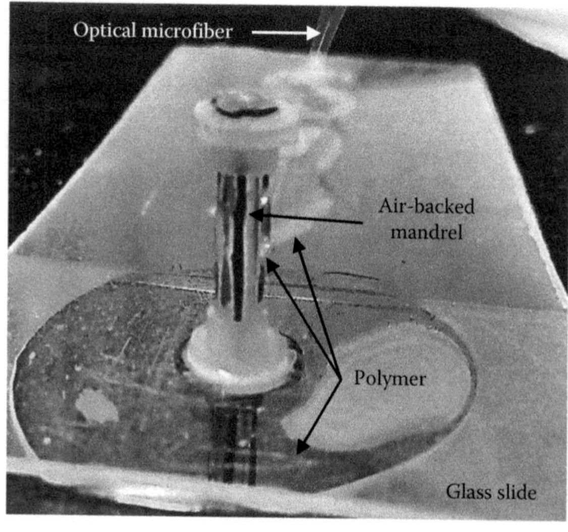

FIGURE 8.7 Photograph of the MNF-based acoustic sensor.

8.4.4 Bend/Curvature

Bend or curvature sensors have various applications in the areas of mechanical engineering, robotic arms, structural health monitoring, and turbulence sensing. With immunity to EMI and compact size, the advantages of MNF-based devices are clear.

A vector bend sensor (as shown in Figure 8.8) was proposed by Zhang et al. [36], consisting of a lateral-offset splicing joint and an up-taper formed through an excessive fusion splicing method. A broadband light source was used to excite the modal MZI and the change in the transmission spectrum was monitored by an optical spectrum analyzer (OSA). The diameter and length of the expanded section are 168 and 280 μm, respectively, with a lateral offset of 6.5 μm. The lateral-offset splicing breaks the center symmetry of the fiber, and thus, the input light splits into fundamental core and cladding modes. The cladding modes couple back into the core when propagating through the up-taper, forming an MZI due to the phase difference between the core and cladding modes. For concave bending, the RI experienced by the mode propagating in the fiber increases in the region close to the inner side of fiber axis, while the RI decreases in the region close to the outer side. For convex bending, the cladding modes will experience a longer propagation path than the core mode. Hence, the OPD between the core mode and cladding modes increases, resulting in a red shift of the interference pattern. On the other hand, with convex bending, the OPD between the core mode and cladding modes decreases, resulting in a blue shift of the interference pattern. Inverse bending directions lead to opposite differential phase change, which makes it possible to recognize the direction of bending.

For curvatures ranging from –3 to 3 m^{-1}, the bend sensitivity at 1463.86 and 1548.41 nm is 11.987 and 8.697 nm/m, respectively.

The authors have also tested the bending response along the transverse axis and find no directional sensitivity. This is due to the MZI being symmetric along the transverse axis, and thus, the same differential phase change occurs for convex or concave bending.

To increase the sensitivity, the length of the fiber between the lateral-offset joint and the up-taper segment must be elongated. There is no upper limit of the detection bandwidth assuming the bending motion is always unidirectional and never like a sine wave. Otherwise, integer cycles of bending will cause all phase differences to cancel out, suppressing the observed interference fringe shift. To reduce the response time, the length of the fibers between the sensor head and the detection system should be minimized. The ultimate compactness of the sensor head is limited by the length of the offset fiber prior to the up-taper. For any fiber-optic bend sensor, the cross-sensitivity to temperature is a critical issue that needs to be addressed. To solve this problem, much effort has been devoted to fabricating photonic crystal fiber–based bend sensors due to their low thermal dependence.

8.4.5 Current

There are well-known advantages in using fiber-optic current sensors over conventional current transformers. Due to the dielectric nature of optical fibers, its sensors have the robustness to take measurements in high-voltage or high magnetic induction noise fields and immunity from saturation

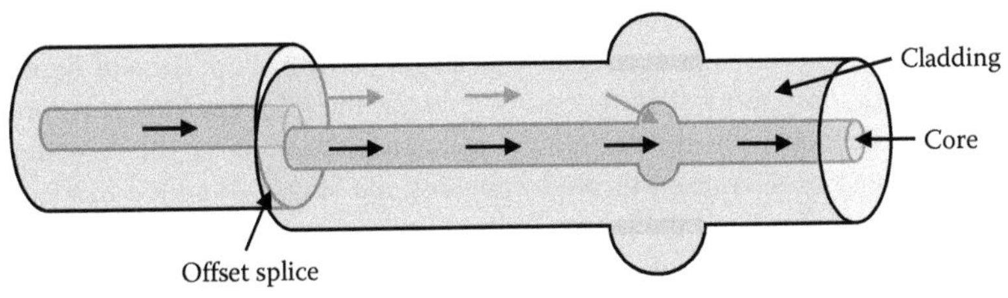

FIGURE 8.8 Schematic diagram of the MNF-based bend sensor.

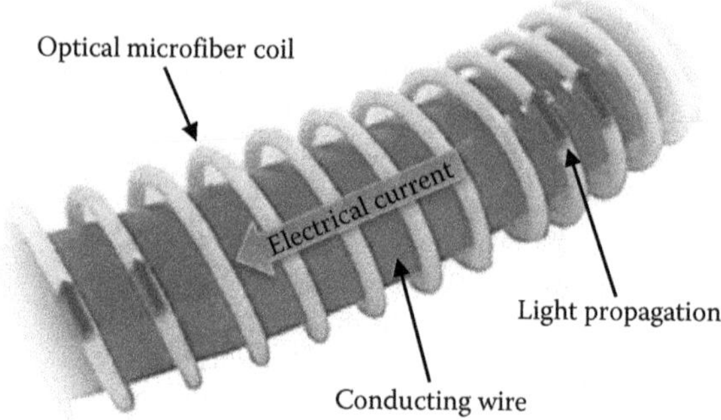

FIGURE 8.9 Schematic diagram of the MNF-based current sensor.

effects that otherwise may limit conventional current transformers. In addition, fiber-optic current sensors have a highly linear response over a wide detection bandwidth that allows them to detect transient electrical faults, inspect noise on direct current (dc) lines for the monitoring of partial discharges during automatic control, and protect high power equipment and vital electrical components. Their compactness, lightweight, and potential availability at low cost make them an attractive choice as sensors. The shorter OPL of coiled MNFs can host a higher detection bandwidth compared to regular-sized fiber coils with the same number of turns (i.e., same sensitivity), which are also much bulkier due to the larger minimum bend radius.

Belal et al. introduced coiled MNFs as compact current sensors, by wrapping an MNF around a conductive wire to form a microfiber coil (MC), as shown in Figure 8.9. In one case, fixed wavelength light from a TLS was used to excite a simple polarimeter and the Faraday rotation/intensity modulation was detected by a photoreceiver [37]. In another case, a broadband light source was used to excite the MI and the change in the transmission spectrum was monitored by an OSA. The former relied on the Faraday effect as the sensing mechanism, while the latter relied on thermally induced phase shifts [38]. The sensor heads all consisted of an MNF coiled around a short length of copper wire rather than a copper wire wound around an MNF. Apart from being more viable in a real measurement environment, this also minimizes the load impedance and heating effects in the current source to permit higher frequency switching with sufficient current. The Faraday rotation of the plane of polarized light is linearly proportional to the current-induced magnetic field and the distance traveled. The change in the polarization azimuth is then translated into a power modulation that provides a measure of current. Thermal effects on the other hand tend to be proportional to the square of the current flow (i.e., electrical power).

For the former configuration using a simple polarimeter, a current sensitivity of 16.8 μrad/A for a dynamic range of 0–19 A and a detection limit of 0.04 A/$\sqrt{\text{Hz}}$ was achieved with 25 MNF turns. The detection bandwidth was predicted to be in the gigahertz regime. For the latter configuration using an MI, the measured sensitivity was 1.28×10^{-4} rad/A^2 at 50 Hz between 0 and 120 A_{rms}. In both cases, the MNF was of 5 μm diameter and 10 cm length, and the sensitivity was expected to increase with the number of MNF turns.

The detection bandwidth is governed by the round-trip time of the MC. Since the direction of Faraday rotation can reverse during the course of propagation by each portion of light, the net rotation could remain static despite local changes. As a result, the total Faraday rotation will drop to zero for the first time beyond a certain modulation frequency. To increase the detection bandwidth, the round-trip time must be decreased through reducing the length of coiled MNF. The response time can be reduced by minimizing the length of the fibers between the sensor head and the detection system. To increase the sensitivity, a larger number of MNF turns must be used, with the limiting factor being the maximum length of MNF that can be fabricated. Hence, a high detection bandwidth can be achieved at the cost of diminishing the sensitivity and vice versa. The rotation and current detection limits can be improved by normalizing the alternating current (ac) component of the detected signal by the dc component in real time to eliminate power fluctuations from the laser source. The ultimate compactness of the sensor head is limited by the diameter of the copper wire to be measured, followed by the minimum bend radius of the MNF. It must be pointed out that MCs are not the ideal candidates for measuring large-diameter wires/cables, because the advantages of small dimensions and high detection bandwidth are lost.

However, linear birefringence played a major role in the quality of fabricated MC sensor heads. The net power transfer between the fast and slow axes due to Faraday rotation is maximum after a quarter of the polarization beat length and reduces to zero over a half of the beat length. In order to improve the sensitivity and reproducibility of the samples, the spun MNF was introduced by Chen et al. [39]. It was found that spinning a side-polished optical fiber during the tapering process packs enough intrinsic linear birefringence to resist bend- and packaging-induced birefringence and, at the same time, sufficient circular birefringence to support efficient Faraday rotation. Later, an alternative to the spun MNF was proposed by Chen et al. [40] in the form of a postfabrication technique for reversing the detrimental effect of linear birefringence in nonideal samples. By means of changing the local birefringence at specific positions along the fiber, the differential phase of light can be progressively corrected to ensure unidirectional Faraday rotation and maximum sensitivity.

8.4.6 Displacement

This category of sensors can be found in a wide variety of industrial applications including semiconductor processing, assembly of disk drives, precision thickness measurements, machine tool metrology, and assembly line testing. The aforementioned characteristics of MNF-based structures enable such optical sensors to operate in confined spaces.

The MLR transmission spectrum is strongly dependent on the loop size. As a result, they have been exploited as displacement sensors by Martinez-Rios et al. [41]. A broadband source was used to excite the MLR and the change in the transmission spectrum was monitored by an OSA. Figure 8.10 shows one of the loop pigtails attached to the moving surface. The principle of operation is based on the interaction between the fundamental cladding mode propagating through the MNF waist and the excited higher-order cladding modes when the MNF is deformed to form a loop. The effect of the curvature is to break the center symmetry of the fiber, thus allowing the coupling of the fundamental mode with cladding modes. As the curvature radius increases, the difference in the propagation constants of the fundamental and higher-order modes grows significantly, thus reducing the magnitude of coupling. In fact, the self-coupling at the touching point of the loop is very small even for a waist diameter of 8.5 μm. Hence, the sensor head does not work as a loop resonator but as a bending-based device. The notch wavelength resonances shift as a function of the loop diameter.

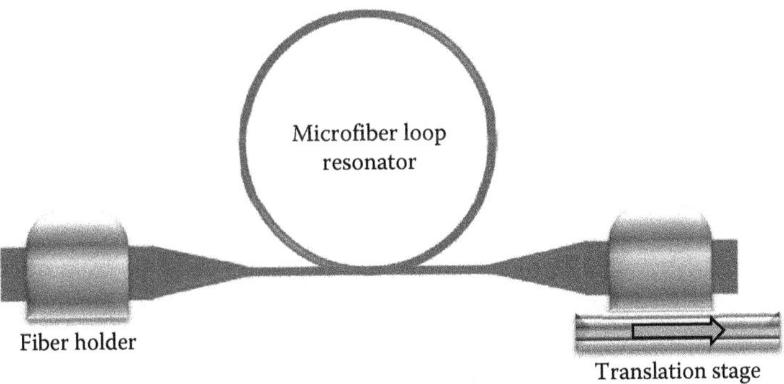

FIGURE 8.10 Schematic diagram of the MNF-based displacement sensor.

A sensitivity of 0.116 nm/μm has been achieved in a displacement range of 0–3.125 mm and the maximum wavelength shift was 360.93 nm. Alternatively, a sensitivity of 2.7 nW/μm was recorded when the measurements were carried out around a center wavelength of $\lambda = 1280$ nm with a low-power light-emitting diode of 100 nm bandwidth and a Ge-based photodetector.

To increase the sensitivity, the taper transition profile and loop length must be optimized to deliver the maximum amount of coupling. The detection limit can be improved by refining the wavelength resolution of the OSA. The detection bandwidth is rather complex to deduce, because the direction and magnitude of coupling are not always proportional to the displacement (unlike phase modulation or polarization rotation). To reduce the response time, the length of the fibers between the sensor head and the detection system should be minimized. The ultimate compactness of the sensor head is limited by the minimum bend radius of the MNF.

8.4.7 Electric Field

Optical electric field sensors have attracted the attention of several industrial segments due to the fact that they can act as dielectric receiver antennas. MNF-based sensor probes have an excellent performance as field receivers compared to metal-based electric field probes due to their immunity to EMI. Such low-invasive and noise-rejecting qualities exhibit the ideal way of detecting localized electric field distributions. Moreover, these sensors offer reduced power consumption and smaller size than their conventional electronic counterparts.

Veilleux et al. [42] have demonstrated control over the transmission of an MNF using a layer of electric-field-tunable liquid crystal (Merck ZLI-1800-100). Fixed wavelength light from a polarized laser diode was used to excite the taper, and the intensity modulation was detected by a photoreceiver. The MNF shown in Figure 8.11 was of 15 μm diameter and 5 mm length. The MNF was placed in a liquid-crystal cell made of two tin oxide–coated glass plates separated by thick spacers. Before the cell was closed, the electrodes and the MNF were optionally treated to ensure a particular orientation of the liquid-crystal molecules at the interfaces. The cell causes the guide modes to become leaky when the RI of the liquid-crystal cladding becomes higher than that of the silica core.

When the voltage applied between two electrodes changed from 0 to 350 V, the external RI changed from $n = 1.33$ to 1.48.

To increase the sensitivity, the MNF length must be increased and the diameter decreased. Since the changes to the guided mode along the length of the MNF are irreversible (unlike phase modulation or polarization rotation), there is no upper limit of the detection bandwidth. To reduce the response time, the length of the fibers between the sensor head and the detection system should be minimized. The ultimate compactness of the sensor head is limited by the length of the taper as well as the cell dimensions.

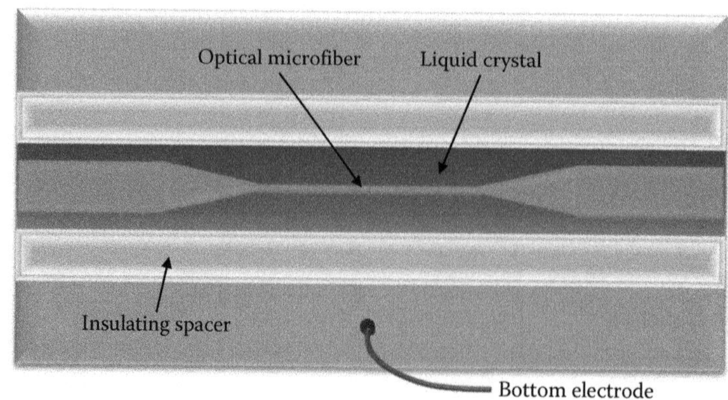

FIGURE 8.11 Schematic diagram of the MNF-based electric field sensor (without top electrode in place).

8.4.8 Force/Pressure

Force sensors have gathered much attention over the years and have established its role in many fields such as biomechanics, civil engineering, fluid-flow measurements, motor sport, process monitoring, and control. For standard optical fiber–based sensors, the sensitivity tends to scale inversely with the fiber cross-sectional area. Therefore, it is possible to increase force sensitivity by using MNFs instead. The small dimensions of MNFs can also minimize the intrusiveness when taking sensitive measurements.

Wang et al. [43] proposed combining MNFs with an optical frequency-domain reflectometry (OFDR) technique (shown in Figure 8.12) to create a high-sensitivity and high-resolution force/strain sensor. In OFDR, a TLS sweeps the wavelength and the resulting interference pattern of the Rayleigh backscattered signal from the fiber taper is monitored before the complex Fourier transform is carried out along the taper length. By calculating the vector sum of the P and S components of the received signal, a polarization-independent signal can be obtained. First, a reference measurement was taken with zero applied force on the taper. Then sensing measurements were taken after each change in the applied force. The applied force changes the effective index of the fundamental mode via the stress-optic effect, and it is revealed as a cross-correlation wavelength shift from the OFDR. The force sensitivity was significantly improved due to its reduced diameter.

A force sensitivity of 620.83 nm/N and detection limit of 6.35 μN were demonstrated with a spatial resolution of 3.85 mm for an MNF of 6 μm diameter, which is about 500 times higher than that of its SMF equivalent.

To increase the sensitivity, the MNF diameter must be reduced. Strong reflections even from distances further than the coherence length of the TLS must be minimized or they will contribute to the overall interference and deteriorate the detection limit. To improve the spatial resolution, the tuning range of the TLS must be broadened. The sensing range can be increased by increasing the coherence length of the TLS. Hence, there is an optimization issue with increasing the coherence length to extend the sensing range and decreasing it to improve the spatial resolution. To increase the detection bandwidth, while there is no fundamental limit associated with the nonreciprocal stress-optic effect, the sampling rate of the data acquisition system can be increased. The response time can be reduced by minimizing the length of the fibers between the sensor head and the detection system. The ultimate compactness of the sensor head is limited by the length of the taper as well as the dimensions of the translation stage setup.

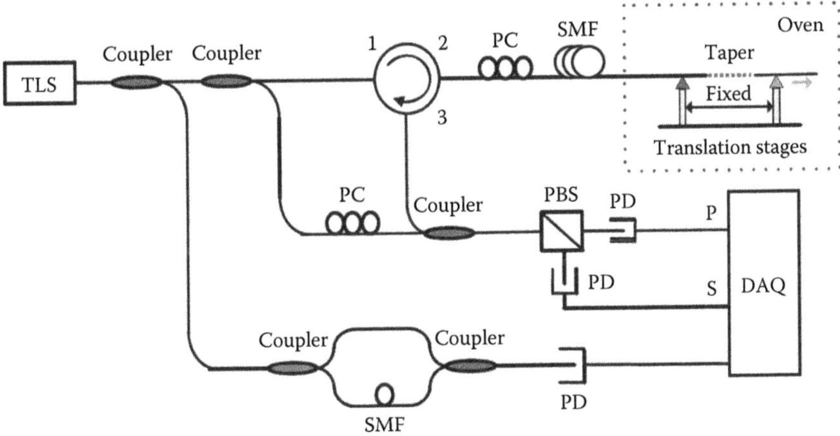

FIGURE 8.12 Schematic diagram of the OFDR interrogation setup. DAQ, data acquisition; PBS, polarization beam splitter; PC, polarization controller; PD, photodetector/photoreceiver; SMF, single-mode fiber; TLS, tunable laser source.

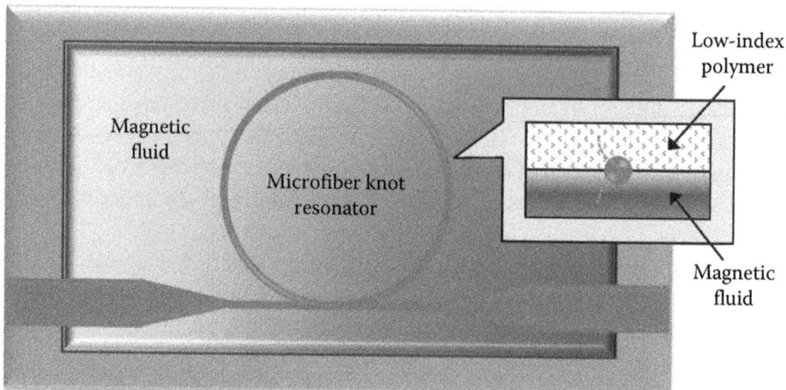

FIGURE 8.13 Schematic diagram of the MNF-based magnetic field sensor.

8.4.9 Magnetic Field

Magnetic field sensors have coexisted alongside electric field sensors due to their duality transformations. Fidelity, compactness, and configurability are the most important parameters for a good magnetic field sensor design. Thus, magnetic field sensing with MNF devices has inherent advantages when it comes to meeting these requirements. The broad range of possible applications includes automotive, navigation, proximity sensors, spatial and geophysical research, transducers for microactuators, traffic counting, and vehicle detection.

An MKR was demonstrated as a magnetic field sensor by Li and Ding [44]. A broadband source was used to excite the MKR and the change in the transmission spectrum was monitored by an OSA. The device shown in Figure 8.13 was submersed in a layer of magnetic fluid (Ferrotec EMG509) in a glass cell between two Helmholtz coils, such that the evanescent field of the MKR penetrates the external cladding. By modulating the applied magnetic field, changes in the RI of the magnetic fluid induce changes in the effective index seen by the guided mode. As a result of the change in OPL, the resonant wavelength will vary as a nonlinear function of the magnetic field signal.

A linear sensitivity of 0.3 pm/Oe up to 300 Oe and a detection limit of 10 Oe were reported.

To increase the sensitivity, the MNF diameter must be reduced. The detection limit can be improved by refining the wavelength resolution of the OSA. The detection bandwidth is governed by the round-trip time of the MKR. Since the effective index change can be positive or negative during the course of a round-trip by each portion of light, the OPL seen by the guided mode could remain constant despite local changes. As a result, the resonant wavelength shift will drop to zero for the first time beyond a certain modulation frequency. Hence, to increase the detection bandwidth, the round-trip time must be decreased by reducing the length of the knotted MNF. The response time can be reduced by minimizing the length of the fibers between the sensor head and the detection system. The ultimate compactness of the sensor head is limited by the minimum bend radius of the MNF, followed by the cell dimensions.

8.4.10 Refractive Index

RI sensing is a prominent subject of optical sensing that has a broad range of uses, including the inspection of concentration levels in aqueous solutions and quality control in the monitoring of food engineering processes. The compact form of MNF-based sensors makes sensing RI in microfluidic channels and humidity environments highly feasible and robust.

Refractometric sensors based on the MCR have been proposed [45] and demonstrated by Xu and Brambilla [46]. A broadband source was used to excite the MCR and the change in the transmission spectrum was monitored by an OSA. Due to the 3D geometry, MCRs have an intrinsic channel

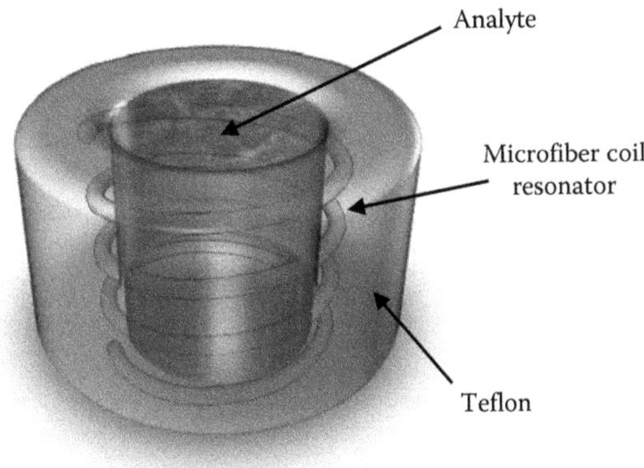

FIGURE 8.14 Schematic diagram of the MNF-based RI sensor.

that can be exploited for microfluidic applications. An MNF with 50 mm length and 2.5 μm diameter of the uniform waist region was fabricated and wrapped five times around a 1 mm diameter poly(methyl methacrylate) rod. The structure was coated with Teflon resin to form a protective layer. The embedded MCR was then carefully treated with acetone to remove the support rod. Finally, a microfluidic channel of ~1 mm diameter was fabricated. Figure 8.14 shows the schematic of the MCR structure. The working principle relates to the overlap between the analyte and the evanescent field of the mode propagating in the MNF. Any change in the analyte RI is reflected as a resonant wavelength shift due to the change in effective index. The wavelength shift is particularly affected by the wavelength/MNF diameter ratio and the coating thickness between the MNF and the fluidic channel. For optimized designs, sensitivities up to 700 nm/RIU and a detection limit of the order of 10^{-7} have been predicted.

Experimental demonstration was carried out by inserting the MCR in solutions of isopropanol and methanol. The resonant wavelength underwent a red shift for increasing analyte RIs and a sensitivity of 40 nm/RIU was reported. The inferior experimental achievement was reportedly due to the lack of smoothness of the internal wall surface and the large MNF diameter used.

To increase the sensitivity, the channel wall thickness must be reduced and the MNF diameter decreased. The detection limit can be improved by refining the wavelength resolution of the OSA. The detection bandwidth is zero because the channel requires cleaning before the next analyte can be injected, and therefore, rapid changes in the RI are not possible. To reduce the response time, the length of fibers between the sensor head and the detection system should be minimized. The ultimate compactness of the sensor head is limited by the minimum channel clearance, followed by the minimum bend radius of the MNF.

8.4.11 Rotation

Rotation sensors have numerous applications, notably in inertial navigation systems and control, stabilization, and positioning systems. Although both mechanical and optical techniques exist for rotation sensing, the latter was found to provide higher sensitivities and lower drift rates. Microresonators are advantageous in the sense that they can enhance the sensitivity without enlarging the size of the device.

Rotation sensors and gyroscopes with enhanced sensitivities have been predicted for MCR-based structures by Scheuer [47]. The combination of slow-light and conventional propagation effects leads to an improvement in sensitivity. The sensitivity enhancement increases rapidly as the loss is decreased. Simulations estimated that for a lossless MCR, the sensitivity to rotation could be improved by four orders of magnitude with respect to the case of nonoptimized standard optical fiber gyroscopes with similar parameters.

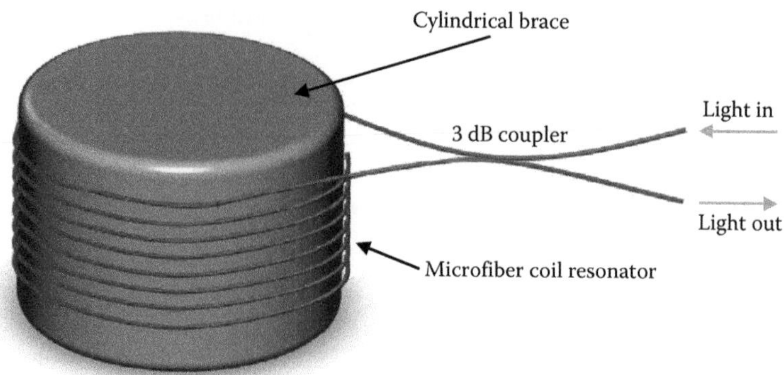

FIGURE 8.15 Schematic diagram of the MNF-based rotation sensor.

However, a later theoretical analysis by Digonnet [48] revealed that gyroscopes made from MCRs (as shown in Figure 8.15) are actually less sensitive than standard resonant fiber-optic gyroscopes consisting of a single loop. The sensitivity is proportional to the total group delay of light through the structure and the maximum achievable group delay is limited in the same manner by loss.

To increase the sensitivity, the MNF length and extinction ratio must be increased while the optical loss is decreased. The detection bandwidth is governed by the transit time of the MCR. Since the relative phase shift can be positive or negative during the course of propagation by each portion of light, the OPL seen by the guided mode could remain constant despite local changes. As a result, the total phase shift will drop to zero for the first time beyond a certain modulation frequency. Hence, the detection bandwidth can be increased by reducing the round-trip time by minimizing the length and/or extinction ratio of the coiled MNF. The ultimate compactness of the sensor head is limited by the minimum bend radius of the MNF.

8.4.12 Roughness

The detection and minimization of surface and bulk nonuniformities of MNFs are of prime importance for reducing their optical loss during fabrication. These sensing techniques can be extended for quality control of standard and special types of optical fibers.

Birks et al. [49] devised and tested a simple way of measuring the diameter uniformity of optical fibers. A broadband source was used to excite the sensing taper and the change in the transmission spectrum was monitored by an OSA. The proposed method shown in Figure 8.16 translates the MNF across small distances along the fiber under test and periodically touches it for discrete measurements. At these points, the MNF transmission spectrum exhibits WGM resonances that shift in wavelength depending on the OPL or the effective radius of the local region of the fiber.

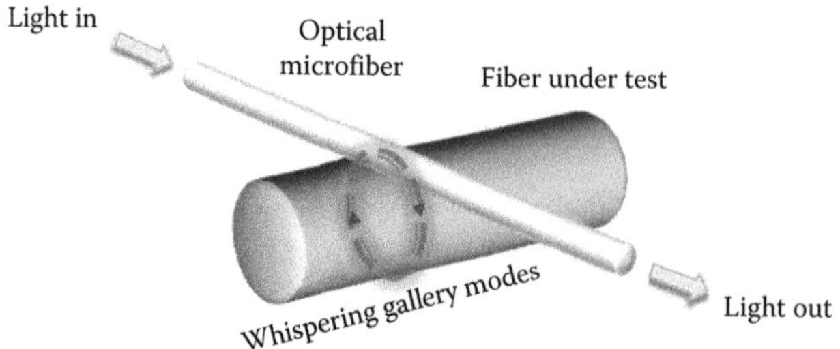

FIGURE 8.16 Schematic diagram of the MNF-based roughness sensor.

A 1.6 μm diameter sensor taper was tested on target fibers that have been tapered to a notional uniform diameter of 19.7 μm over a length of ~10 mm. Diameter variations of less than 1 part in 10^4 (i.e., 2 nm) were reported to be measurable.

To improve the detection limit, the wavelength resolution of the laser source or OSA must be refined. The detection bandwidth is governed by the round-trip time of the WGM. Since the effective index change can be positive or negative during the course of a round-trip by each portion of light, the OPL seen by the guided mode could remain constant despite local changes. As a result, the resonant wavelength shift will drop to zero for the first time beyond a certain modulation frequency. Unfortunately, as the OPL is set by the target fiber, there is no way of increasing the detection bandwidth by means of reducing the round-trip time. The ultimate compactness of the sensor head is limited by the minimum fiber diameter that still guides light [26,27].

8.4.13 Temperature

Over the past few decades, temperature sensing has matured into a profound technology that has been successfully deployed in many industrial sectors, including ecological monitoring, fire detection, leakage detection, oil and gas exploration, plant and process monitoring, power cable and transmission line monitoring, storage tanks and vessels, and structural health monitoring. The compactness of MNF-based structures in conjunction with its inherent immunity to EMI has recently attracted considerable interest.

The application of MCRs as a tool for the rapid inspection of electrical wires for insulation faults and current surges was demonstrated by Chen et al. [50]. A broadband source was used to excite the MCR and the change in the transmission spectrum was monitored by an OSA. The MCR shown in Figure 8.17 was integrated on a sliding probe that maps the local temperature to identify positions with insulation faults that can result in electrical arcing. Defects can be distinguished by an increase in heat signature. The ring-shaped detection area facilitates rotationally symmetric coverage and thus removes the need for radial alignment of the sensor head.

A sensitivity of 95 pm/°C from 26°C to 76°C was demonstrated with an MNF of 2 μm diameter and 15 mm length, coiled around a polymer-coated Teflon capillary of 1.8 mm diameter. The temperature sensitivity of the MCR is 0.21°C when using an OSA with a wavelength resolution of 20 pm.

To increase the sensitivity, a material of higher thermal expansion coefficient should be selected for the support tube/rod. The detection limit can be improved by refining the wavelength resolution of the OSA. The detection bandwidth is governed by the transit time of the MCR. Since the effective

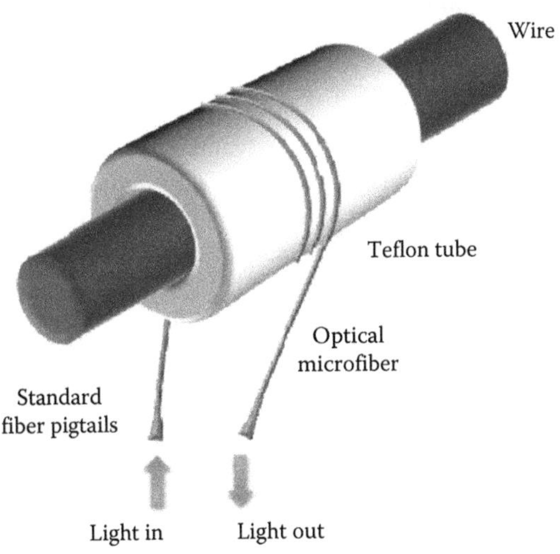

FIGURE 8.17 Schematic diagram of the MNF-based temperature sensor.

index change can be positive or negative during the course of propagation by each portion of light, the OPL seen by the guided mode could remain constant despite local changes. As a result, the resonant wavelength shift will drop to zero for the first time beyond a certain modulation frequency. Hence, to increase the detection bandwidth, the round-trip time must be decreased by reducing the length of the coiled MNF. The response time can be reduced by minimizing the length of the fibers between the sensor head and the detection system. To extend the upper limit of measurable temperature, the polymer packaging must be replaced with Teflon, which has a much higher melting temperature in excess of 300°C. However, Teflon resin is very expensive and large quantities are needed to fully embed the MCR, due to significant shrinkage in volume after drying in air. The ultimate compactness of the sensor head is limited by the diameter of the Teflon tube, which has to be larger than the range of wire diameters to be measured. It is then followed by the minimum bend radius of the coiled MNF.

8.5 DISCUSSIONS

There have been many interesting developments in the various areas of MNF-based sensing. In this section, the numerous advantages of MNF-based sensors over their traditional counterparts employing standard optical fibers are highlighted, with an overview on the challenges ahead and the practical issues that need to be addressed.

8.5.1 Advantages of Optical Microfiber and Nanofiber

By significantly reducing the size of the sensors and their associated electronics with supporting hardware, miniaturization offers the possibility of complete portable microsystems that can carry out many of the operations traditionally performed in a laboratory. The other benefits of high compactness due to bend insensitivity include easier deployment and minimal intrusion. The consequential reduction in weight also draws substantial interest, particularly by the aerospace industry.

It is widely recognized that for sensors responding to a mechanical stimulus, the higher sensitivity associated with the lower stiffness of MNFs grants a notable improvement in the detection of weak signals of the measurand. As for current sensors based on the Faraday effect, the higher detection bandwidth resulting from the shorter OPL and bend insensitivity of MCs enables the measurement of faster-changing current signals.

Although optical nanofibers have the potential to surpass optical microfibers in terms of performance, their fragility currently limits the extent of their use.

8.5.2 Challenges and Practical Issues

To compete with traditional fiber-optic sensors in terms of sensitivity, some MNF-based sensors would require their tapered uniform waist region to be in excess of ~10 cm. Generally, the sensitivity of non-resonator-type MNF-based sensors scales with the MNF length. However, the detection bandwidth associated with certain sensing mechanisms such as the Faraday effect decreases with longer OPLs. Hence, a trade-off must be considered for a good balance between sensitivity and detection bandwidth.

The optical loss of MNFs is one of the key areas that need improvement. Although the theoretical minimum attenuation when embedded in polymers is very low [15], in practice, it is usually not the case. Microbubbles in the polymer cladding cause scattering of the evanescent field [51], and contaminations of the polymer material introduce unwanted absorption. The average loss observed in packaged samples of MNF can range from 0.5 to 3 dB/cm. Therefore, to reach acceptable levels of loss (<0.1 dB/cm), a significant amount of work needs to be done to remove microbubbles from the polymer resin (e.g., vacuum pump) before ultraviolet (UV) curing and to minimize contaminations (e.g., cleanroom conditions).

Another practical issue with MNF-based sensors is the dynamic temperature range, which is determined by the polymer material used for packaging. For the commonly used Efiron PC-373 AP, the polymer experiences significant evaporation in OH content around ~80°C and noticeable deformation by ~120°C. Apart from the physical deterioration of the material, thermal expansion and contraction can induce internal stresses in the embedded MNF, resulting in unpredictable loss and birefringence.

The lifetime of each fabricated sensor head primarily depends on the curing state of the polymer coating. The UV-curing time of the polymer makes a profound impact on the long-term stability of the packaged device. A short duration tends to avoid the buildup of internal stresses that can modify the intended MNF geometry. On the other hand, the protective coating is more volatile and the geometrical stability of the embedded MNF is at risk from external perturbations. Hence, it is to be appreciated that ensuring long-term reliability is likely to be challenging.

An alternative solution is suspended-core fibers, which exhibit the same optical properties of MNFs albeit in a mechanically stable arrangement. However, since the evanescent field is not accessible outside the fiber, they cannot be used for resonators that rely on self-coupling. In addition, suspended-core fibers are far bulkier than MNFs, which eliminates their advantages over standard optical fibers.

8.6 SUMMARY

Optical MNF-based sensors are a rapidly growing field. Optical MNFs exhibit many desirable characteristics such as large evanescent field, strong optical confinement, bend insensitivity, high configurability, high compactness, and the feasibility of extremely high-Q resonators. The resulting sensors hold numerous advantages over their standard optical fiber counterparts, including high sensitivity, high detection bandwidth, fast response, high selectiveness, low intrusiveness, small size, and lightweight. With sensing areas spanning acceleration, acoustic, bend/curvature, current, displacement/strain, electric field, force/pressure, magnetic field, RI, rotation, roughness, and temperature, the future of optical microfiber technology looks exceptionally promising. There is no doubt that new applications will continue to arise from the development of optical microfiber-based sensors.

REFERENCES

1. Tong, L., Gattass, R. R., Ashcom, J. B., He, S., Lou, J., Shen, M., Maxwell, I., and Mazur, E., Subwavelength-diameter silica wires for low-loss optical wave guiding, *Nature*, 426, 816–819, 2003.
2. Brambilla, G., Optical fibre nanowires and microwires: A review, *J. Opt.*, 12(4), 043001, 2010.
3. Stiebeiner, A., Garcia-Fernandez, R., and Rauschenbeutel, A., Design and optimization of broadband tapered optical fibers with a nanofiber waist, *Opt. Express*, 18(22), 22677–22685, 2010.
4. Birks, T. A. and Li, Y. W., The shape of fiber tapers, *J. Lightwave Technol.*, 10(4), 432–438, 1992.
5. Tong, L., Hu, L., Zhang, J., Qiu, J., Yang, Q., Lou, J., Shen, Y., He, J., and Ye, Z., Photonic nanowires directly drawn from bulk glasses, *Opt. Express*, 14(1), 82–87, 2006.
6. Brambilla, G., Koizumi, F., Feng, X., and Richardson, D. J., Compound-glass optical nanowires, *Electron. Lett.*, 41(7), 400–402, 2005.
7. Mägi, E. C., Fu, L. B., Nguyen, H. C., Lamont, M. R. E., Yeom, D. I., and Eggleton, B. J., Enhanced Kerr nonlinearity in sub-wavelength diameter As_2Se_3 chalcogenide fiber tapers, *Opt. Express*, 15(16), 10324–10329, 2007.
8. Harfenist, S. A., Cambron, S. D., Nelson, E. W., Berry, S. M., Isham, A. W., Crain, M. M., Walsh, K. M., Keynton, R. S., and Cohn, R. W., Direct drawing of suspended filamentary micro- and nanostructures from liquid polymers, *Nano Lett.*, 4(10), 1931–1937, 2004.
9. Nain, S., Wong, J. C., Amon, C., and Sitti, M., Drawing suspended polymer micro-/nanofibers using glass micropipettes, *Appl. Phys. Lett.*, 89(18), 183105–183107, 2006.
10. Gu, F., Zhang, L., Yin, X., and Tong, L., Polymer single-nanowire optical sensors, *Nano Lett.*, 8(9), 2757–2761, 2008.

11. Xing, X., Wang, Y., and Li, B., Nanofiber drawing and nanodevice assembly in poly(trimethylene terephthalate), *Opt. Express*, 16(14), 10815–10822, 2008.
12. Love, J. D. and Henry, W. M., Quantifying loss minimisation in single-mode fibre tapers, *Electron. Lett.*, 22(17), 913–914, 1986.
13. Love, J. D., Henry, W. M., Stewart, W. J., Black, R. J., Lacroix, S., and Gonthier, F., Tapered single-mode fibers and devices: 1. Adiabaticity criteria, *IEE Proc.—J. Optoelectron.*, 138(5), 343–354, 1991.
14. Brambilla, G., Xu, F., and Feng, X., Fabrication of optical fibre nanowires and their optical and mechanical characterization, *Electron. Lett.*, 42(9), 517–519, 2006.
15. Brambilla, G., Finazzi, V., and Richardson, D., Ultra-low-loss optical fiber nanotapers, *Opt. Express*, 12(10), 2258–2263, 2004.
16. Sumetsky, M., Optical fiber microcoil resonator, *Opt. Express*, 12(10), 2303–2316, 2004.
17. Tong, L., Lou, J., Ye, Z., Svacha, G. T., and Mazur, E., Self-modulated taper drawing of silica nanowires, *Nanotechnology*, 16(9), 1445–1448, 2005.
18. Qiu, S., Chen, Y., Kou, J., Xu, F., and Lu, Y., Miniature tapered photonic crystal fiber interferometer with enhanced sensitivity by acid microdroplets etching, *Appl. Opt.*, 50(22), 4328–4332, 2011.
19. Naqshbandi, M., Canning, J., Nash, M., and Crossley, M. J., Controlling the fabrication of self-assembled microwires from silica nanoparticles, *OFS*-22, Beijing China, 2012, pp. 842182-1–842182-4.
20. Huang, Z., Zhang, Y., Kotaki, M., and Ramakrishna, S., A review on polymer nanofibers by electrospinning and their applications in nanocomposites, *Compos. Sci. Technol.*, 63(15), 2223–2253, 2003.
21. Praeger, M., Saleh, E., Vaughan, A., Stewart, W. J., and Loh, W. H., Fabrication of nanoscale glass fibers by electrospinning, *Appl. Phys. Lett.*, 100(6), 063114-1–063114-3, 2012.
22. Tong, L., Lou, J., and Mazur, E., Single-mode guiding properties of subwavelength-diameter silica and silicon wire waveguides, *Opt. Express*, 12(6), 1025–1035, 2004.
23. Chen, G. Y., Ding, M., Newson, T. P., and Brambilla, G., A review of microfiber and nanofiber based optical sensors, *Open Opt. J.*, 7, 32–57, 2014.
24. Wu, M., Huang, W., and Wang, L., Propagation characteristics of the silica and silicon subwavelength-diameter hollow wire waveguides, *Chin. Opt. Lett.*, 6(1), 732–735, 2008.
25. Zhai, G., and Tong, L., Roughness-induced radiation losses in optical micro or nanofibers, *Opt. Express*, 15(21), 13805–13816, 2007.
26. Sumetsky, M., Dulashko, Y., Domachuk, P., and Eggleton, B. J., Thinnest optical waveguide: Experimental test, *Opt. Lett.*, 32(7), 754–756, 2007.
27. Sumetsky, M., How thin can a microfiber be and still guide light?, *Opt. Lett.*, 31(7), 870–872, 2006.
28. Chuo, S. and Wang, L., Propagation loss, degradation and protective coating of long drawn microfibers, *Opt. Commun.*, 284(12), 2825–2828, 2011.
29. Xu, F. and Brambilla, G., Embedding optical microfiber coil resonators in Teflon, *Opt. Lett.*, 32(15), 2164–2166, 2007.
30. Tong, L., Lou, J., Gattass, R. R., He, S., Chen, X., Liu, L., and Mazur, E., Assembly of silica nanowires on silica aerogels for microphotonic devices, *Nano Lett.*, 5(2), 259–262, 2005.
31. Teipel, J., Franke, K., Türke, D., Warken, F., Meiser, D., Leuschner, M., and Giessen, H., Characteristics of supercontinuum generation in tapered fibers using femtosecond laser pulses, *Appl. Phys. B*, 77(2), 245–251, 2003.
32. Chen, G. Y., Newson, T. P., and Brambilla, G., Optical microfibers for fast current sensing, *Opt. Fiber Technol.*, 19(6B), 802–807, 2013.
33. Chen, G. Y., Zhang, X., Brambilla, G., and Newson, T. P., Theoretical and experimental demonstrations of a microfiber-based flexural disc accelerometer, *Opt. Lett.*, 36(18), 3669–3671, 2011.
34. Chen, G. Y., Zhang, X., Brambilla, G., and Newson, T. P., Enhanced responsivity of a flexural disc acceleration sensor based on optical microfiber, *Opt. Commun.*, 285(23), 4709–4714, 2012.
35. Chen, G. Y., Brambilla, G., and Newson, T. P., Compact acoustic sensor based on air-backed mandrel coiled with optical microfiber, *Opt. Lett.*, 37(22), 4720–4722, 2012.
36. Zhang, S., Zhang, W., Gao, S., Geng, P., and Xue, X., Fiber-optic bending vector sensor based on Mach–Zehnder interferometer exploiting lateral-offset and up-taper, *Opt. Lett.*, 37(21), 4480–4482, 2012.
37. Belal, M., Song, Z., Jung, Y., Brambilla, G., and Newson, T. P., Optical fiber microwire current sensor, *Opt. Lett.*, 35(11), 3045–3047, 2010.
38. Belal, M., Song, Z., Jung, Y., Brambilla, G., and Newson, T. P., An interferometric current sensor based on optical fiber micro wires, *Opt. Express*, 18(19), 19951–19956, 2010.
39. Chen, G. Y., Brambilla, G., and Newson, T. P., Spun optical microfiber, *IEEE Photon. Technol. Lett.*, 24(19), 1663–1666, 2012.

40. Chen, G. Y., Newson, T. P., and Brambilla, G., Birefringence treatment of non-ideal optical microfibre coils for continuous Faraday rotation, *Electron. Lett.*, 49(11), 714–715, 2013.
41. Martinez-Rios, A., Monzon-Hernandez, D., Torres-Gomez, I., and Salceda-Delgado, G., An intrinsic fiber-optic single loop micro-displacement sensor, *Sensors*, 12(1), 415–428, 2012.
42. Veilleux, C., Lapierre, J., and Bures, J., Liquid-crystal-clad tapered fibers, *Opt. Lett.*, 11(11), 733–735, 1986.
43. Wang, X., Li, W., Chen, L., and Bao, X., Thermal and mechanical properties of tapered single mode fiber measured by OFDR and its application for high-sensitivity force measurement, *Opt. Express*, 20(14), 14779–14788, 2012.
44. Li, X. and Ding, H., All-fiber magnetic field sensor based on microfiber knot resonator and magnetic fluid, *Opt. Lett.*, 37(24), 5187–5189, 2012.
45. Xu, F., Horak, P., and Brambilla, G., Optical microfiber coil resonator refractometric sensor, *Opt. Express*, 15(12), 7888–7893, 2007.
46. Xu, F. and Brambilla, G., Demonstration of a refractometric sensor based on optical microfiber coil resonator, *Appl. Phys. Lett.*, 92(10), 101126-1–101126-3, 2008.
47. Scheuer, J., Fiber microcoil optical gyroscope, *Opt. Lett.*, 34(11), 1630–1632, 2009.
48. Digonnet, M. J. F., Rotation sensitivity of gyroscopes based on distributed-coupling loop resonators, *J. Lightwave Technol.*, 29(20), 3048–3053, 2011.
49. Birks, T. A., Knight, J. C., and Dimmick, T. E., High-resolution measurement of the fiber diameter variations using whispering gallery modes and no optical alignment, *IEEE Photon. Technol. Lett.*, 12(2), 182–183, 2000.
50. Chen, G. Y., Brambilla, G., and Newson, T. P., Inspection of electrical wires for insulation faults and current surges using sliding temperature sensor based on optical microfibre coil resonator, *Electron. Lett.*, 49(1), 46–47, 2013.
51. Wang, S., Pan, X., and Tong, L., Modeling of nanoparticle-induced Rayleigh–Gans scattering for nanofiber optical sensing, *Opt. Commun.*, 276(2), 293–297, 2007.

9 Fiber Bragg Grating Sensors and Interrogation Systems

Dipankar Sengupta

CONTENTS

9.1 INTRODUCTION

Among the spectrally modulated fiber sensors the most promising developments are those based on grating technology, the fiber Bragg grating (FBG). FBG sensors have attracted widespread attention and have been the subject of continuous and rapid development since FBGs were used for the first time for sensing purposes in 1989. Two decades have passed since the use of FBGs began. Various ideas have been proposed and various demodulation techniques have been developed for various measurands and applications. A review of these developments is summarized in this chapter.

The discovery of an FBG made a significant impact on research and development in fiber-optic sensing and telecommunications. FBGs are intrinsic in nature and have the ability to control the properties of light passing within the fiber. FBGs are mainly used in telecommunications as optical filters, dispersion-compensating components, and wavelength division multiplexing (WDM) systems [1]. Also, FBGs are sensitive to any perturbation of the fiber in the grating region and this led to extensive research to make use of it as a sensing element [2].

The measurand information from the FBG is wavelength encoded. FBG is a periodic or a quasiperiodic orthogonal perturbation of the refractive index along the longitudinal axis of the fiber core (Figure 9.1). The grating structure acts like a selective mirror for the wavelength that satisfies the Bragg condition. When FBG is illuminated by a broadband optical source, a narrowband spectral component corresponding to the Bragg resonance wavelength of the grating, λ_B, is reflected, while all other wavelengths outside the narrow reflection band will be transmitted. The Bragg wavelength, λ_B, is given by

$$\lambda_B = 2n_{eff}\Lambda \tag{9.1}$$

where

Λ is the grating periodicity

n_{eff} is the effective refractive index of the waveguide mode

The FBG sensors are one of the most exciting developments in the field of optical fiber sensing. The most important advantages are as follows:

1. FBG sensors are small and rugged passive components having a high lifetime.
2. FBGs can form an intrinsic part of the fiber-optic cable that can transmit the signal over several tens of kilometers.

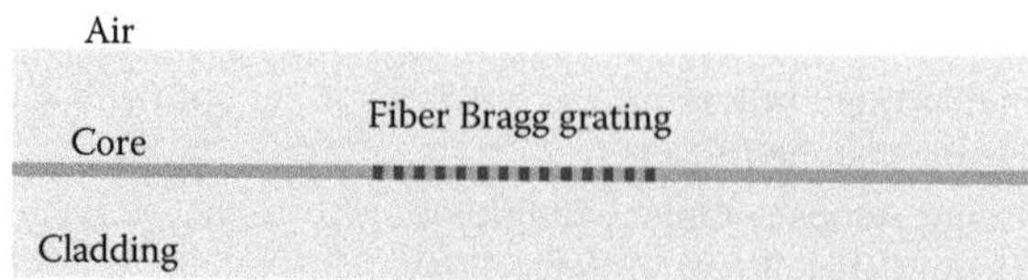

FIGURE 9.1 Fiber Bragg grating.

3. FBGs show no interference with electromagnetic radiation and can function in many hostile environments where conventional sensors fail.
4. FBGs do not make use of electrical signals, which make their explosion safe.
5. Several FBGs can be multiplexed in one optical fiber, driving down the cost of complex control systems.

9.1.1 Bragg Grating: A Brief History

The journey of FBG started at the Communications Research Centre, Canada, with successful demonstration by Kenneth O. Hill et al. in 1978. Initially, the gratings were fabricated using a 488 nm argon ion laser propagating along the fiber core [3,4]. Indirect measurement confirmed the formation of FBG over the entire 1 m length of the fiber, and thereafter, these gratings are called *Hill gratings* as a compliment for their research work on the nonlinear properties of germanium-doped silica fiber. These gratings are also known as internal inscription gratings because of their writing procedure. These gratings are limited to operating at a Bragg wavelength coinciding with the excitation laser wavelength.

In 1981, Lam and Garside reported that the mechanism of the formation of FBGs as reported by Hill was more or less a two-photon process due to the interaction of ultraviolet (UV) light with defect states just below the bandgap of the doped silica forming the core [5]. Later, Julian Stone demonstrated that any change in refractive index could be induced in all germanium (Ge)-doped optical fibers, and this reopened the research activity in FBG [6]. The permanent change in refractive index of the glass material due to light radiation refers to photosensitivity. Until this discovery, the only fiber available, which was photosensitive, was a small core fiber that was developed and made by Bell Northern Research Laboratories.

However, almost 10 years later, a breakthrough in FBG inscription occurred with the demonstration of the side writing technique (transverse holographic technique) by Meltz et al. [7]. It is popularly known as the external inscription technique of writing FBG. In this method, 244 nm radiation was split into two and recombined to form an interference pattern in the core of a photosensitive fiber. The fringe pattern induces a permanent localized index of refraction change. The period of the interference maxima and the index change was set by the angle between the beams, which is useful to form gratings that may reflect any wavelength. The prepared grating was found to be orders of magnitude more efficient than Hill grating.

Hereafter, several external inscription techniques were developed and reported. Bragg gratings in optical fibers have been fabricated using both *amplitude splitting* and *wave front splitting interferometers*. The amplitude splitting method is the one reported by Meltz et al. [7]. In wave front splitting techniques, the prism interferometer [8] and Lloyd interferometer [9] were used. Both of these methods require a UV source with good spatial coherence. The major benefit of this method is that both interferometers have only single optical component, which causes a reduction in the sensitivity to vibrations. The major drawback of the two wave front methods is that the grating length is limited to half the beam's width, and the wavelength tuning is restrained by the interferometer.

The fabrication of FBGs with *phase masks* was first demonstrated in 1993 [10–12]. Phase masks are made up of period patterns usually etched onto fused silica (Figure 9.2). They are made in such a way that when the UV beam is incident on the mask, the zero-order diffracted beam is suppressed to less than a few percent (typically <3%) of the transmitted power, and the diffracted ±1 orders are maximized, each typically carrying >33% of the transmitted power. The rest of the power is divided into the higher-order diffractions, which are always present. For efficient diffraction into the first orders, a near-field fringe pattern with period $\Lambda = \Lambda_{pm}/2$ is produced by the interference of the beams. The interference pattern photo imprints a refractive index modulation in the core of the photosensitive optical fiber placed in close proximity to the phase mask. The period of the grating written in the core of the fiber is one-half of the phase mask period and does not depend on the wavelength of the writing beam or its incident angle on

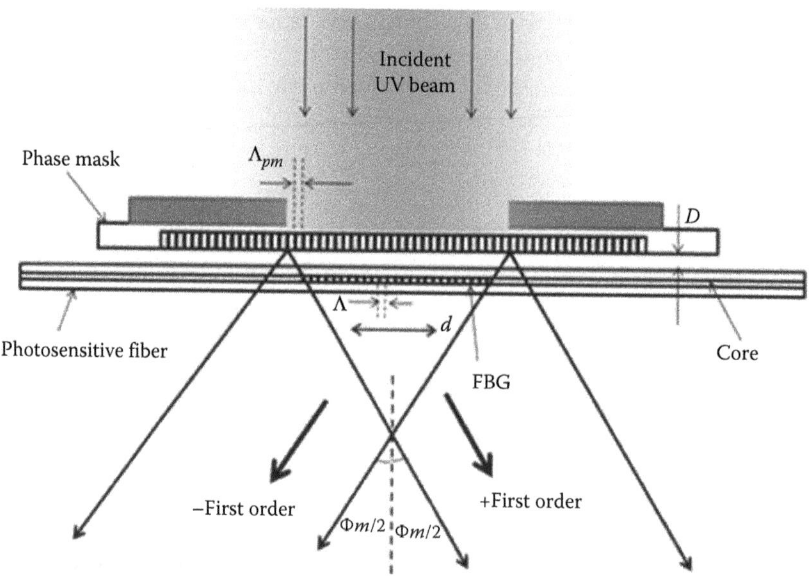

FIGURE 9.2 Fabrication of FBG using a phase mask.

the phase mask: $\Lambda_{mask} = 2\lambda_B$. One of the disadvantages of this technique is that a different phase mask is required for each specific Bragg wavelength.

The *point-by-point* method is one of the more flexible methods in inscribing Bragg gratings into an optical fiber. In this method, a UV source is passed through a slit and then focused down onto the fiber to produce an index change at one point. The fiber is then shifted a distance "Λ," which corresponds to the period of the Bragg condition. Due to submicron translation and tight focusing requirements, this technique is not feasible for first-order gratings. Therefore, the grating's length typically needs to be relatively short and research groups such as Malo et al. [13] have only been able to create second- and third-order gratings.

In the early days, manufacturing the photosensitive fiber and writing grating were performed separately. But as on today, gratings are produced in the industry that typically performs the manufacturing of the photosensitive optical fiber and the *writing* of the FBG all in a single stage. This reduces the time and associated cost of the grating and is generally used to draw a large number of embedded FBGs along a single length of the fiber.

It may become incomplete without mentioning the *writing of FBGs using femtosecond laser.* In 2003, FBG was inscribed on the standard Ge-doped telecom fiber (Corning SMF-28) by the use of 800 nm, 120 fs laser irradiation and a deep-etch silica zero-order nulled phase mask, and the grating was stable at 300°C [14]. Bennion's and coworkers in 2004 reported direct point-by-point inscription of FBGs by the fs laser without using the phase mask. The FBGs obtained were stable at the temperature of up to 900°C [15]. The principal advantage of high-energy pulses is their ability of grating inscription in any material type without preprocessing, such as hydrogenation or special core doping with photosensitive materials—the inscription process is controlled multiphoton absorption, void generation, and subsequent local refractive index changes. Furthermore, the use of an infrared (IR) source or its second harmonic removes the requirement to strip the optical fiber as gratings can readily be written through the buffer layer.

9.1.2 Photosensitivity

Photosensitivity in optical fiber refers to a permanent change in the index of refraction of the fiber core when exposed to light with characteristic wavelength and intensity that depends on the core

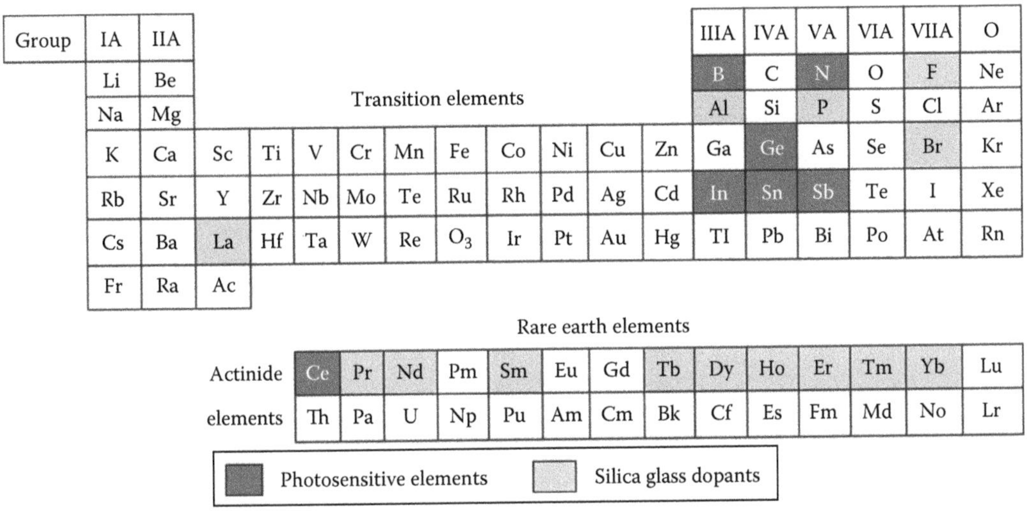

Group	IA	IIA											IIIA	IVA	VA	VIA	VIIA	O
	Li	Be	Transition elements										B	C	N	O	F	Ne
	Na	Mg											Al	Si	P	S	Cl	Ar
	K	Ca	Sc	Ti	V	Cr	Mn	Fe	Co	Ni	Cu	Zn	Ga	Ge	As	Se	Br	Kr
	Rb	Sr	Y	Zr	Nb	Mo	Te	Ru	Rh	Pd	Ag	Cd	In	Sn	Sb	Te	I	Xe
	Cs	Ba	La	Hf	Ta	W	Re	O_3	Ir	Pt	Au	Hg	TI	Pb	Bi	Po	At	Rn
	Fr	Ra	Ac															

Rare earth elements

Actinide elements	Ce	Pr	Nd	Pm	Sm	Eu	Gd	Tb	Dy	Ho	Er	Tm	Yb	Lu
	Th	Pa	U	Np	Pu	Am	Cm	Bk	Cf	Es	Fm	Md	No	Lr

Photosensitive elements   Silica glass dopants

FIGURE 9.3 Photosensitive elements in the periodic table.

material. Initially, photosensitivity has been thought to be a phenomenon only associated with optical fibers having a large concentration of germaniums in the core and photoexcited with 240–250 nm UV light.

However, photosensitivity has been observed with excitation at different UV wavelengths (e.g., 157, 193, 325, and 351 nm) [16–18] in a wide variety of different fibers, many of which have other dopants (e.g., boron, erbium, tin, and phosphorus) [19–22] in addition to germaniums and some of which contain no germanium at all (e.g., rare earth–doped fibers, aluminosilicate fiber, fluorozirconate fiber doped with cerium/erbium) [23–25]. Figure 9.3 shows list of photosensitive elements in the periodic table. There have been efforts to enhance the photosensitivity in optical fibers by introducing techniques such as hydrogen loading [26], flame brushing [27], and codoping [20,21]. A brief summary of photosensitivity of fiber with doped with different elements [1,2] is given in Table 9.1.

It is important to mention here that there have been several efforts, both theoretical and experimental, to understand the mechanisms of UV-induced refractive index change in germanosilicate fibers [28]. Numerous models have also been proposed, such as the color-center model [29], the compaction–densification model [30], the dipole model [31], and the stress relief model [32], for understanding the mechanism of photoinduced refractive index changes in germanium-doped silicate fibers; these models can again be essentially classified under two categories: the electronic models that try to connect the photosensitivity with the defects that are inherently present in the glass and those formed during the fiber drawing process and the structural models that associate the photosensitivity with the local structural changes in the glass that occur during illumination. However, after accumulating all the experimental findings, it becomes obvious that both the electronic defect–oriented changes and structural changes are involved in photosensitivity. More precisely, two main models, namely, the color-center model and the compaction model, have been identified to be the root cause for germanosilicate fibers to exhibit photosensitivity at wavelengths around 240 nm.

9.1.3 Types of Grating

The term *type* in this context refers to the underlying photosensitivity mechanism by which grating fringes are produced in the fiber. The different methods of creating these fringes have a significant effect on physical attributes of the produced grating, particularly the temperature response and ability to withstand elevated temperatures. Thus far, five types of FBG have been reported with different underlying photosensitivity mechanisms. They are Type I, Type IA, Type IIA, Type II, and regenerative grating.

TABLE 9.1
Summary of Different Types of Silica Fiber with Different Dopants

Fiber	Concentration of Elements	Characteristics of Writing Sources	Fiber Treatment	Change in Refractive Index
Silica fiber with germanium	Ge < 10 mol%	240–262 nm (pulse or CW laser)	No treatment	1.5×10^{-4}
			Boron doping	1×10^{-3}
			Hydrogenation	3×10^{-3}
			Flame brushing	1×10^{-3}
		248 nm (single pulse 1 J/cm^2)	No treatment	2×10^{-3}
		193 nm, 400 mJ/cm^2 (two-photon process)	No treatment	10^{-3}
	Ge > 15 mol%	244 nm (pulse or CW laser under long exposure)	No treatment	2.5×10^{-4}
	Ge 30 mol%	334 nm CW	No treatment	0.8×10^{-4}
		Argon ion laser	Boron doping	1×10^{-4}
Germanosilicate fiber	P	248 nm, pulsed	Hydrogenation + heating (400°C)	7×10^{-4}
Silica fiber with Co doping	P	193 nm, pulsed	Hydrogenation	2×10^{-4}
	P + Yb^{3+}, Er^{2+}	193 nm, pulsed	Hydrogenation	10^{-3}
	P, Al, Yb^{3+}, Er^{2+}	248 nm, pulsed	Hydrogenation	2.5×10^{-5}
	P, Sn	248 nm, pulsed	No treatment	5×10^{4}

In *Type I* regime, with the UV exposure, there is a monotonic increase in refractive index modulation of a grating. The growth dynamics of the Type I grating is characterized by a power law with time [33]. This temporal evolution of the index of refraction change is a characteristic behavior of Type I photosensitivity, and the grating formed is referred to as a Type I Bragg grating [34]. The index modulation is typically less than 0.001. Type I index change has a low thermal stability. Writing Type I grating is simple in photosensitive fibers, mainly germanosilicate, including doped photonic crystal fibers [35]. A red shift in Bragg wavelength is observed during the writing process. At temperature below 300°C, the index modulation remains approximately constant and starts decaying afterward (Figure 9.4).

Type IA gratings are subtype of Type I grating. They are formed after prolonged UV exposure of a standard grating in hydrogenated germanosilicate fiber [36,37]. Type IA grating has a red shift in resonance wavelength during the formation and exhibits the lowest temperature coefficient of all

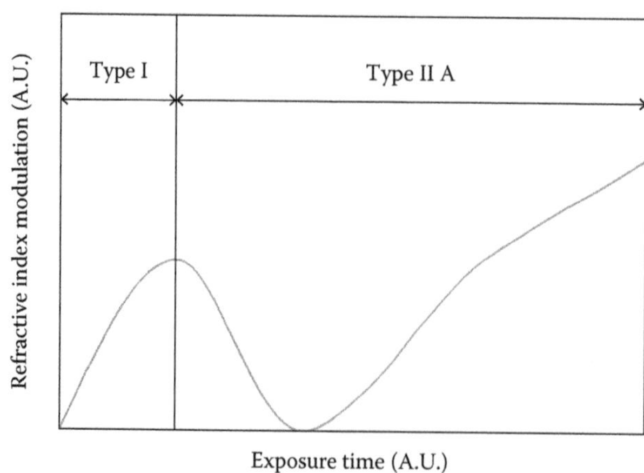

FIGURE 9.4 Change in the refractive index with exposure time.

grating types reported to date, which makes them ideal for use in a temperature compensating in sensor applications.

Type IIA FBGs have the same spectral characteristics as found in Type I gratings. The gratings are inscribed through a long process, following Type I grating inscription [38]. Type IIA grating has a blue shift in resonance wavelength during the formation. They are much more stable than Type I grating at high temperatures around 500°C. A change in the refractive index during prolonged exposure to UV radiation is shown in Figure 9.4.

Type II Bragg gratings are formed under very high, single excimer laser pulse fluence (>0.5 J/cm^2) [39]. As the power of the UV light used for exposure exceeds a threshold value, a sharp increase in the refraction index occurs, which is estimated to be close to 10^{-2}. This type of index change is referred to as Type II. As a result, the Type II index change has the highest stability at elevated temperature (800°C).

Regenerated gratings/chemical composition gratings are produced after annealing the Type I grating. They are regenerated at temperatures > 500 ± °C and usually, though not always, in the presence of hydrogen. They have been interpreted in different ways including dopant diffusion (oxygen being the most popular current interpretation) and glass structural change. Regenerated gratings show similarities with Type II gratings although reflectivity is weak and therefore comparable losses and scattering are not observed. In regenerated gratings, the grating can be repeatedly cycled to very high temperatures with no degradation [40]. Recent work has shown that there exists a regeneration regime beyond diffusion where gratings can be made to operate at temperatures in excess of 1295°C, outperforming even Type II femtosecond gratings. These are extremely attractive for ultrahigh-temperature applications.

9.1.4 Structured Grating

There are eight common structures for FBGs based on the periodicity and refractive index profile. They are shown in Figure 9.5.

The most common FBG is *uniform grating*, which has a constant grating pitch. The grating period is typically 0.25–0.5 μm with the light coupled into the backward propagating direction. This grating is considered to be excellent strain- and temperature-sensing devices because the measurements are wavelength encoded. This eliminates the problems of amplitude or intensity fluctuations that exist in many other types of fiber-based sensor systems [41].

The *blazed grating* has phase fronts tilted with respect to the fiber axis; that is, the angle between the grating planes and fiber axis is <90° [42,43]. In this grating, it is possible to couple light out from the core into backward propagating radiation modes. This grating is used for gain equalization in erbium-doped fiber (EDF) amplifiers.

A *chirped grating* has an aperiodic pitch, displaying a monotonic increase in the spacing between grating planes. This grating is used in specific applications in telecommunications and sensor technology, such as dispersion compensation and the stable synthesis of multiple wavelength sources [44–46].

A *superstructure FBG* (SFBG) is a standard FBG with rapidly varying spatial refractive index of uniform amplitude and pitch, on which is superimposed a slow spatial refractive index modulation [47]. This grating structure has applications in signal processing and tunable fiber lasers.

The structure of the FBG can vary via the refractive index also. The refractive index profile of an FBG can be uniform or apodized. The reflection spectrum of a finite length Bragg grating with uniform modulation of the index of refraction gives rise to a series of side lobes at adjacent wavelengths. It is very important to minimize and if possible eliminate the reflectivity of these side lobes, in devices where high rejection of the nonresonant light is required.

In an *apodized grating*, the refractive index is graded to approach zero at the end of the grating. Apodized gratings offer significant improvement in the suppression of side lobes while maintaining reflectivity and a narrow bandwidth. The two functions typically used to apodize FBGs are Gaussian [48] and raised cosine [49].

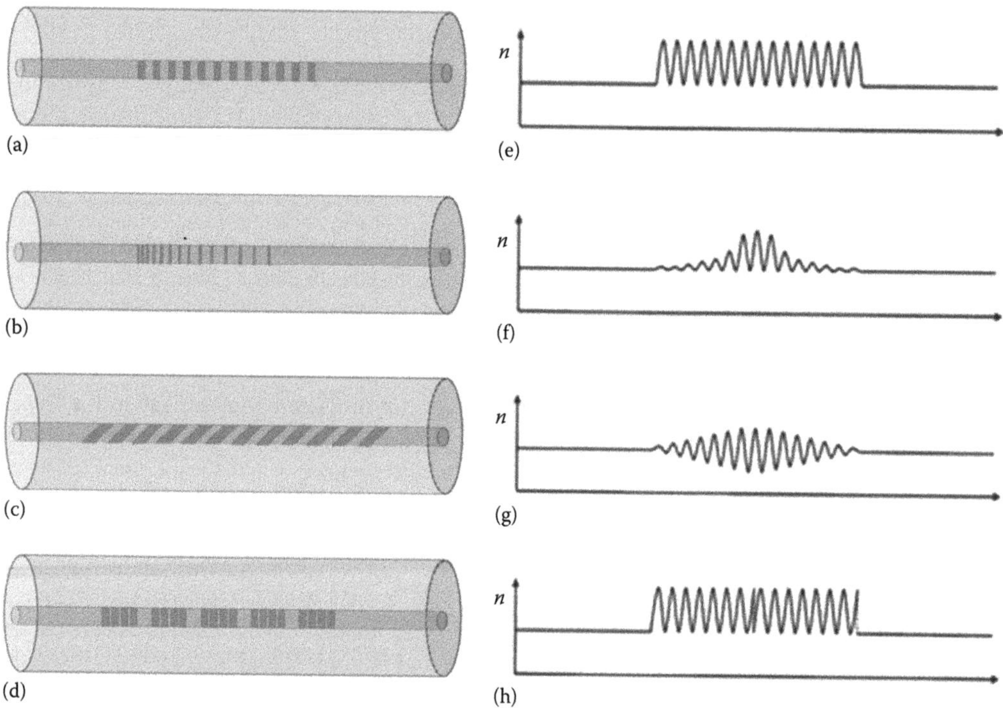

FIGURE 9.5 Uniform FBG with constant index modulation and amplitude. (a) Uniform FBG, (b) chirped FBG, (c) TFBG, (d) SFBG, (e) uniform positive-only index change, (f) Gaussian-apodized index change, (g) raised-cosine-apodized zero-dc index change, and (h) discrete phase shift index change.

The phase of the index modulation is usually set to a constant value. However, introducing a π phase shift in the middle of the grating results in two identical gratings separated by a half wavelength. Such gratings are called *phase-shifted gratings* [50–53]. Phase shifts open up narrowband transmission windows inside the stop band of the Bragg grating and the transmission wavelength can be changed by adjusting the amount of phase shift. Phase-shifted FBGs can be used as narrowband transmission filters and to design an all-fiber demultiplexer capable of demultiplexing more than 20 channels.

9.2 THEORY OF FBG

In its simplest form, an FBG consists of a periodic modulation of the index of refraction in the core of a single-mode optical fiber [54]. Generally, uniform fiber gratings, where the phase fronts are perpendicular to the longitudinal axis of the fiber and the grating planes are of a constant period, are considered as the fundamental building blocks of most Bragg grating structures (Figure 9.6).

Light guided along the core of an optical fiber is scattered by each grating plane. When Bragg condition is not satisfied, the reflected light from each of the subsequent planes becomes progressively out of phase and will eventually cancel out. Additionally, light that is not coincident with the Bragg resonance wavelength will experience very weak reflection at each of the grating planes because of the phase mismatch.

When the Bragg condition is satisfied, the contributions of reflected light from each grating plane add constructively in the backward direction to form a back-reflected peak with the center wavelength defined by the grating parameters. The Bragg grating condition is simply the requirement that satisfies both energy and momentum conservation. Energy conservation ($\hbar\omega_f = \hbar\omega_i$) requires

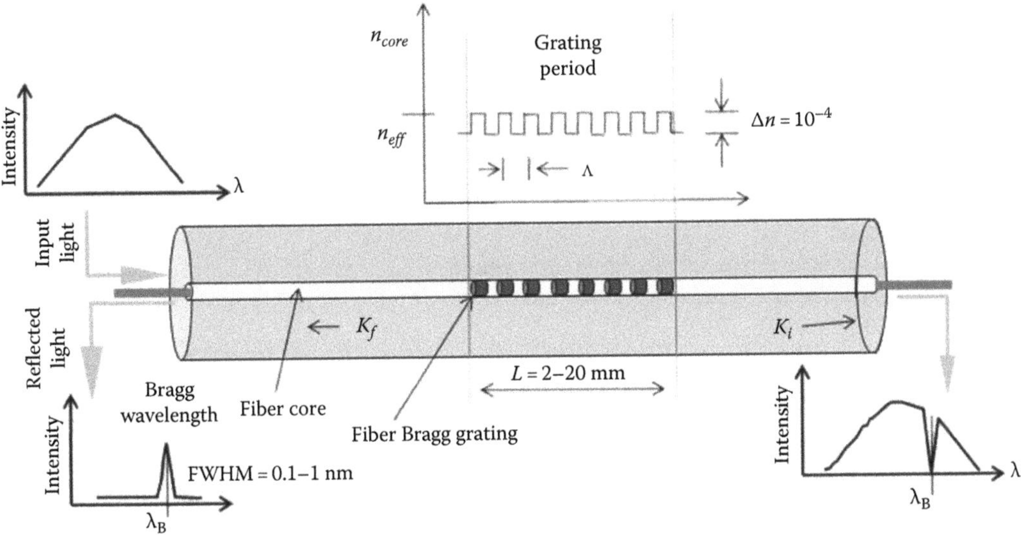

FIGURE 9.6 Illustration of a uniform fiber Bragg grating with constant index modulation amplitude and period.

that the frequency of the incident radiation and the reflected radiation be the same. Momentum conservation requires that the sum of incident wave vector (k_i) and the grating wave vector (K_g) should be equal to the wave vector of the scattered radiation (k_f), expressed as

$$k_i + K_g = k_f \tag{9.2}$$

where

K_g is the grating wave vector with a magnitude $2\pi/\Lambda$ and direction normal to the grating planes
Λ is the grating spacing shown in Figure 9.7

The diffracted wave vector is equal in magnitude but opposite in direction to the incident wave vector. Hence, the momentum conservation condition becomes

$$2\left(\frac{2\pi n_{eff}}{\lambda_B}\right) = \frac{2\pi}{\Lambda} \tag{9.3}$$

This simplifies to the first-order Bragg condition:

$$\lambda_B = 2n_{eff}\Lambda$$

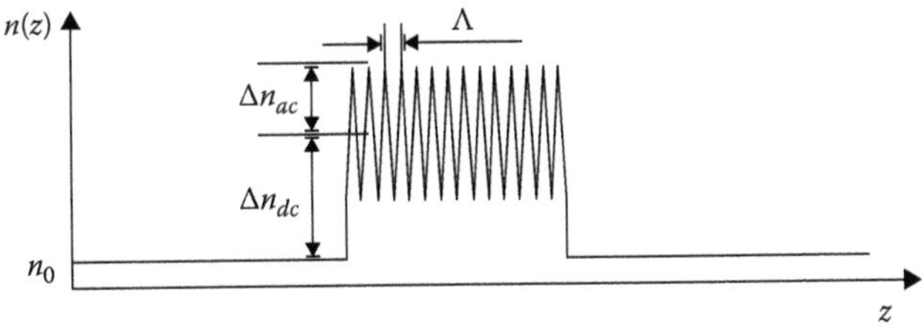

FIGURE 9.7 Refractive index of a uniform FBG.

A Bragg grating is completely characterized by its refractive index distribution along the fiber $n(z)$ [55]:

$$n(z) - n_0 = \Delta n_{dc}(z) + \Delta n_{ac}(z) \cdot \cos\left(\frac{2\pi}{\Lambda} z + \theta(z)\right) \tag{9.4}$$

where
- z is the position
- n_0 is the refractive index prior to grating inscription
- $\Delta n_{ac}(z)$ is the refractive index modulation amplitude
- Λ is the grating period
- $\theta(z)$ is the period chirp (slowly varying with z)
- $\Delta n_{dc}(z)$ is the average change in refractive index

The refractive index modulation amplitude remains sinusoidal until the exposed region reaches the maximal refractive index change (Figure 9.7).

In order to design, fabricate, and use FBGs in various applications, tools for analysis, synthesis, and characterization are crucial. For example, when the gratings are used as sensor elements, synthesis methods can be used to extract any distributed sensing parameters like strain, temperature, and pressure along the fiber. The most widely used technique for modeling the optical properties of FBG is coupled mode theory. The theory can be used to accurately model the optical properties of fiber gratings and is initially developed for uniform gratings. The theory is often used as a technique for obtaining quantitative information about the diffraction efficiency and spectra dependence of fiber gratings.

The derived expression from coupled mode theory for the reflectivity of a uniform grating is [2]

$$R = |r|^2 = \frac{k^2 \sinh^2 \gamma L}{\left(\Gamma^2/4\right)\sinh^2 \gamma L + \gamma^2 \cosh^2 \gamma L} \tag{9.5}$$

where $\gamma^2 = k^2 - (\Gamma^2/4)$. For $\Gamma = 0$ (phase matching), the maximum reflectivity can be expressed as $R_{max} = \tanh^2 kL$, where k is given as

$$k = \frac{\pi \eta \upsilon \Delta n}{\lambda} \tag{9.6}$$

in which υ and η are the fringe visibility and the confinement factors, respectively.

The grating bandwidth $\Delta\lambda_{HM}$ defined as the wavelength range between the first zeros apart from the Bragg peak is given by [2]

$$\Delta\lambda_{HM} = \frac{\lambda_B^2}{\pi n_{eff} L}\left(k^2 L^2 + \pi^2\right)^{1/2} \tag{9.7}$$

Equation 9.7 suggests that the bandwidth of a Bragg grating depends on induced refractive index changes. For strong gratings, the light does not penetrate the full length of the grating, and the bandwidth is thus independent of length and directly proportional to the induced index change. Stronger gratings have a much higher reflection than the weaker gratings, making it possible even to design short gratings and still provide more than 99% reflectivity.

9.3 SENSING PRINCIPLES

This section describes working principle of the FBG as sensors. The Bragg wavelength of FBG sensor strongly depends on applied strain to the FBG and local temperature surrounding it. The shift in Bragg wavelength due to the effect of changes in n_{eff} and/or grating pitch "Λ" of the FBG by the applied strain and temperature is given by [54]

$$\Delta\lambda_B = 2\left(\Lambda\frac{\partial n_{eff}}{\partial l} + n_{eff}\frac{\partial \Lambda}{\partial l}\right)\Delta l + 2\left(\Lambda\frac{\partial n_{eff}}{\partial T} + n_{eff}\frac{\partial \Lambda}{\partial T}\right)\Delta T \tag{9.8}$$

where

- T is the temperature
- l is the grating length
- $\Delta\lambda_B$ is the change in Bragg wavelength

The first term and the second term in Equation 9.8 represent the outcome of strain and temperature on Bragg wavelength. Physical parameters like strain, temperature, surrounding refractive index (SRI), displacement, and magnetic field are sensed by the FBGs due to the their modulation of n_{eff} and/or "Λ."

9.3.1 STRAIN SENSOR

One of the most important potential uses of Bragg gratings in sensors is for strain monitoring, be it static, quasistatic, or dynamic. When an FBG written on the core of a conventional single-mode optical fiber is subjected to strain along its z-axis at constant temperature, the Bragg wavelength varies due to the change in the grating spacing (Λ) and the photoelastic-induced change in the effective refractive index (n_{eff}). A schematic experimental setup for strain measurement using FBG is shown in Figure 9.8.

The fractional change in Bragg wavelength with longitudinal strain ($\Delta T = 0$) is given by

$$\frac{\Delta\lambda_B}{\lambda_B} = (1 - P_e)\Delta\varepsilon_z \tag{9.9}$$

where

- $P_e = (n_{eff}^2/2)\left[p_{12} - \nu(p_{11} + p_{12})\right]$, p_{11} and p_{12} are Pockel's coefficients of the strain-optical tensor, "ν" is the material Poisson's ratio of the optical fiber
- $\Delta\lambda_B$ is the change in Bragg wavelength
- $\Delta\varepsilon_z$ is the applied longitudinal strain

For a typical value of germanosilicate optical fiber $p_{11} = 0.113$, $p_{12} = 0.252$, $\nu = 0.16$, and $n_{eff} = 0.148$, the strain sensitivity at 1550 nm is 1.2 pm/με [54,55].

FBG sensors give a linear strain relationship to the reflected wavelength shift from the sensor within the elastic deformation limit of the fiber. The strain sensitivity of FBG depends on

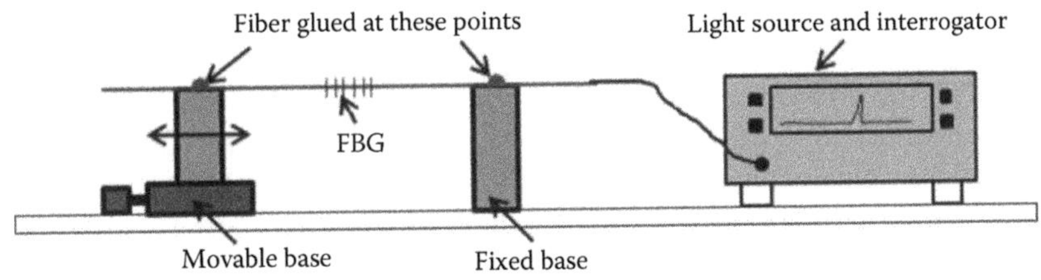

FIGURE 9.8 Schematic experimental setup of an FBG strain sensor.

the type of photosensitive fiber used to fabricate the FBG as well as the diameter of the FBG-containing fiber [56].

It has been found that when a nonisotropic transverse load is applied to an optical fiber, a state called birefringence occurs in the optical fiber. If such load is applied to an FBG written in single-mode fiber, birefringence leads to the separation of the single reflected Bragg peak into two distinct ones, which is a measure of the transverse load. The Bragg wavelength shifts under isothermal condition ($\Delta T = 0$) due to transverse load are given by [57]

$$\frac{\Delta\lambda_{B(x)}}{\lambda_{B(x)}} = \varepsilon_z - \frac{n_{eff}^2}{2}\left[p_{11}\varepsilon_x + p_{12}\left(\varepsilon_y + \varepsilon_z\right)\right] \tag{9.10}$$

$$\frac{\Delta\lambda_{B(y)}}{\lambda_{B(y)}} = \varepsilon_z - \frac{n_{eff}^2}{2}\left[p_{11}\varepsilon_y + p_{12}\left(\varepsilon_x + \varepsilon_z\right)\right] \tag{9.11}$$

where $\varepsilon_x, \varepsilon_y$ are the strains along the axes of the fiber coordinate system. The effect of birefringence can be utilized to sense multiple parameters like strain, force, pressure, and temperature using an FBG [58].

In most cases of practical interest, the FBG is subjected to axisymmetric homogeneous strain, that is, $\varepsilon_x = \varepsilon_y$. Equation 9.1 has been used by several researchers to measure strains in various experimental configurations by monitoring the wavelength shift $\Delta\lambda_B$ since λ_B, P_e are known parameters for a given sensor [59]. Several physical parameters like pressure, force, displacement, vibration, liquid level, and magnetic field can be measured indirectly using strain sensing of the FBG and are discussed in this section.

9.3.2 Temperature Sensor

FBGs are sensitive to temperature and are employed widespread particularly in harsh environments. The thermal-induced changes in Bragg wavelength are due to two major contributions: the first due to the change in effective refractive index of the guided mode and the second owing to the change in the grating period. This fractional wavelength shift for a temperature change (ΔT) can be written as from Equation 9.1 at constant strain [54]

$$\Delta\lambda_B = \lambda_B(\alpha + \delta)\Delta T \tag{9.12}$$

where

$\alpha = (1/\Lambda)(\partial\Lambda/\partial T)$ is the thermal expansion coefficient of the fiber
$\delta = (1/n_{eff})(\partial n_{eff}/\partial T)$ is the thermo-optic coefficient

Typical temperature responses of the Bragg wavelength are 6.8, 10, and 13 pm/°C for the Bragg wavelength of around 830, 1300, and 1550 nm, respectively [60].

Although FBGs are often referring to permanent refractive index modulation in the fiber core, exposure to high-temperature environments usually results in the bleach of the refractive index modulation. The maximum temperature reported for the conventional FBG temperature sensor is around 600°C due to its weak bonds of germanium and oxygen. Figure 9.9 shows a schematic experimental setup for measuring the temperature sensitivity of the FBG.

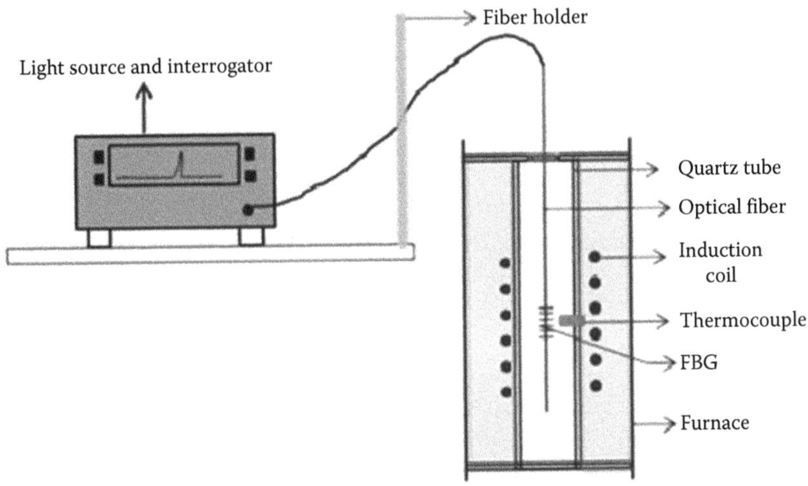

FIGURE 9.9 Schematic experimental setup for the measurement of temperature sensitivity of FBG.

Conventional Type I FBG has been used in various infrastructural health-monitoring applications. However, this type of FBG has poor thermal survivability and accuracy degradation at relative elevated temperature of greater than 300°C. The thermal properties of FBGs such as wavelength stability and the decay in peak reflected power play a vital role in many applications. The thermal decay characteristics of UV-induced FBGs behave very differently from others. Some kind of gratings written into the B- and Ge-codoped silicate fibers can survive only at comparatively low temperatures of about 300°C and disappear after annealing for a few hours at temperatures over 350°C [61]. Compared to these, the FBGs written into fibers codoped with Sn, Sb, or In have been reported to survive much higher temperatures, typically up to 800°C [62,63] or 900°C [64]. It is thus clear that the thermal stability of the gratings is strongly dependent on the glass composition of the fiber into which the gratings are written. In addition, special compositions of dopants incorporated into the fibers such as gratings written into N-doped fibers [65,66], chemical composition gratings in fluorine-doped silica optical fibers [67,68], regenerated gratings [69], and the use of femtosecond IR (fs-IR) lasers for writing the gratings have been reported to create FBGs [14,70] sustainable to temperatures as high as 1000°C. Type II FBGs written by fs-IR are operable up to 1200°C [71]. The reflectivity of this grating is an order of magnitude higher than what is obtainable using Type II UV or regenerated gratings.

At temperatures close to or above 1000°C in air, unpackaged standard silica single-mode fibers lose almost all of their mechanical strength. While the fibers themselves survive hundreds of hours at 1000°C [72] when left untouched, any subsequent handling of the fiber after the test is not possible as the fiber becomes extremely brittle. Obviously, optical fibers experience severe mechanical degradation when tested in oxidizing atmospheres at high temperature [73]. To avoid this, a suitable encapsulation is required to protect the fiber from exposure to oxygen at high temperature. Metallic-coated gratings like gold are the most obvious choice for using them in higher-temperature applications [74]. FBGs written in a single-mode optical fiber with 400 μm cladding using fs-IR laser were reported, which is an another approach to improve the mechanical strength of the fiber at high temperature [75]. Such device can be considered as a self-packaged FBG device and useful in applications operating at or above 1000°C [76]. Experimental results of femtosecond laser–inscribed sapphire FBGs (SFBGs) used for high-temperature measurement were also reported [77,78]. SFBGs can sense temperatures up to 1745°C with no degradation of the grating strength [79].

Not only high-temperature but also low-temperature sensing using FBG is useful in applications related to superconductivity around the liquid nitrogen temperature (77 K) and helium temperature (4 K). In 1996, Gupta et al. demonstrated temperature sensing to as low as 80 K with 1.55 mm FBGs [80]. Mizunami et al. proposed Teflon-embedded FBG to enhance the low-temperature sensitivity [81]. They demonstrated that the use of Teflon-made slab as substrate for FBG can enhance 1.5 times (39 pm/K at 77 K) the temperature sensitivity than that of the PMMA substrate reported earlier. The use of different metal coatings on the FBG sensor for cryogenic temperature measurement with higher low-temperature sensitivity is demonstrated in the work of Lupi et al. [82].

9.3.3 Refractive Index Sensor

Fiber-optic liquid refractive index sensor is vital for designing the optical instruments and also plays a major role in chemical applications. FBG-based refractive index sensors work on the interactions between the evanescent field of the fundamental core mode and the surrounding materials. In principle, the effective refractive index is not influenced by the external one for standard optical fibers and thus no sensitivity to external refractive index [83]. However, if fiber cladding diameter is reduced along the grating region, the effective refractive index, n_{eff}, is significantly affected by the external refractive index. As a consequence, shifts in the Bragg wavelength combined with a modulation of the reflected amplitude occur. The dependence of the sensor responsivity on the cladding diameter can be easily evaluated by resolving numerically the mathematic model of a doubly cladding fiber model [84]. However, if the cladding diameter is reduced, n_{eff} shows a nonlinear dependence on the SRI leading to an increase in sensitivity (Figure 9.10).

Optical fiber diameter can be reduced through wet chemical etching process and hydrofluoric acid is used for the purpose. However, the etching rate of the process depends on concentration of the acid, temperature, and composition of the fiber.

Iadicicco et al. have demonstrated an etched FBG refractive index sensor capable of sensing SRI of both liquid and temperature. In their experiment, half the length of the FBG were etched for SRI sensing. The experimental characterization conducted by them confirms the theoretical and numerical analysis carried out to predict the sensor performances. Resolutions of $\approx 10^{-5}$ and $\approx 10^{-4}$ for the SRI are around 1.45 and 1.33, respectively [85].

Refractive index measurement using an FBG written in the waist of biconically tapered fiber was demonstrated by Grobnic et al. The FBG was fabricated by femtosecond laser technique [86]. Experimental results show that the sensor can measure variation of the refractive index with a sensitivity of 2.4×10^{-5}.

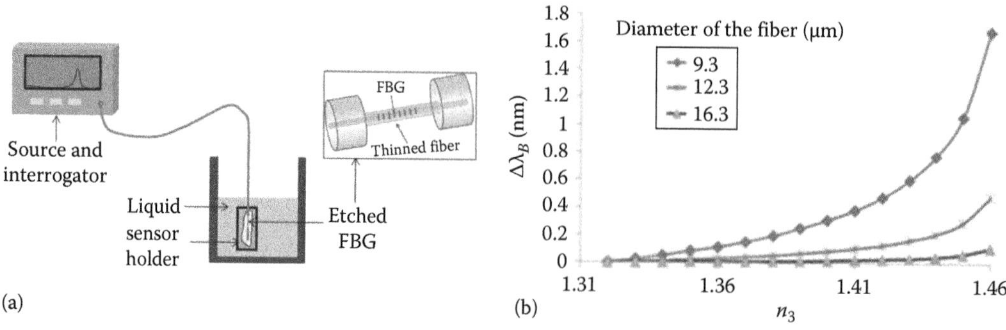

FIGURE 9.10 Schematic (a) experimental setup and (b) response of the FBG in solution with different refractive indices.

Miao et al. proposed a tilted FBG (TFBG)-based SRI sensor where the refractive index of the liquid can be measured by measuring the transmission power of the grating. It is found that the transmission power decreases monotonously from −2.14 to −3.26 dBm when the SRI varies from 1.3723 to 1.4532. Under the aforementioned condition, the resolution of the refractive index of the tilted grating is predicted to be 10^{-4} [87].

9.3.4 Pressure Sensor

Pressure sensors are devices that are designed to accurately detect the magnitude of the pressure applied to any application. The FBG technology has been rapidly applied in the pressure-sensing technology field.

When an FBG is compressed by external pressure ΔP, the fractional change $d\lambda_{BP}/\lambda_B$ is induced in the Bragg wavelength λ_B and is given by

$$\frac{d\lambda_{BP}}{\lambda_B} = \frac{d(n\Lambda)}{n\Lambda} = \left(\frac{1}{\Lambda}\frac{d\Lambda}{dP} + \frac{1}{n}\frac{dn}{dP}\right)\Delta P \tag{9.13}$$

where $d\lambda_{BP}$ represents the change in Bragg wavelength due to pressure.

The fractional change in the physical length of the fiber and the effective refractive index of the fiber core, respectively, is given by

$$\frac{\Delta L}{L} = -\frac{(1-2\upsilon)P}{E} \tag{9.14}$$

$$\frac{\Delta n}{n} = \frac{n^2 P}{2E}(1-2\upsilon)(2\rho_{12}+\rho_{11}) \tag{9.15}$$

where E and υ are Young's modulus and Poisson's ratio of the fiber, respectively. As the fractional change in the spatial period of the grating equals the fractional change in the physical length of the sensing section (i.e., $\Delta L/L = \Delta\Lambda/\Lambda$), the pressure sensitivity is given by

$$d\lambda_{BP} = \lambda_B\left[-\frac{(1-2\upsilon)}{E} + \frac{n^2}{2E}(1-2\upsilon)(2\rho_{12}+\rho_{11})\right]\Delta P \tag{9.16}$$

The experimental result presented by Xu et al. shows that the intrinsic pressure sensitivity of an FBG written in a Ge-doped silica fiber is −3.04 pm/MPa over a pressure range of 70 MPa [88], which is too low for practical pressure measurement (Figure 9.11). The "–" sign indicates that the

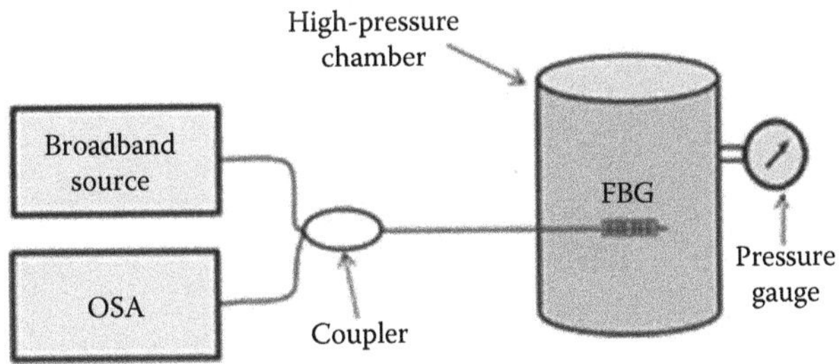

FIGURE 9.11 Experimental layout for studying pressure sensitivity of FBG.

Bragg wavelength is negative. In order to enhance the pressure sensitivity, several indirect techniques are adopted by researchers such as embedding FBG in polymer-filled metal cylinder [89] and metal bellows [90], FBG encapsulated in a polymer half-filled metal cylinder [91], metal-coated FBG [92], etched FBG [93], and using FBGs connected to a diaphragm [94].

9.3.5 Vibration Sensor

Optical fiber vibration sensors have been greatly developed and exploited due to its application for structural health monitoring (SHM) and damage detection. These sensors can be used in any one of the two ways: (a) a single-point sensor that only works in a localized region and (b) a distributed sensor with sensing capability along the length of the optical fiber. Low-frequency response vibration sensors are suited for large structures such as bridges, dams, and skyscrapers due to their low intrinsic frequencies. Vibration sensors with high-frequency response and high spatial resolution are very much desirable in areas like crack detection of materials and anomaly detection of engines where the frequency range of these events could be as high as several kHz to MHz [95]. FBG sensors are used in various arrangement schemes to measure the frequency of vibrations. The basic working principle of an FBG vibration sensor is that vibrations induce high-speed dynamic strain variations in FBG, and consequently monitoring the shift in Bragg wavelength allows measuring those vibrations. In vibration measuring applications, the bandwidth of the interrogation system is a key parameter that strongly limits the application range of the system. Different approaches have been reported to reach high-speed interrogation rates required for FBG-based vibration sensors. The most used optical setups for high-speed interrogation are based in splitting two or more different wavelength components of the reflected light from the FBG, combined with conventional intensity-based optoelectronic detectors.

This passive differential light intensity measurement arrangement can reach interrogation speeds as high as 50 kHz [96]. In another approach, the reflected Bragg signal was allowed to return via a 3 dB coupler to transmit through a wavelength-dependent coupler (Figure 9.12). Since the reflectivity of the FBG is being spectrally altered by the strain, the signals coming from the wavelength-dependent coupler will vary in intensity as the FBG is stretched during each half cycle of vibration. A simple electronic processing circuit is high enough to reveal the voltage directly proportional to the FBG dynamic strain and measure frequency of vibration. Cusano et al. made another approach to detect the FBG reflected signal, where the reflected light from an FBG sensor is selectively divided by an optical filter tuned with the Bragg wavelength of the FBG. The reflected signal from the FGB sensor is then divided into two components by the optical filter, and consequently, when the Bragg peak is displaced due to the strain, the two measured components vary their relation with

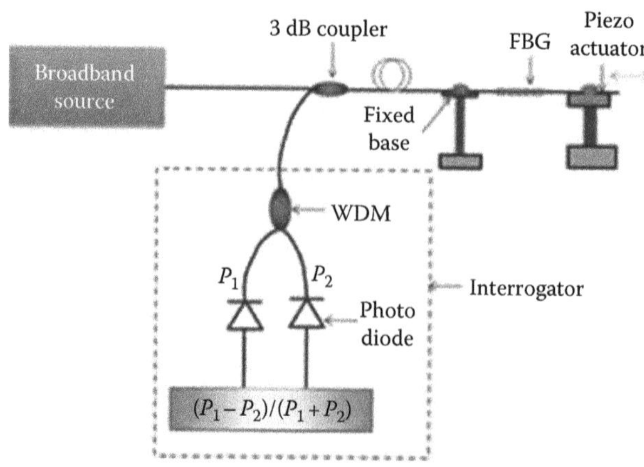

FIGURE 9.12 FBG-based vibration sensor.

respect to the other [97]. This type of passive interrogation can measure dynamic strain at a very high frequency up to 400 kHz. Likewise, many researchers have developed and reported different ways to interrogate the reflected FBG for fast measurement. An elaborate discussion is presented in Section 9.5 of this chapter.

9.3.6 Flow Sensor

Precise measurement of gas and liquid flow is important for many applications in the chemical, aerospace, and medical equipment industries. To measure a fluid flow, an orifice plate can be used, which restricts the fluid flow and causes a pressure drop across itself. An FBG sensor can be used to measure this pressure drop. Lim et al. demonstrated a differential pressure (DP)–based flow sensor [98]. In their sensor design, a steel plate diaphragm and two FBG sensors are used (Figure 9.13). When a DP ($P_1 - P_2$) is applied, the diaphragm is strained. The applied strain is transformed into optical information by using one of the optical FBG mounted at the center of the diaphragm on the side that is in tension. The second grating is mounted on the diaphragm side that is in compression and responds to temperature but not strain. The differential wavelength shift between the reflected signals from the two gratings provides a measurement of the fluid flow rate. Temperature variation causes the variation in the reflected wavelengths to be the same, and therefore, there is no change in differential wavelength.

The DP across an orifice plate is given by

$$Q = \frac{C_d A_0 \sqrt{2\Delta P/\rho}}{\sqrt{1-(A_0/A_1)^2}} \tag{9.17}$$

where

Q is the volumetric flow rate
A_0 is the hole area of the orifice plate
A_1 is the cross-sectional area of the pipe
ρ is the fluid density constant
C_d is the discharge coefficient of the orifice plate
ΔP is the DP across the orifice

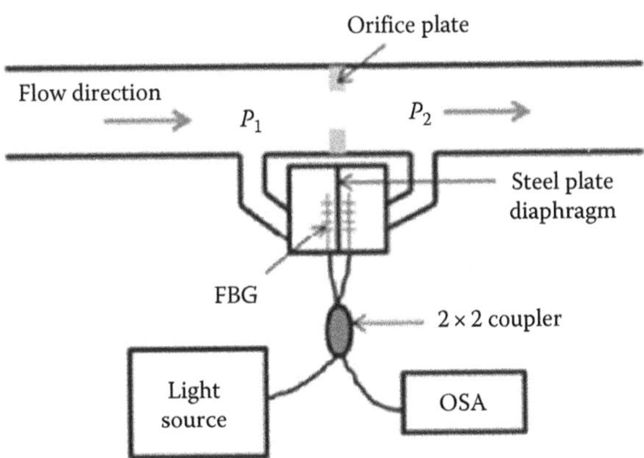

FIGURE 9.13 FBG-based flow sensor. (From Lim, J. et al., *Sens. Actuat. A*, 92, 102, 2001.)

The relationship between the DP across the orifice and the strain on a circular thin plate diaphragm is given by the equation

$$\epsilon_r = \pm\frac{6M_r}{Eh^2} - \nu\frac{6M_\theta}{Eh^2} \tag{9.18}$$

where

$$M_r = \frac{\Delta P}{16}\left[a^2(1-\nu) - r^2(3+\nu)\right]$$

$$M_\theta = \frac{\Delta P}{16}\left[a^2(1+\nu) - r^2(1+3\nu)\right]$$

ϵ_r is the strain in radial direction on the diaphragm center
M_r is the strain in radial direction on the diaphragm center
M_θ is the bending moment in angular direction at the diaphragm center
E is the modulus of elasticity of the diaphragm material
h is the diaphragm thickness
ν is Poisson's ratio of the diaphragm material
a is the diaphragm radial
r is the distance of the point of consideration from the center

The experimental result shows that the arrangement can detect fluid flow rates from 0 to 800 cm^3/s with temperature varying from 25°C to 75°C and also with hydraulic oil with a flow rate up to 6 cm^3/s at a pressure of 700 kPa.

Takashima et al. have reported a cross-correlation flow meter [99]. The sensor has FBGs embedded on metal cantilevers as strain sensor. The experimental results show that the flow meter has a good linearity at a velocity range from 0 to 1.0 m/s and a minimum detectable velocity of 0.05 m/s.

A cantilever-based flow sensor was proposed by Zhao et al. [100]. A couple of FBGs mounted on either side of a cantilever were used. Results show that the sensor can be used within the flow range from 0 to 1000 cm^3/s.

9.3.7 Liquid Level Sensor

Liquid level measurement is one of the most common types of process measurement in chemical, oil, food, beverage industry, biogas plants, fuel tank, and other everyday applications. The liquid level measurement devices can be for either point level or continuous level. Liquid level sensors based on FBG were reported.

Yun et al. [101] have demonstrated continuous liquid level sensing using chemically etched uniform FBG of length 24 mm (Figure 9.14). When the region of FBG of the fiber is chemically etched to a certain extent and beyond, it becomes sensitive to the refractive index of surrounding media. If a portion of chemically etched FBG is immersed in the liquid, the original single transmission dip splits into two transmission dips. This is due to the fraction of the length of the etched FBG that is surrounded by the liquid. The variation of power of transmission dips is a measure of the liquid level surrounding the FBG. A high liquid level sensitivity of 2.56 dB/mm is reported.

A liquid level sensor using FBG embedded on the cantilever beam was proposed by Guo et al. [102]. The bending of cantilever beam induces axial strain gradient along the FBG, resulting in a Bragg bandwidth modulation. The broadening of FBG spectrum bandwidth and the change in reflected optical power are a measure of liquid level. The liquid level sensing range reported was 50 cm. However, it is difficult to determine the direction of the liquid-level variation.

Sohn et al. proposed a liquid level sensor scheme in which the deflection of the cantilever induces a shift in the Bragg wavelength without changing the bandwidth of the spectrum [103]. In the sensor

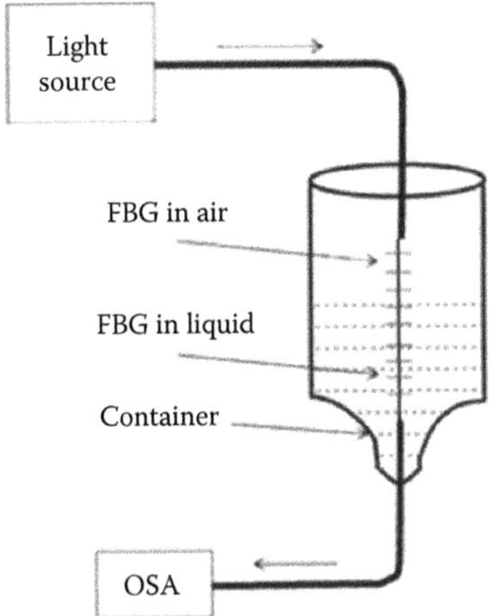

FIGURE 9.14 Etched FBG liquid level sensor. (From Yun, B. et al., *IEEE Photon. Technol. Lett.*, 19, 1747, 2007.)

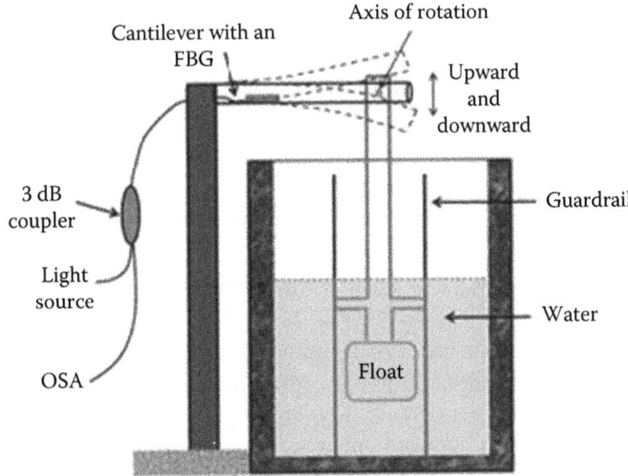

FIGURE 9.15 Buoyancy force–based liquid level sensor. (From Sohn, K.-R. and Shim, J.H., *Sens. Actuat. A*, 152, 248, 2009.)

arrangement, as the liquid level increases, the float attached to the cantilever experiences a buoyancy force and bends the cantilever resulting in a Bragg wavelength shift (Figure 9.15). The sensing range is 36 cm with sensitivity 0.15 nm/cm or 0.1 dB/cm.

A hydrostatic pressure–based liquid level sensor was demonstrated by Fukuchi et al. [104]. The sensor head consists of a diaphragm, a customized bourdon tube, and two FBGs, one for tensile measurement and the other for temperature compensation. The FBG attached to the bourdon tube is strained due to the increase in hydrostatic pressure of water as the water level increases. It causes a shift of the center wavelength of the reflected light from the FBG. The sensing range of 10 m was reported.

Sheng et al. reported one FBG sensor encapsulated in a half silicone rubber filled metal cylinder used for liquid level measurement [105]. The operating mechanism of the design is based on transferring radial pressure into axial strain to induce Bragg wavelength shift (Figure 9.16). The experimental results show a water level sensitivity of 1.526×10^{-5}/cm within a sensing range of 300 cm. With the same operating mechanism, Sengupta and Kishore have presented another liquid level sensor where reduced clad FBG is used [106]. A 23 pm/cm sensitivity and 0–100 cm sensing range are reported.

Of late, another hydrostatic pressure–based liquid level sensor using an FBG and carbon fiber composite diaphragm was proposed by Song et al. [107].

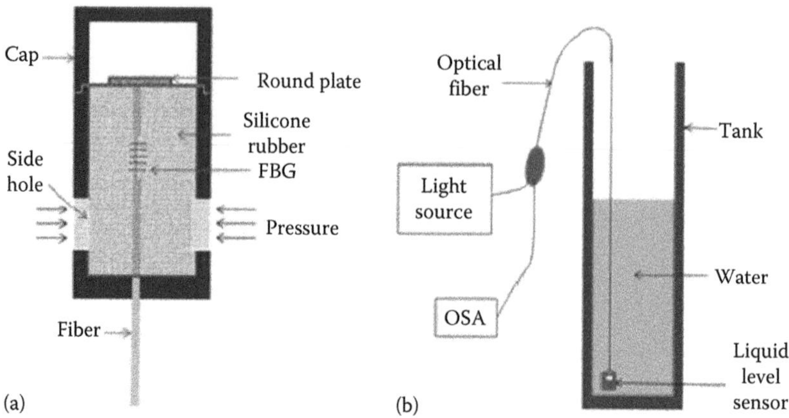

FIGURE 9.16 Hydrostatic pressure–based (a) liquid level sensor and (b) the experimental setup for liquid level measurement. (From Sengupta, D. and Kishore, P., *Opt. Eng.*, 53(1), 017102, 2014.)

9.3.8 Magnetic Field Sensor

Magnetic field measurement has become a fundamental procedure in many industrial and scientific applications. One can find that when a magnetic field is applied to a magnetostrictive alloy, the magnetic domains in the material tend to align along the field direction, and as a result of the magnetoelastic coupling, the material suffers an elastic lengthening in the direction of the magnetic field. If an FBG is embedded on a magnetostrictive alloy and is subjected to magnetic field, the FBG shows a shift in Bragg wavelength due to the straining effect by the alloy. Thus, a magnetic field can be sensed through strain sensing of FBG.

An experiment demonstrated by Cruz et al. used a magnetostrictive rod of composition $Tb_{0.27}$ $Dy_{0.73}$ Fe_2 and dimensions 6 × 100 mm [108] (Figure 9.17). The maximum sensitivity of this alloy, when it operates stress free, is of the order of 10 ppm/mT, and the maximum strain when the material is magnetized to saturation is about 1000 ppm. The embedded FBG on the magnet shows a shift of 1.1 nm of Bragg wavelength by an applied magnetic field of 103 mT. Also, the grating has been chirped up to 0.7 nm by applying a magnetic field gradient of 23 mT/cm as reported.

A side-polished FBG coated with a thin iron film was presented as a magnetic field sensor by Tien et al. The Bragg wavelength shift of 0.08 nm was measured at the distance of 0.38 mm between the fiber sensor and NdFeB magnet with remanent flux density of 1.115 T [109]. The magnitude of interaction depends on the distance between the fiber sensor and the NdFeB magnet. By controlling the interaction length, the diameter of side-polished fiber and thin iron film thickness sensitivity of the sensor can be improved.

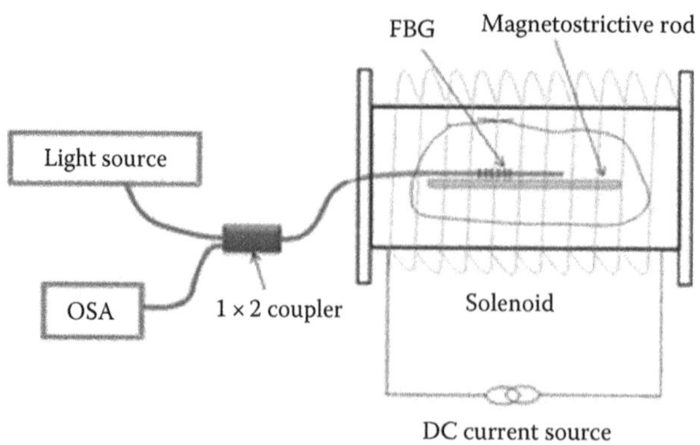

FIGURE 9.17 Magnetic field measurement using FBG.

An FBG embedded on a magnetic shape memory alloy (MSMA) was demonstrated a magnetic field sensor by Cusano and coworkers [110]. MSMA is a new kind of material able to show extremely large strains in response to a magnetic field. Moreover, differently from classical shape memory alloys, which are actuated by temperature, MSMAs normally show quite fast responses. The results show that the sensitivity of the sensor is $(\Delta\lambda_B/\lambda_B)/(\Delta H/H_m) = 1.81\times10^{-4}$ where H_m is the mean magnetic field in the considered range. In their later work, they have shown a Terfenol-D-based FBG magnetic sensor where the magnetostrictive response on prestress is explained. An improved result in terms of magnetic resolution up to 0.0116 A/m has been demonstrated [111,112]. A magnetic field–sensing mechanism based on the polarization-dependent loss (PDL) in the transmission spectrum of FBG is also reported [113].

Quintero et al. investigated an FBG coated with thick layer of a magnetostrictive composite consisting of particles of Terfenol-D (particle size 212–300 μm) dispersed in a polymeric matrix for magnetic field sensing [114]. The designed sensor is small and cylindrical in shape with 1.5 mm diameter and 7 mm length. The sensor was tested at magnetic fields of up to 750 mT under static conditions. The effect of a compressive preload in the sensor was also investigated. The achieved resolution was 0.4 mT without a preload or 0.3 mT with a compressive prestress of 8.6 MPa.

Yang and Dai proposed an idea of depositing magnetostrictive TbDyFe thin films on cladding-etched FBG to sense magnetic field [115]. In order to improve the sensitivity, the diameter of the FBG-containing fiber was reduced by chemical etching. With the same 0.8 μm TbDyFe coating, the sensitivity responses of the FBG wavelength shift for nonetched (125 μm in diameter), 1 h etched (105 μm in diameter), and 2 h etched (85 μm in diameter) are reported as 0.386, 0.563, and 0.950 pm/mT, respectively. However, fibers with thin diameters are mechanically weak and prone to break. With the interest to improve the sensitivity of the magnetic field response, the magnetostrictive multilayer is also investigated. The multilayer coating of TbDyFe/FeNi on FBG written on a 125 μm standard fiber shows a sensitivity of 1.08 pm/mT, which is two times higher than that of the 1 μm TbDyFe single layer.

Another new approach in which the magnetic fluid was used for sensing the magnetic field was demonstrated. The magnetic fluid is a kind of stable colloidal solution of ferromagnetic nanoparticles [115]. The behaviors of ferromagnetic particles in the magnetic fluid are dependent on the external magnetic field, so the refractive index of the magnetic fluid is shown to be magnetic field dependent. An FBG was etched to 11.32, 9.98, and 8.53 μm diameters, respectively. A microtube containing etched FBG and magnetic fluid was prepared as a sensor for magnetic field sensing. As the magnetic field increases, the reflected wavelength of etched FBGs with different diameters shifts to shorter wavelength. The wavelength shift of the etched FBG shows a nonlinear dependence on the magnetic field, which is similar to the theoretical simulation of the etched FBG under different ambient refractive indices. When the magnetic field increases to 24 mT, the wavelength shifts of the FBG with diameters of 11.32, 9.98, and 8.53 μm are 25, 46, and 86 pm, respectively. This concludes higher sensitivity of magnetic field for small diameter of the fiber.

9.3.9 Tilt Sensor

Inclination or tilt angle measurement is an important parameter in terms of civil, mechanical, instrumentation, robotics, and aeronautical engineering applications. An FBG sensor can be configured as a tilt sensor using various schemes in which a strain is induced to shift the Bragg wavelengths of the FBGs. Chen et al. have presented a scheme for a tilt angle measurement, in which only one prestrained FBG is employed to obtain temperature independence in a 1-D measurement [116]. In the arrangement, an FBG is anchored at its two ends (marked by A and B in Figure 9.18) between an iron ball and a polyvinyl chloride (PVC) cylinder (Figure 9.18). The PVC cylinder is fixed at one end onto the inner surface of the aluminum box. Another fiber (dummy fiber) is anchored between one end upon the iron ball (i.e., point C) and the aluminum housing (at point D).

The dummy fiber is then pulled at the other end to create an extensile strain on itself and the FBG, and later, the fiber was fixed at point D using epoxy resin. However, when the aluminum box

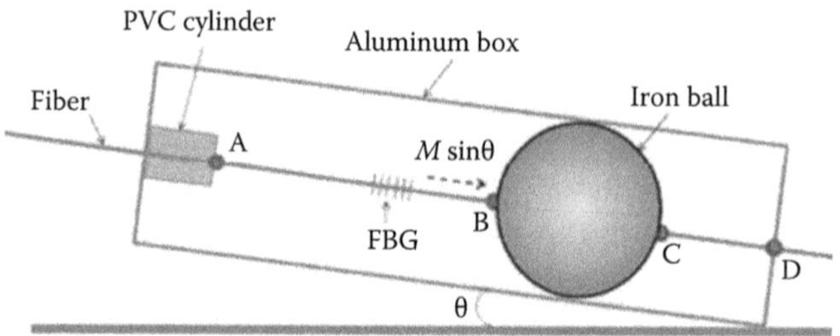

FIGURE 9.18 Tilt sensor using FBG. (From Chen, H.-J. et al., *Appl. Opt.*, 47(4), 556, 2008.)

is tilted by some angle, a pulling force due to the slight gravity-induced movement of the iron ball is produced to strain the FBG. The force would be $M \sin \theta$, where M is the mass of the iron ball and θ is the tilt angle to be measured. The strain on the FBG can thus be detected by monitoring the red shift in the Bragg wavelength of the FBG and is used to calibrate the tilt angle. The strain on the FBG is equal to $M (\sin \theta - \mu \cos \theta)/AE$, where A is the cross-sectional area of the fiber, E is Young's modulus of the fiber, and μ is the frictional coefficient for the movement of the iron ball. The shift in Bragg wavelength, corresponding to tilt angle θ, is given by

$$\Delta\lambda_B = (1 - P_e)\frac{\lambda_B}{AE} M(\sin\theta - \mu\cos\theta) \tag{9.19}$$

However, the accuracy in measuring the tilt angle may strongly be affected by the environmental temperature.

The arrangement described earlier also performs temperature compensation. When the temperature rises, the iron ball, aluminum box, and PVC cylinder expand longitudinally. The expansion of both the iron ball and the PVC cylinder exerts a compressive strain on the FBG, while the expansion of the aluminum box generates an extensile strain.

Because the PVC expands more than the aluminum, the FBG may suffer from a compressive strain and induce a blue shift in its Bragg wavelength. Such a blue shift can be deliberately tailored to match the inherent red shift due to the temperature rise by adjusting the length of the PVC cylinder. An accuracy of ±0.167° in tilt angle measurement and resolution of tilt angle measurement 0.0067° is achieved by the arrangement.

A number of designs of tilt sensors using pendulums and fiber suspension methods were reported by researchers. A vertical-pendulum-based FBG tilt sensor, which can detect the magnitude as well as the direction of the inclination from the horizontal direction, was demonstrated [117,118].

Bao et al. have demonstrated a 2-D tilt sensor by incorporating only two FBGs with a hybrid pendulum transducer. The 2-D tilt angle can be determined by monitoring the wavelength separations of the split peak in two FBGs, which is inherently insensitive to temperature. Experimental results show a sensitivity of 0.054 nm/° over a wide range of 20°, with the accuracy of 0.27° [119].

Aneesh et al. proposed a tilt sensor that consists of four FBGs and a vertical pendulum, which was designed for a dynamic measurement range of ±45°. The sensor is capable of measuring the magnitude as well as the direction of inclination from the horizontal with a complete reversible response. The most important feature of the reported sensor is its inherent enhanced tuning capability for its sensitivity. An excellent sensitivity of the order of ~0.0626 nm/° with a tilt angle resolution better than 0.008° and a tilt accuracy of ~±0.36° within a range of ±0.10° was presented [120].

9.3.10 Humidity Sensor

Humidity is one of the most commonly required physical quantities like temperature and pressure. It has significant importance in air conditioning for human comfort and combating bacterial growth to process control, maintaining product quality, etc. Humidity is a measure of water in gaseous state present in the environment. The ratio of the amount of water vapor present in the atmosphere to the maximum amount the atmosphere can hold is known as relative humidity (RH). The requirements for humidity monitoring vary according to the application and hence various techniques have been employed to perform humidity measurements.

RH sensing using a polyimide-coated FBG was first reported by Kronenberg et al. [121]. The basic working principle of the sensor is that the polyimide polymers are hygroscopic and swell in aqueous media as the water molecules migrate into them. The swelling of the polyimide coating is utilized to strain the FBG-containing fiber (Figure 9.19). This modifies the Bragg condition of the FBG and shifts the Bragg wavelength accordingly. The shift in the Bragg wavelength for the polymer-coated FBG is given by

$$\frac{\Delta\lambda_B}{\lambda_B} = (1-P_e)\alpha_{RH}\cdot\Delta RH + \left[(1-P_e)\alpha_T + \xi\right]\Delta T \tag{9.20}$$

where

α_{RH} and α_T are the moisture expansion coefficient and the thermal expansion coefficient of the coated FBG
ΔRH is the change in RH

The findings from their investigations conclude that an FBG with polyimide coating was able to respond linearly over a wide humidity range. The sensor with sensitivity 2.21 ± (0.10) × 10^{-6} % RH was found with a coating thickness 29 ± 1 μm and can respond well to a humidity range of 10%–90% RH. The sensitivity of the sensor increases with the coating thickness.

Grattan and coworkers have demonstrated FBG samples that are dip coated with commercial-grade polyimide solution (HD MicroSystems, Pyralin PI 2525) to form a sensing element [122,123]. FBG-sensing elements are coated with different thicknesses and are tested at different humidity levels. Results show that for 42, 33, and 10 μm thickness of coating, the RH sensitivities are ~5.6, ~4.5, and ~1.4 pm/%RH, respectively.

Huang et al. have performed a similar attempt [124] to measure humidity, in which FBG is coated with thermoplastic polyimide (moisture-sensitive polymer material) and packaged in a copper tube to protect the polyimide-coated FBG from breakage and pollution. The thickness and length of the polyimide coating are 8 μm and 2.5 cm, respectively. The prepared sensor is tested within a sensing range from 11% to 98%. The sensitivity of the sensor is −0.000266 V/% RH, and the response time is approximately 5 s.

An optical fiber RH sensor based on TFBG, is proposed by Y. Miao et al. utilizing polyvinyl alcohol (PVA) as the sensitive cladding film [125]. The refractive index of the PVA film decreases when it is exposed to a humid environment. Due to the TFBG's sensitivity to ambient refractive

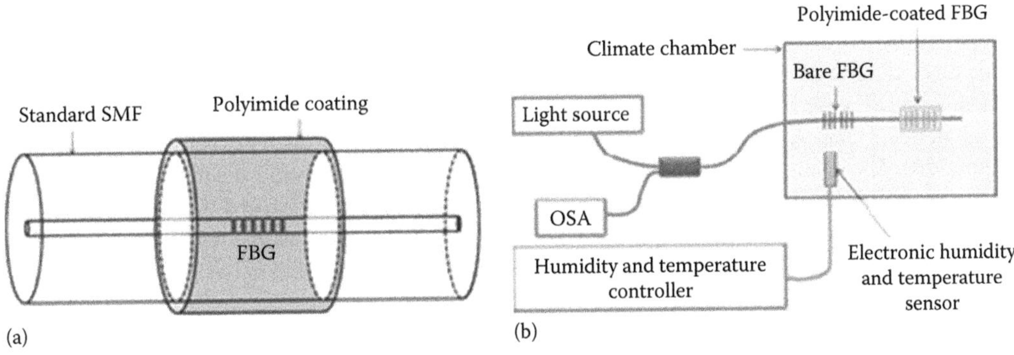

FIGURE 9.19 Humidity sensor (a) polyimide-coated FBG and (b) experimental setup.

index, the spectral properties of PVA-coated TFBG are modified under exposure to different ambient humidity levels ranging from 20% to 98% RH. The reported results show that the transmission power of TFBG has different linear behaviors for two different humidity ranges (20%–74% RH and 74%–98% RH), and the sensitivity for each humidity range reaches as high as 2.52 and 14.947 dBm/%RH, respectively.

Recently, Correia et al. proposed an RH sensor based on FBG associated with a coating of organosilica hybrid material prepared by the solgel method [126]. The organosilica-based coating has a strong adhesion to the optical fiber, and its expansion is reversibly affected by the change in the RH values (15.0%–95.0%) of the surrounding environment, allowing an increased sensitivity (22.2 pm/%RH) and durability due to the presence of a siliceous-based inorganic component.

9.3.11 Gas Sensor

Fiber-optic gas sensors have been designed and constructed to detect hazardous and other gases in the past. Today, hydrogen is becoming an attractive alternative fuel source for use in clean-burning engines and power plants. Some mission-critical applications such as the Space Shuttle main engine already employ liquid hydrogen as fuel. Unfortunately, the use of highly flammable liquid H_2 also introduces a number of safety concerns due to its rapid evaporation rate and low explosive limit. To detect the leaks in hydrogen-fueled systems, many configurations of fiber-optic sensors were developed and reported.

Butler and Ginley [127] recognized the change in the elastic property of Pd during H_2 absorption and invented the fiber-optic sensor based on this property. Their model can be used to determine the axial strain in the fiber core.

FBG sensor has been incorporated to sense hydrogen and reported by researchers. The sensing mechanism is mostly based on mechanical stress that is induced in the palladium coating when it absorbs hydrogen. As the Pd film on the FBG absorbs hydrogen, it expands (depends on the hydrogen concentration) because hydrogen absorption converts Pd to PdHx, which has a lower density and larger volume. The stress in the palladium coating stretches the fiber and causes the grating period, Λ, to expand and shifts the Bragg wavelength of the FBG. The shifts in Bragg wavelengths of the Pd-coated FBGs as a function of hydrogen partial pressure can be expressed as [128]

$$\Delta\lambda_B = \frac{0.026\sqrt{p}}{K}\left[\frac{\left(b^2-a^2\right)Y_{Pd}}{a^2Y_F+\left(b^2-a^2\right)Y_{Pd}}\right]\times 0.78\lambda_B \tag{9.21}$$

where

a and b are the cladding and total sensor diameters
Y_{Pd} and Y_F are Young's modulus of palladium and silica fiber
p is the hydrogen partial pressure (Torr)
K is Sievert's coefficient ($K = 350$ Torr$^{1/2}$)

Sutapun et al. have demonstrated an optical hydrogen sensor with a fiber-optic Bragg grating (FBG) coated with palladium. The FBG had a 35 μm cladding and a 560 nm Pd coating with a Bragg wavelength of 829.73 nm. A simple experimental setup of FBG-based hydrogen gas sensor is presented in Figure 9.20. The results show a linear increase in Bragg wavelength within the hydrogen concentration range of 0.3%–1.8%. In the linear range of its operation, the sensor had 1.95×10^{-2} nm/1%/H_2. However, the sensor becomes irreversible when exposed above 108% of H_2, which reveals that the Pd coating peeled off. A solution of this can be avoided by adding a thin adhesion layer such as Ni or Ti to improve adhesion between the Pd and the fiber [128].

A nanolayer of palladium on an FBG used for hydrogen sensing was reported by Trouillet et al. A slight dependence of the saturation response with the palladium layer's thickness has been observed for the thicker layers (more than 100 nm). It shows that the enhancement of the saturated response

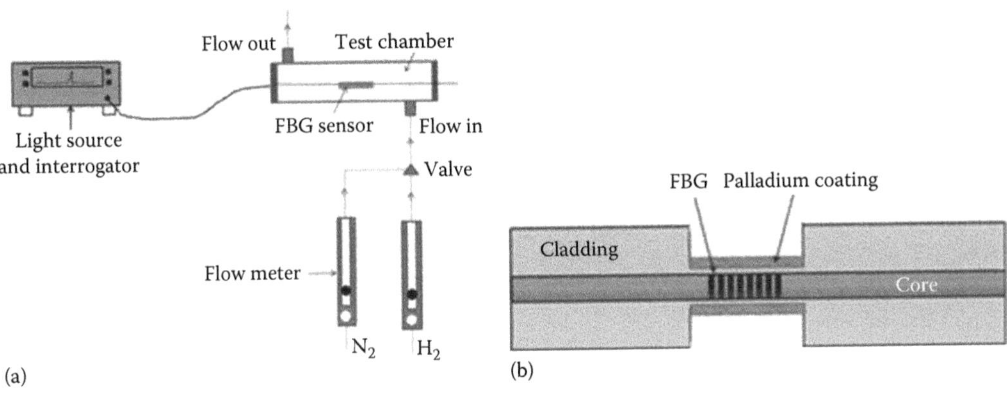

FIGURE 9.20 (a) Schematic experimental setup for hydrogen gas sensor using FBG and (b) Pd-coated FBG.

with the palladium layer's thickness is followed by an increase in the response time. Furthermore, delamination was observed with layers of thicknesses much greater than 100 nm. The FBG sensors appear to be pure strain sensors using the mechanical deformations of the palladium layer when absorbing hydrogen. This was found when compared with the hydrogen-sensing response of a long-period grating (LPG) sensor coated with the same nanolayer thickness of palladium [129].

Caucheteur et al. reported a new technique in which instead of Pd coating on the FBG to detect hydrogen, a sensitive layer made of WO doped with Pt was coated on the FBG [130]. The sensitive layer generates an exothermic reaction in the presence of hydrogen and air, while FBGs act as probes to measure the wavelength shifts induced by temperature changes. Highly reflective long FBGs are required to detect 1% hydrogen concentration in dry air.

Successively, a side-polished FBG sensor coated with thin palladium film of thickness 20 nm for hydrogen sensing was demonstrated by Tien et al. The results shows a linear measuring characteristic in the range of low hydrogen concentrations [131]. A similar working principle used for hydrogen sensing was adopted by J. Dai et al. WO_3 –Pd composite films were deposited on the side-face of side polished FBG and used as sensing elements. The experimental results shows that a maximum Bragg wavelength shifts of 25 pm and 55 pm in 4% and 8% in volume of hydrogen concentration and the sensor response is reversible.

9.3.12 Displacement Sensor

Though there are many commercial displacement sensors available in the market, still displacement sensors used for SHM have been an important topic of research in civil and aerospace fields. In SHM, FBG sensors are routinely used for monitoring strain and temperature. Unlike strain and temperature, displacement is not a directly measurable quantity using bare FBG sensors. Utilizing the strain response of FBG to its equivalent Bragg wavelength, displacement sensors were developed and reported by researchers. A simple schematic displacement sensor arrangement using FBG is shown in Figure 9.21.

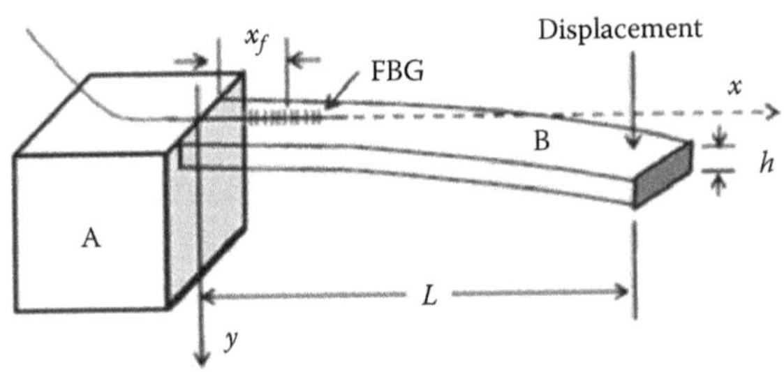

FIGURE 9.21 Schematic arrangement for the displacement measurement using FBG.

A cantilever (B) of length L and thickness h is fixed on a rigid block (A). An FBG is attached on a cantilever structure so that the vertical displacement of the cantilever's free end can be obtained by measuring the wavelength shift of FBG. Using knowledge of material mechanics theory, the axially applied strain can be represented as [133]

$$\epsilon_x = \frac{3}{2L^3}\left(h+\phi_f\right)\cdot\left(L-x_f\right)\cdot D_y \tag{9.22}$$

where

ϕ_f is the diameter of the fiber
x_f is the distance to the center of the FBG to the fixed end of the cantilever
D_y is the vertical displacement at the free end of the cantilever

Thus, the shift in the Bragg wavelength at $\Delta T = 0$ is

$$\frac{\Delta\lambda_B}{\lambda_B} = \frac{3}{2L^3}\left(1-P_e\right)\cdot\left(h+\phi_f\right)\cdot\left(L-x_f\right)\cdot D_y \tag{9.23}$$

where P_e is the photoelastic coefficient of the fiber.

Dong et al. have demonstrated an FBG-based displacement sensor along with temperature compensation using a single FBG [133]. In the scheme, half the length of the FBG is made sensitive to displacement by embedding it on a cantilever beam. As the free end of the cantilever is vertically displaced, this part of the FBG experiences strain and the remaining part of the FBG is unaffected. The measured displacement sensitivity of this arrangement was found to be $8.22\times 10^4\,mm^{-1}$ and the measurement range can be up to 10.5 mm.

Usage of a bilateral cantilever beam (BCB) and an FBG for displacement measurement was reported by Zhang et al. [134]. An FBG is bonded to the surface of the BCB middle that is anchored. A force sensitivity of 1.046 nm/N, a sensitivity of the displacement-based strain of 0.317 nm/mm, and a temperature sensitivity of 0.190 nm/°C between 0°C and 70°C, respectively, were reported.

A differential FBG sensor based on an isosceles triangle cantilever structure is used to measure the displacement along with temperature. The benefit of using the isosceles triangle cantilever is the constant strain along the cantilever, resulting in no chirp of the FBG spectrum. The displacement sensitivity of ~1.75 pm/m is reported [135]. Similarly, Dong et al. have demonstrated a temperature-compensated displacement using a single grating, which is glued at a slant orientation onto the lateral side of a right-angle-triangle cantilever beam [136].

A BCB used for displacement measurement was presented by Guo et al. Linear displacement measurement was done up to 20 mm with displacement resolution of 0.054 mm [137].

Guru Prasad et al. developed a small high sensitive package displacement sensor. The sensor includes a stainless steel cantilever strip over which an FBG sensor has been bonded. A linear shift of 12.12 nm in Bragg wavelength of the FBG sensor is obtained for a displacement of 6 mm with a calibration factor of 0.495 μm/pm [138].

9.4 SIMULTANEOUS MEASUREMENT OF TEMPERATURE AND STRAIN

9.4.1 Introduction

One basic and important issue with FBG sensors is the discrimination between the effects of strain and temperature, because strain and temperature have an identical effect on Bragg wavelength shift. Over the years, many techniques were developed to discriminate the effect of strain and temperature. The temperature-compensating methods fall into two categories, namely, extrinsic and intrinsic methods. In the extrinsic method, the grating is combined with an external material of suitable properties and dimensions to compensate the temperature, whereas the intrinsic method relies on fiber properties.

9.4.2 Intrinsic Strain and Temperature Discrimination

In general, shift in Bragg wavelength of an FBG is not high enough to determine the value of strain/ other physical parameter and temperature acting on the FBG. The most straightforward way is to use an identical, separated, strain-free FBG as the temperature sensor to measure the temperature of the strain sensor directly (Figure 9.22).

The reference FBG (FBG 2) is located in the same thermal environment as the strain sensor (FBG 1). The strain error caused by the temperature variation can be compensated to first order by subtracting the wavelength shift induced by temperature variation from the total wavelength shift obtained with the strain sensor [139]. The advantage of this method is that it is a simple and straightforward technique for temperature compensation; however, the resolution of the technique is low and interrogation of two fibers is cumbersome.

Another method utilizes the two sets of wavelength shift obtained from two superimposed FBGs written at the same location in the fiber. In general, the change in Bragg wavelength of the FBG due to a combination of the strain and temperature effect can be expressed as

$$\Delta\lambda_B = A\Delta\varepsilon + B\Delta T \tag{9.24}$$

where

$\Delta\lambda_B$ is the Bragg wavelength shift, in response to a strain change, $\Delta\varepsilon$, and a temperature change, ΔT

A and B are the strain and temperature sensitivities of the FBG, respectively

This assumes that the strain and thermal response are essentially independent, that is, the related strain–temperature cross-term is negligible, a behavior that has already been found to apply well for small perturbations. As a result for the two Bragg wavelengths to be measured, the following relation holds:

$$\begin{pmatrix} \Delta\lambda_{B1} \\ \Delta\lambda_{B2} \end{pmatrix} = \begin{pmatrix} A1 & B1 \\ A2 & B2 \end{pmatrix} \begin{pmatrix} \Delta\epsilon \\ \Delta T \end{pmatrix} \tag{9.25}$$

where 1 and 2 refer to the two wavelengths of the FBGs. The elements of the matrix ($A1$, $A2$, $B1$, and $B2$) can be determined experimentally by separately measuring the Brag wavelength changes with strain and temperature [140]. As long as the elements of the matrix are known, $\Delta ò$ and ΔT can be determined easily.

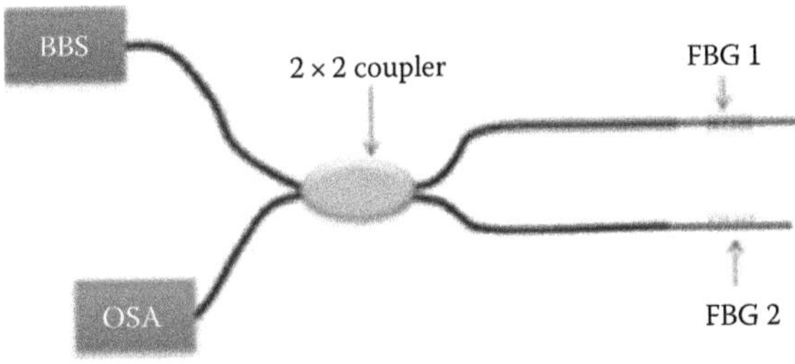

FIGURE 9.22 Discrimination of strain and temperature using reference FBG.

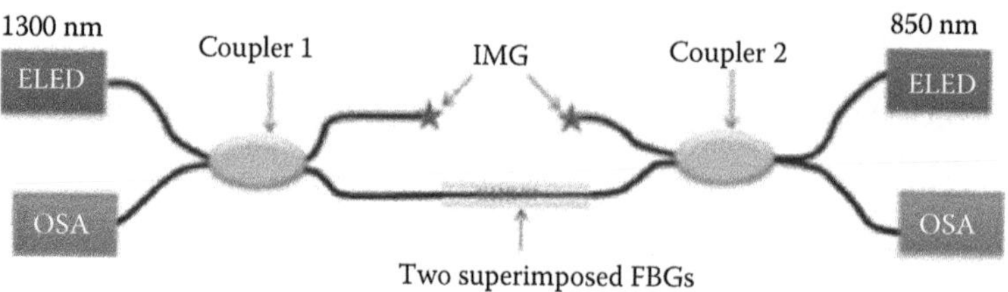

FIGURE 9.23 Discrimination of strain and temperature using dual wavelength.

A schematic experimental setup of this method is shown in Figure 9.23. The sensor probe in this method is small and drawing of two FBGs is easy but needs two sets of light sources and demodulation systems, and the high cost limits the practical applications in the industry fields.

The arrangement of FBG–LPG strain–temperature discrimination setup is shown in Figure 9.24. In this, two FBGs and an LPG are used as sensing elements. The basic idea behind this method is that the LPG has a much larger temperature response than the FBG and a smaller strain response. Therefore, simultaneous strain/temperature measurement could be achieved if the strain and temperature wavelength response coefficients of both the LPG and the Bragg grating were known accurately. The advantage of this approach is that it does not need two sets of independent detection systems for detecting changes in two widely separated wavelengths. A resolution of ±9 με and ±1.5°C from this arrangement was reported [141].

James et al. demonstrated that the strain responses to relative wavelength shifts and the temperature responses to the weighted wavelength shift differences of two FBGs with different cladding diameters like 125 and 80 μm are not the same (Figure 9.25). The discrimination of strain and temperature can be evaluated through the similar process presented by Xu et al. [140]. Here, the Bragg wavelengths of the two FBGs may differ by a few nanometers, allowing them to be measured independently with WDM. But low splice strength and high splice loss between the FBGs require further investigation. In a measurement range of 2500 με and 120°C, the maximum error reported is 17 με and 1°C [56].

Later, Xiaohong et al. have also demonstrated a similar concept of discriminating strain and temperature using a single Bragg grating [142]. In their experiment, they have chemically etched

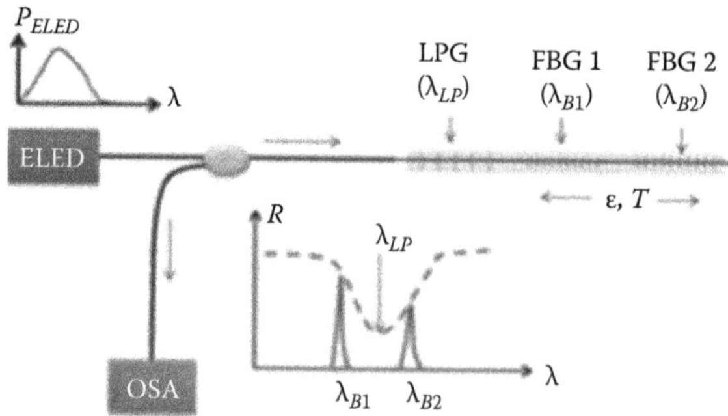

FIGURE 9.24 Discrimination of strain and temperature using FBG and LPG.

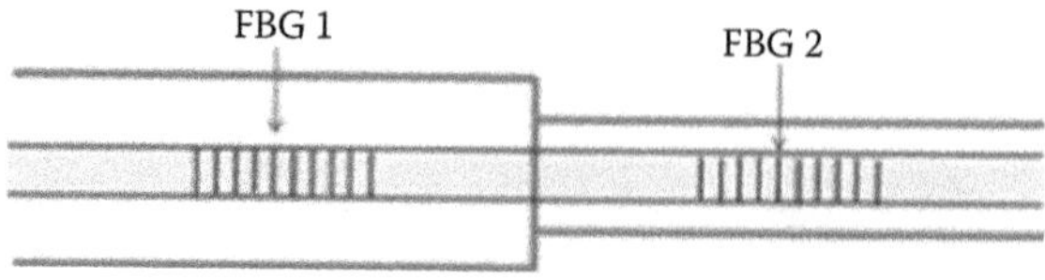

FIGURE 9.25 Discrimination of strain and temperature using FBGs with different cladding diameters.

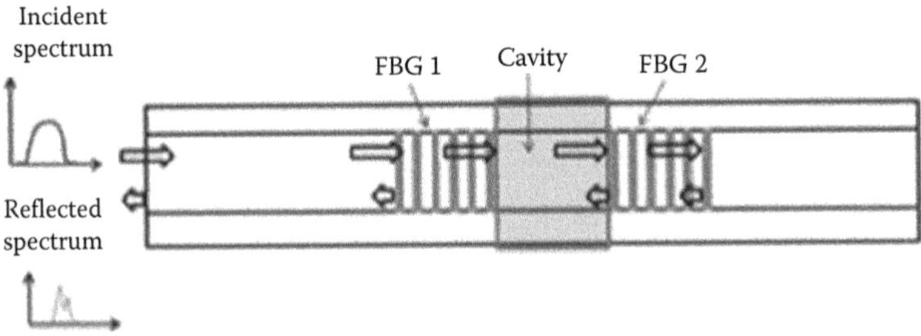

FIGURE 9.26 Discrimination of strain and temperature using FBGs and F–P cavity.

half the length of the FBG to reduce the diameter of that region. This process eliminates the usage of two separate FBGs, but in true sense, the etched region of the FBG is mechanically weak to use it in real time.

The FBG Fabry–Pérot (F–P) cavity method consists of one very short (5 mm) fiber sensor. Strain and temperature are measured simultaneously from the power and the Bragg wavelength shift of the reflected light from the sensor. A schematic picture of the sensor is shown in Figure 9.26.

Two identical fiber gratings and an F–P cavity with a length of 1 mm were used in the sensor. However, it is difficult to fabricate the FBG F–P cavity sensor. The accuracy of this particular sensor in measuring strain and temperature was estimated to be ±30 με in a range from 0 to 3000 με and ±0.4°C from 20°C to 60°C, respectively [143].

Guan et al. proposed a simple fiber sensor based on an SFBG that can measure strain and temperature simultaneously (Figure 9.27). The SFBG is a special type of FBG fabricated using periodically modulated exposure over the length of a phase mask. As a periodically modulated FBG, the SFBG couples the forward-propagating LP_{01} mode to the reverse propagating LP_{01} mode at a series of wavelength and introduces several narrow loss peaks in the transmission spectrum. By measuring the transmitted intensity and wavelength at one of the loss peaks, strain and temperature can be determined simultaneously [144]. The disadvantage of this method is that due to the use of light intensity as one of the signals, measurement results will be influenced by the fluctuation of light source and the loss in the fibers.

A chirped fiber grating in a tapered optical fiber was demonstrated as a temperature-independent strain sensor [145]. A tapered profile was designed such that the grating becomes linearly chirped when tension is applied and it creates a strain gradient along the length of the grating (Figure 9.28).

The effective bandwidth of the grating spectrum only depends on the strain applied and is independent of temperature. In this process, the interrogation simply involves monitoring the back-reflected intensity from the grating. A strain resolution of 4.4 με over a total measurement range of 4066 με has been experimentally demonstrated. A combination of a uniform FBG and an FBG written in the tapered region of FBG (resulting in a chirped grating) was also demonstrated by the researchers [146].

A fiber grating pair written in a bow tie fiber was demonstrated for simultaneous measurement of strain and temperature [147]. The inherent polarization-encoded reflected signals from these gratings were interrogated using a scheme based on the interferometric technique. The performance of

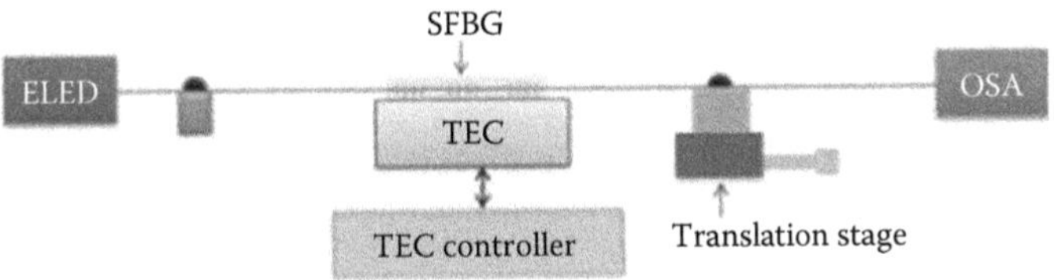

FIGURE 9.27 Discrimination of strain and temperature using SFBG.

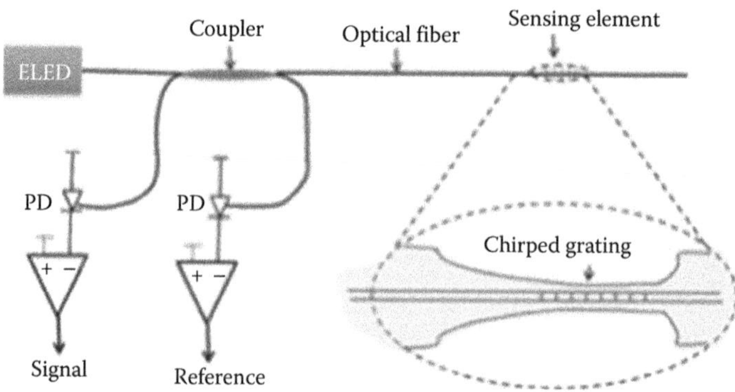

FIGURE 9.28 Discrimination of strain and temperature using chirped grating.

the proposed technique in simultaneous measurement of temperature and strain is experimentally demonstrated and resolutions of ±2.5°C/√Hz and ±26 με/√Hz are obtained for a fiber with birefringence of $B = 5.5 \times 10^{-4}$.

A new scheme of discrimination of strain and temperature was proposed by Guan et al. In the scheme, an FBG written on the splice joint between two different fibers and the fabricated grating has two resonance peaks. This is because of the different refractive indices of the two fibers. One section of the fabricated grating was bonded onto the substrate of large thermal expansion coefficient like aluminum (Figure 9.29). The two resonance peaks exhibit different strain and temperature response and can thus be used for strain/temperature discrimination. The experimental data with the use of linear regression show that the embedded region of FBG has 0.97 pm/με and 10.2 pm/°C, whereas the nonembedded region of FBG shows 0 pm/με and 34.4 pm/°C of strain and temperature sensitivity, respectively [148].

Another configuration suitable for strain and temperature discrimination was demonstrated using a HiBi-fiber loop back mirror (HiBi-FLM) combined with an FBG, whose bandwidth is much narrower than that of the LPG, serves as a sensing head (Figure 9.30). HiBi-FLM shows different strain and temperature sensitivity responses from those of the FBG; therefore, it is feasible to realize simultaneous measurement of the strain and temperature variations based on their combination. By measuring the resonance wavelength shifts of the HiBi-FLM and the FBG accurately and constructing a well-conditioned coefficient matrix based on this configuration, the applied strain and temperature changes can be determined simultaneously with high resolution. The reported experimental result shows a sensing resolution of ±1°C in the temperature and ±21 με in strain over a temperature range of 60°C and strain range of 600 με [149].

An FBG written in a polarization-maintaining fiber connected to a Sagnac fiber loop mirror was used to demonstrate the discrimination between the strain and temperature. The PMFBG acts as a sensor head, and the Sagnac fiber loop mirror serves as a filter to convert wavelength variation to optical power change. By measuring the wavelength difference and power difference of two

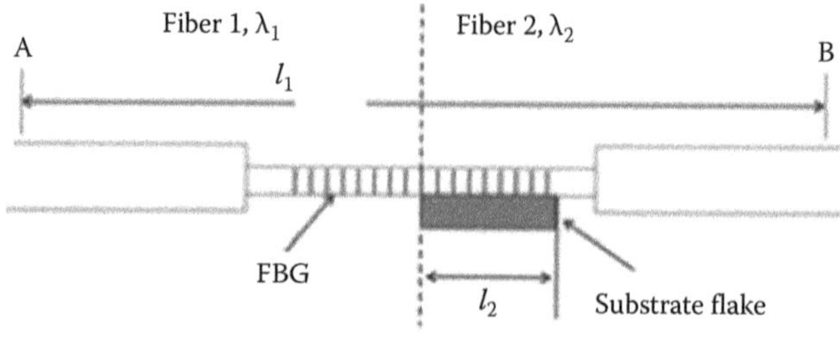

FIGURE 9.29 Discrimination of strain and temperature using FBG written at the splice joint.

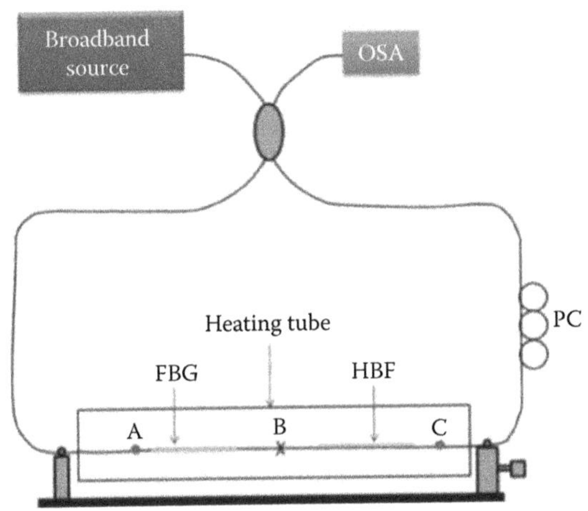

FIGURE 9.30 Discrimination of strain and temperature using FBG and HiBi-FLM.

resonant peaks of the PMFBG, the two parameters can be measured simultaneously. The reported experimental results show maximum errors of ±66.821 με and ±0.001°C and are reported over 480 με and 60°C measurement ranges, respectively. This scheme has the advantage of low cost, compact configuration, and convenience of use [150].

Jung et al. proposed and demonstrated a scheme in which a single-FBG can simultaneously measure the strain and temperature with the aid of an EDF amplifier. Temperature can be determined by using a linear variation in the amplified spontaneous emission power of the EDF amplifier with temperature. And by subtracting the temperature effect from the FBG wavelength shift, the strain can be determined. Experimental result shows rms deviations of 18.2 με and 0.7°C for strain and temperature, respectively [151].

The same group of researcher also demonstrated another scheme capable of the simultaneous measurement of strain and temperature through the measurement of transmitted power and Bragg wavelength shift of an FBG written in an erbium–ytterbium-doped fiber [152].

Here, with the use of a Yb^3-codoped fiber, the size of the sensor head sufficiently decreases (compared with that of Ref. [151]) and can be used as a point sensor (Figure 9.31). This compact fiber-grating-based sensor scheme can be used for synchronous measurement of strain and temperature over ranges of 1100 με and 50°C–180°C with rms errors of 55.8 με and 3°C.

Cavaleiro et al. utilize the effect of boron codoping on the temperature dependence of the refractive index in germanosilicate fibers [153]. By writing gratings with dose wavelengths in undoped acid boron-doped fibers, different temperature sensitivities were obtained while strain sensitivities remained the same. By splicing, these two gratings can be used to measure the strain and temperature simultaneously.

Shu et al. reported an investigation on the dependence of the temperature and strain coefficients on the grating type for FBGs that are UV inscribed in B–Ge-codoped fiber with and without

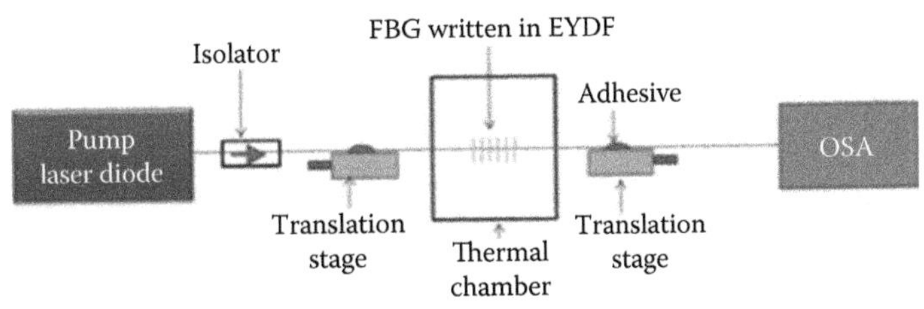

FIGURE 9.31 Discrimination of strain and temperature with FBG written in EYDF.

hydrogenation [154]. The results reveal that all types of gratings exhibit similar strain sensitivities but markedly different temperature sensitivities, greater for gratings inscribed in hydrogen-free rather than hydrogenated fiber and substantially less in Type IA gratings than all others. These characteristics can be implemented for simultaneous measurement of temperature and strain using a dual grating scheme [139].

Lima et al. designed a sensing head and presented the necessary interrogation parameters to perform strain and temperature discrimination [155]. They used a single FBG written in an optical fiber taper with a linear diameter variation. When subjected to a tension and due to the different cross sections of the fiber along its length, different values of strain arise, causing the broadening of the FBG spectrum and allowing the use of the information contained in both peak wavelength and spectral width.

Using a single TFBG, discrimination of these two parameters was demonstrated [156]. The technique exploits the core–cladding mode coupling of a TFBG. The core and cladding modes exhibit different thermal sensitivities, while the strain sensitivities are approximately equal. Monitoring the core–core mode coupling resonance and the core–cladding mode coupling resonance of the TFBG spectrum allows the separation of the temperature- and strain-induced wavelength shifts.

Frazao et al. have presented a scheme for the simultaneous measurement of strain and temperature using a sampled FBG. It is based on a long-period structure written using the electric arc technique. The reported temperature and strain measurement resolutions are estimated to be $\pm 0.50°C/\sqrt{Hz}$ and $\pm 3.38\ \mu\varepsilon/\sqrt{Hz}$, respectively [157].

Zhou et al. reported a sensing head formed by an FBG combined with a section of multimode fiber (MMF), which acts as a Mach–Zehnder interferometer for temperature and strain discrimination. The strain and temperature coefficients of MMFs vary with the core sizes and materials and are useful to improve the strain–temperature resolution. The reported sensor shows for a 10 pm wavelength resolution a resolution of 9:21 $\mu\varepsilon$ in strain and 0.26°C in temperature [158].

A simple and effective technique for the simultaneous measurement of temperature and strain using two closely spaced fiber gratings embedded in series in a silica glass capillary tube was demonstrated by Song et al. In this scheme, one of the two gratings is firmly attached to a carefully designed glass tube and thus responds to heat only, while the other responds to both heat and strain [159].

Haran et al. have demonstrated a construction of a fiber-optic strain gauge rosette (FOSGR) using four in-line FBGs. In this arrangement, three FBGs were used as strain sensors and the fourth FBG was isolated from the strain and acts as a temperature sensor [160].

Brady et al. reported a method of discriminating between temperature and strain effects in fiber sensing using a Bragg grating [161]. The technique uses wavelength information from the first and second diffraction orders of the grating element to determine the wavelength-dependent strain and temperature coefficients, from which independent temperature and strain measurements can be made. The best part of this technique is that it requires neither the writing of a second grating nor any physical alteration of, nor addition to, the fiber. However, the second harmonic (780.5 nm) reflectivity of the FBG was considerably smaller than the primary (1561 nm).

Frazao et al. have presented a fiber-optic-based sensing head for making simultaneous measurements of temperature and strain, which operates over a large temperature range [162]. The configuration is based on the different temperature sensitivities of Type I and Type IIA gratings written in a fiber with high germanium content. Maximal errors of $\pm$ 0.7°C/$\sqrt{Hz}$ and $\pm 3.8\ \mu\varepsilon/\sqrt{Hz}$ are reported over 500°C and 1200 $\mu\varepsilon$ measurement ranges, respectively.

9.4.3 Extrinsic Strain and Temperature Discrimination

Figure 9.32 shows a simple extrinsic strain–temperature discrimination technique [163].

In the experimental setup, one small in length grating was embedded on one specific surface of a bimetal cantilever beam that consists of two metallic materials with different thermal expansion

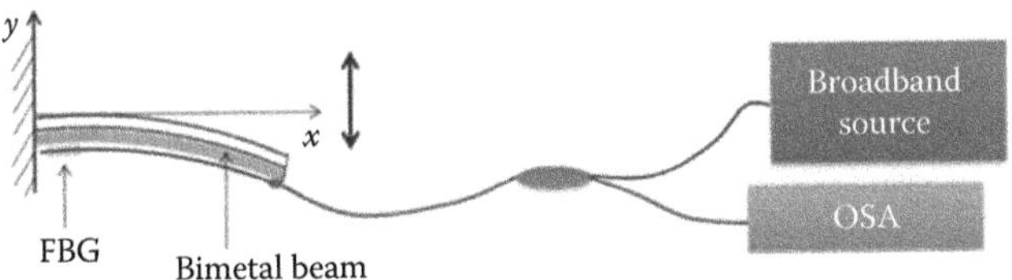

FIGURE 9.32 FBG embedded on a bimetal beam for temperature compensation.

coefficients. The FBG's temperature compensation was realized through the bending of cantilever that bends with temperature and the grating attached to it compressed or stretched accordingly. The grating chirping can be avoided through this scheme. The reported results show the temperature dependence of Bragg wavelength reduced to −0.4 pm/°C over the temperature range from −20°C to 60°C.

Lu et al. [164] proposed a sensor based on two FBGs with distinct polymeric coatings. From the different optical responses that resulted from the gratings of different polymeric coatings, the sensitivity to individual parameters can be exactly determined. The experimental results show that the axial strain and temperature sensitivities are 1.228 pm/με and 11.433 pm/°C for the acrylate-coated FBG and 1.170 pm/με and 11.333 pm/°C for the polyimide-coated FBG, respectively.

Mondal et al. [165] incorporated two nearly identical gratings and embedded them on opposite sides of an arch-shaped steel strip. The compressive and tensile strain effects were explored for thermal and strain discrimination. The results show an improved temperature and load (strain) sensitivity to 28.5 pm/°C and 2.8 pm/g (2.6 pm/με). The sensor can measure temperature and load (strain) accurately with small error of ±1°C and ±1 g, respectively.

9.5 FBG INTERROGATION SYSTEMS

Optical wavelength detection in sensing can be basically categorized into two types: passive detection schemes and active detection schemes. In passive schemes, there are no power-driven components involved. Examples in such cases are those based on linear wavelength-dependent devices. In active detection schemes, the measurement depends on externally powered devices and examples of this scheme include those based on tunable filters and interferometric scanning methods. A detailed review of different wavelength demodulation schemes is presented in the following sections.

9.5.1 Bulk Optics

The earliest reported spectral method of determining the center wavelength of an FBG sensor involved the use of either a monochromator or a spectrometer. Monochromator acts as tunable band-pass filters for light, and they can be used to create tunable light sources and also to take high-precision spectral measurements. Using a monochromator, Meltz et al. in 1989 have performed some of the first studies involving FBG [7]. A spectrometer or optical spectrum analyzer (OSA) is an instrument used to measure light properties, mostly the wavelength and intensity within a certain wavelength range and giving values at every wavelength.

It consists of three major parts: (1) collimating mirror, (2) grating, and (3) focusing mirror. Light from a source is placed in the focus of a collimating mirror that collimates the light to the grating. The light will then be diffracted from the grating to the focusing mirror that will focus the spectra into the exit slit that contains a photodiode (detector) to convert it into an equivalent electrical form. In a conventional spectrometer, the signal in the exit varies with the rotation of the grating (moving part) allowing a scanning of desired spectral range (Figure 9.33). In more advanced spectrometers, the grating is fixed and the single slit is replaced by an array of detectors (CCD).

In order to determine Bragg wavelengths with high-sensitivity, high-density diffraction gratings associated with high-precision rotational stage, large distances between grating, focusing mirror,

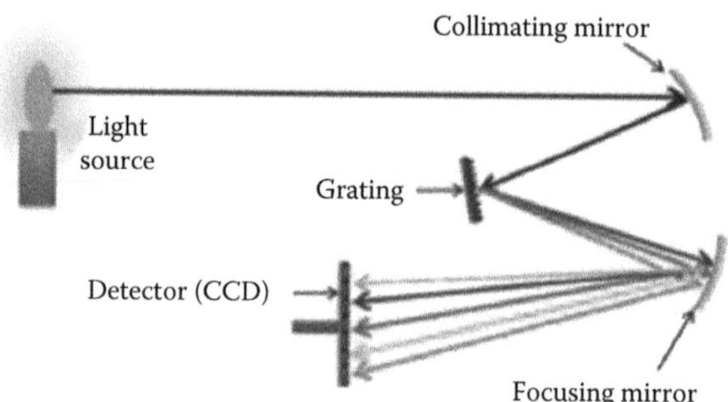

FIGURE 9.33 Schematic diagram of a spectrometer.

and detector are required, which increase the dimension of the spectrometer. These instruments are not attractive in practical application due to their bulk application nature, limited resolution capability, and lack of ruggedness and cost effectiveness. Askins et al. have proposed a spectrometer based on low-resolution diffraction gratings and small CCD arrays utilizing subpixel algorithms. Spectrometer of this kind can be a suitable instrument, where sensitivity is not important [166]. In general, monochromators and OSA are used in the laboratory.

Stephen et al. introduced volume holograms written in photorefractive materials ($BaTiO_3$) as diffractive elements in order to interrogate the Bragg wavelength. The photo refractive material on which multiple volume holograms are written acts as a multiple filters and can be tuned by the application of electrical fields. A strain range of 2500 με, with minimum detectable strain of 4 με, was measured and reported [167]. The insertion loss and bulk in nature are the major drawbacks of this system.

9.5.2 Passive Edge Filters

An edge filter provides a wavelength-dependent transmittance, offering a linear relationship between the Bragg wavelength shift and the output intensity change of the filter. A schematic function of edge filter is shown in Figure 9.34. By measuring the intensity change, the wavelength shift induced by the measurand can be obtained. The measurement range is inversely proportional to the detection resolution. The edge filter method is not suitable where FBG sensors are multiplexed. Several approaches of detecting wavelength are made on this principle.

9.5.2.1 Using Wavelength-Dependent Bulk Edge Filter

Melle et al. have demonstrated a simple interrogation system in which the reflected signal from an FBG sensor was split into two beams by a 2 × 2 coupler [168]. The one beam was fed directly into a photodetector and used as a reference where there may have been fluctuations in the intensity of the source, misalignment of connectors, or microbend losses. The other was fed through a

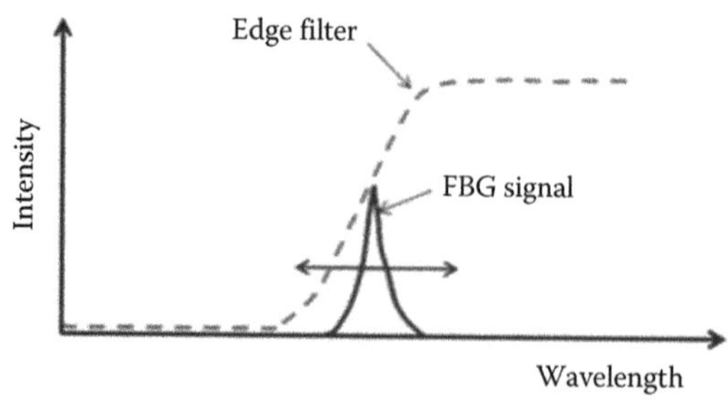

FIGURE 9.34 Schematic function of an edge filter.

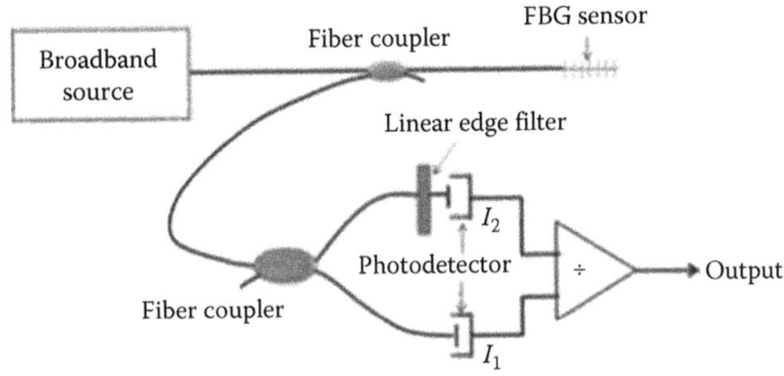

FIGURE 9.35 Transmissive passive edge filter.

wavelength-dependent filter before another photodetector detects it (Figure 9.35). The wavelength-dependent transmittance filter is linear over the full-scale measurement range. Both signals were amplified and the filtered signal was divided by the reference signal to get the final output signal. The ratio of the filtered beam over the nonfiltered beam gives the wavelength information of the reflection peak. In order to eliminate the influences of intensity fluctuations of the light source and fiber link, the ratio between the signal intensity, I_2, and the reference intensity, I_1, is used and is given by

$$\frac{I_2}{I_1} = A\left(\lambda_B - \lambda_0 + \frac{\Delta\lambda}{\sqrt{\pi}}\right) \tag{9.26}$$

where

A and λ_0 are the gradient and the starting value of the edge filter
λ_B and $\Delta\lambda$ are the Bragg wavelength and the line width of the FBG, respectively

It can be seen that this system has several advantages, such as low cost, fast response, and ease of use. Using a wavelength-dependent filter of opposite signs at both beams can double the sensitivity.

9.5.2.2 Using Long-Period Grating

A substantially easy and simpler method to interrogate an FBG sensor is to employ an LPG as an edge filter. LPG consists of a periodic modulation of the modal propagation constant, which allows coupling from the fundamental core mode to some resonant cladding modes. Its transmission spectra show some dips at wavelengths that satisfy the resonant condition. Simply by utilizing the wavelength-dependent transmission loss of an LPG, the realization of an FBG interrogator is possible. An LPG-based interrogation system is shown in Figure 9.36a. In this arrangement, the reflected light from the FBG sensor is transmitted through the LPG and the intensity of the transmitted light is altered when there is a change in the Bragg wavelength. Figure 9.36b plots the reflection and transmission spectra of the FBG and LPG. The slope of transmission dip of the LPG allows the device to be employed as a linear response edge filter and acts as a wavelength to amplitude converter. The reported result shows an 8100 με dynamic range and 0.5 με static strain resolution, respectively [169].

9.5.2.3 Using WDM Filter

An FBG interrogation system consisting of a bulk optical edge filter shows a critical alignment problem and thus reduced portability. In such situation, an all-fiber approach is obviously more attractive. Davis and Kersey have demonstrated a demodulation scheme using a WDM coupler for an FBG sensor (Figure 9.37a). The WDM coupler shows a monotonic change in the coupling ratio between the two output fiber ports. Figure 9.37b shows its transfer function. Taking the ratio

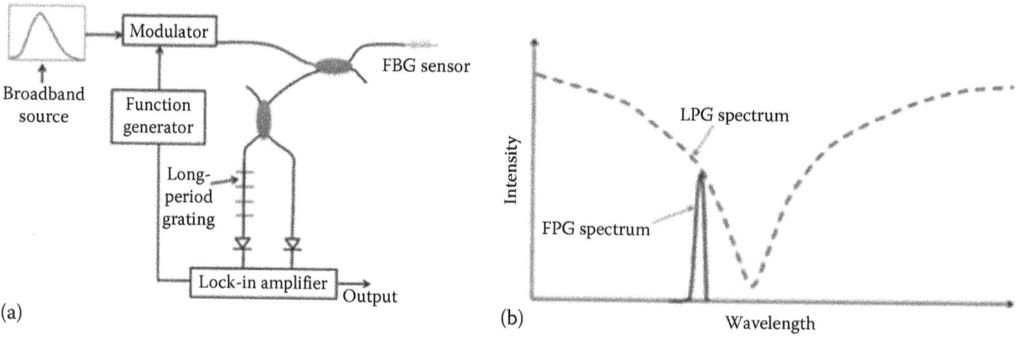

FIGURE 9.36 (a) Schematic arrangement of an LPG-based FBG interrogator and (b) spectrums of LPG and FBG.

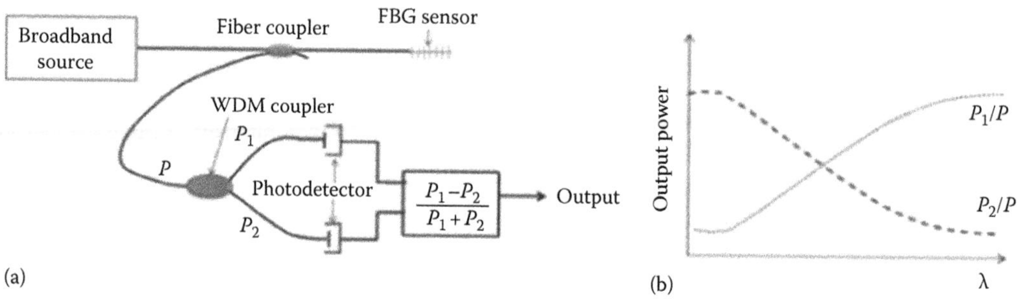

FIGURE 9.37 (a) Schematic FBG interrogator using a WDM coupler and (b) spectral response of a WDM coupler.

of the difference and sum of the two outputs of the WDM coupler gives a drift compensated output for wavelength shift detection. The spectral slope of a WDM coupler is typically less and the slope steepness determines the sensitivity, and the minimum detectable wavelength shifts ±3 με and 0.5 με/√(Hz) are the static and dynamic strain resolution of the system as reported [170]. The system is simple and cost effective and can be implemented where high sensitivity is not a constraint.

9.5.2.4 Using Macrobend Edge Filter

Rajan et al. have demonstrated a low-cost edge filter using an SMF 28 fiber. It is well known that an optical wave guide with an infinite cladding shows an increase in bend loss with the decrease in bending radius. However, a practical fiber contains one or two coating layers outside to offer mechanical protection. Because of the reflection of the radiated field at the interface between the cladding layer and the coating layer, the so-called whispering gallery mode is produced, and the fiber shows significantly different bend loss characteristics as compared with the simple case of infinite cladding outside. A macrobending standard single-mode fiber was developed as an edge filter with an optimal design and simple surface processing. A schematic experimental setup is shown in Figure 9.38. The ratiometric wavelength measurement system employing the developed macrobending fiber filter shows a resolution of ~10 pm in a wavelength range from 1500 to 1560 nm with ease of assembly and calibration [171].

9.5.2.5 Using SMS Fiber Filter

A low-cost intrinsic in-fiber edge filter was demonstrated by Zhang et al. This edge filter is made up of an multimode fiber as MMF (e.g., 100/125 μm fiber) of suitable length, which is spliced between two single-mode fibers.

When light propagates through one end of the single-mode fiber to MMF, then the excitation of a number of guided modes takes place in the MMF (Figure 9.39). This leads to interference between

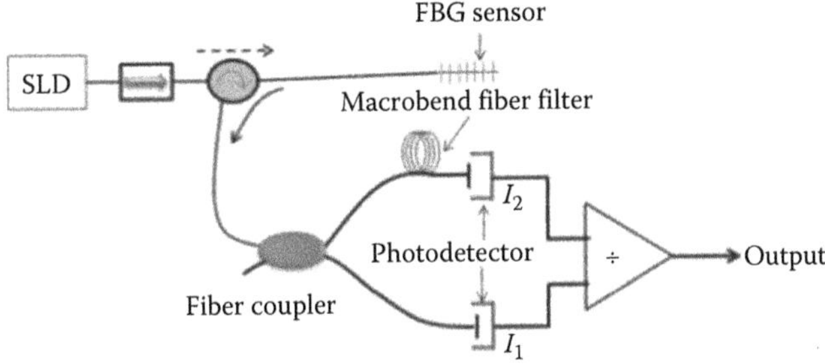

FIGURE 9.38 Schematic experimental setup of an FBG interrogator using a macrobend fiber filter as an edge filter.

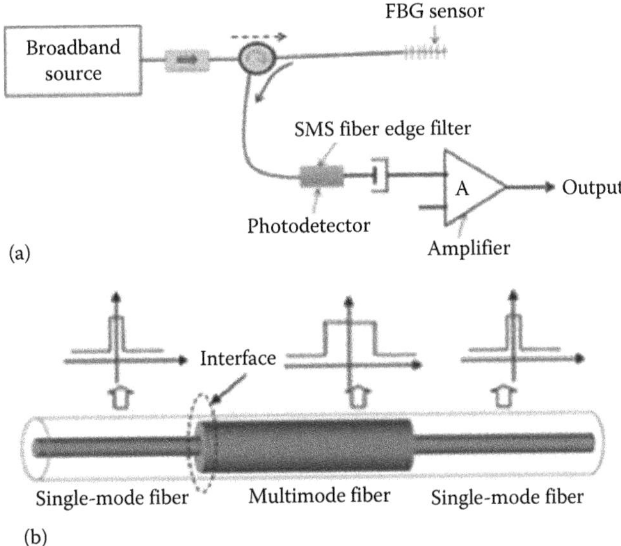

FIGURE 9.39 (a) Schematic FBG interrogator using an SMS fiber edge filter and (b) SMS fiber filter structure.

the different modes during the propagation of light through the MMF section [172]. The excitation of peak wavelength of the SMS edge filter (as shown in Figure 3.39b) can be optimized by varying the length of the MMF. The transmission spectrum of the SMS edge filter for an optimized length shows a band-pass filter response with two edge slopes on either side of the allowed peak wavelength of the filter. Consequently, this device behaves as an edge filter with either a positive or a negative slope for a selected wavelength range [173].

9.5.2.6 Using Narrowband Source and Detector

Graham et al. have implemented a new edge filter interrogation technique where the sensing FBG itself acts as an edge filter. Figure 9.40a shows the reflectivity as a function of wavelength for a typical FBG with maximum reflectivity at λ_B. There is a linear region, $\delta\lambda$, between reflectivities of approximately 20% and 80% centered at λ_0. This linear *edge* of the FBG can be used as an optical filter. A narrowband laser source centered about λ_0 is then intensity modulated by the strain-induced shift in the Bragg wavelength of the FBG [174]. The variation of reflected optical power obtained as the linear edge of the FBG is shifted in the spectrum (Figure 9.40b).

An alternate approach can also give a similar effect. In this case, the selected photodiode should have a narrow spectral bandwidth response. Both the methods are simple and easy to implement.

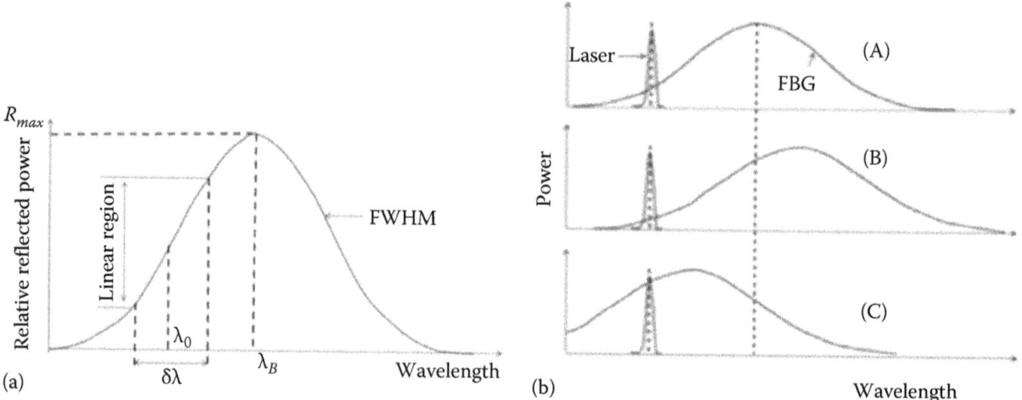

FIGURE 9.40 (a) Spectrum of FBG and (b) operating principle of the TRDS. A—FBG with no change in measurand, B—positive change, and C—negative change.

9.5.2.7 Using Tilted Grating

A TFBG can be used as an edge filter for linear demodulation of the Bragg signal. A system of linear edge filter demodulation based on the TFBG structure was proposed by Zhou et al. In TFBG, the grating planes are parallel to each other but they are not normal to the axis of the fiber. It is found that the radiation mode of TFBG propagates along different directions for different wavelengths and was exploited to achieve TFBG-based wavelength interrogation through side radiation [175]. The advantages of this simple system are an all-fiber design, quasistatic and dynamic operation, high stability, and lower cost.

9.5.2.8 Using Erbium-Doped Fiber

Figure 9.41 shows a simple approach to interrogate a reflecting Bragg wavelength shift. In this approach, a 10 m length of EDF was used as an edge filter [176]. The working principle behind this filter is on the spectral dependence of absorption in EDF and the length of the EDF. The experimental results show a wide filter bandwidth ~10 nm and a slope detection sensitivity of 1.0 dB/nm in the C-band, and the measured strain and temperature resolution are 20 με and 2.0°C, respectively, in terms of minimum change in measurable power from the filter. The resolution of the proposed interrogator can be enhanced by increasing the length of the EDF. The drawback of this scheme is that the power absorbed by the EDF is higher at lower wavelengths, which results in small power corresponding to lower wavelengths.

9.5.3 Active Tunable Edge Filters

9.5.3.1 Using Matched FBG

In general, an FBG interrogator using an edge filter operated under transmissive mode shows smaller power loss with a shift in the Bragg wavelength. An interrogator scheme that consists

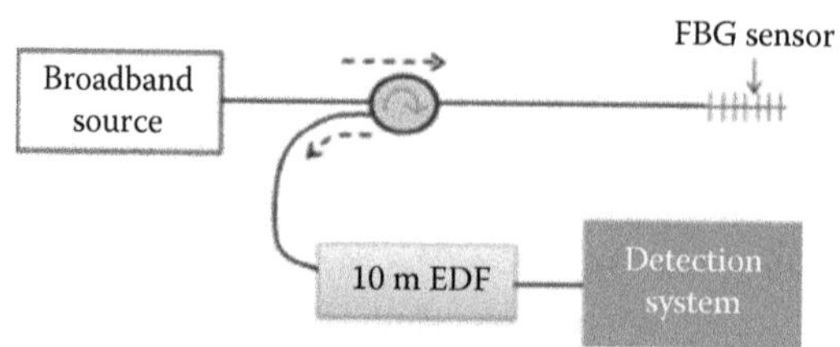

FIGURE 9.41 Schematic experimental setup of FBG interrogation using 10 m EDF.

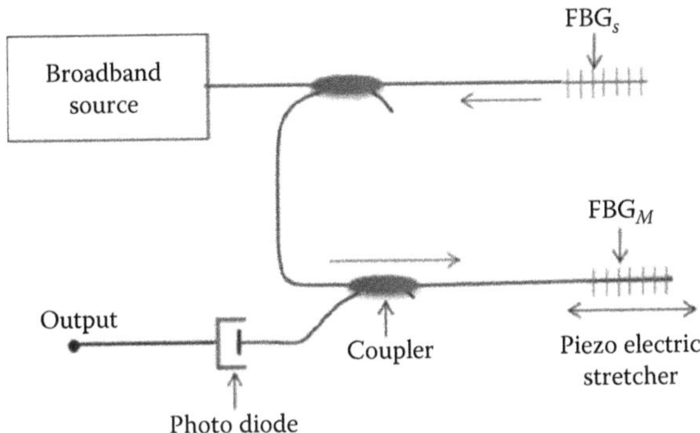

FIGURE 9.42 FBG interrogation using a matched FBG.

of a matched FBG with improved sensitivity was demonstrated by Ribeiro et al. (Figure 9.42). Light from the broadband source is fed to the sensor grating (FBG$_S$) via the fiber coupler. The reflected light from the FBG propagates back to the matched grating (FBG$_M$), which can be strained using a piezoelectric stretcher. As the central reflecting wavelength of the sensor grating varies in direct proportion to the measured the mismatch of the grating peak wavelength is obtained with FBG$_M$.

The central wavelength of the FBG$_M$ can be matched with FBG$_S$ using strain induced by piezoelectric transducer (PZT) stretcher. When the central wavelengths of the FBG$_M$ and FBG$_S$ are exactly matched, then a strong signal will be back reflected from the matched grating and detected by the photodiode. If the relationship between the driving voltage of the PZT and the wavelength of the FBG$_M$ is known, the instantaneous wavelength value of the sensor grating can be determined. The resolution of this sensor arrangement depends on the line width of the gratings [177].

9.5.3.2 Using Acousto-Optic Modulator

An acousto-optic tunable filter (AOTF)-based FBG interrogator was demonstrated by Xu et al. [178]. A schematic setup is shown in Figure 9.43. The filter was driven by a frequency-shift keying of RF drive to track the wavelength shifts in transmission mode of its operation. But the filter can be operated in reflection mode also. The proposed system has a wide tuning range and the ability to recover after transient signal loss. It was demonstrated for FBG temperature sensor integration. The detection system can track a large number of gratings in a quick moving frequency manner.

9.5.3.3 Using Fabry–Pérot Tunable Filter

Kersey et al. have demonstrated the use of a fiber Fabry–Pérot (FFP) wavelength filter for the interrogation of one or more FBG sensors [179]. The bandwidth of a tunable F–P filter is comparable to the FBG. An interrogation scheme using a high-finesse FFP filter tuned by a piezo element is shown in Figure 9.43.

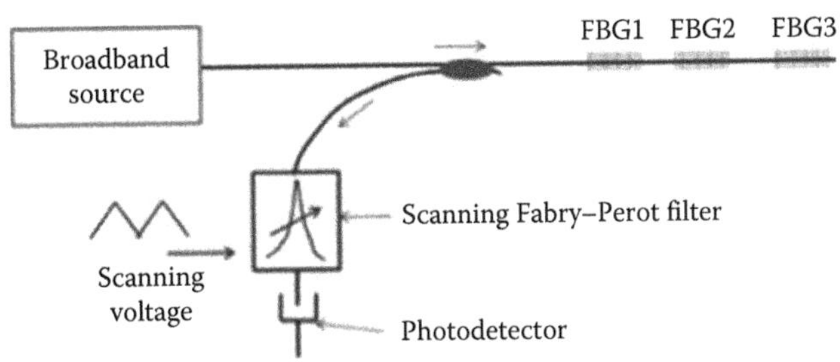

FIGURE 9.43 Tunable FFP filter.

A coupler is used to redirect the reflected signal from the FBG sensor to the F–P filter. In the reported process, the filter was scanned over the wavelength range required for measurement and the peak reflectance wavelength of one or more FBGs was determined. The F–P transmission wavelength was modulated by roughly 1 nm for the interrogation of an FBG. During the process, when the filter wavelength and the sensor Bragg wavelength were aligned, the modulated signal acted as an error signal used to lock the FFP passband wavelength to the sensor return wavelength.

In a multiplexing system, the transmission wavelength of the F–P filter needs to be scanned with a ramp voltage over the entire wavelength range of all the sensors. The return signals from different FBG sensors do not overlap in this process. The frequency of the scanning waveform determined the frequency of measurement updates. The proposed system shows a resolution of <0.3 με for a single sensor setup and better than ±3 με for a setup with four sensors.

9.5.3.4 Using Fiber Laser Source/Analyzer

Ball et al. [180] demonstrated a new interrogation technique using a continuously wavelength-tunable erbium fiber laser as a source for a single FBG temperature sensor or an FBG temperature-sensing array. The designed laser used FBGs for cavity feedback and is intrinsically compatible to the fiber system (Figure 9.44).

When the proposed system is used for sensing a single FBG, it is locked into the central Bragg wavelength of the FBG sensor and tracks the wavelength shift, whereas for sensing multiple FBGs, it is performed through scanning over the total bandwidth of all the sensors, determining the peak reflectivities.

The reported results show a minimum wavelength resolution of approximately 2.3 pm corresponding to 0.2°C and a strain of 2.3 με resolution at 1550 nm.

9.5.3.5 Using Frequency-Modulated Multimode Laser

A signal processing scheme to measure the wavelength shifts in optical FBGs based on the generation of an electrical carrier via ramp modulation of a multimode laser diode (MMLD) was presented by Ferreira et al. Figure 9.45a shows a basic configuration of the proposed interrogation system using the MMLD for illuminating the FBG sensor. In general, if one of the laser modes coincides with the spectrum of the FBG, reflected light associated with that laser mode will be detected at the output (Figure 9.45b). The intensity of the signal at the detector is proportional to the overlap integral of the functions and represents the spectral characteristics of the laser mode and the FBG, respectively.

The scheme described that the phase of the carrier signal is linearly dependent on the difference between the Bragg wavelength and the laser mode wavelength and varies when the sensing FBG is affected by external parameters like temperature.

The change in Bragg wavelength is measured by tracking the phase of the carrier at the detector output in either an open- or a closed-loop scheme. From the experimental result, sensitivities of 0.7 με/√Hz and 0.05°C/√Hz were obtained over a dynamic range of ≈60 dB [181].

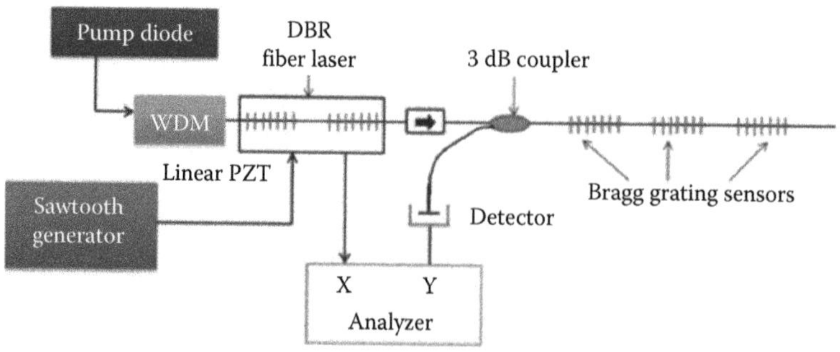

FIGURE 9.44 Source analyzer setup for multiple FBG sensors.

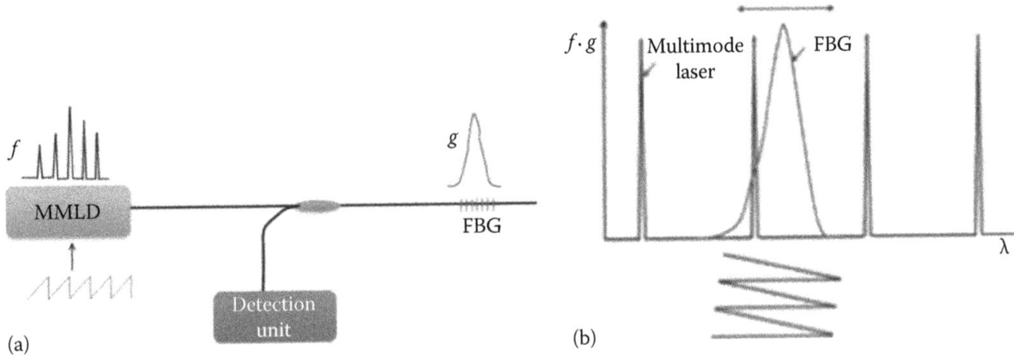

FIGURE 9.45 (a) Basic configuration of FBG interrogator using an MML diode and (b) functions f and g with slight difference in their central wavelengths caused by a sawtooth wave applied to the laser diode.

9.5.3.6 FBG-Based Fiber Laser Using a Mode-Locked Modulator

Kersey and Morey have demonstrated a system where FBG sensors were used as feedback elements for a mode-locked fiber laser cavity [182]. Using the FBGs as optical feedback elements, a cavity was formed. The mode-locked operation of the cavity was used to selectively address different sensors. The working principle of this interrogation is that the system output wavelength is a measure of the perturbation on the sensor because the system will lase at the Bragg wavelength of the FBG element and the system can only interrogate a single FBG. However, the realization of a system to interrogate multiple FBGs is possible when a mode-locked modulator (MLM) is used. An illustration of the scheme is shown in Figure 9.46.

As such, the system can only interrogate a single FBG sensor, but when an MLM is included, the realization of a multiplexed system is possible. A pulse train at frequency f_1 and wavelength λ_{B1} determined by the FBG sensor formed the laser output. In this scheme, a shift in the Bragg wavelength was detected by a shift in the wavelength of the laser output. Another FBG element, FBG2, could be added in the system as shown in Figure 9.46. The cavity was forced to lase in a mode-locked fashion at either of the two FBG wavelengths if the elements were spaced by ΔL and if a certain MLM drive frequency was selected.

9.5.3.7 Using Fiber Loop Reflector Laser

In continuation to the earlier work reported by Kersey and Morey (Figure 9.47), another laser sensor where a wavelength selective filter was located with the laser cavity has been developed [183].

The filter was tuned over the gain bandwidth for selective lasing of the laser at each of the Bragg wavelengths of the FBG sensors. In this scheme, by measuring the wavelength of the laser output wavelength, the Bragg wavelength was determined. Figure 9.47 shows a schematic arrangement of sensing FBGs using this scheme.

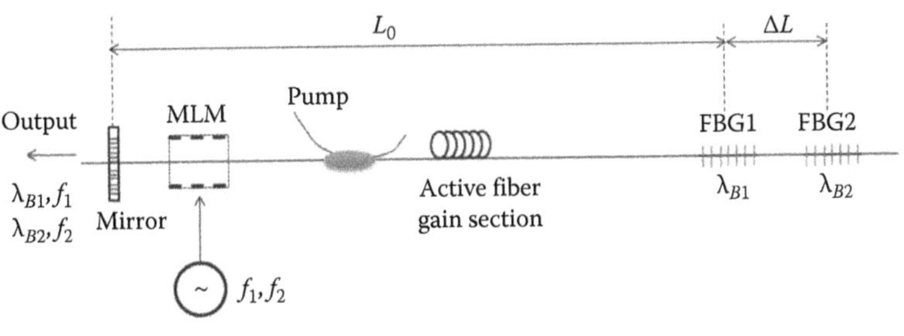

FIGURE 9.46 FBG-based fiber laser using MLM.

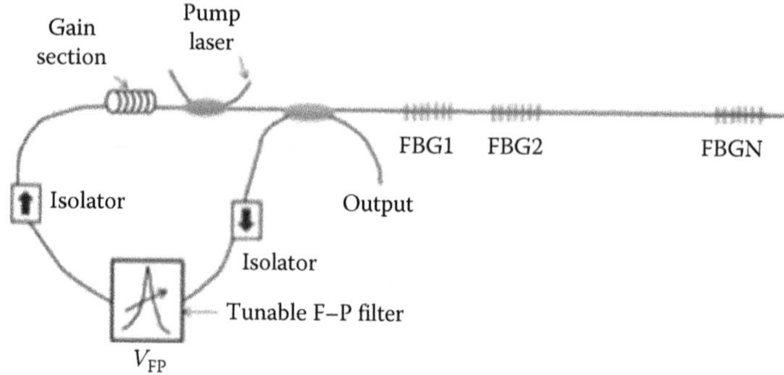

FIGURE 9.47 Fiber loop reflector laser.

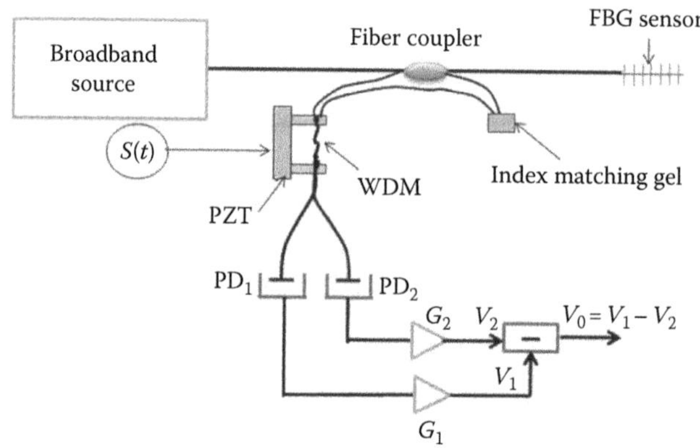

FIGURE 9.48 Tunable WDM.

9.5.3.8 Using Tunable WDM

Misas et al. have demonstrated an all-fiber interrogation system using a tunable WDM filter. Figure 9.48 shows that a WDM coupler is mounted on a PZT stretcher. The spectral response of the WDM depends on its coupling length (Figure 9.48). Figure 9.48 shows a tunable WDM that originated by application of a dc voltage signal to the PZT stretcher [184]. When the light returning from the FBG is directed to the WDM, the amplitude of the transmitted FBG spectral signature changes due to the relative displacement of the WDM spectral response. This behavior indicates a possible time-varying signal obtained at the system output as a consequence of a modulating signal applied to the PZT. If the Bragg wavelength changes due to the action of a certain measurand, a phase term dependent on the Bragg wavelength will be introduced in the output signal. This scheme shows a potential to recover with submicrostrain resolution static and dynamic signals induced in fiber Bragg sensors.

9.5.4 FBG Interrogation Using Mach–Zehnder Interferometer

Kersay et al. have demonstrated an interrogation system using an unbalanced fiber Mach–Zehnder interferometer for dynamic strain measurement. An unbalanced interferometer is an optical fiber element with a transfer function of the form (1+ cos φ) where the phase term depends on the input wavelength. Figure 9.49 shows the general principle of this technique applied to FBG sensors. Light reflected from a grating is directed through an interferometer that has unequal paths.

The shifts in Bragg wavelength are converted into phase shifts due to the inherent wavelength dependence of the phase of an unbalanced interferometer on the input wavelength.

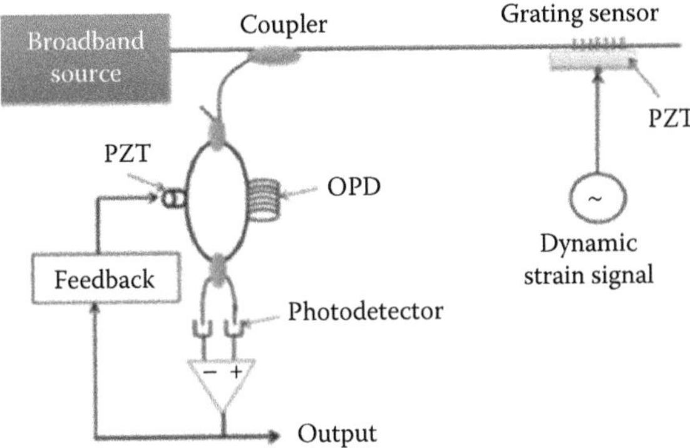

FIGURE 9.49 Grating sensor system with fiber interferometric wavelength.

The interferometer output can be modulated via control of the path imbalance between the interferometer arms. Different phase reading techniques can be applied to determine the phase modulation, $\Delta\varphi$, induced by FBG wavelength shifts $\Delta\lambda$. The dependence of the output phase change on Bragg grating wavelength shift is given by

$$\Delta\varphi = \frac{2\pi nd}{\lambda^2}\Delta\lambda \tag{9.27}$$

where n and d are the effective refractive indices of the core and length imbalance between the fiber arms of the interferometer, respectively.

It is important to note that the interferometer path difference must be kept less than the effective coherence length of the light reflected from the grating. A strain resolution of ~0.6 $n\varepsilon/\sqrt{Hz}$ at frequencies >100 Hz was reported [185].

9.6 CONCLUSIONS

The journey of FBG began with the first report of photorefractive gratings in 1978. It has become a real subject of interest and commercial importance after 10 years of its invention when the phase mask technique for the fabrication of FBG was reported. This chapter reveals a detailed discussion on the history of fiber Bragg granting, measurement of different physical parameters like temperature, strain, and refractive index, and different methods to discriminate strain and temperature sensed by the FBG. A review on active and passive interrogation techniques used for FBG sensors is also presented in this chapter.

REFERENCES

1. A. Othonos and K. Kalli, *Fiber Bragg Gratings: Fundamentals and Applications in Telecommunications and Sensing*, Artech House, Boston, MA, 1999.
2. R. Kashyap, *Fiber Bragg Gratings*, Academic Press, San Diego, CA, 1999.
3. K. O. Hill, Y. Fujii, D. C. Johnson, and B. S. Kawasaki, Photosensitivity in optical fiber waveguides: Application to reflection filter fabrication, *Appl. Phys. Lett.*, 32, 647–649, 1978.
4. B. S. Kawasaki, K. O. Hill, D. C. Johnson, and Y. Fujii, Narrow band Bragg reflectors in optical fibers, *Opt. Lett.*, 3, 66–78, 1978.
5. D. K. W. Lam and B. K. Garside, Characterization of single-mode optical fiber filters, *Appl. Opt.*, 20(3), 440–445, 1981.
6. J. Stone, Photorefractivity in GeO_2 doped silica fibers, *J. Appl. Phys.*, 62, 4371–4374, 1987.
7. G. Meltz, W. W. Morey, and W. H. Glenn, Formation of Bragg gratings in optical fibers by a transverse holographic method, *Opt. Lett.*, 14, 823–825, 1989.

8. R. Kashyap, J. R. Armitage, R. Wyatt, S. T. Davey, and D. L. Williams, All-fiber narrow band reflection grating at 1500 nm, *Electron. Lett.*, 26, 730–732, 1990.
9. B. J. Eggleton, P. A. Krug, and L. Poladian, Experimental demonstration of compression of dispersed optical pulses by reflection from self-chirped optical fiber Bragg gratings, *Opt. Lett.*, 19, 877–880, 1994.
10. K. O. Hill, B. Malo, F. Bilodeau, D. C. Johnson, and J. Albert, Bragg gratings fabricated in monomode photosensitive optical fiber by UV exposure through a phase mask, *Appl. Phys. Lett.*, 62, 1035–1037, 1993.
11. D. Z. Anderson, V. Mizrahi, T. Erdogan, and A. E. White, Production of in-fiber gratings using a diffractive optical element, *Electron. Lett.*, 29, 566–568, 1993.
12. A. Othonos, Novel and improved methods of writing Bragg gratings with phase mask, *IEEE Photon. Technol. Lett.*, 7(10), 1183–1185, 1995.
13. B. Malo, K. O. Hill, F. Bilodeau, D. C. Johnson, and J. Albert, Point-by-point fabrication of micro-Bragg gratings in photosensitive fibre using single excimer pulse refractive index modification techniques, *Electron. Lett.*, 29, 1668–1669, 1993.
14. S. J. Mihailov, C. W. Smelser, P. Lu, R. B. Walker, D. Grobnic, H. Ding, G. Henderson, and J. Unruh, Fiber Bragg gratings made with a phase mask and 800-nm femtosecond radiation, *Opt. Lett.*, 28(12), 995–997, 2003.
15. A. Martinez, M. Dubov, I. Khrushchev, and I. Bennion, Direct writing of fiber Bragg gratings by femtosecond laser, *Electron. Lett.*, 40(19), 1170–1172, 2004.
16. S. Ness and P. R. Herman, 157 nm photosensitivity in germanosilicate planar waveguides, *CLEO 99*, Baltimore, MD, 1999, pp. 76–77.
17. J. Albert, B. Malo, F. Bilodeau, D. C. Johnson, K. O. Hill, Y. Hibino, and M. Kawachi, Photosensitivity in Ge doped silica optical waveguides and fibers with 193 nm light from an ArF excimer laser, *Opt. Lett.*, 19, 387–389, 1994.
18. R. M. Atkins and R. P. Espindola, Photosensitivity and grating writing in hydrogen loaded germanosilicate core optical fibers at 325 and 351 nm, *Appl. Phys. Lett.*, 70, 1068–1069, 1997.
19. F. Bilodeau, D. C. Johnson, B. Malo, K. A. Vineberg, K. O. Hill, T. F. Morse, A. Kilian, and L. Reinhart, Ultraviolet light photosensitivity in Er^{3+}-Ge doped optical fiber, *Opt. Lett.*, 15, 1138–1140, 1990.
20. D. L. Williams, B. J. Ainslie, J. R. Armitage, R. Kashyap, and R. Campbell, Enhanced UV photosensitivity in boron codoped germanosilicate fibers, *Electron. Lett.*, 29, 45–47, 1993.
21. L. Dong, J. L. Cruz, L. Reekie, M. G. Xu, and D. N. Payne, Enhanced photosensitivity in tin codoped germanosilicate optical fibers, *IEEE Photon. Technol. Lett.*, 7, 1048–1050, 1995.
22. L. Dong, L. Reekie, L. Cruz, and D. N. Payne, Grating formation in a phosphorus doped germanosilicate fiber, *Conference on Optical Fiber Communication*, San Jose, CA, 1996, Paper TuO_2, p. 82.
23. M. M. Broer, A. J. Bruce, and W. H. Grodkiewicz, Photoinduced refractive index changes in several Eu^{3+}, Pr^{3+}, and Er^{3+} doped oxide glasses, *Phys. Rev. B*, 45, 7077–7083, 1992.
24. K. O. Hill, B. Malo, F. Bilodeau, D. C. Johnson, and T. F. Morse, Photosensitivity in Eu^{3+}:Al_2O_3 doped core fiber: Preliminary results and applications to mode converters, *Conference on Optical Fiber Communication, Technical Digest Series*, Optical Society of America, Washington, DC, 1991, Vol. 14, p. 14.
25. G. M. Williams, T. E. Tsai, C. I. Merzbacher, and E. Joseph Friebele, Photosensitivity of rare earth doped ZBLAN fluoride glasses, *IEEE J. Lightwave Technol.*, 15, 1357–1362, 1997.
26. P. J. Lemaire, R. M. Atkins, V. Mizrahi, and W. A. Reed, High pressure H_2 loading as a technique for achieving ultrahigh UV photosensitivity and thermal sensitivity in GeO_2 doped optical fibers, *Electron. Lett.*, 29, 1191–1193, 1993.
27. F. Bilodeau, B. Malo, J. Albert, D. C. Johnson, K. O. Hill, Y. Hibino, M. Abe, and M. Kawachi, Photosensitization of optical fiber and silica on silicon/silica waveguides, *Opt. Lett.*, 18, 953–955, 1993.
28. M. Douay, W. X. Xie, T. Tounay, P. Bernage, P. Niay, P. Cordier, B. Poumellec et al., Densification involved in the UV based photosensitivity of silica glasses and optical fibers, *IEEE J. Lightwave Technol.*, 15, 1329–1342, 1997.
29. D. P. Hand and P. St. J. Russell, Photoinduced refractive index change in germanosilicate fibers, *Opt. Lett.*, 15, 102–104, 1990.
30. J. P. Bernardin and N. M. Lawandy, Dynamics of the formation of Bragg gratings in germano optical fibers, *Opt. Commun.*, 79, 194–199, 1990.
31. D. L. Williams, B. J. Ainslie, R. Kashyap, G. D. Maxwell, J. R. Armitage, R. J. Campbell, and R. R. Wyatt, Photosensitive index changes in germania doped silica fibres and waveguides, *Proc. SPIE*, 2044, 55, 1993.
32. M. G. Sceats, G. R. Atkins, and S. B. Poole, Photo induced index changes in optical fibers, *Annu. Rev. Mater. Sci.*, 23, 381–410, 1993.

33. H. Patrick and S. L. Gilbert, Growth of Bragg gratings produced by continuous wave ultraviolet light in optical fiber, *Opt. Lett.*, 18, 1484–1486, 1993.
34. J. Canning, Fibre gratings and devices for sensors and lasers, *Laser Photon. Rev.*, 2, 275–289, 2008.
35. B. J. Eggleton, P. S. Westbrook, R. S. Windeler, S. Spalter, and T. A. Strasser, Grating resonances in air-silica micro structured optical fibers, *Opt. Lett.*, 24, 1460–1462, 1999.
36. Y. Liu, J. A. R. Williams, L. Zhang, and I. Bennion, Abnormal spectral evolution of fibre Bragg gratings in hydrogenated fibres, *Opt. Lett.*, 27, 586–588, 2002.
37. A. G. Simpson, K. Kalli, K. Zhou, L. Zhang, and I. Bennion, Formation of type IA fibre Bragg gratings in germanosilicate optical fibre, *Electron. Lett.*, 40, 163–164, 2004.
38. I. Riant and F. Haller, Study of the photosensitivity at 193 nm and comparison with photosensitivity at 240 nm influence of fiber tension: Type IIA aging, *J. Lightwave Technol.*, 15, 1464–1469, 1997.
39. J. L. Archambault, L. Reekie, and P. St. J. Russle, 100% Reflectivity Bragg reflectors produced in optical fibres by single excimer laser pulse, *Electron. Lett.*, 29, 1484–1486, 1993.
40. S. Bandyopadhyay, J. Canning, M. Stevenson, and K. Cook, Ultrahigh-temperature regenerated gratings in boron-codoped germanosilicate optical fiber using 193 nm, *Opt. Lett.*, 33, 1917–1919, 2008.
41. T. Erdogan, Fiber grating spectra, *J. Lightwave Technol.*, 15, 1277–1294, 1997.
42. T. Erdogan and J. E. Sipe, Tilted fiber phase gratings, *J. Opt. Soc. Am. A*, 13(2), 296–313, 1996.
43. J. Albert, L.-Y. Shao, and C. Caucheteur, Tilted fiber Bragg grating sensors, *Laser Photon. Rev.*, 7(1), 83–108, 2013.
44. F. Ouellette, Dispersion cancellation using linearly chirped Bragg grating filters in optical waveguides, *Opt. Lett.*, 12, 847–849, 1987.
45. K. C. Byron, K. Sugden, T. Bricheno, and I. Bennion, Fabrication of chirped Bragg gratings in photosensitive fibre, *Electron. Lett.*, 29, 1659–1660, 1993.
46. R. Kashyap, P. F. McKee, R. J. Campbell, and D. L. Williams, Novel method of producing all fibre photo induced chirped Gratings, *Electron. Lett.*, 30, 996–997, 1994.
47. B. J. Eggleton, P. A. Krug, L. Poladian, and F. Ouellette, Long periodic superstructure Bragg gratings in optical fibers, *Electron. Lett.*, 30(19), 1620–1622, 1994.
48. J. Albert, K. O. Hill, B. Malo, S. Theriault, F. Bilodeau, D. C. Johnson, and L. E. Erickson, Apodisation of the spectral response of fiber Bragg gratings using a phase mask with variable diffraction efficiency, *Electron. Lett.*, 31, 222–223, 1995.
49. R. Kashyap, A. Swanton, and D. J. Armes, Simple technique for apodising chirped and unchirped fiber Bragg gratings, *Electron. Lett.*, 32, 1226–1228, 1996.
50. J. Canning and M. G. Sceats, π Phase shifted periodic distributed structures in optical fibers by UV post processing, *Electron. Lett.*, 30(16), 1244–1245, 1994.
51. R. Kashyap, P. E. McKee, and D. Armes, UV written reflection grating structures in photosensitive optical fibres using phase shifted phase masks, *Electron. Lett.*, 30(23), 1977–1978, 1994.
52. G. P. Agrawal and S. Radic, Phase shifted fiber Bragg gratings and their applications for wavelength demultiplexing, *IEEE Photon. Technol. Lett.*, 6, 995–997, 1994.
53. D. Uttamchandani and A. Othonos, Phase shifted Bragg gratings formed in optical fibers by UV fabrication thermal processing, *Opt. Commun.*, 127, 200–204, 1996.
54. A. Othonos, Fiber Bragg gratings, *Rev. Sci. Instrum.*, 68, 4309–4341, 1997.
55. K. O. Hill, and G. Meltz, Fiber Bragg grating technology—Fundamentals and overview, *J. Lightwave Technol.*, 15(8), 1263–1276, 1997.
56. S. W. James, M. L. Dockney, and R. P. Tatam, Simultaneous independent temperature and strain measurement using in-fibre Bragg grating sensors, *Electron. Lett.*, 32(12), 1133–1134, 1996.
57. R. Gafsi and M. A. El-Sherif, Analysis of induced-birefringence effects on fiber Bragg gratings, *Opt. Fiber Technol.*, 6, 299–323, 2000.
58. C. Lawrence, D. Nelson, and E. Udd, Multi-parameter sensing with fiber Bragg gratings, *Proc. SPIE*, 2872, 24–31, 1996.
59. B. Lee, Review of the present status of optical fiber sensors, *Opt. Fiber Technol.*, 9, 57–79, 2003.
60. Y.-J. Rao, In-fibre Bragg grating sensors, *Meas. Sci. Technol.*, 8, 355–375, 1997.
61. S. R. Baker, H. N. Rourke, V. Baker, and D. Goodchild, Thermal decay of fiber Bragg gratings written in boron and germanium codoped silica fiber, *J. Lightwave Technol.*, 15, 1470–1477, 1997.
62. G. Brambilla, V. Pruneri, and L. Reekie, Photorefractive index gratings in SnO_2:SiO_2 optical fibers, *Appl. Phys. Lett.*, 76, 807–809, 2000.
63. Y. Shen, T. Sun, K. T. V. Grattan, and M. Sun, Highly photosensitive Sb/Er/Ge codoped silica fiber for fiber Bragg grating (FBG) writing with strong high-temperature sustainability, *Opt. Lett.*, 28, 2025–2027, 2003.

64. Y. Shen, J. He, T. Sun, and K. T. V. Grattan, High temperature sustainability of strong FBGs written into Sb/Ge co-doped photosensitive fiber-decay mechanisms involved during annealing, *Opt. Lett.*, 29, 554–556, 2004.
65. E. M. Dianov, K. M. Golant, R. R. Khrapko, A. S. Kurkov, and A. L. Tomashuk, Low-hydrogen silicon oxynitride optical fibers prepared by SPCVD, *J. Lightwave Technol.*, 13, 1471–1474, 1995.
66. O. V. Butov, E. M. Dianov, and K. M. Golant, Nitrogen-doped silica core fibers for Bragg grating sensors operating at elevated temperatures, *Meas. Sci. Technol.*, 17, 975–979, 2006.
67. M. Fokine, Growth dynamics of chemical composition gratings in fluorine-doped silica optical fibers, *Opt. Lett.*, 27, 1974–1976, 2002.
68. M. Fokine, Formation of thermally stable chemical composition gratings in optical fibers, *J. Opt. Soc. Am. B*, 19, 1759–1765, 2002.
69. J. Canning, K. Sommer, and M. Englund, Fiber gratings for high temperature sensor applications, *Meas. Sci. Technol.*, 12, 824–828, 2001.
14. S. J. Mihailov, C. W. Smelser, P. Lu, R. B. Walker, D. Grobnic, H. Ding, and J. Unruh, Fiber Bragg gratings made with a phase mask and 800 nm femtosecond radiation, *Opt. Lett.*, 28, 995–997, 2003.
70. S. J. Mihailov, C. W. Smelser, D. Grobnic, R. B. Walker, P. Lu, H. Ding, and J. Unruh, Bragg gratings written in all SiO_2 and Ge-doped core fibers with 800-nm femtosecond radiation and a phase mask, *J. Lightwave Technol.*, 22, 94–100, 2004.
71. Y. Li, M. Yang, D. N. Wang, J. Lu, T. Sun, and K. T. V. Grattan, Fiber Bragg gratings with enhanced thermal stability by residual stress relaxation, *Opt. Express*, 17, 19785–19790, 2009,
72. D. Grobnic, C. W. Smelser, S. J. Mihailov, and R. B. Walker, Long-term thermal stability tests at 1000°C of silica fibre Bragg gratings made with ultrafast laser radiation, *Meas. Sci. Technol.*, 17, 1009–1013, 2006.
73. P. S. Reddy, R. L. N. Sai Prasad, D. Sengupta, M. Sai Shankar, K. S. Narayana, and P. Kishore, Encapsulated fiber Bragg grating sensor for high temperature measurements, *Opt. Eng.*, 50(11), 114401, 2011.
74. A. Méndez and T. F. Morse, *Specialty Optical Fibers Handbook*, Elsevier Academic Press, San Diego, CA, 2007, p. 284.
75. D. Grobnic, S. J. Mihailov, R. B. Walker, and C. W. Smelser, Self-packaged Type II femtosecond IR laser induced fiber Bragg grating for temperature applications up to 1000°C, *Proc. SPIE*, 7753, 2011.
76. M. A. Putnam, T. J. Bailey, M. B. Miller, J. M. Sullivan, M. R. Fernald, M. A. Davis, and C. J. Wright, Method and apparatus for forming a tube-encased Bragg grating, US Patent 6298,184, 2001.
77. D. Grobnic, S. J. Mihailov, C. W. Smelser, and H. Ding, Sapphire fiber Bragg grating sensor made using femtosecond laser radiation for ultrahigh temperature applications, *IEEE Photon. Technol. Lett.*, 16, 2505–2507, 2004.
78. D. Grobnic, S. J. Mihailov, H. Ding, F. Bilodeau, and C. W. Smelser, Single and low order mode interrogation of a multimode sapphire fibre Bragg grating sensor with tapered fibres, *Meas. Sci. Technol.*, 17, 980–984, 2006.
79. M. Busch, W. Ecke, I. Latka, D. Fischer, R. Willsch, and H. Bartelt, Inscription and characterization of Bragg gratings in single-crystal sapphire optical fibres for high-temperature sensor applications, *Meas. Sci. Technol.*, 20, 115301, 2009.
80. S. Gupta, T. Mizunami, T. Yamao, and T. Shimomura, Fiber Bragg grating cryogenic temperature sensors, *Appl. Opt.*, 35(25), 5202–5205, 1996.
81. T. Mizunami, H. Tatehata, and H. Kawashima, High-sensitivity cryogenic fibre-Bragg-grating temperature sensors using Teflon substrates, *Meas. Sci. Technol.*, 12, 914–917, 2001.
82. C. Lupi, F. Felli, L. Ippoliti, M. A. Caponero, M. Ciotti, V. Nardelli, and A. Paolozzi, Metal coating for enhancing the sensitivity of fibre Bragg grating sensors at cryogenic temperature, *Smart Mater. Struct.*, 14, N71–N77, 2005.
83. A. Asseh, S. Sandgren, H. Ahlfeldt, B. Sahlgren, R. Stubbe, and G. Edwall, Fiber optical Bragg grating refractometer, *Fiber Integr. Opt.*, 17, 62, 1998.
84. A. Iadicicco, A. Cusano, and S. Campopiano, Thinned fiber Bragg gratings as refractive index sensors, *IEEE Sens. J.*, 5(6), 1288–1295, 2005.
85. A. Iadicicco, S. Campopiano, A. Cutolo, M. Giordano, and A. Cusano, Nonuniform thinned fiber Bragg gratings for simultaneous refractive index and temperature measurements, *IEEE Photon. Technol. Lett.*, 17(7), 1495–1497, 2005.
86. D. Grobnic, S. J. Mihailov, H. Ding, and C. W. Smelser, Bragg grating evanescent field sensor made in biconical tapered fiber with femtosecond IR radiation, *IEEE Photon. Technol. Lett.*, 18(1), 160–162, 2006.

87. Y.-P. Miao, B. Liu, and Q.-D. Zhao, Refractive index sensor based on measuring the transmission power of tilted fiber Bragg grating, *Opt. Fiber Technol.*, 15(3), 233–236, 2009.
88. M. G. Xu, L. Reekie, Y. T. Chow, and J. P. Dakin, Optical in fiber grating high pressure sensor, *Electron. Lett.*, 29(4), 398–399, 1993.
89. Y. Zhang, D. Feng, Z. Liu, Z. Guo, X. Dong, K. S. Chiang, and B. C. B. Chu, High-sensitivity pressure sensor using a shielded polymer-coated fiber Bragg grating, *IEEE Photon. Technol. Lett.*, 13(6), 618–619, 2001.
90. H. Fu, J. Fu, and X. Qiao, High sensitivity fiber Bragg grating pressure difference sensor, *Chin. Opt. Lett.*, 2(11), 621–623, 2004.
91. H.-J. Sheng, M.-Y. Fu, T.-C. Chen, W.-F. Liu, and S.-S. Bor, A lateral pressure sensor using a fiber Bragg grating, *IEEE Photon. Technol. Lett.*,16(4), 1146–1148, 2004.
92. R.-S. Shen, J. Zhang, Y. Wang, R. Teng, B.-Y. Wang, Y.-S. Zhang, W.-P. Yan, J. Zheng, and G.-T. Du, Study on high temperature and high pressure measurement by using metal coated FBG, *Microw. Opt. Technol. Lett.*, 50(5), 1138–1140, 2008.
93. C. R. Dennison and P. M. Wild, Enhanced sensitivity of an in-fibre Bragg grating pressure sensor achieved through fibre diameter reduction, *Meas. Sci. Technol.*, 19, 125301, 2008.
94. H.-J. Sheng, W.-F. Liu, K.-R. Lin, S.-S. Bor, and M.-Y. Fu, High-sensitivity temperature-independent differential pressure sensor using fiber Bragg ratings, *Opt. Express*, 16(20), 16013–16018, 2008.
95. Y. R. García, J. M. Corres, and J. Goicoechea, Vibration detection using optical fiber sensors, *J. Sens.*, 2010, 936487, 2010, 12 pp. doi:10.1155/2010/936487.
96. T. K. Gangopadhyay, Prospects for fibre Bragg gratings and Fabry–Perot interferometers in fibre-optic vibration sensing, *Sens. Actuat. A*, 113(1), 20–38, 2004.
97. A. Cusano, A. Cutolo, J. Nasser, M. Giordano, and A. Calabró, Dynamic strain measurements by fibre Bragg grating sensor, *Sens. Actuat. A*, 110(1–3), 276–281, 2004.
98. J. Lim, Q. P. Yang, B. E. Jones, and P. R. Jackson, DP flow sensor using optical fiber Bragg grating, *Sens. Actuat. A*, 92, 102–108, 2001.
99. S. Takashima, H. Asanuma, and H. Niitsuma, A water flowmeter using dual fiber Bragg grating sensors and cross-correlation technique, *Sens. Actuat. A*, 116, 66–74, 2004.
100. Y. Zhao, K. Chen, and J. Yang, Novel target type flowmeter based on a differential fiber Bragg grating sensor, *Measurement*, 38, 230–235, 2005.
101. B. Yun, N. Chen, and Y. Cui, Highly sensitive liquid-level sensor based on etched fiber Bragg grating, *IEEE Photon. Technol. Lett.*, 19, 1747–1749, 2007.
102. T. Guo, Q. Zhao, Q. Dou, H. Zhang, L. Xue, G. Huang, and X. Dong, Temperature insensitive fiber Bragg grating liquid level sensor based on bending cantilever beam, *IEEE Photon. Technol. Lett.*, 17, 2400–2402, 2005.
103. K.-R. Sohn and J. H. Shim, Liquid level monitoring sensor systems using FBG embedded in cantilever, *Sens. Actuat. A*, 152, 248–251, 2009.
104. K. Fukuchi, S. Kojima, Y. Hishida, and S. Ishi, Optical water level sensors using fiber Bragg grating technology, *ISI, Hitachi Cable Rev.*, 21, 23–28, 2002.
105. H.-J. Sheng, W.-F. Liu, S.-S. Bor, and H.-C. Chang, Fiber-liquid-level sensor based on a fiber Bragg grating, *Jpn. J. Appl. Phys.*, 47(4), 2141–2143, 2008.
106. D. Sengupta and P. Kishore, Continuous liquid level monitoring sensor system using fiber Bragg grating, *Opt. Eng.*, 53(1), 017102, 2014.
107. D. Song, J. Zou, Z. Wei, Z. Chen, and H. Cui, Liquid level sensor using a fiber Bragg grating and carbon fiber composite diaphragm, *Opt. Eng.*, 50, 14401–14405, 2011.
108. J. L. Cruz, A. Díez, M. V. Andrés, A. Segura, B. Ortega, and L. Dong, Fibre Bragg gratings tuned and chirped using magnetic fields, *Electron. Lett.*, 33(3), 235–236, 1997.
109. C.-L. Tien, C.-C. Hwang, H.-W. Chen, W. F. Liu, and S.-W. Lin, Magnetic sensor based on side-polished fiber Bragg grating coated with iron film, *IEEE Trans. Magn.*, 42(10), 3285–3287, 2006.
110. C. Ambrosino, P. Capoluongo, S. Campopiano, A. Cutolo, M. Giordano, D. Davino, C. Visone, and A. Cusano, Fiber Bragg grating and magnetic shape memory alloy: Novel high-sensitivity magnetic sensor, *IEEE Sens. J.*, 7(2), 228–229, 2007.
111. D. Davinoa, C. Visonea, C. Ambrosinoa, S. Campopianob, A. Cusanoa, and A. Cutoloa, Compensation of hysteresis in magnetic field sensors employing fiber Bragg grating and magneto-elastic materials, *Sens. Actuat. A*, 147, 127–136, 2008.
112. C. Ambrosino, S. Campopiano, A. Cutolo, and A. Cusano, Sensitivity tuning in Terfenol-D based fiber Bragg grating magnetic sensors, *IEEE Sens. J.*, 8(9), 1519–1520, 2008.

113. Y. Su, Y. Zhu, B. Zhang, J. Li, and Y. Li, Use of the polarization properties of magneto-optic fiber Bragg gratings for magnetic field sensing purposes, *Opt. Fiber Technol.*, 17, 196–200, 2011.
114. S. M. M. Quintero, A. M. B. Braga, H. I. Weber, A. C. Bruno, and J. F. D. F. Araújo, A magnetostrictive composite-fiber Bragg grating sensor, *Sensors*, 10, 8119–8128, 2010.
115. M. Yang and J. Dai, Review on optical fiber sensors with sensitive thin films, *Photon. Sens.*, 2(1), 14–28, 2012.
116. H.-J. Chen, L. Wang, and W. F. Liu, Temperature-insensitive fiber Bragg grating tilt sensor, *Appl. Opt.*, 47(4), 556–560, 2008.
117. B.-O. Guan, H.-Y. Tam, and S.-Y. Liu, Temperature-independent fiber Bragg grating tilt sensor, *IEEE Photon. Technol. Lett.*, 16(1), 224–226, 2004.
118. X. Dong, C. Zhan, K. Hu, and P. Shum, Temperature-insensitive tilt sensor with strain-chirped fiber Bragg gratings, *IEEE Photon. Technol. Lett.*, 17(11), 2394–2396, 2005.
119. H. Bao, X. Dong, L.-Y. Shao, C.-L. Zhao, and S. Jin, Temperature-insensitive 2-D tilt sensor by incorporating fiber Bragg gratings with a hybrid pendulum, *Opt. Commun.*, 283, 5021–5024, 2010.
120. R. Aneesh, M. Maharana, P. Munendhar, H. Y. Tam, and S. K. Khijwania, Simple temperature insensitive fiber Bragg grating based tilt sensor with enhanced tenability, *Appl. Opt.*, 50(25), E172–E176, 2011.
121. P. Kronenberg, P. K. Rastogi, P. Giaccari, and H. G. Limberger, Relative humidity sensor with optical fiber Bragg grating, *Opt. Lett.*, 27, 1385–1387, 2002.
122. T. L. Yeo, T. Sun, K. T. V. Grattan, D. Parry, R. Lade, and B. D. Powell, Polymer-coated fiber Bragg grating for relative humidity sensing, *IEEE Sens. J.*, 5(5), 1082–1089, 2005.
123. T. L. Yeo, T. Suna, K. T. V. Grattan, D. Parry, R. Lade, and B. D. Powell, Characterisation of a polymer-coated fibre Bragg grating sensor for relative humidity sensing, *Sens. Actuat. B*, 110, 148–155, 2005.
124. X. F. Huang, D. R. Sheng, K. F. Cen, and H. Zhou, Low-cost relative humidity sensor based on thermoplastic polyimide-coated fiber Bragg grating, *Sens. Actuat. B*, 127, 518–524, 2007.
125. Y. Miao, B. Liu, H. Zhang, Y. Li, H. Zhou, H. Sun, W. Zhang, and Q. Zhao, Relative humidity sensor based on tilted fiber Bragg grating with polyvinyl alcohol coating, *IEEE Photon. Technol. Lett.*, 21(7), 441–443, 2009.
126. S. F. H. Correia, P. Antunes, E. Pecoraro, P. P. Lima, H. Varum, L. D. Carlos, R. A. S. Ferreira, and P. S. André, Optical fiber relative humidity sensor based on a FBG with a di-ureasil coating, *Sensors*, 12, 8847–8860, 2012.
127. M. A. Butler and D. S. Ginley, Hydrogen sensing with palladium-coated optical fibers, *J. Appl. Phys.*, 64, 3706–3711, 1988.
128. B. Sutapun, M. Tabib-Azar, and A. Kazemi, Pd-coated elastooptic fiber optic Bragg grating sensors or multiplexed hydrogen sensing, *Sens. Actuat. B*, 60, 27–34, 1999.
129. A. Trouillet, E. Marin, and C. Veillas, Fibre gratings for hydrogen sensing, *Meas. Sci. Technol.*, 17, 1124–1128, 2006.
130. C. Caucheteur, M. Debliquy, D. Lahem, and P. Mégret, Catalytic fiber Bragg grating sensor for hydrogen leak detection in air, *IEEE Photon. Technol. Lett.*, 20(2), 96–98, 2008.
131. C.-L. Tien, H.-W. Chen, W.-F. Liu, S.-S. Jyu, S.-W. Lin, and Y.-S. Lin, Hydrogen sensor based on side-polished fiber Bragg gratings coated with thin palladium film, *Thin Solid Films*, 516, 5360–5363, 2008.
132. J. Dai, M. Yang, Y. Chen, K. Cao, H. Liao, and P. Zhang, Side-polished fiber Bragg grating hydrogen sensor with WO_3-Pd composite film as sensing materials, *Opt. Express*, 19(7), 6141–6148, 2011.
133. X. Dong, Y. Liu, Z. Liu, and X. Dong, Simultaneous displacement and temperature measurement with cantilever-based fiber Bragg grating sensor, *Opt. Commun.*, 192, 213–217, 2001.
134. W. Zhang, X. Dong, Q. Zhao, G. Kai, and S. Yuan, FBG-type sensor for simultaneous measurement of force (or displacement) and temperature based on bilateral cantilever beam, *IEEE Photon. Technol. Lett.*, 13(12), 1340–1342, 2001.
135. Y. Zhao, C. Yu, and Y. Liao, Differential FBG sensor for temperature-compensated high-pressure (or displacement) measurement, *Opt. Laser Technol.*, 36, 39–42, 2004.
136. X. Dong, X. Yang, C.-L. Zhao, L. Ding, P. Shum, and N. Q. Ngo, A novel temperature-insensitive fiber Bragg grating sensor for displacement measurement, *Smart Mater. Struct.*, 14, N7–N10, 2005.
137. T. Guo, Q. Zhao, H. Kang, J. Lv, L. Xue, S. Li, B. Dong, H. Gu, G. Huang, and X. Dong, Temperature-independent FBG displacement measurement based on bandwidth modulation and optical power detection, *Proc. SPIE*, 6595, 65953R, 2007.
138. A. S. Guru Prasad and S. Asokan, Fiber Bragg grating sensor package for submicron level displacement measurements, *Exp. Tech.*, 1–6, 2013.
139. M. G. Xu, J. L. Archmbault, L. Reekie, and J. P. Dakin, Thermally-compensated bending gauge using surface mounted fibre gratings, *Int. J. Optoelectron.*, 9, 281–283, 1994.

140. M. G. Xu, J.-L. Archambault, L. Reekie, and J. P. Dakin, Discrimination between strain and temperature effects using dual wavelength fibre grating sensors, *Electron. Lett.*, 30(13), 1085–1087, 1994.
141. H. J. Patrick, G. M. Williams, A. D. Kersey, and J. R. Pedrazini, Hybrid fiber Bragg grating/long period fiber grating sensor for strain/long period fiber grating sensor for strain/temperature discrimination, *IEEE Photon. Technol. Lett.*, 8, 1223–1225, 1996.
142. L. Xiaohong, W. Dexiang, Z. Fujun, and D. Enguang, Simultaneous independent temperature and strain measurement using one fiber Bragg grating based on the etching technique, *Microw. Opt. Technol. Lett.*, 43(6), 478–481, 2004.
143. W. C. Du, X. M. Tao, and H. Y. Tam, Fiber Bragg grating cavity sensor for simultaneous measurement of strain and temperature, *IEEE Photon. Technol. Lett.*, 11(1), 105–107, 1999.
144. B.-O. Guan, H. Y. Tam, X. M. Tao, and X. Y. Dong, Simultaneous strain and temperature measurement using a superstructure fiber Bragg grating, *IEEE Photon. Technol. Lett.*, 12(6), 675–677, 2000.
145. M. G. Xu, L. Dong, L. Reekie, J. A. Tucknott, and J. L. Cruz, Temperature-independent strain sensor using a chirp Bragg grating in a tapered optical fibre, *Electron. Lett.*, 31(10), 823–825, 1995.
146. O. Frazão, M. Melo, P. V. S. Marques, and J. L. Santos, Chirped Bragg grating fabricated in fused fibre taper for strain–temperature discrimination, *Meas. Sci. Technol.*, 16, 984–988, 2005.
147. L. A. Ferreira, F. M. Araújo, J. L. Santos, and F. Farahi, Simultaneous measurement of strain and temperature using interferometrically interrogated fiber Bragg grating sensors, *Opt. Eng.*, 39(8), 2226–2234, 2000.
148. B.-O. Guan, H.-Y. Tam, H. L. W. Chan, C.-L. Choy, and M. S. Demokan, Discrimination between strain and temperature with a single fiber Bragg grating, *Microw. Opt. Technol. Lett.*, 33(3), 200–202, 2002.
149. D.-P. Zhou, L. Wei, W.-K. Liu, and J. W. Y. Lit, Simultaneous measurement of strain and temperature based on a fiber Bragg grating combined with a high-birefringence fiber loop mirror, *Opt. Commun.*, 281, 4640–4643, 2008.
150. Z. Tong, J. Zhao, and X. Yang, Simultaneous measurement of axial strain and temperature using a PM fiber Bragg grating, *Microw. Opt. Technol. Lett.*, 53(4), 867–870, 2011.
151. J. Jung, H. Nam, N. Park, J. H. Lee, and B. Lee, Simultaneous measurement of strain and temperature by use of a single Bragg grating and an erbium-doped fibre amplifier, *Appl. Opt.*, 38(13), 2749–2751, 1999.
152. J. Jung, N. Park, and B. Lee, Simultaneous measurement of strain and temperature by use of a single Bragg grating written in an erbium:ytterbium-doped fibre, *Appl. Opt.*, 39, 1118–1120, 2000.
153. P. M. Cavaleiro, F. M. Araújo, L. A. Ferreira, J. L. Santos, and F. Farahi, Simultaneous measurement of strain and temperature using Bragg gratings written in germanosilicate and boron-codoped germanosilicate fibers, *IEEE Photon. Technol. Lett.*, 11, 1635–1637, 1999.
154. X. Shu, Y. Liu, D. Zhao, B. Gwandu, F. Floreani, L. Zhang, and I. Bennion, Dependence of temperature and strain coefficients on fiber grating type and its application to simultaneous temperature and strain measurement, *Opt. Lett.*, 27, 701–703, 2002.
155. H. F. Lima, P. F. Antunes, J. de Lemos Pinto, and R. N. Nogueira, Simultaneous measurement of strain and temperature with a single fibre Bragg grating written in a tapered optical fibre, *IEEE Sens. J.*, 10(2), 269–273, 2010.
156. E. Chehura, S. W. James, and R. P. Tatam, Temperature and strain discrimination using a single fibre Bragg grating, *Opt. Commun.*, 275(2), 344–347, 2007.
157. O. Frazao, R. Romero, G. Rego, P. Marques, H. M. Salgado, and J. L. Santos, Sampled fiber Bragg grating sensors for simultaneous strain and temperature measurement, *Electron. Lett.*, 38(14), 693–695, 2002.
158. D.-P. Zhou, L. Wei, W.-K. Liu, Y. Liu, and J. W. Y. Lit, Simultaneous measurement for strain and temperature using fiber Bragg gratings and multimode fibers, *Appl. Opt.*, 47(10), 1668–1672, 2008.
159. M. Song, S. B. Lee, S. S. Choi, and B. Lee, Simultaneous measurement of temperature and strain using two fiber Bragg gratings embedded in a glass tube, *Opt. Fibre Technol.*, 3, 194–196, 1997.
160. F. M. Haran, J. K. Rew, and P. D. Foote, A fibre Bragg grating strain gauge rosette with temperature compensation, *SPIE Proc.*, 3330, 220–230, 1998.
161. G. P. Brady, K. Kalli, D. J. Webb, D. A. Jackson, L. Reekie, and J. L. Archambault, Simultaneous measurement of strain and temperature using the first- and second-order diffraction wavelengths of Bragg gratings, *IEE Proc.-Optoelectron.*, 144(3), 156–161, 1997.
162. O. Frazao, M. J. N. Lima, and J. L. Santos, Simultaneous measurement of strain and temperature using type I and type IIA fibre Bragg gratings, *J. Opt. A-Pure Appl. Opt.*, 5, 183–185, 2003.
163. K. Tian, Y. Liu, and Q. Wang, Temperature-independent fiber Bragg grating strain sensor using bimetal cantilever, *Opt. Fiber Technol.*, 11, 370–377, 2005.
164. P. Lu, L. Men, and Q. Chen, Resolving cross sensitivity of fiber Bragg gratings with different polymeric coatings, *Appl. Phys. Lett.*, 92, 17112–17115, 2008.

165. S. K. Mondal, V. Mishra, U. Tiwari, G. C. Poddar, N. Singh, S. C. Jain, S. N. Sarkar, and P. Kapur, Embedded dual fiber Bragg grating sensor for simultaneous measurement of temperature and load (strain) with enhanced sensitivity, *Microw. Opt. Technol. Lett.*, 51(7), 1621–1624, 2009.
166. C. G. Askins, M. A. Putnam, and E. J. Friebele, Instrumentation for interrogating many-element fiber Bragg grating arrays, *Smart Sensing, Processing and Instrumentation, Proceedings of the SPIE*, San Diego, CA, 1995, Vol. 2444, pp. 257–261.
167. S. W. James, M. L. Dockney, and R. P. Tatan, Photorefractive volume holographic demodulation of in-fiber Bragg grating sensors, *IEEE Photon. Technol. Lett.*, 8(5), 664–666, 1996.
168. S. M. Melle, K. Liu, and R. M. Measures, A passive wavelength demodulation system for guided wave Bragg grating sensors, *IEEE Photon. Technol. Lett.*, 4(5), 512–518, 1992.
169. R. W. Fallon, L. Zhang, L. A. Everall, J. A. R. Williams, and I. Bennion, All-fibre optical sensing system: Bragg grating sensor interrogated by a long-period grating, *Meas. Sci. Technol.*, 9, 1969–1973, 1998.
170. M. A. Davis and A. D. Kersey, All fibre Bragg grating strain sensor demodulation technique using a wavelength division coupler, *Electron. Lett.*, 30(1), 75–77, 1994.
171. Q. Wang, G. Farrell, T. Freir, G. Rajan, and P. Wang, Low-cost wavelength measurement based on a macrobending single-mode fiber, *Opt. Lett.*, 31(12), 1785–1787, 2006.
172. J. Zhang, W. Sun, L. Yuan, and G.-D. Peng, Design of a single-multimode-single-mode filter demodulator for fiber Bragg grating sensors assisted by mode observation, *Appl. Opt.*, 48(30), 5642–5646, 2009.
173. Q. Wang, G. Farrell, and W. Yan, Investigation on single-mode-multimode-single-mode fiber structure, *J. Lightwave Technol.*, 26, 512–519, 2008.
174. G. Wild and S. Hinckley, Distributed optical fibre smart sensors for structural health monitoring: A smart transducer interface module, *Intelligent Sensors, Sensor Networks and Information Processing (ISSNIP)*, Melbourne, Victoria, Australia, 2009, pp. 373–378.
175. K. Zhou, A. G. Simpson, X. Chen, L. Zhang, and I. Bennion, Fiber Bragg grating sensor interrogation system using a CCD side detection method with superimposed blazed gratings, *IEEE Photon. Technol. Lett.*, 16(6), 1549–1551, 2004.
176. U. Tiwari, K. Thyagarajan, M. R. Shenoy, and S. C. Jain, EDF-based edge-filter interrogation scheme for FBG sensors, *IEEE Sens. J.*, 13(4), 1315–1319, 2013.
177. A. B. Lobo Ribeiro, L. A. Ferreira, J. L. Santos, and D. A. Jackson, Analysis of the reflective-matched fiber Bragg grating sensing interrogation scheme, *Appl. Opt.*, 36(4), 934–939, 1997.
178. M. G. Xu, H. Geiger, J. L. Acchambault, L. Reekie, and J. P. Dakin, Novel interrogation system for fibre Bragg grating sensor using an acousto optic tunable filter, *Electron. Lett.*, 29(17), 1510–1511, 1993.
179. A. D. Kersey, T. A. Berkoff, and W. W. Morey, Multiplexed fiber Bragg grating strain-sensor system with a fiber Fabry Perot wavelength filter, *Opt. Lett.*, 18(16), 1370–1372, 1993.
180. G. A. Ball, W. W. Morey, and P. K. Cheo, Fiber laser source/analyser for Bragg grating sensor array integration, *J. Lightwave Technol.*, 12(4), 700–703, 1994.
181. L. A. Ferreira, E. V. Diatzikis, J. L. Santos, and F. Farahi, Frequency-modulated multimode laser diode for fiber Bragg grating sensors, *J. Lightwave Technol.*, 16(9), 1620–1630, 1998.
182. A. D. Kersey and W. W. Morey, Multiplexed Bragg grating fibre laser strain sensor system with mode locked interrogation, *Electron. Lett.*, 29(1), 112–114, 1993.
183. A. D. Kersey and W. W. Morey, Multi element Bragg grating based fibre laser strain sensor, *Electron. Lett.*, 29(11), 964–966, 1993.
184. C. J. Misas, F. M. Molita Araújo, L. A. Ferreira, J. L. Santos, and J. M. López-Higuera, Fiber Bragg sensors interrogation based on carrier generation by modulating the coupling length of a wavelength-division multiplexer, *IEEE J. Quant. Electron.*, 6(5), 750–755, 2000.
185. A. D. Kersey, T. A. Berkoff, and W. W. Morey, High resolution fibre grating based strain sensor with interferometric wavelength shift detection, *Electron. Lett.*, 28, 236–238, 1992.

10 Polymer Fiber Bragg Grating Sensors and Their Applications

David J. Webb

CONTENTS

10.1 INTRODUCTION: HISTORICAL DEVELOPMENT

Interest in grating-based sensors fabricated in polymer optical fibers (POFs) stems from the different material properties of polymers compared to silica. Polymer fiber sensors offer increased stress sensitivity and a larger strain range. When used to monitor compliant structures, they perturb the mechanical behavior of the structure less than the much stiffer silica fiber. Some polymers can provide intrinsic sensitivity to water. Being composed of organic materials, there is the possibility of utilizing a wide range of chemical processing techniques to create sensors with a response to specific chemical or biochemical species. Finally, for in vivo medical applications, polymer fibers may be more attractive when considering the consequences of a fiber breakage.

Historically, the origins of polymer fiber Bragg grating (POFBG) technology can be traced back to early work on the photosensitivity of poly(methyl methacrylate) (PMMA) in the 1970s. Workers at Bell Labs in the United States were able to create Bragg gratings in bulk samples of PMMA and used these to create a laser by diffusing a dye into the polymer [1]. Despite this early start—well before the discovery of photosensitivity in silica fibers—more than 25 years were to pass before the first demonstration of a Bragg grating in a (multimode) POF [2]. Initial work with the fiber devices was undertaken by Gang-Ding Peng, Pak L. Chu, and colleagues at the University of New South Wales, Australia. They were able to demonstrate gratings in single-mode fiber and show high reflectivity (28 dB) [3] and a large (70 nm) tuning range [4]. The first attempts to improve the intrinsic photosensitivity of PMMA through the addition of dopants were reported in 2004 by Tam and coworkers at Hong Kong Polytechnic University [5], and a year later, the first gratings in microstructured (photonic crystal) polymer fiber were demonstrated [6].

Although the first application of a POFBG—as a widely tunable laser cavity mirror—was demonstrated in 2006 [7], their use was initially limited due to the need to butt couple the POF to a low-loss silica fiber. Glued connections between POF and single-mode silica fiber down-leads [8] allowed the POFBGs to be used away from the optical bench, facilitating the development of applications. Further details of applications of POFBGs are given in Section 10.5. Demountable connections were not demonstrated until 2012 [9].

Here, we attempt to briefly identify the key developments in our understanding of the behavior of POFBGs and our attempts to mature the technology, acknowledging that this is necessarily a subjective exercise. Unlike with silica fiber, for polymers, the losses in the near infrared tend to be extremely high (measurements in our lab indicate a typical loss of 1 dB/cm for PMMA-based fibers). Consequently, there is motivation to work at shorter wavelengths and there has been a steady reduction in the reported Bragg wavelengths extending operation down into the 900 nm region, the 800 nm region [10], and the visible [11], the latter device requiring a recording resolution (in terms of lines per millimeter) similar to that reported in bulk PMMA [1].

Soon after the demonstration of the first POFBGs, their temperature dependence was studied [12]. However, early experiments did not control the humidity environment of the gratings and this can lead to problems, since PMMA exhibits an affinity for water, which causes a swelling of the fiber and increases its refractive index, both leading to an increase in the Bragg wavelength. The influence of humidity on the Bragg wavelength of POFBGs in PMMA-based fiber was first studied by Harbach [13]. Later studies revealed that thermal annealing could produce a permanent contraction of the fiber leading to a reduction in Bragg wavelength [14], a process that has also been used to record wavelength multiplexed POFBGs using a single-phase mask [8].

The attractiveness of POFBGs for strain sensing has already been mentioned. Nonlinearities in the relation between the Bragg wavelength and the applied strain that are particularly important at high strain levels were studied by Peters and coworkers [15], and the potential for using POFBGs to detect high-frequency strain was explored by Stefani et al. [16]. Being viscoelastic materials, polymers can suffer from hysteresis in their stress–strain curves, particularly at high strains; means of reducing these effects have been explored, which rely on pre-tensioning or annealing the fiber [17,18].

Although most research has dealt with fiber based on PMMA and related methacrylates (though often with the addition of dopants to modify the refractive index or increase the photosensitivity), there are many other potentially useful polymers. Perfluorinated fiber offers the possibility of much lower losses in the near infrared [19], while Ole Bang and Henrik Rasmussen at the technical University of Denmark have pioneered the use of TOPAS—a cyclic olefin copolymer. Compared to PMMA, this material has several advantages, a key one for POFBGs being insensitivity to water [20]. It is notable that photonic crystal waveguides for the THz region can be constructed from TOPAS and indeed Bragg gratings for this region have been produced by cutting notches in the fiber using a laser [21].

10.2 POF VS. SILICA

In this section, we provide a more detailed comparison of the material properties of silica and POF in order to identify the kind of applications where POF might provide some advantages as well as to understand some of the challenges to be faced when using POF. As a default, we will be discussing PMMA, since this forms the basis of most fibers used in POFBG research to date; any other polymers will be explicitly mentioned. Even when we restrict ourselves to *pure* PMMA, we are still in effect dealing with a range of different materials. Depending on the polymerization process, different samples of PMMA can have very different molecular weight distributions. The situation may be further complicated by the presence of initiators, chain transfer agents, and plasticizers added to control the polymerization process and aid manufacturing. This all goes to explain why for parameters relating to PMMA, a range of values may be found in the literature.

10.2.1 Mechanical Properties

Silica is an isotropic elastic material with a Young's modulus of 73 GPa [22]. PMMA is a viscoelastic material. Its Young's modulus is typically around 3.3 GPa [23] but values quoted in the literature for bulk PMMA range from 1.6 to 3.4 GPa [15,24]. Note that unlike bulk PMMA, POF is not in general an isotropic material. The drawing process leads to preferential alignment of the long molecular chains along the fiber axis causing the material to become transverse isotropic. This reveals itself in the form of birefringence if the fiber is observed from the side [25], and it may be expected that the relevant modulus for axial strains will not be the same as for radial strains [16].

The greater elasticity of POF is an advantage in situations where it is necessary to monitor strain in structures that are themselves highly compliant. In such situations, the stiff silica fiber can locally reinforce the structure, restricting its movement and reporting much lower strains than would occur without the silica fiber in the structure [26].

The failure strain of pristine silica fiber at room temperature is typically in the range 5%–10% [27], though great care must be taken not to introduce any scratches onto the surface of the glass as these can considerably reduce the strain limit [28]. Repetitive loading can also decrease the failure strain. For PMMA-based fibers, failure strains of over 100% have been reported [29], though the value seems to vary significantly depending on the polymer processing and fiber drawing conditions. Annealing can improve the behavior of POF; annealing for several days at 95°C can reduce Young' modulus, yield point, and tensile strength and increase the failure strain [30]. For PMMA, the yield strain—the limit of quasielastic behavior—is about 6% [31].

The viscoelastic nature of polymers considerably complicates their behavior. For example, with PMMA, the yield strength and tensile strength both increase with increasing strain rate [30]. Moreover, as illustrated in Figure 10.1, PMMA-based POF displays hysteresis when the applied

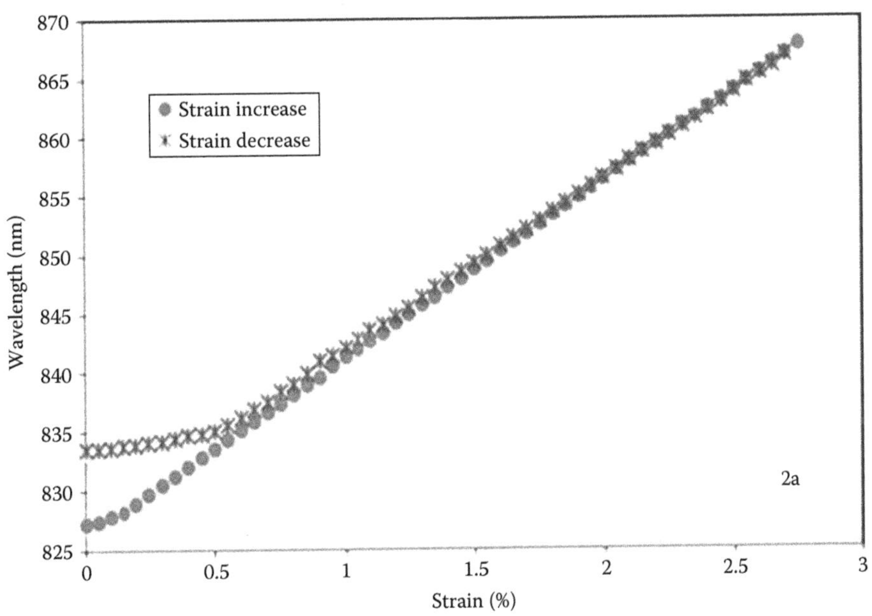

FIGURE 10.1 Hysteresis in the relationship between the Bragg wavelength of a POFBG and the applied strain. (From Abang, A. and Webb, D.J., *Meas. Sci. Technol.*, 25, 015102, 2014.)

strain is cycled. In this particular experiment [18], strain was imposed by stretching a length of fiber containing the POFBG between two posts and the strain was increased every 30 s up to 2.75%. The strain was then reduced at the same rate. Hysteresis is clearly visible at low strain values; in fact, in this region, the fiber was observed to be hanging slack between the two posts and so the applied stress in the fiber was essentially zero. The final Bragg wavelength recovered to close to its initial value over a period of 48 h.

This behavior seems to raise a serious question mark over the use of POFBGs as strain sensors. Fortunately, the situation is not as bad as it seems. First, application of pre-tension or thermal annealing can reduce the hysteresis [17]. Second, in many applications, the POFBG would be glued to, or embedded in, the structure to be monitored rather than stretched between two points on either side of the POFBG. In the former situations, the structure being monitored will effectively force the fiber back to its original length when strain is removed, thus considerably reducing the hysteresis problem [32].

For both silica and POF, even in the elastic region, the relationship between stress and strain is nonlinear, with the nonlinearity for POF being an order of magnitude greater than that of silica [15]. For practical purposes, the nonlinearity may need to be taken into account for strains above about 1% for POF and 3% for silica fiber.

10.2.2 Chemical Properties

POF is an organic material, which can be modified by a wide range of organic chemical techniques. Useful organic species can be incorporated in the POF as copolymers or dopants of else as part of the host polymer in order to enhance nonlinear properties [33], to provide optical amplification [34], to sensitize the fiber to specific chemical species [35], or to improve the photo sensitivity [5].

10.2.3 Optical Properties

A significant difference between POF and silica fiber is the amount of light absorption. Silica fiber has its lowest loss of less than 0.2 dB/km in the 1550 nm spectral region. At longer wavelengths, this increases due to resonances of the molecular bonds, while at lower frequencies, Rayleigh scattering dominates and the loss rises to about 1 dB/km at 800 nm and 5 dB/km at 600 nm [36]. Contrasting this, in the 1550 nm region, the loss of PMMA is extremely high, with values of around 1 dB/cm being typically measured in our laboratory. Losses are much lower in the visible region—though still much higher than for silica fiber. The large-core (usually close to 1 mm) POF used in telecommunications can have a loss of around 0.2 dB/m at 650 nm [37]. The losses of fibers typically used for POFBGs are in practice rather higher at a few dB/m [38]. The difference here may be partly due to the fact that most single or few molded POFs are made by research groups using nonoptimized processes. Additionally, small-core diameters increase the contribution to the loss of scattering due to imperfections at the core–cladding boundary.

The route to improving the losses of PMMA can be understood by noting that most of the loss arises from vibrational resonances of the hydrogen bonds in the material (C–H and O–H). Replacing the hydrogen by a heavier atom will shift the resonances to lower frequencies and hence longer wavelengths, reducing the loss in the visible and near-infrared region. Two hydrogen substitutes are currently used. Deuteration—replacing the hydrogen with the heavier deuterium isotope—can reduce the loss of multimode fiber to less than 30 dB/km in the 600–700 nm range. Alternatively, perfluorination involves replacement with fluorine atoms and this can reduce losses to around 10 dB/km in the 1300 and 1500 nm regions [39].

In the visible and near infrared region, the refractive index of PMMA is approximately 0.04 larger than silica. The refractive index of perfluorinated PMMA is much lower at about 1.33, while for TOPAS cyclic olefin copolymer, the index is about 1.59 in the visible region [36].

10.3 PHOTOSENSITIVITY

Similar to the case with silica fibers, the subject of photosensitivity in polymer fibers is a complex topic, with a variety of possible mechanisms dependent on the inscription wavelength and intensity as well as on the polymerization process used. Unless otherwise noted, in this discussion, we will focus on the photosensitivity of PMMA to UV light at modest power levels (meaning that we can ignore multiphoton processes). Possible mechanisms for photosensitivity include chain scission (the cutting of the polymer *backbone*), photo polymerization, and enhancing or reducing the cross-linking between polymer chains.

The first studies of photosensitivity in bulk PMMA were carried out in the early 1970s by a group at Bell Laboratories in the United States who used 325 nm light from a HeCd laser, which is today the most common source used for POFBG fabrication. Following UV illumination, Tomlinson et al. noted an increase in the refractive index of PMMA, so long as it had been polymerized from methyl methacrylate monomer, which had first been oxidized [40]. Index changes up to 3×10^{-3} were obtained, accompanied by a density change of 0.8%. It is important to note that these index changes actually developed over a period of a few 100 h following illumination [41] and so are not a simple explanation for POFBG fabrication, where gratings appear in tens of minutes *during* UV illumination. This latter paper also noted that the index change was largest in the center of the PMMA slab falling off to zero over 300 μm near the two sides from which the UV beam entered and exited the slab. The group initially speculated that the mechanism might involve photo-cross-linking [40], but after finding no evidence for this eventually concluded that the most likely mechanism was the photopolymerization of unreacted monomers within the PMMA [42], additional evidence for this being provided by electron spin resonance, which indicated that UV illumination increased the concentration of free radicals by over one order of magnitude. The role of residual monomers in the photosensitivity was also confirmed by Marotz [43].

Kopietz et al. studied the evolution of gratings in bulk PMMA during inscription with a mercury UV lamp [44]. They observed an initial decrease in refractive index, which then became positive, reaching as much as 0.01. They explained this behavior as the result of initial monomer creation, followed by later photopolymerization of the monomers.

Later work by Mitsuoka et al. [45] investigated the photodegradation of PMMA in vacuum at a range of wavelengths. They observed main chain scission at wavelengths between 260 and 320 nm, with a peak around 300 nm, and suggested that the photodegradation could produce monomer and radical groups.

The response of PMMA to different wavelengths was also studied by Wochnowski et al. [46], who used mass spectrometry to characterize the ablated products and SPS and FTIR spectroscopy to study the irradiated samples. For 193 nm light, they detected partial scission of the polymer side chains but no significant scission of the polymer main chain. At 248 nm with a fluence of under 15 mJ/cm^2, they observed scission of complete side chains, while at more than 30 mJ/cm^2, side chain fragmentation was detected. At both wavelengths, the refractive index was seen to increase on exposure, possibly due to van der Waals forces pulling adjacent molecules closer together once the side chains were removed; the effect was greater at 248 nm. There was a maximum refractive index as a function of exposure after which the index reduced, possibly due to scission of the main polymer backbone. Experiments were also carried out at 308 nm, but surprisingly, no modification of the polymer was detected.

The first report of a grating in POF by Peng's group at the University of New South Wales in Australia [47] noted that the fiber was produced with a low level of lauryl peroxide initiator and at a temperature below the glass transition temperature, following the method of Tomlinson et al. [40]. No mention was made of the inscription time in that paper; however, later papers by the group discussed the growth behavior of gratings recorded in PMMA POF doped with benzyl methacrylate during exposure to a pulsed 325 nm laser with an average intensity of 60 mW/cm^2 [48]. They observed two regimes. There was an initial linear change in index over a period of about 60 min, leading to an index difference of about 2×10^{-4}, with a negative wavelength shift being observed indicating an index *decrease*. There then followed a rapid increase in index modulation up to a maximum of around 2×10^{-3}, accompanied by losses at wavelengths below the Bragg wavelength. Inspection by microscope revealed apparent damage at the core–cladding boundary.

In a later paper on this topic [49], the authors described a study at a slightly lower intensity level of 45 mW/cm^2. This time, they observed an increase in reflectivity for 28 min after which the reflectivity remained roughly constant up to 48 min. The reflectivity then *decreased*; by 88 min, there was very little reflected signal visible. The UV light was then turned off and the authors observed that the reflectivity then *increased* over about 8 h, after which it remained constant. The authors speculated that the recovery of the grating strength might be due to the relaxation of thermal stresses.

The behavior just described may be indicative of a competitive process with one mechanism leading to an index reduction and another to an increase in index. Saez-Rodriguez et al. [50] have suggested that for pure PMMA, the relevant mechanisms may be photodegradation (main chain scission) for the index reduction and photopolymerization for the index increase. They reported experiments with microstructured POF (mPOF) that showed that inscribing gratings at 325 nm under higher levels of strain led to larger index modulations and linked this to the fact that higher stress enhances photodegradation [51]. By scanning a UV beam over an existing grating, they were able to confirm a net positive wavelength shift due to UV irradiation.

10.4 MEASURAND SENSITIVITY

10.4.1 Temperature

The temperature response of POFBGs can be complicated by the cross-sensitivity to humidity unless care is taken. For example, a common method to determine the temperature sensitivity of a silica fiber sensor is to mount the device on a miniature heating element controlled by a Peltier heat pump. If this is done with POFBGs in the open laboratory environment, our experience is that the obtained sensitivity is much larger than that obtained when the temperature is adjusted with the POFBG in an environmental chamber at constant relative humidity (RH). Perhaps because of this, the literature contains a wide range of values for the temperature sensitivity.

The first reported characterization of a 1579 nm grating in a PMMA-based, step-index fiber resulted in a wavelength shift of 18 nm for a 50°C temperature rise [12]. This corresponds to a temperature sensitivity of −360 pm/°C, though it should be noted that the response was nonlinear, with a greater thermal sensitivity at higher temperatures. The thermal dependence was, however, reversible with no significant hysteresis visible. In a later paper, the authors reported a temperature sensitivity of −149 pm/°C, with this time, a linear response and a working range of 20°C–65°C [52].

It should be noted that the wavelength shift with increasing temperature is negative—the opposite from silica FBGs. This is due to the dominance of the negative thermooptic coefficient of polymers, as will be discussed later.

Experiments on C-band gratings in pure PMMA mPOF revealed an interesting behavior shown in Figure 10.2 [14]. Here, we see three sets of Bragg wavelength data resulting from heating the grating first to 75°C and returning to room temperature, then to 85°C and returning to room temperature, and finally to 95°C. For the first set of data, there is an approximately linear (and it turns out reversible) response up to about 55°C. Thereafter, the wavelength changes much more rapidly with temperature and much of this change is permanent, such that when the device returns to room temperature there is a permanent shift in the Bragg wavelength of about 8 nm. In the second cycle, the linear region is extended, roughly up to the previous maximum temperature, after which there is again a rapid and permanent reduction in Bragg wavelength, such that on returning to room temperature, there is now a reduction in wavelength of almost 20 nm from the initial value. In the final cycle, the linear region has been extended up to 85°C, but thereafter, there is once again a rapid decrease.

The permanent wavelength shifts observed were identified as being due to fiber shrinkage. This comes about because when the temperature is raised there is a relaxation of the molecular

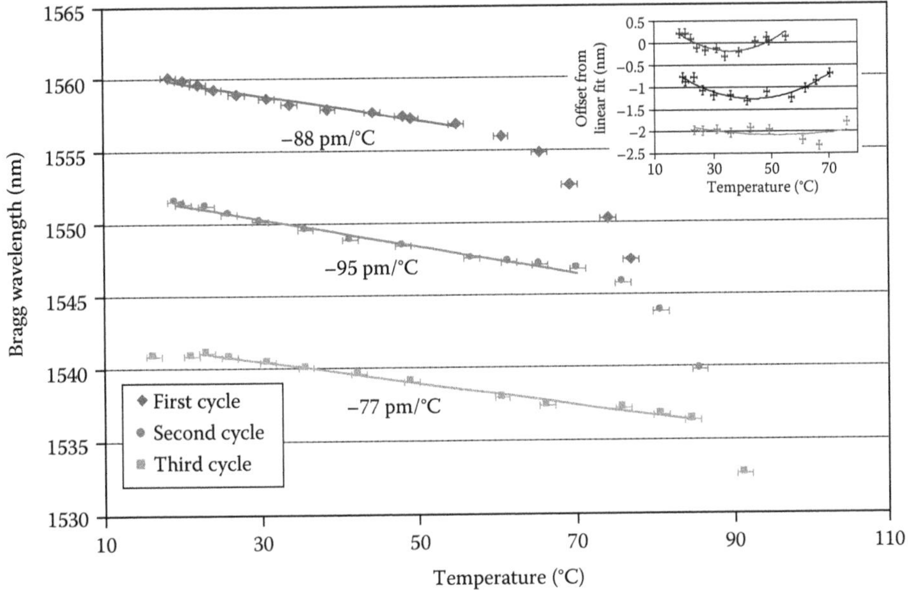

FIGURE 10.2 Hysteresis in the thermal response of a Bragg grating in PMMA mPOF, resulting from permanent fiber shrinkage. (From Carroll, K.E. et al., *Opt. Express*, 15, 8844, 2007.)

alignment that occurs when the fiber is drawn under tension. The hysteresis that arises due to shrinkage can be removed by annealing the fiber before or after the grating is recorded, as was done in this experiment. Note that in the linear regions of the three cycles, the thermal sensitivities vary significantly and are also rather smaller in absolute value from those reported earlier in the literature.

All the thermal characterization experiments described so far were carried out in the open laboratory without control of the humidity. When humidity is held constant, the thermal sensitivity of POFBGs tends to be rather less than the figures reported earlier. The first tests of gratings in the 1550 nm region in step-index PMMA-based fibers revealed that the temperature sensitivity varied from -10 ± 0.5 pm/°C under dry conditions to -36 ± 2 pm/°C with the POFBG submerged in water [13]. The precise value of the thermal sensitivity appears to be fiber dependent. For example, Zhang et al. described a POFBG in the 1550 nm region in a PMMA-based step-index fiber, which had a sensitivity at a constant 50% RH of 55 pm/°C [53]. More intriguingly, Zhang and Tao described a study of a *trans*-4-stilbenemethanol-doped PMMA-based fiber [54]. A grating in the 1550 nm region displayed temperature sensitivities of less than 10 pm/°C at constant humidity, with the sensitivities being positive at 20% RH, negative at 60% RH, and close to 0 at 40% RH.

While there is evidence that the temperature response at constant humidity can be nonlinear [55], a linear response can often be obtained over a temperature range that is big enough to be useful. In this case—and in the absence of cross-sensitivity issues—the Bragg wavelength shift is given by [56]

$$\Delta\lambda_B = \lambda_B(\alpha + \xi)\Delta T \tag{10.1}$$

where

$\Delta\lambda_B$ is the Bragg wavelength shift
ΔT is the temperature change
λ_B is the Bragg wavelength
α is the thermal expansion coefficient
ξ is the thermooptic coefficient

For silica fiber, both α and ξ are positive leading to an increase in wavelength with temperature. PMMA—along with many other polymers—has a negative thermooptic coefficient and so the sign of the wavelength shift depends on the relative magnitudes of α and ξ. Taking typical values for bulk PMMA from the literature [57], we have $\alpha = 7 \times 10^{-5}/°C$ and $\xi = -8.8 \times 10^{-5}/°C$, which lead to a predicted sensitivity of −28 pm/°C for a grating with an initial Bragg wavelength of 1550 nm. To compare this with the measured sensitivity, it is important to realize that α and ξ refer to bulk (isotropic) PMMA, whereas as we have already seen, the fiber material is anisotropic. The anisotropic nature of the fiber means that the thermal expansion along the fiber axis cannot be obtained by simply dividing the bulk volume thermal expansion coefficient by a factor of three [58]. The fiber drawing process will result in an axial thermal expansion less than predicted by this ratio; indeed, under extreme conditions, it is possible that the thermal expansion could even be negative, though no evidence for this occurring with optical fibers has yet been reported.

Much lower temperature sensitivities have been reported for PMMA fiber doped with *trans*-4-stilbenemethanol [54]. Sensitivities of less than 10 pm/°C were obtained at constant humidity, with a negative sensitivity for humidities above 40%, a slight positive sensitivity at 20%, and approximately zero temperature sensitivity at 40% RH.

Gratings have also been fabricated in mPOF fabricated in TOPAS, a cyclic olefin copolymer [59]. This material is insensitive to humidity [20] and displays a temperature sensitivity of −37 pm/°C for a grating at 1568 nm. Furthermore, some variants of this polymer have relatively high glass transition temperatures allowing gratings to be used up to 110°C [60].

10.4.2 Strain

As has already been discussed, polymers are viscoelastic materials. However, in the case of PMMA and for strains applied over relatively short time periods, reversible elastic behavior has been seen for strains up to 13 mε, and in fact, a linear strain tunability of 48.4 mε has been demonstrated [4]. With a total tuning range of 73 nm, this last figure corresponds to a strain sensitivity of 1.5 pm/με, for a grating in PMMA-based fiber with an initial Bragg wavelength of 1560 nm.

Within the literature, there is much more agreement on the measured strain sensitivities of POFBGs than there is for their temperature sensitivities. For gratings in the 1550 nm region, values as low as 1.15 pm/με have been reported [61] (see Figure 10.3), though most measurements provide results in the range of 1.3–1.4 pm/με (e.g., see [17]).

Stefani et al. [16] characterized the dynamic behavior of different polymer fibers (TOPAS, pure PMMA mPOF, and step-index fiber with a core doped with polystyrene) at frequencies up

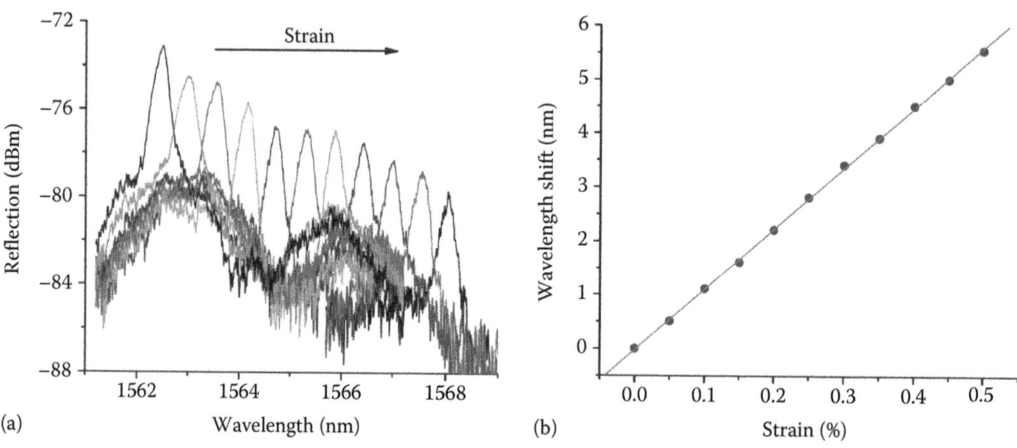

FIGURE 10.3 Strain tuning of a POFBG. (a) Spectra for increasing strain. (b) Wavelength shift as a function of applied strain. (From Chen, C. et al., *Meas. Sci. Technol.*, 21, 094005, 2010.)

to 100 Hz. They found that viscoelastic relaxation introduced nonlinearities that increased with increasing strain and frequency. However, performance sufficient for accelerometer applications could be obtained for strains up to 0.28%.

When Bragg gratings are used to detect high-frequency strains, POF can offer some advantages over silica fiber. Medical applications of ultrasound require the monitoring of MHz frequency strains induced in the fiber by acoustic waves propagating in an aqueous environment. Here, the much smaller Young's modulus of POF is helpful, as is its acoustic impedance, which is closer matched to that of water than is silica (impedance = 13.1 for silica, 1.48 for water, 3.26 for PMMA, all in units of $kg/m^2 \cdot s \times 10^6$). Gallego and Lamela [62] compared the phase sensitivities of POF and silica fiber to 1 MHz ultrasound in water finding the POF to be about 15 times greater.

10.4.3 Water

In contrast to silica FBGs, POFBGs recorded in PMMA-based fiber are water sensitive. PMMA has an affinity for water (absorbing 2% at 23°C [23]) that causes a swelling of the fiber and an increase in its refractive index, both of which contribute to an increase in the Bragg wavelength of any grating recorded in the fiber. This first seems to have been demonstrated by Harbach [13] in experiments investigating the humidity sensitivity of POFBGs. The relationship appears to be fairly linear in the range 40%–90% RH (see Figure 10.4b), though this breaks down at lower humidities. The sensitivity displayed by this device is 38 pm/% RH. The response time (defined here as the time taken to change the Bragg wavelength by 90% of its ultimate change following a step change in humidity) of course depends on the fiber diameter, but even after that is taken into account, there appears to be a significant variation in response time, probably due to the varying molecular weight distributions and degree of cross-linking encountered. The response times tend to be typically a few tens of minutes, that in Figure 10.4b being 30 min.

Some applications would benefit from response times much faster than 30 min. To address this issue, Zhang et al. [63] investigated the use of acetone to etch down the fiber diameter. By reducing the diameter from 190 to 135 μm, the response time of one POFBG was improved from 31 to 12 min. Nominally similar POF samples from different performs displayed different response times, despite being of similar diameters. Zhang et al. noticed a correlation between the response time and the rate at which the diameter was reduced during etching. They speculated that a fast response time and high susceptibility to etching were both symptomatic of a more open molecular structure. Furthermore, they noted that the diffusion constant needed to model the response varied as the fiber was etched, indicating that the molecular structure in the fiber was not homogeneous.

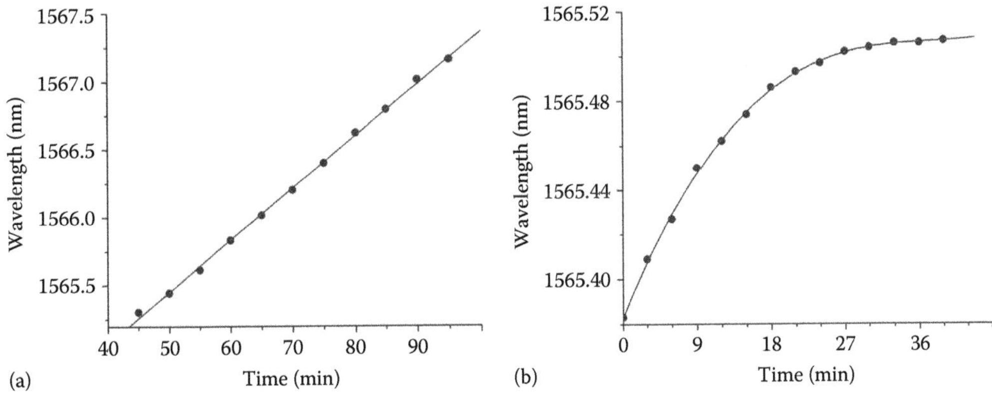

FIGURE 10.4 Humidity response of a PMMA fiber–based POFBG. (a) Dependence of Bragg wavelength on RH at 22°C. (b) Time response for a step change in humidity.

Continuing with this theme, Rajan et al. [64] proposed etching using a 1:1 mixture of acetone and methanol to avoid damaging the surface of the fiber. They reduced the diameter down to 25 μm and claimed a response time of less than 5 s.

Humidity sensitivity is just one consequence of the water affinity of PMMA. In fact, POFBGs in this material are generally sensitive to the concentration of water-based solutions surrounding the fiber; examples range from sugar solution to water in aviation fuel. Zhang and Webb have suggested [65] that the key factor here is not the content of water in the surrounding solution (e.g., expressed as percentage by volume), but the *activity* of the water [66]. Water activity is the fraction of the maximum (saturated) water content of the solution. So for a sugar solution, a water activity of one corresponds to pure water, whereas for aviation fuel, a water activity of one corresponds at room temperature to a water concentration of less than 100 ppm. Despite the difference in total water content, the two solutions will provide similar Bragg wavelength shifts to a POFBG inserted in the liquid.

Of course, where POFBGs are intended to be used as strain or temperature sensors, then a dependency of the Bragg wavelength on humidity is a considerable disadvantage. Fortunately, not all polymers that are suitable for drawing into optical fibers display the same affinity for water as PMMA. Experiments have shown that gratings recorded in TOPAS fiber display a humidity sensitivity at least 50 times less than PMMA [20], where the determination of that factor was actually limited by the stability of the environmental chamber used.

10.4.4 Cross-Sensitivity

A general issue with sensors is cross-sensitivity to undesired measurands. For Bragg gratings in silica fiber, the temperature–strain cross-sensitivity issue is well known and a host of cunning solutions have been proposed, which usually rely on having two gratings that somehow respond differently to temperature and strain. The simplest version of this approach, which is often employed, is to colocate two gratings but shield one of them from the effects of strain.

For POFBGs recorded in PMMA, the intrinsic sensitivity to water also needs to be taken into account. Moreover, the situation is further complicated by the fact that the dependence on temperature and humidity is not independent [65]. The Bragg wavelength shift induced by a change in humidity ΔH and a change in temperature ΔT is given by

$$\Delta\lambda_B = \lambda_B(\eta+\beta)\Delta H + \lambda_B(\xi+\alpha)\Delta T \quad (10.2)$$

where

λ_B is the initial Bragg wavelength
η is the normalized dependence of refractive index on humidity
β is the swelling coefficient related to humidity-induced volumetric change (% RH)$^{-1}$
ξ is the thermooptic coefficient
α is the thermal expansion coefficient

Neither η nor β is constant [67,68]; η is approximately linear but decreases with increasing temperature, while β is independent of temperature, but is only linear in the region 40%–100% RH and decreases for humidities below 40%.

Because of the humidity-dependent expansion of PMMA, the humidity response of the Bragg wavelength also shows a dependency on the fiber strain. However, this dependency can be removed if the fiber is initially strained by an amount greater than any water-induced expansion. For example, Figure 10.5 shows the humidity response of a grating subjected to initial strains of 0, 1500, and 3000 με. Note that the removal of the strain dependency is at the price of a reduction in sensitivity.

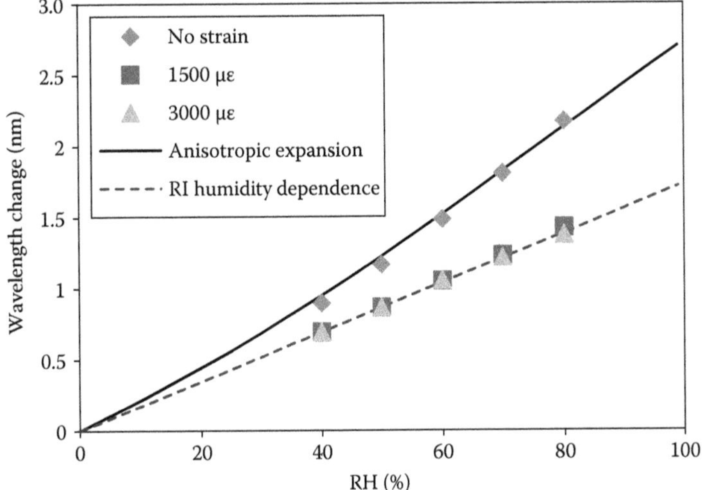

FIGURE 10.5 Humidity response of a PMMA fiber–based POFBG subjected to initial strains of 0, 1500, and 3000 με. (From Zhang, W. and Webb, D.J., *Opt. Lett.*, 39, 10, 2014.)

10.4.5 Other Grating Structures

A number of researchers have also investigated long-period grating (LPG) structures in POF. Because of their large period, typically in the 100 s of μm, LPGs do not require high spatial resolution inscription techniques and are amenable to several simple fabrication processes. LPGs can be produced mechanically compressing the fiber between ridged plates, with the grating structures being made permanent by using heated plates to induce a permanent spatial modulation of the fiber structure [69].

LPGs can also be produced by photoinscription using an amplitude mask and cheap UV lamp [70]. LPGs have also been produced by Saez and Webb in PMMA-based step-index fiber using a point-by-point writing process [71]. These gratings were used to study the water absorption process of PMMA, since the LPG is sensitive to changes in both the core and cladding. Figure 10.6 shows the response of one attenuation band in such an LPG to the fiber being immersed in water. The initial red shift occurs

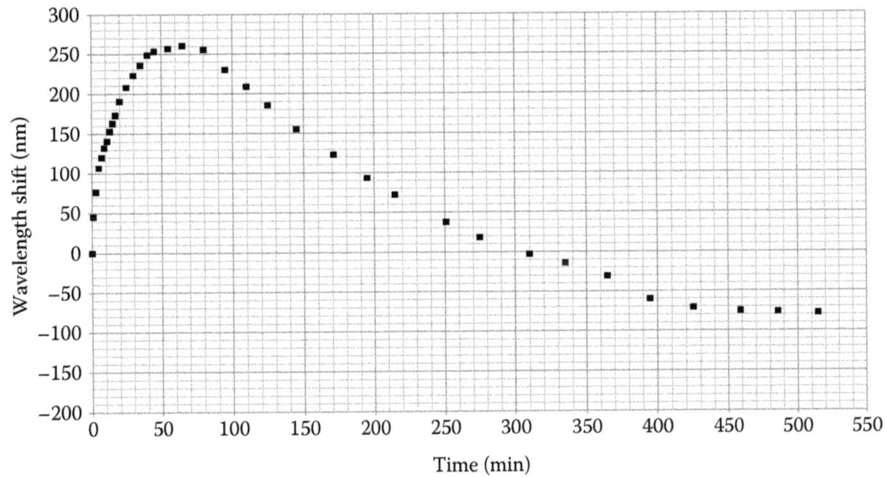

FIGURE 10.6 Wavelength shift in an attenuation band of an LPG written in POF on submersion in water.

as the water diffuses into the cladding, raising its index as well as starting to increase the fiber length. However, once the water enters the core, there is a reduction in the resonant wavelength, leading ultimately to a net blue shift. The process was reversible when the grating was withdrawn from the water.

Fabry–Pérot-interferometer-type structures have also been recorded in POF [72]. However, because of the large losses in the 1550 nm wavelength range used, the finesse of the observed reflection spectrum was as quite low and more reminiscent of fringes were produced by a two-beam interferometer.

10.5 APPLICATIONS

While there were at the time of writing no commercial systems incorporating POFBGs, over the last decade, they have been shown to have potential in a range of applications that seek to take advantage of the rather different properties of polymers compared to silica.

10.5.1 Compliant Structures

The much lower Young's modulus of POF is important when it is necessary to monitor structures that are themselves compliant. This was first brought to light in a project in which silica FBGs were being used to monitor strain in tapestries. It was noticed that the recovered strain was much less than that anticipated and so an experiment was set up to compare the POFBGs and silica FBGs attached to a woven piece of textile using two kinds of glue: Araldite, a relatively stiff epoxy resin, and DMC2, which is a compliant adhesive used in heritage conservation [26].

Digital image correlation was used to provide a visual display of the strain in the textile when it was loaded with 20 N (see Figure 10.7a). It may be seen that the highest strain is seen in the region of the POFBG glued with DMC2, while all other gratings produce a significantly lower strain indicating that the fiber (or in the case of the other POFBG, the glue) is locally reinforcing the material. This is brought out in Figure 10.7b, where the stress–strain curves of the four gratings are plotted. POF-FBG-2 can be seen to display by far the highest strain level for a given stress.

The suitability of POFBGs to monitor compliant structures was brought out further in PHOSFOS, a Seventh Framework project of the European Commission seeking to develop passive sensing skins [73]. One aspect of the project involved the instrumentation of elastic membranes and a comparison was made of polymer and silica fibers embedded in polydimethyl siloxane (PDMS), which has a Young's modulus of around 1 MPa at room temperature [74]. Figure 10.8a shows a 10 cm square skin containing a POFBG sensor being tested for pressure sensitivity, by moving a cylindrical weight across the surface. Figure 10.8b shows the Bragg wavelength shift when the experiment was carried out with both a polymer and a silica grating sensor. It may be seen that the response of the POFBG is approximately 10 times that of the silica FBG, despite their intrinsic strain sensitivities being similar.

The different behavior of POFBGs and silica FBGs in such a situation is even more striking when the skin is strained by clamping and pulling apart the two edges perpendicular to the fiber axis. The results presented in Figure 10.9 clearly show that while the POFBG displays good linearity and minimal hysteresis (which is possibly caused by the skin itself rather than the polymer fiber), the response of the silica FBG is compromised, displaying a high degree of hysteresis probably indicative of the grating pulling away from the surrounding PDMS. It should be noted too that the wavelength shifts experienced by the silica grating are a couple of orders of magnitude smaller than for the POFBG.

10.5.2 Dynamic Sensing

The low Young's modulus of POF compared to silica fiber is potentially attractive in dynamic applications, for example, accelerometry or acoustic sensing. A concern here though is the consequences of the viscoelastic nature of POF. This issue was investigated by Stefani et al. [16],

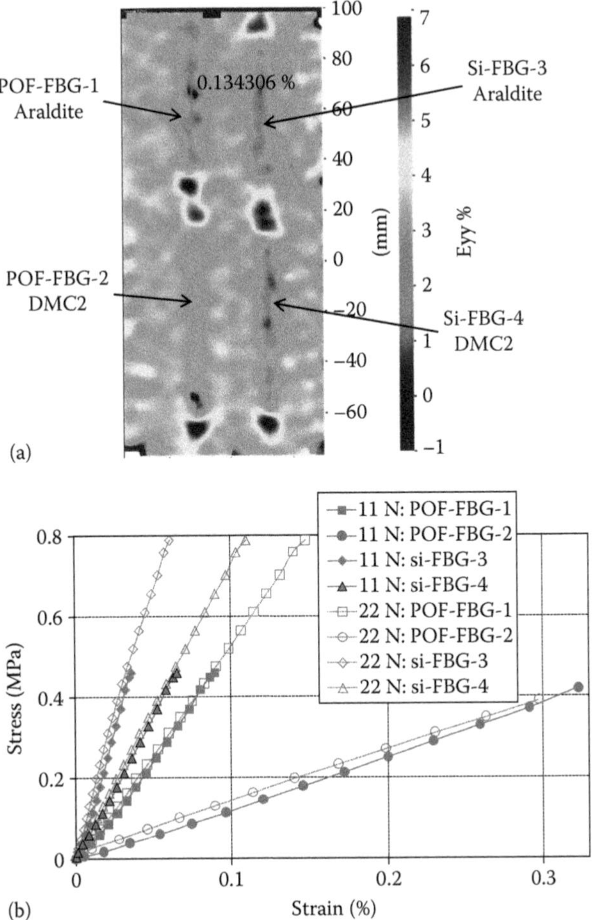

FIGURE 10.7 (a) Strain field revealed by digital image correlation when a textile sample instrumented with FBGs was loaded by 20 N. (b) Stress–strain curves for the four gratings indicated in (a), with the sample loaded to 11 and 22 N. (From Ye, C.C. et al., Applications of polymer optical fibre grating sensors to condition monitoring of textiles, Presented at *the 20th International Conference on Optical Fibre Sensors*, Edinburgh, U.K., 2009.)

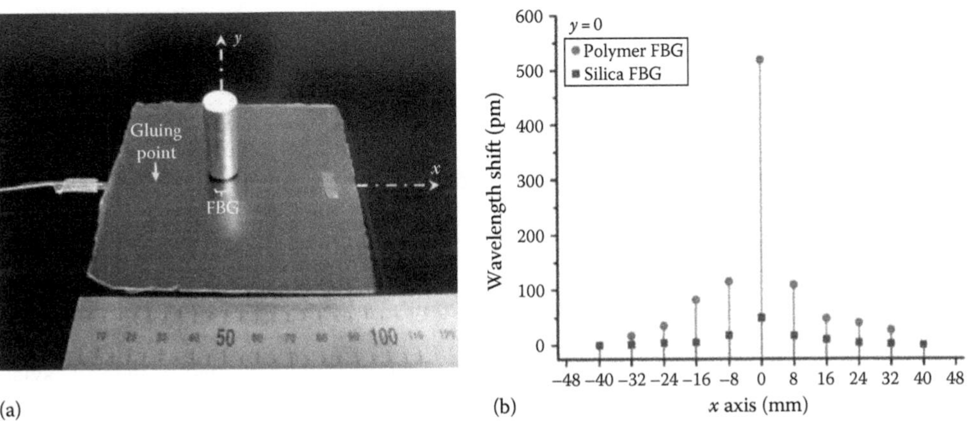

FIGURE 10.8 (a) Polydimethylsiloxane skin instrumented with a POFBG being tested for pressure sensitivity. (b) Comparison of pressure sensitivity of POFBG and silica FBG in the PDMS skin.

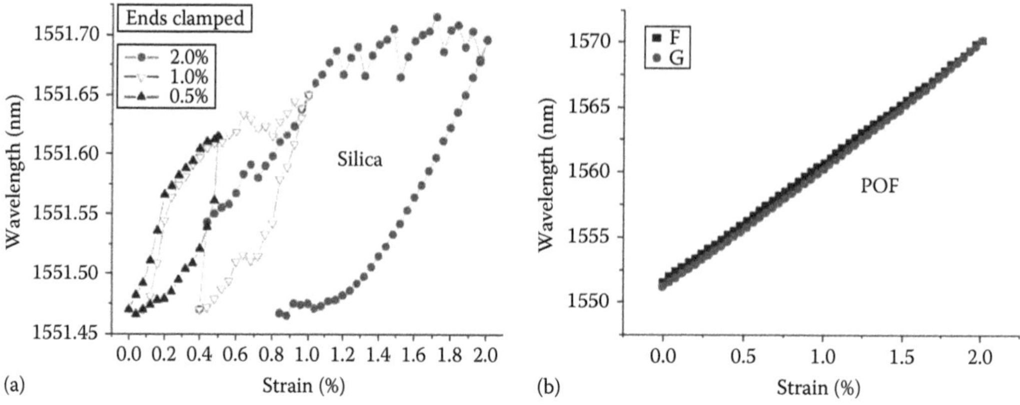

FIGURE 10.9 Wavelength responses of silica (a) and PMMA-based (b) FBGs embedded in a polydimethylsiloxane sheet subject to strain.

who studied fiber samples composed of PMMA and TOPAS and concluded that at low strains of 0.28%, Young's modulus was constant up to about 100 Hz (limited by the measurement system). Under sinusoidal excitation at 1 and 10 Hz, very little effect of viscoelasticity was visible (the phase shift between the force and elongation curves was only a few degrees). For step changes in strain of 2.8% though, significant stress relaxation was observed with a time constant of around 5 s linked to a viscosity of 20 GPa·s. The authors took advantage of these properties to demonstrate a high-sensitivity POF-based accelerometer [75]. The device could respond to accelerations up to 15 g and had a flat frequency response to over 1 kHz with a sensitivity a factor of four higher than an equivalent silica fiber.

The potential of POFBGs for acoustic sensing has not yet been properly explored. However, an indication of the potential has been provided by Marques et al. [76] who investigated control of a FBG's profile by launching acoustic waves with frequencies of hundreds of kHz from a piezoelectric transducer onto a fiber containing the grating via an acoustic horn. They found larger strain levels in mPOF compared to equivalent silica fiber and were able to modify the grating reflection profile for applications in tunable filtering.

Further indication of the potential of POFBGs for acoustic sensing has been provided in work on interferometric ultrasound detection by Gallego and Lamela [62]. They compared fiber interferometers for 1–5 MHz ultrasound detection, finding POF-based devices to be over an order of magnitude more sensitive.

10.5.3 Water Sensing

The affinity of PMMA for water mentioned earlier has a number of potential applications. Significant Bragg wavelength shifts can be obtained for different concentrations of water-soluble substances. For example, Figure 10.10 shows a wavelength shift of almost 1 nm occurring as a result of translation from pure water to a saline concentration of 22% [77].

A high degree of water activity is obtained for very low levels of dissolved water in aviation fuel, in the 10–100 ppm range. Consequently, these modest amounts of water nevertheless produce a large wavelength shift in a PMMA-based POFBG in equilibrium with the fuel. Figure 10.11 shows data from two experiments where a PMMA-based POFBG was used to monitor aviation fuel [78]. In the first experiment (squares), three fuel samples were prepared with known water contents measured using a coulometer. In the second experiment (diamonds), the water content of the fuel was inferred from it being at equilibrium with different humidity levels produced in an environmental chamber.

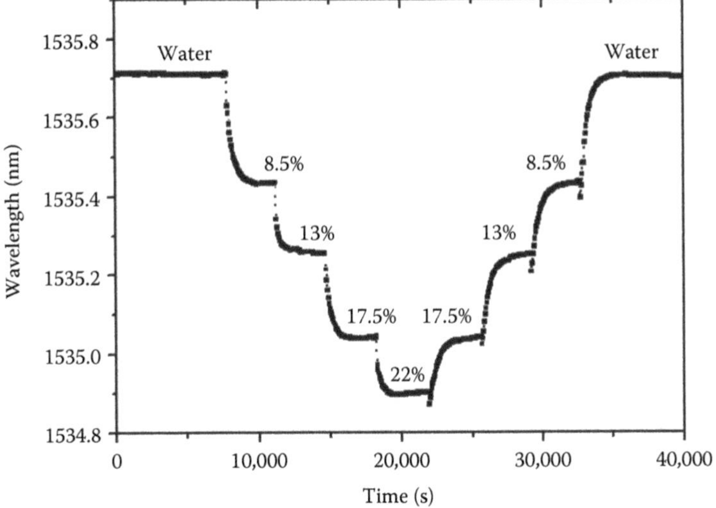

FIGURE 10.10 Shift in Bragg wavelength due to immersion of a PMMA-based POFBG into varying concentrations of saline solution. (From Zhang, W. et al., *Opt. Lett.*, 37, 1370, 2012.)

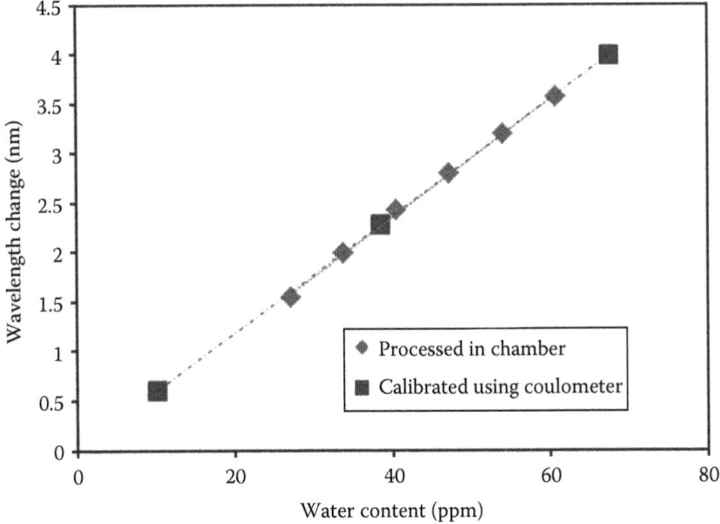

FIGURE 10.11 Shift in Bragg wavelength due to immersion of a PMMA-based POFBG into Jet A-1 fuel containing dissolved water. (From Zhang, W. and Webb, D., Polymer optical fiber grating as water activity sensor, *Proc. SPIE 9128, Micro-structured and Specialty Optical Fibres III*, 91280F, 2 May 2014.)

10.6 CONCLUSIONS

The research presented in this chapter indicates that POFBGs have potential for use in a range of niche applications where the material properties of the constituent polymer(s) provide advantages over silica. Nevertheless, at the time of writing, there remain obstacles to the commercialization of the devices:

- There is only one commercial supplier of single-mode POF, Paradigm Optics [79], with a supplier of mPOF, Kiriama Pty of Sydney Australia, recently ceasing trading.
- There is a lack of single-mode POF components (e.g., couplers and pigtailed devices).

- Sensor properties reported in the literature (and described in this chapter) vary significantly. It seems that full control of the sensor production process (from polymerization to inscription) will be needed to produce repeatable devices.
- The mechanisms of photosensitivity are not fully known, preventing optimization.

The number of research groups tackling these issues has been growing steadily, so there is hope that in the short to medium term the technology may reach a level of maturity compatible with commercial exploitation.

REFERENCES

1. I. P. Kaminow, H. P. Weber, and E. A. Chandross, Poly(methyl methacrylate) dye laser with internal diffraction grating resonator, *Applied Physics Letters*, 18, 497, 1971.
2. G. D. Peng, Z. Xiong, and P. L. Chu, Photosensitivity and gratings in dye-doped polymer optical fibers, *Optical Fiber Technology*, 5, 242–251, 1999.
3. H. Y. Liu, G. D. Peng, and P. L. Chu, Polymer fiber Bragg gratings with 28-dB transmission rejection, *Photonics Technology Letters*, 14, 935–937, 2002.
4. G. D. Peng and P. L. Chu, Polymer optical fiber photosensitivities and highly tunable fiber gratings, *Fiber and Integrated Optics*, 19, 277–293, 2000.
5. J. Yu, X. Tao, and H. Tam, Trans-4-stilbenemethanol-doped photosensitive polymer fibers and gratings, *Optics Letters*, 29, 156–158, 2004.
6. H. Dobb, D. J. Webb, K. Kalli, A. Argyros, M. C. J. Large, and M. A. van Eijkelenborg, Continuous wave ultraviolet light-induced fiber Bragg gratings in few- and single-mode microstructured polymer optical fibers, *Optics Letters*, 30, 3296–3298, December 15, 2005.
7. H. Y. Liu, H. B. Liu, G. D. Peng, and P. L. Chu, Polymer optical fibre Bragg gratings based fibre laser, *Optics Communications*, 266, 132–135, 2006.
8. I. P. Johnson, D. J. Webb, K. Kalli, M. C. Large, and A. Argyros, Multiplexed FBG sensor recorded in multimode microstructured polymer optical fibre, Presented at *the Photonic Crystal Fibres, Brussels—Photonics Europe*, Brussels, Belgium, 2010.
9. A. Abang and D. J. Webb, Demountable connection for polymer optical fiber grating sensors, *Optical Engineering*, 51, 080503, August 1, 2012.
10. I. P. Johnson, K. Kalli, and D. J. Webb, 827 nm Bragg grating sensor in multimode microstructured polymer optical fibre, *Electronics Letters*, 46, 1217–1218, August 19, 2010.
11. C. A. F. Marques, L. B. Bilro, N. J. Alberto, D. J. Webb, and R. N. Nogueira, Narrow bandwidth Bragg gratings imprinted in polymer optical fibers for different spectral windows, *Optics Communications*, 307, 57–61, October 15, 2013.
12. H. Y. Liu, G. D. Peng, and P. L. Chu, Thermal tuning of polymer optical fiber Bragg gratings, *IEEE Photonics Technology Letters*, 13, 824–826, 2001.
13. N. G. Harbach, Fiber Bragg gratings in polymer optical fibers, PhD, EPFL, Lausanne, Switzerland, 2008.
14. K. E. Carroll, C. Zhang, D. J. Webb, K. Kalli, A. Argyros, and M. C. J. Large, Thermal response of Bragg gratings in PMMA microstructured optical fibers, *Optics Express*, 15, 8844–8850, July 9, 2007.
15. S. Kiesel, K. Peters, T. Hassan, and M. Kowalsky, Behaviour of intrinsic polymer optical fibre sensor for large-strain applications, *Measurement Science and Technology*, 18, 3144–3154, 2007.
16. A. Stefani, S. Andresen, W. Yuan, and O. Bang, Dynamic characterization of polymer optical fibers, *IEEE Sensors Journal*, 12, 3047–3053, October 2012.
17. S. Yuan, A. Stefani, M. Bache, T. Jacobsen, B. Rose, N. Herholdt-Rasmussen et al., Improved thermal and strain performance of annealed polymer optical fiber Bragg gratings, *Optics Communications*, 284, 176–182, 2011.
18. A. Abang and D. J. Webb, Effects of annealing, pre-tension and mounting on the hysteresis of polymer strain sensors, *Measurement Science and Technology*, 25, 015102, 2014.
19. H. Y. Liu, G. D. Peng, P. L. Chu, Y. Koike, and Y. Watanabe, Photosensitivity in low-loss perfluoropolymer (CYTOP) fibre material, *Electronics Letters*, 37, 347–348, 2001.
20. W. Yuan, L. Khan, D. J. Webb, K. Kalli, H. K. Rasmussen, A. Stefani et al., Humidity insensitive TOPAS polymer fiber Bragg grating sensor, *Optics Express*, 19, 19731–19739, 2011.
21. S. F. Zhou, L. Reekie, H. P. Chan, Y. T. Chow, P. S. Chung, and K. M. Luk, Characterization and modeling of Bragg gratings written in polymer fiber for use as filters in the THz region, *Optics Express*, 20, 9564–9571, April 2012.

22. G. W. Kaye and T. H. Laby, *Tables of Physical and Chemical Constants*, 16th edn. London, U.K.: Longmann, 1995.
23. J. Brandrup, *Polymer Handbook*, Vols. 1 and 2. New York: Wiley, 1999.
24. D. J. Welker, J. Tostenrude, D. W. Garvey, B. K. Canfield, and M. G. Kuzyk, Fabrication and characterization of single-mode electro-optic polymer optical fiber, *Optics Letters*, 23, 1826–1828, 1998.
25. M. K. Szczurowski, T. Martynkien, G. Statkiewicz-Barabach, W. Urbanczyk, L. Khan, and D. J. Webb, Measurements of stress-optic coefficient in polymer optical fibers, *Optics Letters*, 35, 2013–2015, June 15, 2010.
26. C. C. Ye, J. M. Dulieu-Barton, D. J. Webb, C. Zhang, G. D. Peng, A. R. Chambers et al., Applications of polymer optical fibre grating sensors to condition monitoring of textiles, Presented at *the 20th International Conference on Optical Fibre Sensors*, Edinburgh, U.K., 2009.
27. C. R. Kurkjian, J. T. Krause, and M. J. Matthewson, Strength and fatigue of silica optical fibers, *Journal of Lightwave Technology*, 7, 1360–1370, 1989.
28. Ph. M. Nellen, P. Mauron, A. Frank, U. Sennhauser, K. Bohnert, P. Pequignot et al., Reliability of fiber Bragg grating based sensors for downhole applications, *Sensors and Actuators A: Physical*, 103, 364–376, 2003.
29. M. Aressy, Manufacturing optimisation and mechanical properties of polymer optical fibre, MPhil, Birmingham University, Birmingham, U.K., 2006.
30. C. Jiang, M. Kuzyk, J.-L. Ding, W. Johns, and D. Welker, Fabrication and mechanical behavior of dye-doped polymer optical fiber, *Journal of Applied Physics*, 92, 4, December 2002.
31. D. Yang, J. Yu, X. Tao, and H. Tam, Structural and mechanical properties of polymeric optical fiber, *Materials Science and Engineering*, *A*, 364, 256–259, 2004.
32. A. Abang and D. J. Webb, Influence of mounting on the hysteresis of polymer fiber Bragg grating strain sensors, *Optics Letters*, 38, 1376–1378, May 1, 2013.
33. M. G. Kuzyk, U. C. Paek, and C. W. Dirk, Guest-host polymer fibers for nonlinear optics, *Applied Physics Letters*, 59, 902, 1991.
34. A. Tagaya, Y. Koike, T. Kinoshita, E. Nihei, T. Yamamoto, and K. Sasaki, Polymer optical fiber amplifier, *Applied Physics Letters*, 63, 883–884, 1993.
35. G. Emiliyanov, J. B. Jensen, O. Bang, P. E. Hoiby, L. H. Pedersen, E. M. Kjaer et al., Localized biosensing with Topas microstructured polymer optical fiber: Erratum, *Optics Letters*, 32, 1059, 2007.
36. T. Ishigure, Y. Koike, and J. W. Fleming, Optimum index profile of the perfluorinated polymer based GI polymer optical fiber and its dispersion properties, *Journal of Lightwave Technology*, 18, 178–184, 2000.
37. K. Makino, T. Kado, A. Inoue, and Y. Koike, Low loss graded index polymer optical fiber with high stability under damp heat conditions, *Optics Express*, 20, 12893–12898, June 4, 2012.
38. A. Abang, D. Saez-Rodriguez, K. Nielsen, O. Bang, and D. J. Webb, Connectorisation of fibre Bragg grating sensors recorded in microstructured polymer optical fibre, Presented at *the Fifth European Workshop on Optical Fibre Sensors*, Krakow, Poland, 2013.
39. Y. Koike and M. Asai, The future of plastic optical fiber, *NPG Asia Materials*, 1, 22–28, October 21, 2009 (online).
40. W. J. Tomlinson, I. P. Kaminow, A. Chandross, R. L. Forck, and W. T. Silfvast, Photoinduced refractive index increase in poly(methyl methacrylate) and its applications, *Applied Physics Letters*, 16, 486–489, 1970.
41. J. M. Moran and I. P. Kaminow, Properties of holographic gratings photoinduced in polymethyl methacrylate, *Applied Optics*, 12, 1964–1970, 1973.
42. M. J. Bowden, E. A. Chandross, and I. P. Kaminow, Mechanisms of the photoinduced refractive index increase in polymethyl methacrylate, *Applied Optics*, 13, 112, 1974.
43. J. Marotz, Holographic storage in sensitized polymethyl methacrylate blocks, *Applied Physics B*, 37, 181–187, August 1, 1985.
44. M. Kopietz, M. D. Lechner, D. G. Steinmeier, J. Marotz, H. Franke, and E. Kratzig, Light-induced refractive-index changes in polymethylmethacrylate (PMMA) blocks, *Polymer Photochemistry*, 5, 109–119, 1984.
45. T. Mitsuoka, A. Torikai, and K. Fueki, Wavelength sensitivity of the photodegradation of poly(methyl methacrylate), *Journal of Applied Polymer Science*, 47, 1027–1032, 1993.
46. C. Wochnowski, S. Metev, and G. Sepold, UV-laser-assisted modification of the optical properties of polymethylmethacrylate, *Applied Surface Science*, 154–155, 706–711, 2000.
47. Z. Xiong, G. Peng, B. Wu, and P. Chu, Highly tunable Bragg gratings in single-mode polymer optical fibers, *IEEE Photonics Technology Letters*, 11, 352–354, 1999.
48. H. Y. Liu, H. B. Liu, G. D. Peng, and P. L. Chu, Observation of type I and type II gratings behaviour in polymer optical fiber, *Optics Communications*, 16, 159–161, 2004.

49. H. B. Liu, H. Y. Liu, G. D. Peng, and P. L. Chu, Novel growth behaviours of fiber Bragg gratings in polymer optical fiber under UV irradiation with low power, *IEEE Photonics Technology Letters*, 16, 159–161, 2004.
50. D. Sáez-Rodríguez, K. Nielsen, O. Bang, and D. J. Webb, Increase of the photosensitivity of undoped poly(methylmethacrylate) under UV radiation at 325 nm, in *Micro-Structured and Specialty Optical Fibres*, *Photonics Europe*, Brussels, Belgium, 2014.
51. D. R. Tyler, Mechanistic aspects of the effects of stress on the rates of photochemical degradation reactions in polymers, *Polymer Reviews*, 44, 351–388, 2004.
52. H. B. Liu, H. Y. Liu, G. D. Peng, and P. L. Chu, Strain and temperature sensor using a combination of polymer and silica fibre Bragg gratings, *Optics Communications*, 219, 139–142, 2003.
53. C. Zhang, W. Zhang, D. J. Webb, and G. D. Peng, Optical fibre temperature and humidity sensor, *Electronics Letters*, 46, 643–663, April 29, 2010.
54. Z. F. Zhang and X. M. Tao, Synergetic effects of humidity and temperature on PMMA based fiber Bragg gratings, *Journal of Lightwave Technology*, 30, 841–845, March 2012.
55. Z. F. Zhang and X. M. Tao, Intrinsic temperature sensitivity of fiber Bragg gratings in PMMA-based optical fibers, *IEEE Photonics Technology Letters*, 25, 310–312, February 2013.
56. A. Othonos and K. Kalli, *Fiber Bragg Gratings: Fundamentals and Applications in Telecommunications and Sensing*. Boston, MA: Artech House Publishers, 1996.
57. Z. Zhang, P. Zhao, P. Lin, and F. Sun, Thermo-optic coefficients of polymers for optical waveguide applications, *Polymer*, 47, 4893–4896, 2006.
58. D. R. Salem, *Structure Formation in Polymeric Fibers*. Munich, Germany: Hanser-Gardner Publications, 2001.
59. I. P. Johnson, W. Yuan, A. Stefani, K. Nielsen, H. K. Rasmussen, L. Khan et al., Optical fibre Bragg grating recorded in TOPAS cyclic olefin copolymer, *Electronics Letters*, 47, 271–272, 2011.
60. C. Markos, A. Stefani, K. Nielsen, H. K. Rasmussen, W. Yuan, and O. Bang, High-T-g TOPAS microstructured polymer optical fiber for fiber Bragg grating strain sensing at 110 degrees, *Optics Express*, 21, 4758–4765, February 25, 2013.
61. X. Chen, C. Zhang, D. J. Webb, G.-D. Peng, and K. Kalli, Bragg grating in a polymer optical fibre for strain, bend and temperature sensing, *Measurement Science and Technology*, 21, 094005, 2010.
62. D. Gallego and H. Lamela, High-sensitivity ultrasound interferometric single-mode polymer optical fiber sensors for biomedical applications, *Optics Letters*, 34, 1807–1809, 2009.
63. W. Zhang, D. J. Webb, and G. D. Peng, Investigation into time response of polymer fiber Bragg grating based humidity sensors, *Journal of Lightwave Technology*, 30, 1090–1096, April 15, 2012.
64. S. Katayama, M. Horiike, K. Hirao, and N. Trutsumi, Structures induced by irradiation of femto-second laser pulse in polymeric materials, *Journal of Polymer Science, Part B: Polymer Physics*, 40, 537–544, 2002.
65. W. Zhang and D. J. Webb, Humidity responsivity of polymer optical fiber Bragg grating sensors, *Optics Letters*, 39, 10, 2014.
66. G. V. Barbosa-Cánovas, A. J. Fontana, S. J. Schmidt, and T. P. Labuza, *Water Activity in Foods-Fundamentals and Applications*. Ames, IA: Blackwell Publishing, 2007.
67. T. Watanabe, N. Ooba, Y. Hida, and M. Hikita, Influence of humidity on refractive index of polymers for optical waveguide and its temperature dependence, *Applied Physics Letters*, 72, 1533, 1998.
68. D. T. Turner, Polymethyl methacrylate plus water: Sorption kinetics and volumetric changes, *Polymer*, 23, 197–202, 1982.
69. M. Hiscocks, M. van Eijkelenborg, A. Argyros, and M. Large, Stable imprinting of long-period gratings in microstructured polymer optical fibre, *Optics Express*, 14, 4644–4648, 2006.
70. Z. C. Li, H. Y. Tam, L. X. Xu, and Q. J. Zhang, Fabrication of long-period gratings in poly(methyl methacrylate-co-methyl vinyl ketone-cobenzyl methacrylate)-core polymer optical fiber by use of a mercury lamp, *Optics Letters*, 30, 1117–1119, May 15, 2005.
71. D. Sáez-Rodríguez, J. L. Cruz, I. Johnson, D. J. Webb, M. C. J. Large, and A. Argyros, Water diffusion into UV inscripted long period grating in microstructured polymer fibre, *IEEE Sensors Journal*, 10, 1169–1173, 2010.
72. D. Webb, M. Aressy, A. Argyros, J. Barton, H. Dobb, M. van Eijkelenborg et al., Grating and interferometric devices in POF, Presented at *the Proceedings of XIV International Conference on Polymer Optical Fibre*, Hong Kong, China, 2005.
73. Web site of the EU Framework 7 project PHOSFOS, http://www.phosfos.eu/, accessed 27/9/14.
74. Z. Wang, Polydimethylsiloxane mechanical properties measured by macroscopic compression and nanoindentation techniques, Dissertation, Mechanical Engineering Department, University of South Florida, Tampa, FL, 2011.

75. A. Stefani, S. Andresen, W. Yuan, N. Herholdt-Rasmussen, and O. Bang, High sensitivity polymer optical fiber-Bragg-grating-based accelerometer, *IEEE Photonics Technology Letters*, 24, 763–765, May 1, 2012.
76. C. A. F. Marques, L. Bilro, L. Kahn, R. A. Oliveira, D. J. Webb, and R. N. Nogueira, Acousto-optic effect in microstructured polymer fiber Bragg gratings: simulation and experimental overview, *Journal of Lightwave Technology*, 31, 1551–1558, May 15, 2013.
77. W. Zhang, D. Webb, and G. Peng, Polymer optical fiber Bragg grating acting as an intrinsic biochemical concentration sensor, *Optics Letters*, 37, 1370–1372, April 15, 2012.
78. W. Zhang and D. Webb, Polymer optical fiber grating as water activity sensor, Proc. SPIE 9128, Micro-structured and Specialty Optical Fibres III, 91280F (2 May 2014); doi: 10.1117/12.2054207
79. Paradigm optics. http://www.paradigmoptics.com/.

11 Acousto-Optic Effect and Its Application in Optical Fibers

Alexandre de Almeida Prado Pohl

CONTENTS

11.1 INTRODUCTION

Acoustics is the branch of physics that deals with the interaction of mechanical waves in solids, liquids, and gases. Early observations of vibrating strings date back to the time of Pythagoras in Ancient Greek, and major advances in the understanding of the acoustic processes occurred in the Renaissance period with Gianfrancesco Sagredo and Marin Mersenne [1]. As with acoustics, optics has also a very long and rich history. However, the study of the interaction of light and sound begins with Brillouin in 1922 predicting the diffraction of light by an acoustic wave [2], which has been further observed in 1932 by experiments performed separately by Debye and Sears [3] and Lucas and Biquard [4].

The development of optical fibers in the early 1970s and the enormous potential it presented for the fields of telecommunications and sensors has rapidly led researchers to exploit different physical mechanisms to control the light propagating in the silica waveguide. Following this trend, the acousto-optic mechanism has soon been considered in the design and construction of a variety of optical fiber devices. The first successful use of the effect in optical fibers was demonstrated by Engan and collaborators in 1986 with an all-fiber acousto-optic frequency shifter [5]. Since then, many devices and applications have been devised and demonstrated in conventional and microstructured optical fibers (MOFs) [6,7]. Specially, the fabrication of Bragg and long-period gratings (LPGs) in fibers and their interaction with acoustic waves has greatly expanded the applications, enabling innovative fiber-based devices, such as tunable filters and fiber-based lasers, to be developed. In this regard, Liu, Russel, and Dong have pioneered the application of the acousto-optic mechanism in fibers using fiber Bragg gratings (FBGs) [8].

The acousto-optic effect is based on the change in the refractive index of a medium due to the presence of a sound wave. Such a wave deforms the medium generating a mechanical strain that leads to the change in the optical permittivity through the mechanism known as photoelasticity [9]. The study of the acousto-optic interaction in fibers begins with the understanding of the propagation of acoustic waves in rodlike structures. And the basic device for generating such waves in fibers makes use of a modulator, whose design and principle of operation was first proposed and patented by Zemon and Dakss [10].

This chapter aims to provide a brief overview of the field covering the basic development and an application in the sensoric domain. The chapter is divided as follows: Section 11.2 describes the main parameters of acoustic waves in glass rods and the different types of acoustic modes that are of interest for the acousto-optic interaction. Section 11.3 provides the reader with the knowledge on the two possible configurations used in acousto-optic fiber-based modulators and their operation employing piezoelectric elements. Section 11.4 gives a description of the acousto-optic interaction in fibers, which leads to the coupling of modes in the optical waveguide. Section 11.5 presents the interaction of acoustic waves with Bragg grating and LPG inscribed in optical fibers. Section 11.6 describes the use of photonic crystal fibers (PCFs) in acousto-optic devices. Section 11.7 discusses the development of a fiber-based viscometer whose operating principle relies on the acousto-optic mechanism, and Section 11.8 brings the conclusion and gives some perspectives for the future work in the field.

11.2 ACOUSTIC WAVES IN OPTICAL FIBERS

In order to explain the effects observed by the interaction of acoustic waves with optical fibers (such as the coupling between optical modes and the modulation of fiber grating parameters), one needs to understand the propagation of acoustic waves in rodlike structures. Textbooks on acoustics approach the basic solutions of mechanical waves propagating in slender bars [11,12]. Thurston [13], for instance, has investigated acoustic modes propagating in linear, elastic, isotropic cylindrical structures, named clad rods (structures consisting of a core of one material surrounded by a cladding of a material having different properties). However, the analysis in this chapter is restricted to optical fibers, which can be considered, for the sake of propagation of mechanical waves, as a solid, circular rod consisting of an isotropic, homogeneous material.

The analysis is performed by solving the vector equation of motion that describes the space-time-dependent displacement function $u(r, \theta, z, t)$ in the rodlike structure. One is faced with the solution of a set of three scalar equations in the cylindrical coordinate system from which the displacement field components in the radial (r), circular (θ), and axial (z) directions are obtained [11,13]. Considering the propagation of a harmonic field, the solution of this set of equations is given as

$$\begin{aligned} u_r &= U(r)\exp j(\omega_a t-\beta_a z)\begin{cases}\sin(n\theta)\\ \cos(n\theta)\end{cases} \\ u_\theta &= V(r)\exp j(\omega_a t-\beta_a z)\begin{cases}\cos(n\theta)\\ -\sin(n\theta)\end{cases} \\ u_z &= W(r)\exp j(\omega_a t-\beta_a z)\begin{cases}\sin(n\theta)\\ \cos(n\theta)\end{cases} \end{aligned} \tag{11.1}$$

where

$\omega_a = 2\pi f_a$ and β_a are the angular frequency and the propagation constant, respectively, of an acoustic mode in the z direction

n is an integer describing the circumferential field behavior

U, V, and W represent the radial variation, described by Bessel functions [13], given as

$$U(r) = A k_d J_n'(k_d r) + B\beta_a J_n'(k_t r) + C\frac{n}{r}J_n(k_t r)$$

$$V(r) = A\frac{n}{r}J_n(k_d r) + B\frac{\beta_a}{k_t}\frac{n}{r}J_n(k_t r) + C k_t J_n'(k_t r) \tag{11.2}$$

$$W(r) = -i\left[A\beta_a J_n(k_d r) + B k_t J_n(k_t r)\right],$$

where

J_n is the Bessel function of the first kind and of order n

J_n' is its first derivative with respect to the argument

k_d and k_t are given as $k_d^2 = \omega_a^2/c_d^2 - \beta_a^2$ and $k_t^2 = \omega^2/c_t^2 - \beta_a^2$ respectively

The parameters c_d and c_t are known as the bulk dilatational and transverse (shear) wave velocities, given as $c_d^2 = (\lambda + 2\mu)/\rho$ and $c_t^2 = \mu/\rho$, respectively. λ and μ are Lamé's constants, and ρ is the material density. For fused silica, $\rho = 2200$ kg/m^3, and Lamé's constants are $\mu = 31.26$ GPa and $\lambda = 15.87$ GPa [14], from which the transverse and dilatational phase velocities are estimated as $c_t = 3769$ m/s and $c_d = 5969$ m/s, respectively.

Either the upper or lower set of trigonometric functions in (11.1) is used to find the desired solutions. The dispersion relation determining the value of β_a, as well as the values of the magnitude constants A, B, and C, is obtained requiring the stresses to be zero at the cylindrical surface ($r = a$, with a as the radius of the rod) and choosing an integer value for n in (11.1) and (11.2). The solution of the resulting characteristic equation delivers the dispersion curves from which the propagation parameters of an acoustic mode and the behavior of its phase velocity (the dispersion relationship) as a function of frequency are assessed. A detailed description of the dispersion curves for the cylindrical rod can be found in Thurston [13].

Different values of n provide solutions for the different modes propagating in the guide. By using the lower set of trigonometric functions in (11.1) and choosing $n = 0$, one obtains longitudinal waves, which show no angular dependence in θ and are axially symmetric, with displacement components in the radial and axial directions only. By choosing the upper set of trigonometric functions and still keeping $n = 0$, one obtains torsional waves, for which the displacement components u_r and u_z vanish and the remaining component, u_θ, is independent of θ. Finally, flexural waves that depend on the circumferential angle θ are obtained, setting $n \neq 0$. In this case, either set of trigonometric functions in (11.1) can be chosen. Figure 11.1 sketches the outlook of the rod when each one of these modes propagates.

The analysis of the dispersion curves brings relevant information on the acoustic phase velocity of each propagating acoustic mode (longitudinal, torsional, and flexural [15]), which is important for the operation of the acousto-optic modulator in order to obtain the desired optical effect in the fiber. In using such dispersion curves, it is convenient to work with the normalized frequency, defined as $f_a a/c_t$, where f_a is the acoustic frequency, a is the fiber radius, and c_t is the transverse wave velocity in bulk silica, as described earlier. Figure 11.2 shows the behavior of the phase velocity of the lowest-order longitudinal and flexural modes plus the phase velocity curves of the five lowest-order torsional modes, according to Engan et al. [15]. The plots show the behavior of the ratio c/c_{ext} over the normalized frequency $f_a a/c_t$, in which the parameter c is the phase velocity of the mode under analysis and $c_{ext} = (E/\rho)^{1/2}$ is the phase (extensional) velocity of the fundamental longitudinal mode in the low-frequency regime, with E as Young's modulus (in fused silica, $c_{ext} = 5760$ m/s). The phase velocity shows a characteristic behavior according to the frequency regime. In the low-frequency regime, when $f_a a/c_t \ll 1$, only few acoustic modes propagate in the rod. For the lowest guided longitudinal mode, the phase velocity approaches that given by the extensional wave in silica (c_{ext}) when $f_a a/c_t \ll 1$ (as the ratio c/c_{ext} approaches the unit value), and it asymptotically assumes the velocity of

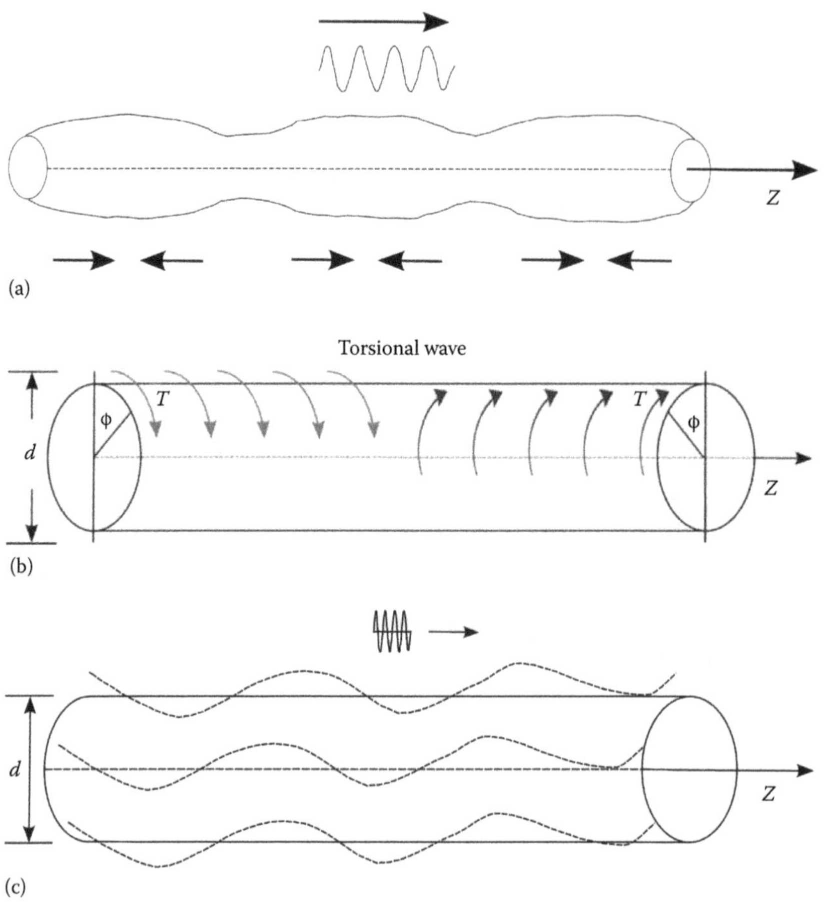

FIGURE 11.1 Outlook of the rod when either (a) the longitudinal, (b) the torsional, or (c) the flexural mode propagates.

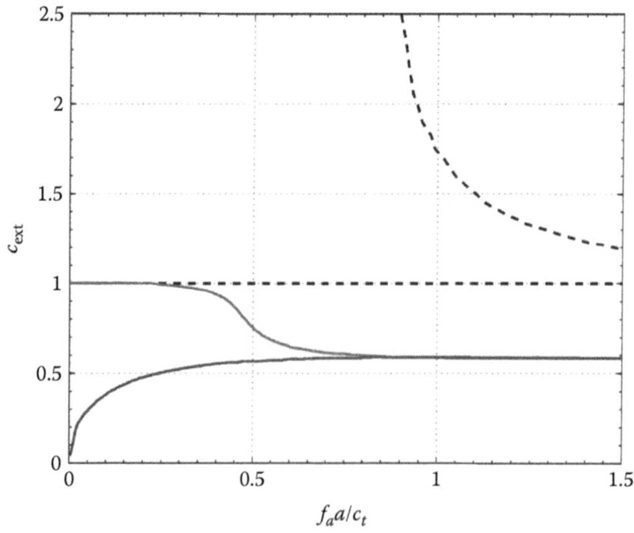

FIGURE 11.2 Behavior of the phase velocities of the lowest-order longitudinal (upper solid curve), two lowest-order torsional (dashed curves) and flexural mode (lower solid curve) as a function of the frequency–radius product, $f_a a/c_t$. (According to Engan, H.E. et al., *J. Lightw. Technol.*, 6(3), 428, 1988.)

Rayleigh waves (surface waves) when $f_a a/c_t \gg 1$. In this last case, the wave energy is confined to a thin layer within the rod surface, a situation also experienced by the torsional and flexural modes. In the high-frequency regime, that is, when $f_a a/c_t \gg 1$, the silica rod becomes highly multimode, which is not interesting for the acousto-optic interaction, as the existence of many modes makes it difficult to have a proper control of the acousto-optic device. Studies also point out that for the acousto-optic interaction to take place, there must be a significant overlap between the acoustic and the optical fields in the fiber, which means that the regime $f_a a/c_t \ll 1$ should be chosen as the most relevant for the acousto-optic operation [15].

Considering only the lowest-order flexural mode, Blake and coworkers [16] have classified the dispersion behavior according to three different acoustic propagation regimes in the silica fiber. For higher frequencies, in which case $f_a a/c_t \gg 1$, the acoustic energy concentrates on the surface of the fiber and the acoustic wavelength, λ_a, varies as $1/f_a$. For lower frequencies, where $f_a a/c_t \ll 1$, the acoustic energy is uniformly distributed over the fiber cross section and λ_a varies as

$$\lambda_a = \left(\frac{\pi a c_{\text{ext}}}{f_a} \right)^{1/2}, \tag{11.3}$$

as it can be inferred from the plot of the dispersion relationship of the fundamental flexural mode in Figure 11.2. Between these two extremes, where $\lambda_a \approx 2\pi a$, there is a transition region. Given that the acousto-optic interaction must take place in the core of the guide, it is important that the flexural mode is excited in the low-frequency region. Considering the dispersion curves given in Engan et al. [15] for flexural modes with one circumferential period ($n = 1$ in (11.1)), the cutoff for higher modes is achieved when $f_a a/c_t < 0.3$. Applying the same condition for longitudinal and torsional modes, the cutoff is achieved when $f_a a/c_t < 0.5$ for longitudinal and $f_a a/c_t < 0.8$ for torsional modes, respectively. Considering these different cutoff limits, a simple calculation shows that in untapered conventional single-mode fibers ($2a = 125$ μm), there exist only three types of modes (the lowest-order longitudinal mode, the lowest-order torsional mode, and the lowest-order flexural mode) when $f_a < 18$ MHz (given that the flexural case is the critical limit). It is important to note that the real dispersion curves of each type of acoustic mode should be taken into account for an exact analysis through the solution of the characteristic equation.

11.3 ACOUSTO-OPTIC MODULATORS

The basic device for generating acoustic waves in fibers makes use of an acousto-optic modulator. In practice, the modulator is composed of a piezoelectric transducer (PZT) (generally a disk made of, for instance, lead zirconate titanate), an acoustic horn, and a short length of fiber. The PZT [17] serves as an electromechanical driver for the acoustic excitation. By applying a harmonic time-variable tension, the PZT is periodically expanded or contracted in a privileged direction (for instance, shear or longitudinal). This movement is transferred to the horn and further coupled to the fiber, setting the conditions for the propagation of the acoustic modes in the optical waveguide. The horn is made of glass or metal and has a decreasing diameter from its base to a sharp tip of about the same diameter of the optical fiber, therefore acting as an acoustic amplitude amplifier [18]. In this way, the amplitudes generated at the larger end will be converted into higher amplitudes of the acoustic field at the other end because of the much smaller area at the horn tip.

There are two possible configurations for such a modulator. Figure 11.3a illustrates the configuration, in which the piezo horn arrangement is placed transversally to the fiber. We call it the transversal configuration. The PZT is bonded to the larger diameter of the horn, and when operated at higher frequencies, it aims at launching longitudinal waves at its narrower end (horn tip), whose

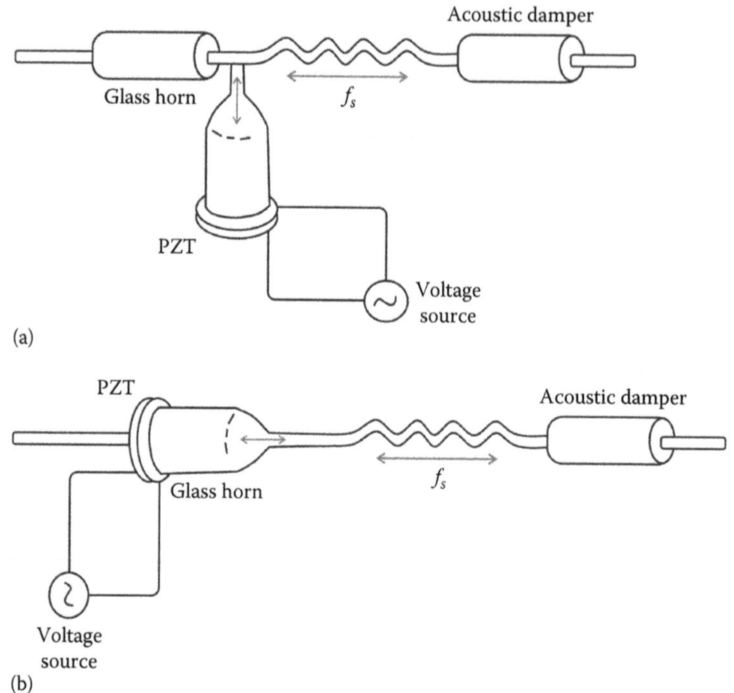

FIGURE 11.3 (a) Transversal and (b) longitudinal configuration of the acousto-optic modulator.

back and forth movement will create a flexural wave of larger amplitude in the optical fiber [15,16]. Figure 11.3b shows the configuration in which the optical fiber is axially aligned with the horn. This configuration, named longitudinal, is versatile and suitable for the excitation of different acoustic modes (flexural, longitudinal, or torsional), the excitation being dependent on the type and way (shear or longitudinal) the PZT is driven. Due to its compactness, it also presents the advantage of saving space in the optical assembly.

The commonly used PZTs do not respond to any excitation frequency but only to narrow bands of resonances in which the vibration modes manifest larger and specific mechanical displacements [19,20]. As the efficiency of the acousto-optic effect is strongly dependent on the applied frequency, and the excitation of a particular acoustic mode is also determined by the directional deformation of the piezo, knowledge on the transducer response and its resonance modes is essential for the efficient control and operation of fiber-based acousto-optic devices.

Piezoceramic disks present a distinct modal behavior that depends on the excitation frequency, material properties, polarization, and geometric dimensions [21]. Piezo resonances can be measured by impedance-phase methods [22]; however, the combined analysis of the piezo spectral response and the modal displacement is useful and required for the identification of the acoustic modes coupled to the fiber through the horn. Figure 11.4a shows the whole device and Figure 11.4b shows only the piezo disk with the axis along which the element is set to vibrate. Due to the material's high dielectric constant and the electric polarization of the electrodes, the PZT can be compared to a capacitor. The element can be represented by an equivalent circuit known as Butterworth–van Dyke and is illustrated in Figure 11.4c [23]. The circuit is composed of an *RLC* series and a parallel circuit ($RLC//C_o$), where the resistor R, the inductor L, and the capacitor C are related to the PZT damping, mass, and elastic constant, respectively. C_o is the electric capacitance between the electrodes to which external connection wires are bonded. Based on this equivalent circuit, a resonator with angular frequency $\omega = 2\pi f$ can be modeled using electrical

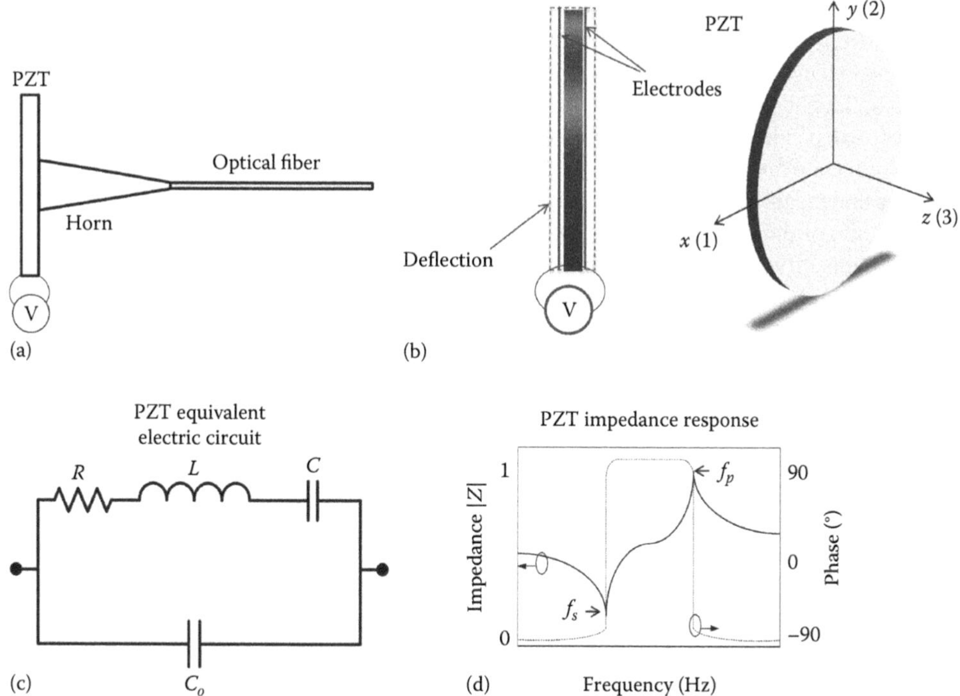

FIGURE 11.4 The acousto-optic modulator and its parts: (a) diagram of the acousto-optic modulator in the longitudinal configuration, (b) 2D and 3D view of the piezo disk, (c) Butterworth–van Dyke equivalent circuit, and (d) piezo impedance-phase response.

parameters only. The frequency response of the transducer is then obtained solving the *RLC* circuit in terms of the impedance Z, written as

$$Z = R + j\left(\omega L - (\omega C)^{-1}\right). \tag{11.4}$$

For the situation where the losses are negligible and the impedance is null, the series resonance f_s is obtained from (11.4) as

$$f_s = \left(2\pi\sqrt{LC}\right)^{-1}. \tag{11.5}$$

A similar deduction applied to the parallel circuit ($RLC//C_o$) in Figure 11.4c allows the parallel or antiresonance f_p to be obtained:

$$f_p = \left(2\pi\sqrt{\frac{LCC_o}{(C + C_o)}}\right)^{-1}. \tag{11.6}$$

Figure 11.4d illustrates the transducer spectral response in terms of the impedance magnitude and phase. The piezo resonance, f_s, and antiresonance, f_p, frequencies correspond to the condition of minimum and maximum impedance, respectively, within a short frequency range. At these frequencies, the inductance L and the capacitance C cancel each other making the phase null. As the electrical current flowing across the piezo is related to the mechanical deformation, the reduction in impedance Z causes a maximum current flow I, which is then responsible for the maximum PZT deflection at the resonance f_s. It is important to observe that the parameters R, L, and C are frequency dependent; therefore, many resonance and antiresonance peaks will appear over an extended frequency range.

The type and magnitude of the deformation manifested by the modes of the piezo element depend on the direction of the electric field polarization, dimensions and geometric shape, and mainly on the material anisotropy. Once the parameters R, L, and C are known or calculated from the material constants, Equations 11.5 and 11.6 are used to calculate the piezo resonances and antiresonances. A detailed analysis of the coupled electromechanical relations and material constants for the piezo behavior can be performed using the finite element method (FEM). The method requires the discretization of the device geometry in small subdomains (elements) and builds a matrix equation system representing the piezo behavior. Solving the matrix equation allows one to obtain the resonant frequencies and, particularly, the mechanical displacements of the piezo disk in its transversal (z) and radial directions (x- or y-axis in Figure 11.4b).

Figure 11.5a shows, as an example, the measured and simulated PZT responses in terms of the impedance magnitude as a function of the excitation frequency (for piezo type P26 from Ferroperm manufacturer). The FEM simulation presents larger impedance-phase amplitudes compared to experimental values. This difference is related to distinct losses that each mode presents in the resonance. Since the anisotropic constants used in simulations do not include the PZT losses, the simulated amplitudes become higher and sharper in the resonances. The losses reduce the impedance-phase amplitude, which affects the localization of some resonances by the condition of minimum impedance and null phase. A better distinction of the resonances as compared to the simulation is observed at low frequencies, since much attenuated and closer resonances at higher frequencies result in narrower bandwidths. Besides the frequency response, the FEM simulation can

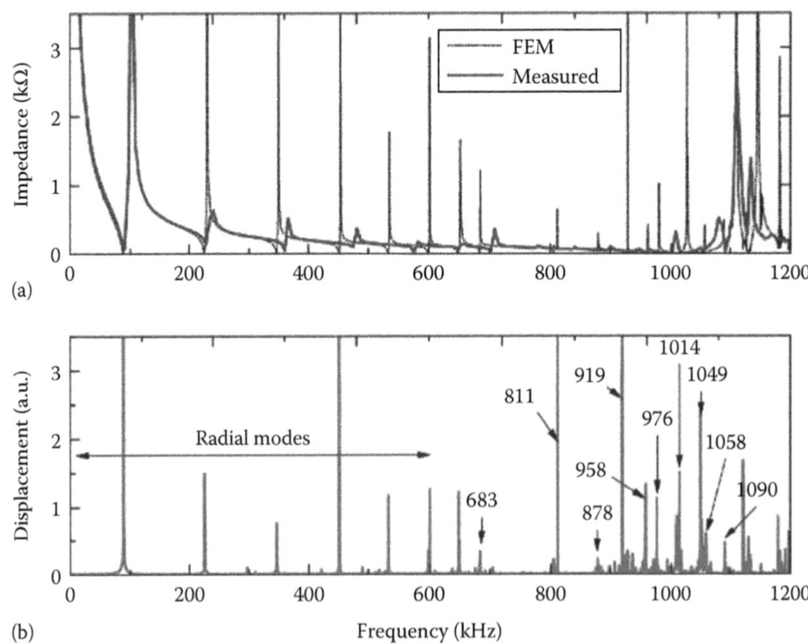

FIGURE 11.5 (a) Measured and simulated frequency responses of the piezoelectric disk (Ferroperm P26) and (b) simulated piezo displacement in the frequency range up to 1200 KHz. (From Silva, R.E. et al., *Opt. Express*, 21(6), 6997, 2013.)

also provide the piezo mechanical response, such that its displacement can be analyzed along the axial (z) and transversal (x–y) directions. The PZT response in terms of displacements indicates also the localization of the vibration modes by the maximum values as shown in Figure 11.5b. A good agreement between simulated and experimental resonances is obtained, mainly at the frequency ranges from 1 to 400 kHz and from 650 to 1200 kHz, respectively [24,25]. The transducer responses in Figure 11.5a and b are used to locate the useful vibration modes, considering the correspondent measured and simulated resonances. A more detailed description of the piezo resonance modes along the axial and transversal displacements can be found in Silva et al. [25].

Generally, acousto-optic applications utilize either flexural or longitudinal acoustic modes to control the properties of guided light in the fiber. Applications using torsional modes have also been demonstrated [26,27]. As described earlier, depending on the vibration direction of the piezo disk, either one or the other type of mode is particularly excited. However, the frequency response of the piezo disk itself is not sufficient to determine the type of acoustic modes excited in the fiber. The quest on which mode (or modes) is (are) excited depends on the physical arrangement and on the highest displacement of the piezo along the radial (x–y) or transversal (z) directions at a certain frequency. The FEM technique can be of further help and be extended to perform a 3D simulation of the whole modulator, which includes the piezo, the horn, and the fiber. This indeed helps the better identification of the acoustic modes excited in the fiber at the piezo resonances. For instance, Figure 11.6a and b shows the simulated acousto-optic modulator for the configuration in which the optical fiber is axially aligned with the horn at the resonance frequencies f = 86 and 960.3 kHz, respectively. For instance, the resonance f = 86 kHz causes the excitation of a flexural mode, which is characterized by pronounced bendings in the fiber, while at f = 906.3 kHz, bendings are no longer seen and the result is a longitudinal mode, where the displacement component is highest along the fiber axial direction. Moreover, from simulations and experiments, it has been observed that the acousto-optic modulator can also excite acoustic waves of complex oscillations, which can be generally termed as hybrid, once they are a composition of flexural, torsional, and longitudinal modes.

Measurement of flexural and longitudinal acoustic modes can be accomplished using an extrinsic Fabry–Perot interferometer [24,28,29]. The technique allows the analysis of oscillating surfaces in the space and time domains. For instance, the cleaved end of a fiber serves as a probe for carrying light from a laser to the vibrating surface under analysis. Light from the probe is back reflected by the probe end face itself and by the external surface (vibrating surface), both forming a low-finesse Fabry–Perot interferometer. The reflected light is detected and analyzed by a digital oscilloscope employing the fast Fourier transform (FFT) function. It can be shown that the detected signal contains harmonics of the vibration frequency of the surface, the relative amplitudes of which can be

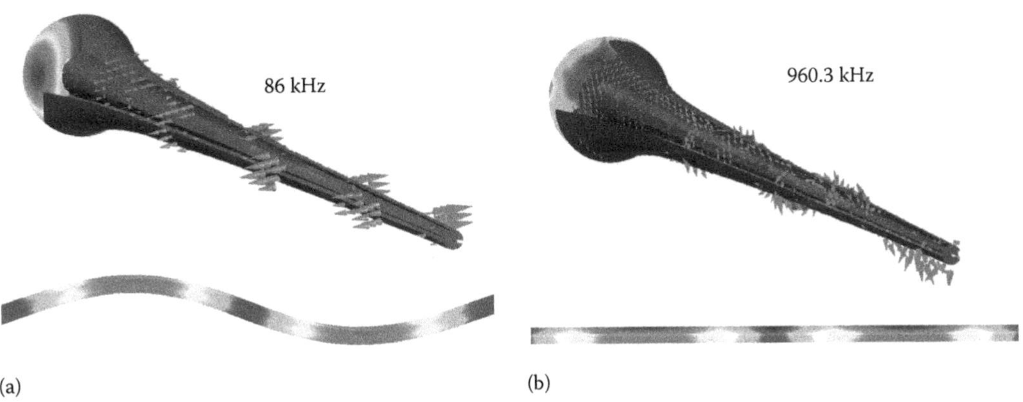

FIGURE 11.6 FEM simulation of the acousto-optic modulator: (a) when excited by a flexural mode and (b) when excited by a longitudinal mode. (From Pohl, A.A.P. et al., *Photon. Sens.*, 3(1), 1, 2013.)

related to the amplitude of the acoustic vibration. The measurement of flexural acoustic modes and the characterization of bendings are more accurately performed by using the standing acoustic mode formed when the vibrating surface of the fiber under test is fixed on its two ends (at the horn tip on one side and, for instance, at the fiber holder on the other). The fiber probe is then set to measure the temporal harmonic oscillation at an arbitrary position along the fiber (with the exception of points located at the wave nodes, where the oscillation is zero!) or is swept along the fiber under test to measure the troughs and crests of the flexural mode. In the case of longitudinal modes, the measurement is accomplished by sensing the stretching of the fiber tip, which is then related to the change in the interferometer cavity length d. Details of such measurements are found in Silva and Pohl [29].

11.4 ACOUSTO-OPTIC INTERACTION IN FIBERS

The employment of acoustic waves in fibers was initially proposed by Engan et al. [5], Kim et al. [16], and Blake et al. [30]. In their work, flexural acoustic modes were used to create microbendings and achieve optical mode conversion in a two-mode optical fiber using the acousto-optic modulator in the transversal configuration. The optical mode conversion mechanism is a well-known phenomenon in fibers, through which propagating modes exchange power due to the change in the effective refractive index in the region where these modes propagate. Coupled-mode theory has been widely employed to describe quantitatively the coupling phenomenon in several situations [31]. A thorough study of mode coupling between the LP_{01} and radiation modes or between the LP_{01} and the LP_{11} modes in optical fibers due to very small bends has been accomplished by Taylor [32]. In his work, Taylor proposes the generation of microbendings using mechanical transducers to produce a periodic perturbation along the fiber axis, which can be used as an $LP_{01} \leftrightarrow LP_{11}$ mode converter. In a similar manner, flexural acoustic modes of a proper frequency produce microbends in the fiber, which cause the change in the effective refractive index of the optical modes. This effective change is due to the contribution of two mechanisms: one concerning the material index modification through the elasto-optic effect and the other concerning the path length change caused by the microbendings. The net change is mainly influenced by the geometrical effect, which predominates over the elasto-optic mechanism [33]. The net change Δn is described as

$$\Delta n(x,y,z,t) = n_0(1-\chi)S_z(x,y,z,t) \tag{11.7}$$

in which S_z represents the strain distribution along the z-axis (fiber axis). In the general case, S_z is a function of the propagating acoustic mode as described in (11.1), but in a more simple case, it can be represented by a harmonic displacement function, which gives Δn as [33]

$$\begin{aligned}\Delta n(x,y,z,t) &= n_0(1-\chi)\beta_a^2\, y\, u_0 \cos(\omega_a t - \beta_a z)\\ &= \Delta n(x,y)\cos(\omega_a t - \beta_a z)\end{aligned} \tag{11.8}$$

where

- n_0 is the refractive index of the undisturbed fiber
- $\chi = 0.22$ is the elasto-optic coefficient for silica
- u_0 is the amplitude of the acoustic field
- $\omega_a = 2\pi f_a$ is the angular frequency
- $\beta_a = 2\pi/\lambda_a$ is the propagation constant of the acoustic mode
- y is the magnitude of the displacement in the fiber radial direction

Considering now the coupling between two optical modes, and denoting mode 1 as the fundamental mode and mode 2 as the one that carries a portion of the optical power after the interaction, the exchange of power between these modes can be described using the well-known coupled-mode equations as follows [33]:

$$\frac{dE_1}{dz} + je^{-j\omega_a t} C e^{-2j\delta z} E_2 = 0, \tag{11.9}$$

$$\frac{dE_2}{dz} + je^{j\omega_a t} C e^{2j\delta z} E_1 = 0,$$

in which δ is the detuning parameter given as

$$\delta = \frac{1}{2}(\Delta\beta - \beta_a) = \pi\left(\frac{1}{L_B(\lambda)} - \frac{1}{\lambda_a(f_a)}\right), \tag{11.10}$$

where

λ_a is the wavelength of the acoustic mode

$\Delta\beta = \beta_2 - \beta_1$ is the difference in the effective propagation constants of optical modes 1 and 2

$L_B = 2\pi/\Delta\beta$ is the intermodal beat length, defined as the necessary length for mode 1 to acquire a phase shift of $\Delta\varphi = 2\pi$ in relation to mode 2

The fraction of light coupled from mode 1 to mode 2 over the interaction length L is given as [33]

$$\eta = \frac{C^2}{\delta^2 + C^2}\sin^2\left(L\sqrt{\delta^2 + C^2}\right), \tag{11.11}$$

with C as the acousto-optic coupling coefficient calculated as

$$C = \frac{k}{2}\int_A \Delta n(x,y) E_1(x,y) E_2(x,y)\,dxdy, \tag{11.12}$$

where

k is the optical wave number

E_1 and E_2 are the transversal field distributions of the optical modes 1 and 2, respectively

$\Delta n(x, y)$ is the refractive index change over the fiber cross section as given in (11.8)

From (11.11), if δ is much larger than C in the denominator, almost no light is coupled from mode 1 to mode 2, while if $\delta = 0$, all light is coupled to mode 2 (if also the product $LC = \pi/2$ in (11.11) is satisfied). For this particular case, the phase matching is satisfied, and the following condition holds:

$$L_B(\lambda) \equiv \frac{\lambda}{n_1(\lambda) - n_2(\lambda)} = \lambda_a(f_a). \tag{11.13}$$

This means that for obtaining the total power conversion from one mode to the other, the acoustic wavelength must be equal to the beat length between the interacting modes. The optical coupling is thus resonant in wavelength; it takes place at the optical wavelength that verifies the phase-matching condition between the beat length of the two optical modes and the acoustic wavelength. At the output of the modulator, only light that remains guided by the core mode is transmitted. Thus, the coupling of power from the fundamental mode to cladding modes results in the appearance of attenuation notches in the spectrum. From (11.13), one obtains the resonance wavelength, λ_R, as

$$\lambda_R = \left[n_1(\lambda) - n_2(\lambda)\right]\lambda_a, \tag{11.14}$$

which corresponds to the wavelength of the attenuation notch. For standard single-mode fibers operating in the range from 1300 to 1600 nm, estimation of beat lengths at a specific optical wavelength gives values for L_B varying from 450 to 650 μm [34]. In this way, for optimal mode coupling, the flexural acoustic wavelength, λ_a, must also lie in this range. However, given that the beat length shows considerable wavelength dispersion, the resonance wavelength, λ_R, will vary as the optical frequency changes, which demands a new matching value for λ_a.

This acousto-optic interaction can be seen as the dynamic counterpart of an LPG, which will be described later in this chapter. For instance, in LPGs, the periodic perturbation of the refractive index is created by UV radiation and their spectral characteristics are fixed during the fabrication. LPGs also show a period in the range of hundreds of micrometers and the resonance wavelength is calculated by the same equation as (11.14). Due to the similarity in optical response as compared to LPGs, the devices obtained using the acousto-optic mechanism are called acoustic gratings.

Since the first studies of acousto-optics in fibers, flexural modes have been excited using the configuration where the horn is transversally placed in relation to the fiber axis and used to generate microbendings [35–37]. However, the generation of microbendings can also be achieved using the acousto-optic modulator with the longitudinal configuration [38]. This fundamentally depends on the type and the way the piezoelectric element attached to the fiber horn is driven. Figure 11.7 brings an example of the coupling between the fundamental LP_{01} mode to three cladding modes ($LP_{11}^{(cl)}$, $LP_{12}^{(cl)}$, $LP_{13}^{(cl)}$) [38], where the modulator employs the longitudinal configuration, a conventional telecommunications fiber with a nominal core diameter of 8.5 μm, a cladding diameter of 125 μm, and a normalized index difference of 0.37%.

Yet, an additional degree of freedom in the design of all-fiber acousto-optic modulators is achieved when the fiber is adiabatically tapered along its length, making it possible to modify the propagation properties of acoustic and optical modes. For instance, a single-mode optical fiber

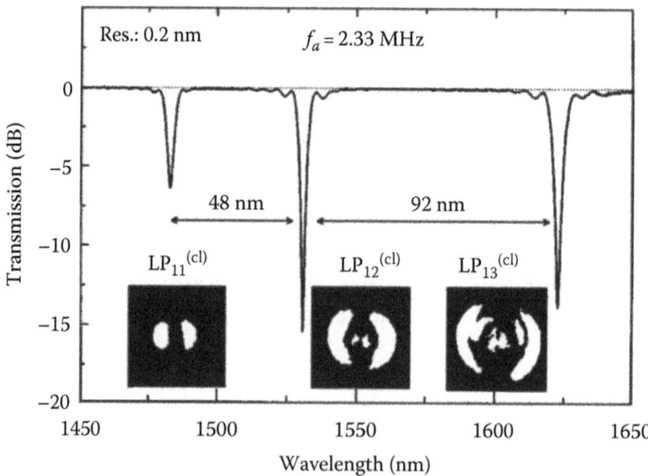

FIGURE 11.7 Transmission spectrum for an unpolarized broadband source, showing coupling to three different cladding modes ($LP_{11}^{(cl)}$, $LP_{12}^{(cl)}$, $LP_{13}^{(cl)}$) at an acoustic frequency of 2.33 MHz. Insets show the far-field radiation patterns of the cladding modes. (Reproduced with permission from Kim, H. S. et al., *Opt. Lett.*, 22(19), 1476, 1997.)

with finite cladding supports cladding modes, which are primarily guided by the entire cladding-air structure. In a tapered single-mode fiber, the guidance will be provided by the thin silica waist with a radius of few micrometers, so that the residual core will not play any role in the propagation. The effective index for these modes has the refractive index of the cladding as the upper bound and the refractive index of air as the lower bound. The coupling due to the acousto-optic interaction takes place between the fundamental LP_{01} mode and several low-order odd cladding modes LP_{1m} (for $m = 1, 2, 3, \ldots$). One of the advantages of using tapers lies on the need for low RF driving power to create the acousto-optic effect, once the acoustic wave is optimally conveyed to the silica waist. Feced et al. [36] have early shown the dependence of the acoustic resonance wavelength as a function of the taper radius in a silica fiber. From the behavior of the beat length dispersion as a function of the *V*-parameter ($V = [(2\pi/\lambda)a(NA)]$, in which a is the fiber radius and NA is its numerical aperture), they calculated a double-branched curve for the dependence of the resonance wavelength, λ_R, over the taper radius, once the condition in (11.13) is obeyed. This means that the resonant condition for the optimal mode coupling involving a specific mode pair (such as the LP_{01}–LP_{11}) can be achieved at two different taper radii [36].

Another important condition that has been observed in the interaction of acoustic waves with light is the fact that the frequency of the optical modes will slightly change by an amount corresponding to the frequency of the acoustic wave. This is known as the energy conservation requirement [16,33], which is expressed by the condition

$$\omega_m - \omega_0 = \pm 2\pi f_a \tag{11.15}$$

where

ω_m is the angular frequency of the optical mode of order m

ω_0 is the angular frequency of the fundamental mode of the optical fiber

The fast mode (for instance, LP_{11}) will be downshifted in frequency by an amount equal to the acoustic frequency when the acoustic flexural mode is traveling in the same direction as the optical signal. Frequency upshift will take place for the coupling from the fast mode to the slow mode for the same acoustic wave. The sign of the frequency shift will be reversed when the acoustic mode is traveling in the opposite direction in relation to that of the optical signal.

The bandwidth of acousto-optic modulators is important in several applications, particularly where such devices are employed as filters. Birks, Russel, and Culverhouse [33] derived an expression, which for silica-based waveguides is given by

$$\Delta\lambda \cong 0.8\lambda \frac{\lambda_a}{L}, \tag{11.16}$$

where

$\Delta\lambda$ is the acoustic filter bandwidth

λ_a is the acoustic wavelength on resonance, given by (11.13)

L is the acousto-optic interaction length

In the derivation of (11.16), it is assumed that modal dispersion is dominant (i.e., the effects of material dispersion and waveguide dispersion are neglected) [33]. From (11.16), one sees that narrow bandwidths are achieved by either increasing the interaction length or reducing the resonant acoustic wavelength. Generally, the reduction of the acoustic wavelength is preferred since this leads to devices of practical size [39]. For devices based on standard fiber diameters (not tapered), narrow bandwidths can only be achieved with very long interaction lengths, once the requirement that the acoustic energy couples efficiently to the core of the fiber limits the minimum acoustic wavelength.

However, by tapering the optical fiber, shorter acoustic wavelengths can be obtained and narrow bandwidth devices of practical length are possible. This points out to another advantage of using tapers in fiber-based acousto-optic modulators.

11.5 ACOUSTO-OPTIC INTERACTION IN FIBER GRATINGS

Fiber gratings are versatile and useful devices employed in several sensing and telecommunications applications [40]. A fiber grating is simply an optical diffraction grating formed by the periodic permanent change in the refractive index along the fiber core, whose effect causes the coupling between propagating optical modes. Fiber gratings are broadly classified into two types: Bragg gratings (also called reflection or short-period gratings), in which the coupling occurs between modes traveling in opposite directions, and LPGs (also called transmission gratings), in which the coupling is between modes traveling in the same direction. For first-order diffraction, which is usually dominant in an FBG, coupling takes place between the fundamental forward propagating and its correspondent backward propagating mode. The fundamental relationship that describes this particular interaction is expressed as

$$\lambda_{FBG} = 2n_{eff}\,\Lambda_B \tag{11.17}$$

where

λ_B is the Bragg or resonance wavelength
n_{eff} is the effective index of the propagating mode
Λ_{FBG} is the grating period

On the other hand, an LPG is a fiber grating, whose period is chosen in order to couple light from the fundamental guided mode to forward propagating cladding modes. Given that the energy of the cladding modes is lost due to absorption and scattering in the surrounding environment, a rejection band is measured in the transmission spectrum [41]. The wavelength-dependent phase-matching condition in an LPG is governed by the relationship

$$\lambda_{LPG}^{m} = \left(n_{LP_{01}} - n_{LP_{1m}}\right)\Lambda_{LPG} \tag{11.18}$$

where

λ_{LPG}^{m} is the peak wavelength (or resonance wavelength) of the mth attenuation band
$n_{\text{LP}_{01}}$ and $n_{\text{LP}_{1m}}$ represent the effective indices of the fundamental guided mode, LP_{01}, and the mth LP_{1m} cladding mode, respectively [42]
Λ_{LPG} is the grating period, which, in contrast to FBGs, ranges in the hundreds of micrometers due to the small differences between the effective refractive indices of the core and cladding modes

The minimum transmission of the rejection band at λ_{LPG}^{m} is calculated as

$$T^{m} = 1 - \sin^2\left(\kappa_m L\right), \tag{11.19}$$

where

L is the length of the LPG
κ_m is the coupling coefficient for the mth cladding mode, which is determined by the overlap integral of the core and cladding modes and by the amplitude of the periodic modulation of the mode propagation constants [42]

Gratings can be inscribed in fibers employing several methods [40]. As seen earlier, they are characterized by three main parameters: reflectivity/transmissivity, period (or pitch), and length, which can be dynamically controlled by the strain field generated by the acoustic wave in the fiber. Indeed, the acousto-optic mechanism provides another degree of flexibility, offering new modulating and control alternatives and extending the range of applications of all-fiber grating devices.

The following sections describe the interaction of flexural and longitudinal acoustic modes with the different types of gratings written in fibers. Usually, such interaction depends on the ratio between the acoustic wavelength and the grating length. For instance, when the acoustic wavelength is much higher than the grating length ($\lambda_a \gg L_{\text{grat}}$), the effective index and the grating pitch can be taken as homogeneously perturbed along the length. However, when the acoustic wavelength is much lower than the grating length ($\lambda_a \ll L_{\text{grat}}$), the acoustic mode generates many short-range compressed and expanded strain sections within the grating, which leads to a change in its period.

The first experimental application of acoustic waves in fiber gratings was reported by Liu et al. [8,43], in which the grating is excited by a longitudinal mode. In this case, the spectral response of the original FBG shows new and narrow reflection bands symmetrically located at both sides of the original Bragg wavelength. The effect is equivalent to a superlattice modulation of the grating, and the structure formed in the interaction is known as super grating or sampled grating. The following sections provide an understanding of the interaction of longitudinal and flexural modes with Bragg grating and LPG through the acousto-optic mechanism.

11.5.1 Interaction of Fiber Bragg Gratings with Flexural Acoustic Modes

Section 11.2 described the excitation of acoustic modes. The rodlike structure represented by the standard silica fiber supports the lowest-order flexural, torsional, and longitudinal modes in the frequency range < 18 MHz for untapered fibers given the limit $f_a a/c_t < 0.3$. However, the existence of a certain mode in the guide depends on the excitation condition. This condition is particularly determined by the type and the way the piezoelectric element is driven (in the radial or transversal direction).

Literature is rich on reports concerning the interaction of flexural modes with gratings. Two cases can be distinguished here: the excitation in the frequency range below 1 MHz and the frequency range above 1 MHz. For f_a < 1 MHz, the acoustic wavelength is not able to provide the phase-matching condition observed in Equation 11.13, so that no coupling to cladding modes takes place. However, reports using the modulator in the longitudinal configuration have demonstrated that flexural modes are generated and cause the appearance of strong bendings in the fiber, whose periods lie in the range of millimeters [24,44]. The FEM modal analysis has been very helpful in the identification of such acoustic modes in the structure [25,45]. For instance, for the frequency regime (up to 650 kHz), most piezo resonances generate flexural acoustic modes that cause strong bendings in the fiber. The observed frequency range depends mostly on the PZT characteristics. Simulation shows that even though the flexural modes are dominant, the lowest-order longitudinal mode is simultaneously excited. The result is the propagation of a hybrid wave. The joint effect of the resulting strain due to both modes leads to the generation of a chirp in the Bragg grating reflection response as seen in Figure 11.8. This mechanism has been early observed in gratings inscribed in microstructured fibers [46] but has been later observed in gratings written in conventional ones [47].

Previous works have shown that fibers with S-shape bending allow the chirping of Bragg gratings without the shift of the grating central wavelength [48,49]. For instance, a fiber fixed or embedded on a thin and flexible metal beam is subjected to two antisymmetric curvatures when an S shape is imposed to the beam. Similarly, a flexural acoustic mode of amplitude A_F and wavelength $\lambda_F(f_a)$ imposes the same displacement to the fiber, as shown in Figure 11.9a. Such a mode causes a fiber bending in which the curvature radius is approximated by Jones et al. [48]

$$R_C = \frac{\lambda_F^2}{(16A_F) + A_F^{-4}} \tag{11.20}$$

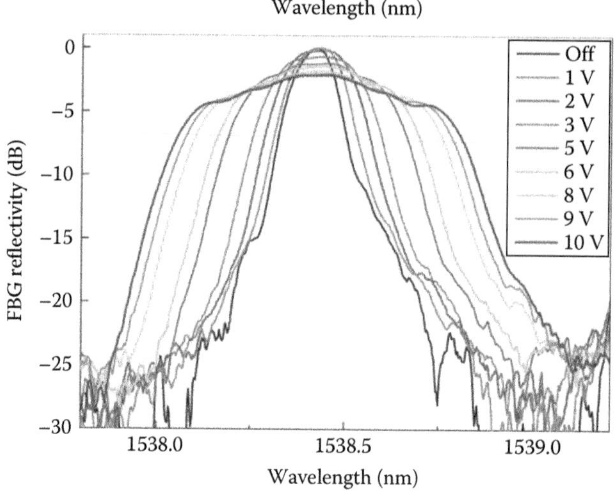

FIGURE 11.8 Chirp generated by a flexural mode at $f = 227.4$ kHz in a 50 mm long Bragg grating.

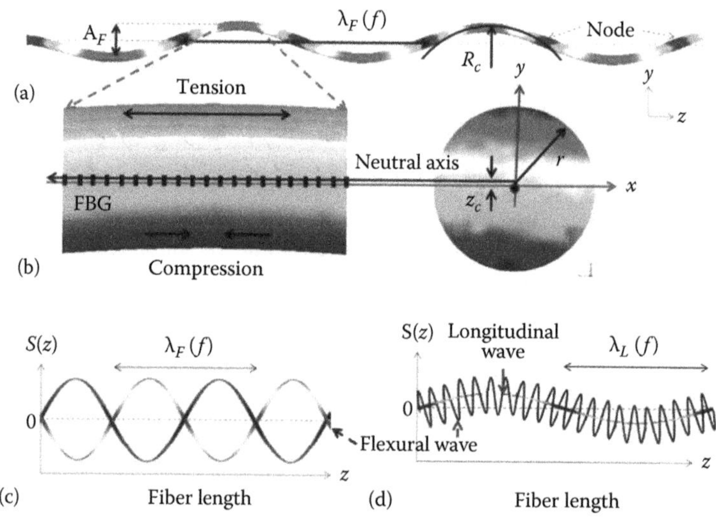

FIGURE 11.9 (a) Flexural acoustic wave parameters and (b) induced strain at curvature. (c) Flexural and (d) combined flexural–longitudinal strains.

At the cross section, the fiber experiences tension strain in the upper side and compression strain in the bottom, as shown in Figure 11.9b. At the neutral axis [48] and at the wave nodes, the resulting strain is null. Since the distance between the neutral axis and the fiber central axis, given by Z_C in Figure 11.9b, is very small, one can assume that the fiber is submitted to antisymmetric strains at its cross section. Similarly, a grating written in such a fiber will be subjected to tension strain in the upper side and compression strain in the bottom side of the cross section (see Figure 11.9b). In this way, the resulting average strain due to the flexural mode presents a small contribution, as seen in Figure 11.9c. However, if a longitudinal acoustic mode of period $\lambda_L(f)$ is simultaneously present in the fiber, as seen in Figure 11.9d, it causes a resulting nonzero strain, which is then responsible to induce the chirp in the grating. The relationship between the wavelength and the frequency of the flexural and longitudinal acoustic modes is $\lambda_F = (\pi r c_{ext}/f_a)^{1/2}$ and $c_{ext} = \lambda_L f_a$, respectively, where r is the fiber radius, c_{ext} is the extensional velocity in the silica ($c_{ext} = 5740$ m/s), and f_a is the excitation frequency of the acoustic field. Results shown in Figure 11.9 considered the modulator excited at $f = 227.4$ kHz, in which the

PZT radial and thickness deformations allow simultaneously the excitation of flexural and longitudinal acoustic modes. The wavelengths of flexural and longitudinal modes in this particular case are estimated to be $\lambda_F = 2.22$ mm and $\lambda_L = 2.52$ cm, respectively (see Figure 11.9c and d).

As described in Section 11.4, flexural modes excited above 1 MHz lead to the creation of microbends that serve equivalently as an acoustic long period grating for coupling the core mode to high-order cladding modes. For obtaining the highest coupling efficiency from the fundamental mode to cladding modes, the acoustic wavelength must equal the beat length of the interacting modes. Thus, the coupling of power from the fundamental to cladding modes results in the appearance of attenuation notches in the transmission spectrum, whose resonance wavelength λ_R is determined by the relationship in (11.14). The preferable configuration employed to generate microbendings uses the modulator in the transversal configuration [16], although the configuration with the modulator in the longitudinal position has been also employed [38]. Now, if a Bragg grating is inscribed in the fiber and the period Λ_{FBG} of the grating matches the acoustic wavelength λ_a in the fiber, three basic types of coupling mechanisms simultaneously take place, which are described as follows [50–54]:

1. Contradirectional coupling between the forward and backward propagating fundamental modes (LP_{01}) resulting from the phase-matching condition provided by the short-period Bragg grating
2. Contradirectional coupling between the propagating core mode (LP_{01}) and higher-order contrapropagating cladding modes (LP_{11}, LP_{21}, LP_{31}, LP_{41}, etc.), whose coupling is still given by the phase-matching condition provided by the short-period Bragg grating (the effect is enhanced if tilted gratings are employed)
3. Codirectional coupling between the propagating backward fundamental mode and copropagating backward cladding modes resulting from the acoustically generated LPG with period λ_a through either the first- or *n*th-order diffraction

The result of these interactions is the appearance of acoustic-driven reflection bands in the reflection spectrum, corresponding to the transfer of power from the backward LP_{01} mode to higher-order backward cladding modes. Figure 11.10 pictorially shows this situation for the coupling between the LP_{01} and the LP_{11} modes.

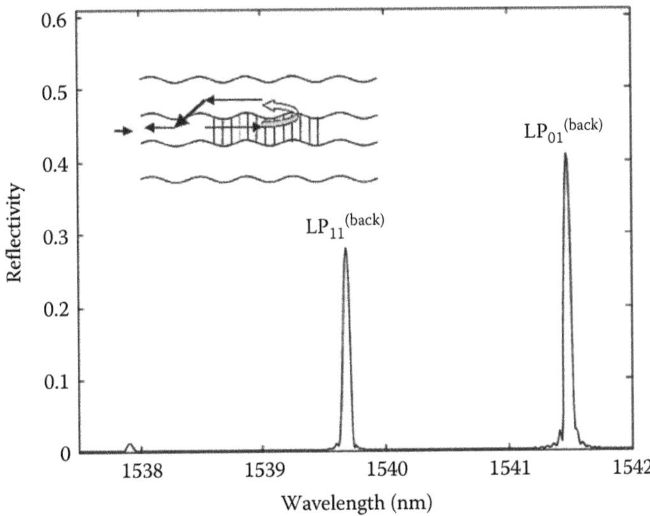

FIGURE 11.10 Reflection bands resulting from the transfer of power from the forward to the backward propagating LP_{01} mode due to the short-period Bragg grating and from this mode to the LP_{11} backward cladding mode due to the long-period acoustic grating. (Adapted from Liu, W.F. et al., *Opt. Lett.*, 25(18), 1319, 2000; Abrishamian, F. et al., *Opt. Fiber Technol.*, 13, 32, 2007.)

Usually, the mechanism described in item 2 above for ordinary situations does not come into play, as the Bragg condition privileges the coupling of the first diffraction order (in the case no tilted gratings are used). A set of multimode-coupled equations can be used to simulate numerically the power exchange between modes. Luo et al. [53] and Abrishamian et al. [54] have successfully calculated such mode coupling. The calculated reflectivity curves are in good agreement with those obtained experimentally in Liu et al. [50]. Moreover, they have also shown that the influence of the higher-order optical modes, such as the LP_{31} and LP_{51}, on the reflective spectra is very small and the coupling occurs mostly between the LP_{01} and LP_{11} modes, so that the analysis can be simplified and the reflective spectra of higher modes can be neglected. In this case, the two-mode coupled equations are sufficient to obtain the accurate spectral characteristics of an FBG subjected to the flexural acoustic mode. This sort of acousto-optic modulation can be used to switch and tune the wavelength of an optical carrier and can be employed as an add–drop in wavelength-division multiplexing (WDM) systems and in fiber sensor devices.

11.5.2 Interaction of Fiber Bragg Gratings with Longitudinal Acoustic Modes

Acoustic longitudinal modes are described by the displacement vector field using the lower set of trigonometric functions in (11.1) and choosing $n = 0$, giving a solution for the displacement components that has no angular dependence in θ and is axially symmetric. That is, the displacement function presents components in the radial and axial directions only, which is translated into a deformation of the fiber as shown in Figure 11.11c. The periodic strain field of the longitudinal acoustic mode perturbs the short-period grating in two ways: (1) by increasing the average refractive index due to the elasto-optic mechanism and (2) by periodically modulating the grating pitch, causing spatial-frequency modulation.

This is particularly relevant when the wavelength of the longitudinal acoustic mode is short as compared to the grating length ($\lambda_a \ll L_{grat}$). Therefore, when such a mode propagates along the fiber, the displacement field generates many periodic compression and rarefaction zones related to the strain field in the material. The result is the appearance of additional bands on both sides of the grating reflection spectrum with regularly spaced peaks that originate from the phase matching of the counterpropagating optical mode with the modulated grating planes [8]. Russell and Liu [55]

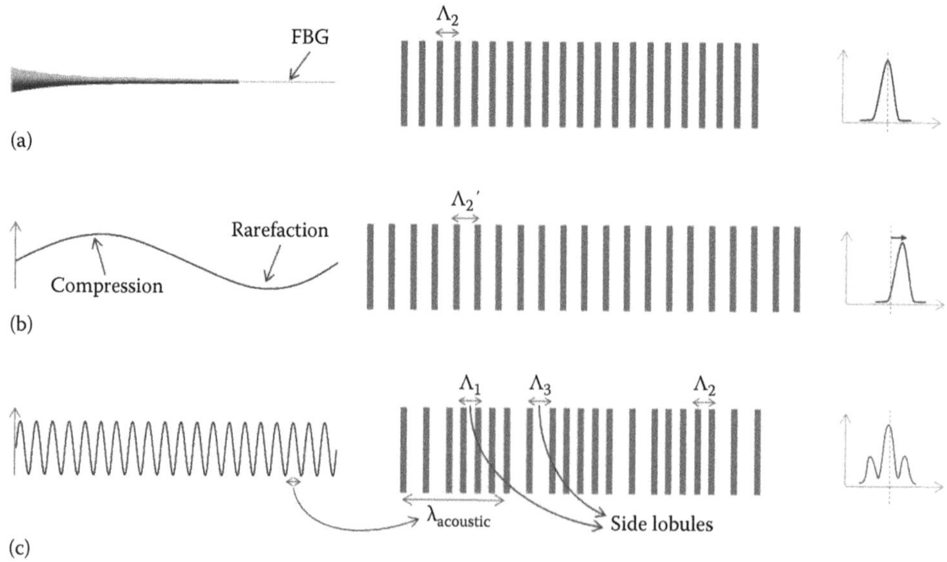

FIGURE 11.11 Representation of the sampled modulation produced by the longitudinal acoustic mode upon interaction with the Bragg grating. (From Pohl, A.A.P. et al., *Photon. Sens.*, 3(1), 1, 2013.)

developed a simple mathematical model to describe the appearance of these side bands. The total change in the effective refractive index, experienced by the grating, is given by the strain field and the contribution of the pitch modulation and can be calculated as

$$\Delta n_{\text{eff}} = (1-\chi)\varepsilon(z,t) + n_0\left\{1 + C\cos\left[K\left(z - \int \varepsilon(z,t)dz\right)\right]\right\}, \tag{11.21}$$

where

$$\varepsilon = \varepsilon_0 \cos(\omega_a t - k_a z) \tag{11.22}$$

is the periodic strain associated with the propagation of the longitudinal acoustic mode. In these equations, $k_a = 2\pi/\lambda_a$ is the acoustic wave vector, λ_a is the acoustic wavelength, and ω_a is the angular acoustic frequency

χ is the magnetic susceptibility of the fiber
C is the modulation depth of the effective refractive index
n_0 is the refractive index of the unperturbed FBG
$K = 2\pi/\Lambda_{\text{FBG}}$ is the grating vector

For an acoustic beam of area A, carrying power P_a in a medium with Young's modulus E, and acoustic group velocity υ_{ga}, the peak strain has the value [55]

$$\varepsilon_0 = \sqrt{\frac{2P_a}{EA\upsilon_{ga}}} \tag{11.23}$$

The pitch modulation produces a sequence of sidebands in the spatial-frequency domain whose amplitudes are described by a standard Bessel function expansion. This formulation leads to a sequence of ghosts of the original fiber grating forming at spatial frequencies given by successive spatial sidebands of K. This situation is depicted in Figure 11.12.

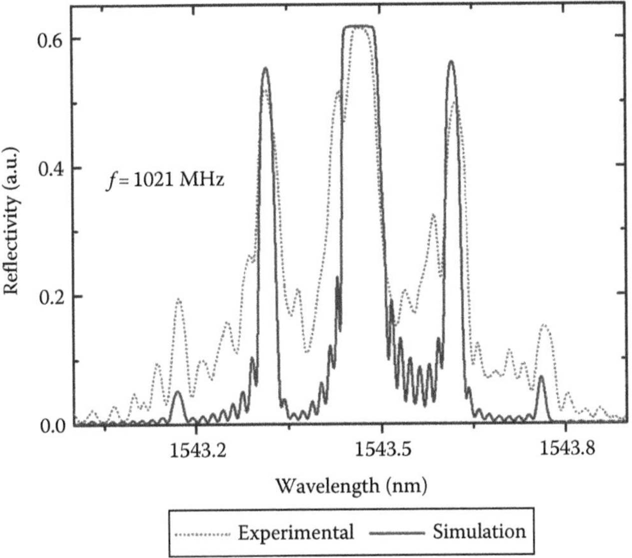

FIGURE 11.12 Experimental and simulated results of the excitation of a longitudinal acoustic mode with an FBG at $f = 1021$ MHz. (From Oliveira, R.A., Characterization and new applications of the acousto-optic effect in fiber gratings, Ph.D. Thesis, Federal University of Technology, Paraná, Brazil, 2011.)

A rigorous approach between the interaction of the longitudinal acoustic mode and the FBG can be performed using the FEM and the transfer matrix (TMM) [45] methods. The FEM approach allows the strain field caused by the acoustic longitudinal mode to be completely characterized along the structure, while the TMM is used to obtain the optical spectrum of the corresponding strained grating. Moreover, the FEM can also take into account the strain in the horn fiber setup, considering the real dimensions of the device. The FEM and TMM approaches build an adequate and accurate method, particularly if the structure under analysis presents an arbitrary cross-sectional shape. The methodology consists of two steps. First, the strain field in the whole structure (horn, taper, and FBG) is obtained by using the FEM. Second, the calculated strain field is used in the TMM algorithm to obtain the FBG spectrum. Oliveira et al. [45] discuss discretization of the acousto-optic modulator for the 1D case. The structure is decomposed in N 1D elements of length $\Delta z = L_D/N$ separated by nodes, with L_D as the total length of the modulator, including the horn. Each element is associated with a value that represents the area of the structure at that section. The differential equation of motion that represents the propagation of the acoustic mode in the structure is given as

$$E \cdot \frac{\partial}{\partial z}\left(A(z)\frac{\partial u(z,t)}{\partial z} \right) - \rho \cdot A(z)\frac{\partial^2 u(z,t)}{\partial t^2} = 0, \tag{11.24}$$

where

u is the axial displacement, which is dependent on the position z and on the time t

$\partial u/\partial z$ is the longitudinal strain ε

$A(z)$ accounts for the variable size of the structure along the z-axis

E is Young's modulus

ρ is the silica density, assumed to be 72.5 GPa and 2200 kg/m^3 for the silica, respectively

The external excitation applied to the modulator is expressed as the combination of a constant load and a harmonic load of frequency ω and amplitude P_0 generated by the piezo element. In this way, the solution for the displacement function $u(z, t)$ is a combination of a fixed (preload) and time-varying load. Once the displacement field is obtained, the strain field in each finite element of the structure is found by differentiation and given as

$$\varepsilon^e = \frac{u^{e+1} - u^e}{\Delta z}, \tag{11.25}$$

in which the finite element is linear and u^{e+1} and u^e are the displacements in the local nodes $e + 1$ and e, respectively [45]. The TMM [56] is a well-known method derived from the coupled-mode theory. The fiber grating is divided into piecewise-uniform sections and the final solutions for each section are combined multiplying the matrices associated with each section, from which the amplitude and power reflection coefficients of the grating response are calculated. When the TMM approach is applied to a uniform grating, the average refraction index change is constant. In such case, the effective refractive index perturbation in the core is described by Erdogan [56]

$$\delta n_{\text{eff}}(z) = \delta n_0(z)\left[1 + \nu \cos\left(\frac{2\pi}{\Lambda}z\right)\right], \tag{11.26}$$

where

ν is the fringe visibility

Λ is the grating nominal pitch

The presence of the grating imposes a dielectric perturbation to the waveguide and forces the coupling between the forward and backward propagating modes. However, the propagating acoustic mode turns the grating pitch nonuniform and changes the path of the optical mode. Taking into account this nonuniformity, the reflection and transmission spectra can still be calculated by considering the same piecewise approach, whereby the grating is divided into discrete uniform sections that are individually represented by a matrix. The solution is found by multiplying the matrices associated with each of the sections. When the characteristic equation is solved by making the matrix determinant equal to zero, the resulting polynomial enables the eigenvalues to be found [57]. The connection between the FEM and TMM approaches takes place, considering the relationship that computes the variation of the design wavelength along the fiber axis (z-axis) in the TMM as a function of the strain field. This relationship is given as

$$\lambda_D(z) = \lambda_{D0} + [1 + (1 - p_e)\varepsilon(z)], \tag{11.27}$$

where
p_e is the photoelastic coefficient
$\varepsilon(z)$ is the strain field calculated using (11.25) in the FEM

This way, once the strain field is known along the fiber and, particularly, along the grating, one can calculate the grating optical response (reflected and transmitted spectra) [45]. Figure 11.12 shows a comparison of the experimental and simulated response (using the FEM and TMM) of an FBG to which an acoustic excitation at $f = 1021$ MHz and $V_{PZT} = 10$ V (corresponding to a load amplitude of 1 N applied to the base of the silica horn) was applied. In the simulation, the damping of the acoustic wave along the structure is neglected.

11.5.3 Interaction of LPGs with Flexural Acoustic Modes

As described earlier, the LPG is an optical fiber grating, whose period is chosen in order to couple light from the fundamental guided mode to forward propagating cladding modes. Given that the energy of the cladding modes is lost due to absorption and scattering in the surrounding environment, a rejection band is measured in the transmission spectrum. The LPG period ranges in the hundreds of micrometers due to the small differences between the effective refractive indices of the core and cladding modes. The strain caused by an acoustic flexural mode in the fiber will shift the wavelength of the LPG rejection band λ_m. This shift is computed as

$$\frac{\Delta\lambda_m}{\lambda_m} = (1 + p_e)\varepsilon(z), \tag{11.28}$$

where
$\Delta\lambda_m$ is the wavelength shift experienced by the grating
λ_m is the peak wavelength of the attenuation band when the fiber is at rest
The parameter p_e is the strain-optic coefficient, assumed to be $(-1.191 \times 10^{-6})\ \varepsilon^{-1}$ for an LPG [58]

The numerical approach using both the FEM and TMM can be employed in this case. An alternative technique, known as the method of assumed modes, can also be used to calculate the strain field [59]. The calculated strain is passed over to the TMM via the *design wavelength* in (11.27). This is achieved by using the relationship in (11.28) and substituting λ_m by the design wavelength, λ_{D0}, in the TMM, considering an infinitesimally weak grating with period Λ and making $\Delta\lambda_m = \Delta\lambda_D$ and $\lambda_m = \lambda_{D0}$, where λ_{D0} is the peak wavelength when the fiber is at rest. In this way, both the effect of the period variation and the change in the LPG effective refractive index can be accounted for at each point inside the grating.

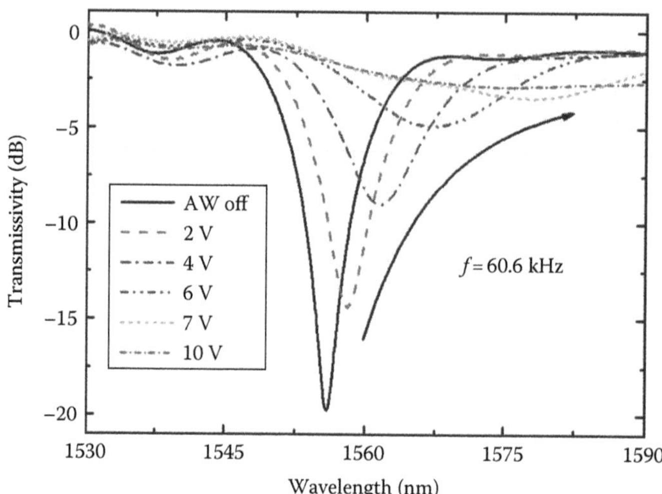

FIGURE 11.13 LPG spectrum behavior when the PZT load varies from 0 to 10 V at $f = 60.6$ kHz in a grating of 40 mm length. (From Oliveira, R.A. et al., *Meas. Sci. Technol.*, 22(045205), 1, 2011.)

Figure 11.13 shows experimental results for the excitation of an LPG of 40 mm length at $f = 60.6$ kHz. The plot shows curves for piezo loads varying from $V_{PZT} = 0$ to 10 V. The increase in the amplitude of the acoustic flexural mode reduces the peak transmissivity of the LPG attenuation band and induces a shift to longer wavelengths (*red shift*). The periodic bending induced by the acoustic wave modifies the effective refractive indices of the core and cladding modes and modulates the intermodal overlap integral, consequently reducing the coupling coefficient between them. A similar behavior is observed for acoustic waves generated at other low piezo resonances.

Excitation of LPGs at higher frequencies, where the predominant acoustic mode is longitudinal [44], leads to no effects in the LPG spectrum. This behavior indicates that the changes in the LPG characteristics are due mainly to the geometric contribution (bendings) of the flexural wave to changes in the effective refractive index through the elasto-optic mechanism.

11.5.4 Switching Time of the Acousto-Optic Effect in Gratings

The time taken by the acoustic wave to change the grating spectrum is called switching time. This time basically depends on a variety of factors, such as the mechanical setup, temperature, and fiber tension. For example, a pure longitudinal mode travels at the acoustic speed of $c_{ext} = 5760$ m/s, which corresponds to the speed of sound in the silica optical fiber. However, a pure flexural mode propagates at a lower speed ($c_t = 3769$ m/s). If a Bragg grating is inscribed the fiber and the modulator is set to excite a longitudinal mode at a certain acoustic frequency, lobes will appear on the side of the grating reflection spectrum. For measuring the switching time, one can use a narrowband optical signal (such as one emitted by a DFB laser) positioned at the same wavelength of one of the side lobes generated by the corresponding acoustic frequency. The time response is assessed using a fast photodetector connected to an oscilloscope. When the acousto-optic effect is off, the photodetector measures only the average power (CW) reflected by the grating. However, when the effect is turned on, the acoustic mode propagates in the fiber and creates the side lobes. The optical signal is therefore reflected by the grating side lobe at the moment it appears enabling a change in the previously measured optical power using the photodetector. Measurements taken with the modulator in the longitudinal configuration at an excitation frequency of $f = 617$ kHz are plotted in Figure 11.14. The switching time $t_s = 17$ μs corresponds to the time the longitudinal acoustic mode takes to settle in the grating. The time $t_s = 42$ μs corresponds approximately to the delay the acoustic wave takes to propagate along the horn and silica fiber until the point where the grating is located

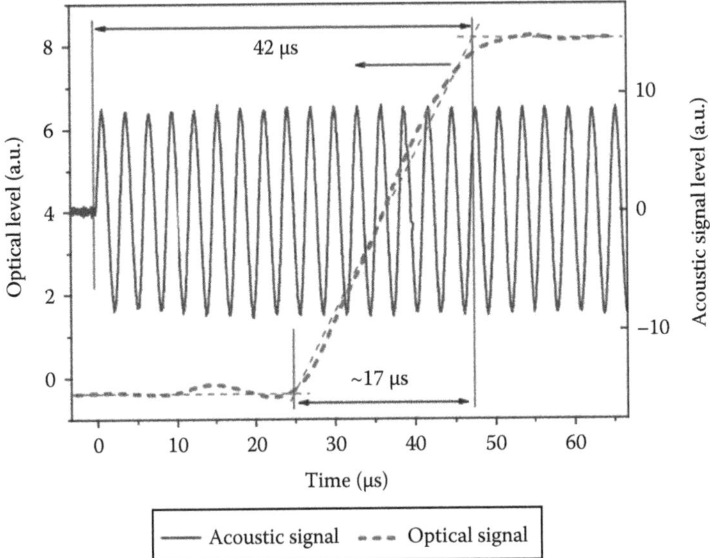

FIGURE 11.14 Characterization of the switching time of the acousto-optic effect in an FBG. (From Marques, C.A.F. et al., *Opt. Commun*, 284, 1228, 2011.)

[60]. Optimizing the physical arrangement makes it possible to achieve lower delay times. However, the switching time of 17 μs is the lowest one can achieve for the propagation of acoustic modes in silica in the linear regime [33].

11.6 ACOUSTO-OPTIC INTERACTION IN PHOTONIC CRYSTAL FIBERS

The PCF, also known as microstructured optical fiber (MOF), is a new class of optical waveguides that provide a high degree of functionality not achieved by their conventional counterparts in silica- or polymer-based materials. For instance, conventional optical fibers consist of a solid glass core with higher refractive index encircled by a cladding with a relatively lower refractive index. PCFs, however, consist of a glass cladding with a number of air holes running along its length and a core consisting of either glass or air. In the first case, light is guided due to the total internal reflection mechanism caused by the lower effective refractive index in the cladding due to the air holes. This class of PCF is known as solid-core or index-guiding fiber. In the latter case, light is confined to a central defect due to the coherent Bragg scattering off the periodic array of air holes in the cladding, where light within well-defined wavelength stop bands is prohibited from propagating.

The PCF was first demonstrated experimentally by Knight et al. [61,62]. Due to their peculiar composite structure, the properties of a PCF can be tailored, such as by adding specific materials into the holes. The refractive index step in index-guiding PCFs is far higher compared with that of the conventional optical fiber, which is typically 1%–2%. The design parameters of the PCF are the air hole diameter, d; the interhole spacing or pitch, Λ; and the number of layers of air holes in the cladding. Such fibers are usually described by the air-filling fraction or the ratio d/Λ, a ratio that ranges from a few percent up to 90%, whereas the interhole spacing values vary typically from 1.0 to 20 μm. By manipulating these parameters, one can easily change the propagation constant of the guided modes and, consequently, tailor the fiber modal, dispersion, and nonlinear properties. Since their invention, the PCF has been used as a platform for developing new concepts in sensors and devices for signal processing. In this way, the acousto-optic effect has also been employed to PCFs and to gratings written in such fibers [46]. Basically, the same mechanisms described in the previous sections apply to acoustic modes propagating in PCFs. For instance, Diez et al. [63] first reported the excitation of cladding modes in a solid-core PCF with a $\Lambda = 3$ μm interhole spacing and

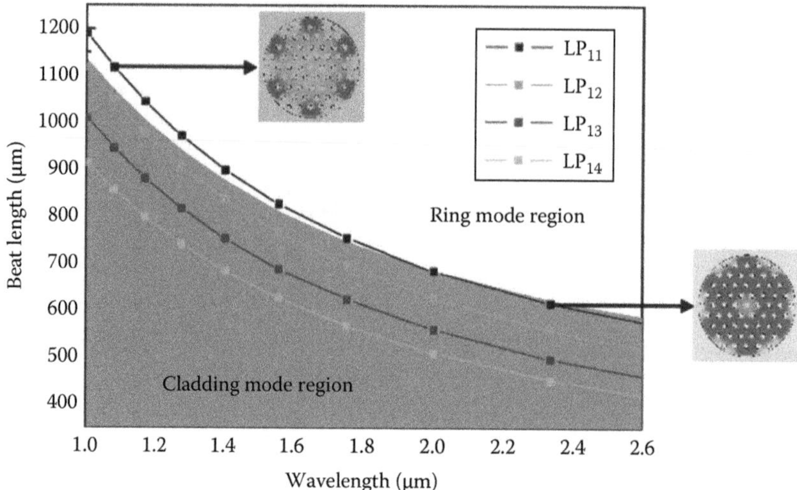

FIGURE 11.15 Beat length between the core and four cladding modes in a four-ring PCF with pitch Λ = 9.7 μm and $d/\Lambda = 0.42$. (Reproduced with permission from Park, H.C. et al., *Opt. Express,* 15(23), 15154, 2007.)

a 0.5 μm hole diameter, producing an acousto-optic filter that could be tuned from 0.4 to 1.8 MHz. Given the condition stated in (11.13), for which the optical beat length between the modes, L_B, equals the acoustic wavelength, λ_a, the coupling takes place between the fundamental core mode and a cladding mode in the PCF (Figure 11.15).

The acousto-optic properties in PCFs, such as the acoustic velocity dispersion, are influenced by the periodicity of the cladding structure. For the acousto-optic interaction to take place, there must be a significant overlap area between the acoustic wave and the optical modes. This is achieved for the limit $f_a a/c_t < 0.3$ in standard untapered silica fibers, in which only the fundamental flexural, torsional, and longitudinal acoustic modes can propagate. For excitation with flexural modes, the displacement of the lowest-order flexural mode resembles that of a purely bent rod. Under this condition, the acoustic dispersion of a silica-based PCF can be approximately calculated from the Euler–Bernoulli (E–B) theory [64], which holds for fibers of arbitrary cross section, and is stated as

$$c = k_a \sqrt{\frac{I_e}{\rho A_{\text{eff}}}} \tag{11.29}$$

where

- ρ is the density of the fiber material
- A_{eff} is the effective area of the fiber cross section
- k_a is the wavenumber of the acoustic wave
- I_e is the flexural rigidity defined by the following integral equation calculated over the fiber cross section

$$I_e = \int_S E(x,y)(x\cos\theta + y\sin\theta)^2\, dxdy, \tag{11.30}$$

where

- $E(x, y)$ is Young's modulus
- θ is the angle that the polarization of the acoustic mode makes with the x-axis

The evaluation of such integral along with the definition of c_{ext} gives

$$\frac{c}{c_t} = \alpha\left(\frac{\pi c_{\text{ext}}}{c_t}\right)^{1/2}\left(\frac{f_a a}{c_t}\right)^{1/2}, \tag{11.31}$$

where
f_a is the acoustic frequency
a is the fiber radius
α is a parameter that depends on the geometry of the fiber

$\alpha = 1$ for a homogeneous cylinder, while α is slightly above 1 for PCFs (the authors in Haakestad and Engan [65] calculate $\alpha = 1.013$ for a solid-core PCF with an outer diameter of 125.3 μm, an average interhole spacing of 9.97 μm, and an average hole diameter of 4.79 μm). One sees that the influence of the holes in the PCF on the phase velocity of the lowest-order flexural mode is rather small, though still significant for the accurate determination of the intermodal beat length when the acousto-optic effect takes place. Equation 11.31 can be rewritten to give a more useful relationship in terms of the acoustic wavelength, λ_a, which is further used to verify the matching condition provided by (11.13). Yeom et al. [66] have derived such a relationship, which also accounts for the influence of a hollow-core fiber structure and is expressed as

$$\lambda_a = \left[\left(\frac{f_a a}{c_t}\right)\frac{c_t}{c_{\text{ext}}}\frac{1}{a a_0 \pi}\right]^{1/2} \tag{11.32}$$

in which $a_0 = [(a^4 - 0.92r^4)/(a^2 - 0.92r^2)]^{1/2}$. In this relationship, the microstructured region of the hollow-core fiber is replaced by a cylindrical shell of radius r representing the silica-based material with a strongly reduced density and Young's modulus, $E(x, y)$, as compared to that of pure silica. The equation shows that, for a given acoustic frequency, the acoustic wavelength becomes larger in the hollow core than in the solid fiber with the same cladding radius. The difference becomes larger as the inner radius increases. A similar approach for taking into account the mechanical properties of the microstructured region in simulations (particularly if the FEM is used) employs the theory of homogenization or the rule of mixture of composite materials [67].

The condition for which the optical beat length must equal the acoustic wavelength requires a systematic analysis of the effective refractive indices of the cladding modes in the PCF structure. As an example, such analysis has been performed in a four-ring solid-core PCF [68] and in a PCF presenting inversion and sixfold rotation symmetry in the air-hole structure [69]. The work shows that the degeneracy and also the field profiles of HE_{12} and the mode group formed by TE_{01}, HE_{21}, and TM_{01} are equivalent to those found in conventional step-index fibers. In this way, the nearly or completely degenerate modes can also be grouped and labeled as LP_{nm} modes for the calculated cladding modes, as in the case of conventional single-mode fibers. Moreover, results in Park et al. [68] show that the slopes of the beat length curves, given by $L_B = \lambda/\left(n_{\text{LP}_{01}} - n_{\text{LP}_{1x}}\right)$, are always negative in this type of PCF, in contrast to conventional single-mode fibers, which show a positive slope. For instance, guided modes in the analyzed PCF tend to have larger field distributions in the air holes at longer wavelengths; therefore, the mode effective index becomes lower. The reduction of the effective refractive index of the fundamental mode at longer wavelengths is smaller than those for higher-order modes because that mode is most tightly confined in the silica core region.

The inscription of Bragg gratings in PCFs adds spectral selectivity to devices and therefore enhances many applications. There are several challenges facing the writing of gratings in structured compared to that in conventional fibers. These include high levels of scattered light arising from multiple interface reflections in the cladding structure and, more critically, rotationally variant symmetry. Nevertheless, the writing of gratings in structured fibers has been successfully achieved and dominated over the past years [70,71]. The acousto-optic effect applied to gratings written in microstructured fibers follows the same basic principles ruled in previous sections. Two cases can be distinguished: the excitation in the low-frequency (<1 MHz) and in the high-frequency (>1 MHz) range. In the low-frequency range, the acoustic wavelength is not able to provide the mode-matching condition, so that no coupling to cladding modes takes place. In the low-frequency range, one observes only the broadening of the grating reflection spectrum. In the high-frequency range, the periodic strain field of the acoustic wave generates shorter compression and rarefaction zones in the material, inducing a refractive index change in the propagating mode but also producing a modulation of the grating pitch. In this latter case, one sees regularly spaced additional bands on both sides of the grating reflection spectrum, as explained earlier.

The interaction of flexural waves in the low-frequency acousto-optic range with a grating written in a PCF has been reported in Pohl et al. [46], in which a triangular core PCF of diameter $d = (2.1 \pm 0.3)$ μm surrounded by 12 rings of holes with a relatively high hole diameter to pitch ratio ($d/\Lambda = 0.58$) was used [72]. A 12 mm long FBG written into the PCF by means of an ArF exciplex (193 nm, 15 ns, 10–200 Hz) through direct writing with a phase mask was employed. In the experiments, the horn fiber system using a modulator in the longitudinal configuration was excited at $f = 82.3$ kHz. Simulation results show that both longitudinal and flexural modes are excited. The wavelength of the flexural mode is estimated as 3.5 mm and the wavelength of the longitudinal mode is estimated as 70 mm, which is much longer than the grating length. Figure 11.16 shows the grating transmission and reflection spectra before and after the acoustic excitation is turned on at 82.3 kHz. Once the acoustic wave is established, both the transmitted and reflected spectra show a reduction in strength while also broadening. The broadening is explained by the interaction of both flexural and longitudinal modes with the grating, as already described earlier. Considering the bending caused by the flexural mode, one observes that the PCF experiences tension strain in the upper side and compression strain in the bottom side of its cross section. In this case, the resulting average strain is small once the distance between the fiber neutral axis and the fiber central axis (see Figure 11.9) is very small. However, the presence of the longitudinal acoustic mode causes a larger strain, which is then responsible for the observed chirping of the grating.

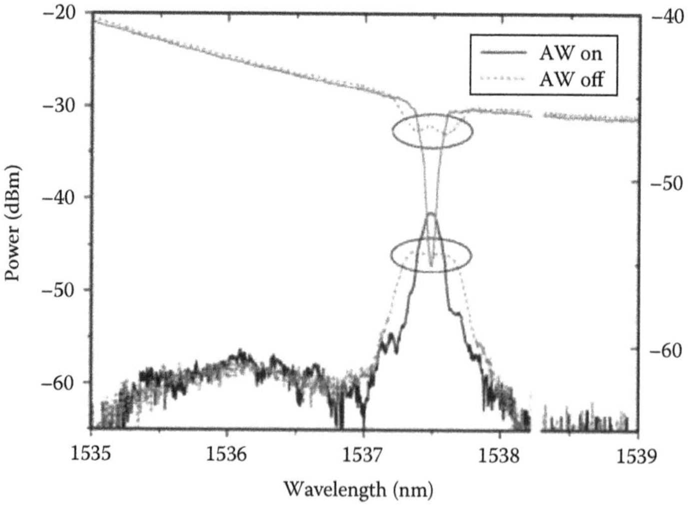

FIGURE 11.16 Broadening of the FBG transmission and reflection spectra due to the interaction of a hybrid acoustic mode in the grating written in the PCF. (From Pohl, A.A.P. et al., Acoustic-induced modulation of photonic crystal fibre Bragg gratings, in *Proceedings of the 10th International Conference on Transparent Optical Networks*, Athen, Greece, vol. 2, pp. 51–54, 2008.)

11.7 APPLICATION IN SENSORS

The acousto-optic effect in fibers has been largely applied to devices such as mode converters [16], tunable filters [38,39,73], Q-switching and mode-locking modulators in fiber lasers [74,75], gain equalizers for erbium-doped fiber amplifiers [76,77], and add–drop multiplexers [52] used in optical communication systems. However, few applications using the acousto-optic modulator in sensors have been reported. The example given in the following describes the use of the acousto-optic modulator to a fiber-based viscometer.

Viscosity is the property of materials characterized by their resistance to flow. Viscometers are important and practical instruments used to measure viscosity and are particularly employed in the food and pharmaceutical industry to monitor and control processes during the fabrication of food and drugs. The most common viscometers used in industry are based on electromechanical techniques having the rotation and vibration of a probe inside the fluid as the operating principle. An interesting alternative to this conventional approach is offered by an optical viscometer based on the acousto-optic effect [78]. An LPG in combination with an acousto-optic modulator can be used as the sensing element. For instance, LPGs are able to measure changes in the refractive index due to the optical interaction of cladding modes with the material in its surrounding. Those changes are manifested through a modification in the resonance dip wavelength and minimum transmissivity. Though viscosity can be related and measured by the induced wavelength ($\Delta\lambda$) and transmittance (ΔT) shift imparted to the LPG as a function of concentration, the complex resulting behavior turns these parameters impractical for sensing. However, the measurement of viscosity can also be accomplished using the LPG temporal response. This is achieved using the acousto-optic modulator to excite a flexural acoustic mode, which forms bends along the fiber and the grating. The effect modifies the coupling coefficients of propagating modes in the LPG, shifting the peak wavelength, and changing the maximum attenuation coefficient, depending on the employed acoustic frequency and intensity. The viscosity is obtained by recording the optical response that critically depends on the damping of the acoustic mode by the surrounding medium. The damped response is a direct measure of the relaxation time associated with the viscous flow—the more viscous the flow the slower the relaxation after the impulse signal is applied to the LPG. Figure 11.17 shows the schematic diagram showing how the viscosity damps the acoustic mode. Figure 11.17a shows the silica horn fiber at rest, and Figure 11.17b shows the attenuation of the flexural mode in the medium after excitation. From bottom to up Figure 11.17b shows how the acoustic wave damps, considering increasing in the viscosity (η) of the environment where the fiber is immersed.

The measurement of viscosity requires the switching of the acoustic flexural mode in the solution for short time periods. This can be accomplished using the burst mode of a signal generator. The idea

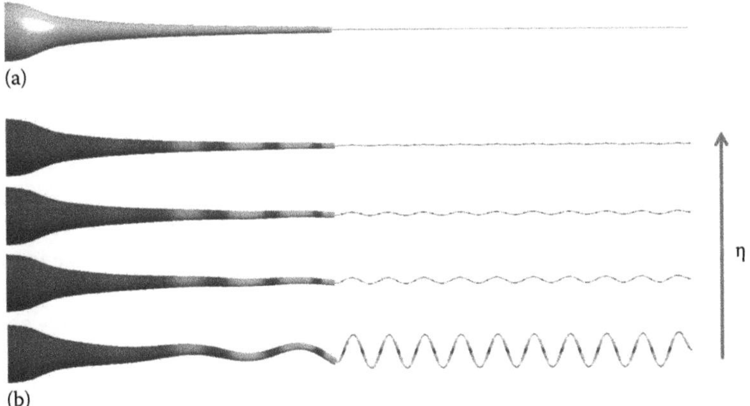

FIGURE 11.17 Illustration of the acoustic behavior of the flexural mode when (a) the fiber is at rest and (b) when the acousto-optic effect is turned on and soon after off. (From Oliveira, R.A. et al., *Sens. Actuat. B Chem.*, 1, 1, 2011.)

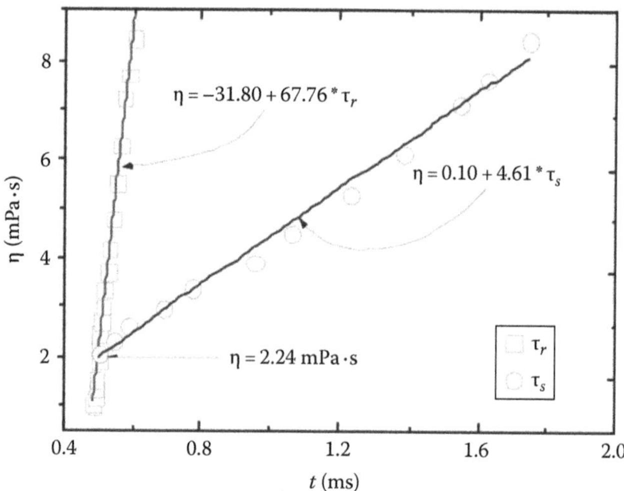

FIGURE 11.18 Viscosity versus time behavior in the acousto-optic viscometer. (From Oliveira, R.A. et al., *Sens. Actuat.. B Chem.*, 1, 1, 2011.)

is to probe the LPG response within a short interaction time using a photodetector. When this happens, the LPG spectrum is shifted, and a maximum signal level in the photodetector is quickly reached. The analysis is based on observing the grating temporal response. For testing the viscometer, measurement is performed in anhydrous D(+)-glucose ($C_6H_{12}O_6$) in deionized, distilled water. This is a solution with a well-known refractive index and viscosity as a function of concentration and is often used to characterize and calibrate commercial devices [79]. First, the behavior of the viscosity and the rise time, depending on the glucose concentration, is assessed. Results show that the viscosity is directly related to [$C_6H_{12}O_6$] through a quadratic dependence [78]. In this case, given that the rise time and concentration both have a quadratic dependency, this means that the viscosity has a linear dependence with the rise time. Figure 11.18 shows the viscosity linear dependence for both the rise time and the total relaxation time, τ_s, after the acoustic burst is turned off. Both parameters show an increase when η increases, which is consistent with the damping of the acoustic mode in the solution. By confining the measurements to the rising time, τ_r, a substantially larger viscosity range becomes possible.

11.8 CONCLUSION

Almost three decades have passed since the first application of the acousto-optic effect in optical fibers. The analysis of the acoustic dispersion curves in cylindrical waveguides of the size of the optical fiber brings relevant information on the phase velocity of acoustic propagating modes (longitudinal, torsional, and flexural), which is important for the operation of the acousto-optic modulator. In conventional untapered single-mode fibers ($2a$ = 125 μm), when $f_a a/c_t$ < 0.3 MHz, the solution of the acoustic propagation equation reveals that there exist only three types of modes: the lowest-order longitudinal mode, the lowest-order torsional mode, and the lowest-order flexural mode. In practice, however, hybrid modes, for example, modes that are a combination of the fundamental ones, play also a role in the operation of the modulator. In this way, the adequate excitation of a particular acoustic mode, achieved by the proper choice of the modulator configuration plus the setting of the correct acoustic frequency and amplitude, is a very important step for obtaining the desired effect in the fiber and grating. Additionally, the previous characterization of piezo resonances is also very helpful in order to identify the relevant resonant modes that are further employed in the acoustic excitation. Though the transversal modulator configuration has been exclusively applied to the excitation of flexural modes, particularly for achieving the coupling between optical modes, the longitudinal configuration is more versatile and compact as it provides a wider range of effects once one chooses the proper piezo element (and the way it is driven) in the modulator.

The interaction of the acoustic modes with Bragg grating and LPG inscribed in the fiber extends the range of applications, which has led to the development of tunable filters, dynamic gain equalizers, the control of laser parameters, and new fiber sensors. The perturbation introduced by an acoustic mode, through the frequency and amplitude of the acoustic wave, respectively, makes it possible to control the period and strength of the grating. Thus, the spectral properties of the optical device can be controlled dynamically through the characteristics of the acoustic perturbation. For instance, in the low-frequency range (<1 MHz), when flexural modes dominate, the interaction of such a mode with an FBG causes the chirp of the grating spectrum. Moreover, the interaction of flexural modes with LPGs makes it possible, by increasing the acoustic intensity, to reduce the peak transmissivity of the LPG peak attenuation band and to induce a shift of the attenuation band to longer wavelengths. On the other hand, when longitudinal modes are excited, the period of the compression and rarefaction strain field in the fiber becomes shorter. Such a strain field will not only cause an additional change of the effective indices of the optical forward and backward propagating modes of a Bragg grating but will also modulate its pitch, leading to the appearance of additional bands on both sides of the grating reflection spectrum.

In recent years, new applications of the acousto-optic mechanism have been reported, such as the technique for writing gratings under the acousto-optic excitation [80,81], the control of the phase shift in gratings using acoustic bursts [82–84], and the application of the effect to polymer fibers [85]. Several other devices and applications were also proposed and demonstrated using the acousto-optic effect over the past years, either using conventional or photonic crystal fibers [86–91]. Although the mechanism is well known, the study and design of acousto-optic modulators is still a topic that deserves attention, particularly if one wishes to achieve high reproducibility. Moreover, the experimental characterization of acoustic modes is also an area that requires further investigation, as the excitation of a specific mode impacts the effect one wishes to obtain.

ACKNOWLEDGMENTS

The author expresses his sincere gratitude to the many collaborators, who contributed to several ideas and results presented in this chapter. The following people deserve special thanks: Ricardo Ezequiel da Silva (PhD student at UTFPR, Brazil), Dr. Roberson Assis de Oliveira (Volvo do Brasil), Dr. Carlos Alberto Ferreira Marques (Universidade de Aveiro, Portugal), Dr. Kevin Cook (The University of Sydney, Australia), Prof. Paulo de Tarso Neves Jr. (UTFPR, Brazil), Prof. Marcos Antonio Ruggieri Franco (IEAv/CTA, Brazil), Prof. Rogerio Nunes Nogueira (Universidade de Aveiro, Portugal), and Prof. John Canning (The University of Sydney, Australia). The author is also grateful to Dr. Wyllian Bezerra da Silva, who carefully assisted in the preparation of several figures for this chapter.

REFERENCES

1. Beyer, R. T., *Sounds of Our Times: Two Hundred Years of Acoustics*, Springer-Verlag, New York, 1998.
2. Brillouin, L., Diffusion of light and x-rays by a transparent homogeneous body, *Annales de Physique* **17**: 88–122, 1922.
3. Debye, P. and Sears, F. W., On the scattering of light by supersonic waves, *Proceedings of the National Academy of Sciences of the United States of America* **18**(6): 409–414, 1932.
4. Lucas, R. and Biquard, P., Optical properties of solid and liquid medias subjected to high-frequency elastic vibrations, *Journal de Physique* **71**: 464–477, 1932.
5. Engan, H. E., Kim, B. Y., Blake, J. N., and Shaw, H. J., Optical frequency shifting in two-mode optical fibers by flexural acoustic waves, *IEEE Ultrasonic Symposium*, Williamsburg, VA, 1986, pp. 435–438.
6. Pohl, A. A. P. et al., Advances and new applications using the acousto-optic effect in optical fibers, *Photonic Sensors* **3**(1), 1–25, 2013.
7. Cuadrado-Laborde, C. et al., Applications of in-fiber acousto-optic devices, in *Acoustic Waves—From Microdevices to Helioseismology*, Beghi, M. G. (ed.), InTech Europe, Rijeka, Croatia, 2011.
8. Liu, W. F., Russell, P. St. J., and Dong, L., Acousto-optic superlattice modulator using a fiber Bragg grating, *Optics Letters* **22**(19), 1515–1517, 1997.
9. Yariv, A. and Yeh, P., *Optical Waves in Crystals*, John Wiley & Sons, Inc., New York, 1984.

10. Zemon, S. A. and Dakss, M. L., Acoustooptic modulator for optical fiber waveguides, US Patent, 4.068.191, 1978.
11. Achenbach, J. D., *Wave Propagation in Elastic Solids*, North-Holland, Amsterdam, the Netherlands, 1973.
12. Rao, J. S., *Advanced Theory of Vibration*, John Wiley & Sons, New Delhi, India, 1992.
13. Thurston, R. N., Elastic waves in rods and clad rods, *Journal of the Acoustic Society of America* **64**(1), 1–37, 1978.
14. De Jong, B. H. W. S., Beerkens, R. G. C. and van Nijnatten, P. A., Glass. *Ullmann's Encyclopedia of Industrial Chemistry*, Wiley-VCH Verlag, 2000.
15. Engan, H. E., Kim, B. Y., Blake, J. N., and Shaw, H. J., Propagation and optical interaction of guided acoustic waves in two-mode optical fibers, *Journal of Lightwave Technology* **6**(3), 428–436, 1988.
16. Kim, B. Y., Blake, J. N., Engan, H. E., and Shaw, H. J., All-fiber acousto-optic frequency shifter, *Optics Letters* **11**(6), 389–391, 1986.
17. Jaffe, H. and Berlincourt, D. A., Piezoelectric transducer materials, *Proceedings of the IEEE* **53**(10), 1372–1386, October 1965.
18. Lee, S. S., Kim, H. S., Hwang, I. K., and Yun, S. H., Highly-efficient broadband acoustic transducer for all-fibre acousto-optic devices, *Electronics Letters* **39**(18), 1309, 2003.
19. Meitzler, A. H., Jr. O'Bryan, H. M., and Tiersten, H. F., Definition and measurement of radial mode coupling factors in piezoelectric ceramic materials with large variations in Poisson's ratio, *IEEE Transactions on Ultrasonics, Ferroelectrics and Frequency Control* **20**(3), 233–239, July 1973.
20. Brissaud, M., Characterization of piezoceramics, *IEEE Transactions on Ultrasonics, Ferroelectrics and Frequency Control* **38**(6), 603–617, November 1991.
21. Brissaud, M., Three-dimensional modeling of piezoelectric materials, *IEEE Transactions on Ultrasonics, Ferroelectrics and Frequency Control* **57**(9), 2051–2065, September 2010.
22. Hong Du, X., Ming Wang, Q., and Uchino, K., An accurate method for the determination of complex coefficients of single crystal piezoelectric resonators II: Design of measurement and experiments, *IEEE Transactions on Ultrasonics, Ferroelectrics and Frequency Control* **51**(2), 238–248, February 2004.
23. Baliato, A., Modeling piezoelectric and piezomagnetic devices and structures via equivalent networks, *IEEE Transactions on Ultrasonics, Ferroelectrics and Frequency Control* **48**(5), 1189–1240, September 2001.
24. Silva, R. E. and Pohl, A. A. P., Characterization of flexural acoustic waves in optical fibers using an extrinsic Fabry–Perot interferometer, *Measurement Science and Technology* **23**(5), 055296, 2012.
25. Silva, R. E., Franco, M. A. R., Neves, P. T., Jr., Bartelt, H., and Pohl, A. A. P., Detailed analysis of the longitudinal acousto-optical resonances in a fiber Bragg modulator, *Optics Express* **21**(6), 6997–7007, 2013.
26. Berwick, M., Pannell, C. N., Russell, P. St. J., and Jackson, D. A., Demonstration of birefringent optical fibre frequency shifter employing torsional acoustic waves, *Electronics Letters* **27**, 713–715, 1991.
27. Lee, K. J., Hwang, I.-K., Park, H. C., and Kim, B. Y., Sidelobe suppression in all-fiber acousto-optic tunable filter using torsional acoustic wave, *Optics Express* **18**(12), 12059–12064, 2010.
28. Andres, M. V., Tudor, M. J., and Foulds, K. W. H., Analysis of an interferometric optical fibre detection technique applied to silicon vibrating sensors, *Electronics Letters* **23**, 774, 1987.
29. Silva, R. E. and Pohl, A. A. P., Characterization of longitudinal acoustic waves in a fiber using an extrinsic Fabry-Perot interferometer, *22st International Conference on Optical Fiber Sensors (OFS)*, Beijing, China, 2012.
30. Blake, J. N., Kim, B. Y., Engan, H. E., and Shaw, H. J., Analysis of intermodal coupling in a two-mode fiber with periodic microbends, *Optics Letters*, **12**(4), 281–283, 1987.
31. Snyder, A. W. and Love, J. D., *Optical Waveguide Theory*, Chapman and Hall, New York, 1983, p. 542.
32. Taylor, H. F., Bending effects in optical fibers, *Journal of Lightwave Technology* **LT-2**(5), 617–627, 1984.
33. Birks, T. A., Russel, P. St. J., and Culverhouse, D. O., The acousto-optic effect in single-mode fiber tapers and couplers, *Journal of Lightwave Technology* **14**(11), 2519–2529, 1996.
34. Matsui, T., Nakajima, K., Shiraki, K., and Kurashima, T., Ultra-broadband mode conversion with acousto-optic coupling in hole-assisted fiber, *Journal of Lightwave Technology* **27**(13), 2183–2188, 2009.
35. Birks, T. A., Russel, P. St. J., and Pannell, C. N., Low power acousto-optic device based on a tapered single mode fiber, *IEEE Photonics Technology Letters*, 6(6), 725–727, 1994.
36. Feced, R., Alegria, C., Zervas, M. N., and Laming, R. I., Acousto-optic attenuation filters based on tapered optical fibers, *IEEE Journal of Selected Topics in Quantum Electronics* **5**(5), 1278–1288, 1999.
37. Zhao, J. and Liu, X., Fiber acousto-optic mode coupling between the higher-order modes with adjacent azimuthal numbers, *Optics Letters*, **31**(11), 1609–1611, 2006.
38. Kim, H. S., Yun, S. H., Kwang, I.-K., and Kim, B. Y., All-fiber acousto-optic tunable notch filter with electronically controllable spectral profile, *Optics Letters*, **22**(19), 1476–1478, 1997.

39. Dimmick, T. E., Kakarantzas, G., Birks, T. A., Diez, A., and Russel, P. St. J., Compact all-fiber acoustooptic tunable filters with small bandwidth-length product, *IEEE Photonics Technology Letters* 12(9), 1210–1212, 2000.
40. Othonos, A. and Kalli, K., *Fiber Bragg Gratings: Fundamentals and Applications in Telecommunications and Sensing*, Artech House, Boston, MA, 1999.
41. Vengsarkar, M., Lemaire, P. J., Judkins, J. B., Bhatia, V., Erdogan, T., and Sipe, J. E., Long-period fiber gratings as band-rejection filter, *Journal of Lightwave Technology* **14**, 58–65, 1996.
42. James, S. W. and Tatam, R. P., Optical fibre long-period grating sensors: Characteristics and application, *Measurement Science and Technology* **14**, R49–R61, 2003.
43. Liu, W. F., Russel, P. St. J., Culverhouse, D. O., and Reekie, L., Acousto-optic superlattice modulator using fiber Bragg grating, *Conference on Lasers and Electro-Optics*, Vol. 9 of 1996 OSA Technical Digest Series, Optical Society of America, Washington, DC, 1996, pp. 243–244.
44. Oliveira, R. A., Neves, P. T., Jr., Pereira, J. T., Canning, J., and Pohl, A. A. P., Vibration mode analysis of a silica horn fiber Bragg grating device, *Optics Communications* **283**, 1296–1302, 2010.
45. Oliveira, R. A., Neves, P. T., Jr., Pereira, J. T., and Pohl, A. P. P., Numerical approach for designing a Bragg grating acousto-optics modulator using finite element and transfer matrix methods, *Optics Communications* **281**, 4899–4905, 2008.
46. Pohl, A. A. P., Cook, K., and Canning, J., Acoustic-induced modulation of photonic crystal fibre Bragg gratings, in *Proceedings of the 10th International Conference on Transparent Optical Networks*, Athens, Greece, 2008, Vol. 2, pp. 51–54.
47. Oliveira, R. A., Characterization and new applications of the acousto-optic effect in fiber gratings, PhD thesis, Federal University of Technology of Paraná, Paraná, Brazil, 2011.
48. Jones, S. L., Murtaza, G., Senior, J. M., and Haigh, N., Single-mode optical fiber microbend loss modeling using the finite difference beam propagation method, *Optics Fiber Technology* **4**(4), 471–479, 1998.
49. Suhir, E., Elastic stability, free vibrations, and bending of optical glassfibers: Effect of the nonlinear stress—Strain relationship, *Applied Optics* **31**(24), 5080–5085, 1992.
50. Liu, W. F., Liu, I. M., Chung, L. W., Huang, D. W., and Yang, C. C., Acoustic-induced switching of the reflection wavelength in a fiber Bragg grating, *Optics Letters* **25**(18), 1319–1321, 2000.
51. Sun, N.-H. et al., Analysis of phase-matching conditions in flexural-wave modulated fiber Bragg grating, *Journal of Lightwave Technology* **20**(2), 311–315, 2002.
52. Diez, A., Delgado-Pinar, M., Mora, J., Cruz, J. L., and Andrés, M. V., Dynamic fiber-optic add-drop multiplexer using Bragg gratings and acousto-optic-induced coupling, *IEEE Photonics Technology Letters* **15**(1), 84–86, 2003.
53. Luo, Z., Ye, C., Cai, Z., Dai, X., Kang, Y., and Xu, H., Numerical analysis and optimization of optical spectral characteristics of fiber Bragg gratings modulated by a transverse acoustic wave, *Applied Optics* **46**(28), 6959–6965, 2007.
54. Abrishamian, F., Nakai, Y., Sato, S., and Imai, M., An efficient approach for calculating the reflection and transmission spectra of fiber Bragg gratings with acousticly induced microbending, *Optical Fiber Technology* **13**, 32–38, 2007.
55. Russel, P. St. J. and Liu, W. F., Acousto-optic superlattice modulation in fiber Bragg gratings, *Journal of the Optical Society of America A* **17**(8), 1421–1429, 2000.
56. Erdogan, T., Fiber grating spectra, *Journal of Lightwave Technology* **15**(8), 1277–1294, 1997.
57. Yamada, M. and Sakoda, K., Analysis of almost-periodic distributed feedback slab waveguides via a fundamental matrix approach, *Applied Optics* **26**(16), 3474–3478, 1987.
58. Kamikawachi, R. C., Possetti, G. R., Falate, R., Muller, M., and Fabris, J. L., Influence of surrounding media refractive index on the thermal and strain sensitivities of long period gratings, *Applied Optics* **16**, 2831–2837, 2007.
59. Oliveira, R. A. et al., Control of the long period grating spectrum through low frequency flexural acoustic waves, *Measurement Science and Technology* **22**(045205), 1–6, 2011.
60. Marques, C. A. F., Oliveira, R. A., Pohl, A. P. P., Canning, J., and Nogueira, R. N., Dynamic control of a phase-shifted FBG through acousto-optic modulation, *Optics Communications* **284**, 1228–1231, 2011.
61. Knight, J. C., Birks, T. A., Russel, P. St. J., and Atkin, D. M., Pure silica single-mode fiber with hexagonal photonic crystal cladding, *Optical Fiber Communication Conference (OFC)*, San Jose, CA, March 1996, Postdeadline Paper PD3.
62. Russel, P. St. J., Photonic crystal fibers, *Journal of Lightwave Technology* **24**(12), 4729–4749, 2006.
63. Diez, A., Birks, T. A., Reeves, W. H., Mangan, B. J., and Russel, P. St. J., Excitation of cladding modes in photonic crystal fibers by flexural acoustic waves, *Optics Letters* **25**(20), 1499–1501, 2000.
64. Graff, K. F., *Wave Motion in Elastic Solids*, Clarendon Press, Oxford, U.K., 1975.

65. Haakestad, M. W. and Engan, H. E., Acoustooptic properties of a weakly multimode solid core photonic crystal fiber, *Journal of Lightwave Technology* **24**(2), 838–844, 2006.
66. Yeom, D.-I., Park, H. C., Hwang, I.-K., and Kim, B. Y., Tunable gratings in a hollow-core photonic bandgap fiber based on acousto-optic interaction, *Optics Express* **17**(12), 9933–9939, 2009.
67. Oliveira, R. A., Neves, P. T., Jr., Pereira, J. T., and Pohl, A. A. P., Analysis of mechanical properties of a photonic crystal fiber Bragg grating acousto-optic modulator, in *Proceedings of the First Workshop on Specialty Optical Fibers and Their Applications*, São Pedro, Brazil, 2008, pp. 1–4.
68. Park, H. C., Hwang, I.-K., Yeom, D.-I., and Kim, B. Y., Analyses of cladding modes in photonic crystal fiber, *Optics Express* **15**(23), 15154–15160, 2007.
69. Do Lim, S., Park, H. C., Hwang, I.-K., Lee, S. B., and Kim, B. Y., Experimental excitation and characterization of cladding modes in photonic crystal fiber, *Optics Express* **18**(3), 1833–1840, 2010.
70. Canning, J. et al., Gratings in photonic crystal and other structured optical fibres, *Trends in Photonics*, J. Canning (ed.), Research Signpost, Kerala, India, 2010. ISBN 978-81-7895-441-7.
71. Cusano, A., Paladino, D., and Iadicicco, A., Microstructured fiber Bragg gratings, *Journal of Lightwave Technology* **27**(11), 1663–1697, 2009.
72. Cook, K., Pohl, A. A. P., and Canning, J., High-temperature type IIa gratings in 12-ring photonic crystal fibre with germanosilicate core, *Journal of the European Optical Society—Rapid Publications* **3**, 08031, 2008.
73. Yeom, D. I., Park, H. S., and Kim, B. Y., Tunable narrow-bandwidth optical filter based on acousticly modulated fiber Bragg grating, *IEEE Photonics Technology Letters* **16**(5), 1313–1315, 2004.
74. Delgado-Pinar, M., Zalvidea, D., Díez, A., Pérez-Millán, P., and Andrés, M. V., Q-switching of an all-fiber laser by acousto-optic modulation of a fiber Bragg grating, *Optics Express* **14**(3), 1106–1112, 2006.
75. Cuadrado-Laborde, C., Diez, A., Delgado-Pinar, M., Cruz, J. L., and Andrés, M. V., Mode-locking of an all-fiber laser by acousto-optic superlattice modulation, *Optics Letters* **34**(7), 1111–1113, 2009.
76. Kim, H. S., Yun, S. H., Kim, H. K., Park, N., and Kim, B. Y., Dynamic erbium-doped fiber amplifier based on active gain-flattening with fiber acousto-optic tunable filters, *IEEE Photonics Technology Letters* **10**, 790, 1998.
77. Marques, C. A. F., Oliveira, R. A., Pohl, A. A. P., and Nogueira, R. N., Adjustable EDFA gain equalization filter for DWDM channels based on a single LPG excited by flexural acoustic waves, *Optics Communications*, **285**, 3770–3774, 2012.
78. Oliveira, R. A., Canning, J., Cook, K., Naqshbandi, M., and Pohl, A. A. P., Compact dip-style viscometer based on the acousto-optic effect in a long period fibre grating, *Sensors and Actuators B: Chemical* **1**, 1–20, 2011.
79. Lide, D. R., *CRC Handbook of Chemistry and Physics*, CRC Press, Boca Raton, FL, 2008.
80. Oliveira, R. A., Cook, K., Canning, J., and Pohl, A. A. P., Bragg grating writing in acousticly excited optical fiber, *Applied Physics Letters* **97**(1), 041101, 2010.
81. Oliveira, R. A., Cook, K., Marques, C. A. F., Canning, J., Nogueira, R. N., and Pohl, A. A. P., Complex Bragg grating writing using direct modulation of the optical fibre with flexural acoustic waves, *Applied Physics Letters* **99**, 161111, 2011.
82. Marques, C. A. F., Oliveira, R. A., Pohl, A. A. P., Canning, J., and Nogueira, R. N., Dynamic control of a phase-shifted FBG through acousto-optic modulation, *Optics Communication* **284**(5), 1228–1231, 2011.
83. Marques, C. A. F., Oliveira, R. A., Pohl, A. A. P., and Nogueira, R. N., Tunable acoustic bursts for customized tapered fiber Bragg structures, *Journal of the Optical Society of America B* **29**(12), 3367–3370, 2012.
84. Marques, C. A. F., Oliveira, R. A., Pohl, A. A. P., and Nogueira, R. N., Tunability of the FBG group delay through acousto-optic modulation, *Optical Fiber Technology* **19**, 121–125, 2013.
85. Marques, C. A. F., Bilro, L., Kahn, L., Oliveira, R. A., Webb, D. J., and Nogueira, R. N., Acousto-optic effect in microstructured polymer fiber Bragg gratings: Simulation and experimental overview, *Journal of Lightwave Technology* **31**(10), 1551–1558, 2013.
86. Zhou, B., Guan, Z., Yan, C., and He, S., Interrogation technique for a fiber Bragg grating sensing array based on a Sagnac interferometer and an acousto-optic modulator, *Optics Letters* **33**(21), 2485–2487, 2008.
87. Hong, K. S., Park, H. C., and Kim, B. Y., 1000 nm tunable acousto-optic filter based on photonic crystal fiber, *Applied Physics Letters* **92**, 031110, 2008.
88. Barmenkov, Y. O., Cruz, J. L., Díez, A., and Andrés, M. V., Electrically tunable photonic true-time-delay line, *Optics Express* **18**(17), 17859–17864, 2010.
89. Zhang, H., Qiu, M., Miao, Y., Liu, B., Gao, S., and Dong, H., Acoustic-birefringence-induced orthogonal acousto-optic gratings in grapefruit microstructured fibers, *IEEE Photonics Journal* **5**(3), 2201610, 2013.
90. Zhang, W., Gao, F., Bo, F., Wu, Q., Zhang, G., and Xu, J., All-fiber acousto-optic tunable notch filter with a fiber winding driven by a cuneal acoustic transducer, *Optics Letters* **36**(2), 271–273, 2011.
91. Zhang, W., Huang, L., Gao, F., Bo, F., Xuan, L., Zhang, G., and Xu, J., Tunable add/drop channel coupler based on an acousto-optic tunable filter and a tapered fiber, *Optics Letters* **37**(7), 1241–1243, 2012.

12 Distributed Fiber-Optic Sensors and Their Applications

Balaji Srinivasan and Deepa Venkitesh

CONTENTS

12.1 INTRODUCTION

Optical fibers are ubiquitous in communication primarily because of its high information-carrying capacity, low loss, small size, negligible electromagnetic interference, low maintenance, and relatively smaller operating expense. Though the deployment of optical communication was initiated for long-distance applications, the increasing bandwidth requirements have resulted in a great reduction in these distant ranges of installation, to include metropolitan and access networks, data centers, and possibly even within a semiconductor processor chip. Along with the increasing utilization of optical fibers for communication, there has been a significant progress in its use for sensing physical parameters. Jacketed fibers with different layers of protection are used for communication applications in order to minimize the influence of environmental perturbations such as temperature, stress, strain, vibrations, and humidity on the propagating signals. In contrast, a sensing application is focused around precisely measuring the changes in amplitude, frequency, phase, or polarization of the propagating optical signal to quantify the desired physical parameter that has created the change. The other parameters that could be sensed in general through optical sensing include rotation, torque, acceleration, voltage, current, pH concentration, and nuclear radiation—this list is definitely not exhaustive. A detailed review of optical fiber sensors can be found in [1]. Distributed sensing refers to such an application where the fiber is itself the sensing element [2]. The desired measurand is monitored continuously along a given length of the fiber and as a function of time. Like in communication, there are some unique advantages presented by the optical fibers as compared to the electronic counterparts, when used for sensing. The 1D nature of the fiber is probably the biggest advantage, since the parameter to be sensed can now be estimated in the desired direction, without getting influenced by the changes in other directions. The other key advantages due to its electrically passive operation include the possibility of a safe operation in a chemically or electrically hostile environment, lighter weight, amenability to bending, and minimal susceptibility to electromagnetic interference.

There are a wide variety of sensing applications that would benefit from distributed sensing due to the potential of spanning six orders of magnitude in the spatial scale [3,4]. Distributed sensing of temperature is one of the most sought-after applications in this domain. A detailed review of the various applications of distributed temperature sensing (DTS) with the methodologies adopted can be found in [5,6]. With the advent of renewable energy sources the consequent deployment of smart grid systems where the power distribution systems become extremely dynamic, it is particularly important to monitor the health of the power cables [7]. This is done by embedding fibers in the inner conductors to measure the temperature profile that in turn allows a prediction of ampacity of the circuit that allows an optimization of network performance. The same principle applies to the monitoring of submarine power cables. The high-temperature regions referred to as the hot spots typically limit the power transfer capacity of underground power cables. The DTS in this case would aid in the dynamic optimization of the power distribution with a real-time monitoring of the temperature profile of the cables [8]. DTS for fire detection in different utility tunnels, including those in mines, bore holes, and industrial plants, was proposed and demonstrated with a temperature resolution of 5°C and a distance resolution of about 5 m, about 30 years ago [9]. In each of these applications, the alarm thresholds are typically set depending on the type of deployment and the typical ambient conditions. DTS finds a significant application in the oil and gas industry, specifically for downhole monitoring to indicate geothermal changes through temperature measurements [10].

Distributed sensing of strain has attracted significant attention in research and development ever since its first demonstration by Horiguchi's group in the late 1990s [11–13]. Reliable structural monitoring using fiber-based distributed sensing is possible for large civil structures such as dams and bridges for preventive maintenance. One such implementation is done in the Gotaalv bridge in Sweden, constructed in 1939. The fatigue in the steel girders of the bridge due to aging was monitored through distributed strain sensing to detect cracks that were wider than 0.5 mm [14]. Real-time information of structural integrity in concrete structures, submarine installations, oil pipelines, and aircraft wings help in the timely fault diagnostics and evaluation of stress [15–17]. Ballastless railway track is recently proposed and demonstrated with an active distributed monitoring of strain [16]. Multiparameter sensing is an even more attractive alternative where the same sensing fiber, when operated under different configurations, can be used for extracting multiple parameters. The requirements on the resolution range and sensitivity of measurement are specific to the application, which makes it impossible to find a universal prescription for the design of these sensors.

From the aforementioned discussion, it is quite apparent that there exists a strong need for distributed sensing of physical parameters such as strain, temperature, and vibrations. Optical-fiber-based distributed sensing is uniquely positioned in this respect due to its unique combination of low loss and sensitivity to the aforementioned physical parameters. However, there are significant challenges to be overcome in order to provide a cost-effective, rugged, and practical solution. These challenges include (1) instrumentation issues such as spatial resolution or dynamic range [18], (2) discrimination between physical parameters, for example, strain and temperature [19], and (3) deployment of the sensing fiber with appropriate encapsulation such that they not only maintain intimate contact with physical structures for high sensitivity but are also protected from harsh environments and may reduce cross-sensitivity through ingenuous engineering [20]. Such issues are discussed in further detail in the following sections.

12.2 BACKGROUND

In this section, we will attempt to uncover some of the fundamental aspects of distributed sensing. We begin with a basic understanding of the different light scattering mechanisms and how they could be potentially exploited for sensing applications. Distributed sensing is carried out by amalgamating the scattering mechanisms with reflectometry techniques in the time domain or in the frequency domain. The former technique is commonly referred to as optical time domain reflectometry (OTDR), and the latter is referred to as optical frequency domain reflectometry (OFDR). The OTDR technique is more popular for the long-range distributed sensing applications outlined in Section 12.1 and hence is discussed in much detail. Finally, we discuss the distributed sensing of physical parameters such as strain and temperature using the aforementioned techniques.

12.2.1 Fundamentals of Light Scattering in Optical Fibers

When light enters an optically transparent dielectric medium, several different scattering mechanisms are possible. These may be commonly classified as either elastic scattering or inelastic scattering. The key difference between them is the exchange of energy with the dielectric medium. In case of the former, no such energy transfer occurs and the scattered light retains the frequency of the incident light. A classic example of elastic scattering mechanism is Rayleigh scattering, which is due to microscopic density variations in the dielectric medium. As such, Rayleigh scattering offers important insight on the background loss information in a given material.

On the other hand, inelastic scattering provides a rich variety of information about the medium and the surrounding regions as there is a specific energy exchange with the medium. Brillouin scattering and Raman scattering are two prime examples for the inelastic scattering mechanism. The former refers to the interaction of the light wave with lattice vibrations of the medium (also known as *acoustic phonons*), whereas the latter is due to the light interaction with molecular vibrations or rotations (also known as *optical phonons*). An important aspect of such scattering mechanisms is their unique spectral signature, which is quite different from the frequency of the incident light. The frequency shift for Brillouin scattering is in the order of 10 GHz and the same for Raman scattering is in the order of 10 THz. The relatively large frequency shift for Raman scattering is attributed to the fast timescales within which the molecular vibrations cease to exist after excitation. As shown in Figure 12.1, the spectral bands of Raman scattered light are much easily identified compared to those of Brillouin scattered light. Note that both these mechanisms have a Stokes band (corresponding to the loss of light energy) and anti-Stokes band (corresponding to an increase in light energy).

In fused silica optical fibers, Rayleigh scattering is 3–4 orders of magnitude weaker compared to the incident radiation owing to the high transmission quality of telecommunication grade fibers. Brillouin scattering is typically two orders of magnitude weaker than Rayleigh scattering, although stimulated Brillouin scattering (SBS) can occur at higher incident power levels making the backscattered power comparable with the Rayleigh scattered radiation. In comparison with Brillouin scattering, Raman scattering is even weaker—roughly 3–4 orders of magnitude lower than Rayleigh scattered radiation.

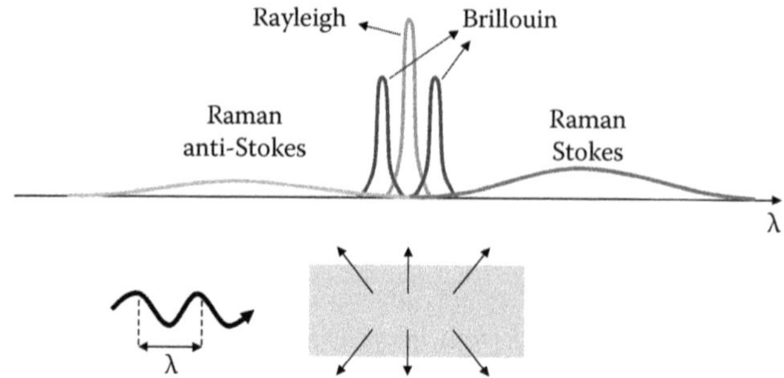

FIGURE 12.1 Conceptual depiction of the elastic and inelastic scattering mechanisms when light interacts with a dielectric medium.

The attractive aspect of inelastic scattering processes such as Raman and Brillouin scattering is the interaction with the propagating medium and the resultant sensitivity to changes in strain and temperature. This aspect is discussed in further detail later in this section. Before that, we will try to understand the basic principles of OTDR and its application to distributed sensing.

12.2.2 Basics of OTDR

When an electromagnetic wave propagates through an optical fiber, it may be subjected to perturbations that alter its amplitude or phase or frequency or polarization. By observing such changes in the propagation characteristics of the wave, one can potentially measure the perturbations that could be due to several physical mechanisms such as strain–temperature–pressure–rotation. As discussed in the other chapters, this is the basis for most of the optical fiber sensors. Invariably, efficient demodulation of the information with low level of uncertainty is a key challenge in all these sensors.

A typical configuration that provides relatively high rejection of unperturbed excitation radiation is the backscatter configuration illustrated in Figure 12.2. An added advantage from the instrumentation perspective is the colocation of the optical source and detector electronics. By launching a pulse of light through a directional fiber coupler into the sensing fiber and by observing the backscattered signal through the other port of the coupler, the perturbations at different sections of the fiber could be identified. This technique is very similar to the time of flight measurements employed in radars over several decades. In several ways, it is much less challenging compared to radars [21], since the propagation direction is confined to 1D in optical fibers. The directional coupler could be replaced by a circulator to reduce the inherent 3 dB loss incurred every time light passes across it. The design principles involved in such OTDRs are discussed in detail in the following.

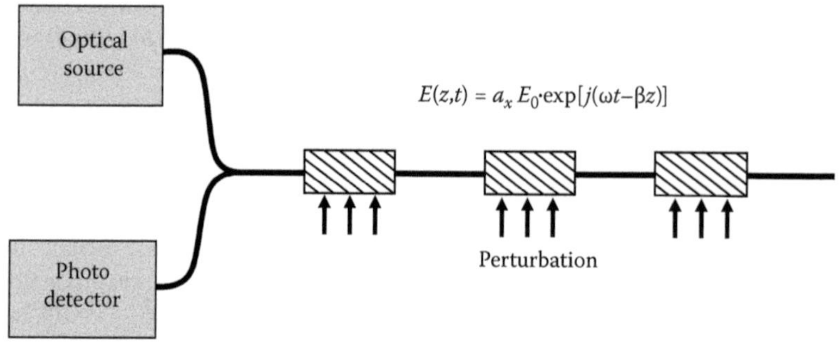

FIGURE 12.2 Schematic diagram of the backscatter configuration typically employed in distributed sensors. The local perturbation information is obtained through the OTDR technique.

An important factor that influences the design of the OTDR receiver is the low intensity level of the backscattered radiation. The receiver design is driven by the OTDR maker's formula:

$$P_L - P_D + \text{SNIR} = R + C + 2L \tag{12.1}$$

where

P_L denotes *the launched laser peak power* (dBm)
P_D denotes the *receiver sensitivity* (dBm)
SNIR denotes the *signal-to-noise-improvement ratio* (dB)
R denotes the *Rayleigh backscatter coefficient* (dB)
C denotes the *coupler and connector losses* (dB)
L is the *single-pass loss* in fiber (dB)

This OTDR equation may be physically interpreted with the help of Figure 12.3. The launched power P_L gets attenuated in the optical fiber as it propagates in the fiber. If there is a break in the fiber, there will be a corresponding 4% Fresnel reflection from the glass–air interface. However, the reflected light will undergo attenuation once again in the fiber before it reaches the receiver. Hence, the slope of the reflected power variation along the length of fiber is twice that of the launched power variation. Similarly, any Rayleigh backscattered light will also undergo twice the loss before reaching the optical receiver.

The receiver will be able to detect backscattered signals that are above the noise floor, that is, power level defined by the receiver sensitivity P_D corresponding to signal-to-noise ratio (SNR) = 1. However, by employing SNR improvement (signal-to-noise-improvement ratio [SNIR]) techniques such as filtering or averaging or correlation, one can effectively lower the noise floor. Such reduction in the noise floor will allow us to interrogate a longer span of the optical fiber. The effective range of signals that could be detected is bounded by the highest level of Rayleigh backscattered power at the initial end of the fiber and the effective noise floor after signal processing. This range is known as the dynamic range of the OTDR. Note that the single-pass dynamic range (L) is half this value as we are dealing with the backscattered configuration.

The Rayleigh backscatter coefficient is of the order of –52 dB/μs of pulse width at 1550 nm. Note that the backscatter coefficient is expressed in terms of dB/μs as the amount of backscattered light gets integrated within the section of fiber that the pulse occupies physically. Supposing we launch a laser

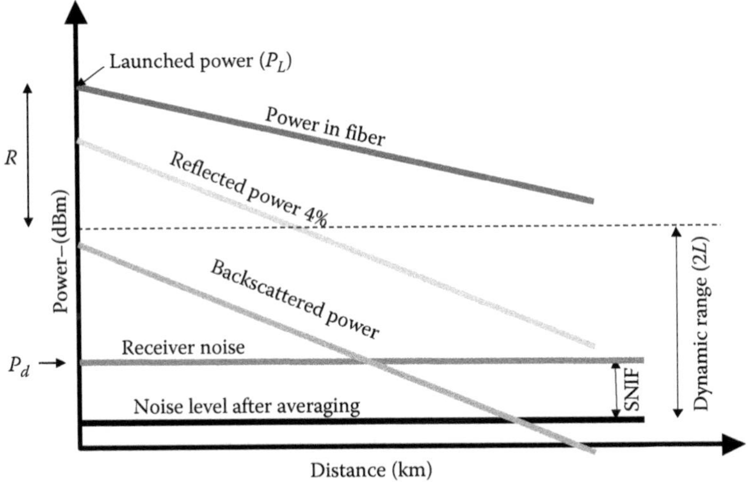

FIGURE 12.3 Typical plot of optical power as a function of distance obtained from an OTDR.

power of (P_L) 20 dBm with a pulse width of 100 ns, the backscattered power is expected to be 62 dB below the launched power level, that is, −42 dBm at the beginning of the test fiber. The receiver sensitivity required to achieve a single-pass dynamic range of 20 dB is of the order of −82 dBm, that is, less than 10 pW of power. It is quite challenging to detect such a low signal in the presence of receiver noise.

One has to take the assistance of signal processing techniques (averaging, Savitzky–Golay filtering) to alleviate the aforementioned requirement. One way to meet the SNR requirement on the receiver design is to average the OTDR trace for various laser pulses. Averaging improves SNR by a factor of $\sqrt{N}$ times for N number of averages. This results in SNIR of 1.5 $\log_2(N)$. Therefore, by averaging the OTDR trace $N = 64$ k times, one can get SNR improvement of 24 dB. Hence, to achieve a dynamic range of 20 dB corresponding to 100 km of standard single-mode sensing fiber, the receiver needs to have sensitivity of the order of −58 dBm.

From these calculations, our goal is to design a receiver system with sufficient transimpedance gain such that it can convert the photocurrent corresponding to a backscattered power of −58 dBm to hundreds of mV. Another important parameter that needs to be taken care of while designing the receiver is the bandwidth, which decides the spatial resolution of the system.

Trade-off between spatial resolution and dynamic range: The key performance parameters for OTDR are the *spatial resolution* (δL) and the *dynamic range* [22]. Spatial resolution is defined as the smallest length of the fiber over which any sensible change in the measurand can be detected. If the total length of the fiber is L, then the system is equivalent to ($L/\delta L$) sensors multiplexed in a linear array. As discussed earlier, dynamic range is defined as the fiber loss range over which the backscattered signal could be reliably measured. The backscattered power is directly proportional to the pulse width of the laser. In general, there is a strong trade-off between dynamic range and spatial resolution in OTDRs.

For example, to achieve minimum spatial resolution, one has to launch the shortest possible light pulse into the fiber. But the short duration of the laser pulse effectively covers only a short length of fiber at any instant and the corresponding backscattered power is low. This in turn limits the dynamic range or, in other words, the distance over which the backscattered signal is above the noise level. Conversely, one can choose pulse widths of tens of microseconds, which results in large SNR and dynamic range but compromises the spatial resolution.

Trade-off between spatial resolution and accuracy: Apart from dynamic range trade-off, there exists another strong trade-off between accuracy and spatial resolution. To achieve minimum spatial resolution, one has to launch short-duration laser pulse, but short-duration pulse effectively covers only a small length of the fiber that results in small backscattered power. If the Gaussian noise level of the receiver is comparable to backscattered signal level, then the OTDR trace looks noisy and the measurement inaccuracy is relatively high. In general, the measurement inaccuracy increases as we reach the farther sections of the fiber due to the inherent attenuation in the fiber. Hence, to arrive at an optimized design, we need a strong theoretical model incorporating the transfer functions of each of the system components, which is capable of predicting the output in an accurate and rugged manner.

12.2.2.1 Design of the OTDR Receiver

The previous discussion may be extended to the design of an OTDR receiver. In order to quantify the previous constraints, one needs to understand the noise sources associated with PIN- and avalanche photodiode (APD)-based optical receivers. In this section, we also discuss the principles associated with the design of low-noise PIN and APD-based receivers and the merits of APD receiver over PIN-based receiver.

By using the OTDR maker's formula as mentioned in (12.1), we have estimated the backscattered signal reaching the receiver end to be of the order of −58 dBm for 100 ns pulse width. The receiver should be sensitive enough to capture such low-intensity signals. There is a constraint on the upper limit of the receiver gain that comes into picture when one uses long-duration laser pulse widths of the order of tens of microseconds, since the output voltage of the low-noise amplifier (LNA) section should not exceed the input voltage swing level of the analog-to-digital converter (ADC). Supposing the voltage swing is limited to 500 mV, the gain of the receiver must be of the order of several megaohms.

One more important parameter that needs to be considered while designing the receiver is the required bandwidth of the LNA section; this in turn depends on the rise time of the system. For a transform-limited pulse, the relation between bandwidth (B) and rise time (t_r) is

$$B \approx \frac{0.35}{t_r} \tag{12.2}$$

where the rise time is given by

$$t_r = \frac{2.\delta L \cdot n_{eff}}{c} \tag{12.3}$$

The effective refractive index (n_{eff}) is approximately 1.46 for a standard fused silica optical fiber. The factor 2 in (12.3) accounts for the to and fro travel of the pulse inside the fiber. Hence, in order to achieve a spatial resolution of 1 m, the rise time of the system needs to be of the order of 10 ns. Using the estimated rise time value, the bandwidth of the system comes out to be 35 MHz, which can be estimated by using (12.2).

LNA design principles: Backscattered anti-Stokes radiation reaching the receiver end is converted from optical signal to photocurrent by using a PIN photodiode or APD. The conversion efficiency is determined from the responsivity (A/W) parameter of the diode, which is defined as the ratio of output photocurrent to the optical power incident on the detector. The transimpedance amplifier (TIA) configuration is mostly preferred for converting the generated photocurrent to a corresponding voltage [23]. In our design, we have used the differential TIA stage so as to filter out the common noise in the receiver. The output signals from the TIA stage are further amplified by using a two-stage differential amplifier (diff. amp.).

While designing a cascaded receiver section, the thumb rule is to keep the noise generated in the first stage as small as possible, since noise gets amplified by the amount of gains of subsequent stages. So in order to design an LNA receiver, the first stage should contribute as less noise as possible. TIA configuration is an attractive choice to use as the first stage of the receiver design, since it scales the output noise voltage by a factor of $\sqrt{R_f}$ (feedback resistance) and scales the signal by a factor of R_f. Therefore, the TIA stage improves the SNR of the receiver circuit by a factor of $\sqrt{R_f}$. In addition, the TIA configuration scales down the bandwidth only by $\sqrt{R_f}$ (whereas a typical high impedance load will scale bandwidth down by a factor R_f) and thus provides relatively high bandwidth for a given gain value. As mentioned earlier, we used two stages of the diff. amp. configuration immediately after the TIA stage for further amplification. The main reason for choosing the diff. amp. configuration is to eliminate the common-mode noise of the operational amplifier.

12.2.2.2 Noise Sources in the OTDR Receiver

Noise sources in the receiver effectively degrade the performance of OTDR because it limits the dynamic range of the system. A detailed study of noise sources that affect the performance of the system is presented as follows.

Shot (quantum) noise: Shot noise arises from the statistical nature of the photon arrival time at the receiver and the resultant generation of photogenerated carriers. It can be mathematically modeled as a stationary random process with Poisson statistics. The root mean square (RMS) value of the shot noise current density is given by

$$\sigma_s^2 = \left\langle i_s^2(t) \right\rangle = 2qI_pB \tag{12.4}$$

For an APD-based receiver, there exists an extra noise due to the process of generation of secondary electron (e) and hole (h) pairs by impact ionization:

$$\sigma_s^2 = \left\langle i_s^2(t) \right\rangle = 2qI_p M^2 F(M) B \tag{12.5}$$

where
- q is the charge of the electron
- B is the effective bandwidth of the receiver
- M is the multiplication factor
- I_P is the photocurrent
- $F(M)$ is the excess noise factor

Bulk dark current noise: Dark current noise arises due to thermally generated electron (e) and hole (h) pairs in the *pn* junction diode even in the absence of light illumination. The dark current noise variance can be estimated by

$$\sigma_d^2 = \left\langle i_d^2(t) \right\rangle = 2qI_{dark} B \tag{12.6}$$

where I_{dark} is the dark current of the photodiode specified by the manufacturer. For APD, there exists an excess noise due to multiplication process. This is expressed as

$$\sigma_d^2 = \left\langle i_d^2(t) \right\rangle = 2qI_{dark} M^2 F(M) B \tag{12.7}$$

Surface dark current noise: This noise arises due to surface defects, bias voltage, and surface area of the APD device. Since avalanche multiplication is a bulk effect, the surface dark current is not affected by the avalanche gain. Surface dark current noise variance is given by

$$\sigma_{SD}^2 = \left\langle i_{SD}^2(t) \right\rangle = 2qI_L B \tag{12.8}$$

where I_L is the surface leakage current.

Since the quantum noise, bulk dark current noise, and surface leakage current noise are uncorrelated, the noise variance of photodiode is given by

$$\sigma_n^2 = \sigma_s^2 + \sigma_d^2 + \sigma_{SD}^2 \tag{12.9}$$

The noise generated by the TIA is common to both PIN- and APD-based receivers. The TIA stage is typically designed using FET input operational LNA, having a gain bandwidth product in the order of 1 GHz. As this operational amplifier is built with JFETs at the input stage, it will have low input noise voltage density (1 nV/√Hz) and virtually no input current noise density (1 fA/√Hz) [24].

Thermal noise: At finite temperature, electrons move randomly in any conductor. Random thermal motion of electrons in a resistor manifests as a fluctuating current even in the absence of applied voltage. Thermal noise in TIA stage is due to the feedback resistance of the operational amplifier, which arises due to thermal agitation of the charge carriers. The input thermal noise variance contributed by feedback resistance is given by

$$\sigma_T^2 = \left\langle i_T^2(t) \right\rangle = \frac{4k_B T}{R_F} B \tag{12.10}$$

where
- k_B is the Boltzmann constant
- R_F is the load resistance
- T is the effective noise temperature in K

The input-referred current noise spectral density of operational amplifier due to input-referred voltage noise (E_N) and current noise (I_N) is given by

$$i_{iv}^2 = I_N^2 + \left(\frac{E_N}{R_F}\right)^2 + \left(\frac{\left(E_N \cdot 2\pi C_D F\right)^2}{3}\right) \tag{12.11}$$

where

C_D is the diode capacitance
R_F is the feedback resistance of the TIA stage
F is the effective noise bandwidth in Hz

The total output-referred noise of the TIA stage using APD for a band limiting frequency B and a feedback resistance R_F is given by

$$v_{rms} = \sqrt{\left(i_{iv}^2 + \frac{4k_BT}{R_F} + 2q\left(I_p + I_d\right)M^2F\left(M\right) + 2qI_L\right)BR_F^2} \tag{12.12}$$

Note that the RMS noise voltage at the TIA output scales with the feedback resistance value as well as the square root of the bandwidth. Hence, it is prudent to choose R_F as high as possible within the allowed bandwidth constraint. Subsequent amplifier stages in the receiver will only further degrade the SNR, although the degradation could be minimized by a careful choice of the op-amps and maintaining a differential signal path.

12.2.2.3 Design of the PIN-Based Receiver

Using the design principles mentioned earlier, a low-noise PIN-based receiver with a gain of 10 MΩ and bandwidth in the order of 1 MHz may be designed. The estimated noise voltage at the output of the TIA stage with a photodiode resistance of 1000 MΩ, capacitance of 0.5 pF, feedback resistance of 470 kΩ, and feedback capacitance of 0.2 pF is of the order of 52.7 mV (RMS). The schematic diagram of a PIN-based receiver is shown in Figure 12.4. The received optical signals are converted to photocurrent using PIN photodiode, which in turn are converted to corresponding voltage signals using TIA.

Since the captured anti-Stokes signals are of the order of tens of nanowatts, further amplification of the signals needs to be performed. Care needs to be taken to maintain the bandwidth to be in the order of megahertz. Since the op-amp gain bandwidth is typically in the order of several 100 MHz, one might have to use two diff. amp. stages as shown in Figure 12.4.

12.2.2.4 Design of the APD-Based Receiver

The schematic of the APD-based receiver is illustrated in Figure 12.5. As with the PIN-based receiver, we used the three-stage receiver design, where the TIA stage is followed by two diff. amp. stages.

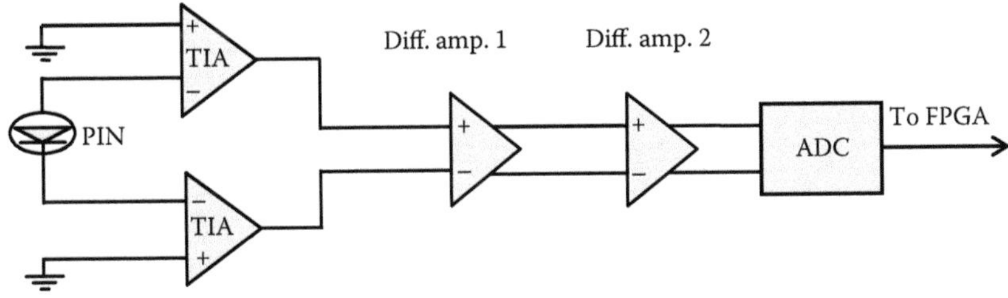

FIGURE 12.4 Schematic diagram of PIN-based differential configuration optical receiver.

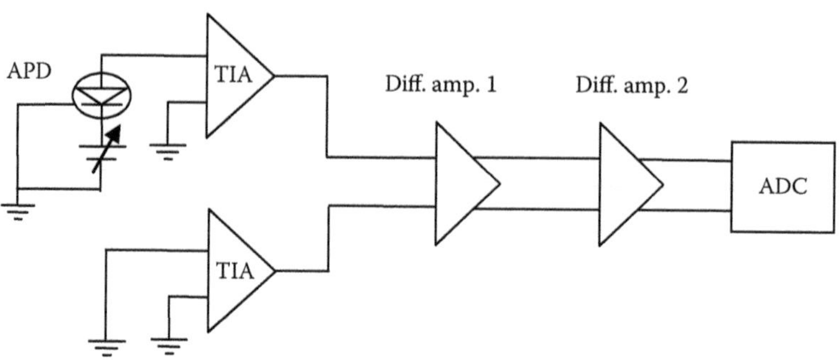

FIGURE 12.5 Schematic diagram of an APD-based optical receiver. A dummy TIA is used to cancel common-mode noise in the receiver.

Unlike PIN diode, APD requires high reverse bias voltage of the order of tens of volts for its operation. Hence, we cannot use the differential TIA configuration at the front end of the receiver. As we are using only one TIA stage, the total gain from TIA section is decreased by a factor of two compared to PIN receiver. However, the common-mode noise of TIA may be canceled out using a dummy TIA op-amp. The main advantage of the APD-based receiver is that it provides amplification of the signal before getting corrupted by the optical receiver noise. This APD gain also helps one to improve the bandwidth of the receiver by decreasing the gain requirement of the TIA and diff. amp. sections.

Although the InGaAs APD has a typical breakdown voltage of around 60 V, it is operated at a lower bias voltage so that one can achieve maximum SNR. If one operates the diode beyond such an optimal operating voltage, the noise contributed by APD dominates the signal amplification. This is illustrated in Figure 12.6, where the RMS noise voltage is plotted as a function of APD bias voltage. From this plot, we see that the optimum bias voltage is around 48 V beyond which the APD noise starts to flare up.

Using the design principles mentioned earlier, one can design an APD-based receiver with a transimpedance gain of 10 MΩ and achieve a bandwidth in the order of 10 MHz. The estimated noise voltage of the TIA stage with a photodiode resistance of 1000 MΩ, capacitance of 0.5 pF, feedback resistance of 47 kΩ, and feedback capacitance of 0.2 pF is of the order of 5 mV (RMS). Figure 12.7 shows a plot comparing the relative performance of the PIN-based and APD-based optical receivers for an 80 km long standard single-mode optical fiber span. Due to a gain of ~12 in the InGaAs APD, the single-pass dynamic range is improved by ~5.6 dB for the APD-based optical receiver design.

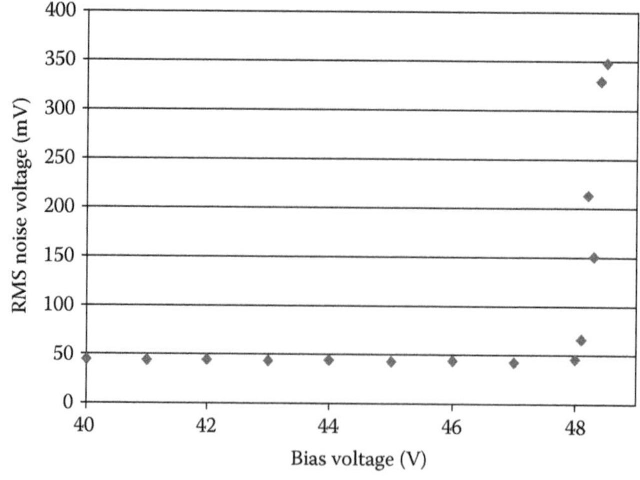

FIGURE 12.6 Typical noise voltage plot as a function of the APD bias voltage showing the drastic increase in noise beyond an optimum bias voltage.

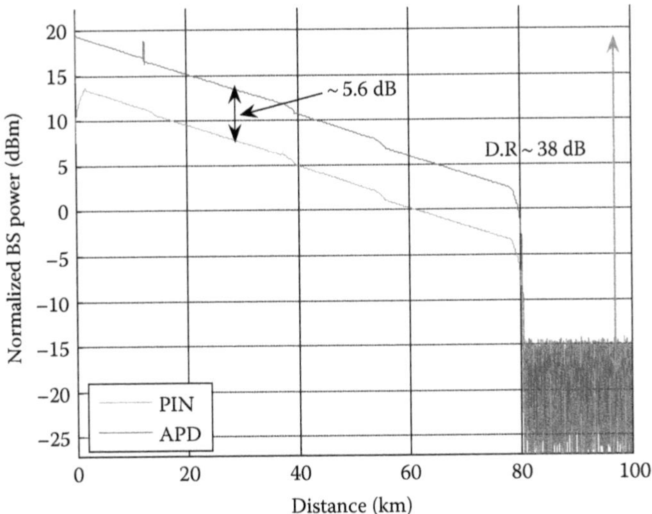

FIGURE 12.7 Comparison of OTDR traces obtained for a PIN- and APD-based receiver. The APD-based receiver demonstrates a single-pass dynamic range improvement of 5.6 dB over the PIN-based optical receiver.

12.2.2.5 SNR Improvement through Averaging

As mentioned previously, by averaging the OTDR trace over several laser pulses, one can improve SNR by a factor of $\sqrt{N}$ for N number of averages, and this results in SNIR by an amount of 1.5 $\log_2(N)$ when expressed in base 2 logarithmic form. Here, we will present the derivation like how one will achieve an improvement of $\sqrt{N}$ by averaging N number of OTDR traces. Since noise is a random variable, let us denote the noise present in the first trace as X_1, the noise in the second trace as X_2, and similarly the noise in the Nth trace as X_N. Since we are averaging all the N captured traces, the averaged noise of "N" traces is given as

$$X_{avg} = \frac{\left(X_1 + X_2 + \cdots + X_N\right)}{N} \tag{12.13}$$

and the corresponding expected value and variance are given by

$$E\left(X_{avg}\right) = \frac{\left(E\left(X_1\right) + E\left(X_2\right) + \cdots + E\left(X_N\right)\right)}{N} \tag{12.14}$$

$$\text{Var}\left(X_{avg}\right) = \text{Var}\left(\frac{\left(X_1 + X_2 + \cdots + X_N\right)}{N}\right) \tag{12.15}$$

As we know, Var(aX) (where a is some number) is equivalent to a^2Var(X) and Var($X + Y$) = Var(X) + Var(Y) for two independent random processes. By applying these two relations, the result comes out as follows:

$$\text{Var}\left(X_{avg}\right) = \frac{\text{Var}\left(X_1 + X_2 + \cdots + X_N\right)}{N^2} \tag{12.16}$$

$$\text{Var}\left(X_{avg}\right) = \frac{\text{Var}\left(X_1\right) + \text{Var}\left(X_2\right) + \cdots + \text{Var}\left(X_N\right)}{N^2} \tag{12.17}$$

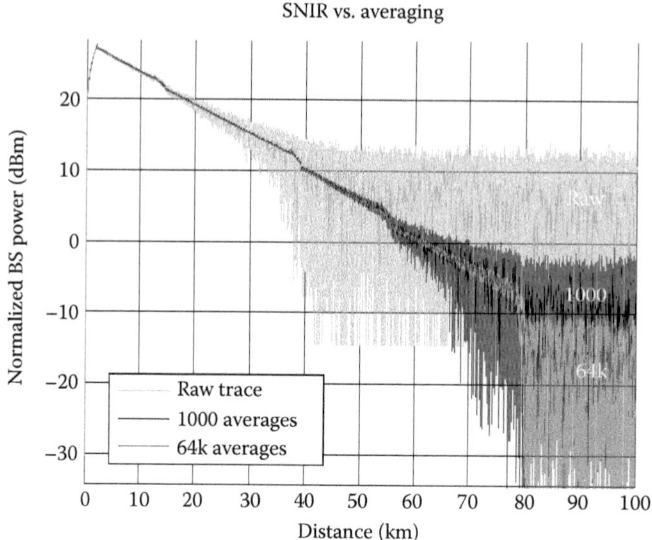

FIGURE 12.8 Plot of the Rayleigh backscattered power as a function of distance for an 80 km standard single-mode fiber span. As the number of averages increases, we are able to detect signals from the farther end of the fiber span.

If $\text{Var}(X_1) = \text{Var}(X_2) = \text{Var}(X_N) = \text{Var}(X)$, then the equation comes out to be

$$\frac{\text{Var}\left(X_1 + X_2 + \cdots + X_N\right)}{N^2} = N\text{Var}\left(X\right)/N^2 \tag{12.18}$$

Therefore, the noise variance comes out to be $\text{Var}(X)/N$ and the standard deviation comes out to be $\sqrt{(\text{Var}(X)/N)}$. Therefore, noise is increased by $\sqrt{N}$ times, whereas the corresponding signal improves by N times resulting in an SNR improvement by $\sqrt{N}$ times.

The effect of averaging on an OTDR trace is illustrated in Figure 12.8. The raw trace corresponding to a single optical pulse launched into an 80 km long standard single-mode optical fiber hardly shows any signal beyond 30 km. As we shoot more numbers of optical pulses and average the backscattered traces, we are able to reduce the noise floor by 24 dB, corresponding to 2^{16} averages. Thus, the entire fiber span becomes visible after the aforementioned averaging process.

12.2.2.6 SNR Improvement through Savitzky–Golay Filtering

Another technique to reduce the overall noise is to use Savitzky–Golay filtering, which is a time-varying finite impulse response (FIR) filter. The fundamental idea is to fit a different polynomial to the data surrounding each data point. The smoothed points are computed by replacing each data point with the value of the fitted polynomial by using $2n + 1$ neighboring points with n being equal to or greater than the order of the polynomial. A sample polynomial fit is shown in Figure 12.9.

Polynomial fitting is performed using the least square error (LSE) method. This filter is slightly more effective than averaging since it reduces noise without adding much of penalty to the pulse shape by selectively applying the order of the polynomial. Using such Savitzky–Golay filtering, one can reduce the noise effectively by choosing a lower-order polynomial (more smoothened response) at the expense of increasing the rise time to the system. Increase in rise time degrades the spatial resolution of the system, which is not desirable. On the other hand, the choice of a higher-order polynomial preserves the rise time and also allows sharp noise features that compromise the noise floor. Hence, there exists a trade-off between the polynomial order and the measurement accuracy. Similarly, if the window size is larger, the noise floor is reduced at the expense of slower rise time.

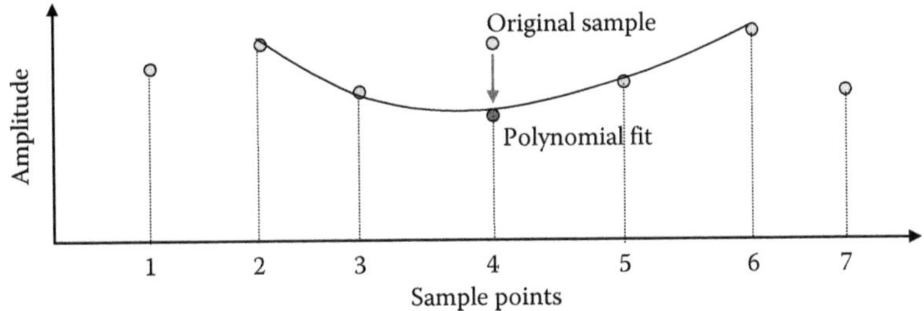

FIGURE 12.9 Illustration of the polynomial fitting implemented as part of the Savitzky–Golay filtering algorithm.

The effectiveness of the Savitzky–Golay filtering is illustrated by the comparison of the noise histogram at different sections of the fiber in Figure 12.10. A Savitzky–Golay filter with an order of one and a window size of 19 is shown to reduce the noise by a factor of 3. Such results may be extended to an OTDR trace and optimized to lower the noise floor without compromising the spatial resolution in a significant manner.

12.2.2.7 SNR Improvement through Correlation Coding

In an earlier discussion, we mentioned that there is a trade-off between dynamic range and spatial resolution for an OTDR system. Such a trade-off may be overcome by adopting a commonly used technique in radar systems known as the correlation technique [25]. The simplest approach of applying correlation to OTDR signal is to correlate the output signal $V(t)$ with the probe signal $p(t)$ such that

$$p(t)*V(t) = [p(t)*p(t)] \otimes f(t) \otimes r(t)$$

As long as the autocorrelation of the probe signal approximates delta function, the system response $f(t)$ can be accurately estimated, subject to the response of the receiver $r(t)$. Thus, the duration of

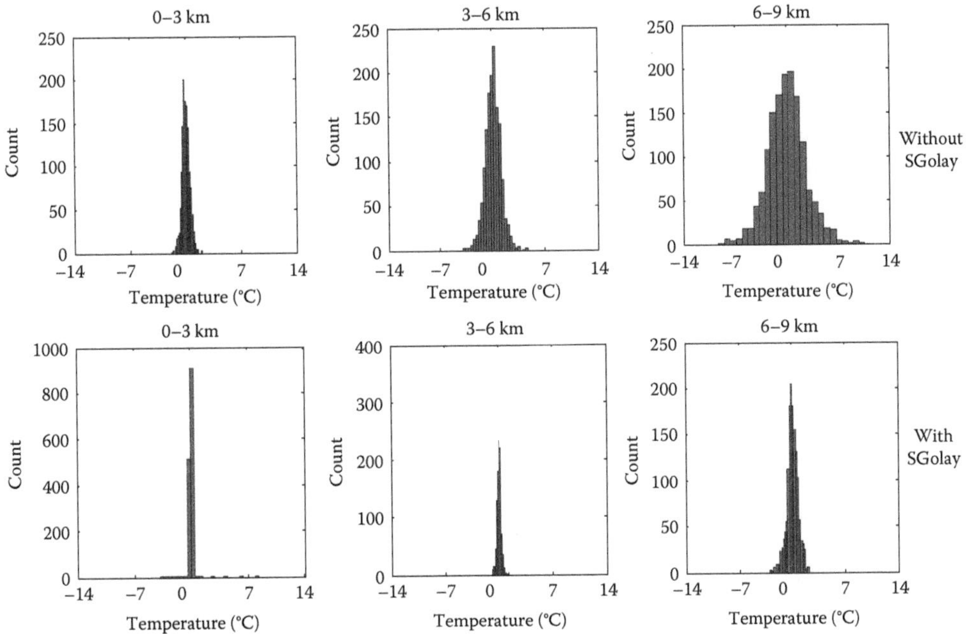

FIGURE 12.10 Histogram of the noise data observed at different locations of a 10 km long fiber span. The use of Savitzky–Golay filter (bottom) reduces the noise variance significantly.

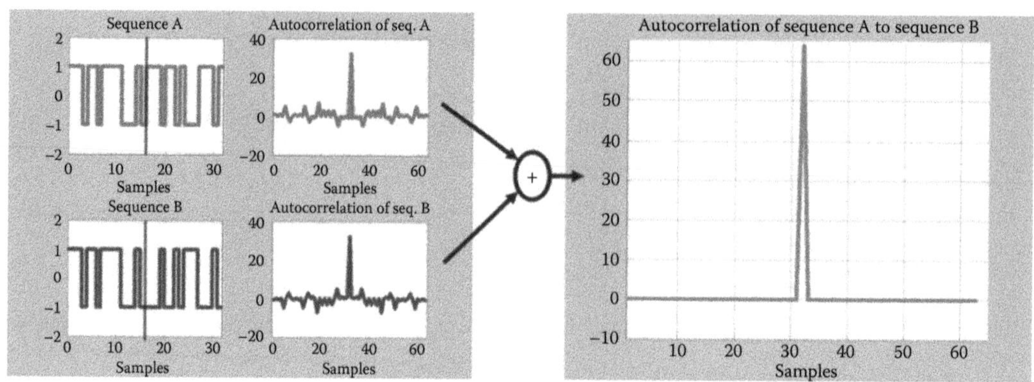

FIGURE 12.11 Schematic diagram showing the two complementary Golay sequences and their corresponding autocorrelation plot. When the two responses are added, they cancel the side lobes and provide a sharp autocorrelation.

autocorrelation and receiver response determines the resolution for the correlation OTDR rather than the probe pulse width as in the case of conventional OTDR. This implies the use of correlation codes for improving the SNR without compromising the resolution.

There are several unipolar aperiodic codes (combination of "1"s and "0"s) with low autocorrelation side lobes, but they are typically not sufficient to satisfy stringent OTDR requirements. Fortunately, there exist bipolar code pairs ("1"s and "–1"s) with complementary autocorrelation property. These codes were first introduced by M.J.E. Golay and are called Golay codes. Due to its complementary autocorrelation property (shown in Figure 12.11), the main lobe of the sum adds up and the side lobes cancel out with each other. Effectively, the sum of the autocorrelation outputs approximate to a delta function.

The key thought to use the Golay sequence for OTDR is that the resolution is determined by the autocorrelation width of the main lobe, whereas the backscattered power (i.e., SNR) is proportional to the sequence length. Hence, the SNR and resolution are decoupled and can be improved independently by choosing an optimum code length and width of a single bit, respectively.

The constraint in using bipolar sequence in OTDR application is that the light intensity is unipolar. The solution is to transmit two unipolar versions of a bipolar sequence. This is achieved by sending the sequence and its ones' complement on a bias (equal to half of the peak power) and subtracting the outputs. For example, instead of sending the code –1,1, one could always send a pair of unipolar code 0,1 and 1,0 and subtract them to get back the original bipolar sequence. The complete algorithm is as follows:

- Send four unipolar versions of a complementary code pair and subtract them.
- Correlate the subtracted outputs with the original bipolar version of sequence.
- Add the resulting output to cancel out the side lobes and add the main lobes.

In order to calculate the SNR improvement, let us assume a sequence of length L bits. The autocorrelation main lobe amplitude would be proportional to the sequence length L and noise amplitude scales as $\sqrt{L}$ [25]. Adding the autocorrelation outputs of the two bipolar sequences would result in net signal improvement of $2L$. We also know that noise adds on an RMS basis, for every addition or subtraction. Therefore, one subtraction and addition involved in the aforementioned algorithm would increase noise by a factor of 2. Effectively, a net SNR improvement of $\sqrt{L}$ can be achieved using the Golay code without sacrificing resolution. However, when conventional method is used, one could have done four parallel averages, in the time that is taken to probe the system with four unipolar sequences, and thereby reduce noise by a factor of 2. Therefore, the actual SNR improvement that can be achieved with respect to the conventional method is $\sqrt{L}/2$ (Figure 12.12).

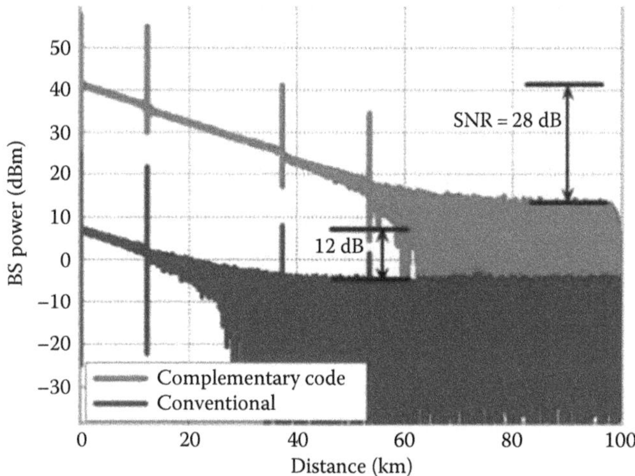

FIGURE 12.12 OTDR traces for the conventional single pulse scheme in comparison with a 1024 bit Golay coded sequence showing the expected 16 dB improvement in SNR.

From this discussion, we gather that the OTDR dynamic range and accuracy could be improved using techniques such as averaging, low-pass filtering, or correlation coding. Of these, the low-pass filtering technique is the least preferred as it compromises the spatial resolution of the OTDR while improving the SNR. It also adds a computational cost to the measurement. Averaging is much better in providing an SNR improvement without affecting the spatial resolution. However, it significantly increases the measurement time (typically in the order of minutes for 216 averages) and is not an option for applications requiring fast measurements. With respect to these, the correlation coding improves the SNR significantly while preserving the spatial resolution within the least measurement time. As such, it is a popular technique for improving the SNR in OTDR instruments.

12.2.3 Physical Parameter Sensing Using OTDR

We started Section 12.2 by discussing the scattering mechanisms in an optical fiber and followed it by a discussion on the basic principles of OTDR. As mentioned earlier, the scattering mechanisms in conjunction with OTDR may be exploited for sensing physical parameters such as strain, temperature, force, pressure, and vibration.

The previous discussion on OTDR is based on Rayleigh backscattered light from a long length of optical fiber. Unfortunately, Rayleigh backscattering itself is quite insensitive to most of the physical parameters mentioned previously. There have been a few reports of using modified versions based on the measurement of polarization changes (polarization OTDR) [26] or phase changes (coherent OTDR) [27]. However, they are not as attractive for static sensing applications compared to sensing using inelastic scattering processes such as Raman and Brillouin scattering.

Raman scattering is well suited for distributed sensing of temperature through the measurement of the anti-Stokes scattered radiation [28]. More importantly, it has very little sensitivity to other physical parameters such as strain or pressure and hence is well suited for industrial applications. Distributed sensing is achieved by slightly modifying the OTDR receiver to incorporate a band-pass filter (BPF) capable of separating out the Stokes and the anti-Stokes components. Since the temperature measurement is dependent on the ratio of the Stokes and anti-Stokes components, simultaneous OTDR traces need to be obtained for both. This necessitates a dual-channel receiver design building on the concepts discussed earlier. The other important aspect in the design of such receivers is the high transimpedance gain requirement for such receivers and the associated constraint on the bandwidth due to the low level of Raman backscattered signals. Such issues are discussed in further detail later.

As discussed earlier, Brillouin scattering is based on the interaction of light with acoustic modes propagating in the optical fiber [29]. Since the acoustic wave propagation is dependent on the density of the material, which in turn is dependent on physical parameters such as strain and temperature, Brillouin scattering may be used to sense such parameters. Sensing is achieved by monitoring the backscattered Brillouin frequency as the optical excitation pulse propagates along the fiber. One of the key challenges is to extract the Brillouin backscattered frequency shift with an accuracy of a few kHz from an optical carrier whose frequency is in the order of 200 THz. The other challenge is to discriminate between strain and temperature as the Brillouin frequency shift is sensitive to both. Such aspects are discussed in more detail in Section 12.4.

12.3 DISTRIBUTED TEMPERATURE SENSING USING RAMAN SCATTERING

As discussed in the previous section, when an electromagnetic wave is coupled into a silica waveguide, several different scattering mechanisms are invoked. One of those scattering mechanisms, which is capable of providing a precise measurement of temperature, is Raman scattering [28]. A key aspect of such a scattering process is that it has very little cross-sensitivity to other physical parameters such as strain, pressure, and vibration and hence is an attractive candidate for DTS.

Raman scattering is due to the interaction of an electromagnetic wave with molecular vibrations in a given material. Such molecular vibrations, also known as optical phonons, are characteristic of the molecules present in the system and provide a unique signature. If the molecule is at an unperturbed state, the incoming electromagnetic wave or photon may transfer some of its energy to the molecular vibrations resulting in a loss of energy ($h\Delta\nu_{ph}$) for the scattered photon. This process resulting in a lower energy photon or a photon with longer wavelength (λ_S) relative to the incident photon is called the Stokes process. The Bose–Einstein probability for such a Stokes process is given by

$$\rho_S(E) = \frac{1}{1 - \exp\left(-\frac{h\Delta\nu_{ph}}{k_B T}\right)} \tag{12.19}$$

On the other hand, if the molecule is already in a vibrational state, the incident photon could annihilate the optical phonon and steal its energy, resulting in a higher energy for the scattered photon. Such a higher-energy photon, corresponding to a shorter wavelength (λ_{AS}) relative to the incident photon, is called the anti-Stokes process. The Bose–Einstein probability for such an anti-Stokes process is given by

$$\rho_{AS}(E) = \frac{\exp\left(-\frac{h\Delta\nu_{ph}}{k_B T}\right)}{1 - \exp\left(-\frac{h\Delta\nu_{ph}}{k_B T}\right)} \tag{12.20}$$

These processes are more clearly explained in the energy level diagram of Figure 12.13. It is to be noted that the scattering process is a spontaneous process involving virtual energy levels and does not correspond to any real energy levels. In the Stokes process, energy is lost by the incoming photon in favor of the phonon energy of the molecule, and the anti-Stokes process corresponds to the energy gain by the incoming photon due to transfer of the phonon energy. A typical reason for the molecule to be in a vibrational state during the anti-Stokes process is that it is experiencing an elevated temperature. Hence, the anti-Stokes process is highly dependent on temperature, whereas the Stokes process does not have much sensitivity to temperature changes. In fact, when we observe the ratio of the optical power in the anti-Stokes signal to that in the Stokes signal, we see that it has an exponential dependence on temperature:

$$R(T) = \frac{P_{AS}}{P_S} = \left(\frac{\lambda_S}{\lambda_{AS}}\right)^4 \exp\left(-\frac{h\Delta\nu_{ph}}{k_B T}\right) \tag{12.21}$$

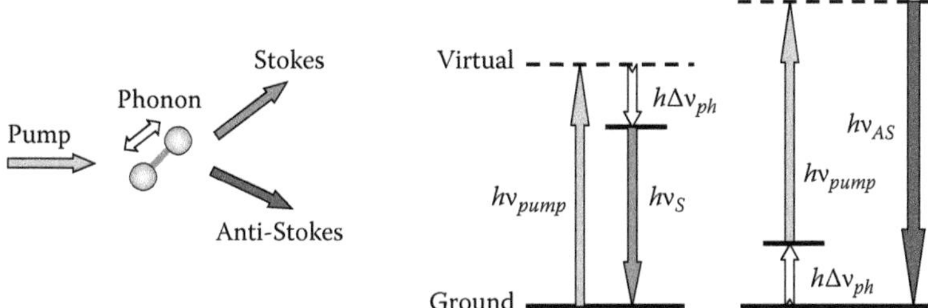

FIGURE 12.13 Illustration of the interaction between the excitation photon (pump) with the molecular vibration modes (phonon) resulting in either Stokes or anti-Stokes scattered photons. The corresponding energy level diagram is shown on the right side.

As mentioned earlier, the phonon energy spectrum is characteristic of the constituent molecules and their arrangement in a given material. Since the Raman scattering process is dependent on the phonon energy spectrum, the corresponding gain spectrum will be similar. For fused silica, the measured Raman gain coefficient normalized to the peak gain value as a function of frequency is as shown in Figure 12.14 [30]. The peak is found to occur at a frequency of 13 THz with respect to the pump frequency. This corresponds to a wavelength shift of ~100 nm when the pump wavelength is at 1550 nm. Thus, we expect to see the Stokes spectrum peak at 1650 nm and the anti-Stokes spectrum peak would be at 1450 nm.

12.3.1 Typical Configurations

In the earlier discussion, we learnt that the anti-Stokes scattered light provides a good measure of the temperature in any given material. Moreover, by normalizing the optical power in the anti-Stokes band with that in the Stokes band, one can eliminate any common artifacts such as background attenuation in the sensing fiber, light coupling losses, or any microbending losses. DTS may be achieved by marrying the OTDR concept described in the previous section with the observation of the ratio of the optical power in the Stokes and anti-Stokes spectral bands. Such a technique is commonly referred to as distributed anti-Stokes Raman thermometry (DART).

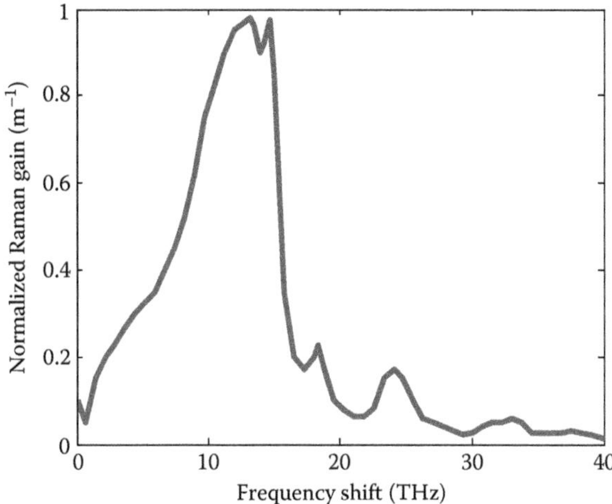

FIGURE 12.14 The Raman gain of a fused silica optical fiber normalized with respect to its peak value shown as a function of the frequency shift with respect to the excitation frequency.

A couple of important points worth noting at this point is the following: (1) The Raman scattering process referred earlier is the spontaneous process (not stimulated) and (2) the sensing fiber could either be a single-mode optical fiber or a multimode optical fiber. Stimulated Raman scattering (SRS) is not preferred for such sensing applications as it converts most of the pump power into the Stokes band and is typically noisy due to the nonlinear nature of the process [31]. So, all the practical DART systems operate at power levels lower than the threshold for SRS in the sensing fiber. A related issue is the use of large area multimode fibers since the SRS threshold is higher owing to smaller power density in such fibers. For example, the SRS threshold for a 10 km long sensing fiber is in the order of 1 W in standard single-mode fibers, whereas it is an order of magnitude higher in multimode fibers. Graded-index multimode fibers are typically preferred as they have much lower intermodal dispersion compared to step-index multimode fibers [32]. However, certain applications such as power line monitoring [5] that rely on existing optical fiber communication infrastructure use single-mode fibers for the DTS.

Another important issue is the pump wavelength used in the DART system. Although the easy and relatively cheap availability of components for the optical communications market makes 1550 nm a good choice of wavelength, commercial systems typically use 1064 nm lasers as they are available with several Watts of power. A key consideration in commercial systems is the availability of fiber-pigtailed lasers based on compact, efficient semiconductor laser diodes as they could be TEC-cooled and readily integrated with other fiber-optic components in a rugged package. The pump lasers are typically pulsed at 10 ns pulse width and several kHz repetition rate using a microcontroller-triggered electronic pulse driver. In certain semiconductor lasers, the spectrum could have a tail extending into the Stokes or the anti-Stokes bands and would need to be trimmed using appropriate in-line filters. Such implementation issues are discussed in further detail in Section 12.3.3.

The schematic diagram of a typical DART system is shown in Figure 12.15. One of the key challenges in the development of a DART system is the precise measurement of the optical power in the Stokes and anti-Stokes spectral bands in the presence of a strong Rayleigh backscattered signal. As mentioned previously, it is even more daunting if the excitation laser itself has spectral components overlapping with either of the Stokes or anti-Stokes spectral bands. The temperature is determined by observing the ratio of the anti-Stokes and Stokes components:

$$R(z,T)=\left(\frac{\lambda_S}{\lambda_{AS}}\right)^4 \exp\left(-\frac{h\Delta\nu_{ph}}{k_B T(z)}\right)\exp\left[-\int_0^z \left(\alpha_{AS}(z')-\alpha_S(z')\right)dz'\right] \tag{12.22}$$

where α_{AS} and α_S are attenuation coefficients of the sensing fiber at the anti-Stokes and Stokes wavelength, respectively.

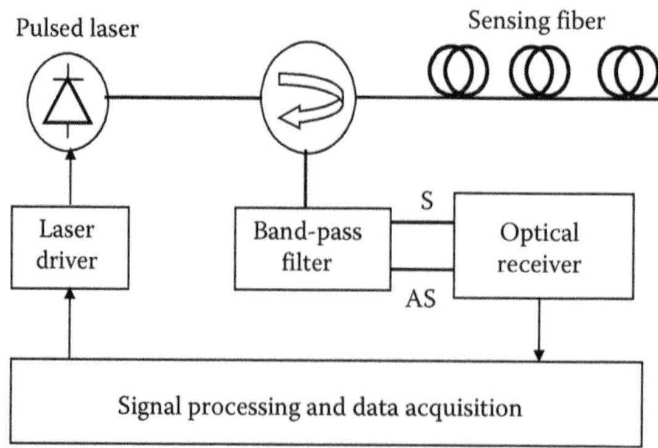

FIGURE 12.15 Schematic diagram of a typical DART system based on measuring the ratio of the Stokes and anti-Stokes backscattered optical power.

The Rayleigh backscattered signal is 3–4 orders of magnitude greater than the Raman backscattered signals. This necessitates the use of dielectric thin-film filters with sharp roll-off, high extinction, and high sideband suppression capable of separating the Stokes, Rayleigh, and anti-Stokes backscattered components before they reach the optical receiver. In addition, one might have to incorporate appropriate BPFs at the output of the pulsed laser to shape the source spectrum. Such custom filters tend to be quite expensive, which increases the cost of the DART system. On the other hand, one may use off-the-shelf components for much lower cost while incurring a performance penalty in terms of the measured temperature uncertainty.

As shown in Figure 12.15, the heart of the DART system is a microcontroller that not only triggers the optical source but also synchronizes the data acquisition from the optical receiver. Electronic pulses with the user-defined pulse width and repetition rate are generated from the microcontroller, which are converted to corresponding optical pulses by directly modulating the laser source. The optical pulses are launched into the sensing fiber through an optical circulator so that any backscattered light can be directed toward an optical receiver. As discussed earlier, the Stokes and anti-Stokes spectral bands are separated using a BPF before capturing them using a dual-channel optical receiver. The electronic signals from the optical receiver are digitized using an ADC and acquired by the microcontroller for further processing and display. The optical receiver itself is typically based on APDs, although in some instances a photon counting detection scheme has been employed to improve the receiver sensitivity, and hence, the spatial resolution of the measurement can be within 1 m [33,34].

One of the major drawbacks of the aforementioned scheme is that the optical filter would have to separate out the various spectral components precisely, which imposes a lot of constraints on the filter design. Typically, there is a finite amount of Rayleigh backscattered component that leaks into the Stokes/anti-Stokes ports, which makes the calibration of the DART system a bit tricky. Moreover, the wavelength dependence of the sensing fiber as well as the other components over the 200 nm spectral band occupied by the Stokes as well as the anti-Stokes signals is another issue. Such constraints may be alleviated through an alternative design with the use of only the anti-Stokes component measured from either ends of the sensing fiber ($P_{AS}{}^{For/Back}$) when the laser pulses are launched from the corresponding end [35]:

$$P_{AS}^{For}(z) = C_{For}\rho_{AS}\exp\left[-\int_{For}^{Back}\left(\alpha_R(z)+\alpha_{AS}(z)\right)dz\right] \tag{12.23}$$

$$P_{AS}^{Back}(z) = C_{Back}\rho_{AS}\exp\left[-\int_{Back}^{For}\left(\alpha_R(z)+\alpha_{AS}(z)\right)dz\right] \tag{12.24}$$

where

C_{For} and C_{Back} are constants for the two launch conditions
α_R and α_{AS} are the Rayleigh and anti-Stokes loss coefficients, respectively

The resultant anti-Stokes signal that has been compensated for the background fiber attenuation is now given by

$$P_{AS}(z) = \left[P_{AS}^{For}(z)\,P_{AS}^{Back}(z)\right]^{1/2} \tag{12.25}$$

The schematic diagram for such a scheme is illustrated in Figure 12.16. Note that the dual-channel BPF is replaced by a filter that passes only the anti-Stokes backscattered light. The launch condition is switched between the forward and backward ports using an optical switch and the corresponding traces are collected. The final anti-Stokes trace is computed as shown previously and

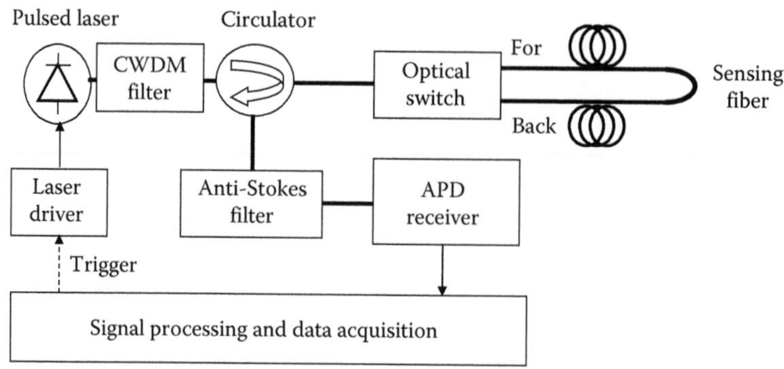

FIGURE 12.16 Schematic diagram of a DART system employing a loop scheme for temperature measurement based on only the anti-Stokes backscattered power.

appropriately calibrated to obtain the temperature map. Although such a scheme is much simpler, it is susceptible to background loss variations such as splice or bend losses. An internal calibration scheme has recently been proposed to counter such loss variations [36].

12.3.2 Simulation Results and Design

The schematic diagram of the setup used for DART is illustrated in Figure 12.15. As described previously, optical power from a pulsed laser source is launched into the sensing fiber through an optical circulator. The backscattered anti-Stokes, Stokes, and Rayleigh components may be optically separated using a BPF. The temperature assessment is usually performed by normalizing the anti-Stokes component with Stokes or Rayleigh scattered light intensities, since both components are independent of temperature. Using an optical switch, the filtered components are selectively allowed to reach the low-noise receiver for amplification of signals. The electrical signals at the receiver output are digitized using an ADC. The digitized signals are averaged in a microcontroller unit or an field programmable gate array (FPGA), which works in tandem with two banks of SRAM to accomplish this task. Finally, the averaged traces are typically extracted through a Universal Serial Bus (USB) port.

An important factor that influences the design of the Raman optical time domain reflectometer (ROTDR) receiver is the low intensity level of the backscattered anti-Stokes radiation because the anti-Stokes capture coefficient for single-mode fibers is of the order of 6e-10 m^{-1} over the anti-Stokes spectral band [37]. Depending on the level of backscattered anti-Stokes component, one has to set the appropriate gain of the receiver. The receiver design is driven by the OTDR equation:

$$P_L - P_D + \text{SNIR} = R_i + C + 2L \tag{12.26}$$

where

P_L denotes *the peak laser output power* (dBm)
P_D denotes the *receiver sensitivity* (dBm)
SNIR denotes the *signal-to-noise-improvement ratio* (dB)
R denotes the *Raman backscattered coefficient* (dB)
i denotes the *Stokes* or *anti-Stokes component*
C denotes the *circulator* and *connector losses* (dB)
L is the *single-pass loss* in fiber (dB)

12.3.2.1 Model for Distributed Temperature Sensing System

In ROTDR system, we measure the Stokes and anti-Stokes components as a function of time that has to be converted into a temperature versus distance map. To do this conversion precisely, we need a forward model, that is, for a given input temperature map, we should be able to estimate the

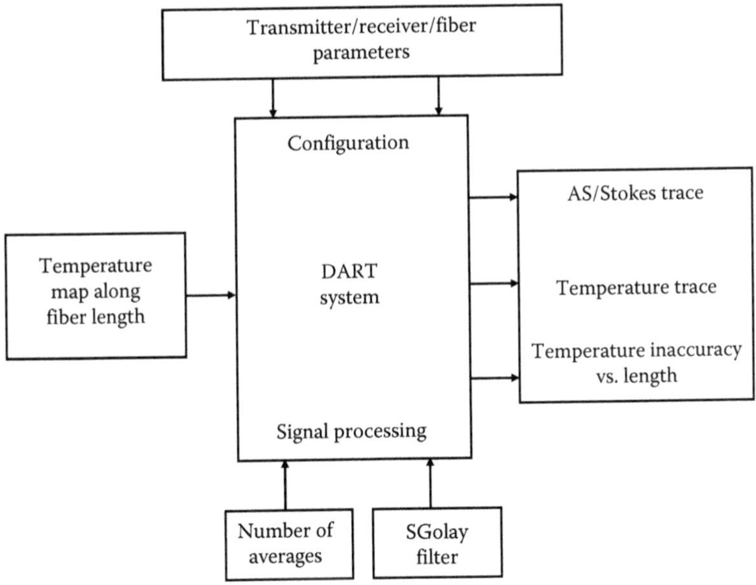

FIGURE 12.17 Block diagram of the simulation model for a DART system. The model computes the expected temperature inaccuracy for a given set of fiber and other component parameters while using different signal processing functions.

backscattered Stokes and anti-Stokes components as a function of time [38]. The block diagram of the model is illustrated in Figure 12.17, where the model takes the input parameters and computes the Stokes and anti-Stokes traces by taking into account the signal processing techniques, namely, Savitzky–Golay filtering and trace averaging.

The flowchart used for modeling the DTS system is illustrated in Figure 12.18. The model takes the following input parameters for computing the Stokes and anti-Stokes backscattered power:

1. Attenuation of the chosen fiber as a function of frequency
2. Normalized Raman gain for the fiber as a function of the frequency shift
3. Pass-band and stop-band cutoff frequencies of the Stokes and anti-Stokes filter
4. Temperature profile along the fiber length
5. Transmitter characteristics like power launched into the fiber and pulse width of the laser source
6. Receiver circuit elements like TIA feedback resistance, diode capacitance, and responsivity for computing noise and bandwidth of the receiver circuit

By using these parameters, the model will proceed to estimate the backscattered spectral components for a section of fiber length covered by the transmitted pulse width. The estimated backscattered power that is calculated for a particular section of a fiber over a fiber length of pulse width is applied to a photodetector model to convert the optical power to the corresponding photocurrent. This in turn is converted to voltage by multiplying it with TIA feedback resistance. The output of the TIA stage is then multiplied with the gain of the diff. amp. This process is repeated for each and every section of the fiber in the lengths of laser pulse width at a sampling frequency of 48 MHz. Finally, the traces are formed at the output of the receiver circuit as a function of the fiber length. In the model, photodetector and receiver circuit random noise sources are taken into account for estimating the Stokes and anti-Stokes traces. These are averaged for $N = 2^{16}$ times to improve the SNR. Averaging of traces in the model is taken into account by dividing the noise with $\sqrt{N}$, because as mentioned earlier, averaging the traces by N times improves the SNR by $\sqrt{N}$ times.

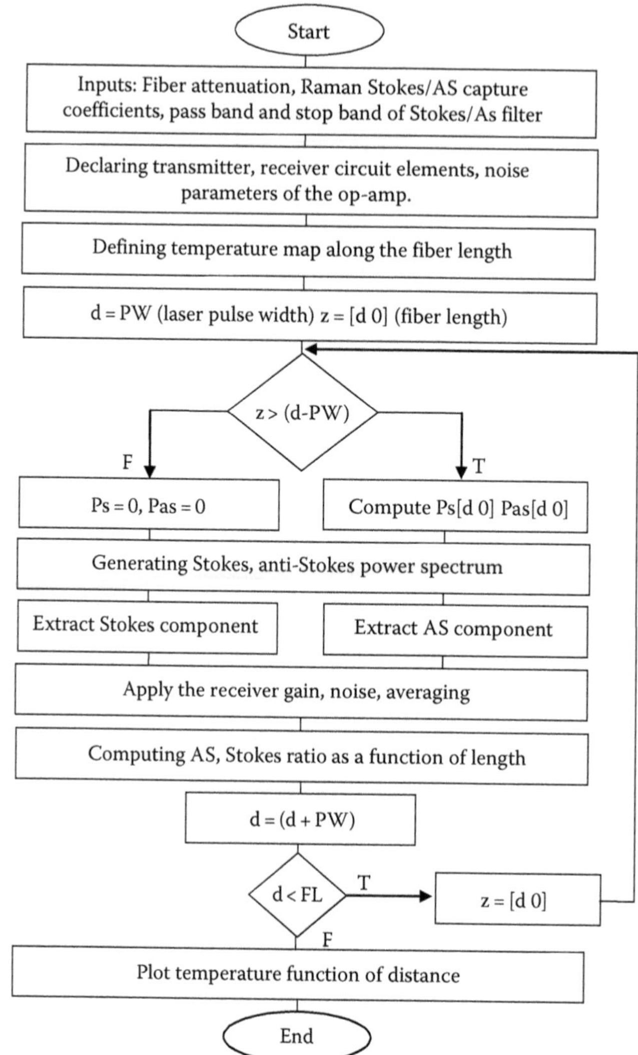

FIGURE 12.18 Flowchart for computing the expected temperature as a function of distance for a given temperature map.

12.3.3 Practical Implementation and Characterization

Using the aforementioned simulation model, one can design an ROTDR that meets the required performance specifications in a cost-effective manner. The critical issue in DART systems is the accurate determination of the anti-Stokes and Stokes power at a given location based on the measured data at the receiver. This is challenging due to the fact that the Stokes and anti-Stokes signals span a spectral width of more than 300 nm around the excitation wavelength of around 1550 nm (preferred due to the relatively easy availability of fiber-optic components). Specifically, elements such as the sensing fiber, circulator, BPF, and APD through which the aforementioned signals propagate have their own spectral response. In this section, we describe the efforts undertaken at IIT Madras to implement a Raman OTDR system.

One of the key aspects is the spectrum of the pulsed laser source, which is illustrated in Figure 12.19. The measurement was performed by continuously firing laser pulses at a repetition rate of 1 kHz and setting the optical spectrum analyzer (OSA) in averaging mode for 100 number of averages. As seen from

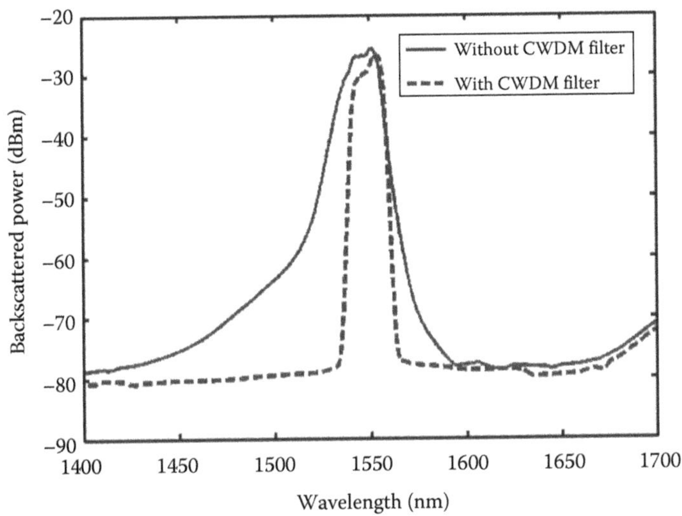

FIGURE 12.19 Averaged spectrum of the pulsed semiconductor laser source at 1550 nm. A CWDM filter inserted at the output of the laser helps to remove any spectral components that are overlapping with the Stokes/anti-Stokes spectrum.

Figure 12.19, the pulsed laser spectrum is primarily concentrated over a 20 nm band around 1550 nm, but there is a long tail seen extending into the anticipated anti-Stokes band (centered around 1450 nm). Even though the relative power in this region is more than 40 dB below the power level at the center wavelength, it is significant since the anti-Stokes power is typically more than 50 dB lower compared to the peak power and hence can potentially corrupt the measurement of the Raman scattered signals.

In order to trim down the source spectrum, a coarse wavelength division multiplexing (CWDM) filter centered at 1550 nm and having a bandwidth of 20 nm may be used. As seen from Figure 12.19, the introduction of the CWDM filter at the output of the pulsed laser source has significantly reduced the source laser spectrum leakage into the anti-Stokes band. The filtered pulse laser is launched into the sensing fiber (SMF-28, Corning Inc.) through a wideband circulator, which directs the backscattered light from the sensing fiber toward port three of the circulator. A BPF is used to separate the anti-Stokes spectrum (1450–1490 nm) from the Stokes/Rayleigh spectrum (1530–1580 nm). It should be noted that all these components have their own loss spectrum and need to be accounted so that the final temperature measurement is accurate. As explained earlier, we have used our Agilent OSA in an averaging mode and high-sensitivity setting (−85 dBm) to accurately measure the loss spectrum for all the aforementioned components.

The BPF has a sharp cutoff near 1520 nm and is able to prevent the anti-Stokes scattered spectrum from leaking into this port. Initially, we were planning to extract the Stokes scattered radiation and use it for normalizing the temperature-dependent anti-Stokes signal. However, since the Rayleigh component dominates the power collected from this port, we decided to use the Rayleigh signal itself to normalize the anti-Stokes signal.

The Rayleigh and anti-Stokes signals are subsequently detected using a low-noise receiver based on an InGaAs APD. The APD is found to have a relatively flat response over both the backscattered bands of interest. We used a bias voltage optimization program that monitors the noise floor as a function of bias voltage and ensures that the voltage is fixed at a level that does not shift the noise floor. The corresponding gain of the APD is measured to be ~13. The resulting photocurrent was converted to a corresponding voltage using a TIA with a gain of 47 k and subsequently amplified using two diff. amp. stages with a total gain of ~50, providing a bandwidth of ~5 MHz. The amplified signal is then digitized at a rate of 48 MSa/s using an ADC and finally acquired using our FPGA-based data acquisition board. The board also provided synchronization pulses such that the

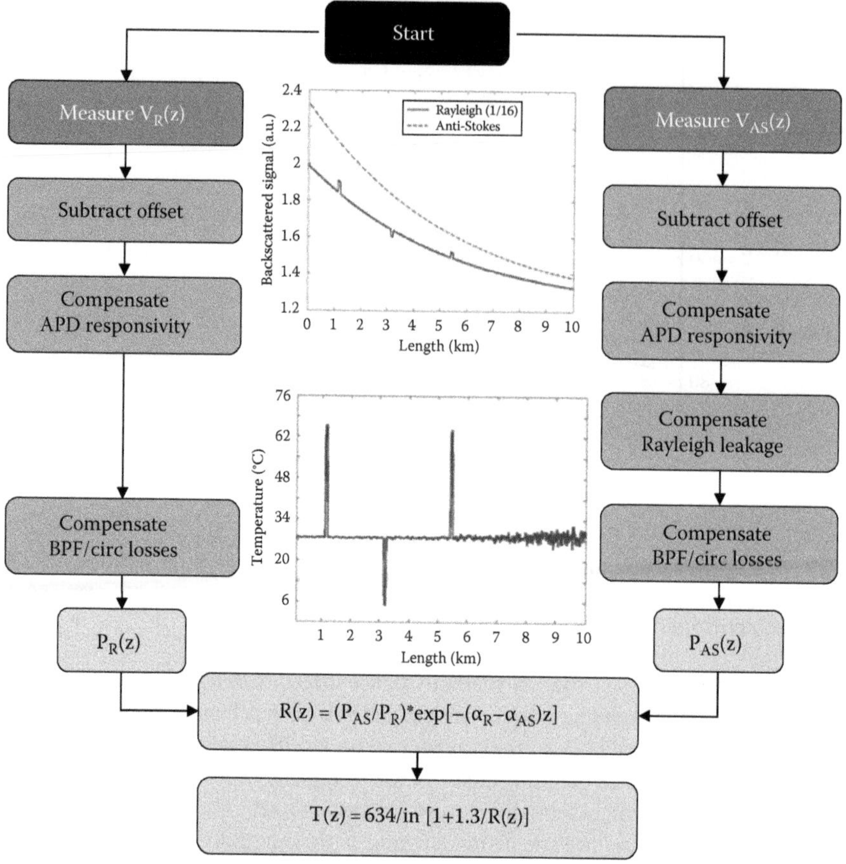

FIGURE 12.20 Flowchart representing the calibration algorithm used to determine the temperature map from experimental observations of the anti-Stokes and Rayleigh components.

laser pulses were fired at 1 kHz repetition rate and the resultant backscattered signals were averaged up to 64 k averages, yielding an SNR improvement of 24 dB.

Based on the previous spectrum measurements, we have developed a calibration algorithm that is represented as a flowchart in Figure 12.20. As explained earlier, we first measured the anti-Stokes and Rayleigh traces for a given temperature setting experienced by a short length (200 m) of the sensing fiber kept inside a box oven. The traces typically are riding a circuit-induced offset of ~20 mV, which was calculated by averaging the last 1000 samples of the respective traces and subtracted from the measured traces. The traces are subsequently corrected for the APD responsivity and the respective losses from the BPF and circulator. In addition to these, the anti-Stokes signal was also corrected for the Rayleigh leakage power. The final corrected Rayleigh and anti-Stokes traces are then used to calculate the ratio, which included a correction factor for the differential attenuation in the sensing fiber. Finally, the measured temperature may be deduced using such data. Thus, our calibration algorithm takes into account the detailed spectral measurements outlined earlier and appropriately uses them to obtain a precise measure of the temperature sensed by the optical fiber [39].

The traces of the anti-Stokes signal are normalized with Rayleigh signal after applying all appropriate correction factors for the losses in the different optical components. A typical plot of the RMS value of temperature error as a function of the distance measured based on noise observed at room temperature condition is shown in Figure 12.21. As expected, the error value increases exponentially with the length as the pump power is depleted at the farther sections of the fiber. The baseline error value observed at the initial section of the fiber is due to the receiver noise in our case.

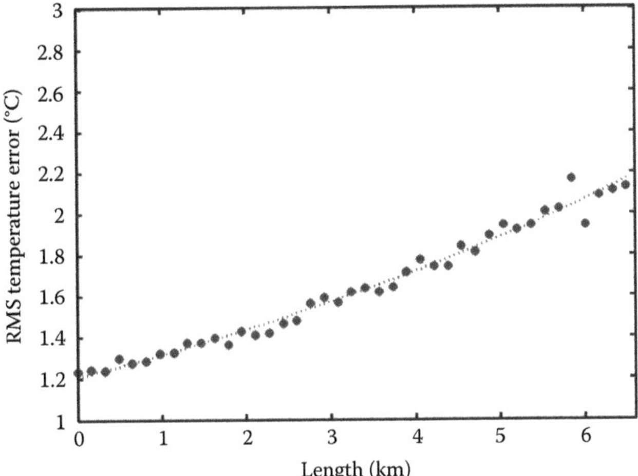

FIGURE 12.21 Typical plot of the measured RMS temperature error as a function of distance for a conventional single-ended ROTDR.

To verify the repeatability of the temperature map, we carried out the measurement of the anti-Stokes and Rayleigh traces for 100 times. Since the sensing fiber is kept at room temperature, we expect the ratio to be ideally constant across the entire length of fiber. However, due to the presence of the receiver noise, there is a temperature error over the fiber span. This could be quantified by subtracting the measured temperature from the expected temperature. The temperature error is plotted as a function of the sensing fiber length in Figure 12.22, along with the RMS temperature error calculated from the previous 100 measurements for a bin containing 4 data points (which corresponds to 8 m spatial resolution).

Based on the previous measurements, one could plot histograms to determine the temperature uncertainty at different sections of the fiber. Figure 12.23 illustrates such histogram data observed at four different locations along the sensing fiber. The histograms fit well to a Gaussian function, as expected for the noise sources. One can also observe that the RMS temperature uncertainty increases as we reach the farther sections of the sensing fiber.

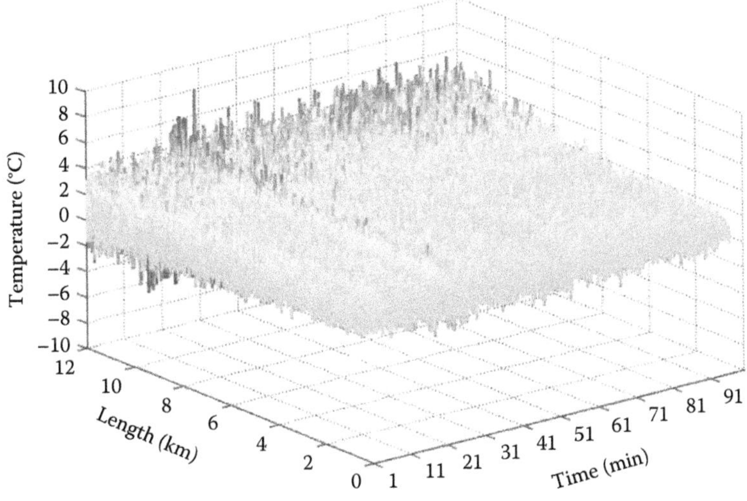

FIGURE 12.22 The RMS temperature error computed over a length of 12 km for 100 measurements when the test fiber is maintained at room temperature.

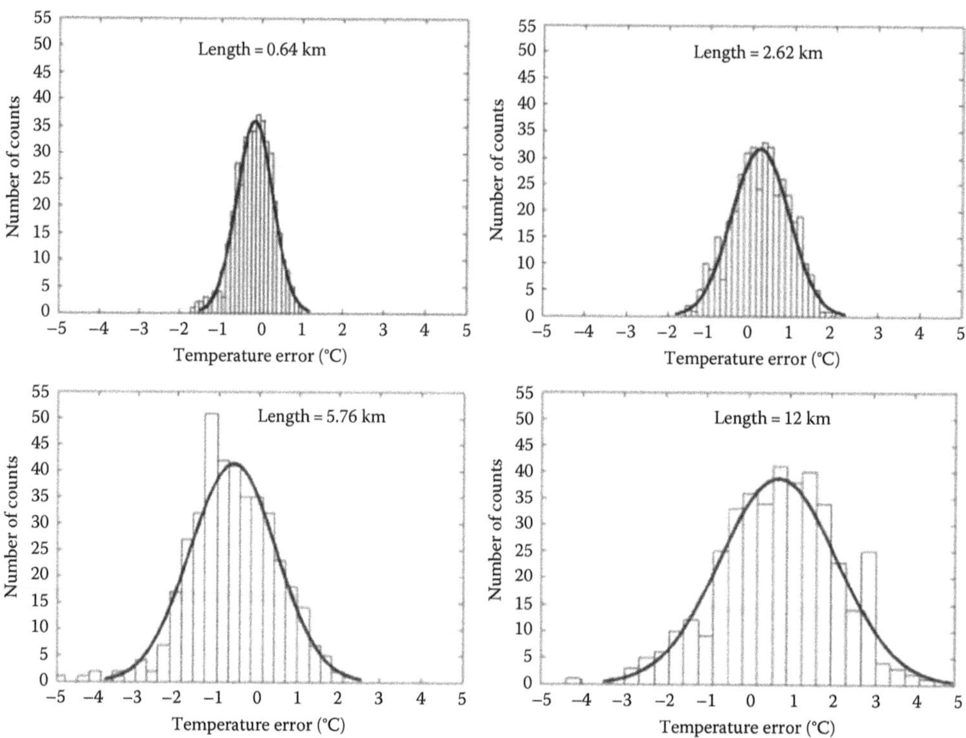

FIGURE 12.23 Histogram of the data obtained at four different locations along the sensing fiber. The envelope of the histogram fits well with a Gaussian function.

In contrast to the single-ended configuration, the loop configuration has a different dependence of the RMS temperature error as a function of distance [35]. The higher error values at the either extremes are due to the low power levels at the farther end of the fiber when the pump is launched from the forward or the backward port. A typical plot of the temperature error is shown in Figure 12.24.

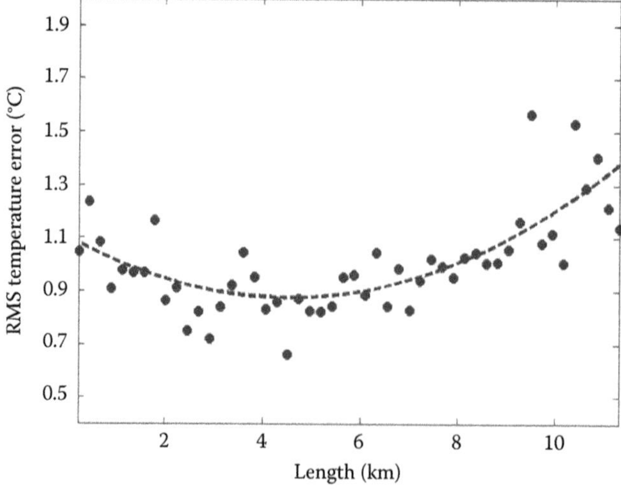

FIGURE 12.24 Typical plot of the measured RMS temperature error as a function of distance for an ROTDR operated in a loop configuration.

12.3.4 Applications

DART is a three-decade-old technology, which has seen several exciting advancements and innovations to reach commercialization. Over the past decade, the technology has matured well enough to support several on-field applications spanning from downhole monitoring to power line monitoring to environmental temperate sensing. The key challenge in several of these applications is the customized design of the cable to cater to the particular needs of any given application. Such aspects are discussed in further detail next.

Downhole monitoring: Downhole monitoring of oil and gas reservoirs is one of the earliest and leading applications of distributed sensors [40–43]. By monitoring the geothermal gradient along the depth of a well at different times, one can understand several physical processes such as flow diagnostics, water injection profiles, and cross-flow between different zones [40]. In such applications, the typical distances are in the order of 2–3 km and the temperatures may rise up to 250°C. As such, DART systems using graded-index multimode sensing fibers will be quite apt for these applications. However, special hermetic encapsulation of the cable has to be designed to protect the fiber from the harsh environment of the bore well, especially from hydrogen penetration as it increases the background loss of the fiber [42,44,45].

Power line monitoring: It is well known that the temperature of power cables is enhanced due to Joule heating effect. As such, by monitoring the thermal profile along the power cable, one can estimate the current at any particular point along the cable. This is particularly useful in determining the current carrying capacity or ampacity of the cable such that the corresponding temperature is below the critical temperature at which the insulation is damaged [46]. Since the recent designs of power cables include single-mode optical fibers embedded in its core for communication and signaling applications [47], the same may be used for power line monitoring as well. In addition to this, several other power system applications such as transformer monitoring, switchgear monitoring, rotating machine monitoring, and pipework monitoring are also enabled by DART technology [5].

Environmental temperature sensing: Distributed fiber sensors potentially have a major role in studying hydrologic processes such as interaction between surface water and groundwater resources, snow hydrology, and land surface energy exchanges [48,49]. In such applications, DART systems with multimode sensing fibers suitably coated with water-resistant materials are typically used.

Pipeline monitoring: One of the applications where the unique capability of distributed sensors to reach ultralong (tens of kilometers) with submeter resolution is quite useful is pipeline monitoring [50–52]. There are primarily two different aspects of pipeline monitoring, namely, leakage detection and blockages in gas pipelines due to wax or hydrate formation. While the former measures a drop in temperature due to Joule Thompson effect on gas leakage, the latter is looking for an increase in the temperature due to crystallization of the hydrates. The DART technology is useful in detection of temperature over distances in the order of 10 km, whereas longer distances require Brillouin sensors.

Sewage monitoring: Sewage monitoring is one of the upcoming applications of DART sensors [53], wherein the temperature of wastewater is monitored to monitor the flow along the primary channel or from individual house connections. The expected temperature changes are only in the order of a few degrees, and hence, the measurement accuracy of the instrument needs to be quite good (0.1°C). Such systems can also be used to monitor the flow during storms and could be useful to precisely locate blockages.

12.4 DISTRIBUTED SENSING USING BRILLOUIN SCATTERING

Interaction of monochromatic light waves with the propagating acoustic waves and its subsequent scattering, with a frequency shift, was first proposed by Leon Brillouin in 1922 [54]. The experimental observation of this theoretical prediction of frequency shift was first observed by Chiao et al. [55]

in quartz and sapphire. Brillouin scattering was well predicted and demonstrated in the early days of evolution of optical fiber. Ippen and Stolen reported the observation of Brillouin scattering in an optical fiber for the first time from Bell Labs in 1972 in a long length of a lossy fiber and using green light [56]. SBS was looked upon as a major deterrent in optical fiber communication systems. A large part of research related to Brillouin scattering is devoted to mitigating this effect in optical fibers. On the other hand, SBS started getting attention for its application in the design of narrow-line lasers, frequency shifters, optical buffers, and spectroscopy in addition to its application in sensors. In this section, we explain the underlying processes in an optical fiber that results in SBS, the possible configurations through which Brillouin scattering is utilized for distributed sensing and some results from the experimental implementation of these schemes.

12.4.1 Background: Basics of Brillouin Scattering

Brillouin scattering is a process resulting from the interaction between an incident optical field and the acoustic waves in the material [54]. The thermoelastic motion of the molecules in the material leads to random density fluctuations within the medium. These density fluctuations propagate through the medium with the velocity of sound as acoustic waves in random directions. The periodic density modulation results in a refractive index modulation, albeit small, and causes diffraction of the incident waves. Since the density modulation and the consequent refractive index grating moves with an acoustic velocity v_A relative to the incident waves, the scattered wave experiences a frequency shift due to Doppler effect, resulting in spontaneous Brillouin scattering. When the spontaneous scattering occurs at sufficiently high intensities, the incident and the scattered waves interfere and in turn intensify the acoustic waves through electrostriction. The increased strength of the moving grating now results in a positive feedback causing a large intensity in the scattered waves, thus causing SBS. The scattered wave in the backward direction is downshifted in frequency and hence called the Stokes wave. Quantum mechanically, the phenomenon is described as the annihilation of a pump photon and the creation of a new photon with a downshifted frequency (Stokes) and a phonon. Thus, the process can be interpreted as a parametric acousto-optic interaction between the pump photon, the Stokes photon, and the acoustic waves.

The conservation of energy and momentum in the acousto-optic interaction leading to the scattering process demands

$$\Omega_B = \omega_p \pm \omega_s \text{ and } k_A = k_p \pm k_s \tag{12.27}$$

where

ω_p, ω_s, and Ω_B represent the frequencies

k_p, k_s, and k_A represent the wave vectors of the incident pump, the Stokes, and the acoustic waves, respectively

Equation 12.27 is written for extreme cases corresponding to the conditions where the pump and the acoustic waves propagate along the same direction (corresponding to the positive sign) or the pump and the acoustic waves move in opposite directions (corresponding to the negative sign).

In general, the phase matching condition for the Stokes wave, corresponding to the negative sign in Equation 12.27, is shown in Figure 12.25.

From this figure, it is deduced that the acoustic wave satisfies the standard dispersion relation:

$$\Omega_B = v_A \left|k_A\right| \approx 2 v_A k_p \sin\left(\theta/2\right) \tag{12.28}$$

where θ is the angle between the pump and Stokes fields. Equation 12.28 shows that the frequency shift of the Stokes wave depends on the angle of scattering and is a maximum in the backward

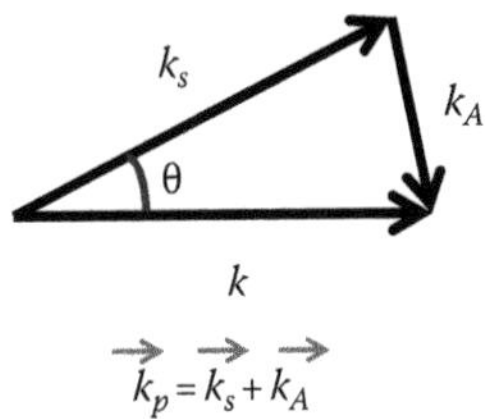

FIGURE 12.25 Representation of the momentum conservation condition for acousto-optic scattering of the Stokes wave.

direction, ($\theta = \pi$), corresponding to the case where the acoustic waves and the pump wave move in opposite directions. It can also be seen that though spontaneous scattering can occur in the forward direction, there is no stimulated scattering in the forward direction. In the case of SBS in single-mode optical fibers, the relevant direction corresponds to the case where $\theta = \pi$, since guidance is not possible in any other directions. Hence, for optical fibers, the Brillouin frequency shift can be written as

$$\nu_B = \frac{2n\nu_A}{\lambda_p} \tag{12.29}$$

where n is the effective modal index for the pump wavelength, λ_p. Thus, the frequency shift depends on the wavelength of the pump, velocity of the acoustic wave, and refractive index. For silica fibers, the typical frequency shift at a pump wavelength of 1.55 μm is about 11.1 GHz.

The spectral width of the Brillouin gain spectrum (BGS) depends on the lifetime of the acoustic phonons. For a quasi-monochromatic pump, assumption of an exponential decay in the time domain results in a Lorentzian function for the gain spectrum in the frequency domain, given as

$$g = g_0 \frac{(\Gamma_B/2)^2}{(\Gamma_B/2)^2 + (\Omega - \Omega_B)^2} \tag{12.30}$$

where

g_0 is the peak Brillouin gain
Γ_B represents the reciprocal of the phonon lifetime

The peak value of gain coefficient is given as [54]

$$g_0 = \left(\frac{4\pi n^8 p_{12}^2}{c\lambda_p^3 \rho_0 \nu_B \Delta\nu_B}\right)\left(\frac{\Delta\nu_B}{\Delta\nu_B \otimes \Delta\nu_p}\right) \tag{12.31}$$

where

p_{12} is the longitudinal elasto-optic coefficient
$\Delta\nu_B$ is the full width at half maximum (FWHM) of the gain spectrum, which corresponds to the reciprocal of the phonon lifetime
$\Delta\nu_p$ is the linewidth of the pump at wavelength λ_p
ρ_0 is the material density

For silica fibers, with continuous wave (CW) pump, the gain coefficient is estimated to be 5×10^{-11} mW^{-1}. For Lorentzian gain profile, the convolution in the denominator of Equation 12.31 can be calculated as $\Delta\nu_B + \Delta\nu_p$. The gain curve is reported to narrow with the increase in pump power [57]. In the case of temporally short pump pulses, the spectrum is correspondingly broad, and the gain curve converges to a Gaussian function as a result of the superposition of a number of Lorentzian functions. The width of the gain spectrum in this case is decided by the duration of the pump pulse, and the gain spectrum gets narrowed when the pump pulse is shorter than the phonon lifetime. For pulses shorter than 5 ns, the gain spectrum becomes identical to the CW value [58].

SBS is one of the prominent deterrents in optical fiber communications. When the pump power exceeds a certain distinct value, the throughput power is found to saturate, indicating the onset of SBS. Beyond this threshold, the Stokes power is also found to increase. SBS threshold is defined differently in the literature [59]. Following ITU recommendations, the threshold power is calculated for CW pump using the following relation [54]:

$$P_{th} \cong 19 \frac{KA_{eff}}{L_{eff} g_0} \tag{12.32}$$

where K is the polarization factor, which varies from 1 to 2 depending on the polarization state of the pump and the signal. Thus, larger lengths of the fiber would indicate lower threshold powers. SBS threshold in communication systems is found to be in the range of 1–10 mW, due to this reason.

The properties of acoustic wave such as its velocity are dependent on temperature and strain in the fiber, and hence, the center frequency of Brillouin gain is expected to vary with the temperature and strain in the medium. This dependence, which is primarily due to the changes in the acoustic velocity, is found to be linear and is represented as

$$\nu_B(T,\varepsilon) = C_\varepsilon \Delta\varepsilon + C_T \Delta T + \upsilon_B(T_0,\varepsilon_0) \tag{12.33}$$

where

C_ε and C_T represent the strain and temperature coefficients, respectively
T_0 and ε_0 represent the reference values of temperature and strain, respectively

For standard single-mode fibers, C_ε and C_T are experimentally measured to be ~50 kHz/με and 1 MHz/K, respectively [3,60,61], with minor variations around these values as determined by the composition of the fiber and its coatings. Note that while the fiber is the most sensitive to tensile strain, lateral strain does not induce a significant change in the frequency. The change in frequency shift of SBS due to temperature was first used for a distributed sensing of temperature by Culverhouse et al. [62] while distributed strain sensing was first demonstrated by Horiguchi et al. [12]. The typical linewidth of Brillouin gain profile for single-mode fiber germanium-doped silica fiber is in the order of 20 MHz [63]. The theory suggests that Brillouin gain linewidth does not change with temperature–strain variations. However, the linewidth is observed to change with temperature. The temperature dependence of the Brillouin linewidth is found experimentally to be 0.25 MHz/°C with higher temperatures having narrower Brillouin spectra [64].

Since the frequency shift due to Brillouin scattering is sensitive to changes in both temperature and strain, one of the primary drawbacks of the Brillouin-based scheme is the cross-sensitivity. One of the earliest methods proposed to tackle this issue was to use two separate sets of fibers—one subjected to strain and the other fiber running parallel to the test fiber to work as a reference fiber to record temperature changes, if any. The use of a second fiber was later eliminated by measuring the intensity in the Stokes, anti-Stokes, and Rayleigh scattering when operated in the spontaneous scattering domain [19,65]. The temperature is measured through the intensity changes in the anti-Stokes wave while the strain is measured through the intensity changes in the Stokes wave. The ratio of the intensities of the Rayleigh scattering to the sum of the Stokes and the anti-Stokes Brillouin scattered light—known as the Landau–Placzek ratio—for an optical fiber is found to be predominantly dependent only on temperature [30]. Hence, when this ratio is measured in conjunction with the frequency dependence of the strain, it was possible to uniquely discriminate the temperature and strain in the fiber, as demonstrated by Newson and coworkers [19]. This technique had only a limited sensitivity and spatial resolution since the operation had to be limited to the spontaneous regime to maintain the linearity in the generated Stokes power, resulting in very low intensity levels. The use of Rayleigh scattering as the reference, however, helped to cancel any fluctuations in the input power or in propagation losses in the fiber. Parker et al. [67] further countered this limitation by mathematically deriving an equivalent linear relation for the Stokes and the anti-Stokes component in the stimulated scattering domain, but the noise associated in their technique limited the performance of the sensor. It was found that the SNR of the anti-Stokes component increases with the input power, while that of the Stokes decreases with the input power. Hence, for a given detection bandwidth, there is an optimal operating point for gain, which in turn limited the resolution and range of measurement.

The approach to tackle the cross-sensitivity issue was to use fibers with different temperature coefficients in the core and alternate refractive index profiles and, hence, multiple peaks in the BGS. Large effective area nondispersion-shifted fiber, which was commercially used to counter the dispersion in a standard single-mode fiber in communication systems, is one such fiber used by Lee et al. [68], who demonstrated that the strain coefficients of the multiple peaks were identical while their temperature coefficients were different. This differential response of the two gain peaks to temperature and strain made the simultaneous and the discriminated measurements. Photonic crystal fibers were further used for improved sensitivity and resolution [69]. Fluorine-doped depressed clad fibers and polarization-maintaining fibers were also reported, which relied on the differential response of the multiple Brillouin gain peaks to temperature and strain [70]. The different dependencies of the first two peaks on strain and temperature have been previously used for simultaneous measurements along the fiber. Bao et al. recently reported the use of four peaks in the BGS of a large effective area fiber (LEAF) for discriminating strain and temperature [71,72]. With an increase in strain, the power and linewidth in the first two peaks were found to decrease, while with temperature, the power in all the peaks was found to have a quadratic relation. The linewidths were found to decrease with the increase in temperature. These temperature and strain coefficients were used to discriminate these parameters, and an improved strain resolution of 37 με and a temperature resolution of 1.8°C with a spatial resolution of 4 m were reported in this work. The beat spectrum of these gain peaks was further used to result in intensity-based measurement by Bao and coworkers [73]. Robust techniques of distinguishing temperature and strain with the best measurement resolutions are thus still a topic of active research. We now proceed to discuss the typical measurement configurations used for distributed sensing.

12.4.2 Typical Configurations for Distributed Sensing Using Brillouin Scattering

Brillouin sensors are primarily classified based on whether spontaneous Brillouin scattering or SBS is used as the mechanism for sensing. In those configurations that use spontaneous scattering, incident light—referred to as pump—is launched into the fiber under test (FUT) and the spontaneous scattering from the fiber is detected and measured. The pump is usually pulsed so that the location of the intensity of this spontaneously scattered light can be detected. The scattered power is typically small, and hence, coherent detection is typically used to increase the dynamic range. The measurement of spontaneous scattering is one of the key techniques typically used for temperature–strain discrimination, as discussed in the previous section. The schematic of such a system—also referred to as Brillouin optical time domain reflectometer (BOTDR)—is shown in Figure 12.26.

In the BOTDR system, the light from the output of a narrow-line CW laser is pulsed externally to act as the pump for creating spontaneous scattering in fiber. External optical amplifiers are often used at the output of the modulator to increase the detection range. The backscattered light is frequency downshifted from ν_p, depending on the temperature and strain conditions in the FUT. The scattered light is allowed to beat with CW light from laser two, whose frequency is typically adjusted to be close

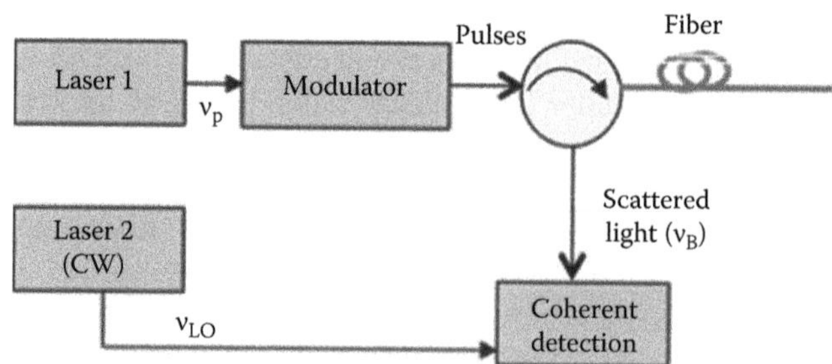

FIGURE 12.26 Schematic of a typical Brillouin OTDR.

to the frequency corresponding to the scattered light (ν_s). The electrical beat signal at the output of the coherent receiver contains frequency components with much lower frequencies, which is further analyzed using a tunable BPF. The center frequency of the BPF is tuned, and the time domain signal is recorded for each of the center frequency in order to reconstruct the frequency distribution of the backscattered signal at each position of the fiber. Alternately, the electrical beat signal is mixed with a microwave oscillator and the frequency of the oscillator is tuned to obtain the frequency map of the fiber. For a simultaneous detection of strain and temperature such as the Landau–Placzek ratio method, care is taken to ensure that system operates in the spontaneous regime. The power in the spontaneous backscattered light is smaller, and though coherent detection improves the dynamic range to a great extent, the spatial resolution of the system is limited to about 1 m. This limitation is primarily posed by the phonon lifetime in the fiber. If the pulse covers a distance larger than the corresponding phonon lifetime (~10 ns), the observed frequency distribution would be a convolution of the natural linewidth of the Brillouin spectrum with the pulse spectrum, resulting in a poorer measurement contrast. BOTDR configuration is particularly attractive in situations where only one end of the fiber is accessible.

A significant improvement in dynamic range can be achieved if SBS is utilized, where the additional seed photons at the Brillouin shifted frequency are launched from the farther end of the fiber, thus leading to a Brillouin optical time domain amplifier (BOTDA) configuration. The schematic of a BOTDA configuration is shown in Figure 12.27.

In the BOTDA system, light from a narrow-linewidth laser is split into two arms, one of which is pulse modulated, which acts as pump to create SBS in the fiber. Light in the other arm is amplitude modulated typically using a Mach–Zehnder-based intensity modulator using an radio frequency (RF) tone at frequencies corresponding to the Brillouin frequency shift in the optical fiber. The modulator is operated at the carrier-suppressed configuration by biasing it at its null, resulting the generation of frequencies at $\nu_p \pm \nu_B$. These frequencies are fed through the farther end of the fiber as the probe/seed photons, which undergo amplification whenever the frequency difference between the seed and pump frequency corresponds to exactly the Brillouin shift. The frequency

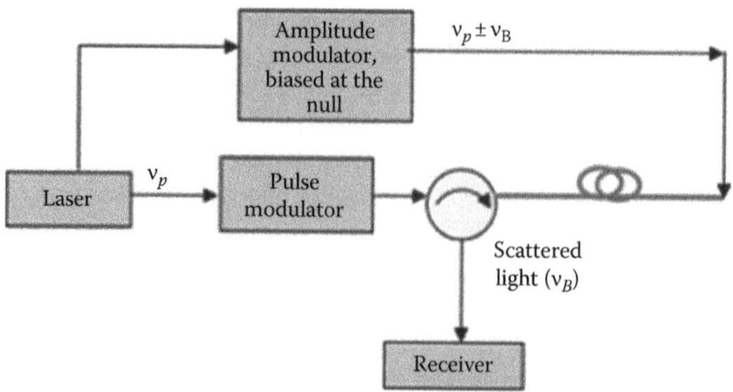

FIGURE 12.27 Schematic of a general Brillouin optical time domain analysis setup.

of modulation is swept and the time domain signal of the amplified probe corresponding to the Stokes frequency is observed for each modulation frequency using a direct detection detector. Time information about amplification of seed at respective instants helps in estimating distance information, and the corresponding frequency shift helps in estimating the temperature–strain difference. The distance resolution is half the distance that pump pulse can cover during its ON time. Since the probe is amplified in this case, its intensity is much larger than the Rayleigh component, which results in a better SNR. The dynamic range for BOTDA is larger than that for BOTDR for the same reason. Resolution is limited by the SNR at the receiver and the spatial resolution is limited by the phonon lifetime. The technique also demands the need of a high bandwidth receiver for detection. BOTDA system is more preferred than BOTDR only when both ends of fiber are accessible. Since each sensing cycle involves the tuning of RF frequency at a desired wavelength and the capture of an averaged time domain trace for each frequency, the measurement time required for a traditional BOTDA scheme could be of the order of minutes. Many health-monitoring applications would require dynamic detection, and hence, the BOTDA scheme is modified for such applications.

The distance resolution can be considerably improved by using the Brillouin optical correlation domain amplification (BOCDA) scheme, first proposed by Hotate and He [74], where the pump and the probe are phase modulated. Coherent interaction between the pump and the probe waves takes place only in those regions in the fiber where they are highly phase correlated. The phase delay between the pump and the probe is tuned by scanning the phase modulation frequency and consequently the region of interaction throughout the length of the fiber. Changing the probe frequency with respect to that of the pump scans the gain spectrum. When the spectral width of the pulse exceeds the Brillouin linewidth, the resolution of correlation increases in this case, as opposed to a deteriorated measurement in the case of BOTDA. The correlation being a periodic function, care should be taken to ensure that there is only one spatial location in the fiber at which the correlation is maximum. The constraints, the measurement length, and the post processing are some limitations of this technique. This method is successfully demonstrated for fiber lengths of up to 1 km, with a spatial resolution of even as small as 1.6 mm [75–77]. In the following sections, we pick one scheme, BOTDA, and analyze the design as well as the experimental results in detail.

12.4.3 Simulation Results and Design of BOTDA

The propagation of pump and signal in the fiber can be derived from the coupled mode theory governing the interaction of light with acoustic waves, and the steady-state equations describing the evolution of the pump and Stokes waves with position z along the fiber length are [78]

$$\frac{dI_p}{dz} = -g_0 I_p I_s - \alpha I_p \tag{12.34}$$

$$\frac{dI_s}{dz} = -g_0 I_s I_p + \alpha I_s \tag{12.35}$$

where

I_p and I_s represent the pump and the Stokes intensities, respectively
α is the propagation loss in the fiber

These coupled differential equations can be solved numerically using the standard Runge–Kutta algorithm. An optical fiber has a cylindrical symmetry, and hence, all three types of acoustic modes corresponding to longitudinal, torsional, and flexural modes are supported in the fiber. Though each of these modes can cause independent, the lowest longitudinal acoustic mode interacts with the input pump photon and shows the strongest influence on Brillouin scattering. The theory of SBS discussed earlier assumes that the optical mode interacts with the lowest-order acoustic mode and the area of interaction is approximated as the effective area of the optical mode. This approximation is not true and should be modified in the case of fibers with multicomposition and modified refractive index profile, such as in LEAF.

The coupled differential equations are solved for the boundary value problem where the pump and probe powers are known at either ends of the fiber. An adaptive technique was used to handle this boundary value problem, and the evolution of the pump and probe power is shown in Figure 12.28.

One of the critical limitations of the BOTDA scheme is the nonuniform distribution of pump power through the length of the fiber due to gain depletion. Equations 12.34 and 12.35 are solved numerically when the pump is pulsed to understand the effect of gain depletion in the pump, as it propagates along the length of the fiber. The probe is assumed to be CW, with a frequency corresponding to the gain peak of Brillouin scattering. The pulse width, pump power, and probe power are varied in this simulation. Figures 12.29 and 12.30 show the evolution of pump and probe powers along the length of the fiber for different values of pulse widths of the pump.

A comparison of Figures 12.29 and 12.30 indicates strong pump depletion for a larger pulse width. A larger pulse width would result in increased interaction length in the fiber and hence a stronger pump depletion. It is also found that the increase in probe power also leads to a larger depletion. Hence, in the context of sensing, it is optimal to choose large pump power, small probe power, and short pulse widths.

Since the Brillouin scattering provides a shift in the frequency whenever there is a strain or temperature perturbation of the sensing fiber, the backscattered power has to be analyzed in the frequency domain as well. The probe signal gets amplified within the sensing fiber when its frequency matches the Brillouin scattered frequency. Thus, we can analyze the strain–temperature changes along the entire length of the fiber. The simulated temporal response of the scattered power from a fiber of length 10 km corresponding to two hot spot events—at 3 km and at 8 km, respectively—at different tuning frequencies is shown in the following. These hot spot events correspond to temperature events of 100°C and 200°C. The same results are obtained when the hot spot events correspond to strain events of 2.3% and 4.6%, respectively.

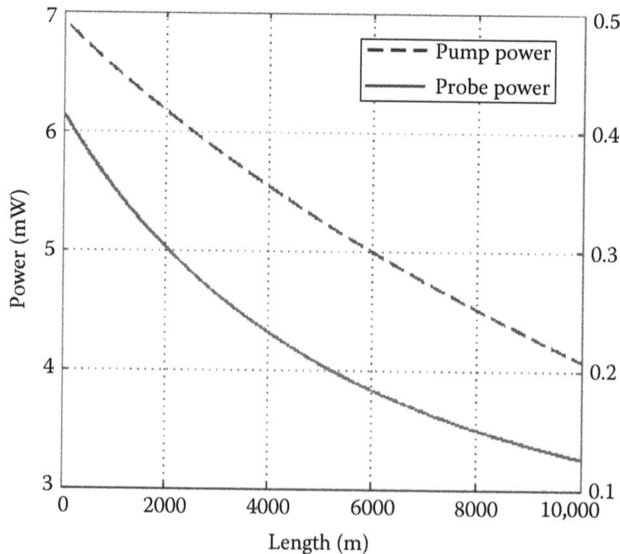

FIGURE 12.28 Evolution of pump and probe powers along the length of the fiber; the pump and the probe are CW. The probe power is found to increase in the counterpropagating direction.

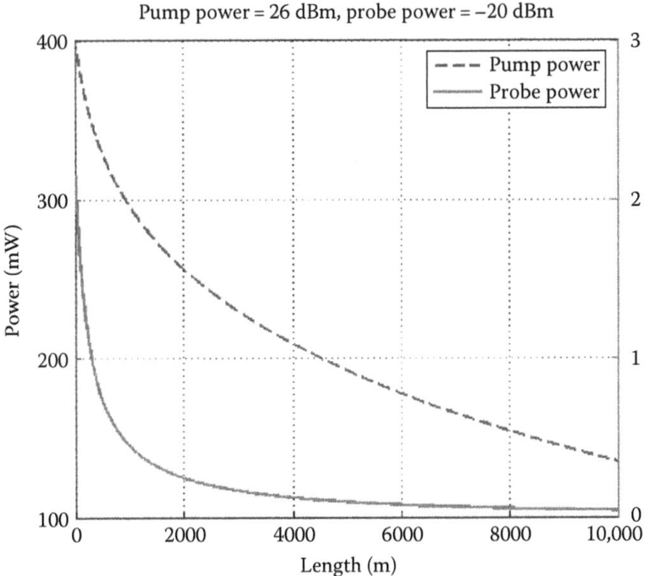

FIGURE 12.29 Evolution of the pump and probe powers along the length of the fiber for a pulse width of 250 ns.

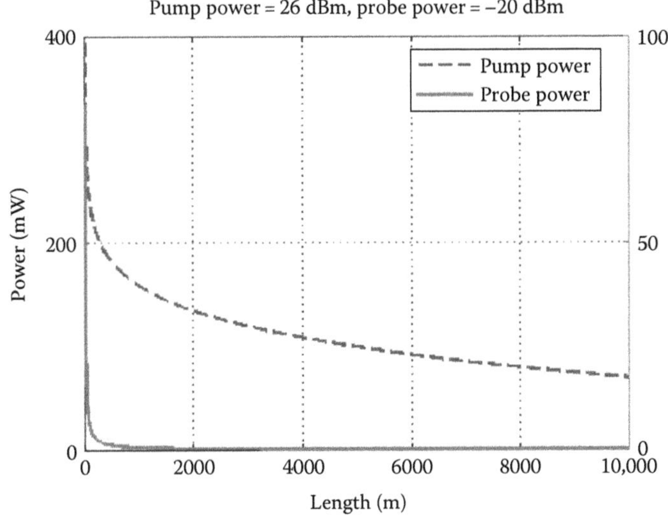

FIGURE 12.30 Evolution of the pump and probe powers along the length of the fiber for a pulse width of 500 ns.

Standard telecommunication grade fiber is expected to have a strain coefficient of 0.05 MHz/με and a temperature coefficient of ~1 MHz/K. Figure 12.31 indicates higher powers at specific locations in the fiber, corresponding to specific frequencies of the probe that correspond to the temperature–strain at that location. An algorithm is further used to extract the location and the frequency information from the 3D trace given in Figure 12.31 to estimate the temperature–strain distribution along the length of the fiber.

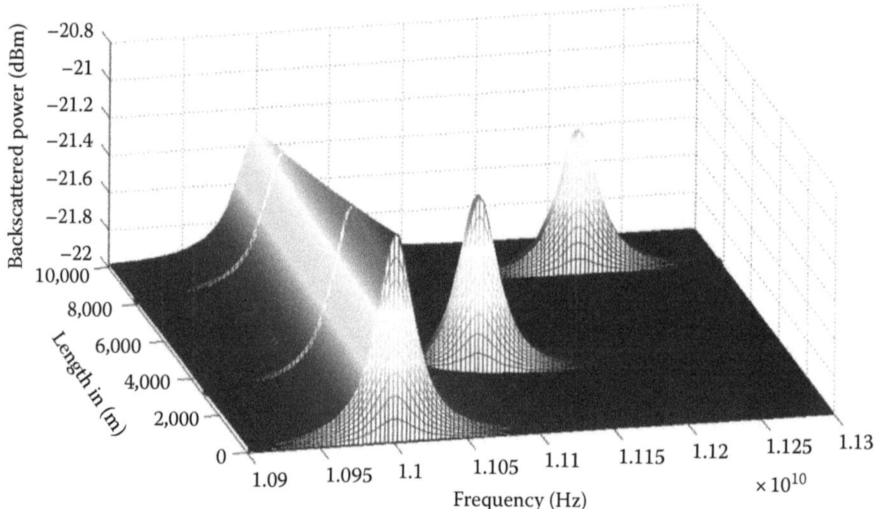

FIGURE 12.31 The simulated output of the BOTDA at different frequencies of the probe for pump pulses of width 50 ns.

12.4.4 Practical Implementation and Characterization

All applications associated with Brillouin scattering rely on the nature and response of BGS under different excitation conditions. It is therefore extremely important to understand the finer features of the BGS under such conditions. The inhomogeneous broadening of BGS as a function of the core-cladding doping concentrations and the state of polarization of the light have been studied earlier [79,80]. The evolution of BGS from a spontaneous to stimulated regime, with an associated linewidth narrowing, has been reported [81,82]. The presence of a weak external probe is also reported to result in the narrowing of the Stokes spectrum in a fiber Brillouin laser cavity [13]. The BGS is usually measured by observing the gain experienced by the counterpropagating probe frequency as it is detuned from the pump frequency [14].

In this section, we provide experimental evidence to distinguish the gain spectrum due to SBS in the absence of the probe (self-stimulated gain spectrum) from that experienced by the probe (externally stimulated gain spectrum) in a Brillouin fiber amplifier. We specifically study the Brillouin gain depletion as the probe frequency is in resonance with the gain spectrum.

12.4.4.1 Experimental Setup to Extract the Brillouin Gain Spectrum

The schematic diagram of the experimental setup for the Brillouin fiber amplifier is illustrated in Figure 12.32. It consists of a single narrow-linewidth (250 kHz) tunable laser (Newport Corp., Vortex), whose output at 1546 nm wavelength is amplified using a +22 dBm output power erbium-doped fiber amplifier (EDFA). The amplified light is split using a 70/30 coupler to generate both the pump and the probe. The output of the coupler carrying higher power is pulse modulated to form the pump, which is propagated through a standard (G652) single-mode fiber of length 25 km in the forward direction. Light from the 30% port is amplitude modulated in the carrier-suppressed mode using a high-speed lithium niobate-based electrooptic modulator (EOM, Photline Technologies), and the lower sideband of the modulated output is used as the probe. The frequency of the probe wave is tuned over the desired range by appropriately tuning the RF generator (Hittite). The reflected and transmitted powers are extracted using the circulators C1 and C2, respectively, combined using a 50/50 fused coupler, and detected using a high-speed photodetector (Model 1414, New Focus). The spectrum of the corresponding RF signal is observed using an electronic spectrum analyzer (ESA) (FSV 30, Rohde & Schwarz).

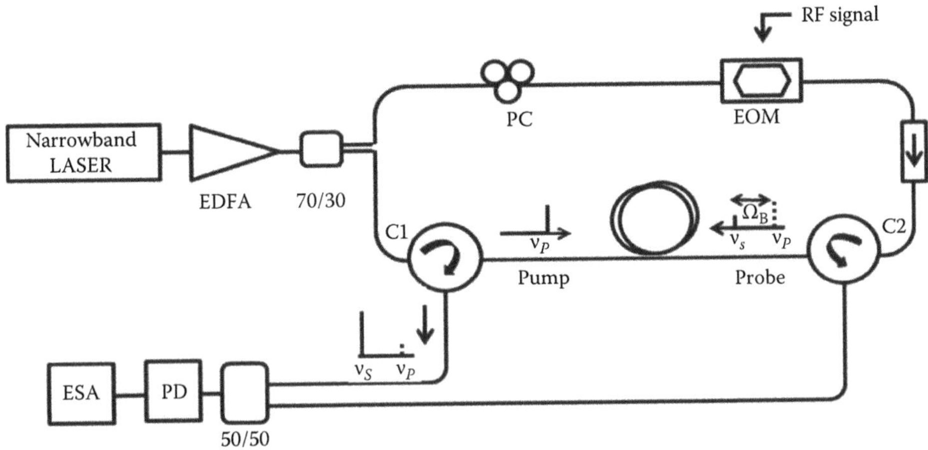

FIGURE 12.32 Schematic of the experimental setup used to extract the BGS through heterodyning.

Tuning of the modulation frequency and the ESA spectrum acquisition are automated using a Python code for easy and repeatable data collection. It is ensured that the ASE levels from the EDFA are at least 30 dB below the signal levels. The experiments are performed for two conditions. In the absence of the probe, the forward-propagating pump initiates SBS for power levels larger than the threshold. The beat spectrum for this condition corresponds to the *self-stimulated gain spectrum*. In the other configuration, the appropriately frequency-shifted probe undergoes amplification and the beat spectrum for this condition corresponds to the BGS of the fiber.

Pump propagating through the fiber typically triggers the onset of a spontaneous Brillouin process. The spectral width of the corresponding gain spectrum is found to be few tens of MHz [81]. With the increase in pump power beyond the threshold of SBS, the corresponding gain spectrum is found to be narrow, and we refer to this as the self-stimulated gain spectrum [81]. Note that the center frequency corresponding to the spontaneous and the self-stimulated gain spectra is identical. The gain spectra typically reported elsewhere correspond to that of the gain experienced by a counterpropagating probe, when the system is operated in the undepleted pump approximation. We refer to this gain spectrum as the stimulated gain spectrum.

We initially study the self-stimulated BGS (in the absence of probe) by launching CW pump light into the single-mode fiber. When the pump power is increased beyond the threshold level for SBS (~7 dBm for the 25 km long test fiber), we observe a significant increase in the backscattered power corresponding to the self-stimulated process. The RF signal obtained as a result of the beating of the transmitted pump as well as the Brillouin backscattered light represents the gain spectrum for the self-stimulated process as shown in Figure 12.33.

The Brillouin frequency shift for the given fiber is found to be centered around 10.884 GHz from the pump frequency, and the FWHM of this gain spectrum is observed to be around 6 MHz. Note that the previous work [81] reports a bandwidth of 11–12 MHz for similar fibers. The relatively lower values observed by us are attributed to the use of a narrower-linewidth (250 kHz) laser source. The FWHM is found to decrease by ~0.6 MHz when the pump power is increased from 10 to 17 dBm as seen from the inset of Figure 12.33. This is in agreement with the spectral narrowing with the increase in pump power reported elsewhere [81].

In order to compare the gain spectrum with that obtained using a counterpropagating probe, the modulation frequency applied to the high-speed modulator is tuned from 10.84 to 10.93 GHz. Consequently, the probe is generated at these frequency shifts with respect to the pump frequency. The tuning resolution is maintained at 500 kHz to capture the finer spectral features. Figure 12.34 shows the spectrum observed on the ESA as the probe frequency is tuned over the range mentioned earlier.

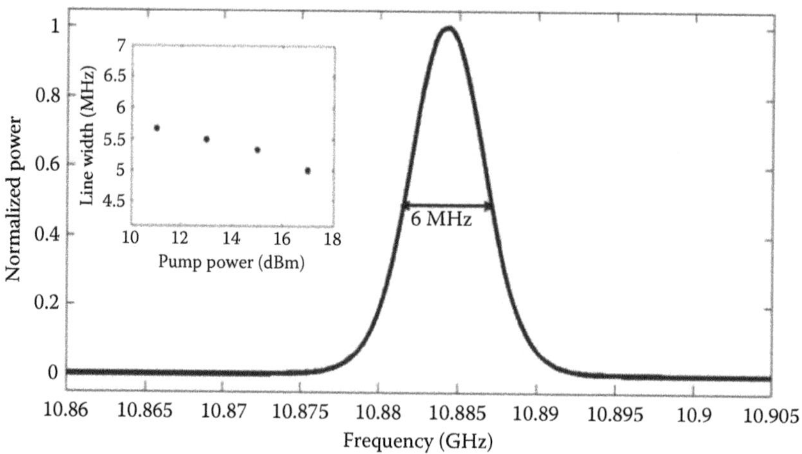

FIGURE 12.33 BGS for a pump power of 10 dBm measured using a fiber of length 25 km in the absence of the counterpropagating probe.

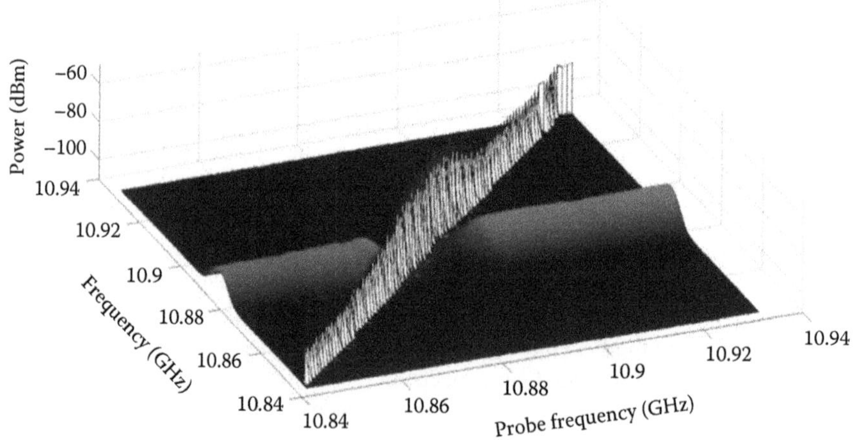

FIGURE 12.34 RF spectrum of the beat between the transmitted pump power and the amplified probe for different probe frequencies. The figure shows the depletion of the self-stimulated BGS as the probe frequency is tuned across its bandwidth.

When the probe frequency is detuned several tens of MHz away from the center of the BGS, the probe does not experience any amplification, and the signature of the self-stimulated gain spectrum is observed with a center frequency of 10.884 GHz. The spectral shape and the FWHM are identical to that observed in Figure 12.32. As the probe is tuned closer to the center frequency, it is found that the probe is amplified at the expense of the self-stimulated BGS. This is corroborated by the observation of depletion in the self-stimulated gain spectrum. By tracking the peak amplitude of the probe frequency in Figure 12.33, we can obtain the BGS as shown in Figure 12.35. It is observed that the FWHM gain bandwidth is around 17 MHz, which is consistent with that reported in the literature [83].

Since the gain spectrum of the self-stimulated Brillouin process (corresponding to the BOTDR configuration) is narrower, the BOTDR should provide finer spectral resolution in sensing applications. However, the BOTDA configuration is typically preferred over the conventional Brillouin

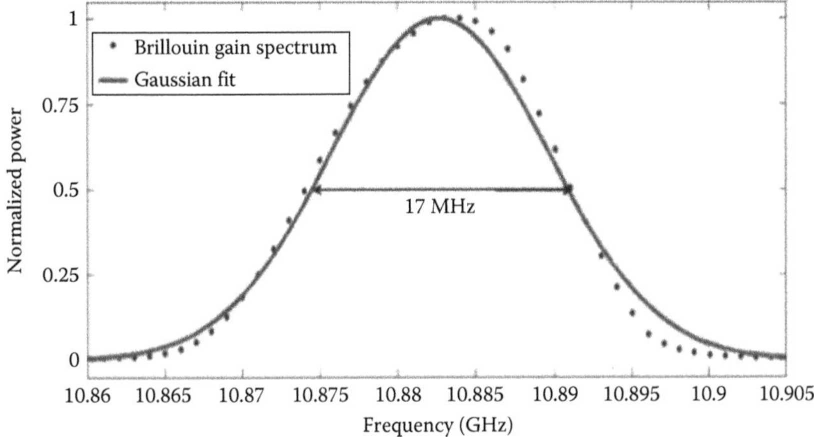

FIGURE 12.35 Experimentally measured gain spectrum and the corresponding Gaussian fit for the counter-propagating probe in the Brillouin fiber amplifier.

OTDR due to the higher SNR of the former. We now present some experimental results of temperature measurements in the BOTDA configuration.

12.4.4.2 Results from Temperature Measurement

The schematic of the BOTDA scheme used in our experiments is shown in Figure 12.36.

Light from the narrow-linewidth external cavity laser is amplified and split using a 30:70 coupler. Light from one of the arms of the coupler is used to generate the probe using the high-speed modulator operating in carrier-suppressed configuration, as discussed in the previous section. Light from the other arm is modulated using a pulsed source and is further amplified using another high-power EDFA. Since the pump is pulsed, the average power at the input of the high-power EDFA is small, and hence, there is a large amplified spontaneous emission built up in the EDFA. An external ASE filter is used to cut off the out-of-band ASE.

The output spectrum is shown in Figure 12.37, indicating a Brillouin gain of >20 dB. The counterpropagating probe power in the absence of pump power is shown in blue, where the carrier is suppressed. In the presence of a strong pump power, the longer wavelengths corresponding to the Stokes frequency are found to amplify. The gain in the probe is also found to increase with the

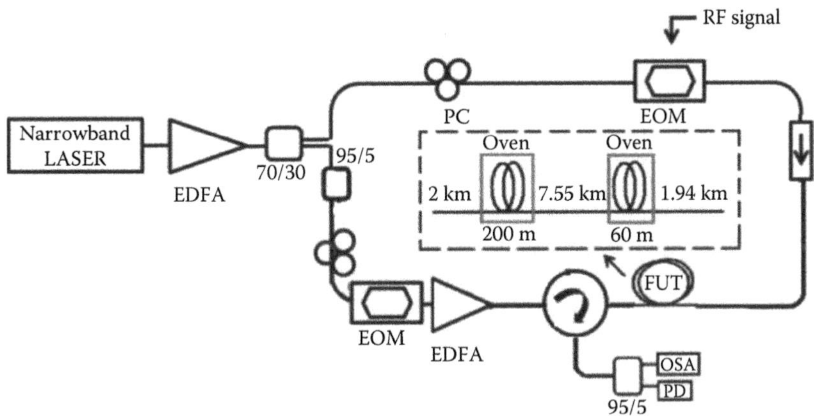

FIGURE 12.36 Experimental setup in the BOTDA configuration; the details of the test bed of the FUT are shown.

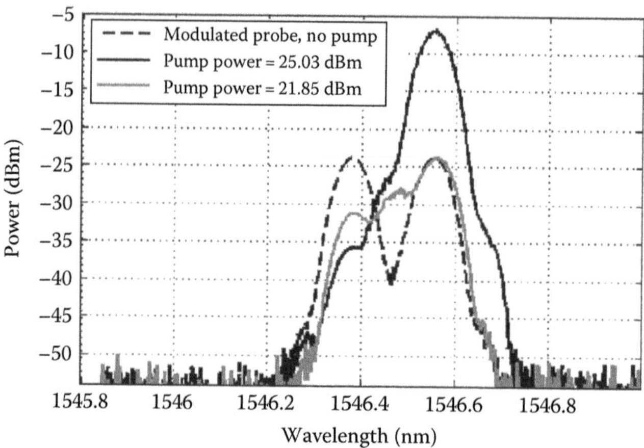

FIGURE 12.37 Output spectrum of the BOTDA, indicating the increase in the amplified probe power with the increase in pump power.

increase in pump power. The amplified probe power is captured on an oscilloscope, and the nature of the time domain traces is discussed next.

The time domain traces of the probe power as a function of the pump power as recorded on an oscilloscope are shown in Figure 12.38. The fiber used for this experiment is 12 km long, with no temperature or strain events. For smaller pump powers, the amplification of the probe along the length of the fiber is insignificant. When the pump power increases, the probe is found to amplify along the length of the fiber. The pulse width of the pump used in these experiments is 500 ns, and hence, the pump is found to significantly deplete within a length of about 4 km. This results in a smaller gain in the probe at longer lengths, as discussed in the simulation. The amplification in the probe at specific lengths of the fiber is now calculated and plotted as function of pump power, as shown in Figure 12.39. The gain estimated from the optical spectrum is also plotted in the same figure. It is evident that the gain at longer lengths of the fiber is smaller due to pump depletion.

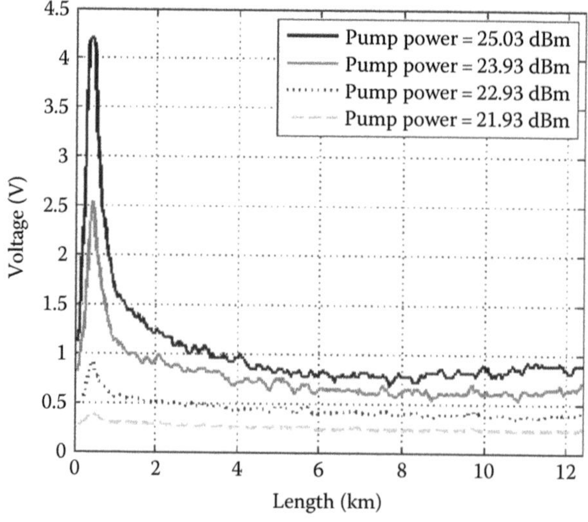

FIGURE 12.38 Time domain traces of the probe when pulsed pump of different peak powers is launched into the 12 km test fiber.

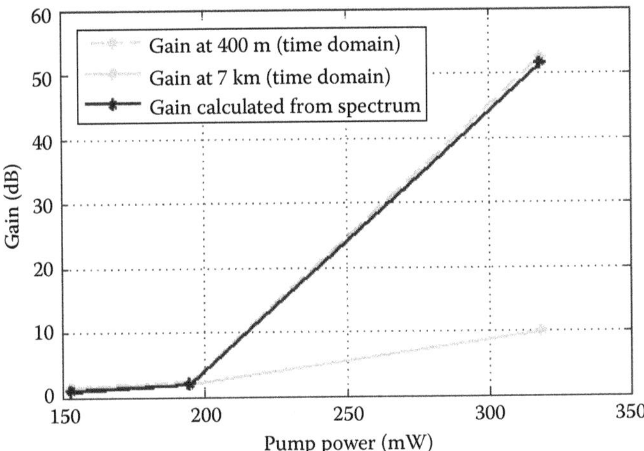

FIGURE 12.39 A comparison between the gain estimated from the experimental results corresponding to the time domain and the spectral domain outputs.

The sensitivity of measurement is improved by beating the BOTDA signal with the pump, and the beat signal is observed on the RF spectrum analyzer, operated in the zero-span mode. The method helps to isolate the beat frequency corresponding to the probe and also helps in automating the process of data capture and analysis. The process of frequency tuning the RF frequency used to drive the modulator in the probe arm and the data acquisition is completely automated. The time domain variations corresponding to the beat frequency are captured on the ESA and the output is shown in Figure 12.40. The entire fiber is kept in room temperature for this experiment. The different frequency peaks seen in different sections of the fiber as seen in Figure 12.40 indicate that the inherent strain on each length of fiber is different. The sections of the fiber are now kept in the oven and the oven is heated to a high temperature. The temperature test bed used for these measurements is indicated in Figure 12.36.

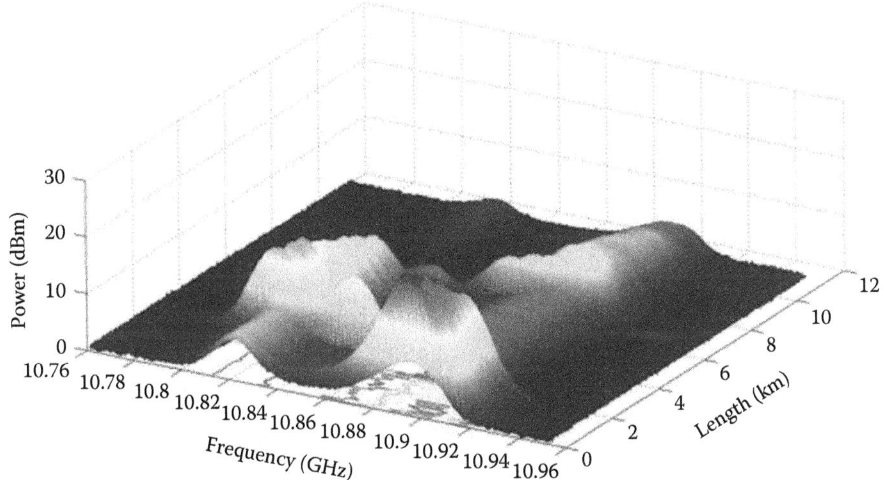

FIGURE 12.40 BOTDA signal as a function of frequency; when the entire length of fiber is kept at room temperature, the difference in frequency peaks in different sections of the fiber indicates that the fiber spools are wound with different strains.

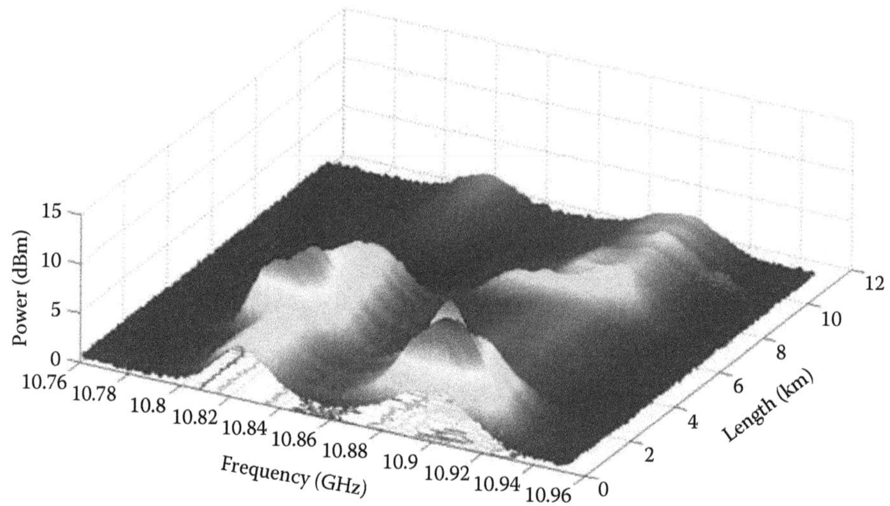

FIGURE 12.41 BOTDA signal as a function of frequency when the section of fiber is kept in the oven.

The probe signal is acquired in the time domain as a function of the RF frequency of the modulator in the probe arm, and the results are shown in Figure 12.41. The difference between the reference at room temperature and the one with the event is shown in Figure 12.42, clearly indicating the temperature events.

The raw data obtained after removing the offset are filtered with optimized window sizes in the time domain and in the frequency domain before finding the difference shown in Figure 12.42. The details of filtering are as discussed in Section 12.2. Peak search algorithms are further used to extract the position of the frequency peaks due to the temperature event. The extracted events are shown in Figures 12.43 and 12.44. The measured temperatures using these frequency shifts are found to match well with the set temperatures. Further optimization of these results are underway. Polarization scrambling of the probe would further improve the correctness of these measurements.

Thus, a distributed measurement of temperature is demonstrated successfully using the Brillouin optical time domain analyzer configuration. The resolution of measurement can be further improved with the use of pulses with shorter widths.

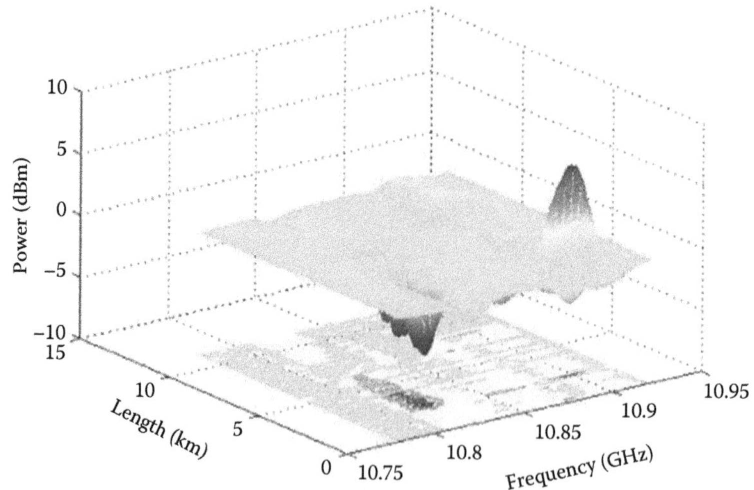

FIGURE 12.42 The temperature event at a distance of 2 km, after subtraction.

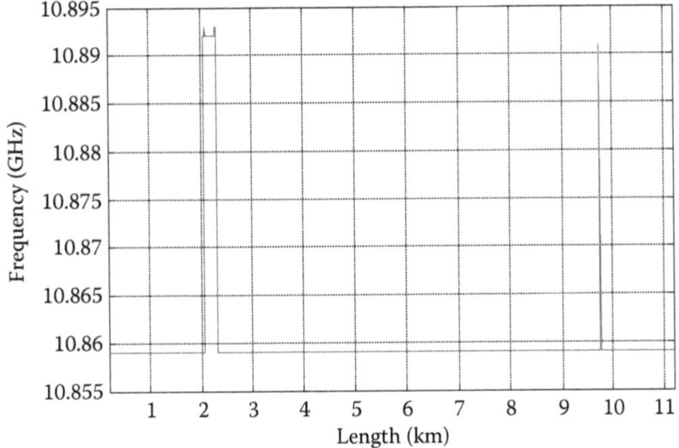

FIGURE 12.43 Temperature events extracted from the BOTDA.

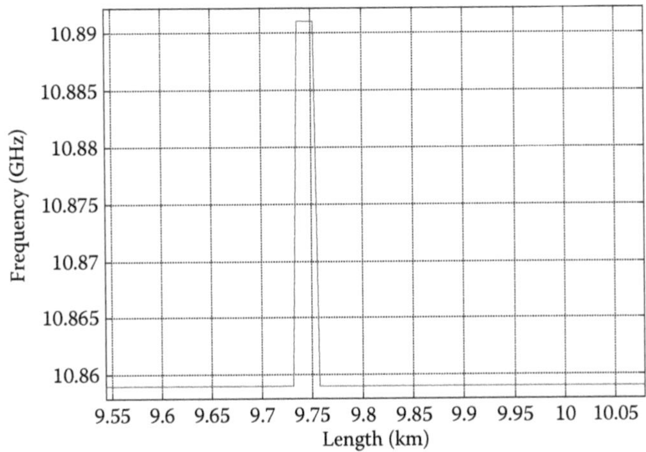

FIGURE 12.44 Temperature events extracted from the BOTDA, with the scale zoomed around the event at 9.75 km.

12.4.4.3 Results from Strain Measurement

An estimate of the applied strain is required in order to assess the strain obtained from BOTDA. In this section, we provide an in situ measurement of strain using fiber Bragg gratings (FBGs). FBGs are optical filters that reflect a narrow spectrum of wavelengths incident on it and transmit the remaining wavelengths. They are fabricated by creating a periodic variation of refractive index within the core of an optical fiber. The period of the index variations determines the reflected wavelength called Bragg wavelength of the FBG:

$$\lambda_B = 2\Lambda n_{eff} \tag{12.36}$$

where

- λ_B is the Bragg wavelength
- Λ is the grating period
- n_{eff} is the effective refractive index of the fiber

Due to strain or change in temperature, the refractive index and grating period changes thereby changing the Bragg wavelength. So, FBG can be used for sensing strain and temperature. The lengths of the FBGs are typically of the order of a few cm, and hence, they are very effective as point sensors. The strain sensitivity of an FBG reported elsewhere is 1.2 pm shifts in Bragg wavelength for 1 με strain applied. The dependence of Bragg wavelength of FBG on strain is utilized for this measurement.

The distributed strain measurement using BOTDA is typically demonstrated in the laboratory with a test bed where specific strain is applied along different lengths of the fiber. The strain applied is either by using a pulley system or by winding the fiber between two posts, one of which is movable. When a test bed for a distributed strain measurement is set up in the laboratory, we require a calibration for the applied distributed strain. The experimental setup used is identical to that discussed in Section 12.4.5. Strain is applied on the 100 m fiber, which is spliced between two spools of length 1 and 10 km shown in Figure 12.45. Three recoated FBGs whose Bragg wavelengths are 1566, 1574, and 1582 nm approximately are spliced after 25, 50, and 75 m within the 100 m fiber.

The pump pulse width is 500 ns and the pump power is 18 dBm. The frequency of the probe is swept from 10.850 to 10.970 GHz in steps of 1 MHz, and time domain traces are captured for unstrained and strained conditions. The transmission spectra of FBGs for unstrained and strained conditions are shown in Figure 12.46.

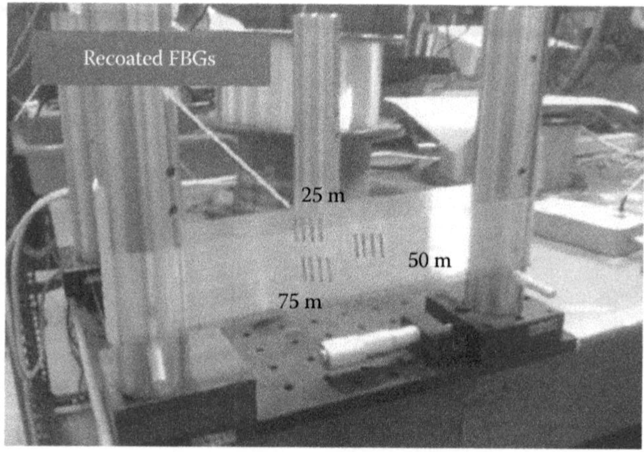

FIGURE 12.45 100 m strain test bed with FBGs incorporated in it.

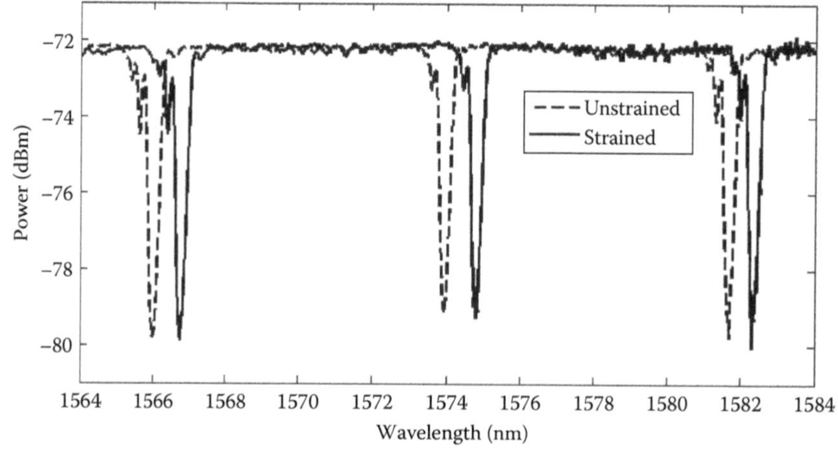

FIGURE 12.46 Transmission spectra of FBGs with and without strain.

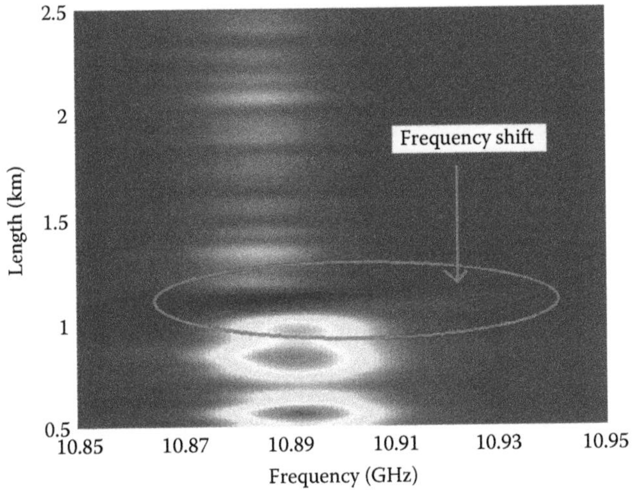

FIGURE 12.47 Top view of 3D plot of strained condition.

The shift in the Bragg wavelength of three FBGs due to the application of strain is almost uniform and is about 800 pm. Using the standard 1.2 pm shift for 1 με strain, the strain applied in this case is approximately 666 με. Figure 12.47 shows the top view of the 3D plot of backscattered signal for strained condition. Figure 12.47 also indicates that the center frequencies of the 1 km and the 10 km spools in the FUT are not identical. This difference could be attributed due to the fiber structures being dissimilar or are wound with dissimilar strains. The measurement contrast can be further improved by scrambling the polarization of the probe. Figure 12.48 shows the spectra for unstrained and strained cases at 1.08 km.

Clearly one can see the shift in Brillouin frequency after 1 km due to the applied strain. The fluctuations in power levels are minimized using an averaging of over 2000 traces. Savitzky–Golay filter is applied for smoothing the data. Correlation with a Gaussian signal with a width of BGS is done to eliminate insignificant fluctuations. In order to observe the strain along the fiber, the Brillouin frequency without strain and with strain at every length is required. Peak search algorithm is implemented for the frequency map along the length of the fiber, corresponding to both

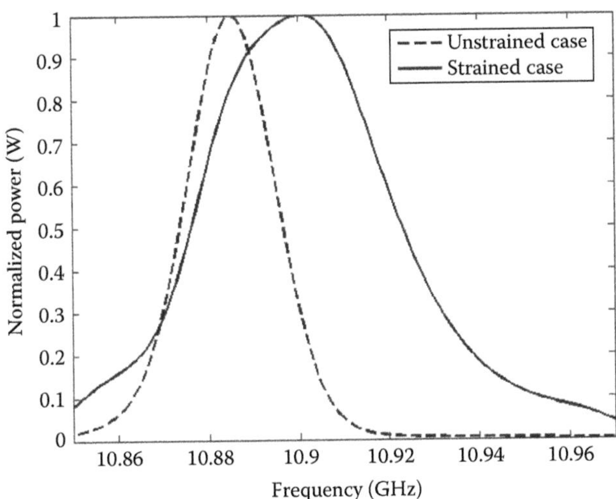

FIGURE 12.48 Spectra at 1.08 km with and without strain.

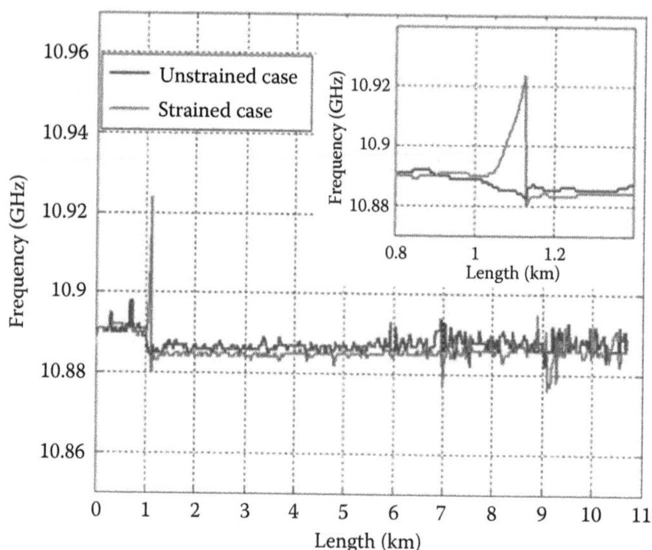

FIGURE 12.49 Frequency at which power is maximum for the strained and unstrained cases.

the unstrained and strained conditions. This information is further used to obtain the frequency at which backscattered power is maximum at every length. The results are shown in Figure 12.49. The inset shown is the zoomed version from 0.8 to 1.4 km. The difference in the frequencies for unstrained and strained cases gives the strain at that particular position.

Significant shift in the Brillouin frequency is noticed at 1 km where strain is applied. The Brillouin frequency shift is not uniform within 100 m. So the strain experienced by the fiber is not uniform. This may be because of uneven strain on winding 100 m fiber on strain test bed, which is done manually. To confirm this, we increased the strain and observed the Brillouin frequency shift, which is found to be of the same pattern as shown earlier. The center frequencies of 1 and 10 km spools are found to be different. The 100 m fiber is taken from the spool of 10 km fiber. So the center frequency of 100 m fiber in unstrained position is 10.885 GHz. So the shift in frequency due to strain is 10.925 − 10.885 GHz = 40 MHz approximately. Using the standard 50 kHz shift in Brillouin frequency for 1 με strain, the strain obtained from the BOTDA technique is 800 με, which is closer to the strain obtained from FBGs. In order to confirm the repeatability of the results, we captured the data thrice at the same strain conditions. The shift in the frequency is almost the same in all the cases. As indicated earlier, the measurement resolution can be improved with the use of pulses with shorter widths, and the SNR can be further improved by employing polarization scrambling the probe frequency. Efforts in these directions are underway.

12.4.5 Applications

Several of the applications requiring DTS (mentioned for Raman scattering in Section 12.3.4) may also be met using Brillouin scattering–based sensors. In fact, due to the ability to support distances in the order of several tens of kilometers, Brillouin sensors are the only choice for ultralong-distance distributed sensing applications such as pipeline monitoring.

In addition to the previous, Brillouin scattering–based sensors are also quite attractive for distributed strain sensing. A specific modality of distributed strain sensing including FBGs for reference in the BOTDA configuration performed by us in our lab is discussed in the previous section. Of the different configurations that we discussed in the previous sections, a particular choice is decided by the application for which the sensing is used for. For instance, in the case of structural health monitoring, a spatial resolution of approximately few centimeters or less is preferred, which necessitates the use of correlation coding. In any case, care must be taken for the suitable placement of the sensing fiber to detect the appropriate parameter. Since the glass optical fiber typically has polymer protection, the entire applied strain does not get transferred to the fiber—the magnitude of this transfer is decided by the mechanical properties of the polymer. The strain transfer coefficient between the structure under analysis and the optical fiber can be estimated exactly once the shear modulus of the coating and the modulus of elasticity of the core along with the radii of the core and the coating are known [84].

Certain civil engineering structures prefer installation of bare fiber, specifically suited for small structures such as concrete cylinders or fiber-reinforced concrete patches. Surface-adhered sensors with suitable adhesives and protective coatings are used in cases where the fiber or the structure is not subjected to harsh environmental or loading conditions. One such application is sensing the curvature in reinforced concrete structure in beams. Reliability of sensing is greatly dependent on the placement of the sensors, mechanical protection, long-term stability of the adhesives/splices used, and protection from the ingress of humidity, seepage, and robustness toward other environmental changes.

12.5 SUMMARY

After being a laboratory curiosity for a couple of decades, optical fiber sensors are gaining widespread commercial acceptance and are rapidly becoming the technology of choice for physical sensing. Distributed sensing of physical parameters such as strain and temperature using inelastic scattering process in optical fibers is a unique technology. For DTS, the Brillouin scattering–based technology is emerging as a serious competitor for DART. On the other hand, Brillouin distributed sensors have no competitors for distributed strain sensing.

While the ability to sense strain and temperature over a spatial range spanning six orders of magnitude is quite attractive, the on-field implementation of the technology is a key issue. Especially, the design of rugged optical fiber cables with adequate protection from the harsh environments and without compromising the sensitivity requires ingenuous engineering. There is also significant scope for cheaper, energy-efficient instrumentation leading to wider acceptance as a technology of choice to enable a smarter, greener planet.

ACKNOWLEDGMENTS

The authors acknowledge contributions from several collaborators and students including Anil Prabhakar, Nitin Chandrachoodan, Venkatesh, Bharath Kumar, Amitabha Datta, Uvaraj, Vinayak, Ramalakshmi, Achuth, and Shahna. The authors also thank Bhargav who proofreading the text and offering constructive comments. Finally, most of the experimental work would not have been possible without the funding of the Department of Electronics and Information Technology, Government of India.

REFERENCES

1. B. Lee, Review of the present status of optical fiber sensors, *Opt. Fiber Technol.*, 9(2), 57–79, April 2003.
2. X. Bao and L. Chen, Recent progress in distributed fiber optic sensors, *Sensors*, 12, 8601–8639, 2012.
3. C. A. Galindez-Jamioy and J. M. López-Higuera, Brillouin distributed fiber sensors: An overview and applications, *J. Sens.*, 2012, 1–17, 2012.
4. M. Niklès and F. Ravet, Distributed fibre sensors: Depth and sensitivity, *Nat. Photon.*, 4(7), 431–432, July 2010.
5. A. Ukil, S. Member, H. Braendle, and P. Krippner, Distributed temperature sensing: Review of technology and applications, *IEEE Sens. J.*, 12(5), 885–892, 2012.
6. S. M. Aminossadati, N. M. Mohammed, and J. Shemshad, Distributed temperature measurements using optical fibre technology in an underground mine environment, *Tunn. Undergr. Sp. Technol.*, 25(3), 220–229, May 2010.
7. G. Bolognini and A. Hartog, Raman-based fibre sensors: Trends and applications, *Opt. Fiber Technol.*, 19(6), 678–688, December 2013.

8. S. Randel, R. Ryf, A. Gnauck, M. A. Mestre, C. Schmidt, R. Essiambre, P. Winzer et al., Mode-multiplexed 6 × 20-GBd QPSK transmission over 1200-km DGD-compensated few-mode fiber, in *National Fiber Optic Engineers Conference*, Los Angeles, CA, 2012, no. I, paper PDP5C.5.
9. H. Ishii, K. Kawamura, T. Ono, H. Megumi, and A. Kikkawa, A fire detection system using optical fibres for utility tunnels, *Fire Saf. J.*, 29(2–3), 87–98, 1997.
10. S. J. Kimminau, M. Rupawalla, K. Rashid, and D. M. Hargreaves, Well characterisation method, US 7778780 B2, August 2010.
11. T. Horiguchi and M. Tateda, Optical-fiber-attenuation investigation using stimulated Brillouin scattering between a pulse and a continuous wave, *Opt. Lett.*, 14, 408–410, 1989.
12. T. Horiguchi, T. Kurashima, and M. Tateda, Tensile strain dependence of Brillouin frequency shift in silica optical fibers, *IEEE Photon. Technol. Lett.*, 1, 5, 1989.
13. T. Horiguchi and M. Tateda, BOTDA-nondestructive measurement of single-mode optical fiber attenuation characteristics using Brillouin interaction: Theory, *J. Lightwave Technol.*, 7(8), 1170–1176, 1989.
14. M. Enckell, B. Glisic, F. Myrvoll, and B. Bergstrand, Evaluation of a large-scale bridge strain, temperature and crack monitoring with distributed fibre optic sensors, *J. Civ. Struct. Heal. Monit.*, 1(1–2), 37–46, March 2011.
15. J.-M. Henault, G. Moreau, S. Blairon, J. Salin, J.-R. Courivaud, F. Taillade, E. Merliot et al., Truly distributed optical fiber sensors for structural health monitoring: From the telecommunication optical fiber drawling tower to water leakage detection in dikes and concrete structure strain monitoring, *Adv. Civ. Eng.*, 2010, 1–13, 2010.
16. X. Chapeleau, T. Sedran, L.-M. Cottineau, J. Cailliau, F. Taillade, I. Gueguen, and J.-M. Henault, Study of ballastless track structure monitoring by distributed optical fiber sensors on a real-scale mockup in laboratory, *Eng. Struct.*, 56, 1751–1757, November 2013.
17. M. Nikles, L. Thevenaz, and P. A. Robert, Simple distributed fiber sensor based on Brillouin gain spectrum analysis, *Opt. Lett.*, 21(10), 758, May 1996.
18. P. Healey, Instrumentation principles for optical time domain reflectometry, *J. Phys. E*, 19(5), 334, 1986.
19. M. N. Alahbabi, Y. T. Cho, and T. P. Newson, Simultaneous temperature and strain measurement with combined spontaneous Raman and Brillouin scattering, *Opt. Lett.*, 30, 1276–1278, 2005.
20. D. Inaudi and B. Glisic, Development of distributed strain and temperature sensing cables, *Proc. SPIE*, 5855, 222–225, 2005.
21. M. I. Skolnik, *Radar Handbook*, Chapter 7, McGraw-Hill Professional, 1990, 1328 pp.
22. P. Healey, Review of long wavelength single-mode optical fiber reflectometry techniques, *J. Lightwave Technol.*, 3, 876–886, 1985.
23. C. Ciofi, F. Crupi, C. Pace, and G. Scandurra, How to enlarge the bandwidth without increasing the noise in OP-AMP-based transimpedance amplifier, *IEEE Trans. Instrum. Measure.*, 55, 814–819, 2006.
24. 1.6 GHz, low-noise, FET-input operational amplifier. http://www.ti.com/lit/ds/symlink/opa657.pdf. Datasheet, Texas Instruments, 2001.
25. M. Nazarathy, S. A. Newton, R. P. Giffard, D. S. Moberly, F. Sischka, W. R. J. Trutna, and S. Foster, Real-time long range complementary correlation optical time domain reflectometer, *J. Lightwave Technol.*, 7, 24–38, 1989.
26. A. J. Rogers, Polarization-optical time domain reflectometry: A technique for the measurement of field distributions, *Appl. Opt.*, 20(6), 1060–1074, March 1981.
27. J. P. King, D. Smith, K. Richards, P. Timson, R. E. Epworth, and S. Wright, Development of a coherent OTDR instrument, *J. Lightwave Technol.*, 5(4), 616–624, 1987.
28. A. Hartog, A distributed temperature sensor based on liquid-core optical fibers, *Lightwave Technol. J.*, 1, 498–509, 1983.
29. T. Horiguchi, K. Shimizu, T. Kurashima, M. Tateda, and Y. Koyamada, Development of a distributed sensing technique using Brillouin scattering, J. Lightwave Technol., 13, 1296–1302, 1995.
30. R. H. Stolen, Raman gain in glass optical waveguides, *Appl. Phys. Lett.*, 22(6), 276, 1973.
31. R. H. Stolen, E. P. Ippen, and A. R. Tynes, Raman oscillation in glass optical waveguide, *Appl. Phys. Lett.*, 20, 62, 1972.
32. A. Signorini, S. Faralli, M. A. Soto, G. Sacchi, F. Baronti, R. Barsacchi, A. Lazzeri, R. Roncella, G. Bolognini, and F. Di Pasquale, 40 km long-range Raman-based distributed temperature sensor with meter-scale spatial resolution, *Optical Fiber Communication (OFC)*, collocated *National Fiber Optic Engineering Conference 2010*, San Diego, CA, 2010.
33. M. Höbel, J. Ricka, M. Wüthrich, and T. Binkert, High-resolution distributed temperature sensing with the multiphoton-timing technique, *Appl. Opt.*, 34(16), 2955–2967, June 1995.
34. P. Eraerds, M. Legre, J. Z. J. Zhang, H. Zbinden, and N. Gisin, Photon counting OTDR: Advantages and limitations, *J. Lightwave Technol.*, 28, 952–964, 2010.

35. M. A. Soto, A. Signorini, T. Nannipieri, S. Faralli, and G. Bolognini, High-performance Raman-based distributed fiber-optic sensing under a loop scheme, *IEEE Photon. Technol. Lett.*, 23(9), 534–536, 2011.
36. M. A. Soto, A. Signorini, T. Nannipieri, S. Faralli, G. Bolognini, and F. Di Pasquale, Impact of loss variations on double-ended distributed temperature sensors based on Raman anti-Stokes signal only, *J. Lightwave Technol.*, 30, 1215–1222, 2012.
37. M. A. Farahani and T. Gogolla, Spontaneous Raman scattering in optical fibers with modulated probe light for distributed temperature Raman remote sensing, *J. Lightwave Technol.*, 17(8), 1379–1391, 1999.
38. A. Datta, B. K. Lagishetty, and B. Srinivasan, Performance evaluation of temperature sensing system based on distributed anti-Stokes Raman thermometry, *Adv. Photon. Renew. Energy*, JThA4, 2010. DOI:http://dx.doi.org/10.1364/SENSORS.2010.JThA4.
39. A. Datta, U. Gajendran, V. Srimal, D. Venkitesh, and B. Srinivasan, Precise, rugged spectrum-based calibration of distributed anti-Stokes Raman thermometry systems, *Proceedings of the SPIE 8311, Optical Sensors Biophotonics III*, 83110E, 8311, 83110E1–8, November 2011. DOI:10.1117/12.904327.
40. G. Brown, Downhole temperatures from optical fiber, *Oilf. Rev.*, 20(4), 34–39, 2009.
41. K. Eriksson, Fibre optic sensing—Case of 'Solutions Looking for Problems,' in *Society of Petroleum Engineers*, Aberdeen, UK, 2001.
42. C. Wang, G. Drenzek, I. Majid, K. Wei, D. Bolte, and A. Soufiane, High-performance hermetic optical fiber for downhole applications, in *Society of Petroleum Engineers*, Houston, TX, 2004.
43. D. J. Hill, M. M. Molenaar, and E. Fidan, Real-time downhole monitoring of hydraulic fracturing treatments using fibre optic distributed temperature and acoustic sensing, in *SPE/EAGE European Unconventional Resources Conference & Exhibition-From Potential to Production*, Vienna, Austria, 2012.
44. C. Smithpeter, R. Normann, J. Krumhansl, D. Benoit, S. Thompson, S. N. Laboratories, and A. Nm, Evaluation of a distributed fiber-optic temperature sensor for logging wellbore temperature at the Beowawe and Dixie Valley geothermal fields, in *Twenty-Fourth Workshop on Geothermal Reservoir Engineering*, Stanford, CA, 1999.
45. T. Reinsch, J. Henninges, and R. Ásmundsson, Thermal, mechanical and chemical influences on the performance of optical fibres for distributed temperature sensing in a hot geothermal well, *Environ. Earth Sci.*, 70(8), 3465–3480, March 2013.
46. G. Yilmaz and S. E. Karlik, A distributed optical fiber sensor for temperature detection in power cables, *Sens. Actuators A: Phys.*, 125, 148–155, 2006.
47. Sterlite Technologies, Optical fiber composite ground wire (OPGW) cables. Available [Online]: http://www.sterlitetechnologies.com/power_products/9.
48. S. W. Tyler, J. S. Selker, M. B. Hausner, C. E. Hatch, T. Torgersen, C. E. Thodal, and S. G. Schladow, Environmental temperature sensing using Raman spectra DTS fiber-optic methods, *Water Resour. Res.*, 45, W00D23, 2009.
49. J. S. Selker, L. Thévenaz, H. Huwald, A. Mallet, W. Luxemburg, N. van de Giesen, M. Stejskal, J. Zeman, M. Westhoff, and M. B. Parlange, Distributed fiber-optic temperature sensing for hydrologic systems, *Water Resour. Res.*, 42(12), 1–8, December 2006.
50. M. Nikles, F. Briffod, R. Burke, and G. Lyons, Greatly extended distance pipeline monitoring using fibre optics, in *Proceedings of OMAE05 24th International Conference on Offshore Mechanics and Arctic Engineering*, Halkidiki, Greece, 2005, pp. 1–8.
51. E. Tapanes, Fiber optic sensing solutions for real-time pipeline integrity monitoring, in *Australian Pipeline Industry Association National Convention*, Gold Coast, Australia, 2001, pp. 27–30.
52. F. Tanimola and D. Hill, Distributed fibre optic sensors for pipeline protection, *J. Nat. Gas Sci. Eng.*, 1(4–5), 134–143, November 2009.
53. R. P. S. Schilperoort and F. H. L. R. Clemens, Fibre-optic distributed temperature sensing in combined sewer systems, *Water Sci. Technol.*, 60(5), 1127–1134, January 2009.
54. G. Agrawal, *Nonlinear Fiber Optics*. Academic Press, San Diego, CA, 2001, 467 pp.
55. R. Y. Chiao, C. H. Townes and B. P. Stoicheff, Stimulated Brillouin scattering and coherent generation of intense hypersonic waves, *Phys. Rev. Lett.*, 12(21), 592–596, 1964.
56. E. P. Ippen and R. H. Stolen, Stimulated Brillouin scattering in optical fibers, *Appl. Phys. Lett.*, 21(11), 539, 1972.
57. A. Gaeta and R. Boyd, Stochastic dynamics of stimulated Brillouin scattering in an optical fiber, *Phys. Rev. A*, 44(5), 3205–3209, September 1991.
58. X. Bao, A. Brown, M. DeMerchant, and J. Smith, Characterization of the Brillouin-loss spectrum of single-mode fibers by use of very short (<10-ns) pulses, *Opt. Lett.*, 24(8), 510, April 1999.

59. R. Billington, Measurement methods for stimulated Raman and Brillouin scattering in optical fibres, NPL report COEM. Available [Online]: http://www.npl.co.uk/publications/measurement-methods-for-stimulated-raman-and-brillouin-scattering-in-optical-fibres. Accessed on March 22, 2014.
60. T. Kurashima, T. Horiguchi, H. Ohno, and H. Izumita, Strain and temperature characteristics of Brillouin spectra in optical fibers for distributed sensing techniques, *24th European Conference on Optical Communication (ECOC'98)*, Madrid, Spain, 1998, vol. 1 (IEEE Cat. No. 98TH8398).
61. T. Kurashima, T. Horiguchi, and M. Tateda, Distributed-temperature sensing using stimulated Brillouin scattering in optical silica fibers, *Opt. Lett.*, 15, 1038–1040, 1990.
62. D. Culverhouse, F. Farahi, C. N. Pannel and D. A. Jackson, Potential of stimulated Brillouin scattering as sensing mechanism for distributed temperature sensors, *Electron. Lett.*, 25(14), 913–915, 1989.
63. Y. Koyamada, S. Sato, S. Nakamura, H. Sotobayashi, and W. Chujo, Simulating and designing Brillouin gain spectrum in single-mode fibers, *J. Lightwave Technol.*, 22, 631–639, 2004.
64. M. Nikles, L. Thevenaz and P. A. Robert, Brillouin gain spectrum characterization in single-mode optical fibers, *IEEE J. Lightwave Technol.*, 15(10), 1842–1851, 1997.
65. H. H. Kee, G. P. Lees, and T. P. Newson, All-fiber system for simultaneous interrogation of distributed strain and temperature sensing by spontaneous Brillouin scattering, *Opt. Lett.*, 25, 695–697, 2000.
66. P. C. Wait and T. P. Newson, Landau Placzek ratio applied to distributed fibre sensing, *Opt. Commun.*, 122, 141–146, January 1996.
67. T. R. Parker, M. Farhadiroushan, R. Feced, V. A. Handerek, and A. J. Rogers, Simultaneous distributed measurement of strain and temperature from noise-initiated Brillouin scattering in optical fibers, *IEEE J. Quant. Electron.*, 34, 645–659, 1998.
68. C. C. Lee, P. W. Chiang, and S. Chi, Utilization of a dispersion-shifted fiber for simultaneous measurement of distributed strain and temperature through Brillouin frequency shift, *IEEE Photon. Technol. Lett.*, 13(10), 1094–1096, 2001.
69. H. Liang, J. Li, B. Han, Y. Chang, L. Cheng, and B.-O. Guan, Potential for simultaneous strain and temperature sensing based on Brillouin scattering in an all-solid photonic bandgap fiber, *Opt. Lett.*, 38, 465–467, 2013.
70. W. Z. W. Zou, Z. H. Z. He, and K. Hotate, Stimulated Brillouin scattering in F-doped optical fibers and its dependences on strain and temperature, *OFC/NFOEC 2008—2008 Conference on Optical Fiber Communication/National Fiber Optic Engineers Conference*, San Diego, CA, 2008.
71. X. Liu, X. Bao, S. Member, and I. Paper, Brillouin spectrum in LEAF and simultaneous temperature and strain measurement, *J. Lightwave Technol.*, 30(8), 1053–1059, 2012.
72. D.-P. Zhou, W. Li, L. Chen, and X. Bao, Distributed temperature and strain discrimination with stimulated Brillouin scattering and Rayleigh backscatter in an optical fiber, *Sensors*, 13, 1836–1845, 2013.
73. Y. Lu, Z. Qin, P. Lu, D. Zhou, L. Chen, and X. Bao, Distributed strain and temperature measurement by Brillouin beat spectrum, *IEEE Photon. Technol. Lett.*, 25(11), 1050–1053, June 2013.
74. K. Hotate and Z. H. Z. He, Synthesis of optical-coherence function and its applications in distributed and multiplexed optical sensing, *J. Lightwave Technol.*, 24, 2541–2557, 2006.
75. W. Z. W. Zou, Z. H. Z. He, and K. Hotate, Single-end-access correlation-domain distributed fiber-optic sensor based on stimulated Brillouin scattering, *J. Lightwave Technol.*, 28, 2736–2742, 2010.
76. M. Belal, Y. T. Cho, M. Ibsen, and T. P. Newson, A temperature-compensated high spatial resolution distributed strain sensor, *Meas. Sci. Technol.*, 21(1), 015204, January 2010.
77. A. Masoudi, M. Belal, and T. P. Newson, Distributed dynamic large strain optical fiber sensor based on the detection of spontaneous Brillouin scattering, *Opt. Lett.*, 38, 3312–3315, 2013.
78. R. Boyd, *Nonlinear Optics*, 3rd edn. Academic Press, Burlington, MA, 2008, 640 pp.
79. N. Shibata, R. G. Waarts, and R. P. Braun, Brillouin-gain spectra for single-mode fibers having pure-silica, *Opt. Lett.*, 12(4), 269–271, 1987.
80. S. Xie, M. Pang, X. Bao, and L. Chen, Polarization dependence of Brillouin linewidth and peak frequency due to fiber inhomogeneity in single mode fiber and its impact on distributed fiber Brillouin sensing, *Opt. Express*, 20(6), 6385–6399, March 2012.
81. A. Yeniay, J.-M. Delavaux, and J. Toulouse, Spontaneous and stimulated Brillouin scattering gain spectra in optical fibers, *J. Lightwave Technol.*, 20, 1425–1432, 2002.
82. R. W. Boyd, K. Rzazewski, and P. Narum, Noise initiation of stimulated Brillouin scattering, *Phys. Rev. A*, 42(9), 5514–5521, 1990.
83. N. Shibata, Y. Azuma, T. Horiguchi, and M. Tateda, Identification of longitudinal acoustic modes guided in the core region of a single-mode optical fiber by Brillouin gain spectra measurements, *Opt. Lett.*, 13(7), 595, July 1988.
84. F. Ansari, Practical implementation of optical fiber sensors in civil structural health monitoring, *J. Intell. Mater. Syst. Struct.*, 18(8), 879–889, March 2007.

13 Fiber Laser–Based Sensing Technologies

Asrul Izam Azmi, Muhammad Yusof Mohd Noor, Haifeng Qi, Kun Liu, and Gang-Ding Peng

CONTENTS

13.1 INTRODUCTION

Lasers have become a very important branch of photonics that play an essential role in modern science and technology since the first demonstrations in 1960. Evidently, optical fibers are now a crucial part for the success of lasers—as the high efficiency gain medium for fiber lasers and/or low loss transmission medium for laser power delivery. The invention of fiber laser can also be dated back to early 1960s when Snitzer demonstrated lasing in Nd^{3+}-doped glass and then in fibers [1,2]. Fiber lasers have attracted intensive industrial and research efforts for a great number of applications—ranging from high-power fiber laser–based industrial manufacturing and optical fiber communications to low-power fiber laser–based sensing, because of their superior properties in terms of power, efficiency, beam quality, compactness, delivery flexibility, thermal stability, and cost. The development of erbium-doped fibers, initially by Poole et al. in mid-1980s [3], leads to successful industrial application of erbium-doped fiber amplifier and lasers. Through continuing improvements in performances in terms of wavelength coverage, output power, slope efficiency, beam quality, pulse shape and duration, and noise properties in recent years, fiber lasers are now valid alternatives replacing their bulk solid-state and gas counterparts in many applications.

Fiber sensors have also made great progress following a similarly successful period. Fiber sensors have a number of attractive advantages over conventional sensors, such as immunity to electromagnetic interference, capability of highly multiplexing, small size, lightweight, and high sensitivity. Over the years, several core technologies for fiber sensing devices, systems, and networks including fiber Bragg grating (FBG) [4], fiber FP [5], Brillouin [6], spectrally coded multiplexing techniques [7], as well as interrogation, multiplexing, sensor packaging, and system integration techniques, have been developed. As a result, a range of fiber-optic sensors have been commercially deployed for industry applications, for example, for structure health monitoring of large-scale civil infrastructures such as bridges, dams, and tunnels and performance monitoring of key equipment and machinery.

In recent years, fiber laser–based sensor (FLS) has been developed aiming at high performance such as fast system response, high channel count, and better long-term performance with low cost. Fiber lasers are being widely used in fiber sensor systems where they are used as light sources or sensing elements. Here, we refer the latter cases as FLSs since fiber lasers are employed as both the optical light source and the sensor element in these systems. Fiber lasers have many features desirable for sensing applications, for example, increased effective interaction length and enhanced sensitivity due to cavity resonance, high power, small size, inherit fiber compatibility, single polarization, and single longitudinal mode operation. In addition, fiber lasers could have very narrow linewidth, low phase noise, and low relative intensity noise (RIN) that are essential for high-sensitivity interferometric type of fiber sensors. They could readily achieve high spectral resolution detection needed for gas and chemical sensing or monitoring. These attributes make fiber lasers appropriate for various sensor applications.

With the rapid development in application of fiber-optic sensors, more advanced and sophisticated requirements on performance and size are emerging. This leads to the significantly intensified research and development on a variety of FLSs including distributed feedback (DFB), distributed Bragg reflector (DBR), composite cavity, and ring FLSs.

In this chapter, we cover briefly both fundamental concepts and practical applications of FLS. By focusing on a few key techniques and application examples, we try to capture some of the most important and innovative aspects of the latest research and development on FLSs.

13.2 FUNDAMENTALS OF FIBER LASERS

Fiber lasers are the lasers developed with special doped optical fibers. They can be used as purely optical power sources in many applications or both power source and fiber sensing elements in FLS systems. On one hand, high-power fiber lasers can supply up to tens of kilowatts currently [8,9], making them useful for laser processing and machining. On the other hand, lower-power fiber lasers can be the sensing elements such those detecting the feeble sound signal that is even weaker than the deep sea noise in the submarine sensing application fields [10–16].

13.2.1 Fiber Laser Types

Fiber lasers can be classified into different types according to different criteria. In terms of cavity structures, fiber lasers can be classified into ring cavity fiber lasers, that is, ring fiber laser (RFL) and linear cavity fiber lasers, which include DBR fiber laser (DBRFL), DFB fiber laser (DFBFL), and composite cavity fiber laser (CCFL). In general, fiber lasers may emit different wavelengths with different doped rare-earth ions or other active materials and under different pump and cavity conditions. There are also many fiber lasers of special designs such as double-cladding fiber lasers, large-core fiber lasers, multimode fiber lasers, and photon crystal fiber lasers.

Here, we focus on the low-power, short-cavity fiber lasers, for example, DBRFL, CCFL, DFBFL, and RFL. These fiber lasers are of great importance in sensing for their compact size, robust single mode operation, narrow linewidth, and low noise.

A DBRFL is made upon a linear cavity consisting of a section of doped fiber sandwiched between two FBGs, as shown in Figure 13.1a. The doped fiber is often an erbium-doped fiber pumped by a semiconductor laser at either 980 or 1480 nm, through a wavelength-division multiplexer. DBRFLs can be used when a longer interaction length is preferred. Usually, a DBRFL can only operate in single longitudinal mode with cavity lengths up to 10 cm. The poor stability induced by mode hopping and spatial-hole burning would prove problematic for DBR multilongitudinal mode lasers for FLS. When a stable single frequency output is desired, fiber laser should operate in single longitudinal mode. Since the longitudinal mode spacing $\Delta\lambda = \lambda^2/(2L_c)$ with λ is the operating wavelength, the effective cavity length L_c must be short enough to ensure $\Delta\lambda$ greater than or nearly equal to the grating bandwidth.

A CCFL is a linear cavity consisting of a section of doped fiber with three or more FBGs as shown in Figure 13.1b. As will be discussed later in this chapter, the composite cavity structure could be used to enhance sensitivity in some applications.

A fiber ring laser, as shown in Figure 13.2a, can be easily constructed with a fiber ring including gain fiber and a fiber coupler.

A DFBFL, as shown in Figure 13.2b, is usually made upon a single continuous FBG with a π-phase shift inside the grating written in an active fiber. The cavity length of a DBFFL ranges from a few millimeters to centimeters. The phase shift in the grating opens an ultra-narrow

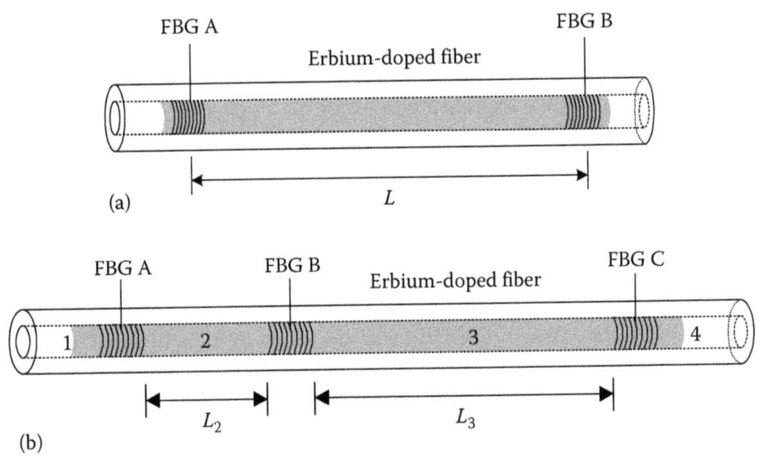

FIGURE 13.1 Schematic diagrams of (a) DBRFL and (b) CCFL.

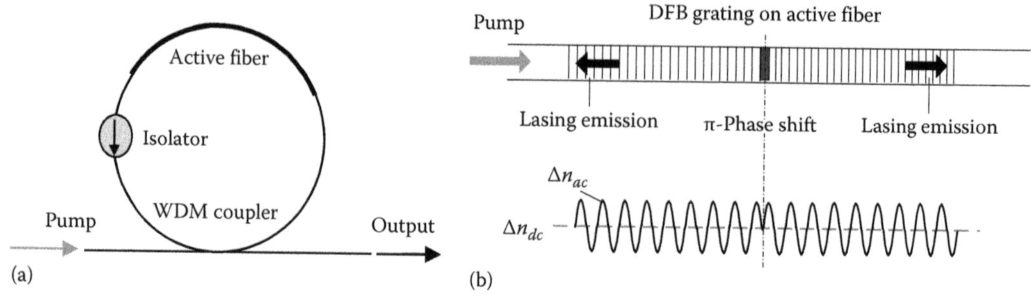

FIGURE 13.2 Schematic diagrams of (a) RFL and (b) DFB fiber laser.

transmission band in the Bragg reflection region, making for the DFBFL laser an extremely narrow linewidth. The effective cavity length is inversely proportional to the grating strength. Hence in good DFBFLs, the laser mode is tightly confined around the phase shift, free from mode hops, and exhibiting extremely narrow linewidth, long coherent length, and excellent environmental stability.

The choice of a particular laser structure depends mainly on specific application condition and requirement. The erbium-doped DBRFL and DFBFL emit 1.55 μm wavelength and relatively low output power; however, they can operate robustly in single longitudinal mode without special cooling control. Although they can provide high output power, the erbium–ytterbium co-doped fiber lasers have an intense heating effect inside fiber and need precise cooling control to keep the stable operation. Hence, Er-doped DBR and DFBFLs are extensively investigated and widely used as the sensing elements in sensing applications such as fiber hydrophone and geophones.

One main task in fabricating linear cavity fiber lasers is grating writing. Grating writing is usually accomplished by direct exposure with phase mask or interference exposure. In fabricating DFBFL, the key is the accuracy in phase shift that will crucially affect the laser performance. Hence, one must pay great attention to the phase shift in fabrication. There are several techniques to introduce the phase shift reported, such as local heating [17], partial shielding exposure [18], local postexposure [19], and fiber/phase mask dithering [20].

13.2.2 Laser Phase and Gain Conditions

The wavelength and intensity responses of fiber lasers affect both the laser output and the sensing performance in FLS. So it is very important to analyze and optimize these responses for specific FLS applications.

For DBRFL and CCFL, simple approximation can be made based on multiple reflection phenomena, leading to simple analytical model. Distinguished intensity and wavelength responses can be observed between single cavity and composite cavity laser. However, for DFBFL, the optical field is spatially distributed within the grating, making it a bit more complex to analyze wavelength and intensity responses.

13.2.2.1 DFBFL

DFBFL can be a stable light source with very narrow linewidth (between 1 and 10 kHz) [21,22] and low phase noise (about 0.3 μrad/√Hz·m) [23]. A stable single longitudinal mode is accomplished from the design of laser with a very short cavity (of π-phase shift). The comparison of field mode distribution of DFBFL and DBRFL obtained from numerical calculation is shown in Figure 13.3a. Such concentrated mode around π-phase shift makes DFBFL less affected by external

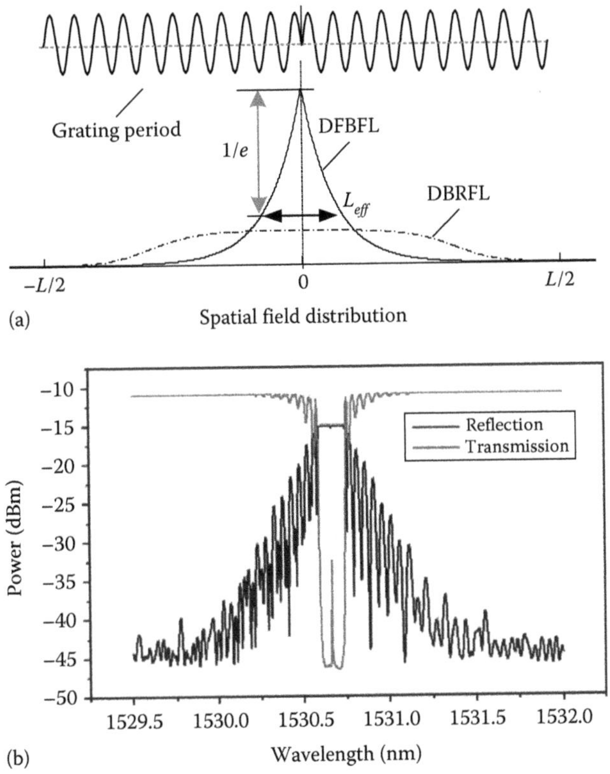

FIGURE 13.3 (a) Spatial field distribution of a DFBFL as compared to DBRFL and (b) transmission and reflection spectra of a fabricated 4 cm DFBFL.

perturbations such as temperature gradient and physical actions (bending or twisting), but in return, it has lower slope efficiency, compared to DBRFL. Figure 13.3b shows the typical spatial field distribution and spectral transmission of a DFBFL. Typically, DFBFL has length of several centimeters up to 10 cm [24]. Longer cavity has advantages in terms of achieving narrower linewidth [25] and provides better stability against external reflection [26]. However, from the sensing perspective, it is more advantageous to have shorter cavity to keep the sensor head as compact as possible and reduce pump absorption, which is critical for arrayed system.

There are a few cavity designs for different applications that have been demonstrated over the past years: (1) apodization for suppression of higher-order lasing mode [27,28] and optimizing output power [29], with previous work obtained about 50 dB suppression using Gaussian apodization; (2) dual phase-shifted DFB to create stable dual-wavelength operation [30,31]; and (3) smoothed phase-shift transition to improve resistance to cross coupling that increases phase noise [32].

Coupled mode theory (CMT) [33] is the most common method in analyzing spatially distributed field of periodic structure such as DFBFL. This treatment is different from CCFL and DBRFL since we assume that the field is spatially distributed within the gratings for DFBFL. For a uniform structure of DFBFL, the phase and gain condition expression derived from CMT [34] can be written as

$$\left[\frac{S-(\gamma-i\Delta\beta)}{S+(\gamma-i\Delta\beta)}\right]e^{2SL}=-1. \tag{13.1}$$

In (13.1), S is given by

$$S = \sqrt{\kappa^2 - (\Delta\beta + i\gamma)^2} \tag{13.2}$$

and $\Delta\beta$ is the detuning of propagation constant from the designed value given by

$$\Delta\beta = 2\pi n_{eff}\left(\frac{1}{\lambda} - \frac{1}{\lambda_B}\right) \tag{13.3}$$

The remaining symbols are defined as follows: γ is the linear gain, κ is the coupling coefficient, n_{eff} is the effective refractive index, L is the total cavity length, λ_B is designed Bragg's wavelength, and λ is the lasing wavelength. Lasing wavelength and threshold gain of DFBFL are obtained by solving Equation 13.1.

Although CMT provides good insights into laser physical properties, the analysis will be limited only for uniform structure. For spatially nonuniform structures, for example, DFBFL with apodization, chirped grating, or multiple phase-shifted grating, the numerical method becomes an essential tool. For example, transfer matrix method (TMM) derived from CMT is required for analysis as in the case with apodization [28].

13.2.2.2 DBRFL and CCFL

The cavity of DBRFL is simply formed by two FBGs inscribed on a continuous EDF. DBRFL is also a single-cavity laser that has a similar type as DFBFL, but it has significantly longer effective cavity length. This results in a more efficient pump-lasing power conversion. Previously, incorporation of an external reflector to DBR lasers has been studied for several purposes including linewidth narrowing [35,36], reduction of low-frequency wavelength instability [37], and continuous wavelength tuning [38,39]. It will be demonstrated by inclusion of an active feedback (hence composite cavity) that phase condition of laser can be significantly altered, which is useful for wavelength-type sensing. The typical spectral transmission of a composite cavity for CCFL is shown in Figure 13.4.

We follow the approach given by Refs. [40,41] to solve the boundary value problem for the fields in the CCFL. Taking into consideration the effects of coherent multiple interferences of the composite cavity structure, the potential of net optical gain in both cavities, and the

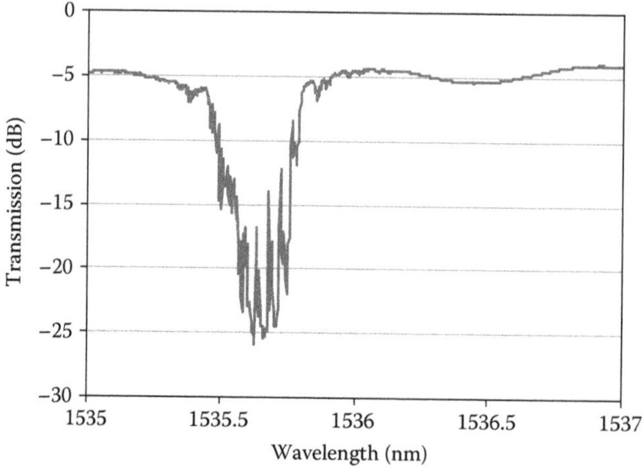

FIGURE 13.4 Measured transmission spectrum of a 2 cm/8 cm CCFL with 6 mm FBGs. (From Azmi, A.I. et al., *Photon. Sens.*, 1(3), 210, 2011.)

symmetrical properties of an FBG, the general condition for the phase and gain conditions for the CCFL can be written as

$$r_A r_B e^{i\phi_2} g_2 - r_B r_C e^{i\phi_3} g_3 + r_A r_C (2r_B^2 - 1) e^{i(\phi_2+\phi_3)} g_2 g_3 = 1 \tag{13.4}$$

where $\phi_j = 2n_j k L_j$ and $g_j = \exp\left[2(\gamma_j - \alpha_j) L_j\right]$ (j = 2 and 3 denotes the respective region) are the round-trip phase and round-trip gain, respectively, with n_j being the effective refractive index, k_j the propagation constant given by $k = 2\pi/\lambda$, λ the lasing wavelength of the CCFL, γ_j the linear gain, and α_j the linear loss. If region two is considered being the primary cavity, the general condition for oscillation can be written as

$$g_2 = \frac{1}{r_A r_B e^{i\phi_2}} Z_C \tag{13.5}$$

with the complex feedback parameter Z_C defined as

$$Z_C = \frac{1 - r_B r_C e^{i\phi_3} g_3}{1 + \frac{r_C g_3}{r_B}(1 - 2r_B^2) e^{i\phi_3}} \tag{13.6}$$

The additional terms that relate the phase and magnitude of Z_C are defined as $G_C = \ln |Z_C|$ and $\phi_c \leq \angle Z_C$. The threshold gain and phase conditions for the CCFL are obtained by rearranging Equation 13.5 into a real and imaginary term:

$$2\gamma_{th} L_2 = 2\alpha_2 L_2 - \ln(r_A r_B) + G_C \tag{13.7}$$

$$\phi_C = \phi_2 + p2\pi, \quad p \text{ integer} \tag{13.8}$$

The final term in Equation 13.7 is the additional term of the Fabry–Perot threshold that accounts the effect of the active feedback. It is apparent that the phase condition in Equation 13.8 has multiple solutions due to the definition of p as an integer. However, the lasing mode is selected at the wavelength with the lowest threshold gain.

13.2.3 Laser Characteristics

The performance of a fiber laser can be specified by several basic parameters, such as output power, pump threshold, directivity, linewidth, RIN, frequency noise, and polarization characteristics. For DBRFL and DFBFL, the pump threshold and output power can be obtained by the previously mentioned theoretical analysis and directly experimentally tested. However, other parameters such as laser linewidth, RIN, frequency noise, and polarization are relatively easy to test but more complicated to analyze.

13.2.3.1 Linewidth

The linewidth denotes the monochrome characteristic and the coherence characteristic of the fiber laser. A stable and narrow linewidth is always desired for laser source and sensing applications. Laser linewidths of <10 kHz have been reported in the erbium-doped DBRFL and DFBFL. The laser linewidth can be measured by using a delayed self-homodyne or a delayed self-heterodyne technique. The optical fiber delay line is usually needed to be longer than the coherence length of the fiber laser to be measured. However, the laser linewidth has the great relation to the cavity perturbation caused by ambient noise. The measured linewidths of the fiber lasers are usually much wider than the theoretical limit base on the Schawlow–Townes linewidth formula:

$$\Delta \nu = \frac{2\pi h\nu(\Delta \nu_g)^2}{P} \tag{13.9}$$

where

ν is the lasing frequency
P is the laser output power
$\Delta\nu_g$ is the passive grating linewidth

A sound and vibration isolating package with temperature control can reduce the linewidth by over one order in magnitude compared to the free-running fiber lasers. Several commercial fiber laser products with a very narrow linewidth that is even several tens of hertz have been provided by American and Danish companies [42,43].

13.2.3.2 Intensity Noise and Frequency Noise

Ronnekleiv investigated the frequency noise and the RIN theoretically and experimentally and considered that the laser frequency and intensity noise are to some degree correlated due to thermal effects in the gain medium [22]. The RIN of the laser describes the intensity stability of the laser output. It is defined as the mean square fluctuation of the power, in unit frequency range, divided by the square of the average optical power, $\bar{P}$:

$$\text{RIN} = \left[\frac{\langle \Delta P^2 \rangle}{\bar{P}^2}\right] \text{Hz}^{-1} \tag{13.10}$$

where $\langle \Delta P^2 \rangle$ is the mean square optical intensity fluctuations in a 1 Hz bandwidth at a specified frequency. In general, the RIN has a peak at the relaxation oscillation frequency (ROF), that is, the oscillation of intensity in the cavity around their steady-state values, caused either when the laser is first turned on or when the laser is suddenly perturbed by a small fluctuation in gain or cavity loss. For erbium-doped DBRFL and DFBFL, the ROF increases with the increasing pump power while the RIN peak decreases in contrary. Cranch has theoretically and experimentally analyzed the RIN characteristics of the erbium-doped DFBFL in details [44]. The optical frequency noise $\nu^2_{rms}(f)$ is defined by the relation $\nu^2_{rms}(f) = < |\Delta\nu^2(f)| >$, which expresses the power spectral density of $\Delta\nu(t)$. The term $\Delta\nu(t) = d\phi_s/(2\pi dt)$ represents the time-dependent frequency fluctuations, where $\phi_s(t)$ is the time-dependent optical phase fluctuations. The frequency noise is usually increased at low frequencies due to ambient temperature drift, acoustical vibrations, etc. The active stabilization by feedback to the pump laser can effectively reduce the intensity and frequency noise level and remove the relaxation oscillation peaks from the noise spectra.

13.2.3.3 Polarization

As for the polarization characteristics of the fiber lasers concerned, they are closely related to the birefringence of the gain medium fiber and the cavity structure. Birefringence intrinsic to the fiber or generated during the fabrication process can cause simultaneous lasing in orthogonal polarization modes in a normal DFBFL or DBRFL. Although the beat frequency mechanism by the two laser polarization mode can be used as a sensing technique, a single polarization lasing is always required in the interferometry sensing to avoid the signal fading. Rønnekleiv et al. studied the polarization mechanism in DFBFL and deduced the conditions to get single polarization laser [45]. There are several methods to obtain single polarization mode output, such as utilizing a birefringent phase shift or a birefringent Bragg grating, by which the threshold for one polarization mode is increased while that for another mode is reduced or constant. By introducing a polarization dependence of the UV-induced refractive index change in the Bragg grating or its phase shift, the detuning of one polarization mode can be changed relative to another, and in extreme case, one can be prevented from reaching threshold. The UV-induced birefringence increases with increasing angle between the UV writing beam polarization state and the fiber axis. Stress can also be applied to generate birefringence during fabrication so that twisting the fiber laser can extinguish one polarization mode.

13.3 SENSING CHARACTERISTICS OF FIBER LASERS

The aspects described earlier are general characteristics of fiber lasers. They are closely related but not the same characteristics of our interest for FLS. For general purposes, fiber lasers of stable characteristics robust to external disturbances are normally desired.

However, for FLS purposes fiber lasers need to have high sensitivity to external disturbance, preferably only to the particular measurand of interest. Fiber lasers operate usually under resonance with high-quality factor and naturally sensitive to many internal and external factors. Nevertheless, it is useless for FLS if it is simultaneously sensitive to too many factors and conditions. Hence for FLS purposes, fiber lasers are required to have both good sensitivity and selectivity at least.

Sensitivity: Sensitivity or responsivity to a particular measurand can be manifested in changes in output parameters—intensity, phase, frequency, and/or polarization—of fiber lasers. In general, all the output parameters, intensity, phase, frequency (wavelength), and polarization of fiber laser, will change to different extents under an external influence. In practice, only one sensitivity, to either intensity, phase, wavelength, or polarization, is to be selected and utilized for FLS purpose. Hence, one may classify the FLS into intensity type, phase type, wavelength or frequency type, and polarization type.

As an example, a wavelength-type FLS for acoustic pressure or hydrophone is discussed in the following. When exposed to an acoustic field, varying pressure (ΔP) induces dynamic strain (ε) onto a fiber laser according to

$$\varepsilon = \frac{2\nu - 1}{E}\Delta P \tag{13.11}$$

where

ν is Poisson's ratio, $\nu \approx 0.23$ for silica glass

E is Young's modulus, $E \approx 72 \times 10^9$ for silica glass

Assuming the surrounding temperature to be stable, the shift in the fiber laser's wavelength can be expressed as [11]

$$\frac{\Delta\lambda_{FL}}{\lambda_{FL}} = \left(1 + \frac{1}{n_{eff}}\frac{\partial n_{eff}}{\varepsilon}\right)\varepsilon = \left(\varepsilon_z - \frac{n_{eff}^{\;2}}{2}((\rho_{11} + \rho_{12})\varepsilon_r + \rho_{12}\varepsilon_z)\right) \tag{13.12}$$

where

ε_z and ε_r are the in-fiber pressure-induced longitudinal and radial strains

ρ_{11} and ρ_{12} are the elasto-optic coefficients, $\rho_{11} \approx 0.121$ and $\rho_{12} \approx 0.265$ for silica glass

n_{eff} is the effective index of the fiber

Under hydrostatic conditions, $\varepsilon_z = \varepsilon_r = \varepsilon$. Substituting Equation 13.11 into Equation 13.12 gives the wavelength response of a fiber laser:

$$\frac{\Delta\lambda_{FL}}{\Delta P} = \left(1 - \frac{n_{eff}^2}{2}(\rho_{11} + 2\rho_{12})\right)\frac{2\nu - 1}{E}\lambda_{FL} \tag{13.13}$$

This is the intrinsic wavelength selectivity of bare fiber laser that is similar to that of FBG or fiber. Taking special sensor head design and packaging considerations and procedures would be able to increase the effective sensitivity.

Selectivity: As mentioned earlier, the key issue in FLS is not only to have high sensitivity to the measurand of interest but also to have high selectivity; thus, other factors do not produce significant effect interfering the sensing process. The high selectivity is usually achieved through special sensor head design and packaging considerations and procedures.

13.3.1 Optimization of Sensing Characteristics

In general, both sensitivity and selectivity in FLS can be optimized, with regard to one particular laser parameter and one particular measurand, by fiber laser design and operation, sensor head design and packaging, system conditions, and interrogation techniques. Hence, there are many ways to optimize or tailor the sensing characteristics for FLS purposes.

The optimization of wavelength sensitivity for FLS using CCFL is discussed here as a practical example [15]. Figure 13.4 shows the transmission spectrum of a CCFL fabricated in UNSW [15]. If the CCFL is under varying pressure as in the case previously discussed, it would have the same wavelength sensitivity as given by Equation 13.13. However, if only one cavity is made susceptible, and the other one nonresponsive, different response would be expected. This partial cavity sensing format is illustrated by Figure 13.5 where only one cavity is utilized for sensing while the other one is mechanically isolated.

To investigate the response difference between the total and partial cavity sensing, a numerical simulation is carried out based on the CCFL that was fabricated and tested. The parameters are set with the following values: $r_B = 0.7$, $r_C g_3 = 0.8$, $n = 1.465$, $L_2 = 3$ cm, $L_3 = 6$ cm, $\nu = 0.23$, $\rho_{11} = 0.121$, $\rho_{12} = 0.265$, and $E = 72$ GPa. Illustrated in Figure 13.6 is the resulting wavelength response according to the phase-match condition for all sensing schemes when pressure of 0–8 MPa is gradually imposed. It can be seen that the response of total cavity sensing is linear, which is a similar trend to the single cavity laser response. As for both partial cavity schemes, the responses are nonlinear and periodically change between high and low. This period is the same as the period of ϕ_C. The highest sensitivity is achieved by the 3 cm partial cavity, with the corresponding region marked in Figure 13.5. The spectral range for the high-sensitivity region of the 3 cm partial cavity is relatively

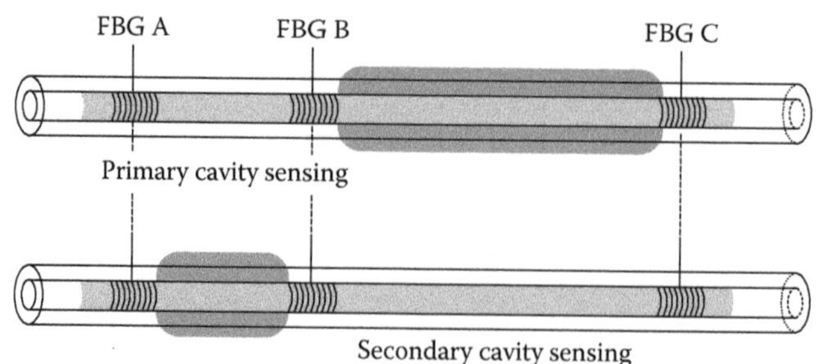

FIGURE 13.5 Partial cavity sensing schemes where the shaded regions indicate the laser cavity area that is mechanically isolated, while the other part is used for sensing. (From Azmi, A.I. et al., *J. Lightwave Technol.*, 28(12), 1844, 2010.)

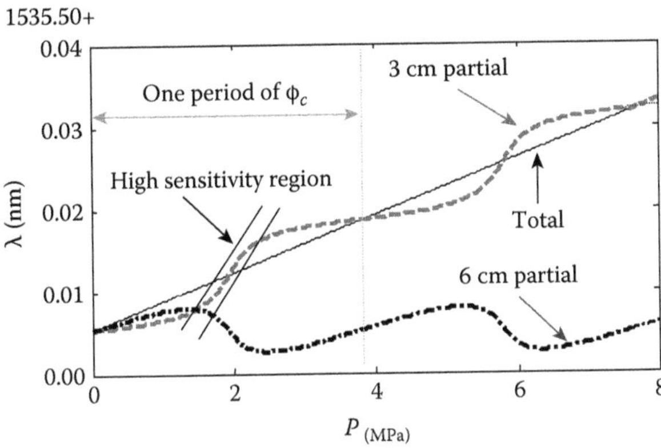

FIGURE 13.6 The resulting response for all sensing formats when pressure is gradually imposed from 0 to 8 MPa. (From Azmi, A.I. et al., *J. Lightwave Technol.*, 28(12), 1844, 2010.)

narrow, approximately 0.01 nm. Such narrow spectral response suggests that this sensing scheme is only suitable for small amplitude signal detection such as acoustic and vibration. To demonstrate how these responses are obtained, the phase-matching conditions at two different pressures are plotted as shown in Figure 13.7. At zero pressure, the same solution of 1535.505 nm is obtained for all sensing schemes. This condition is represented by the solid line and the solution is marked by "+" symbol. After the application of 3 MPa pressure, the change in the phase condition is plotted by the dash line

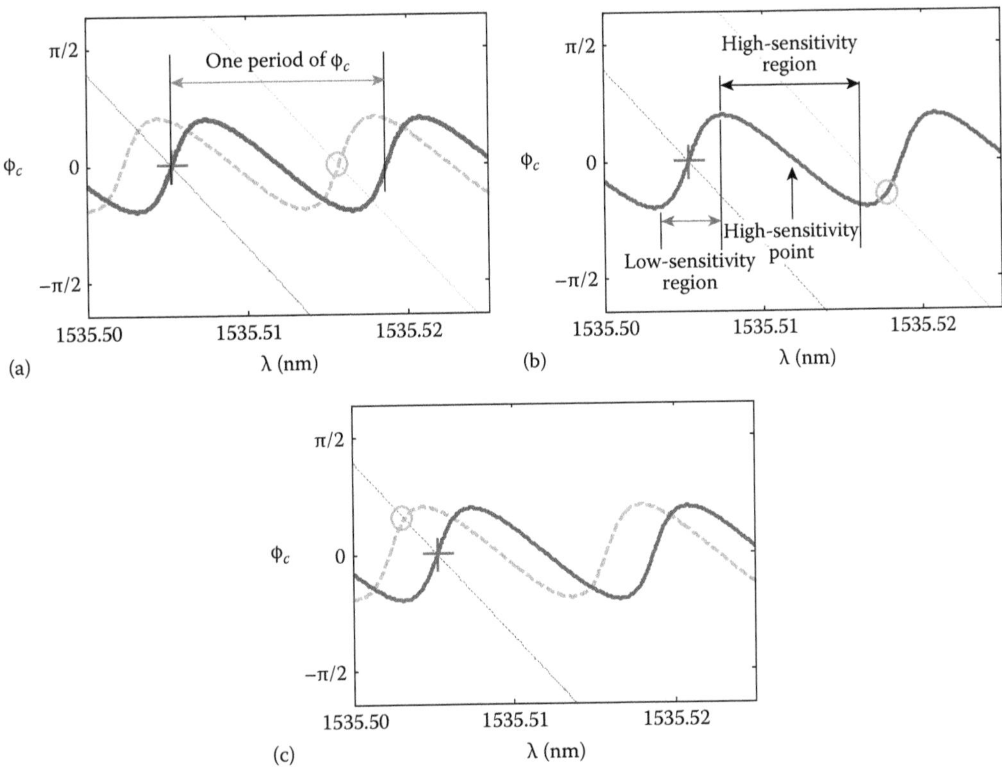

FIGURE 13.7 Phase-matched conditions for (a) total cavity sensing, (b) 3 cm partial cavity sensing, and (c) 6 cm partial cavity sensing. Solution at zero loading is marked by "+" and at 3 MPa loading is marked by "O." (From Azmi, A.I. et al., *J. Lightwave Technol.*, 28(12), 1844, 2010.)

and the solution is marked by "O" symbol. If the phase condition is not altered, the plot remains in solid line. For the total cavity sensing, both sides of Equation 13.8 are affected during sensing, and the resulting wavelength after 3 MPa pressure is 1535.515 nm. This corresponds to the standard response of 3.47×10^{-9} nm/Pa. While for the both partial cavity sensing, either one side of Equation 13.8 is affected, resulting in rather nonlinear response. In the following, we only focus on the 3 cm (primary) cavity sensing scheme since it demonstrates the highest response.

Further assessment on Figures 13.6 and 13.7 yields several important points that can be interpreted in respect to the wavelength response of the primary partial cavity sensing:

1. From Figure 13.7b, when the right-hand side (RHS) of Equation 13.8 is swept across the static left-hand side (LHS) of Equation 13.8, the intersection point changes more drastically when slope difference is minimum, and hence, a higher sensitivity can be achieved at this condition. By having the slope value of RHS of Equation 13.8 always smaller (more negative) than LHS of Equation 13.8, the multiple intersection that led to multiple longitudinal mode can be avoided as mentioned in [41].
2. Parameters L_3 and n_3 control the period of ϕ_C; thus, they also control the period of the high-sensitivity region. Here, we define the high-sensitivity region as the region with response higher than the typical response of a single-cavity laser. The high-sensitivity region in one cycle is located between the maximum and minimum points of ϕ_C (or at the negative-slope region). However, the effect of L_3 and n_3 to the sensitivity should also be considered.
3. As illustrated in Figure 13.6, the initial response of CCFL is commenced at the low-sensitivity region, which is practically undesired. In general case, the phase-matching point of CCFL is unknown since the location of the intersection is spectrally arbitrary. Adjusting the initial solution of Equation 13.8 is possible by adding the phase of ϕ_C, for example, by applying an initial static pressure. A sufficient initial static loading makes operation of the CCFL at the high-sensitivity region possible.

According to the point made in (1), the zero-slope difference condition of Equation 13.8 signifies the highest sensitivity condition and the boundary between single and multiple longitudinal modes. By setting the slope relation as $\phi_c' > \phi_2'$, the zero-slope difference condition can be derived. Since the slope of ϕ_c is a periodic function, the minimum slope location is set at $\phi_3 = m\pi$ (for most cases), where m is an odd integer. The resulting zero-slope difference condition is given by

$$\frac{n_2L_2}{n_3L_3} < \frac{r_B + \left(1-2r_B^{\,2}\right)/r_B}{r_B - \left(1-2r_B\right)^2/r_B - \left(1-2r_B^{\,2}\right)r_Cg_3 + 1/(r_Cg_3)} \tag{13.14}$$

For the operation flexibility, the CCFL response can be optimized with regard to r_cg_3 (controlling pump power and hence gain) rather than n_2L_2/n_3L_3. Figure 13.8a shows the wavelength responses of (3 cm, 6 cm) CCFL at selected r_cg_3 values. Since the phase-matched condition is spectrally arbitrary, the pressure of the x-axis in Figure 13.8a can be assumed a relative value. There are three different characteristics that can be observed when an increased pressure is applied: a positive response only ($r_cg_3 = 1.00$, 1.24), a negative response only ($r_cg_3 = 1.46$, 1.60), and a combination of positive and negative responses ($r_cg_3 = 1.40$). Using Equation 13.14, we solve r_cg_3 using the previous parameters setting and yield $r_cg_3 < 1.24$. This relation agrees with the results obtained in Figure 13.8a, where the highest sensitivity is produced at $r_cg_3 = 1.24$, and further increment of r_cg_3 will cause multimode operation. Indicated in Figure 13.8b is the corresponding relative sensitivity of the plot shown in Figure 13.8a for $r_cg_3 = 1.24$.

Sensitivity enhancement of 40 dB can be achieved within narrow pressure range approximately in order of a tenth of an MPa (thus narrow spectral range). Shown in Figure 13.8c is the peak enhancement against r_cg_3 (y-axis on LHS) and cavity length ratio, L_3/L_2, against r_cg_3 that produces a zero-slope difference condition for the highest-sensitivity operation (y-axis on RHS). The shaded area

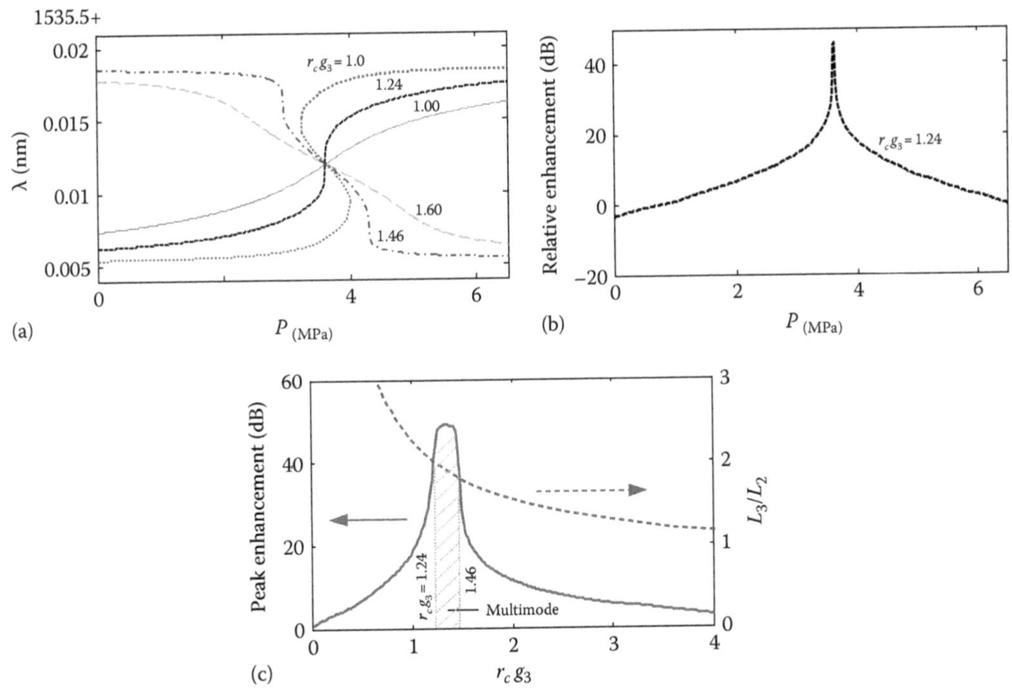

FIGURE 13.8 (a) Lasing wavelength against pressure at different r_cg_3 values for the primary partial cavity sensing, (b) the enhancement relative to the sensitivity of the single-cavity laser for r_cg_3 = 1.24, and (c) the peak enhancement against r_cg_3, where the shaded area under the graph indicates the region where r_cg_3 value produces multimode operation of CCFL (*y*-axis on LHS) and the cavity length ratio L_3/L_2 against r_cg_3 that produces zero-slope difference condition for the highest-sensitivity operation (*y*-axis on RHS). (From Azmi, A.I. et al., *J. Lightwave Technol.*, 28(12), 1844, 2010.)

under the enhancement plot indicates the undesired operating region where r_cg_3 value produces multimode operation of CCFL, which is bounded by r_cg_3 values of 1.24 and 1.46. It is shown that the length of the secondary cavity compensates the feedback requirement for the highest sensitivity condition, with the $L_3/L_2 = 2$ ratio (or [3 cm, 6 cm] CCFL) providing a comparatively feasible r_cg_3 of 1.24.

13.4 INTERROGATION TECHNIQUES FOR FLS

Whether using DBRFL, DFBFL, CCFL, or RFL, FLS is often a wavelength type or intensity type that is based on the wavelength or intensity sensitivity of fiber laser to a measurand. Here, we describe several most frequently used interrogation techniques for FLS applications.

13.4.1 Intensity Interrogation

The simplest FLS would be an intensity type that the output power is directly sensitive to a measurand. Due to its resonance-enhanced effective interaction length and low RIN, an FLS could be made very compact yet significantly more sensitive than its passive intensity-type fiber sensors. In particular, intensity-type fiber laser sensors can be very simple and low cost because the intensity/power changes in laser output can be directly interrogated with an optical detector. So it could be useful and feasible for many industrial applications.

A schematic diagram of intensity-type FLS with direct intensity interrogation is shown in Figure 13.9. Here, the photodetector directly converts optical signal from fiber laser to electrical signal, which passes through a low-pass filter to prevent aliasing in digital data acquisition.

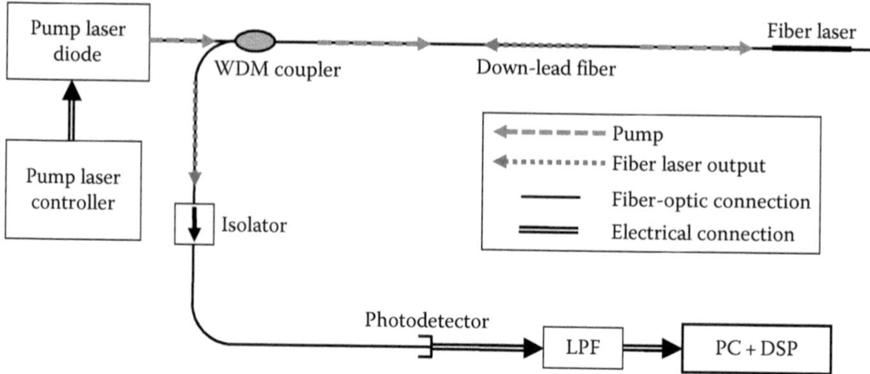

FIGURE 13.9 Intensity-type FLS with direct intensity interrogation.

13.4.2 Wavelength Interrogation

For wavelength-type FLS, there are many ways to interrogate the optical signal that its wavelength is changed with a measurand, for example, using optical spectrum analyzer, optical tunable filter, and interferometer. However, the interrogation by optical interferometer is considered as a technique producing superior sensing performance in terms of wavelength resolution (wavelength sensitivity) and time resolution (frequency range). So interferometric interrogation, rather than the direct intensity interrogation, has been most widely used for high-performance FLS systems.

13.4.2.1 Wavelength Interrogation by MZI

A schematic diagram of the interrogation by interferometer for a wavelength-type FLS is shown in Figure 13.10. In this system, the pump power is delivered to the fiber laser via a WDM coupler

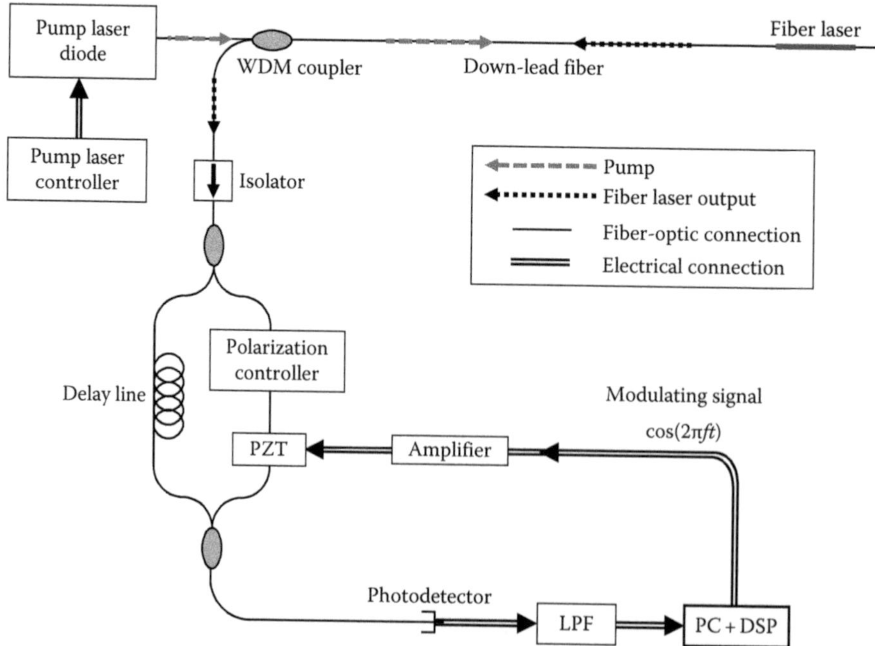

FIGURE 13.10 Wavelength-type FLS with interrogation by MZI.

and a length of down-lead fiber. The fiber laser acts as a very compact sensing element and an array of fiber lasers can be easily used to build a WDM hydrophone system. The operation using backward emission is much more desirable over forward emission for its spectral purity as no residue pump wavelength is present, as well as its simplicity in allowing all the try-end system on one end only.

Here, a fiber unbalanced Mach–Zehnder interferometer (MZI) is used for wavelength interrogation. The wavelength change $\Delta\lambda$ in optical sensing signal from the fiber laser is converted to a change in optical phase difference between the two unequal arms, $\Delta\phi$ given by

$$\frac{\Delta\phi}{\Delta\lambda} = \frac{2\pi n_{eff} d}{\lambda^2} \tag{13.15}$$

where d denotes the path difference related to the length of the delay line. The change in phase difference is again converted to a change in optical intensity in the output arm of the interferometer and detected by the optical detector. A PZT phase modulator is introduced into one arm of the MZI, so that the optical signal phase in that arm can be modulated at the carrier frequency f_c, and the sensing information $\phi(t)$ is carried as the sidebands of the even and odd harmonics of the carrier signals. According to the Nyquist theorem, f_c needs to be at least double the maximum frequency of interest in $\phi(t)$. The modulating signal $\cos(2\pi f_c t)$ can be first generated digitally and then realized to a voltage waveform by a digital-to-analog converter. The modulating signal is then amplified in order to drive the PZT. The amplifier also performs low-pass filtering (LPF) on the signal to reduce the finite step noise caused by the digital-to-analog conversion process. Different demodulation methods may be used to extract the sensing information encoded in the spectral change of the laser output.

Since the interferometer output is nonlinear (in sinusoidal form), different demodulation methods have been introduced to extract the output linearly. Commonly known demodulation techniques include active homodyne with phase tracker [47], passive homodyne [48–50], pseudoheterodyne [51], and synthetic heterodyne [52]. Passive homodyne is preferred compared to others as it requires no feedback loop and no reference source for interrogation.

The most successful technique for passive homodyne is based on phase modulation of carrier signal, known as phase-generated carrier (PGC). In a pioneering work [48], the demodulation is accomplished by computation on quadrature pair from Bessel's expansion of phase-modulated signal. Novel approaches in restraining PGC signal instability caused by changing parameters such as path delay, direct current (DC) bias, and environmental disturbance have been experimented and proposed recently. The time domain orthogonal sampling approach commercially developed by Optiphase Inc. [49] demonstrates a significant leap of demodulation performance from previous methods with main merits of very high dynamic range (from microradians to thousands of radians using fringe counting) and minimizing self-cross talk or noise (since without using harmonics pair). A recent quadrature sampling method reported was based on the phase-shifting interferometry that relies on the determination of an arbitrary phase step in order to minimize phase measurement errors [50]. A symmetrical 3×3 coupler can be used in passive homodyne to assist the demodulation as it generates a phase difference between signals by 120° which allowed the signal to be retrieved using the trigonometry operations [53–55]. Several improvements of the original scheme have been reported, notably: a new algorithm that minimized distortion [54] and a new passive stabilizing technique to eliminate the 3×3 coupler output fluctuation [55]. Based on these developments of passive homodyne PGC, a digital demodulation technique has been implemented in an FLS system [46]. The benefit of using digital demodulation is very obvious because, by taking advantage of modern of DSP techniques, it allows the performance and characteristics of the systems to be continuingly improved by timely updating the program codes. It is much easier and a lot less expensive than updating analog demodulation systems.

13.4.2.2 Digital PGC Demodulation

The variation in the light intensity detected by the photodetector can be written as

$$I_{DET}(t) = A + B\cos\left(C\cos(2\pi f_c t) + \phi(t)\right) \tag{13.16}$$

where

A and B are related to the optical power and mixing efficiency of the two arms

C is the amplitude of the sinusoidal signal modulated onto the optical signal

Prior to digitizing, an LPF is required in order to prevent alias signals. According to the Nyquist theorem, the relationship between the cutoff frequency f_{co} and sampling f_S is $f_{co} \leq f_S/2$. Next, the detector signal gets digitized by the same PCI 6014 card, so that $I_{DET}(t)$ becomes

$$I_{DET}[n] = A + B\cos\left(C\cos\left(2\pi n\frac{f_C}{f_S}\right) + \phi[n]\right) \tag{13.17}$$

Due to the digitizing process, I_{DET} is now expressed as the function of data number $n = tf_s$, which is dimensionless. As both the fundamental and first-harmonic carrier signals are required for the demodulation process, again according to the Nyquist theorem, the sampling frequency must be $f_s \geq 4f_c$. Thus, the sampling frequency must be at least eight times the maximum acoustic frequency of interest. The digital demodulation algorithm applied to the digitized signal is shown in Figure 13.11.

First, two signals of frequencies f_c and $2f_c$ are required, which are of the form

$$S_1[n] = D\cos\left(2\pi n\frac{f_C}{f_S} + \theta_1\right) \tag{13.18}$$

$$S_2[n] = E\cos\left(4\pi n\frac{f_C}{f_S} + \theta_2\right) \tag{13.19}$$

where

D and E are the amplitudes of $S_1[n]$ and $S_2[n]$, respectively

θ_1 and θ_2 are introduced as the phase differences between S_1 and S_2, respectively, against the modulating signal

To reduce complicated synchronization issues, the two mixing signals are derived by sampling the modulating signal on a separate input channel on the PCI card. Since $S_1[n]$ has the same

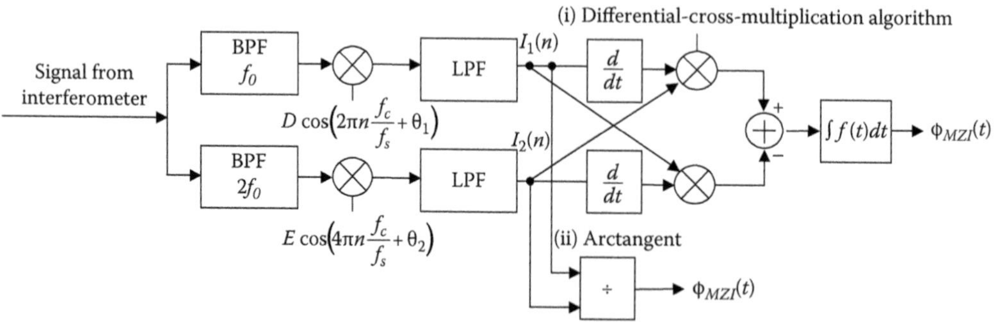

FIGURE 13.11 PGC demodulation scheme using (i) differential cross-multiplication algorithm and (ii) arctangent algorithm.

frequency as the modulating carrier signal, it is readily obtainable. $S_2[n]$ is then obtained through the trigonometric identity "cos(2x) = 2cos²(x)−1." By first squaring $S_1[n]$ to obtain

$$\left(S_1[n]\right)^2 = \frac{D^2}{2}\left(1+\cos\left(4\pi n\frac{f_C}{f_S}+2\theta_1\right)\right) \tag{13.20}$$

and then removing the DC term will result in $S_2[n]$ with $E = D^2/2$ and $\theta_2 = 2\theta_1$. Second, the two signals $S_1[n]$ and $S_2[n]$ are mixed with $I_{DET}[n]$. In order to make it easier to comprehend the process, Equation 13.17 is expanded using Bessel functions to become

$$\begin{aligned} I_{DET}(t) = A + B\Bigg\{&\left[J_0(C)+2\sum_{k=1}^{\infty}(-1)^k J_{2k}(C)\cos\left(4k\pi f_C t\right)\right]\cos\left(\phi(t)\right) \\ &-\left[2\sum_{k=0}^{\infty}(-1)^k J_{2k+1}(C)\cos\left((2k+1)2\pi f_C t\right)\right]\sin\left(\phi(t)\right)\Bigg\} \end{aligned} \tag{13.21}$$

It can be determined that only the components associated with $J_1(C)$ and $J_2(C)$ in Equation 13.21 will produce a signal in the base band when mixed with $S_1[n]$ and $S_2[n]$, respectively, so that for the intended purpose of this demodulation process, $I_{DET}[n]$ for the two mixing signals may be simplified as

$$I_{DET1}[n] = -2BJ_1(C)\cos\left(2\pi n\frac{f_C}{f_S}\right)\sin\left(\phi(t)\right) \tag{13.22}$$

$$I_{DET2}[n] = -2BJ_2(C)\cos\left(4\pi n\frac{f_C}{f_S}\right)\cos\left(\phi(t)\right) \tag{13.23}$$

and the signals after mixing can be written as

$$I_{DET1}[n]S_1[n] \approx -BDJ_1(C)\sin\left(\phi[n]\right)\times\left(\cos\left(4\pi n\frac{f_C}{f_S}+\theta_1\right)+\cos(\theta_1)\right) \tag{13.24}$$

$$I_{DET2}[n]S_2[n] \approx -BEJ_2(C)\cos\left(\phi[n]\right)\times\left(\cos\left(8\pi n\frac{f_C}{f_S}+\theta_2\right)+\cos(\theta_2)\right) \tag{13.25}$$

and finally low-pass digital filtering to extract the base band will recover two orthogonal signals:

$$I_1[n] = -BDJ_1(C)\sin\left(\phi[n]\right)\cos(\theta_1) \tag{13.26}$$

$$I_2[n] = -BEJ_2(C)\cos\left(\phi[n]\right)\cos(\theta_2) \tag{13.27}$$

By careful buffering, $S_1[n]$ and $S_2[n]$ can be delayed independently to make θ_1 and θ_2 close to the integer multiple of π, so that the cosine of these terms can be maximized. The buffering process will also remove the effects other phase shifts introduced throughout the demodulating process that were not accounted for in the derivations.

The two orthogonal signals $I_1[n]$ and $I_2[n]$ can be applied in two methods to extract the desired signal $\phi[n]$. The first is the cross-multiplying option as shown in Figure 13.11. The time derivatives of Equations 13.26 and 13.27 are

$$\frac{\partial I_1[n]}{\partial t} = -BDJ_1(C)\cos(\theta_1)\frac{\partial \phi[n]}{\partial t}\cos\left(\phi[n]\right) \tag{13.28}$$

$$\frac{\partial I_2[n]}{\partial t} = BEJ_2(C)\cos(\theta_2)\frac{\partial \phi[n]}{\partial t}\sin\left(\phi[n]\right) \tag{13.29}$$

Cross multiplying $I_1[n]$ and $I_2[n]$ with their time derivatives gives

$$I_1[n]\frac{\partial I_2[n]}{\partial t} = -B^2DEJ_1(C)J_2(C)\times\cos(\theta_1)\cos(\theta_2)\frac{\partial \phi[n]}{\partial t}\sin^2\left(\phi[n]\right) \tag{13.30}$$

$$I_2[n]\frac{\partial I_1[n]}{\partial t} = B^2DEJ_1(C)J_2(C)\times\cos(\theta_1)\cos(\theta_2)\frac{\partial \phi[n]}{\partial t}\cos^2\left(\phi[n]\right) \tag{13.31}$$

Then subtracting Equation 13.30 from Equation 13.31 gives

$$I_2[n]\frac{\partial I_1[n]}{\partial t} - I_1[n]\frac{\partial I_2[n]}{\partial t} = B^2DEJ_1(C)J_2(C)\cos(\theta_1)\cos(\theta_2)\frac{\partial \phi[n]}{\partial t} \tag{13.32}$$

And finally, integrating Equation 13.32 gives

$$I_{OUT}[n] = B^2DEJ_1(C)J_2(C)\cos(\theta_1)\cos(\theta_2)\phi[n] \tag{13.33}$$

which is the desired signal $\phi[n]$ multiplied by a constant. For this cross-multiplying option, the integration process will suppress the high-frequency noise but emphasize the low-frequency noise.

Further, due to the differentiation and integration processes, no absolute DC level can be determined for $\phi[n]$, and the signal-to-noise ratio (SNR) will be poor for very low frequencies. Slow variations in B due to drift in the interferometer's mixing efficiency can be measured and corrected, but fast variations will add extra noise to the signal due to its squaring effect in (13.28). If θ_1 and θ_2 of the multiplying signals are optimized as previously suggested, then $\cos(\theta_1)\cos(\theta_2)$ should be close to unity. The product of the Bessel functions $J_1(C)J_2(C)$ can be maximized by choosing $C \approx 2.375$ to be approximately 0.22.

The second option that can be applied to the orthogonal signals to extract $\phi[n]$ is the arctangent method. Dividing Equation 13.26 by Equation 13.27 gives

$$\frac{I_1[n]}{I_2[n]} = \frac{D}{E}\frac{J_1(C)}{J_2(C)}\frac{\cos(\theta_1)}{\cos(\theta_2)}\tan\left(\phi[n]\right) \tag{13.34}$$

which can be rearranged to recover $\phi[n]$ according to

$$\phi[n] = \tan^{-1}\left(\frac{I_1[n]}{I_2[n]}\frac{E}{D}\frac{J_2(C)}{J_1(C)}\frac{\cos(\theta_2)}{\cos(\theta_1)}\right) \tag{13.35}$$

If the amplitudes D and E are equal, the phases θ_1 and θ_2 are properly optimized and $C \approx 2.63$ so that $J_1(C) \approx J_2(C) \approx 0.46$, and Equation 13.35 will be simplified to

$$\phi_{MZI}(t) = \tan^{-1}\left(\frac{I_1(t)}{I_2(t)}\right) \tag{13.36}$$

and the display of $I_1[n]$ and $I_2[n]$ on an x–y graph will show a circle (or part of a circle depending on the amplitude of $\phi[n]$). For this arctangent option, variations in B will not affect the result, and the absolute value of $\phi[n]$ can be obtained. However, care must be taken with the amplitudes D and E, phases θ_1 and θ_2, and choice of C in order to avoid distortions. After $\phi[n]$ is recovered, Fourier transform is to be performed to the time signal to calculate the spectral components. The resolution of the calculated spectrum is dependent on the sampling rate and buffer size used to sample the output from the photodetector.

It was found by simulation that large amplitude sinusoidal and square wave $\phi[n]$ inputs showed the arctangent option to be better at tracking the correct amplitude. Also, simulations with Gaussian white noise on the input signal indicated the arctangent option to have lower noise at low frequencies but higher noise at high frequencies when compared with the cross-multiplying option.

13.4.2.3 Wavelength Interrogation by MI

Interrogation by Michelson interferometer (MI), as shown in Figure 13.12, is essentially similar to the interrogation by MZI, in terms of its system design, operation, and demodulation. However, it is usually preferred for better resilience to polarization fading when the MZI is replaced with MI using two Faraday rotation mirrors (FRMs) as reflectors.

13.4.3 Intensity and Wavelength Interrogation

Many different interrogation techniques have been applied in FLS. In addition to the direct intensity interrogation and wavelength interrogation discussed earlier, simultaneous intensity and wavelength interrogation techniques are also developed and applied in FLS. By simultaneous intensity and wavelength interrogation, the spectroscopic information encoded in both intensity and wavelength of sensing laser signals can be extracted. The spectroscopic information is essential for sensing gas, chemical, humidity, etc. In the following, we briefly introduce the two most often useful simultaneous intensity and wavelength interrogation techniques: wavelength modulation spectroscopy (WMS) and wavelength sweep technique (WST).

13.4.3.1 Wavelength Modulation Spectroscopy

WMS is a useful technique for FLS with DBRFL or RFL for gas, chemical, and humidity sensing. WMS shifts the signal detection to higher frequencies well above the strong low-frequency

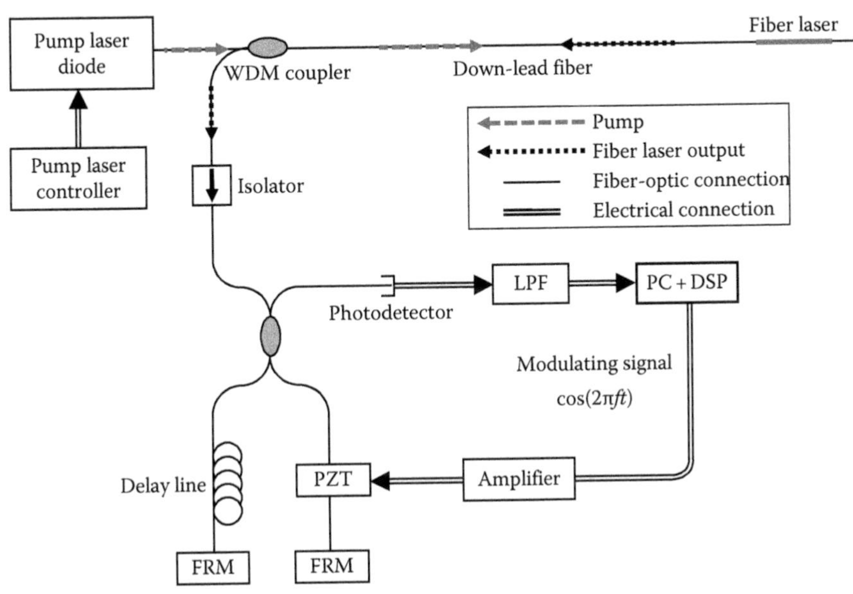

FIGURE 13.12 Wavelength-type FLS with interrogation by MI.

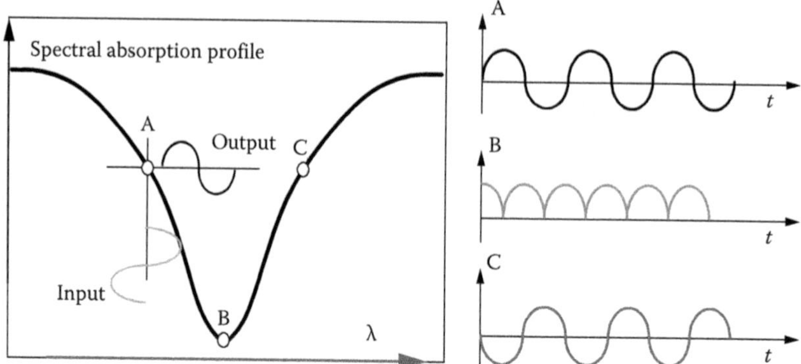

FIGURE 13.13 Wavelength to amplitude modulation conversion in WMS. A, B, and C are special operating points of modulation. B is at the center of the absorption peak where the modulation produces the minimum fundamental amplitude. A and C are points that produce the maximum fundamental amplitude but with a phase difference of π between them.

noises presented in the measurement systems and environments. It is able to achieve significant noise suppression and sensitivity enhancement, offering great advantages over direct spectroscopic absorption in terms of noise resistance and sensitivity.

As shown in Figure 13.13, in WMS, the wavelength of the laser is modulated by a slowly sweeping (sawtooth) signal at a low frequency through a specific absorption line/profile and by a sinusoidal signal at a higher frequency, f, which is superimposed onto the slowly sweeping signal. When the laser wavelength slowly scans through the absorption profile, the wavelength modulation results in an amplitude modulation of the transmitted signal. The highest amplitude of f component in the transmitted output occurs when it passes in the highest slope points of the absorption profile. This method shifts the detection bandwidth to higher frequency; therefore, it reduces $1/f$ noise and consequently increases the SNR.

WMS is sensitive to line-shape curvature rather than absorption magnitude [56], which is helpful for highly pressure broadened spectra. In addition, normalizing the WMS second-harmonic signal by the WMS fundamental harmonic signal [57] can account for the fluctuations in the laser source intensity without the need to scan across the spectral feature to nonabsorbing wings and fit a nonabsorption baseline across the feature. These benefits make WMS an attractive choice for many sensor applications.

The fiber ring laser using WMS has been demonstrated for humidity sensing. Laser source should have a linewidth significantly smaller than the absorption line of water vapor to apply this modulation technique. Assuming the linewidth is much narrower than the absorption linewidth of water vapor, and taking a Lorentzian line-shape absorption profile (under atmospheric pressure), the optical power output can be approximated as [58]

$$P = P_0 \left[1 + \eta \sin \omega t - \frac{\alpha CL}{1 + \left[\dfrac{v_o - v_g + v_m \sin \omega t}{\delta v} \right]^2} \right] \tag{13.37}$$

where

η is an intensity modulation index
v_o represents the light frequency
v_m is the amplitude of the frequency modulation
v_g and δv are center frequency and the half width of the absorption line, respectively

Equation 13.37 can be expanded into a Fourier series, and the power output of the first-harmonic (f) and second-harmonic (2f) signal can be written as [58,59]

$$P = P_0\left[1+\eta\sin\omega t-\frac{\alpha CL}{1+\left[V_m/\delta_v\right]^2\sin^2\omega t}\right] \tag{13.38}$$

$$P_f = P_0\eta \tag{13.39}$$

$$P_{2f} = -k\alpha_0 CLP_0 \tag{13.40}$$

with

$$k=\frac{2\left[2+x^2-2\left[1+x^2\right]^{1/2}\right]}{x^2\left[1+x^2\right]^{1/2}} \tag{13.41}$$

where x is equal to V_m/δ. The output of WMS can be retrieved by means of lock-in amplifiers using f signal (fundamental or first-harmonic signal) and 2f signal (second-harmonic signal) as depicted in Figure 13.14. In Figure 13.14, the f component is at zero and the 2f component is at the maximum,

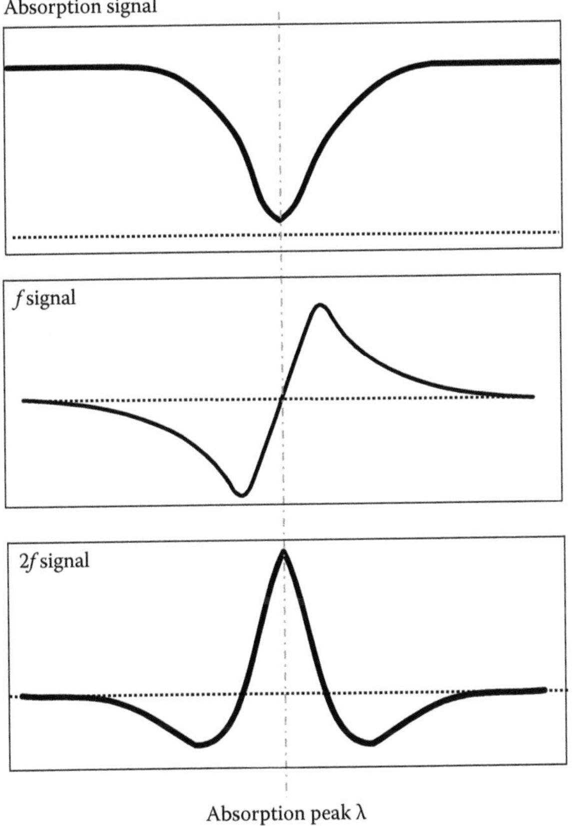

FIGURE 13.14 WMS output obtained with a lock-in amplifier.

when the source wavelength is right at the center of the absorption peak. However, the laser source is in fact both frequency and intensity modulated. Hence, we will always have *f* component offset due to the intensity modulation, even at the center of the absorption peak. By disabling the slowly sweeping modulation and locking the laser source emission wavelength at the center of the absorption profile of water vapor, the ratio 2*f* to *f* signal can be related to the absorption level only. This is because the 2*f* signal is proportional to the gas absorption level as well as to the light power, while the *f* signal (mainly from the intensity modulation) is proportional to the light power, as shown in Equations 13.39 and 13.40. The ratio is not dependent on light power, and therefore, the concentration can be determined free from the interference owing to various disturbances, such as laser power fluctuations, attenuation variations of the sensing element, and fiber [60]. In order to lock the central wavelength properly at the absorption peak, the reference cell is used. The 2*f* signal from a reference cell is used since it always has a prominent maximum at the center of the absorption line. Normally, the WMS is applied in single-pass absorption. However, the WMS can be applied in the ring laser to further improve the performance of the ring laser sensor, for example, in terms of minimum detection characteristic of the ring laser sensor.

13.4.3.2 Wavelength Sweep Technique

WST is based on scanning the laser wavelength around the absorption line of the intended gas. Depending on the center wavelength of the laser, tuning across some tens of nm is feasible. A laser can be wavelength tuned by changing its temperature and the injection current. Laser sources must have a very narrow linewidth, typically smaller than 50 MHz, which is much less than the linewidth of a single rotational gas absorption line, to use this wavelength scanning approach [61].

With the scanned wavelength scheme, a narrow linewidth laser is typically tuned over the absorption feature, and the resultant peak of spectral absorbance over the measured wavelength region is used to determine the gas absorption peak as shown in Figure 13.15. By using a broad range of scanning wavelength, a narrow linewidth of laser is tuned not only the entire absorption feature but also the nonabsorption wing of water vapor using sawtooth modulation of the current, which is injected

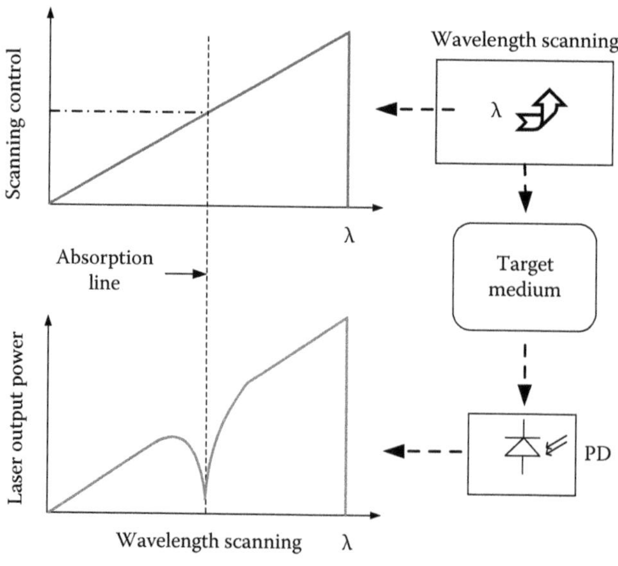

FIGURE 13.15 Wavelength scanning scheme.

into a laser. Voltage changes at the absorption peak of intended gas vary when the gas concentration changes. Normally, the WST scans more than one gas absorption lines so that the position of each absorption lines is near to each other.

13.5 FIBER LASER–BASED SENSING

FLS has been developed for many potential applications [62–65]. The most active and important part of FLS has been for acoustic sensing applications such as fiber laser hydrophones and gas sensors, which will be briefly discussed later. Moreover, FLS has also been developed for applications such as strain [66,67] and magnetic field [68]. Beverini et al. [66] reported a fiber laser-based strain sensor with a DBRFL. They used an imbalanced MZI that converts wavelength variations into phase-amplitude variations. Cranch et al. reported a fiber laser magnetometer recently [68]. In their study, a metalized fiber laser attached to a current-carrying nonmagnetic bridge measures the induced strain. Less than 2 nT/√Hz resolution at frequencies from 0.002 to 10 Hz is achieved with only 75 mA-rms of current. The very short physical gauge length allowed for fiber lasers makes possible a wide range of packaging and transduction mechanisms not available for more traditional fiber-optic sensors. For example, Foster et al. reported experimental results for a micro-engineered silicon fiber laser hydrophone [69]. The hydrophone has a flat pressure responsivity of 107 dB re Hz/Pa over a bandwidth exceeding 5 kHz, corresponding to ocean noise limited acoustic sensitivity. The first structural resonance of the hydrophone is 9 kHz in water and the acceleration rejection is in excess of 0 dB re ms^{-2}/Pa up to 5 kHz.

For its merits such as high sensitivity and compact size, optical fiber hydrophone using DFBFL [70–75] and CCFL [13,76,77] as sensing element has been intensively studied for underwater acoustic sensors in recent years. Conventional interferometric fiber sensors have been considered as the most favorable type of sensors in terms of sensitivity [78]. But they usually need to have long interaction lengths, for example, up to several hundred meters in the case of fiber hydrophones, to achieve the desirable performance. For an unbalanced fiber MZI or MI, the longer the path length, the higher the resolution. For this interferometric type of sensors, a reduction in size could drastically lower the sensitivity. Hence, the requirement on high performance and miniaturization may be hard to satisfy with the interferometric type of fiber sensors. FLSs are capable of resolving the length–sensitivity dilemma. Laser cavity resonance with high-quality factor can significantly enhance the effective interaction length and achieve very high sensitivity. The high sensitivity per unit length achieved with FLSs makes it possible to achieve a compact yet high-performance sensor head design. For smaller fiber sensor heads with reduced inertial mass, another obvious advantage is lower acceleration sensitivities. This is an additional benefit for most of fiber sensors except for an accelerometer.

13.5.1 FLS with DFBFL

Compared with conventional ceramic hydrophones, optical fiber hydrophones have some advantages such as immunity to electromagnetic interference, down-lead insensitivity to disturbances, electrically passive, small size, light weight, and relative simplicity in multiplexing designs.

Conventional fiber-optic interferometric hydrophones generally achieve a satisfactory level of performance, but the required sensitivity imposes a restriction on the minimum length of fiber and hence a lower limit on the physical volume needed to contain the fiber coil. Hence, one must take advantage of FLS by significantly increasing the sensitivity per unit length through the high-quality factor resonance possible in fiber lasers, thus achieving further miniaturization without compromising performance. Fiber lasers could be used for very compact and highly sensitive hydrophones. One of the partly packaged DFBFL hydrophone sensor heads is shown in Figure 13.16.

When used in deep sea applications, the sensitivity of the hydrophone system should be the same as the background acoustic noise level of the quiet ocean, described as deep sea state zero (DSS0),

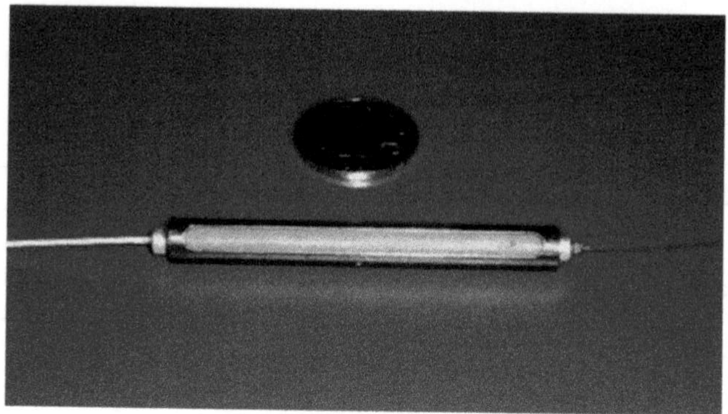

FIGURE 13.16 A hydrophone head of a 5 cm DFBFL with sensitivity enhancement and supporting case.

which is 100 μPa at 1 kHz, which is 11 orders of magnitude smaller than the hydrostatic pressure 1 km below sea surface. Hence, the development of fiber-optic hydrophone systems must consider high sensitivity, operational depth, dynamic range, applicable frequency range, and operating costs. Practical large multichannel systems deploy hydrophone in arrays and have further considerations such as the ease of repair, simplicity of manufacturing, ruggedness to survive in hostile environments, and robustness for handling during deployment and recovery.

13.5.2 Intensity Type of DFB Hydrophone

FLS using DFBFL for hydrophone with the intensity interrogation scheme has been tested. Employing the system configuration as shown in Figure 13.9, we have also tested several intensity-type fiber laser hydrophones. In this system, the sampling frequency needs to be at least double to that of the maximum acoustic frequency of interest, which is a quarter of the sampling frequency required by an interferometric system. This factor gives a big advantage from reduced hardware requirements.

High-sensitivity and high-frequency responses have been achieved. In Figure 13.17, we show the acoustic response of a DFBFL-based intensity-type hydrophone at high frequencies between 10 and 20 kHz.

Implementing the interferometric-type system, homodyne digital demodulation scheme with PGC, and improved packaging schemes, we have achieved pressure sensitivity of 58.0 dB re $\mu\text{Pa/Hz}^{1/2}$ at 1 kHz or minimum detectable acoustic pressure below 800 μPa during field test. With the intensity-type fiber laser hydrophone system, it is possible to obtain high-frequency response with benefits of reduced components, relaxed hardware requirements, simplicity, and increased robustness. This shows that the intensity-type fiber laser could be suitable to low-cost engineering and industrial applications that are not critical. Software-based demodulation also contributes in terms of development flexibility, which is not offered by the traditional analog circuit.

13.5.3 Wavelength Type of DFBFL Hydrophone

FLS systems as advanced hydrophones have been developed mostly for military and geophysical applications. We have worked on DFBFL-based hydrophone system in recent years [46].

The DFBFL-based hydrophone system is tested with an interferometric interrogation scheme as shown in Figure 13.12. From Equations 13.13 and 13.15, the interferometric system phase sensitivity is given by

$$\frac{\Delta\phi}{\Delta P} = -217 + 20\log(d) \text{ dB re rad}/\mu\text{Pa} \tag{13.42}$$

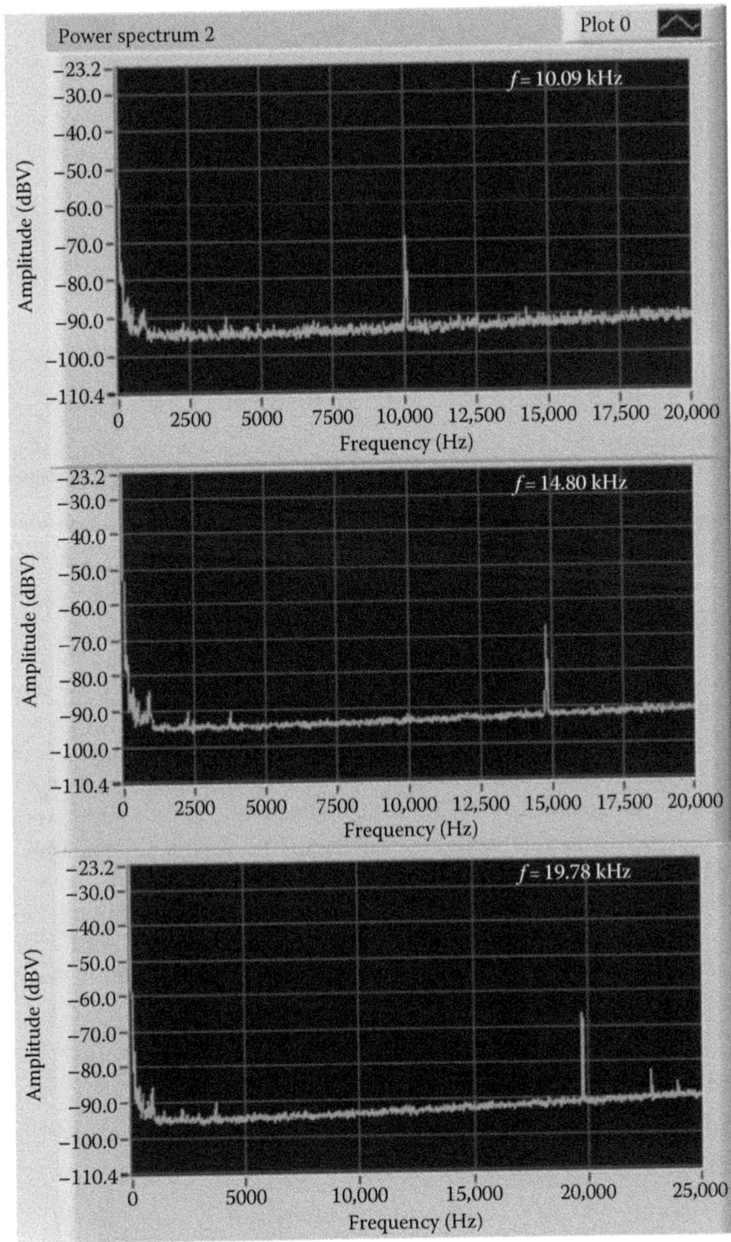

FIGURE 13.17 Acoustic response of DFBFL-based intensity-type hydrophone. (From Azmi, A.I. et al., *Photon. Sens.*, 1(3), 210, 2011.)

The passive homodyne demodulation technique that uses a PGC is employed to demodulate the phase signal. The demodulation technique using MI can detect small phase shifts in the μrad range with a linear dynamic range of approximately 107 and is immune to signal fading from environmental drifts.

Extensive laboratory experiments and field tests in lakes have been successfully carried out. The PGC demodulator discussed earlier with path imbalance was utilized to examine the hydrophone performance. The benchmark performance is the sensitivity of bare fiber that gives pressure sensitivity of −168.2 dB re rad/μPa at 1 kHz. With special DFBFL hydrophone packaging designs,

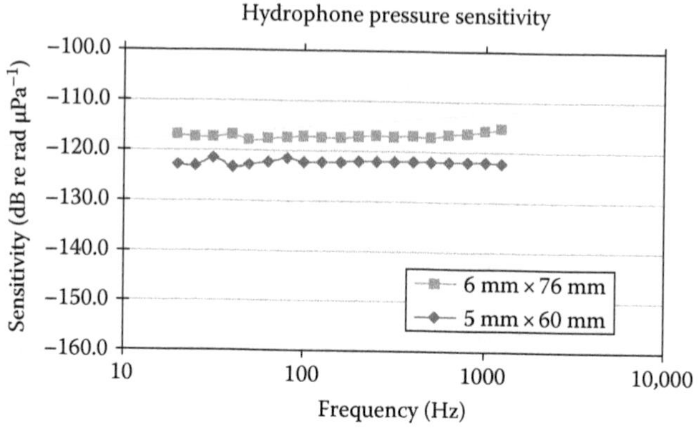

FIGURE 13.18 Pressure sensitivity results of fiber laser hydrophones with two different packaging designs. (From Azmi, A.I. et al., *Photon. Sens.*, 1(3), 210, 2011.)

we are able to vastly improve the pressure sensitivity of a bare fiber as well as flatten the frequency response. As indicated in Figure 13.18, the average pressure sensitivity from two different packaging designs are −117 and −122 dB re rad/μPa, respectively, implying enhancement of more than 40 dB compared to bare fiber.

Taking into account average pressure sensitivity and on-field phase noise level, noise equivalent pressure (NEP) of hydrophone systems has been determined. The NEP resulted from four tests at 1 kHz are 57.8, 58.8, 57.8, and 57.5 dB re μPa/√Hz, respectively. This produces a very low averaged NEP at 58.0 dB re μPa/√Hz, implying the minimum detectable acoustic pressure of 794 μPa/√Hz. The NEP of hydrophone system obtained from one of the measurements is shown in Figure 13.19. Since environment noises during field test are much higher than DSS0, actual minimum detectable pressure should be lower than what we have measured. The ideal NEP supposedly can be determined from anechoic chamber measurement.

With the interferometric-type fiber laser hydrophone system, homodyne digital demodulation with PGC, and improved packaging schemes, it is able to achieve very high sensitivity and low minimum detectable pressure. The focus is now shifted toward arrayed sensors that are more practical in real-world cases [74].

13.5.4 FLS with CCFL

The CCFL-based hydrophone system is also tested with an MZI interrogation scheme [15]. The experimental system is shown in Figure 13.20. The CCFL used in the test was pumped by a 975 nm

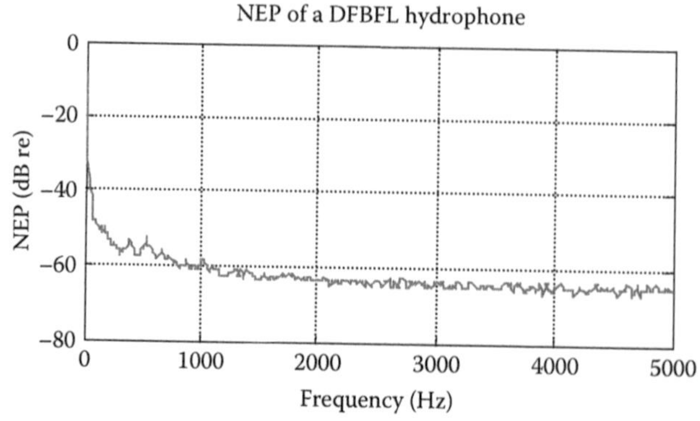

FIGURE 13.19 NEP of the system. (From Azmi, A.I. et al., *Photon. Sens.*, 1(3), 210, 2011.)

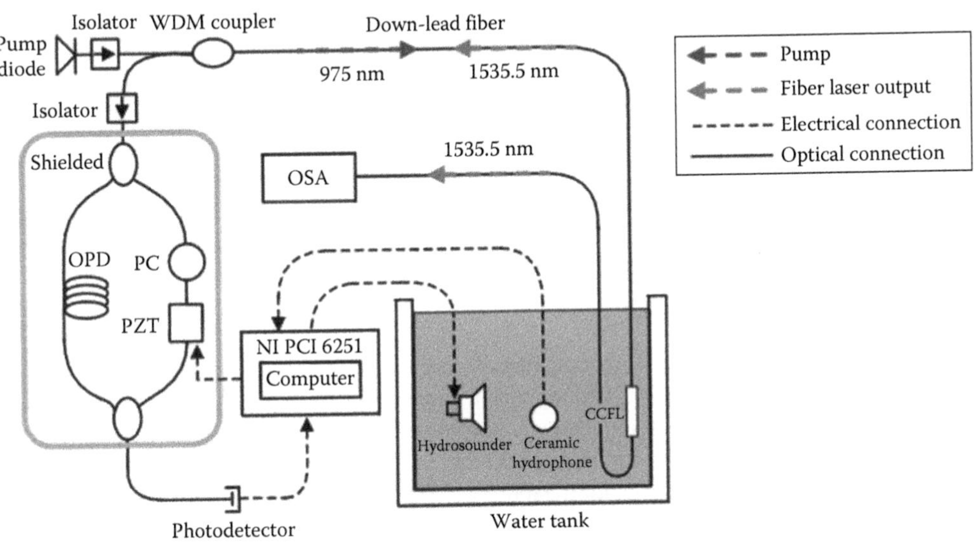

FIGURE 13.20 Experimental setup of FLS using CCFL for hydrophone.

diode laser and produced an output power of −8.9 dBm with 80 mW pump power. Figure 13.21 shows the spectral output of the 2 cm/8 cm CCFL at 37 mW pump power as recorded with the OSA.

A large optical path difference (OPD) of 150 m has been used to further enhance the phase responsivity. With such large path imbalance, the MZI was highly sensitive to external perturbation that can induce phase noise, and therefore, it was shielded and sealed in a thick polystyrene enclosure during measurement.

For partial cavity sensing for higher sensitivity, a steel plate was mounted and glued to the cavity section to reduce pressure sensitivity. When 1 kHz sine signal was insonified, the rms voltage detected by the ceramic hydrophone was 195.7 μV; hence, the signal pressure that arrived at the

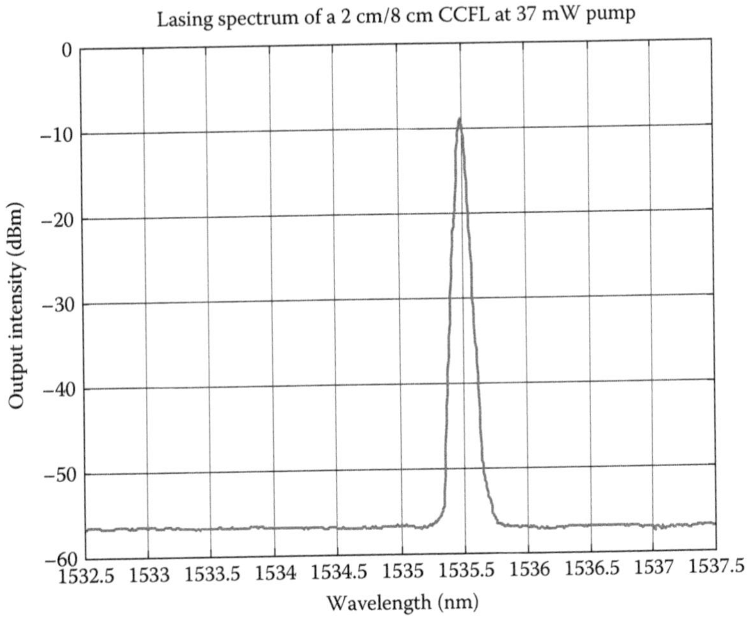

FIGURE 13.21 Lasing spectrum of the CCFL.

close proximity to the CCFL was 0.62 Pa. The amplitude of the demodulated phase excursion for uncoated CCFL was 1.52 mrad, resulting in a phase responsivity of −52.2 dB re rad/Pa. For the steel-mounted CCFL, the phase excursion was 0.12 mrad, resulting in a phase responsivity of −74.2 dB re rad/Pa. Therefore, the responsivity was reduced by 92% when the whole cavity was mounted to the steel plate, which is adequate for realizing the partial cavity format. Results at other frequencies show consistencies in isolating the cavity from the external pressure field.

Responses between 100 and 2000 Hz with spacing of 10 Hz were measured for total and partial cavity sensing schemes. Demodulated signals of both sensing schemes with 1000 Hz signal are shown in Figure 13.22. The resulting phase excursions were 1.52 and 3.05 mrad for total and partial cavity scheme for a 0.62 Pa test signal. Hence, the resulting phase responsivities were −52.2 and −46.2 dB re rad/Pa, respectively.

Responsivity for the entire frequency range was measured using similar approach, and the obtained results are presented in Figure 13.23. Although the responsivities for both schemes were found varied nearly ±10 dB from the average responsivity, a similar trend of responsivity can be observed from these two schemes, which demonstrated some correlations. The resulting calculation yielded an average responsivity of −55.2 dB re rad/Pa for the total cavity and −49.1 dB re rad/Pa for the partial cavity. This implies that the enhancement of average phase responsivity of 6 dB was achieved when partial cavity sensing was utilized. The measured phase responsivity of the total cavity was also close to the theoretical value for bare fiber of −53.8 dB re rad/Pa for 150 m path imbalance.

Operation under prestatic loading is also tested. The CCFL may occur at an arbitrary initial operation point, and therefore, the optimum pressure sensitivity may not be optimized. To change the initial phase-match spectral location and consequently change the wavelength response, a prestatic loading to the primary cavity was experimented. The CCFL phase responsivity for each added load was measured from 100 Hz to 2 kHz with spacing of 10 Hz and quantified by the average responsivity.

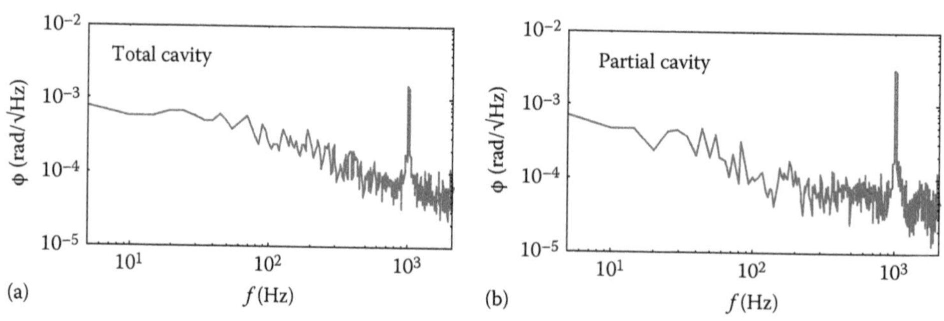

FIGURE 13.22 Detected phase spectrum under a 1000 Hz signal under (a) total cavity sensing and (b) partial cavity sensing. (From Azmi, A.I. et al., *J. Lightwave Technol.*, 28(12), 1844, 2010.)

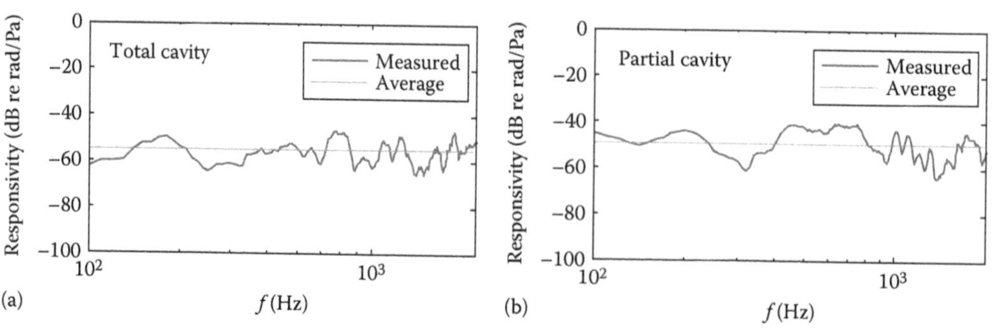

FIGURE 13.23 Measured responsivity for (a) total cavity sensing and (b) 3 cm partial cavity sensing. (From Azmi, A.I. et al., *J. Lightwave Technol.*, 28(12), 1844, 2010.)

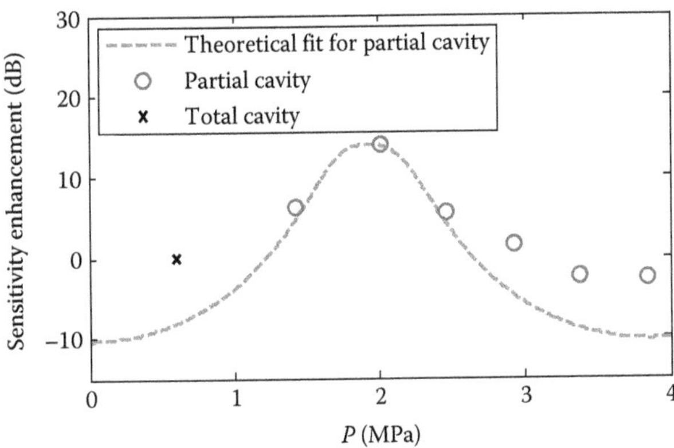

FIGURE 13.24 The relative sensitivity enhancement under loading for partial cavity sensing with regard to total cavity sensing (reference). The dotted line represents the predicted enhancement for one cycle of ϕ_c with $r_c g_3 = 0.86$. (From Azmi, A.I. et al., *J. Lightwave Technol.*, 28(12), 1844, 2010.)

Figure 13.24 is the measured average responsivity of the partial cavity normalized by that of the total cavity. The highest average responsivity obtained is −41.1 dB re rad/Pa, showing an increase of 14 dB from the total cavity sensing, as seen from Figure 13.24. This response trend of the prestrained CCFL agrees to the theoretical expectation and confirms that the CCFL sensitivity enhancement scheme can be implemented in conjunction with the conventional packaging technique.

13.5.5 FLS with RFL

One area of great potential for FLS applications is chemical and gas sensing. Typically, gases such as methane, acetylene, CO_2, and O_2 are often to be monitored where they play an important role in industry and living environment.

Various RFL-based sensing systems have been reported for gas, chemical, and humidity detection in recent years [79–86]. The implementation of intracavity spectroscopy would allow a detection system with very high sensitivity of intracavity detection and the advantages of fiber-optic sensors including remote detection and multiplexing capability [61]. FLS systems have great potential for high-sensitivity gas and chemical detection. Fiber lasers could have very narrow linewidth, low phase noise, and low RIN, and they could readily achieve high spectral resolution detection needed for gas and chemical sensing or monitoring [83].

13.5.6 TFS with RFL: Gas Sensing

A fiber ring laser sensor [61,83] for intracavity absorption spectroscopic detection of gases or chemicals is shown in Figure 13.25. The system contains an erbium-doped fiber and 980 nm optical pump source, a Fabry–Perot-type tunable optical filter (TOF), optical circulator, optical isolator, and a photodetector. The EDFA is pumped by a 980 nm pump laser via a 980/1550 wavelength-division multiplexing coupler. The bandwidth and the free-spectral range of the TOF are 1.25 GHz and 11.23 THz, respectively. The introduction of FBG or FBG array as the wavelength reference allows the system for self-calibration. The gas cell is made with collimated fiber connector using SELFOC lens, as shown in Figure 13.26. Multichannel gas detection is realized in this system using an optical switch. With the gas cells placed within the closed path of a ring cavity via an optical circulator, a large number of passes through the cell can make the effective interaction length multiple of actual of physical length of a short gas cell and thus achieve high sensitivity with small sensor size.

By using the WST discussed earlier and without aligning the passband of the TOF to a particular absorption line center of the gas, multiple absorption lines with different absorption cross

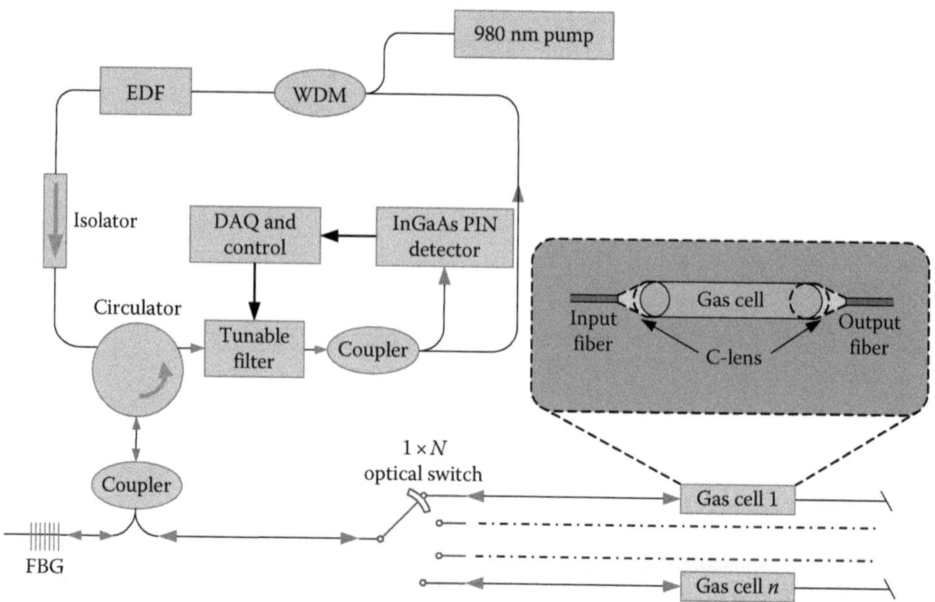

FIGURE 13.25 Schematic diagram of an FLS system with RFL. (From Peng, G.D. et al., Development of fibre laser based gas and chemical sensors, invited paper, *Technical Digest of the 13th International Meeting on Chemical Sensors (IMCS-13)*, Perth, Australia, July 11–14, 2010, vol. 40.)

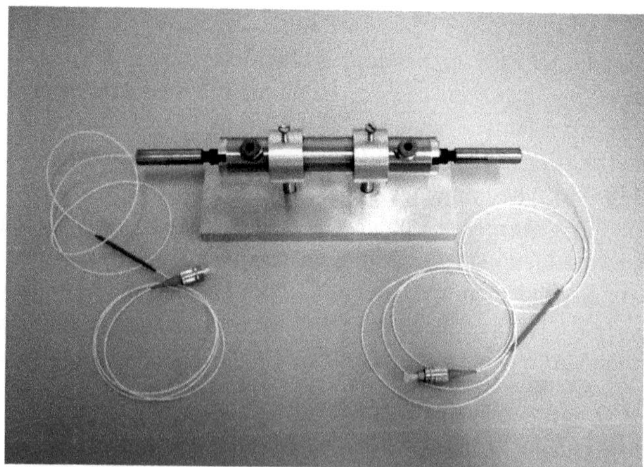

FIGURE 13.26 The sensor cell is made of a pair of fiber pigtailed C-lens (SELFOC lens collimators).

sections can be obtained as shown in Figure 13.27, where the driver voltage of the TOF is swept to detect the output signal of the detector. The original signal illustrates the overall absorption spectra while the peaks in the second derivative curve can be used to identify the absorption lines automatically.

With the WST, the characteristic of the intracavity fiber ring laser sensor can be described according to known gas absorption spectra. Normally, the absorption line with the largest attenuation enhancement factor is to be selected if it has the appropriate absorption cross section so it operates close to the lasing threshold of the laser. Then the absorption spectrum data are logged and processed with the Lorentzian profile fitting, as shown in Figure 13.28. The concentration of the gas can be determined by digitally processing the scanned spectral data.

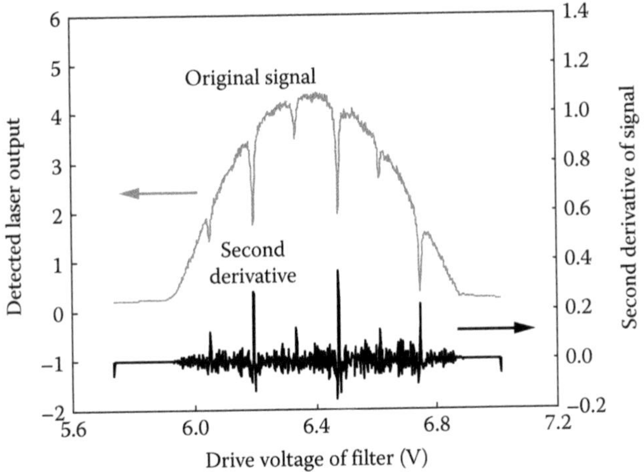

FIGURE 13.27 Typical RFL output signal of across an absorption line. (From Peng, G.D. et al., Development of fibre laser based gas and chemical sensors, invited paper, *Technical Digest of the 13th International Meeting on Chemical Sensors (IMCS-13)*, Perth, Australia, July 11–14, 2010, Technical Digest, p. 40, 2010.)

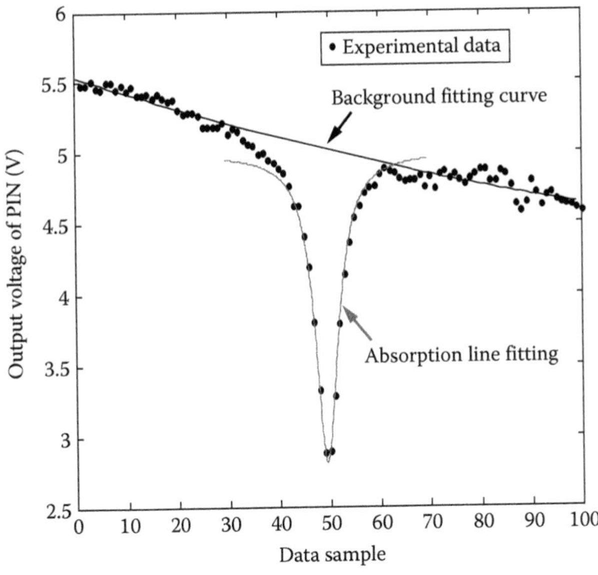

FIGURE 13.28 Experimental data are processed using the Lorentzian profile fitting. (From Peng, G.D. et al., Development of fibre laser based gas and chemical sensors, invited paper, *Technical Digest of the 13th International Meeting on Chemical Sensors (IMCS-13)*, Perth, Australia, July 11–14, 2010, Technical Digest, p. 40, 2010.)

Both the wavelength modulation technique (WMT) and WST can be useful for TLS using RFL for gas and chemical sensing [84]. The principles of using WMT and WST for FLS systems have been introduced earlier in this chapter. When both WMT and WST are applied to FLS using RFL, both gas concentration sensing and absorption wavelength detection are possible. The second-harmonic spectra of gas absorption lines and FBG array can be obtained in one scanning period. FLS using RFL and combining both WMT and WST have been demonstrated to be effective in achieving good system performance and also realizing both concentration sensing and gas-type recognition. In their experiments, the concentration calibration has been realized by using the amplitude of the second-harmonic spectra and applying Lorentzian profile fitting. Take the absorption line at 1530.371 nm

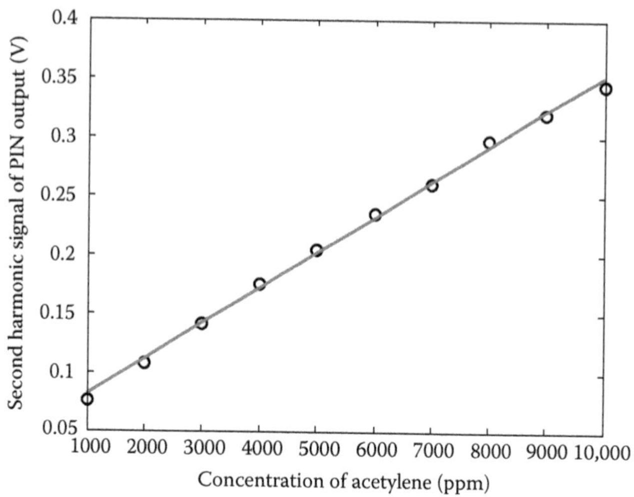

FIGURE 13.29 Acetylene concentration detection. (From Peng, G.D. et al., Development of fibre laser based gas and chemical sensors, invited paper, *Technical Digest of the 13th International Meeting on Chemical Sensors (IMCS-13)*, Perth, Australia, July 11–14, 2010, Technical Digest, p. 40, 2010.)

as an example; the acetylene concentration in the gas cell was varied and tested between 1000 and 10,000 ppm with an interval of 1000 ppm. The different concentrations are obtained by mixing acetylene and nitrogen. From the experimental system, very good linear relationship between the amplitude of the second-harmonic spectra and acetylene concentration has been obtained and shown in Figure 13.29. The amplitude of the gas second-harmonic spectra can be used for gas calibration and concentration detection. The absolute detected error of the system is less than 75 ppm for acetylene sensing when using WST and average algorithms. The absolute error of the measured results is mainly lower than 0.445 nm, which is one order of magnitude smaller than the absorption wavelength ranges space of different gases. So this method can also be used to realize gas-type recognition. This system is available for other gases, whose absorption wavelengths fall in the broad gain bandwidth of EDFA. When the wavelength tuning range of erbium-doped FLS is extended to L-band, gases such as CO, CO_2, and H_2S can be detected. The peak absorbance of absorption line is used for concentration calibration because it increases linearly with the gas concentration, as shown in Figure 13.30. The measurement errors corresponding to gas concentration and absorption wavelengths of CO_2 are less than 400 ppm and 30 pm, respectively [85]. If the rare-earth ions doped in the gain fiber are changed, the wavelength lasing range can be extended further and more kinds of gases can be detected.

13.5.7 TFS with RFL: Humidity Sensing

FLS with RFL for humidity sensor has been reported recently [87]. This FLS with RFL for humidity sensing, as shown in Figure 13.31, is in principle similar to the previous techniques—by means of direct intercavity infrared absorption spectroscopy. However, due to quite different operation wavelengths, both the wavelength scanning technique, which is achieved by an intercavity TOF, and the gain medium, which is an EDFA, cannot be used here for humidity sensing. Here, WMS is achieved by injecting a seeding optical signal from an electrically modulated DFB laser diode. In addition, the optical gain for the RFL is provided by a semiconductor optical amplifier (SOA). In addition, a section of 5 cm PCF, not a traditional SELFOC cell within the fiber ring laser cavity, is used where a small gap between one end of PCF and SMF is used as air diffusion hole to allow air in and out of the air holes inside the PCF.

In the experimental demonstration, the wavelength scanning output from a DFB laser is fed into the fiber ring laser that its output power is sensitive to the humidity absorption. As can be seen at the

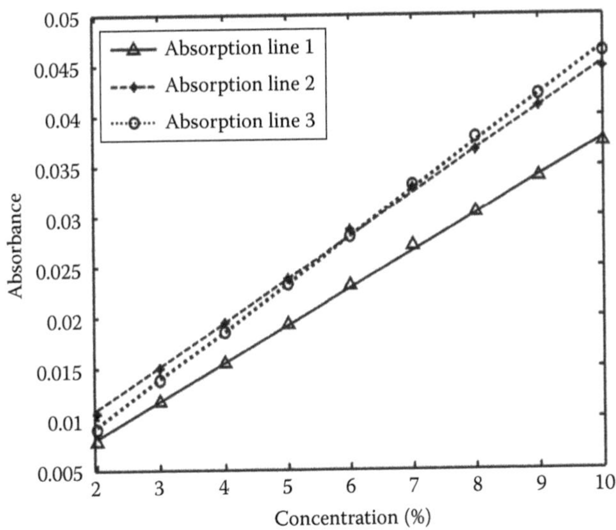

FIGURE 13.30 Carbon dioxide concentration detection. (From Yu, L. et al., *Sens. Actuat. B Chem.*, 193, 356, 2014.)

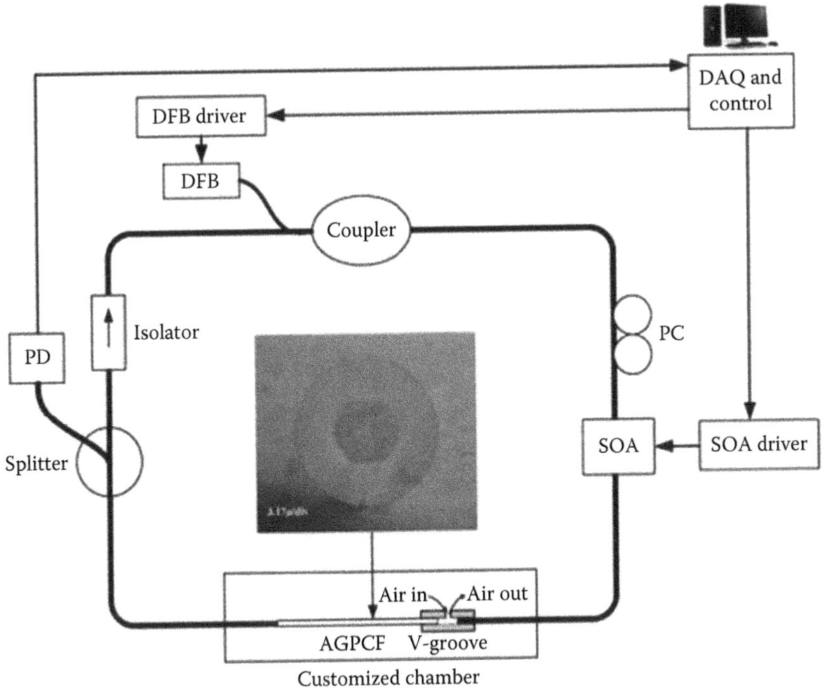

FIGURE 13.31 Experimental setup of the ring laser–based sensor. (From Noor, M.Y.M. et al., *Meas. Sci. Technol.*, 24(8), 085203, 2013.)

ring output in Figure 13.32, the sensitivity of the sensor is estimated around 10.93/1% RH over the range from 0% to 90% RH with good stability.

The fiber laser system performance in terms of the sensitivity enhancement factor, the laser output power, and the SNR to link them to the EDF parameters and operation conditions have also been systematically investigated [83], based on theoretical modeling and analysis established on

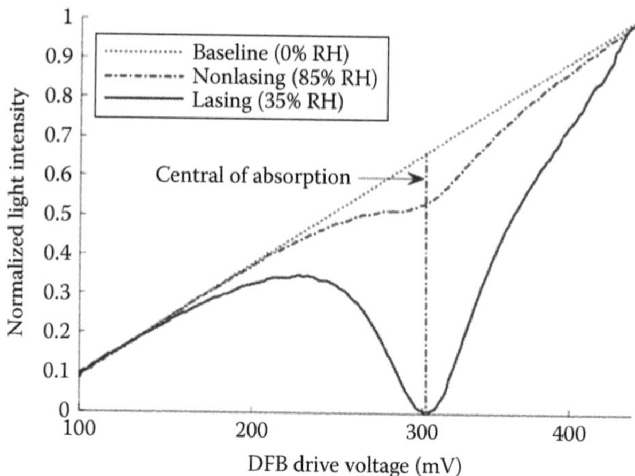

FIGURE 13.32 Comparison of absorption signals in one scanning period for nonlasing mode at 85% RH and lasing mode at 35% RH. Baseline is used as a reference for 0% RH.

coupled rate equations and propagation equations. The relationship between optimal parameters and desirable performances has been established and sensitivity enhancement factors of 50–65 were achieved [83]. This indicates that FLS using RFL has the ability to realize high-performance trace gas detection. When the new gain fibers become available, by combining with WST and WMS [83] and by optimizing system design and operation conditions for best sensitivity enhancement based on theoretical modeling and analysis [86], it is possible for FLS system using RFL to achieve good system performance for low concentration and large dynamic range gas detection. Moreover, the recent research and development on new gain fibers, bismuth-doped fibers for 1250 nm–1450 nm wavelengths [88] and, especially, bismuth/erbium co-doped fibers for 1250 nm–1600 nm wavelengths [89–91], could open up opportunities for new fiber lasers that offer operation over very broad wavelength range that is of great interest for FLS systems covering a wide range of gas, chemical, and humidity applications in the future.

13.6 CONCLUSIONS

Because of their very narrow linewidth, long effective length, and very high enhancement of sensitivity offered by cavity resonance, it is possible to measure extremely small changes in absorption, cavity length, or effective refractive index with FLSs. Hence, high-performance and compact FLSs have many beneficial attributes for a wide range of applications including acoustic, hydraulic, chemical, and biological sensors.

While the FLSs have been demonstrated to provide miniature design and high performance desirable for many applications, there remain many challenges, and there is a need for further intense research and development in converting them into real-world practical systems. We hope that this chapter will provide a useful reference and that it will encourage further research and development of novel technologies for high-performance laser-based sensors.

REFERENCES

1. E. Snitzer, Optical maser action of Nd^{3+} in a barium crown glass, *Phys. Rev. Lett.* 7, 444–446, 1961.
2. C. J. Koester and E. Snitzer, Amplification in a fiber laser, *Appl. Opt.* 3, 1182–1186, 1964.
3. S. B. Poole, D. N. Payne, and M. E. Fermann, Fabrication of low-loss optical fibres containing rare-earth ions, *Electron. Lett.* 17, 730–732, 1985.
4. A. D. Keresy, M. A. Davis, H. J. Patrick, M. LeBlanc, K. P. Koo, C. G. Askins, and M. A. E. J. Friebele, Fiber grating sensors, *J. Lightwave Technol.* 15, 1442–1463, 1997.

5. K. A. Murphy, M. A. Gunther, A. M. Vengsarkar, and R. O. Claus, Quadrature phase shifted extrinsic Fabry-Perot optical fiber sensors, *Opt. Lett.* 16, 273–275, 1991.
6. D. Culverhouse, F. Farahi, C. N. Pannell, and D. A. Jackson, Potential of stimulated Brillouin scattering as sensing mechanism for distributed temperature sensors, *Electron. Lett.* 25, 913–915, 1989.
7. P. Childs, T. Whitbread, and G. D. Peng, Spectrally coded multiplexing in strain sensor system based on carrier-modulated fibre Bragg gratings, in *Photonics Asia Conference on Information Optics and Photonics Technology*, Beijing, China, Nov. 8–12, 2004, *Proc. SPIE* 5634, 204–210, 2005.
8. A. Galvanauskas, High power fiber lasers, *Opt. Photon. News*, 15(7), 42–47, 2004.
9. A. Heidt, Z. Li, J. Sahu, P. Shardlow, M. Becker, M. Rothhardt, M. Ibsen, R. Phelan, B. Kelly, S. Alam, and D. Richardson, 100 kW peak power picosecond thulium-doped fiber amplifier system seeded by a gain-switched diode laser at 2 μm, *Opt. Lett.* 38(10), 1615–1617, 2013.
10. Y. Zhang, M. Zhang, and W. Jin, Sensitivity enhancement in erbium-doped fiber laser intra-cavity absorption sensor, *Sensor. Actuators A: Phys.*, 104, 183–187, Apr. 2003.
11. G. A. Cranch, G. A. Miller, and C. K. Kirkendall, Fiber laser sensors: Enabling the next generation of miniaturized, wideband marine sensors, *Proc. SPIE* 8028, 80280I, 2011.
12. S. Foster, A. Tikhomirov, M. Milnes, J. van Velzen, and G. Hardy, A fiber laser hydrophone, *17th International Conference on Optical Fibre Sensors*, *Proc. SPIE* 5855, 627, 2005. DOI:10.1117/12.623273.
13. I. Leung, Development of composite cavity fibre lasers for fibre laser hydrophone systems, Doctoral thesis, The University of New South Wales, Sydney, New South Wales, Australia, 2008.
14. L. Ma, Y. Hu, H. Luo, and Z. Hu, DFB fiber laser hydrophone with flat frequency response and enhanced acoustic pressure sensitivity, *IEEE Photon. Technol. Lett.* 20(7), 1280–1282, 2009.
15. A. I. Azmi, D. Sen, and G. Peng, Sensitivity enhancement in composite cavity fiber laser hydrophone, *J. Lightwave Technol.* 28(12), 1844–1890, 2010.
16. A. Wong, D. Chen, H. Wang, W. Chung, H. Tam, C. Lu, and B. Guan, Extremely short distributed Bragg reflector fibre lasers with sub-kilohertz linewidth and ultra-low polarization beat frequency for sensing applications, *Measure. Sci. Technol.* 22, 045202, 2011.
17. J. Kringlebotn, J. Archambault, L. Reekie, and D.Payne, Er^{3+}:Yb^{3+}-codoped fiber distributed-feedback laser, *Opt. Lett.* 19(24), 2101–2103, 1994.
18. L. Wang, B. Chen, J. Chen, G. Li, L. Chang, A. Sun, and Z. Lin, Single mode Er-doped π-phase-shifted distributed feedback fibre grating laser, *Chin. Phys. Lett.* 24(3), 721–723, 2007.
19. M. Sejka, P. Varming, J. Hubner, and M. Kristensen, Distributed feedback Er^{3+}-doped fibre laser, *Electron. Lett.* 31(17), 1445–1446, 1995.
20. W. Loh, M. Cole, M. Zervas, S. Barcelos, and R. Laming, Complex grating structures with uniform phase masks based on the moving fiber-scanning beam technique, *Opt. Lett.* 20(20), 2051–2053, 1995.
21. W. H. Loh and R. I. Laming, 1.55 μm phase-shifted distributed feedback fibre laser, *Electron. Lett.* 31(17), 1440–1442, 1995.
22. E. Ronnekleiv, Frequency and intensity noise of single frequency fiber Bragg grating lasers, *Opt. Fiber Technol.* 7(3), 206–235, 2001.
23. G. A. Cranch and P. J. Nash, Optical fiber hydrophones, in *Handbook of Laser Technology and Applications*. Bristol, U.K.: IOP Publishing Ltd., 2003, sec. D2.5.
24. P. Varming, V. C. Lauridsen, J. H. Povlsen, J. B. Jensen, and M. Kristensen, Design and fabrication of Bragg grating based DFB fiber lasers operating above 1610 nm, in *Proceedings of the Optical Fiber Communication Conference*, Baltimore, MD, 2000, vol. 3, pp. 17–19.
25. S. Ogita, Y. Kotaki, K. Kihara, M. Matsuda, H. Ishikawa, and H. Imai, Dependence of spectral linewidth on cavity length and coupling coefficient in DFB laser, *Electron. Lett.* 24(10), 613–614, 1988.
26. E. Rønnekleiv, O. Hadeler, and G. Vienne, Stability of an Er-Yb-doped fiber distributed-feedback laser with external reflections, *Opt. Lett.* 24(9), 617–619, 1999.
27. D. Yu. Stepanov, J. Canning, L. Poladian, R. Wyatt, G. Maxwell, R. Smith, and R. Kashyap, Apodized distributed-feedback fiber laser, *Opt. Fiber Technol.* 5(2), 209–214, 1999.
28. A. I. Azmi, D. Sen, and G. D. Peng, Output power and threshold gain of apodized DFB fiber laser, *Proc. SPIE* 7386, 73860K.1–73860K.11, 2009.
29. A. I. Azmi and G. D. Peng, Performance analysis of apodized DFB fiber laser, in *Proceedings of Photonics Global Conference*, Singapore, 2008, vol. 1, pp. 1–4.
30. Y. Dai, X. Chen, J. Sun, Y. Yao, and S. Xie, Dual-wavelength DFB fiber laser based on a chirped structure and the equivalent phase shift method, *IEEE Photon. Technol. Lett.* 18(18), 1964–1966, 2006.
31. X. Liu, A novel dual-wavelength DFB fiber laser based on symmetrical FBG structure, *IEEE Photon. Technol. Lett.*, 19(9), 632–634, 2007.

32. A. Tikhomirov and S. Foster, DFBFL sensor cross-coupling reduction, *IEEE J. Lightwave Technol.* 25(2), 533–538, 2007.
33. H. Kogelnik and C. V. Shank, Coupled-wave theory of distributed feedback lasers, *J. Appl. Phys.* 43(5), 2327–2335, May 1972.
34. A. Yariv, *Optical Electronics in Modern Communications*, 5th edn. New York: Oxford University Press Inc., 1997, pp. 619–626.
35. N. Schunk and K. Petermann, Numerical analysis of the feedback regimes for a single-mode semiconductor laser with external feedback, *IEEE J. Quant. Electron.* 24(7), 1242–1247, Jul. 1988.
36. S. Saito and Y. Yamamoto, Direct observation of Lorentzian lineshape of semiconductor laser and linewidth reduction with external grating feedback, *Electron. Lett.* 17(9), 325–327, Apr. 1981.
37. L. Goldberg, A. Dandridge, R. O. Miles, T. G. Giallorenzi, and J. F. Weller, Noise characteristics in line-narrowed semiconductor lasers with optical feedback, *Electron. Lett.* 17(19), 677–678, Sept. 1981.
38. R. Wyatt and W. J. Devlin, 10 kHz linewidth 1.5 μm InGaAsP external cavity laser with 55 nm tuning range, *Electron. Lett.* 19(3), 110–112, Feb. 1983.
39. D. Mehuys, M. Mittelstein, and A. Yariv, Optimised Fabry-Perot (AlGa)As quantum-well lasers tunable over 105 nm, *Electron. Lett.* 25(2), 143–145, Jan. 1989.
40. A. Olsson and C. Tang, Coherent optical interference effects in external-cavity semiconductor lasers, *IEEE J. Quantum Electron.* 17(8), 1320–1323, Aug. 1981.
41. J. Osmundsen and N. Gade, Influence of optical feedback on laser frequency spectrum and threshold conditions, *IEEE J. Quant. Electron.* 19(3), 465–469, Mar. 1983.
42. NP Photonics Inc. Datasheet: Rock module compact single-frequency fiber laser, 2013. URL: http://www.npphotonics.com/images/pdfs/products/9_NPP_RockMod.pdf.
43. NKT Photonics A/S Inc. Datasheet: Koheras Basik E15-Ultra low noise, single frequency fiber laser modules in the 1.5μm range, 2014. URL: http://www.nktphotonics.com/files/files/Koheras_BasiK-E15.pdf.
44. G. Cranch, M. Englund, and C. Kirkendall, Intensity noise characteristics of Erbium-doped distributed-feedback fiber lasers, *IEEE J. Quant. Electron.* 39(12), 1579–1586, 2003.
45. E. Rønnekleiv, M. Ibsen, and G. J. Cowle, Polarization characteristics of fiber DFB lasers related to sensing applications, *IEEE J. Quant. Electron.* 36, 656, 2000.
46. A. I. Azmi, I. Leung, X. Chen, S. Zhou, Q. Zhu, K. Gao, P. Childs, and G. Peng, Fiber laser based hydrophone systems, *Photon. Sens.* 1(3), 210–221, 2011.
47. D. A. Jackson, R. Priest, A. Dandridge, and A. B. Tveten, Elimination of drift in a single-node optical fiber interferometer using a piezoelectrically stretched coiled fiber, *Appl. Opt.* 19(17), 2926–2929, Sept. 1980.
48. A. Dandridge, A. Tveten, and T. Giallorenzi, Homodyne demodulation scheme for fiber optic sensors using phase generated carrier, *IEEE J. Quant. Electron.* 18(10), 1647–1653, Oct. 1982.
49. J. Bush, A. Cekorich, and C. K. Kirkendall, Multichannel interferometric demodulator, *Proc. SPIE* 3180, 19–29, May 1997.
50. M. Milnes, A. Tikhomirov, S. Foster, and S. Goodman, Fast four step digital demodulation for multiplexed fibre laser sensors, *Proc. SPIE* 7004, 700422.1–700422.5, Apr. 2008.
51. D. A. Jackson, A. D. Kersey, M. Corke, and J. D. C. Jones, Pseudoheterodyne detection scheme for optical interferometers, *Electron. Lett.* 18(25), 1081–1083, Dec. 1982.
52. J. Cole, B. Danver, and J. Bucaro, Synthetic-heterodyne interferometric demodulation, *IEEE J. Quantum Electron.* 18(4), 694–697, Apr. 1982.
53. D. A. Brown, C. B. Cameron, R. M. Keolian, D. L. Gardner, and S. L. Garrett, A symmetric 3 × 3 coupler based demodulator for fiber optic interferometric sensors, *Proc. SPIE*, 1584, 328–335, Sept. 1991.
54. M. D. Todd, M. Seaver, and F. Bucholtz, Improved, operationally-passive interferometric demodulation method using 3 × 3 coupler, *Electron. Lett.* 38(15), 784–786, Jul. 2002.
55. K. P. Koo, A. B. Tveten, and A. Dandridge, Passive stabilization scheme for fiber interferometers using 3 × 3 fiber directional couplers, *Appl. Phys. Lett.* 41(7), 616–618, Oct. 1982.
56. J. T. C. Liu, J. B. Jeffries, and R. K. Hanson, Wavelength modulation absorption spectroscopy with 2f detection using multiplexed diode lasers for rapid temperature measurements in gaseous flows, *Appl. Phys. B* 78(3–4), 503–511, 2004.
57. G. B. Rieker, H. Li, X. Liu, and J. B. Jeffries, A diode laser sensor for rapid, sensitive measurements of gas temperature and water vapour concentration at high temperatures and pressures, *Measure. Sci. Technol.* 18(5), 1195–1204, 2007.
58. L. Jin, Y. Yan, and Y. Haima, A novel optical fiber gas sensor for on-line measurement of carbon monoxide, *Proceedings of International Workshop on Information Security and Application*, Qingdao, China, 2009.

59. T. Iseki, H. Tai, and K. Kimura, A portable remote methane sensor using a tunable diode laser, *Measure. Sci. Technol.* 11(6), 594, 2000.
60. H. Tai, K. Yamamoto, and M. Uchida, Long-distance simultaneous detection of methane and acetylene by using diode lasers coupled with optical fibers, *IEEE Photon. Technol. Lett.* 4(7), 804–807, 1992.
61. K. Liu, W. C. Jing, G. D. Peng, J. Z. Zhang, Y. Wang, T. G. Liu, D. Jia, H. Zhang, and Y. Zhang, Wavelength sweep of intra-cavity fiber laser for low concentration gas detection, *IEEE Photon. Technol. Lett.* 20(18), 1515–1517, 2008.
62. G. D. Peng, I. Leung, X. B. Chen, S. L. Zhou, Q. Zhu, P. Childs, and K. Gao, Applications of fiber laser based sensors, in *the 14th Opto-Electronics and Communications Conference, OECC'09*, Hong Kong, July 13–17, 2009, Paper MWB5.
63. A. I. Azmi, I. Leung, P. Childs, and G. D. Peng, Composite cavity fiber laser for sensor applications, *Sens. Transducers J.* 116(5), 1–12, May 2010.
64. Y. Liu, W. Zhang, T. Xu, J. He, F. Zhang, and F. Li, Fiber laser sensing system and its applications, *Photon. Sens.* 1(1), 43–53, 2011.
65. G. A. Miller, G. A. Cranch, and C. K. Kirkendall, High performance sensing using fiber lasers, *Opt. Photon. News* 23(2), 30–36, 2012.
66. N. Beverini, E. Maccioni, M. Morganti, F. Stefani, R. Falciai, and C. Trono, Fiber laser strain sensor device, *J. Opt. A: Pure Appl. Opt.* 9, 958–962, 2007.
67. I. Leung, A. C. L. Wong, and G. D. Peng, Strain related characteristics of composite cavity fibre lasers, *Proc. SPIE* 6830, 68301W, 2007.
68. G. A. Cranch, G. A. Miller, C. G. Askins, R. E. Bartolo, and C. K. Kirkendall, Remotely-interrogated three-axis fiber laser magnetometer, *IEEE Sens. J.* 12, 890–899, 2012.
69. S. Foster, G. Cranch, G. Miller, J. Harrison, and A. Tikhomirov, Ultra-low frequency noise DBR fibre laser for sensing strain in the femto-ε regime, *Proceedings of the 37th Australian Conference on Optical Fibre Technology*, Sydney, Australia, Dec. 2012.
70. D. J. Hill, P. J. Nash, D. A. Jackson, D. J. Webb, S. F. O'Neill, I. Bennion, and L. Zhang, A fiber laser hydrophone array, in *SPIE Conference on Fiber Optic Sensor Technology and Applications*, Boston, MA, Sept. 1999, vol. 3860, pp. 55–66.
71. A. Frank, K. Bohnert, K. Haroud, H. Brändle, C. V. Poulsen, J. E. Pedersen, and J. Patscheider, Distributed feedback fiber laser sensor for hydrostatic pressure, *IEEE Photon. Technol. Lett.* 15(12), 1758–1760, Dec. 2003.
72. I. Leung, Z. Brodzeli, T. Whitbread, X. B. Chen, and G. D. Peng, A distributed-feedback fibre laser based optical fibre hydrophone system with very high-sensitivity, in *Photonics Asia Conference on Information Optics and Photonics Technology*, Beijing, China, Nov. 8–12, 2004, *Proc. SPIE*, 5634, 434–443, 2005.
73. B. O. Guan, H. Y. Tam, S. T. Lau, and H. L. W. Chan, Ultrasonic hydrophone based on distributed Bragg reflector fiber laser, *IEEE Photon. Technol. Lett.* 17(1), 169–171, Jan. 2005.
74. S. Foster, A. Tikhomirov, and J. van Velzen, Towards a high performance fiber laser hydrophone, *J. Lightwave Technol.* 29, 1335–1342, 2011.
75. S. Foster, A. Tikhomirov, J. Harrison, and J. van Velzen, A fibre laser sensor seabed array, *Proceedings of Acoustics*, Victor Harbor, Australia, 2013.
76. I. Leung and G. D. Peng, Composite cavity fibre laser with internal active feedback, *Eng. Trans.* 10(2), 23, 2007.
77. J. Z. Zhang, Q. Chai, Q. Q. Hao, Y. B. Ge, X. L. Li, Q. Li, W. M. Sun, L. B. Yuan, and G. D. Peng, Composite cavity fiber laser sensors based on weak feedback, *Appl. Opt.* 50, 5059–5063, Sept. 2011.
78. C. K. Kirkendall and A. Dandridge, Overview of high performance fibre-optic sensing, *J. Phys. D: Appl. Phys.* 37(18), R197–R216, Sept. 2004.
79. G. Stewart, K. Antherton, H. Yu, and B. Culshaw, An investigation of an optical fibre amplifier loop for intra-cavity and ring-down cavity loss measurements, *Meas. Sci. Technol.* 12, 843–849, 2001.
80. Y. Zhang, W. Jin, H. B. Yu, M. Zhang, Y. B Liao, H. L. Ho, M. S. Demokan, G. Stewart, B. Culshaw, and Y.-H. Li, Novel intracavity sensing network based on mode-locked fiber laser, *IEEE Photonics Technol. Lett.* 14(9), 1336–1338, Sept. 2002.
81. M. Zhang, D. N. Wang, W. Jin, and Y. B. Liao, Wavelength modulation technique for intra-cavity absorption gas sensor, *IEEE Trans. Instrum. Meas.* 53(1), 136–139, Feb. 2004.
82. H. Y. Ryu, W. Lee, H. S. Moon, and H. S. Suh, Tunable erbium-doped fiber ring laser for applications of infrared absorption spectroscopy, *Opt. Commun.* 275, 379–384, Jul. 2007.
83. G. D. Peng, K. Liu, M. Li, and W. C. Jing, Development of fibre laser based gas and chemical sensors, invited paper, in *Technical Digest of the 13th International Meeting on Chemical Sensors (IMCS-13)*, Perth, Australia, Jul. 11–14, 2010, Technical Digest, p. 40, 2010.

84. K. Liu, T. Liu, J. Jiang, G. D. Peng, H. Zhang, D. Jia, Y. Wang, W. Jing, and Y. Zhang, Investigation of wavelength modulation and wavelength sweep techniques in intracavity fiber laser for gas detection, *IEEE/OSA J. Lightwave Technol.* 29(1), 15–21, 2011.
85. L. Yu, T. G. Liu, K. Liu, J. F. Jiang, L. Zhang, Y. W. Jia, and T. Wang, Development of an intra-cavity gas detection system based on L-band erbium-doped fiber ring laser, *Sens. Actuat. B Chem.* 193, 356–362, 2014.
86. M. Li, J. M. Dai, K. Liu, and G. D. Peng, Performance analysis and design optimization of intra-cavity absorption gas sensor based on fiber ring laser, *J. Lightwave Technol.* 29(24), 3748–3756, Dec. 2011. doi:10.1109/JLT.2011.2172187.
87. M. Y. M. Noor, N. Khalili, and G. D. Peng, SOA-based fiber ring laser with seed of DFB wavelength scanning for relative humidity measurement using an air-guided photonic crystal fiber, *Meas. Sci. Technol.* 24(8), 085203, 2013.
88. E. M. Dianov, Bismuth-doped optical fibres: A new breakthrough in near-IR lasing media, *Quant. Electron.* 42(9), 754–761, 2012.
89. Y. Luo, J. Wen, J. Zhang, J. Canning, and G. D. Peng, Bismuth and erbium codoped optical fiber with ultrabroadband luminescence across O-, E-, S-, C-, and L-bands, *Opt. Lett.* 37(16), 3447–3449, Aug. 2012.
90. J. Zhang, Y. Luo, Z. M. Sathi, N. Azadpeyma, and G. D. Peng, Test of spectral emission and absorption characteristics of active optical fibers by direct side pumping, *Opt. Express* 20(18), 20623–20628, Aug. 2012.
91. J. Zhang, Z. M. Sathi, Y. Luo, and G. D. Peng, Toward an ultra-broadband emission source based on the Bismuth and Erbium co-doped optical fiber and a single 830 nm laser diode pump, *Opt. Express* 21(6), 7786–7792, Mar. 25, 2013.

14 Active Core Optical Fiber Chemical Sensors and Applications

Shiquan Tao

CONTENTS

14.1 INTRODUCTION

The broad acceptance of fiber-optic technologies in communication industry not only changed information transformation and how we communicate with each other but also brought important impacts in science and engineering. For example, it brings many low-cost, easy-to-use optic parts to the field of optical spectroscopy, which is a critical part of chemical/biochemical measurement. The fast growth of fiber-optic sensors for chemical/biochemical sensing during the past two decades has largely expanded chemical/biochemical measurement capability, increased the measurement technology's sensitivity, and made it possible to carry out the measurements not only in scientific laboratories but also in real-world field applications. The active core optical fiber chemical sensor (AC-OFCS) techniques presented in this chapter are examples demonstrating the high sensitive nature of fiber-optic sensors.

14.2 PRINCIPLE OF AC-OFCS AND EW-OFCS

An optical fiber chemical sensor (OFCS) detects the existence of and measures the concentration of a compound in a sample through detecting the interaction of the compound with light propagating inside an optical fiber.[1,2] Depending on the location at which the interaction occurs, OFCS can be divided into two classes: AC-OFCS[3,4] and evanescent wave optical fiber chemical sensor (EW-OFCS).[5–7] The principles of AC-OFCS and EW-OFCS that detect analyte's optical absorption as sensing signals are diagrammatically shown in Figure 14.1. In an AC-OFCS, the interaction of an analyte compound with the light occurs inside an optical fiber core, while in an EW-OFCS, the interaction of an analyte compound with the light occurs in the cladding layer of an optical fiber.

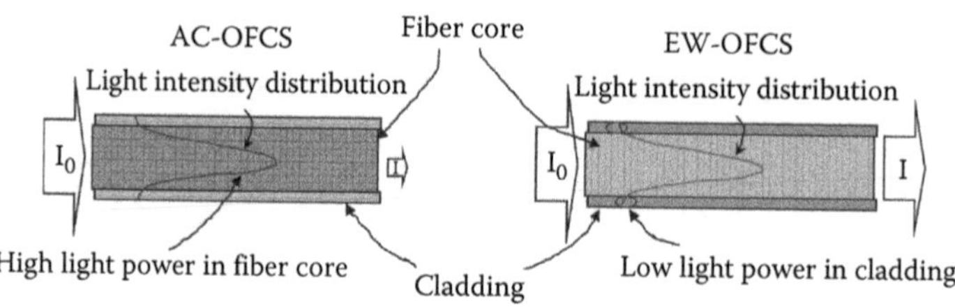

FIGURE 14.1 Diagrammatic graphs show the principles of AC-OFCS and EW-OFCS. The horizontal-lined parts in the fibers are where an analyte (or reaction product(s) of an analyte with a sensing reagent) interacts with light in the optical fiber. The vertical-lined parts in the fibers are not involved in the sensing but only for light guiding. Optical absorption spectrometry is used as an example in the graphs. I_0 is the light intensity injected to the fiber, and I is the intensity of light transmitted out of the fiber. In AC-OFCS, most of the light is absorbed, while in EW-OFCS, only small part of the light is absorbed.

A light beam traveling down an optical fiber can be scattered or absorbed by a compound existing inside the fiber core or inside the cladding. The light propagating in an optical fiber can also excite a compound in the fiber to a higher energy level and causes the emission of fluorescence (FL). All these interactions can be used in designing OFCS. Therefore, analytical spectroscopic techniques, such as ultraviolet/visible (UV/Vis) absorption spectrometry, near-infrared (NIR) and mid-infrared (IR) absorption spectrometry, Raman scattering spectrometry, and FL spectrometry, have been used in OFCS design.[1–11] In addition, the existence of an analyte compound in the fiber core or cladding can change the refractive index of the core or cladding materials, which causes the change in light intensity guided through the fiber. This phenomenon has also been used in designing OFCS.[12–14] The characteristics, including sensitivity, precision, reversibility, response time, and selectivity, of an OFCS are decided by the properties of the compound to be detected, the chemical reactions involved in the sensing process, the analyte/light interaction used for the detection, the location in which analyte/light interaction occurs, and the microstructure of the optical fiber core and the cladding.

14.2.1 OFCS with an Optical Fiber Core as a Transducer

When a light beam is injected into an optical fiber, the light travels down the fiber through a series of total internal reflections (TIRs) at the interface of optical fiber core and cladding layer. If an analyte exists inside the fiber core, the analyte molecules can interact with light propagating inside the fiber core. In this case, the optical fiber core also acts as an optical cell for detecting analyte/light interaction. Theories established in conventional analytical spectroscopy can be used to describe the interaction of an analyte with light inside the fiber core. For example, the absorption of light at a specific wavelength by an analyte inside the fiber core can be described by using the Lambert–Beer's law[15]:

$$A = \text{Log}\,(1/T) = \varepsilon CL \tag{14.1}$$

where

A is the absorbance
T is the transmittance
ε is the absorption coefficient
C is the concentration of the analyte inside the fiber core
L is the length of the analyte/light interaction, which is decided by the following equation[4]:

$$L = \boldsymbol{\ell}/(1 - \sin^2\theta)^{1/2} \tag{14.2}$$

where

$\boldsymbol{\ell}$ is the length of the optical fiber transducer
θ is the incident angle of light beam to the optical fiber

Similarly, a molecule inside an optical fiber core can also be excited by the light guided inside the fiber and emits FL. If the concentration of fluorescent molecule inside the fiber core is very low, the intensity of FL can be expressed as[16]:

$$F = K\phi I_0 \varepsilon CL \tag{14.3}$$

where

K is a constant

ϕ is the quantum efficiency of the fluorescent molecule

I_0 is the incident light intensity

The interaction of light guided in an optical fiber core with an analyte or reaction product(s) of an analyte with a sensing reagent inside the optical fiber core is the basis of AC-OFCS using porous solid optical fibers (PSOFs), liquid-core waveguides (LCWs), or hollow waveguides (HWGs) as transducers. In optical fiber industry, the interactions of light guided in an optical fiber with the fiber core materials have also been used for fiber quality control, optical fiber amplifier design, etc. For example, laser-induced optical fiber FL spectroscopy can be used to monitor possible contamination of fiber core material.[17] Optical fiber core Raman scattering is another technique having been used to monitor the quality of optical fiber cable.[18] The in-fiber amplification technologies significantly improved optical fiber communication capacity. Present in-fiber amplification technologies used in optical fiber communication industry are based on NIR laser-induced FL of erbium ions doped in a silica fiber core.[19,20] Fiber core optical absorption techniques have also been proposed for monitoring high-energy radiation (x-ray, γ-ray) and ionization particles (α-particle, neutron) for applications in nuclear facilities. The irradiation of silica optical fibers by the high-energy radiation and ionization particles breaks down chemical bounds, which causes the generation of radicals, nonbinding oxygen species inside the fiber core. The formed radical species can be detected through monitoring the absorption spectrum of the optical fiber core.[21]

14.2.2 OFCS with Tailored Cladding as a Transducer

When a light beam is guided through an optical fiber, a standing wave, called evanescent wave (EW), is formed at each point of TIRs. The existence of EW distributes part of the light power into the cladding layer adjunct to the fiber core. If a compound inside the cladding layer can interact with the electromagnetic wave distributed in the cladding, this compound can be sensed through detecting the interaction of the compound with EW in the cladding. The interactions of a compound in the cladding with EW can be optical absorption, FL emission, or scattering. Almost all of the analytical spectroscopic techniques used in conventional analytical spectroscopy can be used for detecting the interaction of an analyte with the EW in the cladding layer. Taking optical absorption as an example, the EW absorption by a compound in the cladding layer can be expressed as follows[4]:

$$A_{EW} = \mathrm{Log}(1/T) = \gamma\varepsilon C[d_p \ell n_2 \sin\theta / a(n_1^2 - n_2^2 \sin^2\theta)^{1/2}] \tag{14.4}$$

where

γ is the ratio of light power distributed in the cladding over total light power guided through the optical fiber

n_1 and n_2 are the refractive index of the fiber core and cladding materials

ε is the absorption coefficient of the analyte

$d_p \ell n_2 \sin\theta / a(n_1^2 - n_2^2 \sin^2\theta)^{1/2}$ is the absorption path length, which equals the penetration depth (d_p) times the number of TIRs

The number of TIR is calculated as $\ell n_2 \sin\theta / a(n_1^2 - n_2^2 \sin^2\theta)^{1/2}$, as light travels a length of ℓ inside an optical fiber (diameter equals a).

Similarly, EW-excited FL,[22–24] EW Raman scattering,[25,26] and EW scattering[27] can be used in designing OFCS to detect the interaction of an analyte in the cladding with the EW penetrated into the cladding.

14.3 COMPARISON OF AC-OFCS AND EW-OFCS

Comparing Equation 14.1 with Equation 14.4, it is clear that the sensitivity of an optical fiber core absorption–based chemical sensor is much higher than that of an EW absorption–based chemical sensor. Two factors, the light intensity ($I_{cladding} = \gamma I_{total}$, γ value is usually smaller than 0.05 for multimode optical fibers) and the interaction path length ($= d_p \ell n_2 \sin\theta / a(n_1^2 - n_2^2 \sin^2\theta)^{1/2}$, d_p is only in μm level), limit the sensitivity of an EW absorption–based sensor. For example, Klunder and Russo calculated the absorption path length of an EW-based OFCS using a $\ell = 12$ m optical fiber as a transducer to be only 3 mm.[28] In FL spectroscopy and scattering spectroscopy, the intensity of emitted FL light and scattered light is proportional to the intensity of excitation light and the path length of the interaction. Therefore, the sensitivity of an AC-OFCS using FL emission or Raman scattering as a sensing mechanism is also higher than that of EW-excited FL sensor or EW scattering sensor.

Most reported OFCSs are based on optical fiber EW spectrometry.[1,2,8–11] An EW-OFCS can use conventional optical fibers made for communication industry to construct the sensor. These fibers are inexpensive and easy to handle and are compatible with all kinds of tools and instruments developed for the communication industry. Therefore, these EW-OFCSs are easy to construct and low cost. On the other hand, commercially available optical fibers used by communication industry are solid fibers. It is almost impossible to introduce a sample into fiber core of such a fiber to detect the interaction of an analyte in the sample with light guided inside the fiber core. Therefore, in order to make an AC-OFCS, a specially tailored optical fiber core has to be developed. This fiber core should be able to guide light and more importantly allows the introduction of samples into the fiber core. Presently, three types of special optical fibers, PSOF, LCW, and HWG, have been used in designing AC-OFCS. Some very sensitive chemical sensors have been developed using these special optical fibers. Following are examples of such AC-OFCS.

14.4 AC-OFCS AND APPLICATIONS

14.4.1 AC-OFCS Using Tailor-Made PSOF as Transducers

Several techniques have been reported for making PSOF. Shahriari et al.[29] reported a high-temperature glass fiber pulling process followed by wet chemical etching to make a porous glass optical fiber. A sodium borate–doped glass rod is first made by normal high-temperature glass-making process. The glass rod is then pulled by a fiber-pulling device to make a glass optical fiber. The glass optical fiber is then soaked in a hot concentrated hydrochloric acid solution for more than 12 h to etch out the sodium borate. A porous glass optical fiber was obtained after the etching process. This porous optical fiber can then be soaked into a solution of a chemical agent to impregnate the chemical agent into the fiber core. A short piece of this reagent-impregnated porous glass optical fiber can be used as a transducer of an AC-OFCS. An optical fiber ammonia sensor has been developed by using a short piece of such a porous glass fiber impregnated with a pH indicator as a transducer. This sensor can detect trace ammonia in gas samples with a detection limit of 0.7 ppm.

A significant progress in designing AC-OFCS using PSOF is reported by Tao et al.[3,4] They reported a wet chemical process to make porous silica optical fibers. The starting material for making such porous silica fibers is an ester of silicic acid. The ester is first hydrolyzed in the presence of a mineral acid as a catalyst. The obtained silicic acid is not stable. They dehydrate through hydroxide condensation to form silicon dioxide nanoparticles, which distributed in water to form a colloidal solution (called silica sol solution). The formed sol solution is then mixed with a solution of a sensing agent. The mixed solution is injected into a tube of small diameter and kept in the tube at room temperature until the solution inside the tube is gelatinized. During the gelatinizing process, part of the solvent separates out of the gel and forms a film between the formed silica gel and the wall of the tube. The liquid solvent film prevents the contact of silica gel with the tube and avoids adhesion of the gel to the tube wall. As gelatinizing proceeding, part of the solvent permeates out of the tube and the formed silica gel shrinks. Finally, a silica gel monolith in the shape of a fiber with diameter smaller than that of the tube is formed inside the tube. This gel fiber can be pushed out of the tube by injecting a liquid (water or other liquid) through the tube from one end. The gel fiber just pushed out of the tube is rigid but gradually hardens after exposed to air. Finally, a sensing agent–doped porous silica optical fiber is obtained after the fiber is hardened. This process for making porous silica optical fibers is simple and low cost. In addition, the porous silica optical fiber is made from a wet chemical process; any chemical or biochemical agent, which can be dissolved in water or an appropriate organic solvent, can be doped into the fiber for designing sensors for different application purposes.

Figure 14.2 is a scanning electron microscope (SEM) imaging of such an as-made porous silica optical fiber. The rough structure on the surface of the optical fiber was believed to be originated from the inner surface microstructure of the Tygon tube, which was used as the model in making the porous silica optical fiber. This rough structure scatters light out of the fiber. A chemical polishing procedure by using a hydrofluoric acid solution was developed to remove the surface structure and improve the fiber's light-guiding efficiency. The SEM has also been used to study the inner microstructure of the porous silica optical fiber. However, it was found that the pore size inside the fiber is smaller than the resolution limit (2.5 nm) of the used SEM. The small pore size inside the porous silica optical fiber is one of the reasons the fiber can efficiently guide light, because scattering loss is very low with such small pores. However, the porous silica optical fibers have a high loss level in transmitting UV light below 350 nm.

Several AC-OFCSs using reagent-doped porous silica optical fibers have been reported.[3,4,30,31] Figure 14.3 shows the transducer structure of such an AC-OFCS. A $CoCl_2$-doped porous silica optical fiber has been used to design a moisture sensor having a structure similar to that shown in Figure 14.3.[3,4] This sensor is based on the reversible reaction of $CoCl_2$ with water vapor to form $CoCl_2(H_2O)_x$ complex inside the porous silica optical fiber core when it is exposed to a water vapor containing gas sample. The concentration of formed $CoCl_2(H_2O)_x$ complex is in equilibrium with the concentration of water vapor in the gas sample. The formation of $CoCl_2(H_2O)_x$ complex reduces

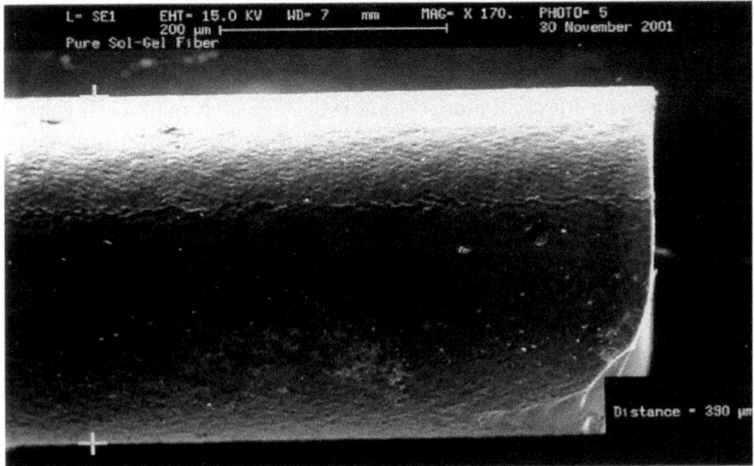

FIGURE 14.2 An SEM picture of a porous silica optical fiber made from a wet chemical method.

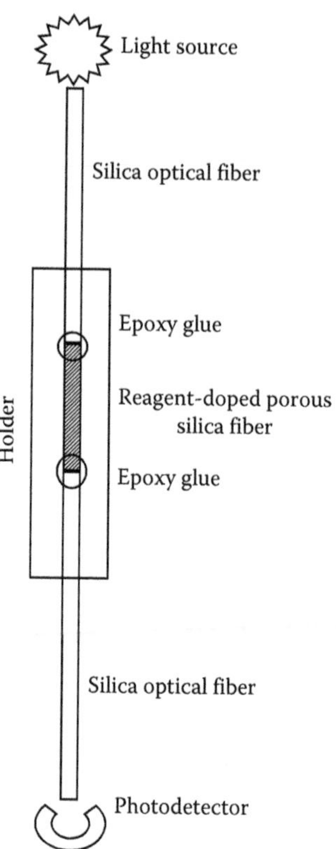

FIGURE 14.3 A diagrammatic graph shows the structure of an AC-OFCS using a reagent-doped porous silica optical fiber as a transducer.

$CoCl_2$ concentration in the porous silica optical fiber. $CoCl_2$ absorbs light at 632 nm. Therefore, water vapor concentration in a gas sample can be detected through detecting the fiber's optical absorbance signal by using a 632 nm diode laser as a light source. This sensor has high sensitivity and can easily detect water vapor in air to ppm level. However, the sensor has a slow response for monitoring water vapor in air at room temperature because the diffusion of water molecules into the porous structure inside the fiber takes a long time at ambient temperature.

An AC-OFCS using a $CuCl_2$-doped porous silica optical fiber has been reported for monitoring trace NH_3 in high-temperature gases, such as stack gas in coal-fired power plant.[30] In recent years, ammonia has been added to the exhaust gas of some combustion systems in order to reduce NO_x emission. Therefore, a sensor for continuous monitoring trace NH_3 in exhaust gas of such combustion system is significant for process control as well as for monitoring air pollutant emission. The AC-OFCS for NH_3 monitoring is based on a simple reversible chemical reaction:

$$CuCl_2 + NH_3 \leftrightarrow Cu(NH_3)_xCl_2$$

The formed $Cu(NH_3)_x^{2+}$ complexes inside the porous silica optical fiber are in equilibrium with the concentration of NH_3 in the gas sample. The formed $Cu(NH_3)_x^{2+}$ absorbs light with peak absorption wavelength at around 550 nm. Therefore, the concentration of NH_3 in gas phase can be monitored through monitoring the fiber-optic absorption signal at around 550 nm. The sensor was tested for monitoring trace ammonia in air gas sample at a temperature of 450°C. Figure 14.4 is the recorded absorption spectra of the AC-OFCS exposed to high-temperature air samples containing NH_3 of different concentrations. With the increase in NH_3 concentration in air sample, the absorbance at 540 nm increased. The recorded absorbance at 540 nm has a linear relationship with the

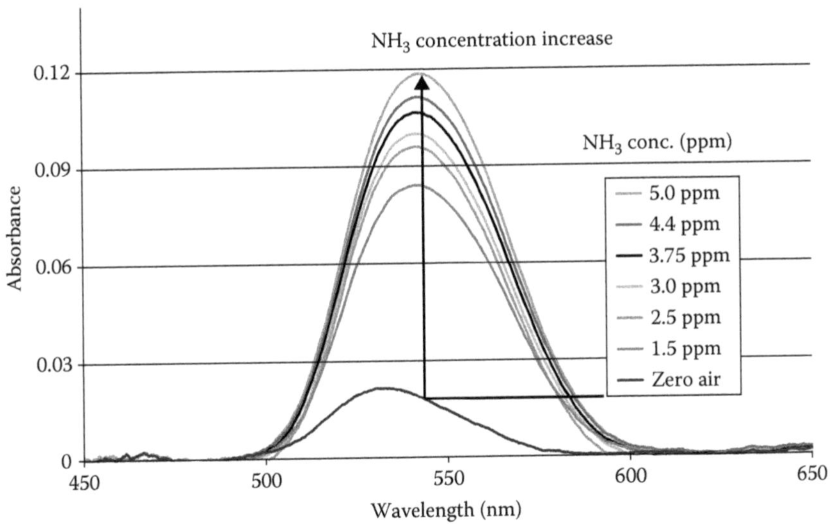

FIGURE 14.4 The recorded UV/Vis absorption spectra of an AC-OFCS using a $CuCl_2$-doped porous silica optical fiber (fiber length, 3 cm) as a transducer exposed to high-temperature air samples containing ammonia of different concentrations.

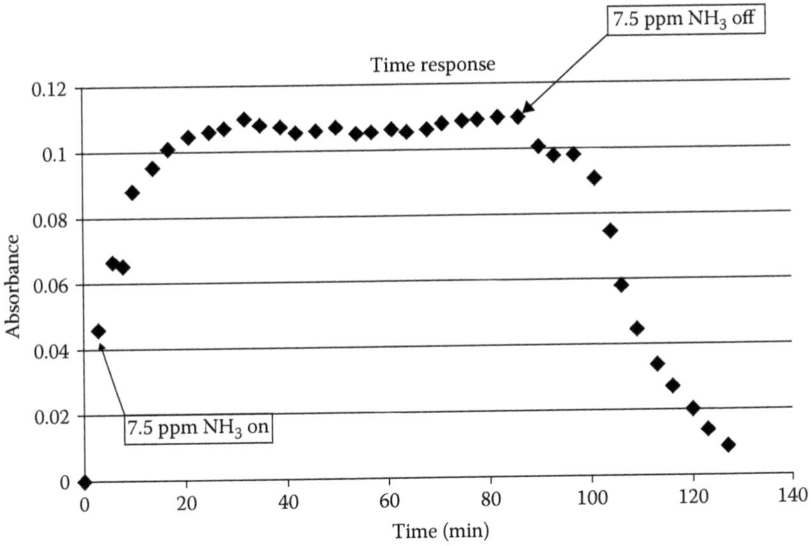

FIGURE 14.5 Time response of the $CuCl_2$-doped porous silica optical fiber ammonia sensor alternatively exposed to an ammonia-containing air sample and a blank air sample.

NH_3 concentration in air gas sample. Figure 14.5 shows the sensor's time response to the change in NH_3 concentration in an air gas sample. This result shows that the sensor is reversible and can be used for continuous monitoring trace NH_3 in a gas sample. The response time is <20 min. It has to be indicated that the response time of such a sensor can be much shorter when a porous silica optical fiber of small diameter will be used to design the sensor.

AC-OFCSs using polydimethylsiloxane (PDMS) polymer optical fibers have also been reported. PDMS has been used in manufacturing organic polymer optical fibers as well as a cladding material for conventional optical fibers, which are used by communication industry. PDMS is a hydrophobic polymer and can be made by polymerizing dimethylsiloxane with some cross-linking agents in the existence of special catalysts. The reagent kits for making this polymer are available from commercial sources. Klunder et al. made PDMS fibers by using silicone sealing products, such as RTV-732 and RTV-3145, available from department stores. They investigated the application of these polymer fibers for sensing trichloroethene (TCE) in environmental samples.[28,32] TCE absorbs NIR light with peak absorption wavelength at around 1.64 μm. A PDMS optical fiber of 10 mm length was connected with conventional optical fibers as a transducer. This sensor was reported for detecting TCE in water samples down to 1.1 ppm.

14.4.2 AC-OFCS Using LCWs as Transducers

LCW is an optical fiber using a liquid confined inside a tube (capillary) or another liquid sheath as light-guiding medium. LCW development was originally targeted for delivering high-intensive laser energy. Liquid materials having high refractive index and polymers of low refractive index have been studied for constructing LCW.[33–36] For example, carbon disulfide (CS_2) is an organic solvent used in organic chemistry. The refractive index of this solvent is 1.63, which is higher than the refractive index of glass and fused silica. Therefore, a glass capillary filled with CS_2 can be used as an LCW. The CS_2 LCW has been used as a sample cell for optical spectroscopic detection of chemical compounds extracted into CS_2.[37] However, due to the compound's strong unpleasant odor, the application of CS_2-based LCW in optical spectroscopy as a sample cell is not widely accepted.

The most significant progress in LCW and the application of LCW in AC-OFCS is the development of special amorphous polyfluoropolymers.[38–46] These polymers are transparent materials and do not absorb light of broad wavelength range from UV to mid-IR. The refractive index of these fluoropolymers is in the range from 1.29 to 1.31, which is lower than the refractive index of water (1.33) and most organic solvents.[47] Therefore, a tube or a capillary made from one of these polymer materials filled with water can act as an optical fiber with water or an organic solvent inside the tube as a light-guiding medium. In addition, these fluoropolymers are chemically stable and do not react with normal chemical reagent. Therefore, an LCW with water or an organic solvent–filled fluoropolymer capillary is a perfect long-path sample cell for spectroscopic detection/monitoring of a chemical/biochemical species in the water or organic solution. Moreover, the fluoropolymers are also highly permeable to many gas molecules.[48,49] Molecules in a gas sample can diffuse into the solution inside the tube and interact with light guided through the solution. This makes the fluoropolymer-based LCW also useful in gas sensing.[38,50,51] The LCWs presently most widely used are made from Teflon AF amorphous fluoropolymer resin developed by DuPont.[52] Two fluoropolymer resins, Teflon AF 1600 and Teflon AF 2400, are available from DuPont. The resin can be made to the form of tube or capillary to form LCW. The resin can also be coated on the inner surface of a silica capillary to form an LCW.[39] Presently, Teflon AF capillary products are available from commercial sources.

An LCW can be used as an optical cell in optical spectrometry. Compared with traditional optical spectrometers, which have optical cells with path length in centimeter range, an LCW optical cell can be as long as meters to hundreds of meters. In optical spectrometric methods, the sensitivity of the methods is usually proportional to the path length of the optical sample cell. Therefore, the application of LCW in optical spectrometry can significantly improve sensitivity. This is demonstrated in several LCW-based sensors.

An LCW-based Cr(VI) sensor has been reported.[45] This simple sensor uses a 2 m LCW as a transducer, a UV light-emitting diode (LED, 375 nm peak wavelength) as a light source, and a photodiode as a photodetector. Light from the UV LED is coupled into the LCW by using a short piece of conventional silica optical fiber and a three-way connector as shown in Figure 14.6. Light guided through the LCW is delivered to the photodiode via another piece of conventional optical fiber. A water sample is delivered into the LCW with a pump. This sensor monitors the optical absorption signal by chromate ion itself, and therefore, no chemical reagent is needed. Figure 14.7 shows the time response of this sensor when water samples having Cr(VI) of different concentrations were

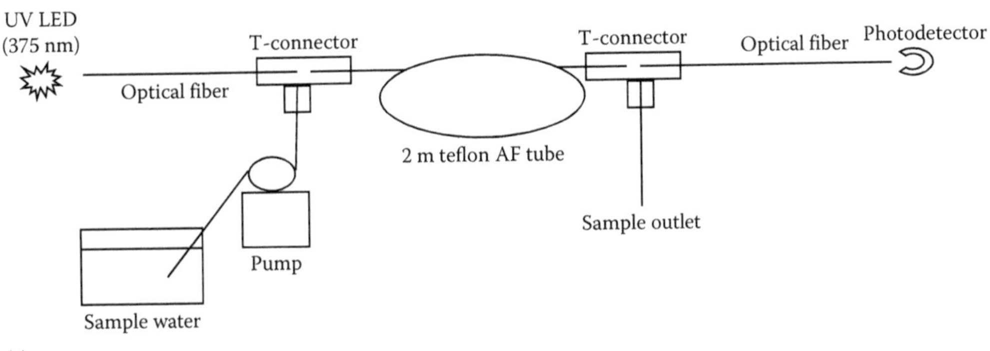

(a)

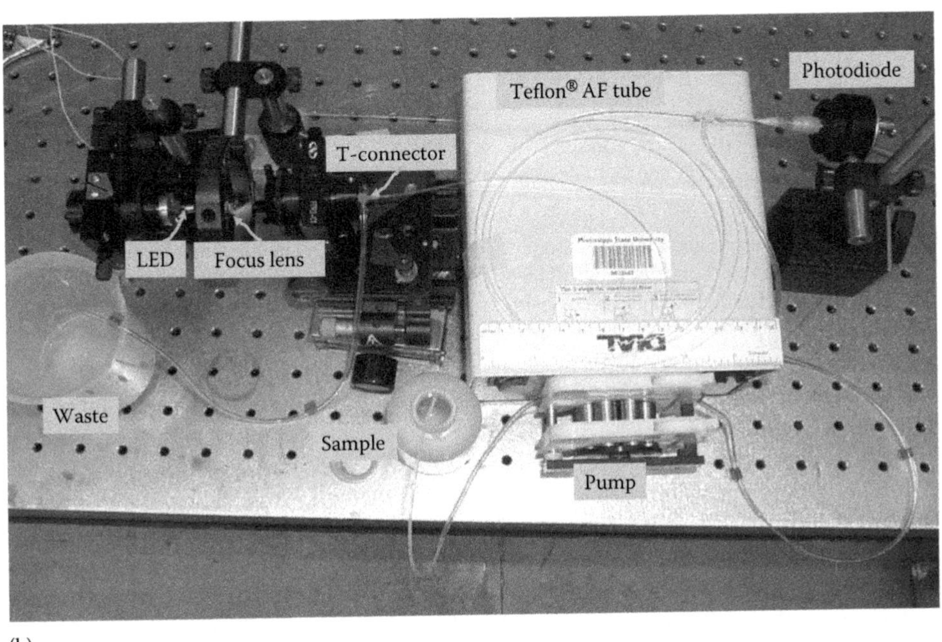

(b)

FIGURE 14.6 (a) A diagrammatic structure of an AC-OFCS using an LCW for monitoring trace Cr(VI) in water samples and (b) a picture of a laboratory-made LCW Cr(VI) sensor for testing the sensing principle.

pumped through the LCW. As demonstrated in the test results, this simple structured sensor can detect Cr(VI) in water sample with a detection limit of 0.10 ng/mL. Cr(VI) has been broadly used as a chemical reagent, and inappropriate discharge of Cr(VI)-containing waste caused water contamination worldwide. Because Cr(VI) is cancerogenic, the existence of Cr(VI) in drinking water is a serious concern today. This simple and easy-to-use sensor was proposed for testing Cr(VI) level in tap water at home.

Another interesting work involving LCW AC-OFCS is the detection of mercury atoms in water by using LCW atomic absorption spectrometry.[46] In traditional analytical chemistry, atomic absorption spectrometry is used to detect the interaction of free atoms in gas phase with light. Conventional samples (liquid or solid) must be atomized to generate free atoms in high-temperature flames or plasmas. Mercury is a special element. Mercury ions in a sample can be reduced to elementary mercury in an aqueous solution. The elementary mercury can be purged out of the aqueous solution by burbling the solution with an inert gas. Mercury in the gas phase exists as mercury atoms, which can be detected by atomic absorption spectrometry. However, what is the existing form of mercury after chemical reduction in aqueous solution is unclear.

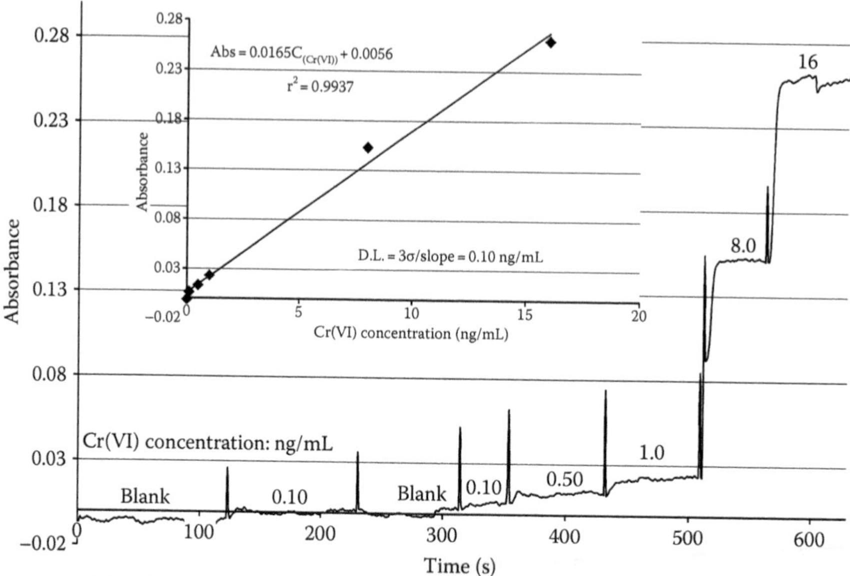

FIGURE 14.7 Time response and calibration curve of the LCW AC-OFCS for monitoring water samples containing Cr(VI) of different concentrations. The sensor achieved a 0.10 ng/mL detection limit, which is much lower than the detection limit achieved by using expensive instruments, such as AAS and ICP-AES.

Tao et al. investigated the state of reduced mercury in water by using LCW optical absorption spectrometry. Mercury in ionic form (Hg^{2+}) was reduced by mixing a $HgCl_2$ solution with the solution of a strong reducing agent ($NaBH_4$). The mixed solution was pumped into a 1.6 m LCW. Light from a UV light source was coupled into the LCW. Light emerged from the LCW was coupled into an optical fiber–compatible UV/Vis spectrometer with a structure similar to that shown in Figure 14.6. The recorded absorption spectra are shown in Figure 14.8. These results indicate that the reduced mercury in the aqueous solution absorbs light with peak absorption wavelength exactly the same as that of atomic absorption spectrum by mercury atoms in gas phase. It can be concluded from these results that the reduced mercury exists in aqueous solutions as mercury atoms. However, the mercury atoms in aqueous solutions are hydrated and encounter high-frequency collisions from molecules surrounding the atoms. This is reflected in the width of the atomic absorption spectrum in Figure 14.8. The half-height width of atomic absorption spectrum by gas-phase mercury atoms is around 10 pm, while the half-height width of the recorded atomic absorption spectrum by mercury atoms in the aqueous solution is more than 20 nm. The LCW atomic absorption spectrometric method can also be used for monitoring mercury in water samples.

LCW can also be used as sample cells for FL spectrometry. A compound inside the solution filled in a Teflon AF capillary can be perpendicularly excited. The emitted FL light is collected by the LCW and guided to the capillary's end, which is connected to an optical fiber–compatible spectrometer or a photodetector via a conventional optical fiber. For example, a Teflon AF 2400 tube has been helically wounded round a quartz tube.[40] Two UV lamps (370 nm) were inserted into the quartz tube. Quinine sulfate and chlorophyll-a in the solution filled inside the LCW have been excited by the UV light from the UV lamps. The emitted FL photons are detected with optical fiber–compatible spectrometer. This sensor can easily detect the analytes to sub-ppb concentration. LCW FL spectrometry using a UV LED as an excitation source has also been used in detecting atmospheric formaldehyde.[43] This fieldable device can be used to detect formaldehyde in air sample in sub-ppb level. A Teflon 2400 capillary has been used as both a separation column and a light-guiding device. Proteins in a sample filled inside the LCW are separated by

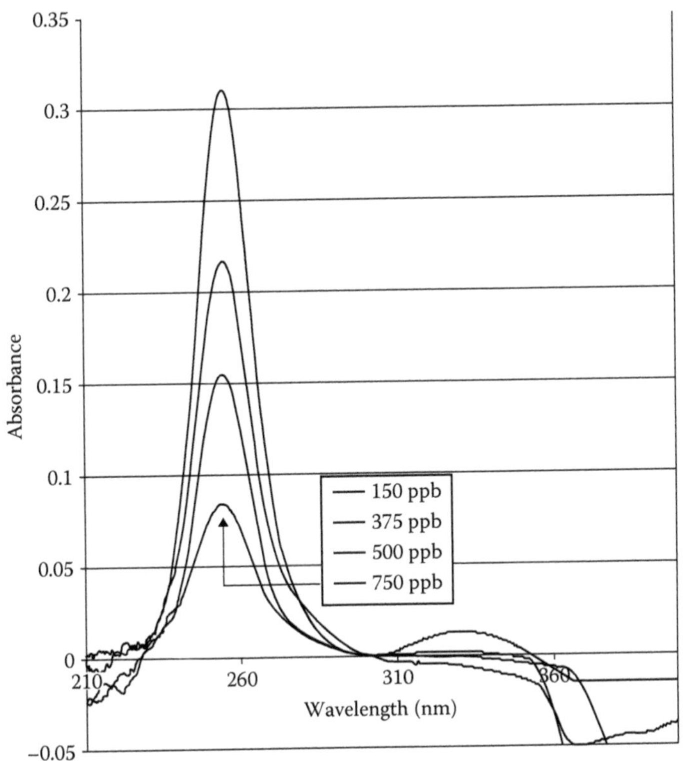

FIGURE 14.8 Atomic absorption spectra of mercury atoms in water recorded by using an LCW as a long-path-length optical cell. The half-height width of the absorption spectra is around 20 nm, which is thousands of times broader than atomic absorption spectrum of mercury atoms in gas phase. The mercury atoms in water encounter high-frequency collisions, which broadens the absorption spectrum according to quantum mechanical principles.

isoelectric focusing technique. A laser beam has been injected into the LCW to excite the separated proteins. A CCD imager was used to detect the distribution of FL from the separated proteins. When compared to a commercially available instrument with UV detection, the separation efficiency and peak capacity were similar, while the detection sensitivity was enhanced by 3–5 orders of magnitude.[53]

Laser-induced Raman spectrometry can be considered as a convenient detection technique, because the excitation is not wavelength dependent, and a laser can excite many compounds. However, Raman spectrometry has an inherited low sensitivity compared with FL spectrometry. The application of a long-path-length sample cell in laser-induced Raman spectrometry can be a solution for improving the sensitivity of Raman spectrometry. LCWs have been investigated as sample cells in laser-induced Raman spectrometry, and significant sensitivity improvement has been achieved.[39,54] LCW as sample cell for laser-induced Raman is especially attractive as a detection technique for separation technologies, such as HPLC and CE, because Raman spectrometry can be considered as a general detection technique and is well suited for detecting multiple analytes from separation eluents.

The gas permeation property of Teflon AF material makes LCW very useful for gas sensing. Teflon AF materials are highly permeable to H_2O, CO_2, O_3, H_2, and N_2. LCW gas sensors can be developed either by directly monitoring the intrinsic optical property of the gas compound or by monitoring reaction products of gas compounds permeated into the solution with sensing reagents in the solution filled inside the LCW. Tao and Le reported an LCW ozone sensor for monitoring water ozonation process.[55] Water ozonation is presently used in industry for water sanitation.

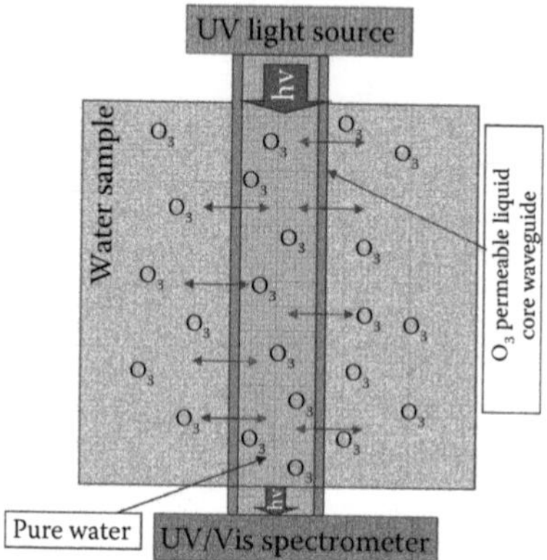

FIGURE 14.9 A diagrammatic graph shows the principle of an LCW AC-OFCS for monitoring trace ozone in water samples. The LCW acts as a light-guiding device as well as a permeation separator, which prevents interference species in water sample from entering the LCW.

A sensor with capability of continuous monitoring ozone concentration in water is significant in the ozonation process control. The principle of the reported sensor is diagrammatically shown in Figure 14.9. This sensor is simple structured and consists of a Teflon AF 2400 capillary filled with pure water, a UV light source (254 nm), and a photodetector. The pure water–filled Teflon AF 2400 capillary is deployed into a water sample. Ozone in the water sample permeated through the Teflon AF 2400 tube and dissolved in the pure water filled in the capillary. Light from the UV light source guided through the LCW is absorbed by ozone molecules dissolved in water, which is monitored with a photodetector as a sensing signal. This sensor is reversible with response time shorter than 4 min. It can be used for monitoring trace ozone in water to sub-ppb level. An LCW-based sensor for monitoring trace SO_2 in gas and liquid samples has also been reported.[56] Similar to the ozone sensor, this sensor also uses a pure water–filled Teflon AF 2400 capillary as a transducer. The intrinsic absorption signal by SO_2 molecule permeated to water filled in the Teflon AF capillary at 254 nm has been monitored as a sensing signal. LCW gas sensors have also been reported for monitoring CO_2.[57] In such a sensor, a pH-indicating reagent is dissolved in water filled in a Teflon AF 2400 capillary. Carbon dioxide permeated into the solution filled inside the capillary changes the pH and thus changes the color of the solution. This color change is monitored by using fiber-optic UV/Vis absorption spectrometry. This sensor can be used for continuous monitoring of CO_2 in air or for industrial process control.

14.4.3 AC-OFCS Using HWGs as Transducers

HWG is originally developed for the purpose of delivering high-energy laser power, such as high-intensive light from CO_2 lasers and Er:YAG lasers.[58–60] An HWG uses a gas (air or specially prepared gas) filled in a special capillary or tube as light-guiding media. Compared with solid and liquid, it is much easier to obtain high-purity gases, which do not absorb laser light of specific wavelength. Therefore, there is much less chance of fiber damage when an HWG is used for guiding high-intensive laser power. Depending on the refractive index of capillary wall material, HWG can be grouped into two categories: (1) those whose inner wall materials have refractive indices greater than 1 (leaky guide) and (2) those whose inner wall materials have refractive index less than 1. The HWG with $n < 1$ wall material is similar to normal optical fibers in that the refractive index of fiber core is higher than that of cladding. Hollow sapphire fiber operating at 10.6 μm ($n = 0.67$) is an example of this class of HWG.[51] Presently, the most popular HWG is a leak guide having a dielectric silver iodide (AgI) film deposited on the surface of a silver film, which is coated on the inner surface of a glass tube or capillary.[60–63] The smooth surface of glass tube reduced the possible scattering loss. With present technique, AgI film–based HWG can work in the wavelength range from 2 to 20 μm with loss level at 0.02 dB/m.[64]

Photonic bandgap HWG is a recent development of optical fibers. Light guiding in this fiber is based on Bragg grating. The transmittance of a photonic bandgap hollow optical fiber depends on wavelength. Light of specific wavelength ranges can be guided through such a fiber, while light of other wavelengths will have a high loss level. The transmitted wavelength range of such a fiber can be tuned through controlling the diameter of holes surrounding the central light-guiding hole.[65]

Early HWG gas sensors use leaky guides or dielectric AgI membrane–coated capillaries as transducers.[58–63] In recent developments, photonic bandgap HWG gas sensors have been reported for monitoring organic compounds (ethane, dichloromethane, trichloromethane, ethyl chloride, etc.).[66–68] A high-resolution quantum cascade laser has been used as a light source in photonic bandgap HWG gas sensors.[65] The stable quantum cascade laser can provide stable light beams of precisely tuned wavelength. These sensors do not require an IR monochromator and largely simplify the sensor structure. Fourier transform infrared spectroscopy and mid-IR optical spectroscopy using cascade lasers as light sources are powerful for identifying/detecting individual organic compounds, and they have been proposed as detectors for gas chromatography. However, the sensitivity of these mid-IR spectrometric methods is very limited for gas detection. Although HWG can be used for increasing the path length of interaction, the reported detection limit of these sensors is still limited in ppm level. As detection techniques, these detectors are not as good as broadly used mass spectrometers.

In viewing the three types of AC-OFCS, LCW can be used for monitoring both liquid and gas samples and can achieve high sensitivity. PSOF-based AC-OFCSs are especially attractive in monitoring high-temperature gas samples. However, PSOF-based sensor's response can be too slow for many industrial process control applications if operated at ambient temperature. HWG-based AC-OFCSs have the potential to achieve high sensitivity and fast response detection if high-resolution lasers are available as light sources.

14.5 CONCLUSION

AC-OFCS is different from traditional EW-OFCS in that it uses the interaction of an analyte in the fiber core with light guided inside the fiber as a sensing signal. Due to its long analyte–light interaction path length, AC-OFCS can have sensitivity thousands of times higher when compared to EW-OFCS. The sensor technologies presented in this chapter demonstrated AC-OFCS's high sensitivity nature. As discussed in this chapter, further development of AC-OFCS technologies largely depends on advances in other fields, such as material science, HWG, laser, and optoelectronics.

REFERENCES

1. R. Narayanaswamy and O. S. Wolfbeis, eds., *Optical Sensors for Industrial, Environmental and Clinical Applications*, Springer-Verlag, Berlin, Germany, 2003.
2. K. T. V. Grattan and B. T. Meggitt, eds., *Optical Fiber Sensor Technology*, vol. 4: Chemical and environmental sensing, Kluwer Academic Publishers, Boston, MA, 1999.
3. S. Tao, C. B. Winstead, J. P. Singh, and R. Jindal, *Opt. Lett.*, **27**, 1382 (2002).
4. S. Tao, C. B. Winstead, R. Jindal, and J. P. Singh, *IEEE Sens. J.*, **4**, 322 (2004).
5. B. D. MacCraith, *Sens. Actuators*, **B11**, 29 (1993).
6. G. Stewarg and W. Johnstone, *Opt. Fiber Sensors*, **3**, 69 (1996).
7. W. Jin, H. L. Ho, G. Stewart, and B. Culshaw, *Trends Anal. Spectrosc.*, **4**, 155 (2002).
8. V. Ruddy, B. D. MacCraith, and J. A. Murphy, *J. Appl. Phys.*, **67**, 8070 (1990).
9. O. S. Wolfbeis, *Anal. Chem.*, **76**, 3269 (2004).
10. O. S. Wolfbeis, *Anal. Chem.*, **74**, 2663 (2002).
11. L. Su, T. H. Lee, and S. R. Elliott, *Opt. Lett.*, **34**, 2685 (2009).
12. M. Chomat, D. Berkova, V. Matejec, I. Kasik, G. Kuncova, and M. Hayer, *Sen. Actuators*, **B87**, 258 (2002).
13. R. G. Heideman, R. P. H. Kooyman, J. Greve, and B. S. F. Altenburg, *Appl. Opt.*, **30**, 1474 (1991).

14. S. Korposh, S. W. James, S. W. Lee, S. Topliss, S. C. Cheung, W. J. Batty, and R. P. Tatam, *Opt. Express*, **18**, 13227 (2010).
15. D. A. Skoog, D. M. West, and F. J. Holler, eds., *Fundamentals of Analytical Chemistry*, 7th edn., Saunders College Publishing, New York, 1996.
16. J. R. Lakowicz, ed., *Topics in Fluorescence Spectroscopy*, vol. 4: Probe design and chemical sensing, Plenum Press, New York, 1991.
17. W. J. Miniscalco and B. A. Thompson, *Materials Research Society Symposium Proceedings*, **88** (Optical Fiber Materials and Proceedings), 127 (1987).
18. B. E. Hubbard, N. I. Agladze, J. J. Tu, and A. J. Sievers, *Phys. B: Condens. Matter*, **316–317**, 531 (2002).
19. J. M. P. Delavaux and J. A. Nagel, *J. Lightwave Technol.*, **13**, 703 (1995).
20. M. Artiglia, P. Di Vita, and M. Potenza, *Opt. Quant. Electron.*, **26**, 585 (1994).
21. D. W. Cooke, B. L. Bennett, and E. H. Farnum, *J. Nucl. Mater.*, **232**, 214 (1996).
22. B. D. MacCraith, V. Ruddy, C. Potter, B. O'Kelly, and J. F. McGilp, *Electron. Lett.*, **27**, 1247 (1991).
23. L. C. Shriver-Lake, K. A. Breslin, P. T. Charles, D. W. Conrad, J. P. Golden, and F. S. Ligler, *Anal. Chem.*, **67**, 2431 (1995).
24. B. D. MacCraith, C. M. McDonagh, G. O'Keeffe, E. T. Keyes, J. G. Vos, B. O'Kelly, and J. F. McGilp, *Analyst*, **118**, 385 (1993).
25. B. Mizaikoff, M. Karlowatz, and M. Kraft, *Proceedings of SPIE—The International Society for Optical Engineering*, **4202**, 263 (2001).
26. J. Baldwin, N. Schuehler, I. S. Butler, and M. P. Andrews, *Langmuir*, **12**, 6389 (1996).
27. L. Xu, J. C. Fanguy, K. Soni, and S. Tao, *Opt. Lett.*, **29**, 1191 (2004).
28. G. L. Klunder and R. E. Russo, *Appl. Spectrosc.*, **49**, 379 (1995).
29. M. R. Shahriari, Q. Zhou, and G. H. Sigel, *Opt. Lett.*, **13**, 407 (1988).
30. S. Tao, J. C. Fanguy, and T. V. S. Sarma, *IEEE Sens. J.*, **8**, 2000 (2008).
31. T. V. S. Sarma and S. Tao, *Sens. Actuators*, **B127**, 471 (2007).
32. G. L. Klunder, R. J. Silva, and R. E. Russo, *Proceedings of SPIE—The International Society for Optical Engineering*, **2068** (Chemical, Biochemical, and Environmental Fiber Sensors V), 186 (1994).
33. J. Stone, *Appl. Phys. Lett.*, **20**, 239 (1972).
34. G. J. Oglivie, R. J. Esdaile, and G. P. Kidd, *Electron. Lett.*, **8**, 533 (1972).
35. I. Pinnau and L. G. Toy, *J. Mater. Sci.*, **109**, 125 (1996).
36. R. Altkorn, I. Koev, R. P. Van Duyne, and M. Litorja, *Appl. Opt.*, **36**, 8992 (1997).
37. K. Fujiwara and K. Fuwa, *Anal. Chem.*, **57**, 1012 (1985).
38. M. Belz, P. Dress, A. Sukhitskiy, and S. Y. Liu, *Proceedings of SPIE—The International Society for Optical Engineering*, **3856** (Internal Standardization and Calibration Architectures for Chemical Sensors), 271 (1999).
39. M. Holtz, P. K. Dasgupta, and G. Zhang, *Anal. Chem.*, **71**, 2934 (1999).
40. J. Li, P. K. Dasgupta, and G, Zhang, *Talanta*, **50**, 617 (1999).
41. R. Altkorn, I. Koev, and M. J. Pelletier, *Appl. Spectrosc.*, **53**, 1169 (1999).
42. B. J. Marquardt, P. G. Vahey, R. E. Synovec, and L. W. Burgess, *Anal. Chem.*, **71**, 4808 (1999).
43. J. Li and P. K. Dasgupta, *Anal. Chem.*, **72**, 5338 (2000).
44. P. K. Dasgupta, G. Zhang, J. Li, C. B. Boring, S. Jambunathan, and R. Al-Horr, *Anal. Chem.*, **71**, 1400 (1999).
45. S. Tao, C. B. Winstead, H. Xia, and K. Soni, *J. Environ. Monit.*, **4**, 815 (2002).
46. S. Tao, S. Gong, L. Xu, and J. C. Fanguy, *Analyst*, **129**, 342 (2004).
47. R. C. Weast, M. J. Astle, and W. H. Beyer, eds., *CRC Handbook of Chemistry and Physics*, 65th edn., CRC Press, Inc., Boca Raton, FL, 1984.
48. A. Yu. Alentiev, Yu. P. Yampolskii, V. P. Shantarovich, S. M. Nemser, and N. A. Plate, *J. Mater. Sci.*, **126**, 123 (1997).
49. P. R. Resnick and W. H. Buck, *Fluoropolymers*, **2**, 25 (1999).
50. M. R. Milani and P. K. Dasgupta, *Anal. Chim. Acta*, **431**, 169 (2001).
51. Z. A. Wang, W. J. Cai, Y. Wang, and B. L. Upchurch, *Anal. Chem.*, **84**, 73 (2003).
52. H. X. Teng, *Appl. Sci.*, 2, 496–512 (2012).
53. Z. Liu and J. Pawliszyn, *Anal. Chem.*, **75**, 4887 (2003).
54. M. J. Pelletier and R. Altkorn, *Anal. Chem.*, **73**, 1393 (2001).
55. L. Le and S. Tao, *Analyst*, **136**, 3335 (2011).
56. S. Gong, J. C. Fanguy, and S. Tao, *226th ACS National Meeting Paper Abstract*, New York, September 7–11, 2003 (2003), ANYL-111.
57. Z. A Wang, Y. C. Wang, W. J. Cai, and S. Y. Liu, *Talanta*, **57**, 69 (2002).

58. J. A. Harrington, *Fiber Integrated Opt.*, **19**, 211 (2000).
59. R. K. Nubling and J. A. Harrington, *Appl. Opt.*, **34**, 372 (1996).
60. R. L. Kozodoy, A. T. Pagkalinawan, and J. A. Harrington, *Appl. Opt.*, **35**, 1077 (1996).
61. M. Alaluf, J. Dror, R. Dahan, and N. Croitoru, *J. Appl. Phys.*, **72**, 3878 (1992).
62. Y. Matsuura, T. Abel, and J. A. Harrington, *Appl. Opt.*, **34**, 6842 (1995).
63. C. D. Rabii and J. A. Harrington, *Appl. Opt.*, **35**, 6249 (1996).
64. R. George and J. A. Harrington, *Appl. Opt.*, **44**, 6449–6455 (2005).
65. J. C. Knight, *Nature*, **424**, 847 (2003).
66. O. Frazao, J. L. Santos, F. M. Araujo, and L. A. Ferreira, *Laser Photon. Rev.*, **2**, 449 (2008).
67. C. Charlton, *Appl. Phys. Lett.*, **86**, 194102 (2005).
68. N. Gayraud, L. W. Kornaszewski, J. M. Stone, J. C. Knight, D. T. Reid, D. P. Hand, and W. N. MacPherson, *Appl. Opt.*, **47**, 1269 (2008).

15 Optical Fiber Humidity Sensors

Muhammad Yusof Mohd Noor, Gang-Ding Peng, and Ginu Rajan

CONTENTS

15.1 INTRODUCTION

Optical fiber sensors (OFS) have evolved mainly from the fiber communications technology and have experienced remarkable *growth* since the 1970s. OFS is suitable for use as the sensing instrument for applications in extraordinary and harsh environments, such as in civil engineering structures, high-voltage and biomedical instrumentation, mining environments, and agriculture ecosystems. Humidity sensors are among the applications of OFS that have been intensively studied by many researchers to cater to the high specific demand of the sensing industry. This chapter first provides humidity measurement–related parameters and the advantages of the fiber sensor for humidity sensing over the conventional electronic sensors. This is followed by comprehensive reviews of the development of various fiber humidity sensors in accordance with coated and noncoated fiber humidity sensors. Their operating principles, various sensing schemes, and related optical components are described. In addition, the performance and applications of the sensor schemes are discussed to investigate which are the most effective. Finally, the gap in the literature is identified and a possible solution is provided. All the sensors reported in the literature are presented in designated tables to allow a broad evaluation and cross comparison.

15.2 HUMIDITY MEASUREMENT PARAMETERS

Water is a substance with the chemical formula H_2O. A water molecule is made up of one oxygen and two hydrogen atoms connected by covalent bonds. Water molecules can exist in three states—solid, liquid, or gaseous. Meanwhile, humidity is defined as a water molecule in gaseous form and is also known as water vapor in the atmosphere. Humidity is present everywhere in the atmosphere of the earth. Humidity is more difficult to measure than other physical properties, such as temperature, strain, and pressure, because it has an extremely broad dynamic range and the measurements have to be made in open atmosphere in different temperature conditions. Many different units or parameters are generally used to express humidity measurements. The parameters most commonly used for humidity measurement are absolute humidity, mixing ratio of parts per million (PPM), dew or frost point temperature (D/F PT), relative humidity (RH), and wet-bulb temperature. The term absolute humidity is defined as the amount of water vapor present in a unit volume of air and is expressed in grams per cubic meter volume of air (g/m^3). Absolute humidity does not fluctuate with the temperature of the air. The higher the amount of water vapor, the higher the absolute humidity. RH also measures water vapor but relative to the temperature of the air. The RH measurement is defined as a measure of the actual amount of water vapor in the air compared to the total amount of water vapor that can exist in the air at a current temperature. It is expressed as the ratio of the actual partial water vapor pressure to the saturation water vapor pressure of the air, multiplied by 100% at the prevailing ambient temperature. For actual water vapor pressure, P_w, and saturation water vapor pressure, P_{ws}, the RH is

$$\%\,\mathrm{RH} = \frac{P_w}{P_{ws}} \times 100 \tag{15.1}$$

Meteorologists usually consider the dew point temperature ($T = T_d$; above 0°C) to evaluate humidity, which is independent of temperature. The dew point temperature, which provides a measure of the actual amount of water vapor in the air, is defined as the temperature to which a volume of air must be cooled, such that it becomes saturated with respect to liquid water. Similarly, frost point ($T = T_f$; below 0°C) is also defined as the temperature to which the air must be cooled in order for that air to be saturated but with respect to ice. PPM parameters, ppm_v and ppm_w, have also been introduced to represent water vapor content fraction by volume or weight, respectively. ppm_v is a volume of water vapor per total volume of air or in terms of vapor pressures:

$$\mathrm{ppm}_v = \frac{\text{Partial pressure of } H_2O}{\text{Total pressure} - \text{Partial pressure of } H_2O} \times 10^6 \tag{15.2}$$

Meanwhile, ppm_w is used to express the amount (mass) of water vapor relative to the total dry air (mixing ratio times one million) as follows:

$$\mathrm{ppm}_w = \frac{\text{Mass of } H_2O}{\text{Mass of carrier air}} \times 10^6 \tag{15.3}$$

When unsaturated air flows over a wet thermometer bulb, water evaporates from the wet surface and the temperature of the wetted surface and enclosed thermometer bulb falls. The wet-bulb temperature is the lowest temperature an object can be cooled through the process of evaporation. The humidity of the air determines the rate of evaporation from the wet-bulb thermometer. It is measured by a psychrometer that has two bulb thermometers (dry-bulb and wet-bulb thermometer).

TABLE 15.1
Correlation among Humidity Parameters

Absolute Humidity (g/m³)	RH (%RH) at 20°C	Dew Point (°C)	PPM at 1 atm (ppm_v)	(ppm_w)	Wet-Bulb Temperature (°C)
0.0009	0.005	−80	1	0.7	6.006
0.003	0.02	−70	5	3	6.009
0.01	0.08	−60	19	12	6.02
0.05	0.2	−50	64	40	6.05
0.1	0.8	−40	189	118	6.1
0.2	1	−38	231	144	6.2
1.7	10	−12.6	2321	1443	7.7
3.5	20	−3.6	4652	2894	9.4
5.2	30	1.9	6994	4350	10.9
6.9	40	6	9348	5814	12.4
8.7	50	9.3	11172	7285	13.8
10.4	60	12	14087	8762	15.2
12.1	70	14.5	16474	10246	16.5
13.9	80	16.4	18872	11738	17.7
15.6	90	18.3	21281	13236	18.9
17.3	100	20	23702	14742	20

The wet-bulb temperature is always lower than the dry-bulb temperature, unless the RH is 100% when the two temperatures are identical. The relationship between wet-bulb temperature T_w, dry-bulb temperature T_a, and the humidity ratio W is described as

$$W = \frac{(2501 - 2.381T_w)W_w - (T_a - T_w)}{2501 - 1.805T_a - 4.186T_w} \tag{15.4}$$

where W_w is the humidity ratio at T_w. Table 15.1 shows the parameters correlation among absolute humidity, RH, PPM by volume and weight, dew point temperature, and wet-bulb temperature.

From Table 15.1, it can be observed that the RH, wet-bulb temperature, and absolute humidity measurement cover the higher humidity range while ppm covers the lower humidity range. Therefore, the RH, wet-bulb temperature, and absolute humidity measurement are suitable to apply in the daily routine life. In practice, RH is constantly used for easy understanding. In contrast, ppm is used extensively in applications related to the industry, especially for the trace humidity measurement due to high coverage in very low humidity conditions. On the other hand, the dew point parameter covers the whole humidity range, which shows that this parameter is available for use in any condition of the measurement.

15.2.1 Electronic versus Optical Fiber Humidity Sensor

The demand for reliable, low-cost, and compact sensors has driven the development of the electronic humidity sensors. Many transduction techniques of various state-of-the-art electronic humidity sensors have been reported, including gravimetric, capacitive, resistive, piezoresistive, and magnetoelastic sensors. These technologies have revolutionized the design of the humidity sensor into a smaller, faster, and reliable one.

Nowadays, the electronic humidity sensors are typically available for less than $10, which are the most low-cost sensors in the market. Despite the low cost, these sensors are quite sophisticated

TABLE 15.2
Examples of Performance of Commercially Available Sensors

Sensor	Response Time (s)		Range (%RH)		Hysteresis (%)
	Rise	Fall	Min	Max	
HMT330 Grid Filter	8	17	0	100	1
HMT330 Steel Netting Filter	20	50	0	100	1
HMT330 Sintered Filter	40	60	0	100	1
HC2-C	15	NA	0	100	NA
Fluke 971	60	NA	0	100	NA

that an integration of microprocessor and temperature sensors is required for valid measurement data. In addition, these commercial electronic sensors also have other common problems as follows:

- Poor accuracy of around 2% at low and high humidity
- Lengthy response time
- Only viable in short-term stability
- Inadequate temperature operational range

The accuracy of these sensors is up to few percent RH (%RH) and become worse in the extreme low and high humidity ranges at 0%–10% RH and 90%–100% RH, respectively. A long response time of the sensors, typically in the range from 5 to 60 s, is too long in some crucial areas such as for a respiratory or breathing sensor. Furthermore, the maximum operating temperature of the sensors is in the range from 80°C to 120°C, which renders them unsuitable for some industrial processing applications. Finally, the electronic sensors commonly have bad long-term stability. Several examples of performance of commercially available sensors in the current market [1] are shown in Table 15.2.

With the advent of optical fiber technology, a substantial level of research has focused on humidity sensors based on fiber-optic approach to tackle the common problems of the electronic sensor. Fiber humidity sensors have already proven suitable in various application areas where electronic humidity sensors have been found to be inappropriate in some areas such as for fuel engine application [2], high-voltage transformer in electrical substations [3], thermal insulation [4], structural health monitoring [5], and biomedical engineering [6]. Therefore, in the future, it is possible that fiber humidity sensors will replace their electronic counterparts to address the increasing demands from industry. However, it still remains challenging to develop a fiber humidity sensor that can provide a specific set of required characteristics in a particular industry area. Table 15.3 shows a cross comparison between a fiber humidity sensor and an electronic humidity sensor, illustrating the superiority of fiber sensor to electronic sensor and their different characteristics.

15.3 CLASSIFICATION OF FIBER-OPTIC HUMIDITY SENSOR

15.3.1 Hygroscopic Material–Coated Sensors

15.3.1.1 Evanescent Wave

In optics, evanescent waves (EWs) are waves that extend beyond the core of the fiber. This wave is formed when light travels in the optical fiber as a result of total internal reflection according to Snell's law as shown in Figure 15.1. An incident and reflected light between the surface of the core and cladding of the fiber must follow the critical angle criterion in order for the light to be total internally reflected into the core. The interference between the incident and reflected signals at the core/cladding creates an EW in the surrounding core of the fiber, as shown in Figure 15.2.

TABLE 15.3
Cross Comparison of Characteristics between Fiber Humidity Sensor and Electronic Humidity Sensor

Requirement Traits of Sensor	Fiber Humidity Sensor	Electronic Humidity Sensor
Sensitivity	Very high sensitivity	High sensitivity; limited by the electronic part
Range	Narrow to standard range	Normal range
Response time	Available for fast response in milliseconds	Generally in seconds
Accuracy	Excellent	Not accurate in extreme ranges
Temperature dependency	Need a compensation scheme	Suffer from temperature dependency but have the ability to compensate it by integrating with the temperature sensor
Low cost	Do not have a proper commercial production line	Well-established mass production of the fabrication technology
EMI influence	Immune to EMI	Not immune to EMI
Size	Very small size (micrometer or millimeter); depends on the required length of the fiber	Available in a miniature size (millimeter to centimeter in size)
Multiplexing	Allow to multiplex large number of sensors	Does not have the ability to multiplex multiple sensors in one sensor's system
Robustness	Able to withstand high-temperature environment	Only able to operate in normal temperature condition
Corrosion resistance	High resistance to hazardous gaseous environments	Erodes in harsh environment
Stability	Long-term stability	Short-term stability
Application	Industrial processes, biomedical and smart structures	Automation, agriculture, household appliances, and domestic applications

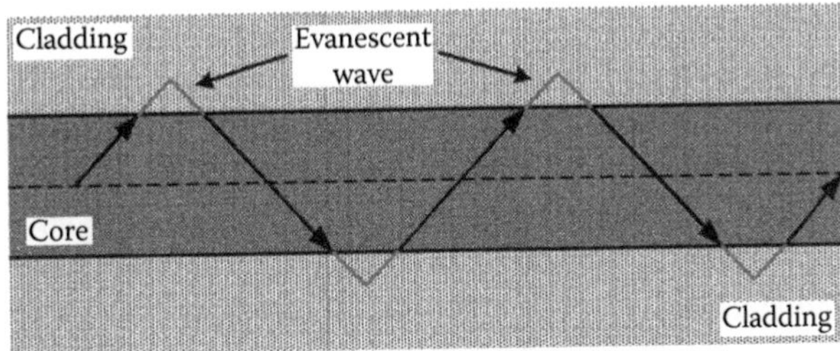

FIGURE 15.1 Total internal reflection in the optical fiber.

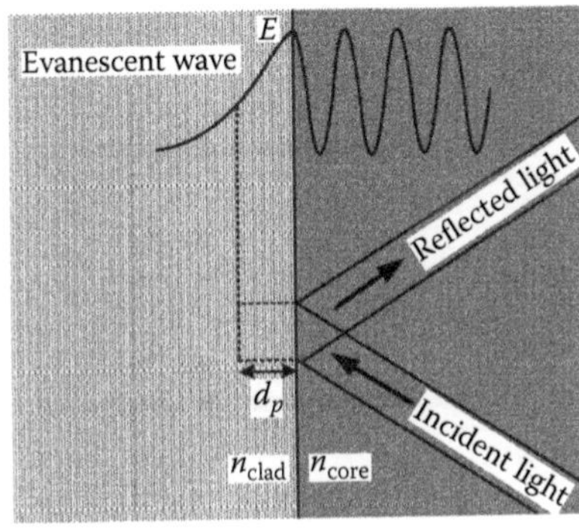

FIGURE 15.2 EW generated at the interface of cladding and core of the fiber.

The amplitude of the EW decays exponentially with the distance away from core/cladding interface and can be represented as

$$E(x) = E_0 \exp\left(\frac{x}{d_p}\right) \tag{15.5}$$

where

E_0 is the initial value evanescent field

$E(x)$ is the intensity of the exponentially decaying evanescent field with the distance x perpendicular to the interface

d_p is the penetration depth of the evanescent field, which is given by

$$d_p = \frac{\lambda}{2\pi n_{\text{core}}[\sin^2\Phi - (n_{\text{clad}}/n_{\text{core}})^2]^{1/2}} \tag{15.6}$$

where

λ is the wavelength of propagating light in the optical fiber

Φ is the incident angle

n_{core} and n_{clad} are the refractive indices of core and cladding of the fiber, respectively

The EW sensing technique enables the light field generated at the core/cladding interface to interact with the target analyte surrounding the core of the fiber and give information due to the optical scattering, light absorption, and change in refractive index. Many optical fiber structure configurations can be used for EW sensing. Generally, the majority of the EW techniques reported is based on the unclad and bend fiber configuration. Optical fibers with a part of the cladding removed to gain access to the evanescent field can be further improved by heating or bending the fiber to form a bent shape. This bending effect will extend the evanescent field further away from the core/cladding interface, thus increasing the interaction between the light and the target analyte. The coating or thin film, which is coated on top of this kind of fiber structure, is used to interact with the water vapor to realize the EW humidity sensor.

Examples of such sensor include those from Vijayan et al. [7], who have demonstrated the use of nanostructure as a humidity-sensitive clad with a bent structure of the plastic optical fiber (POF), as shown in Figure 15.3. This figure shows a transmitter as a light source, bent sensing element, and a photodetector (PD), which is also a common setup configuration for other EW bent sensors. To create the nanostructure cladding, spherical co-nanoparticles with polyaniline are chemically synthesized and dip coated onto the unclad POF. The sensor exhibits a detection range of 20%–100% RH and has a response time of 8 s. The length and thickness of the nanostructure are needed to optimize for better performance. The sensor configuration discussed shows good repeatability and is fully reversible.

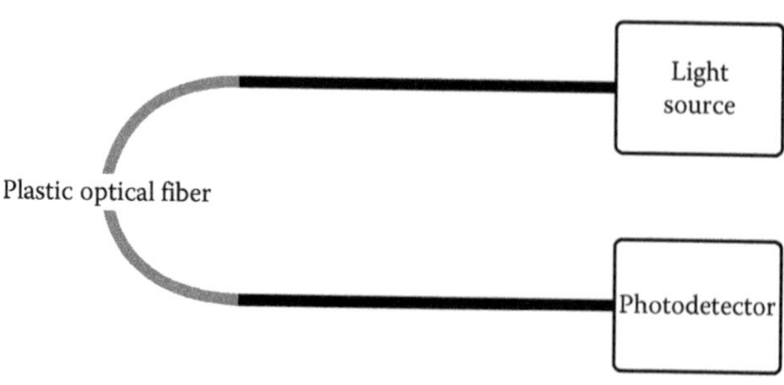

FIGURE 15.3 Generic setup of bent evanescence wave sensor using humidity-sensitive film. (From Vijayan, A. et al., *Sens. Actuat. B Chem.*, 129(1), 106, 2008.)

Mathew et al. [8] recently discussed a humidity sensor based on an agarose-coated single-mode fiber (SMF) bend. For this sensor, agarose is chosen as a hygroscopic coating material due to its simple coating procedure and because it can offer a wide operating humidity range. In addition, agarose has low material degradation compared with the coating used in [9] and has a good adhesion to silica. The agarose solution is prepared by dissolving in distilled water. Then, the sensor head is fabricated by coating the stripped portion of the SMF (1060XP fiber) with hot agarose solution. After fabrication, the sensor head is kept for 1 day at room temperature to make it partially dehydrated and reach equilibrium with the surrounding environment. The fiber with agarose coating is bent using a stable and translating surface without removing the cladding of the fiber. This method offers more flexibility and it can be easily tailored to meet specific bend sensor requirements but with a trade-off of the complicated setup of the sensor head. The sensor with a bend radius of 16.75 mm shows a wide humidity range of between 25% and 90% RH with linear response. It has a fast response time of around 50 ms and shows high sensitivity at higher wavelengths.

A similar sensor configuration using a U-bend optical fiber coated with sensitive film was also investigated by Zhao et al. [10]. The humidity sensing principle of the sensor is based on EW scattering and EW absorption. This low-cost fiber-optic humidity sensor uses a silica film as a coating through a solgel process and enhances the sensitivity of the sensor by doping the silica film with a methylene blue. The developed sensor exhibits a logarithmic response in the range from 1.1% to 70% RH. However, in a region where the range is from 1.1% to 4.1% RH, the sensor shows a good linear response. From the experimental results, the sensitivity and detection limit of the sensor are 0.0087% and 0.062% in a low RH environment. The response time of the sensor is around 20 s to 3 min depending on the RH variation and exhibits a good reversibility. In addition, the sensor seems unaffected by a series of tested non-air gases, such as acetone, chloroform, ethanol, methanol, and argon vapor. The high sensitivity of the sensor in low RH surroundings makes it suitable for use in the manufacturing industry, which demands a high-precision sensor in a low RH environment. The main disadvantage of this sensor is the doping process onto the sensitive coating, which further complicates the fabrication process of the sensor.

Employing an unclad fiber structure without any bending is another common configuration for the EW sensing method. An example of a fiber humidity sensor based on this configuration was demonstrated by Hernaez et al. [11] using a silica optical fiber based on surface plasmon resonance (SPR) in the infrared region. Instead of using a single layer, as discussed so far, two layers of polymeric coating are applied for the sensor, which further complicates the depositing coating processes. The first layer, which is an indium tin oxide (ITO) coating, is deposited onto 200 μm core diameter optical fiber. Then, a polycation polyallylamine hydrochloride and polyanion polyacrylic acid (PAH/PAA) coating is deposited onto the ITO layer using the layer-by-layer (LbL) method to realize a fiber humidity sensor. The thickness of the PAH/PAA will vary with the RH surrounding the medium [12,13]. Thus, the changes in thickness will also change the refractive index of the polymeric coating in contact with the SPR supporting layer (ITO). As a result, a variation of the SPR wavelength is observed according to the surrounding RH. The sensor has shown high repeatability and good linearity in the range from 20% to 80% RH with the average sensitivity of the sensor being 1.08 nm/% RH. They [14] extended their ITO-coated fiber humidity sensor by adjusting the thickness of the PAA/PAH coating to operate on the point of maximum variation of the resonance wavelength (1500 nm). The fabricated sensor has shown a sensitivity improvement of a factor of four by the tuned sensor (100 bilayers) compared to the nontuned one (20 bilayers).

A similar approach, using ITO coating as a resonance supporting layer on top of the optical fiber core, was also demonstrated [15] but with a different humidity-sensitive coating. They deposited an agarose layer onto the ITO coating using the boiling water method [16]. The working principle of this fiber humidity sensor is based on electromagnetic resonance (EMR).

In addition to the ITO coating that has been used in humidity sensors based on the lossy mode resonance (LMR) phenomenon, Sanchez et al. [17] demonstrated a fiber humidity sensor based on LMR but with two different material supporting layers—ITO and In_2O_3. From their experimental results, the In_2O_3-based humidity sensors have shown a 12% improvement in the sensitivity when compared with the ITO-based sensor in the RH changes between 20% and 80% RH. Moreover, the sensor is able to carry out dual-reference measurements due to multiple resonances associated with TE and TM light. Furthermore, the manufacturing time is significantly shorter, to less than a seventh, because the required sensing film bilayer for the In_2O_3-based sensor is reduced to only 13, whereas 100 bilayers are needed for ITO-based sensors to produce a similar performance to the In_2O_3-based humidity sensor. Furthermore, the sensor shows a very good degree of repeatability and reversibility. Aneesh et al. [18,19] demonstrated two RH sensors based on different nanoparticle sensing films over an unclad straight optical fiber. The zinc oxide–doped solgel and TiO_2-nanoparticles-doped nanostructured thin-film sensing material deposited onto the unclad plastic-clad silica multimode fibers and an EW-based intensity modulation technique is applied in both sensor schemes. For sensors based on zinc oxide, two days is required by the sensing probe to dry up at room temperature before it is annealed at high temperature to produce the final sensor device. Meanwhile, the TiO_2-based sensor requires long stirring processes and other related chemical processes to develop the nanostructured thin film. The responses of both sensors are observed to be linear with a wide range of detection with quick time response in milliseconds.

The use of the LMR mechanism for humidity sensing has been explored by Rivero et al. [20] based on silver nanoparticles (Ag-NPs) inside a polymeric coating. The coating is deposited onto a 4 cm exposed part of the optical fiber core by means of the LbL self-assembly. The response of the sensor has been tested in a cycle of exhalation and inhalation, and the result shows that the sensor exhibits a very fast response time of 692 and 839 ms for the rise and fall time, respectively. Therefore, the implementation of nanostructured coating and unclad fiber can be used in practical RH monitoring applications, such as human breathing monitoring devices due to the fast response of this type of sensor. Morisawa et al. [21] investigated a POF humidity sensor using a swelling polyvinylpyrrolidone (PVP) polymer cladding. The attachment of water molecules causes swelling of the polymer cladding that leads to a reduction in the refractive index of the cladding and becomes lower than the core of the POF. Therefore, humid air changes the propagation mode of the POF from leaky to guided mode, causing a remarkable increase in the output light intensity of the POF. They improved the sensor by dye doping the PVP polymer cladding with a phthalocyanine and found that the sensor is three times more sensitive compared to the undoped polymer cladding. The proposed sensor is able to operate in the range from 20% to 90% RH with a response time shorter than 1 s and has a good reproducibility. Furthermore, this sensor has proven that it has high sensitivity at high RH level through testing on a few cycles of human breath.

Employing a tapered optical fiber with a humidity-sensitive thin film represents another fiber configuration for EW humidity sensing. The evanescent field is exposed for such a sensor by tapering the fiber through a heat-and-pull process. This process causes the tapered region to have a smaller diameter compared to the original fiber, and thus, a high fraction of the evanescent field is exposed to the surrounding medium of the fiber. The advantage of using this fiber configuration is that the sensing element can be fabricated without involving any hazardous chemicals. Moreover, different tapered waist diameters can be easily fabricated by controlling the heat temperature, pulling speed, and distance during the fabrication process. Zhang et al. [22] presented a humidity sensor based on standard SMF (SMF-28, Corning), subwavelength-diameter fiber taper coated with a gelatin layer with 80 nm and 8 mm thickness and length, respectively. The humidity response was examined using a wavelength of 1550 nm. When the sensor is exposed to different humidity levels, a change in refractive index of the gelatin thin layer changes the mode field of the guided mode of the coated fiber taper. Then a portion power from the guided mode is converted to radiation mode, resulting in RH-dependent loss at the output of the coated fiber. The fabricated sensor operates within a wide humidity range from 9% to 94% RH with a good reversibility. The sensor has a fast response time of around 70 ms.

A combination of tapered PVA-based sensor and a fiber Bragg grating (FBG) design for humidity sensing with a temperature insensitive characteristic was demonstrated by Li et al. [23]. The FBG acts as a reflective filter, which enables the sensor to operate in reflection mode. The working principle of the sensor is based on the interaction between the optical EW and the PVA coating along the tapered region. The FBG enhances the sensitivity of the sensor due to the light that passes through the taper twice. Although the sensor has a good thermal stability, the transmission loss (i.e., bent fiber) and fluctuations of the laser source may change the sensor output and cause error readings. In addition to SMF, MMF, and POF, which from the literature seem to be the common choices of fiber used in the EW-based fiber humidity sensors discussed so far, other types of fibers are used as a sensor head, such as polarization maintaining fiber (PMF), small-core silica fiber, and no-core fiber with different types of polymer-sensitive films [24–28].

15.3.1.2 In-Fiber Grating

The in-fiber grating technique has become popular recently in various sensor applications due to its inherent sensitivity characteristic to strain, temperature, and refractive index change [29,30]. The grating structure within the fiber is fabricated by exposure to ultraviolet (UV) light, which creates a periodic refractive index modulation of the fiber core. The fiber must have a photosensitivity characteristic to form the grating structure through the UV exposure. The concept of in-fiber grating devices for humidity sensing can be realized by selecting a material that is selectively sensitive to humidity measurand and coating that material in the region where the grating is inscribed at the fiber. The selection of the sensing material for an in-fiber grating type of sensor therefore plays a crucial role in determining the performance of the humidity sensor. The grating-based sensor needs this transducer to react to a humid environment because the grating fiber is not inherently sensitive to humidity. The grating structure determines which category the grating would be classified as, which is dependent on the grating period and position. Generally, the grating can be classified into three main categories, namely, the FBG [29], long-period fiber grating (LPFG) [31], and tilted FBG (TFBG) [32]. Figure 15.4 depicts the structure of the FBG. The sensing principles of the FBG are based on the detection of the shift of wavelength of the reflected light when exposed to the surrounding climate effect. In FBG, the wavelength of the light, also known as Bragg wavelength λ_B, is given by [29]

$$\lambda_B = 2n_{\text{eff}}\Lambda \tag{15.7}$$

where

n_{eff} is the effective refractive index of the fiber core
Λ is the period of the grating (submicron period)

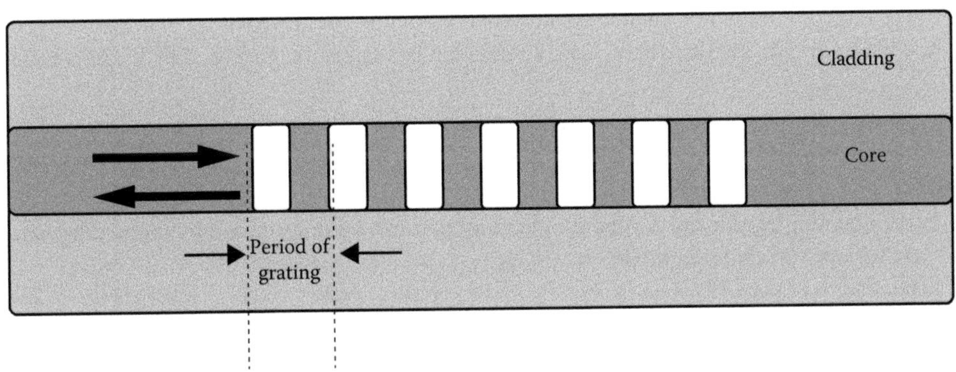

FIGURE 15.4 FBG structure. (From Lee, B., *Opt. Fiber Technol.*, 9(2), 57, 2003.)

Standard FBGs normally have a short period grating around 0.5 μm. The propagating light through the fiber is reflected back when it reaches the grating area in the fiber where the environment element is exposed to the location of this Bragg grating [33]. The surrounding environment changes the properties of the optical fiber: elasto-optic and thermo-optic, which eventually adjusts the period of the grating (Λ) or makes a change to the refractive index [34] and, hence, shifts the Bragg wavelength. To use an FBG as a humidity sensor, the coating material on the Bragg grating structure must not only be sensitive to humidity but also be able to induce mechanical deformation as it interacts in order to induce strain on the FBG, which eventually influences the period of the grating. The FBG represents a class of optical transducer that changes the spectrum of the reflected light.

An LPFG is totally different compared to the FGB concept. Instead of reflecting the light, the LPFG allows the light to go through the grating by the forward-propagating mode to one or several of the forward-propagating cladding modes [36]. Typically, the LPFG has a larger period in the range of 100 μm to 1 mm [37], and the concept of LPFG promotes coupling between the propagating core mode and co-propagating cladding modes. An LPFG is normally fabricated by photoinducing a period refractive index modulation in a fiber core, which results in the transmission spectrum of the fiber containing a series of attenuation bands corresponding to the coupling of a different cladding mode at distinct wavelengths. The forward-propagating cladding modes are guided by the surrounding medium, and therefore, guidance of the cladding modes is affected by the refractive index of the surrounding medium [38]. Like the FBG sensor, in order to build the humidity sensor based on an LPFG, an overlay material that responds to humidity in the local environment is coated surrounding the grating structure. However, the selection of the coating is less stringent than FBG because the selected material species only need to respond to the presence of humidity by changing its refractive index properties [39]. The resonant wavelength and amplitude of the LPFG are very sensitive to the changes in refractive index of the surrounding medium caused by the perturbation of the humidity atmosphere; thus, the LPFGs have been exploited as sensing humidity transducers due to their high sensitivity to variation of the surrounding refractive index. Resonance wavelength shifting is used as the sensing parameter given by the phase matching between the modes propagating in the core and cladding; the expression is given [37] by

$$\lambda = \left(n_{\text{eff}}(\lambda) - n^{i}_{\text{clad}}(\lambda)\right)\Lambda \tag{15.8}$$

where

λ is the resonant wavelength
$n_{\text{eff}}(\lambda)$ is the effective refractive index of the propagating core mode
$n^{i}_{\text{clad}}(\lambda)$ is the refractive index of the ith cladding mode
Λ is the period of the LPFG

while the minimum transmission of the attenuation bands can be governed by the expression

$$T_i = 1 - \sin^2\left(k_i L\right) \tag{15.9}$$

where

k_i is the coupling coefficient for the ith cladding that can be determined by the overlap integral of the core and cladding mode
L is the length of the LPFG

In comparison, a TFBG possesses a similar periodic refractive index modulation along the fiber core axis like a common FBG, but their index modulation pattern has a certain tilt angle with respect to the fiber axis [32]. Due to the existence of the tilt angle, beside the core mode coupling, as in common FBGs, contra-cladding mode coupling also occurs in TFBG, which enhances the sensitivity of the TFBG to the surrounding medium. Therefore, the TFBG, which possesses the merits of the LPFG and FBG, has attracted substantial interest in the sensing area. The Bragg resonance condition for TFBG can be described as [32]

$$\lambda_{\text{Bragg}} = \left(n_{\text{eff}}^{\text{co}} + n_{\text{eff}}^{\text{co}}\right)\frac{\Lambda_g}{\cos\xi} \tag{15.10}$$

where

$n_{\text{eff}}^{\text{co}}$ is the effective refractive index of the propagating core mode
Λ_g represents the grating period along the fiber axis
ξ is the tilt angle of the grating period

A fiber humidity sensor based on TFBG can be realized by coating the section of TFBG with a hygroscopic material similar to the LPFG fiber structure.

Recent studies by Correia et al. [40] have shown the use of di-ureasil as a sensing material for FBG-based humidity detection. The di-ureasil is an organosilica hybrid material prepared by the solgel technique, which has a strong adhesion to the silica fiber and swells reversibly according to the change in the RH values. The RH sensor has a humidity detection range from 5% to 95% RH with a sensitivity of 22.2 pm/%RH. The fabricated FBG sensors were tested in a real structure health application where the coated FBGs were put inside concrete blocks. Two concrete blocks with a different porosity were produced by mixing water and cement at different ratios. The sensors were tested for a time period of 1 year, and the results show that the sensors are feasible for monitoring the civil engineering structures.

Humidity sensing in a concrete environment based on an FBG sensor approach was also demonstrated by Sun et al. [5]. The coated polymer FBG humidity sensor is fabricated with built-in temperature compensation (uncoated FBG as a temperature sensor) to monitor water ingress that can deteriorate the building stone. A polymer coating of 24 μm thickness is applied to the FBG sensor using the dip-coating method. Both FBGs are assembled in one probe, side by side with a careful design of packaging. The first assessment of the probe in the laboratory environment shows that it can operate in RH detection from around 10%–90% RH. The probe sensor is then evaluated in the specially built wall for over several days. It was found that the probe sensor performance is better than the commercial capacitance sensor as the fiber humidity sensor investigated here is not affected by the condensation of moisture inside or on the surface of the sensor itself.

Humidity sensing in the working point of nuclear and subnuclear physics detectors based on the ionization of a gas medium is yet another example of an application in which a polymer-coated FBG sensor approach can be adapted. The recent work by Caponero et al. [41] illustrates the flexibility of the FBG humidity sensor to monitor humidity condition in resistive plate counters (RPCs) packaged in a steel box. This is important, as the humidity influences the working point of the RPC. Due to the RPC operating in a high electromagnetic field environment, the FBG sensor is the best candidate for installation on an RPC due to the fact that FBG sensors are immune to electromagnetic interference (EMI). Polymer material is used as a humidity-sensitive coating for the FBG sensor. The coated FBG sensor is inserted in the gas conduit upstream chamber of the RPC detector through a customized casing to monitor the distribution of the RH value over the RPC. A second uncoated FBG sensor is also included in the casing to compensate for the temperature sensitivity of the FBG. A detection humidity range from 25% to 65% RH and resolution of 2% RH can be achieved for this probe sensor. This shows that the probe sensor is eligible to be installed not only but multiple FBGs to cover the large detectors through multiplexing.

Like FBG, the use of LPFG grating in humidity sensing is also quite popular. Corres et al. [42] have presented a humidity sensor using LPFG sensor by depositing the alumina (Al_2O_3) and poly(sodium 4-styrenesulphonate) on the cladding of the LPFG fiber structure. The deposition of the sensitive layer is done using electrostatic self-assembly (ESA) to produce a thin layer of subwavelength thickness. Moreover, a high refractive index layer can be intercalated to overcome the low refractive index of the sensing material. Details of the ESA layer deposition method can be found in [43]. The sensitivity of the LPFG humidity sensor is enhanced by using an additional intermediate layer, which is PAH+/PAA–. The range of variation of RH demonstrated by this coated LPFG is restricted to 50%–75% RH due to a limitation of the attenuation bands of the LPFG fiber structure. However, a dynamic wavelength shift of 11 nm was obtained with a sensitivity of 0.44 nm/%RH for this LPFG sensor.

Venugopalan et al. [44] developed a grating fiber humidity sensor using LPFG structure in SMF with a setup consisting of a broadband light source, sensor, chamber, and OSA. Salt solution is used as a reference for RH in the small chamber. The sensors are fabricated using a thin layer PVA on the LPFG fiber structure, the refractive index of which changes as a function of the humidity level, thus inducing a wavelength shift in the attenuation bands of the LPFG sensor in the transmission mode. The coating thickness of 800 nm is implemented for the LPFG sensor by using simple dip-coating processes. The sensor performance is evaluated, and it was found that the humidity level detection ranges from 33% to 97% RH with a good repeatability and an average speed of response time of around 1 min. Even though the sensor exhibits a nonlinear response, it has a high sensitivity when the humidity level was changed from 75% to 97% RH with a very large wavelength shift of around 100 nm.

Viegas et al. [45] designed a sensor to measure humidity based on LPFG coated with a thin film of SiO_2 nanospheres. The thin film is deposited on the grating structure using the ESA method. The optical properties change when exposed to a variation of humidity levels, which causes the resonance wavelength of the LPFG to shift. The coated LPFG sensor can produce a wavelength shift of up to 12 nm in the humidity range from 20% to 80% RH. Although the sensor responds to a temperature variation with a temperature coefficient of around 100 pm/°C, the temperature sensitivity of the LPFG sensor can be compensated by using straightforward standard thermal compensation techniques. Viegas et al. [46] extended their work by enhancing the sensitivity of the LPFG-based humidity sensor by using an intermediate layer. The intermediate layer is deposited between the humidity-sensitive film and fiber cladding using the ESA method. Due to a higher refractive index of the intermediate layer compared to sensitive film, the total effective refractive index of the coating is increased and thus improves the sensitivity of the sensor.

Yu et al. [47] introduced cascaded LPFG sensors, where two identical LPFGs are cascaded in series on the same optical fiber separated by a few centimeters distance to produce the interference fringes within the attenuation band of LPFGs. The cascaded structure is coated with a hydrogel overlay using a dip-coating technique to make the cascaded LPFGs function as a humidity sensor. The wavelength shift of the resonance wavelength of the LPFG-based sensor is measured using Fourier analysis. The performance of the cascaded LPFGs sensor shows a good linear response in the upper humidity range from 60% to 100% RH with a sensitivity of 9.9×10^{-3}/%RH, and its accuracy was found to be 0.39% RH.

A common working configuration for coated LPFG structures is in the transmission mode. Therefore, FBGs are a common choice for sensing due to a simple design probe (convenient to work in reflection mode). Nevertheless, the research by Alwis et al. [48] modified the coated LPFG with a period of 330 µm to operate in reflection mode like the FBG and obtain better performance of the sensor. The proposed sensor applies a polymer coating on the cladding of the LPFG at the distal end of the fiber probe. In order to realize a reflectance LPFG-based sensor, a silver mirror was first coated at the tip of the written LPFG fiber with a separation distance of 20 mm between them. Multiple resonance peaks can be observed from the output of the sensor in reflection mode as a complex modal interaction is formed due to the reflections from the silver mirror. By increasing the surrounding humidity, the coated polymer swells accordingly due to the absorption of water vapor, and thus, a longitudinal strain is applied directly to the LPFG. As a result, a ~6 nm blue shift of the resonance LPFG wavelength is observed for range from 20% to 80% RH. The sensitivity of the fabricated sensor was found to be 0.10 nm/%RH with insignificant hysteresis under 1% RH.

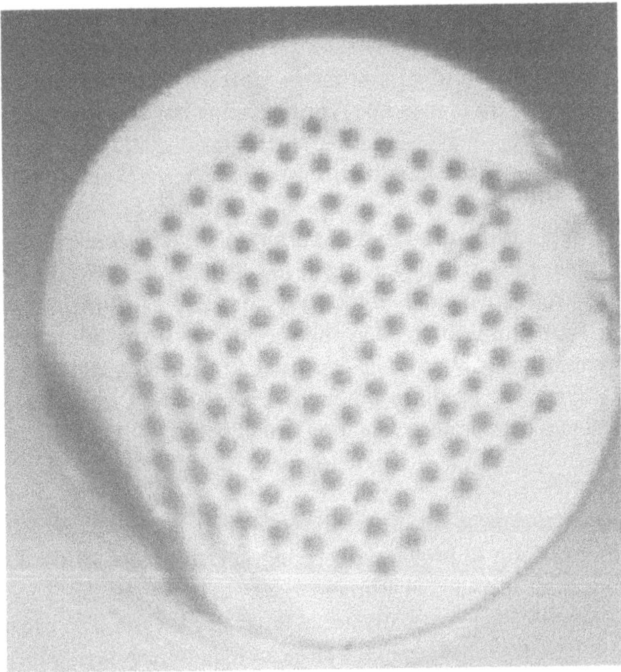

FIGURE 15.5 Cross section of the PCF.

With the advent of fiber technology, Zheng et al. [49] have developed LPFG-based humidity sensor using a new generation of fiber, a photonic crystal fiber (PCF), as shown in Figure 15.5. They proposed two types of nanofilm coating configuration, which are inner and outer PCF coatings. Like a traditional coating design, the outer coating is done by depositing the nanofilm on the cladding of the PCF, while for the inner coating, the film is deposited on the surface of the cladding air channels in the PCF. The ESA technique is employed to deposit the nanofilm for both the coating designs. They used an endlessly single-mode PCF (ESM-PCF) fiber to inscribe the LPFG on it. The core and diameter of the PCF are 10.4 and 125 μm, respectively. The air channels in the cladding of the PCF are arranged in a hexagonal pattern with an average air channel diameter of 3.1 μm. Both of the PCF-LPFG sensors have been tested in the humidity range from 20% to 50% RH. The response of the sensors to the humidity is based on the wavelength shift and resonance intensity change. The experimental results of the sensor show that the inner-coated PCF-LPFG has higher sensitivity compared to the outer-coated PCF-LPFG. The inner-coated PCF-LPFG exhibits a large resonance wavelength shift of 0.0007% pm within the RH region of 22% to 29% RH and has good thermal stability.

TFBG, on the other hand, offers an alternative and additional advantage over the FBG and LPFG as it has a merit from both types of grating. Unlike the FBG and LPFG sensor, the in-fiber grating for the TFBG sensor is not reported much in the literature for humidity sensing, as this grating technology is still fairly new compared to other grating structures. One example of a TFBG sensor that has been used to sense humidity was demonstrated by Miao et al. [50]. These authors proposed a TFBG humidity sensor by utilizing a common humidity-sensitive coating for other types of grating, which is a PVA thin film. As the TFBG is sensitive to ambient refractive index through its cladding mode, the PVA coating on the cladding of TFBG will influence the resonance wavelength of the TFBG due to the change in refractive index of the PVA film. The PVA-coated TFBG has demonstrated a response to various humidity levels ranging from 20% to 98% RH in transmission spectrum. The sensor has a sensitivity of 2.52 and 14.947 dBm/%RH for humidity range of 20%–74% RH and 74%–98% RH. Moreover, the sensor exhibits a good degree of hysteresis and repeatability.

Although the response of this humidity sensor has two different linear behaviors for different RH range detections, it was found to be highly sensitive in the area of high humidity; therefore, this sensor could be used in RH monitoring for ordinary as well as extremely humid atmospheres.

Recently, a combination of more than one type of grating has also attracted interest in humidity sensing. For example, Dong et al. [51] and Li et al. [52] used grating technology combining FBG and TFBG to form a grating-based humidity sensor by superimposing the FBG at the end part of the TFBG. The length of the FBG is much smaller than that of the TFBG. Again, the regular PVA-sensitive film for other types of grating is coated onto the hybrid grating sensor. The coated FBG-TFBG sensor operates by reflecting a part of the cladding mode of the hybrid grating sensor, which is sensitive to the refractive index of the PVA film. Thus, the intensity of the reflection output changes when the PVA film is affected by the variation of surrounding humidity. The range of humidity detection for this coated hybrid sensor is from 30% to 100% RH with a maximum sensitivity of 0.737 nW/%RH. The response time of the sensor is around 2 s, and it has an excellent thermal stability when exposed to a temperature condition from 20°C to 80°C.

15.3.1.3 Interferometry

Interferometry is one of the most sensitive, powerful, and versatile approaches that have been applied in optical fiber sensing to provide a very high performance of the OFS. This method is normally used when a high level of measurement sensitivity is needed. Most of interferometric fiber-optic sensors use an SMF and basically are of the intrinsic type. The use of interferometers in optical measurement is based on changes in the phase of light in response to the physical parameters surrounding the fiber. The phase shift can be detected by combining the intended signal with a reference signal. The mixing of the two optical signals will consequently convert the phase difference between the two signals into an optical intensity change. The phase angle for light with given λ, n_1 is index of the refraction of the core and length L is given by [33]

$$\Phi = \frac{2\pi L}{\lambda} = \frac{2\pi n_1 L}{\lambda} \tag{15.11}$$

A change in length and/or refractive index under perturbation of the external physical environment will cause a phase change, as defined by the following equation:

$$\Delta\Phi = \frac{2\pi}{\lambda_0}\left(n_1 \Delta L + \Delta n_1 L\right) \tag{15.12}$$

Note that very tiny changes in length and refractive index produce a large phase difference, also known as the optical path difference. Various interferometer configurations, such as the Fabry–Perot (FP), Michelson, Mach–Zehnder (MZ), and Sagnac, can be used for sensing.

The most straightforward interferometer configuration is called MZ, which is composed of two fiber beam splitters where one fiber is exposed to a measuring field (sensing fiber) and another fiber is isolated from the surrounding of the sensing element (reference fiber). Both light signals are out of phase when the sensing fiber is perturbed by the external local environment. For example, when the overlay that is sensitive to humidity is coating the sensing fiber, the reaction between the humidity and the coating causes the sensing fiber length to change, which eventually produces changes in the path difference. As a result, the intensity output decreases due to destructive interference. A similar principle is applied in the Michelson interferometer. The only difference is that the light wave is split and recombined in the same splitter [53]. On the other hand, the FP interferometer is a type of multibeam interferometer. This configuration typically consists of two semireflective fiber mirrors. Optical light waves bounce forward and backward between mirrors (cavity) with specific length and wavelength. The mixing of these waves will produce maximum intensity output, also known as resonance. Any deviation contributed by length or wavelength will decrease the output intensity.

To make the humidity detection possible using the FP interferometer configuration, the properties of the cavity must be dependent on humidity, which causes the resonance wavelength to shift in response to the humidity variation, which can be conveniently detected by performing intensity or wavelength shift measurement. The Sagnac interferometer is another type of the interferometer in which two beams from the same point are propagated in opposite directions with different polarization states through the optical fiber loop. Both of the beams are combined again at the source of the beams. The sensing mechanism is dependent on the polarization of the light guided along the fiber loop. Typically, birefringent fibers are used in the sensing part.

A sensor design using an FP-based sensor for humidity sensing has been reported by Corres et al. [54]. In order to fabricate this sensor, the tip of the SMF pigtail is coated with a film of SiO_2 nanoparticles that are highly sensitive to humidity. The coating consists of SiO_2 particles with a low thickness of 760 nm, which is created using the ESA technique. The sensor exhibits two linear regions in the RH range from 40% to 98%RH with a variation of 10 dB reflected optical power. In addition, the sensor has a low hysteresis and fast response time (the rise and fall times are 150 and 100 ms, respectively). The sensor has been tested for the application in breath monitoring, and the results show that the response is fast enough to detect a cycle of human breathing. The thermal sensitivity of the sensor is negligible for temperatures higher than 36°C.

Ruete et al. [55] reported a nano-coated FP dual head in series humidity sensor. The sensor head is coated at the tip of the SMFs by means of the LbL technique with a PDDA/poly R-478 film thickness of 32.5 nm (40 layers). The working principle of the FP sensor is based on the change in the optical properties of a sensing film as a function of humidity. The sensor has a dynamic reflected optical power of 6.71 dB, which is achieved within the RH range from 11% to 85% RH. In addition, the sensor exhibits low hysteresis of under 1%.

Chen et al. [56] demonstrated a miniature FP interferometer humidity sensor. The sensing head consists of an SMF (SMF: 8 nm/125 nm) spliced with a section of hollow-core fiber (Polymicro TSP050150: 75 nm/125 nm) and a thin chitosan sensing film coated over the end face of the hollow fiber. The 50 μm length of the hollow-core fiber is used in this FP sensor configuration. The interfaces of a single-mode, hollow-core fiber and chitosan hollow-core fiber act as two reflective mirrors forming a small FP cavity. The basis of this FP humidity sensor relies on the swelling of the chitosan material, which changes the FP cavity due to the absorption of water molecules by the chitosan film. As a result, the FP interference fringe of the FP sensor is shifted as a function of the humid environment. The experimental results show that the fabricated sensor is able to sense humidity in the range of 20%–95% RH with a good sensitivity of 0.13 nm/%RH. In addition, the sensor has a fast response time of around 380 ms. Although the uncertainty of the chitosan-coated FP sensor is ±1.68% RH, the chitosan sensing film used in this FP sensor configuration offers good mechanical strength, fine stability, and good diaphragm to form the FP cavity.

Yao et al. [57] proposed an FP interferometer humidity sensor with a low cross-temperature sensitivity. The sensor is fabricated by dip coating a highly humidity-sensitive nanocomposite polyacrylamide (PAM) material onto the open cavity of SMF. The cavity is formed using a large offset (~65 μm) splicing method. As the refractive index of the PAM material changes significantly after it absorbs water vapor molecules in the surrounding air, the properties of the FP cavity change and the shifting of FP interference spectrum can be observed. From the experimental results, the sensor shows that the interference spectrum shift is around 4 nm when the environment humidity varies in the range of 38%–78% RH with a sensitivity of 0.1 nm/%RH. However, a huge shift of ~59 nm can be observed for a humidity level range from 88% to 98%RH with a sensitivity of up to around 5.9 nm/%RH. The PAM-coated FP sensor has been proven to have low-temperature dependence with a temperature sensitivity of around 0.6 pm/°C.

Besides the FP sensor, the MZ sensor is one of the most employed in fiber humidity sensing for interferometer-based sensors. As an example, Gu et al. [58] proposed an MZ humidity sensor based on a thin-core fiber with an FBG in between. The surface of the MZ structure is coated with a sensitive film of poly (*N*-ethyl-4-vinylpyridinium chloride) (P4VP·HCl) and poly(vinylsulfonic acid, sodium salt) (PVS) through an LbL ESA technique for humidity sensing. The coated MZ interferometer sensor is used to measure the humidity-induced refractive index change in the nanofilm. On the other hand, the FBG in the middle of the coated MZ interferometer sensor can compensate the cross-temperature sensitivity of the sensor. The sensor has an average sensitivity of 89.8 pm/%RH in a wide range RH detection from 20% to 90% RH. Time response of the sensor is around 2 s with a linear response. The sensor exhibits an estimated detection resolution of 0.78% RH in different temperature conditions. This proves that the FBG in the middle of the MZ interferometer sensor can be utilized to compensate the cross sensitivity to temperature of the sensor.

Mathew et al. [59,60] reported a novel fiber humidity sensor based on a PCF MZ interferometer. Agarose sensing material is coated on the cladding of the PCF interferometer by drawing the interferometer through a hot agarose solution. The interferometer is formed by collapsing both splice sections of the PCF with SMF, which allows the excitation of the core and cladding modes in the PCF section. In principle, the agarose refractive index changes upon the variation of the humidity level. Therefore, when the cladding modes interact with the agarose film, the interference spectrum of the interferometer sensor is shifted due to a change in phase shift over the length of the PCF length. In order to investigate the effect of the thickness of the agarose sensing film on the performance of the PCF interferometer, several identical interferometer sensors are fabricated with a PCF length of ~40 mm. From the test results of all the sensors, the one that has the thickest coating (1250 nm) exhibits the highest sensitivity. It has a good linearity for an RH range of 40%–90% with a detection resolution of 0.07% RH. Above 90% RH, the response of the sensor is not linear. The sensor has a fast response time of 75 ms for an abrupt 40% RH change.

Li et al. [61] also used a similar PCF structured configuration to fabricate an interferometer sensor. However, to turn the interferometer sensor into a humidity sensor, instead of using an agarose material, they used a PVA sensing film, which is a common humidity-sensitive film applied in most of fiber humidity sensors, and deposited the film onto the cladding of the PCF using a dip-coating process. Due to the temperature dependence of the device being 3.1 pm/°C, the sensor is almost temperature insensitive as the sensor has a very high sensitivity to humidity compared to the temperature sensitivity. The main drawback of this sensor is that the sensing element is easily exposed to minor errors, such as fiber bending and induced strain.

Zhang et al. [62] presented a hybrid sensor to measure the humidity and temperature at the same time based on a similar PCF MZ interferometer sensor cascaded with an FBG. Although the PVA-coated interferometer is sensitive to temperature and humidity, the FBG is only temperature sensitive. Therefore, simultaneous measurement of humidity and temperature can be achieved by using the proposed sensor configuration.

Mathew et al. [63] also proposed a novel hybrid fiber sensor to measure humidity and temperature simultaneously. A PCF interferometer sensor in reflection configuration is fabricated by splicing the PCF fiber with SMF. The length of the collapsed region at the splice point is around 250 μm, and the tip of the PCF fiber is cleaved so that it behaves as a mirror in open void air holes. To create a humidity sensor, a length of 1 mm PCF in open void configuration is immersed in a hot agarose solution to infiltrate the air holes of the PCF to improve the humidity sensitivity of this interferometer humidity sensor. The length of the agarose-infiltrated air holes of the PCF is approximately 100 μm long. The FBG is put in line with the agarose-infiltrated PCF interferometer sensor to measure temperature by detecting the shift of the Bragg wavelength. The PCF interferometer infiltrated with agarose, on the other hand, is used to monitor humidity through the intensity change in the reflected spectrum of the sensor. The sensor has shown a dynamic power of more than 7 dB from 20% to 95% RH with a small temperature sensitivity of 9.8 pm/°C.

Chen et al. [64] proposed a fiber interferometer sensor for humidity sensing based on polarization maintaining (PM) the fiber in the Sagnac interferometer configuration, as shown in Figure 15.6. The interferometer PM sensor uses a chitosan as a coating material due to its ability to absorb water molecules from the surrounding environment, which can cause an induced strain on the PM fiber-based

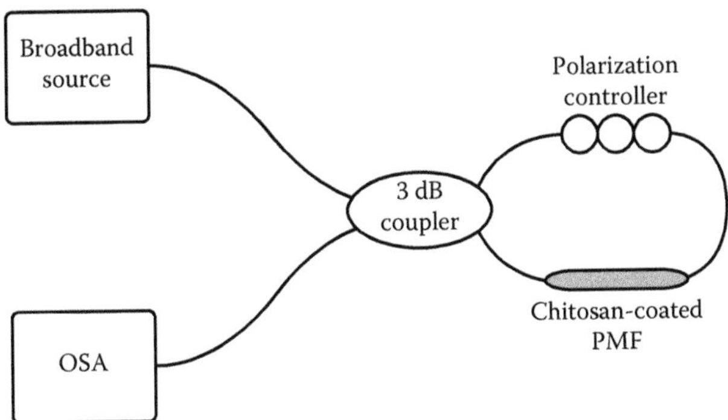

FIGURE 15.6 Chitosan-coated based polarization maintaining fiber based Sagnac interferometer sensor. (From Li Han, C. et al., *IEEE J. Sel. Top. Quant. Electron.*, 18(5), 1597, 2012.)

Sagnac interferometer sensor through the swelling effect of the chitosan. The transmission spectrum of the sensor is in interference pattern. The sensor performance is optimized based on the concentration of the chitosan material and chemical etching of the sensing part of the PM fiber.

15.3.1.4 Spectroscopic

Up to 10 years ago, the spectroscopic method was a popular choice to design a fiber humidity sensor. This method observes the optical power intensities obtained from the measurement and relates them to the concentration of the water vapor through optical absorption or fluorescent spectral properties of the sensing film. Most of the spectroscopic fiber-based humidity sensors used an air gap design [65,66] to detect the characteristic optical spectral properties of the humidity-sensitive film. A variety of fiber humidity spectroscopic sensors have been reported by the researchers in the literature up to 2008 [30]. However, the spectroscopic method for sensing humidity is not the preferred method for most researchers nowadays. This can be seen through the literature in which only a few spectroscopic fiber sensors were reported between 2008 and 2013.

Examples of these sensors are from Ohira et al. [67] who have reported a fiber-optic sensor with an air gap spectroscopic configuration. A transparent plate covered by a thin layer silica gel impregnated with 8-amino-1-naphthol-5-sulfonic acid (ANS) is put in the middle of the gap. The gap is formed by a small distance between the optical fibers and a detector, which are fixed face-to-face with each other. The transmission of the plate at 850 nm wavelength provides the RH value. The transmittance of the silica plate is decreased when the surrounding humidity is increased due to a change in light absorbance of the transparent plate. The sensor device can detect a humidity range from 19% to 99.6% RH with an error of ~±4% RH. The response time of the sensor to humidity is below 5 min. This sensor can also be used to detect other gas species such as NO_2 and O_2 by using a different wavelength operation.

The latest sensor that used the spectroscopic technique for fiber humidity sensing was demonstrated by Wang et al. [68], as shown in Figure 15.7. The spectroscopic sensor configuration

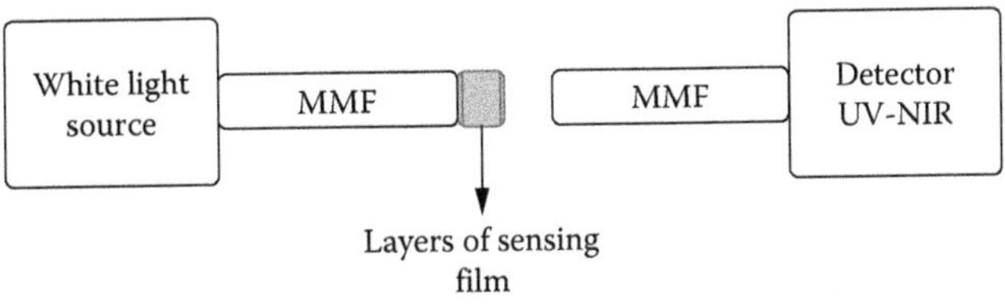

FIGURE 15.7 Fiber humidity sensor based on optical absorption. (From Biao, W., An optical fiber humidity sensor based on optical absorption, In *Communications and Photonics Conference and Exhibition, 2011*, ACP, Asia, 2011.)

is formed by a small gap of bare multimode fibers. First, the end face of the MMF in the air gap configuration is coated with $CoCI_2$-doped PVA/SiO_2-sensitive film. The response of the sensor is observed through the optical absorption characteristic of $CoCI_2$. The wet $CoCI_2$ has a low absorption at the wavelength band around 550–750 nm, whereas it produces a very high absorption in dry conditions. The $CoCI_2$ is doped with PVA/SiO_2 to obtain a good stability at the absorption wavelength band of $CoCI_2$. By detecting the change in absorption of $CoCI_2$, the measurement of humidity can be achieved. From the measurement results, the sensor exhibits a humidity sensing range of 25%–65% RH with a good repetitive response. In addition, the response time of the sensor is less than 2 min. This air gap sensor design needs a proper packaging to make sure the MMFs are aligned well with each other and have a stable design structure to minimize the error reading due to vibration and pollution.

15.3.2 Noncoated Optical Fiber Humidity Sensors

15.3.2.1 Adsorption of Water Molecule

Water molecules adsorb onto virtually all surfaces. For instance, water molecules adsorbed on hydrophilic silicon oxide surface cause great alteration in adhesion and friction in nanoscale contact [69,70]. The mechanism of adsorption of water vapor by the surface of silicon oxide (known as SiO_2@silica) has been reported by Morror et al. [71,72]. They have shown that the adsorption of water vapor creates a nano-thin layer along the surface of silica fiber, and researchers from Switzerland [73] have demonstrated this effect on the waveguide, which alters the effective refractive index of the guided mode along the surface of the waveguide. Asay and Kim [74] reported that the evolution of adsorbed water molecule on silicon oxide at room temperature is a function of RH and the molecular configuration of water molecules on the silica surface when the RH is increased. It was found that below 30% RH the growth rate for water layer is very small, estimated to be below an average of 10 nm thickness. For the RH range from 30% to 60% RH, the average thickness only increases a few nanometers. As the humidity increases around 60% RH, the thickness of the thin layer increases exponentially but is limited to a total thickness below 30 nm. To detect such very small changes, the interferometric configuration is needed to design a fiber humidity sensor based on this working principle due to the interferometric technique being one of the most sensitive approaches available for optical fiber sensing through an optical phase–induced measurement.

The concept of a fiber humidity sensor based on adsorption of water molecules onto the surface of silica fiber is relatively new. To date, only a few studies on this subject have been reported, starting from the year 2010. Fiber humidity sensors using the principle of water molecule adsorption was first reported by a researcher from the DIT Photonics Research Centre, Ireland [75]. The sensor design is based on an in-reflection PCF interferometer, which is composed of a PCF fiber spliced with standard optical fiber. The air holes of the PCF in the splicing region of both fibers are collapsed due to the heat applied during the fusion splicing. The other end of the PCF is cleaved using a fiber cleaver to make the end of the PCF act as a mirror and expose the air holes of the PCF's fiber end to the ambient atmosphere. The diffraction of the SMF mode due to the collapsed section of the PCF enables the excitation of two modes inside the PCF. These modes propagate through the PCF and are reflected back when they reach the cleaved end of the PCF. They are recombined again as an SMF core mode when passing through the same collapsed region. One of the modes in the PCF is propagated through the cladding of the PCF, and the interaction of this mode with the adsorption and desorption of water molecules along the surface of the air holes within the PCF's cladding will shift the interference spectrum of the PCF interferometer. Therefore, humidity can be measured as the adsorption of water molecules as a function of ambient humidity. With the length of 3.4 cm for the PCF interferometer sensor, it has an RH range detection from 40% to 95% RH. The sensitivity of the sensor is high at ~24 pm/%RH above 70% RH while in the region of 40%–70% RH, the sensor shows a low sensitivity of around 5.6 pm/%RH. As the sensor head is fully made of single material silica, the sensor can be used in normal and high-temperature conditions to measure humidity.

Favero et al. [76] introduced a simple novel PCF interferometer sensor based on the same working principle to sense humidity for breathing applications, which operates in reflection mode. The interferometer sensor is formed from a stub of PCF fusion spliced with a standard SMF at one end of the PCF. Instead of the other end of the PCF being left open to the atmosphere, they sealed it by splicing ~10 mm of SMF to the PCF and spliced the short section of SMF with ~5 mm capillary tube with an inner and outer diameter of 40 and 125 μm, respectively. The end of the tube is sealed by means of arc discharges using the splicing machine. This tube structure is used to reflect the light back into the PCF again. The key part for the PCF interferometer is the collapsed air hole region in the PCF, which is what allows the excitation and combination of multiple propagation modes along the PCF. The cladding modes of the PCF are highly sensitive to the adsorption and desorption of water molecules along the PCF cladding surface. As a result, when water molecules are present on the PCF surface due to the variation of humidity level, the interference pattern of the interferometer sensor shifts. In this case, they used a 2 cm long sealed-void interferometer sensor placed in the oxygen mask with an LED central emission at 1550 nm as a light source and the wavelength shift of the interference spectrum is monitored by a commercial FBG interrogator (I-MON E-USB, Ibsen Photonics). From the testing result with the human exhaled air, the sensor device can be used to monitor human breathing at different breathing rates. Moreover, the sensor output is insensitive to fiber bending and movement.

Fu et al. [77] demonstrated a tapered SMF for humidity sensing without any special hygroscopic coating. To fabricate the fiber taper sensor, the SMF is corroded with HF acid for several minutes (20 min for the optimized fiber taper sensor) to decrease the diameters of the core and cladding to form a tapered structure. The sensing mechanism of this sensor is described by analyzing the evanescent field that interacts with the water molecules surrounding the surface of fiber taper. From the experimental data, the sensor exhibits a humidity measurement range from 50% to 95% RH with a measurement resolution of 1% RH.

Another fiber humidity sensor based on the adsorption water molecule effect for applications in breath sensing was demonstrated by Lin et al. [78], as shown in Figure 15.8. In the sensor design discussed, a piece of optical gradient-index fiber (GIF) tip was spliced to the end of a standard single-mode optical fiber. The GIF acts as a collimator lens that expands the light coming from the SMF. When expanded light reaches at the facet of the GIF (interface between the GIF and ambient air), it will interact with the water molecules that attach onto the GIF tip, which is a function of the humidity level, and reflect back into the SMF according to Snell's law. The humidity is determined through the intensity of the reflection light signal. A GIF with a core and cladding of 100 and 140 μm, respectively, is used to fabricate the sensor with a quarter pitch of 446 μm. The results show that the breath sensor has a good sensitivity and that it is able to operate in real-time respiration monitoring applications.

Shao et al. [79] proposed a novel humidity sensor based on in-fiber MZ interferometer, which consists of two arc-induced ellipsoid fiber tapers in between two SMFs. Both of the arc-induced tapers are fabricated on the same SMF. When the fundamental mode of SMF reaches the first taper,

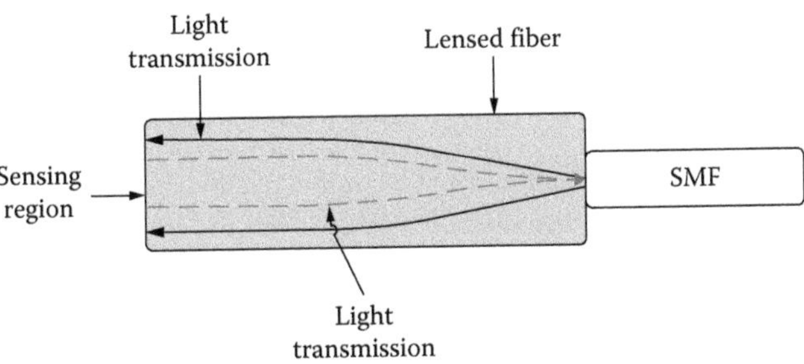

FIGURE 15.8 Schematic working principle of the breathing sensor based on optical lensed fiber. (From Lin, Y.-C., *Microw. Opt. Technol. Lett.*, 55(2), 450, 2013.)

the cladding modes are excited and propagate in the SMF as the interferometer arm, and thus, the modal phase difference between the core and cladding mode is produced. The second taper into the SMF recombines the core and cladding modes. The sensing principle is based mainly on the attachment of water molecules along the cladding surface between the tapers, which gives some force on the fiber with RH variation. In addition, the refractive index of the air, which is a function of RH, also alters the cladding mode of the fiber tapers. As a result, the phase difference due to these effects will change the spectrum of the MZ interferometer.

15.3.2.2 Edlén Theorem and Air Index of Refraction Principle

A series of empirical equations based on the theory of Edlén [80] give the refractive index (n) as a function of the temperature, pressure, and humidity level. The refractive index of the air cavity in the gap is a function of ambient humidity with condition of the other variables (temperature and pressure) being maintained constant. The index at a certain temperature (T) and pressure (P), n_{TP}, can be determined as [81]

$$n_{TP} = 1 + P\frac{n_s - 1}{9609543}\frac{1 + 10^{-8}(0.601 - 0.00972T)P}{1 + 0.0036610T} \tag{15.13}$$

where

T is in °C

P is in Pa

n_s is the index of the standard air (1 atm, 15°C, and 0% RH), as defined by the Edlén equation

When the RH, at a certain partial pressure of water vapor (F), is considered, the index, n_{TPF}, can be expressed as [81]

$$n_{TPF} = n_{TP} - F\left(3.7345 - 0.0401\sigma^2\right) \times 10^{-10} \tag{15.14}$$

with

$$F = P_0 \exp\left(\frac{\alpha(T - T_0)}{b + T}\right) * \%\text{RH} \tag{15.15}$$

where

F is in Pa

T is in K

σ (the wavenumber of the light) is in μm^{-1}

% RH, being the percentage of RH, is valued between 0 and 100 for 0% RH and 100% RH, respectively

$P_0 = 611.213$ Pa

$\alpha = 17.50$

$b = -32$ K

$T_0 = 273.15$ K are parameters relating to the characteristics of water vapor

Therefore, through the Edlén equation, humidity can be calculated based on the refractive index of air.

In addition, the investigation of the humidity effects on the air index of refraction based on the Fresnel relations and the absorption/scattering law can also be used to determine humidity, as described in [82,83]. Although not many sensors that are based on both of the previous principles have been reported for humidity sensing, examples are as follows.

Murty et al. [81] have presented a fiberized low-coherence interferometer humidity sensor based upon the measurement of the refractive index of air. The setup of the sensor probe is based on the Michelson interferometric (MI) configuration. The probe consists of a mirror, a collimator, and a beam splitter.

One arm is a tube containing air at fixed pressure and humidity level that serves as a reference. The second arm is exposed to the humid air, which acts as the sensing region. When there is a change in humidity surrounding the Michelson interferometer sensor, only one arm is affected and changes the optical path difference of the interferometer. A change in humidity due to the optical path difference is measured using a lock-in amplifier with the help of two more subsystems, which are the processing interferometer and FP reference shielded in a thermal enclosure connected through a power distributor. A temperature and pressure electronic sensor is also used in this sensor system to calculate the RH based on the Edlén theorem. The exposed cavity in the second arm of the interferometer has a cavity length of 9.75 cm. The sensor works linearly over the full range of humidity (0%–100% RH) with an accuracy of 1.6% RH and does not suffer from hysteresis. This sensor is designed for industrial applications in which the electronic sensor cannot be used in a harsh environment that can cause material degradation.

Mehrabani et al. [82] designed an optical humidity sensor using a small gap configuration based on the intensity modulation due to the refractive index change. The principle of this sensor is that the light power passing through the gap between two fibers is dependent on the medium refractive index in the gap. The humidity can be determined due to the effect of the humidity on the air refractive index variation. The design of this sensor uses a multimode fiber with 40 cm length and a short multimode fiber at 4 cm in the optical gap arrangement. A liquid cell (thickness of around 0.5 mm) is placed between the gaps in the fibers with an effective gap length of 9 mm. The cladding and core diameters of the fiber are 1 and 0.86 mm. The sensor shows a humidity variation as a function of the intensity power variation from 32.76% to 41.26% RH with a power variation of 58 nW. Although the sensor exhibits a small range of RH detection, the performance of the sensor is reasonable as a refractive index monitoring to determine the humidity condition. Golnabi et al. [83] extended their work to investigate and optimize the fiber gap design humidity sensor based on the air index of refraction. Several fiber-to-fiber gap optical designs are tested as a refractive index monitoring mean, and from the experimental results, the probe with an axial fiber gap distance of 10 mm and using a plastic bottle container between the gap shows the highest sensitivity compared to the other gap designs with an average sensitivity of 1.00% RH/nW in the humidity range from 36.93% to 72% RH.

15.3.2.3 Beer–Lambert Law

In optics, the Beer–Lambert law also known as Beer's law relates the absorption of light to the properties of the fluids or gases through which the light is traveling. The law states that the amount of light emerging from a sample is diminished by three physical phenomena:

1. The amount of absorbing material along its path length (concentration)
2. The distance the light must travel through the material sample (optical path length)
3. The probability that the light of that particular wavelength will be absorbed by the material (absorption coefficient)

The transmittance, T, is expressed as the ratio of the intensity of the light power entering the gas medium, P_0, to the light power exiting the gas medium, P, as shown in Figure 15.9, and the absorbance is calculated as $A = -\log_{10} T = -\log_{10}(P/P_0)$. This relationship may be expressed as

$$A(\lambda) = -\log_{10}\frac{P}{P_0} = -\ln\frac{P(\lambda)}{P_0(\lambda)} = \alpha\rho CL \tag{15.16}$$

where

α is the absorption coefficient of the gas
total pressure ρ (atm)
C is the amount of gas concentration
L (cm) is the distance of absorption path

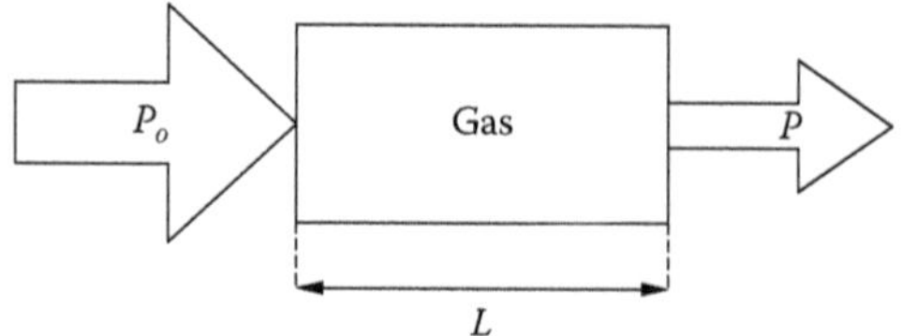

FIGURE 15.9 Principle of the Beer–Lambert law: the absorption of a given gas can be determined by measuring the power of light entering and exiting a certain path length of a gas medium.

As a beam of light passes through an absorbing medium, the amount of light absorbed in any unit volume is proportional to the intensity of incident light times the absorption coefficient. In terms of a gas as the absorbing medium, this is the basis of absorption spectroscopy, which allows the measuring of concentrations of different gases in a gas cell. To develop the sensor based on the optical absorption spectroscopy, the emission wavelength of the light source must be determined accordingly before considering the other factor element in the sensor system. This is because different types of gas have different absorption coefficients, which is stated in Beer's law and is defined as

$$\alpha = S_i \Phi \tag{15.17}$$

where

S_i (cm^{-2} atm^{-1}) is the transition line strength
Φ (cm) is the line shape function

The line strength parameter is wavelength sensitive and represents the gas wavelength identity and absorption sensitivity at a particular wavelength. Every gas has a different absorption wavelength, which can be found in the HITRAN database [84]. The examples of the absorption wavelengths for primary greenhouse gases in the earth atmosphere, such as water vapor, carbon dioxide, and methane, are shown in Table 15.4. In this table, it can be observed that each gas has a different absorption wavelength. However, the overlaps between the absorption lines for different gas species exist, which can be avoided by carefully choosing the wavelength absorption of the gas based on the HITRAN database for each specific application.

In order to develop a humidity sensor based on the principle of the Beer–Lambert law, the water vapor absorption wavelength is used to monitor the condition of the humidity level in the atmosphere. The absorption wavelength for water vapor is not limited to one as there are several absorption wavelengths available in the HITRAN database that can be used to develop a humidity sensor based on the light absorption principle.

An example of the use of this method can be seen in the work reported by Schirmer et al. [87], which describes a fiber-coupled diode laser absorption sensor for high-precision trace humidity detection in the ppm range and was tested in laboratories of the European humidity standards.

TABLE 15.4
Absorption Wavelength of Greenhouse Gases

Gas	Absorption Wavelength (nm)	References
Water vapor (H_2O)	1365	[85]
Carbon dioxide (CO_2)	1573	[85]
Methane (CH_4)	1651	[85]
Ozone (O_3)	254	[86]

The optical gas cell called a white cell is used as a sensor head. The white cell has a reflection of 52 times equivalent to 10.4 m optical path length in 0.751 vol. The wavelength of 1393 nm is used for the absorption line detection and two-tone frequency modulation spectroscopy (TTFMS) is adopted for the detection of the signal. The sensor can operate in the 1–100 ppm_v at atmospheric pressure with good accuracy, stability, and reproducibility. Since the sensor shows no significant cross-spectroscopic sensitivity to other molecule gas species, the sensor is suitable for trace humidity measurement in many other gases. Although it can detect a very low concentration of humid air, the optical alignment of the white cell is very troublesome and sensitive. In addition, the bulky sensor head makes the installation of the sensor in a hard-to-reach, remote location and narrow space for specific engineering applications almost impossible.

Kozlov et al. [88] proposed an optical humidity sensor based on the principle of water vapor absorption in UV radiation. The existence of water vapor absorption in the near-vacuum UV (VUV) region in the wavelength range of 110–190 nm is used in the measurement of gas humidity. Specifically, the favorable absorption wavelength in VUV radiation is at a wavelength of 121.6 nm. The sensor head consists of a VUV source and PD with a small gap between them, magnesium fluoride window, and gas inlet and outlet in the upper and lower parts at the center of the sensor head. From the measurement testing, the sensor has a basic measurement error of ±5% RH. A big disadvantage of the device is the need to carefully select a proper light source and PD before installing it in the sensor head, which makes the sensor less flexible and more expensive. Lauer et al. [89] presented a laser absorption humidity sensor by means of infrared absorption spectroscopy using vertical-cavity surface-emitting lasers (VCSELs) with an emission wavelength of 1.84 μm, as shown in Figure 15.10. The sensing area of the infrared absorption sensor is an open path 30 cm in length through a collimating lens with an effective focal length of 11 mm. The detection of light intensity is performed using the 2*f* harmonic signal data processing technique. The big advantage of the sensor scheme is that the measurement of the humidity is not affected by particle dust or ambient light. The path length of the sensor can be decreased without compromising the sensitivity of the sensor by using a much stronger water vapor absorption feature in the vicinity of 2.7 μm and multiple reflective mirrors to fold the absorption path.

Basu et al. [90] explored the use of the absorption spectroscopic technique to measure the gas humidity within a narrow-channel flow-field noninvasively sensing element. This sensing element is a transparent circular flow that has a 12 cm path length. Tunable diode laser absorption spectroscopy in conjunction with wavelength modulation spectroscopy (WMS) is used to tune the DFB laser source at the water vapor absorption line around the wavelength of 1491.6 nm. With this sensor scheme, range detection in partial water vapor pressure unit from 0.042 to 0.380 atm is achieved. Through the implementation of WMS in this sensor scheme, a very small amplitude optical signal due to the absorption of water vapor is separated from the standard levels of noise that increase the signal-to-noise ratio (SNR) of the sensor. Ke et al. [91] have reported a compact diode laser fiber-optic sensor based on the absorption spectrum of water vapor to measure the humidity by using a distributed feedback (DFB) laser emitting at 1877 nm wavelength.

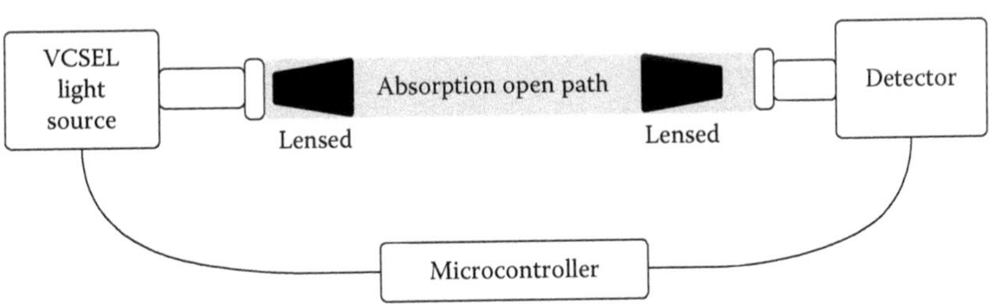

FIGURE 15.10 Block diagram of the experimental setup of laser single-pass absorption-based humidity sensor. (From Lauer, C. et al., *IEEE Trans. Instrum. Meas.*, 54(3), 1214, 2005.)

The integration of a microcontroller unit connected with electronic temperature and pressure sensor makes the laser sensor absorption system able to lock to the absorption peak and calculate the RH. In addition, the laser sensor system can switch functions automatically between direct measurement for high humidity region and second-harmonic detection for low humidity detection. The sensor system has a wide detection range from 0.01% to 100% RH for a temperature range of −20°C to 40°C. The laser sensor has good detection resolution at 0.01% RH but has significant error, which is less than 2%.

Besides the WMS detection scheme, the differential scheme presented by Wang et al. [92] measures gas humidity based on the differential value of two adjacent absorption peaks. The differential values of the two water vapor absorption peaks at 1367.862 and 1368.597 nm produce a linear relationship to water vapor concentration. A DFB laser diode pigtail with central wavelength of around 1370 nm is used as a light source for this sensor scheme. A collimator gas cell with an effective optical path length of 10 cm is employed as a sensing element to detect gas humidity. This method provides a solution to the laser absorption scanning spectra using a single absorption peak, particularly the difficulty in the selection of a reference point caused by random noise and the broadening effect of the water vapor absorption peak. Zhang et al. [93] reported a direct absorption technique by scanning the single absorption peak of water vapor around 1369.59 nm. A narrow DFB laser diode is used in the sensor scheme and transmitted light signal is split into two beams with a ratio of 90:10 intensity. The 90% intensity is transmitted through the sensing element (collimator gas cell) and the rest is used as a reference signal to remove the laser intensity fluctuations. By applying a synchronous superposition average of a LabVIEW-based signal algorithm, the sensor system achieved a range ppm detection from 19.0 to 1865.3 ppm_v. The average measurement error is within a stable level of 10 ppm_v. The use of direct absorption technique simplifies the modulation and demodulation process of the sensor scheme but compromises the sensitivity and SNR of the sensor.

15.3.2.4 Polymer Fiber

One of the most common polymer fiber-based humidity sensors is the polymer optical FBG (POFBG). POFBGs have gained much interest among application engineers and scientists due to their unique advantages compared to their silica counterparts, such as inherent fracture resistance, high flexibility, large strain range, low Young's modulus, humidity sensitivity, and biocompatibility [94]. In recent years, the advancement in the fabrication of single-mode polymer optical fiber (POF) also enhanced the research and development of Bragg grating sensors in POF, and sensors for different applications based on single-mode POFBGs have been reported [2,95]. One of the unique advantages of polymer FBGs recorded in poly(methyl methacrylate) (PMMA)-based fiber is their intrinsic sensitivity to humidity, which originates from the water absorption properties of PMMA.

The intrinsic humidity sensitivity of POFBG can make it superior compared to other techniques if the response time is improved. Recent studies have shown that the diffusion time of water content into the polymer fiber can be improved from around 30 min down to 10 min by reducing the cladding thickness of the polymer fiber [96].

The humidity sensitivity of POFBG arises from the swelling of the fiber and changes in the refractive index due to the water absorption/desorption of the fiber, which leads to a change in peak reflected wavelength of the POFBG. For a strain–stress-free POFBG, the Bragg wavelength then depends on the temperature (T) and water content (w) and can be expressed as

$$\lambda_B = 2n_{eff}\left(T,w\right)\Lambda\left(T,w\right) \tag{15.18}$$

where

n_{eff} is the effective core index
Λ is the grating pitch

RH is defined as the ratio of the amount of water vapor present in atmosphere to the maximum amount that the atmosphere can hold at the existing temperature. The Bragg wavelength in terms of RH can be expressed as [96]

$$\lambda_B = \lambda_{B0} + cH^m \tag{15.19}$$

where
λ_{B0} is the initial Bragg wavelength
c is a coefficient associated with PMMA swelling, humidity dependence of refractive index, and swelling-induced stress
H^m represents the RH with nonlinear water content dependence, which is represented by the factor m

The process of water absorption and desorption in PMMA can be described by the diffusion theory [97]. Assuming the polymer fiber as a homogeneous cylindrical rod, the diffusion equation can be written as

$$\frac{\partial C}{\partial t} = D\frac{\partial^2 C}{\partial r^2} + \frac{D}{r}\frac{\partial C}{\partial r} \tag{15.20}$$

where
D is the diffusion coefficient
$C(t, r)$ is the concentration depending on time t and radial position r

Solving this equation by applying suitable boundary conditions, one can find out the normalized concentration for in-diffusion (sorption) and for out-diffusion (desorption) [98].
For in-diffusion:

$$C(t,r) = 1 - \frac{2}{a}\sum_{n=1}^{\infty}\frac{\exp\left(-D\alpha_n^2 t\right)J_0\left(r\alpha_n\right)}{\alpha_n J_1\left(a\alpha_n\right)} \tag{15.21}$$

For out-diffusion:

$$C(t,r) = \frac{2}{a}\sum_{n=1}^{\infty}\frac{\exp\left(-D\alpha_n^2 t\right)J_0\left(r\alpha_n\right)}{\alpha_n J_1\left(a\alpha_n\right)} \tag{15.22}$$

where
a is the cylinder radius
α_n is the nth positive root of $J_0(\alpha_n) = 0$

From the equation, one can see that time response is influenced by the diffusion coefficient and the fiber geometry. For PMMA, the water desorption is faster than water sorption and the diffusion coefficient is given as 6.4×10^{-9} cm^2/s for sorption and 9×10^{-9} cm^2/s for desorption [99]. Therefore, further control of the response time can be obtained by varying the diameter of the polymer fiber, and fiber with small cladding diameter will help to expedite the water sorption and desorption properties of the polymer fiber.

Zhang et al. [100] demonstrated a POFBG sensor for humidity sensing. They inscribed the grating on a POF SMF of 200 μm diameter. The fiber has a pure PMMA cladding and a doped PMMA core. The polymer FBG sensor can detect a humidity level from 45% to 90% RH with good linearity and a sensitivity of 38.3 pm/%RH (red-shift wavelength). The FBG sensor also responds to temperature with a sensitivity of −43 pm/%RH in the range of 18°C–36°C. The polymer FBG sensor has been used to detect water in jet fuel, and from the initial experiment results, the sensor has adequate sensitivity to detect water molecules in three types of jet fuel.

Instead of using standard polymer fiber, a new type of noncoated fiber sensor can be designed by infiltrating the PCF microholes with the polymer, which has been reported by Tao et al. [101]. These authors have reported an RH fiber sensor using tapered PCF and the air holes of the PCF are filled with a humidity-sensitive polymer. From the simulation results, the loss of the light intensity is from 0.063 to 75.847 dB/cm when RH varies from 0% to 95% RH. The intensity measurement base used in this sensor scheme is simple and less expensive than the standard FBG-based humidity sensor with a wavelength shift detection technique.

Witt et al. [102,103] investigated the single-mode PMMA microstructured POF (MPOF) inscribed with LPFG to sense humidity for the first time. The MPOF was fabricated by the Optical Fibre Technology Centre at the University of Sydney and consists of small air holes along the cladding of the fiber. The mechanism of light propagation in the fiber is through index guiding, and the fiber is purely PMMA without any dopants. The core and outer diameters of the fiber are around 340 and 5.5 μm, respectively. The total length of the fiber is around 2 m and 15 cm length of LPFG with a period of 1 mm is written at the middle of the fiber. When the MPOF LPFG sensor is exposed to humidity, it can detect a humidity cycle from 20% to 90% RH with good repeatability but with a very long response time (up to 32 h) due to the considerable length of the sensing grating. This sensor has a significant temperature cross sensitivity, which means that a temperature compensation scheme is essential for the sensor.

Zhang et al. [104] characterized the FBG sensors inscribed onto the core of POF for humidity sensing. The 10 cm length of PMMA-based polymer fiber is written with a 5 mm long FBG. The humidity sensitivity of the FBG polymer fiber is in the range of 50% to 95%RH. It has a linear response with sensitivity of 35.2 ± 0.4 pm/%RH. The minimum response time of the polymer grating sensor is about 30 min. Due to the inherent temperature sensitivity of the FBG polymer sensor, an FBG silica fiber is used in a dual parameter configuration to compensate for the temperature sensitivity of the FBG polymer fiber sensor that leads to a well-conditioned response.

Zhang et al. [105] investigated the response time of the fiber humidity sensor based on PMMA polymer FBGs by etching the fiber grating with acetone in a controllable manner. The etching process reduces the diameter of the fiber grating sensor and thus speeds up the diffusion time of water molecules into the core of the FBG polymer fiber sensor and eventually decreases the response time of the sensor to the humidity variations. By reducing the diameter of the polymer grating sensor to a diameter of 135 μm, the response time is improved: down to 7 and 12 min for humidity increase and humidity decrease, respectively. The sensor has a sensitivity of 33.6 pm/%RH and minimum detection limit of 0.12% RH.

Another work by Rajan et al. [106] showed that the response time of the polymer FBG-based humidity sensor can be improved by reducing the cladding thickness, and a response time of 4.5 s is observed for a fiber with 25 μm thickness. Figure 15.11 shows the dynamic response of an etched polymer FBG with change in humidity. Figure 15.12 shows the dependence of response time on the cladding thickness of the polymer fiber.

Other than polymer FBGs, other reported work on polymer fibers for humidity sensors includes polymer microfiber. Gu et al. [107] demonstrated a polymer single-nanowire optical sensor for humidity sensing. The polymer nanowire is fabricated by direct drawing of the PAM aqueous solution to form a 410 nm diameter and 250 μm length of polymer nanowire. The nanowire is supported on a MgF_2 substrate and the coupling of light in and out for the nanowire is done by an evanescent coupling method. In this coupling technique, a fiber taper with distal end around 500 nm in diameter is placed in parallel contact with both ends of the nanowire supported by a fluoropolymer bonding; a supercontinuum broadband is used as a the light source. The small footprint nanowire humidity sensor exhibits an RH range from 10% to 88% RH and an incredibly fast response time of around 24 ms when the RH jumps from 10% to 75% RH and 30 ms when RH drops from 88% to 75% RH.

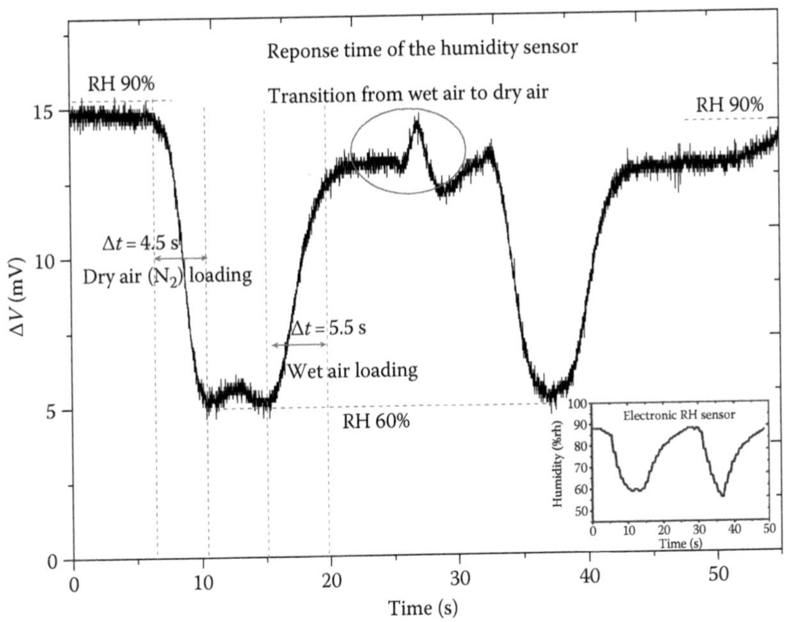

FIGURE 15.11 Plot showing response time of the humidity sensor.

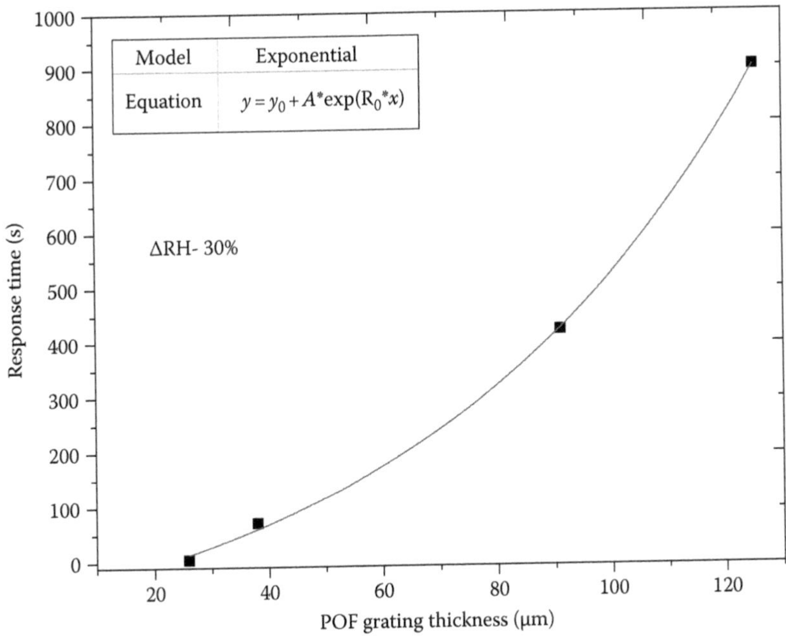

FIGURE 15.12 Response time of humidity sensor with different thicknesses.

15.3.2.5 Recent Development of Microstructured Fiber Humidity Sensor

This section reports several latest novel schemes for noncoated fiber humidity sensing based on a microstructured-type fiber, which is PCF. The first novel scheme [108] exploits the physical parameter for water vapor in air holes of the PCF. A particular point of interest for this humidity sensor is that it can be achieved without the use of any hygroscopic coating onto the fiber with

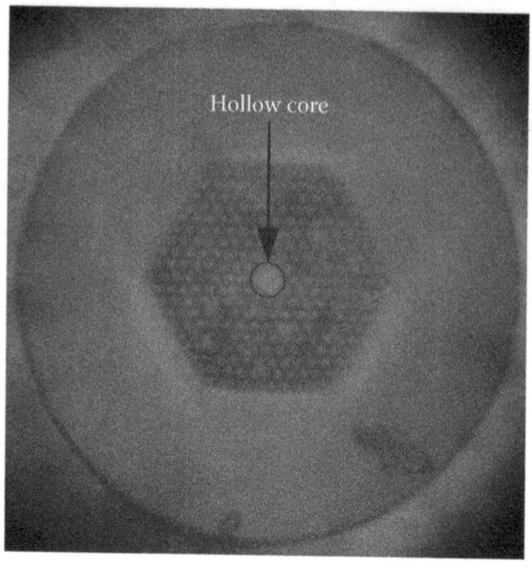

FIGURE 15.13 HC-PCF.

wide humidity range detection and a linear response. PCF, which is hollow-core PCF (HC-PCF), was employed as a sensing element as shown in Figure 15.13.

The wavelength scan and reference modulation technique is experimentally implemented with an HC-PCF sensing head of 5 cm and a small gap between SMF and HC-PCF to allow air diffusion into air holes of HC-PCF. Humidity sensing is based on light absorption caused by water vapor. The laser emitting at 1369 nm is chosen to measure RH because of the strong absorption by water vapor and takes advantages of the mature telecommunication technology in this near-infrared band. A reference wavelength in the modulation technique helps to reduce the outside affection such as fiber loss, fluctuation of DFB laser power, and attenuation by dust in the probed region.

The experimental setup to detect RH in ambient atmospheric pressure is depicted in Figure 15.14. The HC-PCF with total length around 5 cm acts as the sensor head. The splice loss between an SMF and HC-PCF is around 1 dB and reproducible.

The data of scanning modulation signal in one scanning period are shown in Figure 15.15. The absorption in the raw modulation signals is caused by high humidity level at 80% RH and low humidity level at 20% RH. The dip of the absorption at 1368.59 nm can be seen clearly in one scanning period for both humidity levels. A reference signal that acts as a zero absorption baseline is

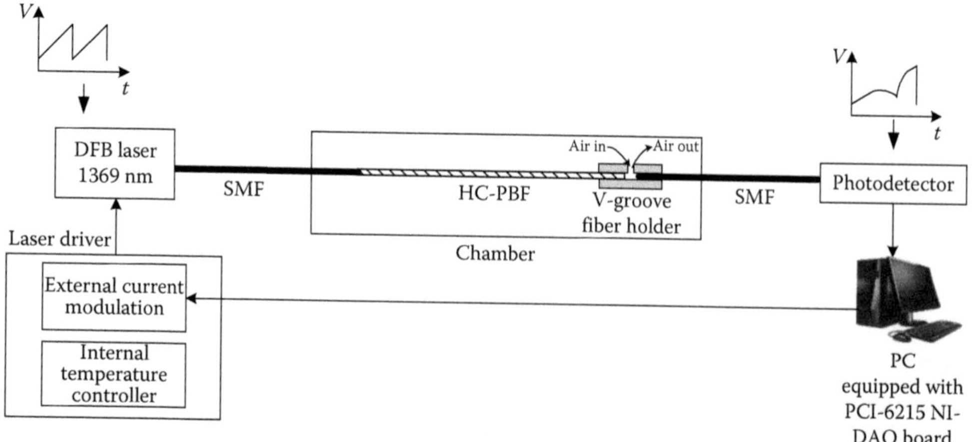

FIGURE 15.14 Experimental setup for humidity detection and HC-PCF (sensor head) cross section.

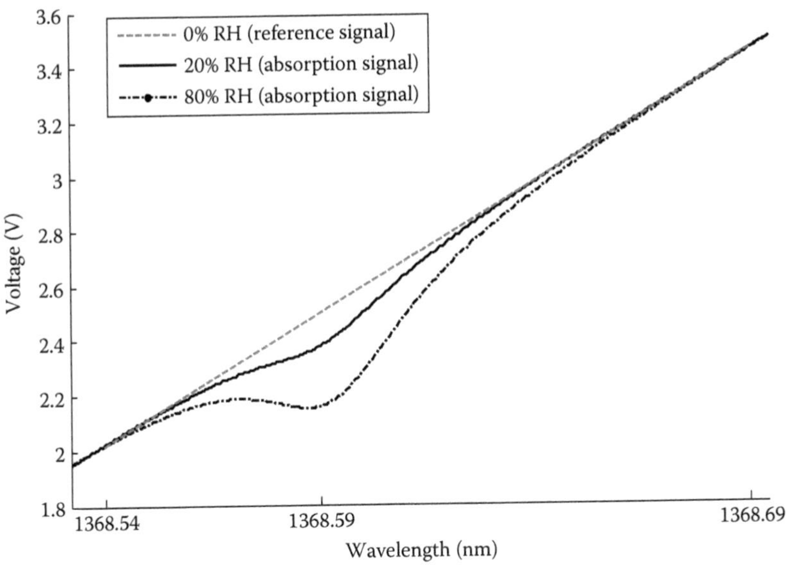

FIGURE 15.15 Raw data of scanning modulation signal for high humidity level (80% RH) and low humidity level (20% RH).

constructed based on linear fit of the two wavelength points at 1368.54 and 1368.69 nm where no absorption of water vapor appeared in both wavelengths by using combinations of waveform measurement and generation function in the LabVIEW.

The sensor demonstrated a humidity sensitivity of 3.02 mV/%RH over the range of 0%–90% RH with linear response, low noise, and good reversibility.

To realize higher sensitivity and more accuracy than by using direct absorption spectroscopy for humidity sensing as shown in previous sensor, a modulation technique can be included to further improve the performance of the sensor. Therefore, we have developed a novel all-fiber digital WMS scheme based on HC-PCF [109]. High-sensitivity detection is achieved by using WMS that shifts the detection bandwidth to a higher frequency, thereby offering a larger SNR and thus higher sensitivity. Wavelength modulation, in conjunction with the ratio between the second harmonic (2*f*) and first harmonic (*f*) of the WMS signal at water vapor absorption line around 1368.59 nm, is used for RH measurement. In addition, an HC-PCF reference cell is introduced in the WMS scheme to lock the DFB laser source to the wavelength absorption peak of water vapor. WMS was performed using a developed compact and flexible digital lock-in detection LabVIEW software system without dedicated signal generation or lock-in hardware. The experimental system based on the WMS modulation technique is shown in Figure 15.16.

With the combination of WMS and fiber reference cell, the disturbances such as laser power fluctuations and attenuation variations of the sensing element and fiber can be suppressed. In addition, the WMS technique takes into account the full-integrated area and avoids the baseline fitting problematic. Therefore, it is more accurate than the absorption technique in the previous sensor, which is based on direct wavelength modulation technique. The digital WMS scheme with an HC-PCF sensing head of 5 cm and an HC-PCF reference cell of 7 cm was experimentally implemented. High humidity resolution of ~0.2% RH over the range 0%–95% RH was achieved.

Both of the previous sensing schemes apply single pass absorption with different modulation techniques. Multipass absorption can be used to significantly increase the sensitivity of the sensor through a ring laser configuration. Another novel scheme is showcased by us based on the multipass absorption [110]. A customized semiconductor optical amplifier (SOA) is used as a gain medium in the ring. Wavelength scanning is introduced into the ring using DFB laser and HC-PCF is included

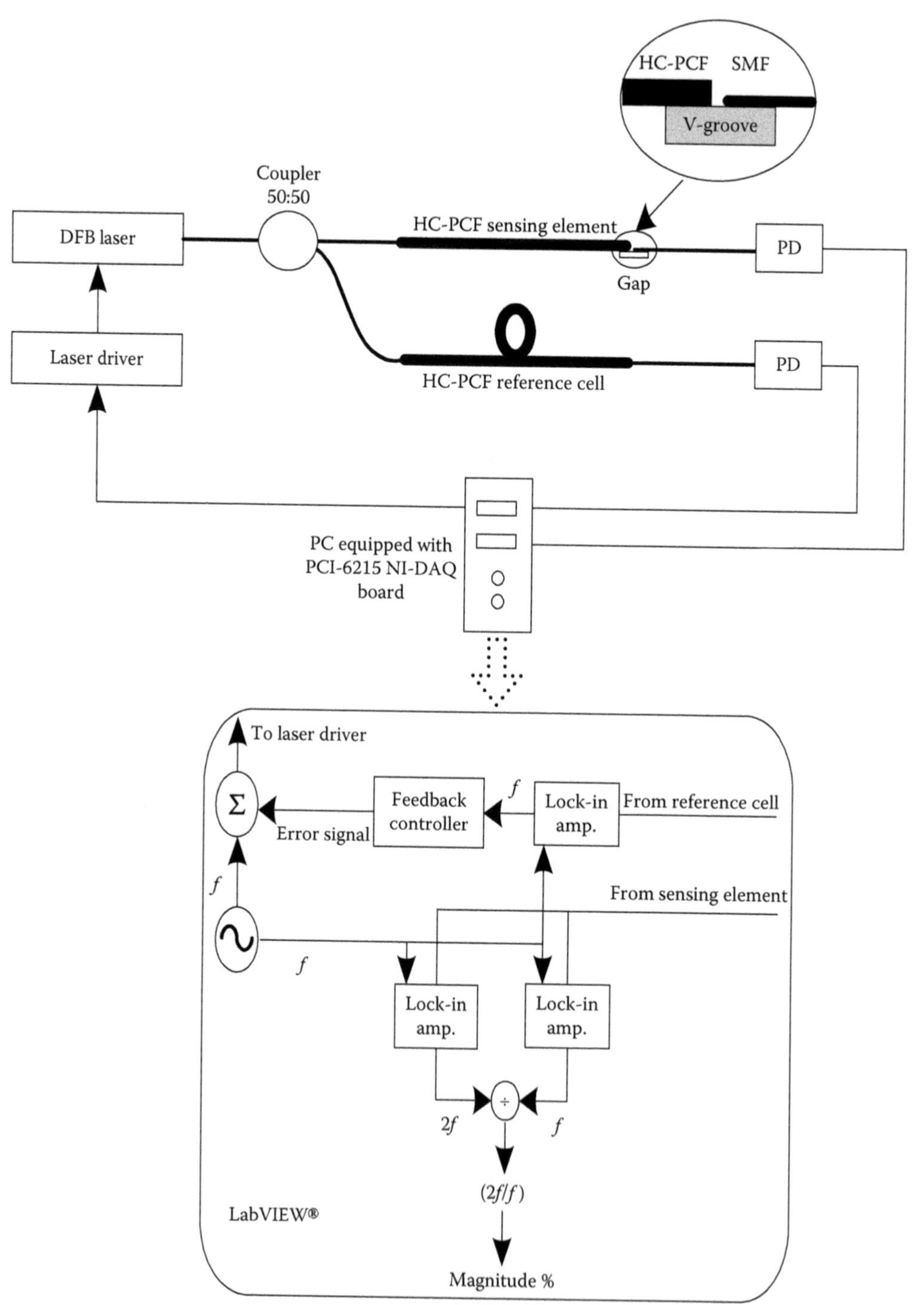

FIGURE 15.16 System design of the all-fiber optic humidity fiber sensor. All the signal generations, lock-in detection, and feedback of error signal are digitally implemented using LabVIEW®.

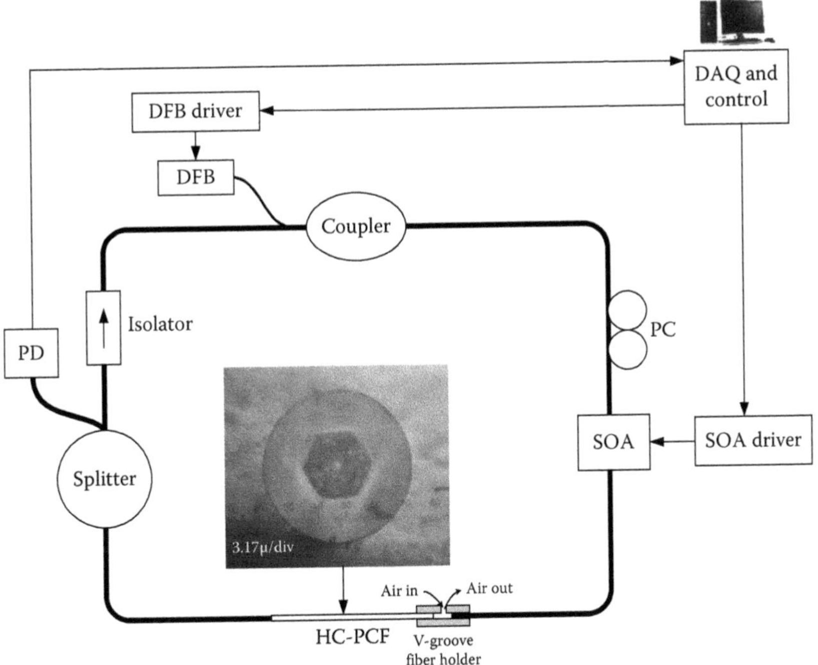

FIGURE 15.17 Experimental setup for humidity detection and HC-PCF (sensor head) cross section.

in the ring cavity that acts as a sensing element. The ring laser operates around the water vapor absorption line at 1368.59 nm that is equal to injected wavelength scanning of the DFB laser to the ring.

The experimental setup of the SOA-based fiber ring laser with DFB wavelength scanning is shown in Figure 15.17. The configuration consists of the SOA (BOA1036S, Covega, THORLABS), a coupler, a polarization controller (PC), a DFB laser source, HC-PCF gas cell, a splitter, an isolator, and a PD.

The SOA fiber ring laser is combined with the DFB laser source to scan the water absorption peak around 1368.59 nm. By applying DFB wavelength scanning scheme, no optical filter is required in the ring cavity as the DFB laser source has a narrow line-width and high peak power. With the HC-PCF sensing element in the SOA ring laser cavity, this sensor scheme demonstrates a high sensitivity performance due to the very large number of light passes through the HC-PCF gas cell, which transform the cell into multipass system in lasing mode. The gain under a driving current of the SOA ring laser based on the lasing and nonlasing mode is 150 and 70 mA, respectively. The sensitivity is improved from 2.47 to 10.93 mV/%RH when the single pass absorption mode is transformed to multipass absorption mode as shown in Figure 15.18.

Although a high sensitivity and broad range of detection can be achieved by the sensors in previous scheme that is based on the absorption in the HC-PCF, these sensors have a slow response time, of the order of several minutes. One solution to this problem is to drill a micro hole along the sensor head but this requires a high cost of femtosecond laser or focus ion beam laser which eventually increase the cost to fabricate the sensor. In addition, the fabrication process becomes more complicated. Therefore, a solid-core PCF (SC-PCF) modal interferometric humidity sensor [111] is proposed by us to overcome the slow response time issue. The sensor is made from a piece of SC-PCF spliced to standard fibers in transmission mode. The cross section of SC-PCF is shown in Figure 15.19. The sensing principle is based on adsorption of water molecules on the surface of the silica; thus, it does not require the use of any special hygroscopic coating or thin layer, as discussed in the previous section, and yet exhibits a quick response time. Moreover, the all-silica nature of the sensing element means the sensor has a very good thermal stability or negligible cross-temperature sensitivity.

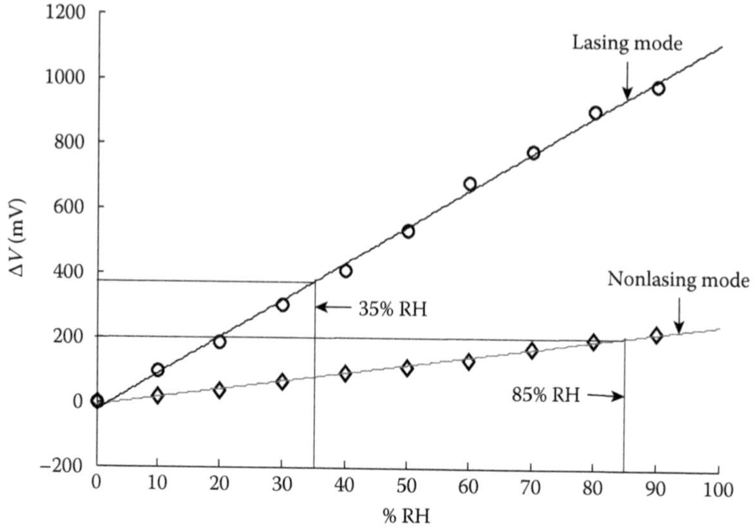

FIGURE 15.18 Voltage difference versus RH for lasing and nonlasing mode of the system.

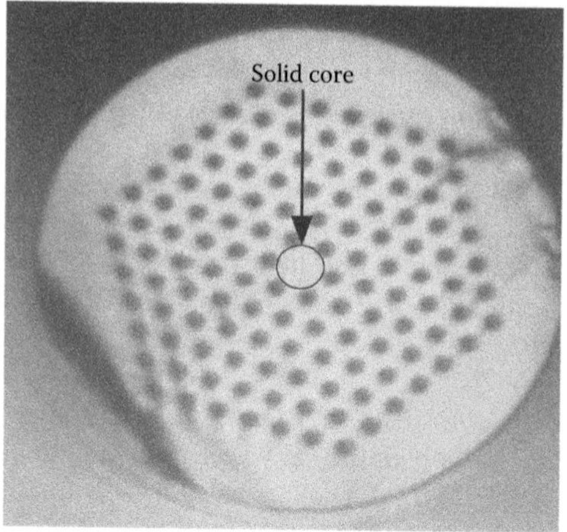

FIGURE 15.19 SC-PCF.

Figure 15.20 shows the experimental setup for characterizing the SC-PCF-MI as a humidity sensor. The spectra of the interference as shown in Figure 15.21 shifted to longer wavelengths when the humidity level inside the chamber was increased. The desorption and adsorption of water molecules from the surface of the SC-PCF changed the effective refractive index of the cladding mode [73]. Then the variation in the effective refractive index caused a phase change between the cladding and core modes of the interferometer. Thus, the interference peak of the interferometer was shifted when the surrounding humidity changed.

The experimental results show that the interference spectrum shift is ~4 nm when the surrounding RH varies from 60% to 95% RH. This SC-PCF-MI optical fiber humidity sensor, with advantages such as small size, fast response, simple structure, and fabrication, has potential applications in real-time monitoring. Moreover, a negligible cross-temperature sensitivity of the SC-PCF-MI humidity sensor makes the temperature compensation not necessary if such a sensor is operated in normal atmosphere.

Mathew et al. [112] reported a miniature fiber humidity sensor based on a polymer-infiltrated SC-PCF interferometer operating in reflection mode. The selected hygroscopic polymer agarose material is

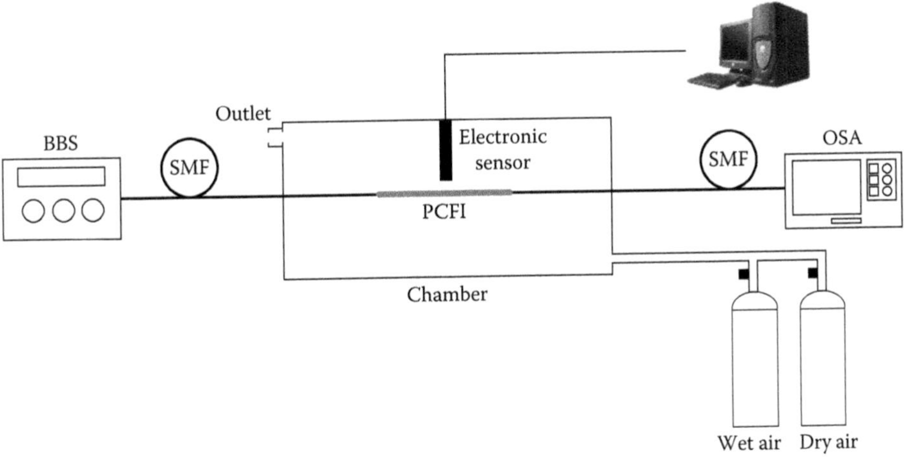

FIGURE 15.20 Experimental setup of the fabricated sensor.

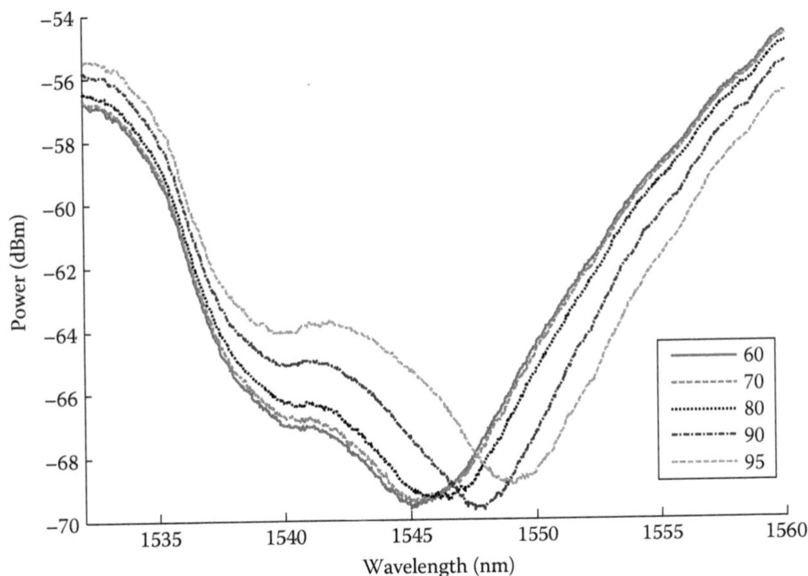

FIGURE 15.21 The sensor detection range was tested from 60% to 95% RH.

infiltrated into the micro air holes of the PCF interferometer. In the interferometer sensor, the collapsed air hole regions of the PCF allow the excitation and recombination of the core and cladding modes. The refractive index of the agarose inside the air holes of the PCF is a function of ambient RH. Therefore, the modal propagation of the cladding mode is influenced by agarose material, and therefore, a phase change induced between the cladding and core modes of the PCF interferometer causes a shift in the interference spectrum. The length of the compact PCF interferometer is 1 mm, and the collapsed air holes of the PCF are about 250 μm. The agarose infiltration length for the PCF sensor is estimated to be around 100 μm using a polarizing microscope, and the infiltration process is done through the capillary forces method. The agarose-infiltrated PCF sensor exhibits an RH detection in the range of 14%–98% RH, which has two different resolution regions. For the first region, which is from 14% to 86% RH, the sensor shows 0.06 dB% RH while 0.6 dB/%RH resolution from 86% to 98% RH for the second region. The sensor response has a good long-term stability and repeatability. In addition, the fast response time of 400 ms (30% RH change) observed for this sensor suggests that the interferometer sensor has a potential application in human breath monitoring. For a practical breath sensor, temperature compensation is required due to the sensor having a thermal sensitivity of 0.066 dB/°C.

15.4 APPLICATION AREAS

The SC-PCF-based sensor as discussed in the previous section has shown a potential to be applied in real-time monitoring application due to its fast response time. Nevertheless, to match the need of real-time monitoring applications such as breath monitoring, the sensor in transmission configuration is not a practical solution. This is due to the complicated packaging setup of the sensor in the compartment of a disposable oxygen mask for breath sensing. In addition, a temperature insensitive feature is a must due to cycles of human breath consist of temperature variation. As a result, we have introduced a simple sealed-void PCF interferometric humidity sensor that operates in reflection mode and is insensitive to temperature condition [113]. The sensor is fabricated by collapsing the holes of the SC-PCF at both ends to form an interferometer with core and cladding mode excitation as shown in Figure 15.22. As a result, the reflection spectrum of the sensor exhibits a sinusoidal interference peak that shifts when the adsorption and desorption of H_2O molecules occur along the surface of SC-PCF. This 1.3 cm long SC-PCF interferometer sensor shows a sensitivity of 20.3 and 61.6 pm/%RH in the range of 60%–80% RH and 80%–95% RH, respectively. The sealed-void configuration gives the sensor a fast response time of ~452 ms and protects the air holes of the SC-PCF from dirt as well. Moreover, the all-silica nature of the sensing element means the sensor has a good thermal stability. The combination of fast response and insensitivity to temperature variation makes this sensor a great candidate for breath sensing device.

To verify the interferometer sensor for breath sensing, a human respiration experiment is performed. The sensor head is carefully fixed in a homemade holey tube as shown in Figure 15.23a to allow a person's exhaled air to enter and exit the sensitive part of the sensor head. After that, the tube is inserted into the compartment of a disposable silicone oxygen mask as shown in Figure 15.23b. Then the package sensor is set in the oxygen mask, approximately at 3 cm away from the person's nose. The full packaging of the sensor into the mask is depicted in Figure 15.23c. A setup consisting of tunable laser source (Agilent 81940A), circulator, PD (V500, Optiphase Inc.) and a PC equipped with LabVIEW card are used for monitoring the respiratory condition in real time. The breath signals are observed for ~15 s as shown in Figure 15.24. From this figure, cycles of human breathing in

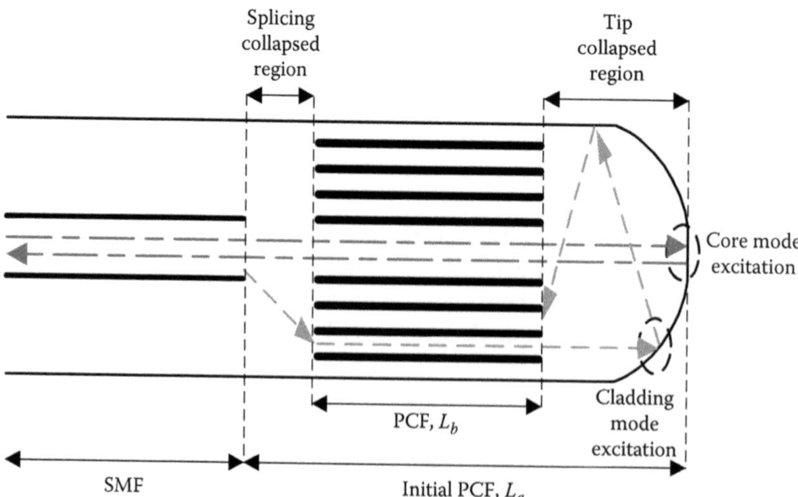

FIGURE 15.22 Schematic diagram for operating mechanism of the in reflection sealed-void SC-PCF interferometric sensor.

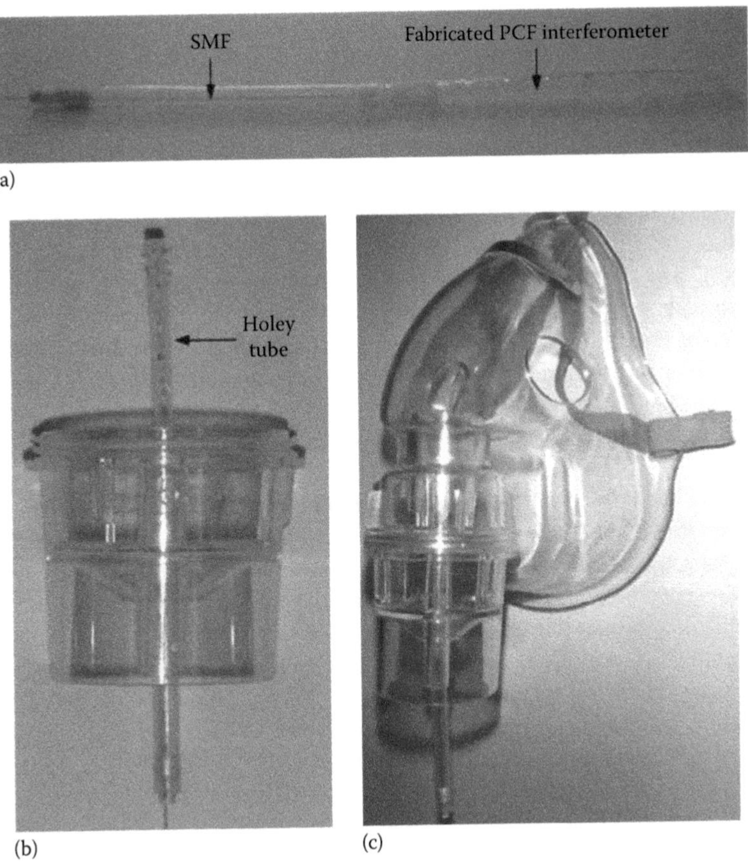

FIGURE 15.23 Photograph of the sensor and the position of the holey tube in the mask. (a) The sensor head is carefully fixed in a home-made holey tube. (b) The tube was inserted into the compartment of a silicone nebulizer mask. (c) The package sensor was set in the nebulizer mask.

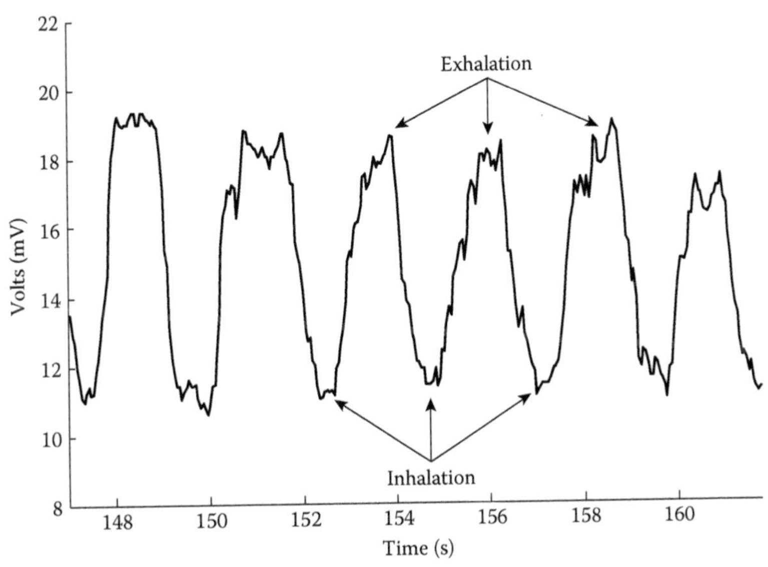

FIGURE 15.24 The human respiratory monitoring data with the package sensor set in oxygen mask for ~15 s.

and out using oxygen mask can be seen clearly. It is evident that this sealed-void PCFI sensor has a potential application for monitoring breathing condition.

In addition, other types of noncoated fiber humidity sensors that have been used as a breath sensors are demonstrated by Lin et al. [78] and Favero et al. [76], which are based on reflective optical lensed fiber and PCF-SMF interferometer.

On the other hand, other researchers have also demonstrated the breath sensing application using the hygroscopic material-coated sensors. For example, a polymer-coated hetero-core optical fiber [114], nanometer-scale-coated FP interferometer [115,116], and dye-doped fluorescent plastic fiber [117] were developed in order to monitor RH in a medial mask to determine the breath condition of a patient.

Despite the breath sensing application, the fiber humidity sensors have also been used in other applications such as control of moisture in power transformers with paper-oil insulation [3], monitoring of corrosion under insulations [4], and structural health monitoring [5,44].

15.5 SUMMARY

In this chapter, the background knowledge of gas humidity was introduced and the advantages of fiber humidity sensors compared to electronic humidity sensors were discussed. An overview was provided of the various optical fiber sensing techniques employed for gas humidity detection, which covered the diverse coated and noncoated fiber humidity sensors reported in the literature. Most of the reported sensors were tested in a controlled laboratory environment and few of them were tested in the field for a specific application, particularly in the biomedical and civil engineering fields. The major findings show that a new generation of sensor technology based on optical fibers is emerging for coated and noncoated fiber. The present trends suggest that electronic humidity sensors could soon be substituted by sensors that are based on photonic fiber structures.

REFERENCES

1. S. A. Kolpakov, N. T. Gordon, C. Mou, and K. Zhou, Toward a new generation of photonic humidity sensors, *Sensors*, 14(3): 3986–4013, 2014.
2. F. Ding, L. Wang, N. Fang, and Z. Huang. *Experimental Study on Humidity Sensing Using a FBG Sensor with Polyimide Coating*. Communications and Photonics Conference and Exhibition (ACP), 2010, Asia, Shanghai. Optical Society of America, 2010.
3. J. H. Rodriguez-Rodriguez, F. Martinez-Pinon, J. A. Alvarez-Chavez, and D. Jaramillo-Vigueras. Polymer optical fiber moisture sensor based on evanescent-wave scattering to measure humidity in oil-paper insulation in electrical apparatus, *Proceedings of IEEE Sensors*, 1052–1055, 2008.
4. H. Cho, Y. Tamura, and T. Matsuo, Monitoring of corrosion under insulations by acoustic emission and humidity measurement, *Journal of Nondestructive Evaluation*, 30(2): 59–63, 2011.
5. S. Tong, K. T. V. Grattan, S. Srinivasan, P. A. M. Basheer, B. J. Smith, and H. A. Viles, Building stone condition monitoring using specially designed compensated optical fiber humidity sensors, *IEEE Sensors Journal*, 12(5): 1011–1017, 2012.
6. J. Mathew, Y. Semenova, and G. Farrell, A miniature optical breathing sensor, *Biomedical Optics Express*, 3(12): 3325–3331, 2012.
7. A. Vijayan, M. Fuke, R. Hawaldar, M. Kulkarni, D. Amalnerkar, and R. C. Aiyer, Optical fibre based humidity sensor using co-polyaniline clad, *Sensors and Actuators B: Chemical*, 129(1): 106–112, 2008.
8. J. Mathew, Y. Semenova, and G. Farrell, A fiber bend based humidity sensor with a wide linear range and fast measurement speed, *Sensors and Actuators A: Physical*, 174: 47–51, 2012.
9. D. C. Bownass, J. S. Barton, and J. D. C. Jones, Detection of high humidity by optical fibre sensing at telecommunications wavelengths, *Optics Communications*, 146(1–6): 90–94, 1998.
10. Z. Zhao and Y. Duan, A low cost fiber-optic humidity sensor based on silica sol–gel film, *Sensors and Actuators B: Chemical*, 160(1): 1340–1345, 2011.
11. M. Hernáez, C. R. Zamarreño, I. R. Matías, and F. J. Arregui, Optical fiber humidity sensor based on surface plasmon resonance in the infra-red region, *Journal of Physics: Conference Series*, 178(1): 012019, 2009.

12. J. Goicoechea, C. R. Zamarreño, I. R. Matias, and F. J. Arregui, Utilization of white light interferometry in pH sensing applications by mean of the fabrication of nanostructured cavities, *Sensors and Actuators B: Chemical*, 138(2): 613–618, 2009.
13. S. S. Shiratori and M. F. Rubner, pH-dependent thickness behavior of sequentially adsorbed layers of weak polyelectrolytes, *Macromolecules*, 33(11): 4213–4219, 2000.
14. C. R. Zamarreño, M. Hernaez, I. Del Villar, I. R. Matias, and F. J. Arregui, Tunable humidity sensor based on ITO-coated optical fiber, *Sensors and Actuators B: Chemical*, 146(1): 414–417, 2010.
15. M. Hernaez, C. R. Zamarreño, C. Fernandez-Valdivielso, I. del Villar, F. J. Arregui, and I. R. Matias, Agarose optical fibre humidity sensor based on electromagnetic resonance in the infra-red region, *Physica Status Solidi (c)*, 7(11–12): 2767–2769, 2010.
16. F. J. Arregui, Z. Ciaurriz, M. Oneca, and I. R. Matías, An experimental study about hydrogels for the fabrication of optical fiber humidity sensors, *Sensors and Actuators B: Chemical*, 96(1–2): 165–172, 2003.
17. P. Sanchez, C. R. Zamarreño, M. Hernaez, I. D. Villar, C. Fernandez-Valdivielso, I. R. Matias, and F. J. Arregui, Lossy mode resonances toward the fabrication of optical fiber humidity sensors, *Measurement Science and Technology*, 23(1): 014002, 2012.
18. R. Aneesh and S. K. Khijwania, Zinc oxide nanoparticle based optical fiber humidity sensor having linear response throughout a large dynamic range, *Applied Optics*, 50(27): 5310–5314, 2011.
19. R. Aneesh and S. K. Khijwania, Titanium dioxide nanoparticle based optical fiber humidity sensor with linear response and enhanced sensitivity, *Applied Optics*, 51(12): 2164–2171, 2012.
20. P. J. Rivero, A. Urrutia, J. Goicoechea, I. R. Matias, and F. J. Arregui, A Lossy Mode Resonance optical sensor using silver nanoparticles-loaded films for monitoring human breathing, *Sensors and Actuators B: Chemical*, 187, 40–44, 2012.
21. M. Morisawa and H. Yokomori, Improvement in the sensitivity of dye-doped POF-type moisture sensor, *Proceedings of IEEE Sensors*, 508–511, 2010.
22. L. Zhang, F. Gu, J. Lou, X. Yin, and L. Tong, Fast detection of humidity with a subwavelength-diameter fiber taper coated with gelatin film, *Optics Express*, 16(17): 13349–13353, 2008.
23. L. Tao, D. Xinyong, C. Chi Chiu, Z. Chun-Liu, and Z. Peng, Humidity sensor based on a multimode-fiber taper coated with polyvinyl alcohol interacting with a fiber Bragg grating, *IEEE Sensors Journal*, 12(6): 2205–2208, 2012.
24. H. Liang, Y. Jin, J. Wang, and X. Dong, Relative humidity sensor based on polarization maintaining fiber loop mirror with polymer coating, *Microwave and Optical Technology Letters*, 54(10): 2364–2366, 2012.
25. S. Akita, H. Sasaki, K. Watanabe, and A. Seki, A humidity sensor based on a hetero-core optical fiber, *Sensors and Actuators B: Chemical*, 147(2): 385–391, 2010.
26. Q. Wu, Y. Semenova, J. Mathew, P. Wang, and G. Farrell, Humidity sensor based on a single-mode hetero-core fiber structure, *Optics Letters*, 36(10): 1752–1754, 2011.
27. Y. Liu, Y. Zhang, H. Lei, J. Song, H. Chen, and B. Li, Growth of well-arrayed ZnO nanorods on thinned silica fiber and application for humidity sensing, *Optics Express*, 20(17): 19404–19411, 2012.
28. L. Xia, L. Li, W. Li, T. Kou, and D. Liu, Novel optical fiber humidity sensor based on a no-core fiber structure, *Sensors and Actuators A: Physical*, 190(0): 1–5, 2013.
29. A. Kersey, M. A. Davis, H. J. Patrick, M. Leblanc, K. P. Koo, C. G. Askins, M. A. Putnam, and E. J. Friebele, Fiber grating sensors, *Journal of Lightwave Technology*, 15(8): 1442–1463, 1997.
30. T. L. Yeo, T. Sun, and K. T. V. Grattan, Fibre-optic sensor technologies for humidity and moisture measurement, *Sensors and Actuators A: Physical*, 144(2): 280–295, 2008.
31. V. Bhatia and A. M. Vengsarkar, Optical fiber long-period grating sensors, *Optics Letters*, 21(9): 692–694, 1996.
32. X. Dong, H. Zhang, B. Liu, and Y. Miao, Tilted fiber Bragg gratings: Principle and sensing applications, *Photonic Sensors*, 1(1): 6–30, 2011.
33. D. Donlagic, *Fiber Optic Sensors: An Introduction and Overview*, University of Maribor, Maribor, Slovenia, 2000.
34. N. Beverini, E. Maccioni, M. Morganti, F. Stefani, R. Falciai, and C. Trono, Fiber laser strain sensor device, *Journal of Optics A: Pure and Applied Optics*, 9(10): 958–962, 2007.
35. B. Lee, Review of the present status of optical fiber sensors, *Optical Fiber Technology*, 9(2): 57–79, 2003.
36. X. Yu, M. Zhang, P. Childs, L. Wang, M. Lei, Y. Liao, J. Ju, and W. Jin, Research on testing the characteristics of hydrogel film by using a long-period fiber grating, *Applied Optics*, 48(11): 2171–2177, 2009.
37. S. W. James and R. P. Tatam, Optical fibre long-period grating sensors: characteristics and application, *Measurement Science and Technology*, 14: R49–R61, 2003.
38. S. Yin, K.-W. Chung, and X. Zhu, A highly sensitive long period grating based tunable filter using a unique double-cladding layer structure, *Optics Communications*, 188(5–6): 301–305, 2001.

39. L. Yang, W. Liwei, Z. Min, T. Dongsheng, M. Xianhui, and L. Yanbiao, Long-period grating relative humidity sensor with hydrogel coating, *IEEE Photonics Technology Letters*, 19(12): 880–882, 2007.
40. S. F. H. Correia, P. Antunes, E. Pecoraro, P. P. Lima, H. Varum, L. D. Carlos, R. A. S. Ferreira, and P. S. André, Optical fiber relative humidity sensor based on a FBG with a di-ureasil coating, *Sensors*, 12(7): 8847–8860, 2012.
41. M. A. Caponero, A. Polimadei, L. Benussi, S. Bianco, S. Colafranceschi, L. Passamonti, D. Piccolo et al., Monitoring relative humidity in RPC detectors by use of fiber optic sensors, *Journal of Instrumentation*, 8(3): T03003, 2013.
42. J. M. Corres, I. del Villar, I. R. Matias, and F. J. Arregui, Two-layer nanocoatings in long-period fiber gratings for improved sensitivity of humidity sensors, *IEEE Transactions on Nanotechnology*, 7(4): 394–400, 2008.
43. G. Decher, Fuzzy nanoassemblies: Toward layered polymeric multicomposites, *Science*, 277(5330): 1232–1237, 1997.
44. T. Venugopalan, T. Sun, and K. T. V. Grattan, Long period grating-based humidity sensor for potential structural health monitoring, *Sensors and Actuators A: Physical*, 148(1): 57–62, 2008.
45. D. Viegas, J. Goicoechea, J. M. Corres, J. L. Santos, L. A. Ferreira, F. M. Araújo, and I. R. Matias, A fibre optic humidity sensor based on a long-period fibre grating coated with a thin film of SiO_2 nanospheres, *Measurement Science and Technology*, 20(3): 034002, 2009.
46. D. Viegas, J. Goicoechea, J. L. Santos, F. M. Araújo, L. A. Ferreira, F. Arregui, and I. Matias, Sensitivity improvement of a humidity sensor based on silica nanospheres on a long-period fiber grating, *Sensors*, 9(1): 519–527, 2009.
47. Y. Xiujuan, P. Childs, Z. Min, L. Yanbiao, J. Jian, and J. Wei, Relative humidity sensor based on cascaded long-period gratings with hydrogel coatings and Fourier demodulation, *IEEE Photonics Technology Letters*, 21(24): 1828–1830, 2009.
48. L. Alwis, T. Sun, and K. Grattan, Analysis of polyimide-coated optical fiber long period grating-based relative humidity sensor, *IEEE Sensors Journal*, PP(99): 1–1, 2012.
49. S. Zheng, Y. Zhu, and S. Krishnaswamy, Fiber humidity sensors with high sensitivity and selectivity based on interior nanofilm-coated photonic crystal fiber long-period gratings, *Sensors and Actuators B: Chemical*, 176: 264–274, 2013.
50. M. Yinping, L. Bo, Z. Hao, L. Yuan, Z. Haibin, S. Hua, Z. Weihua, and Z. Qida, Relative humidity sensor based on tilted fiber Bragg grating with polyvinyl alcohol coating, *IEEE Photonics Technology Letters*, 21(7): 441–443, 2009.
51. X. Dong, T. Li, Y. Liu, Y. Li, C.-L. Zhao, and C. C. Chan, Polyvinyl alcohol–coated hybrid fiber grating for relative humidity sensing, *Journal of Biomedical Optics*, 16(7): 077001, 2011.
52. T. Li, X. Dong, C.-L. Zhao, and Y. Liu, Polymer-coated hybrid fiber grating for relative humidity sensing, *Proceedings of SPIE*, 7853: 78532V, pp. 1–6, 2010.
53. K.-T. Lau, L.-M. Zhou, P.-C. Tse, and L.-B. Yuan, Applications of composites, optical fibre sensors and smart composites for concrete rehabilitation: An overview, *Applied Composite Materials*, 9(4): 221–247, 2002.
54. J. M. Corres, I. R. Matias, M. Hernaez, J. Bravo, and F. J. Arregui, Optical fiber humidity sensors using nanostructured coatings of SiO_2 nanoparticles, *IEEE Sensors Journal*, 8(3): 281–285, 2008.
55. L. Ruete, J. Goicoechea, M. Hernaez, I. R. Matias, and F. J. Arregui. Two nano Fabry-Perot interferometers for humidity sensing, *Proceedings of IEEE Sensors*, 1004–1007, 2008.
56. L. H. Chen, T. Li, C. C. Chan, R. Menon, P. Balamurali, M. Shaillender, B. Neu et al., Chitosan based fiber-optic Fabry–Perot humidity sensor, *Sensors and Actuators B: Chemical*, 169: 167–172, 2012.
57. J. Yao, T. Zhu, D.-W. Duan, and M. Deng, Nanocomposite polyacrylamide based open cavity fiber Fabry & Perot humidity sensor, *Applied Optics*, 51(31): 7643–7647, 2012.
58. B. Gu, M. Yin, A. P. Zhang, J. Qian, and S. He, Optical fiber relative humidity sensor based on FBG incorporated thin-core fiber modal interferometer, *Optics Express*, 19(5): 4140–4146, 2011.
59. J. Mathew, Y. Semenova, G. Rajan, and G. Farrell, Agarose coated single mode fiber bend for monitoring humidity, *Proceedings of SPIE*, 8073: 807317, pp. 1–8, 2011.
60. J. Mathew, Y. Semenova, and G. Farrell, Effect of coating thickness on the sensitivity of a humidity sensor based on an Agarose coated photonic crystal fiber interferometer, *Optics Express*, 21(5): 6313–6320, 2013.
61. T. Li, X. Dong, J. Chan, K. Ni, S. Zhang, and P. Shum, Humidity sensor with a PVA-coated photonic crystal fiber interferometer, *IEEE Sensors Journal*, PP(99): 1–1, 2013.
62. S. Zhang, X. Dong, T. Li, C. C. Chan, and P. P. Shum, Simultaneous measurement of relative humidity and temperature with PCF-MZI cascaded by fiber Bragg grating, *Optics Communications*, 303(11), 42–45, 2013.

63. J. Mathew, Y. Semenova, and G. Farrell, Fiber optic hybrid device for simultaneous measurement of humidity and temperature, *IEEE Sensors Journal*, PP(99): 1–1, 2013.
64. C. Li Han, C. Chi Chiu, L. Tao, M. Shaillender, B. Neu, P. Balamurali, R. Menon et al., Chitosan-coated polarization maintaining fiber-based Sagnac interferometer for relative humidity measurement, *IEEE Journal of Selected Topics in Quantum Electronics*, 18(5): 1597–1604, 2012.
65. S. Otsuki, K. Adachi, and T. Taguchi, A novel fiber-optic gas sensing arrangement based on an air gap design and an application to optical detection of humidity, *Analytical Sciences*, 14(3): 633–635, 1998.
66. S. McMurtry, J. D. Wright, and D. A. Jackson, A multiplexed low coherence interferometric system for humidity sensing, *Sensors and Actuators B: Chemical*, 67(1–2): 52–56, 2000.
67. S.-I. Ohira, P. K. Dasgupta, and K. A. Schug, Fiber optic sensor for simultaneous determination of atmospheric nitrogen dioxide, ozone, and relative humidity, *Analytical Chemistry*, 81(11): 4183–4191, 2009.
68. W. Biao, Z. Fujun, P. Fufei, and W. Tingyun. An optical fiber humidity sensor based on optical absorption. In *Communications and Photonics Conference and Exhibition, 2011*, ACP, Asia, 2011.
69. M. Urbakh, J. Klafter, D. Gourdon, and J. Israelachvili, The nonlinear nature of friction, *Nature*, 430(6999): 525–528, 2004.
70. M. Binggeli and C. M. Mate, Influence of capillary condensation of water on nanotribology studied by force microscopy, *Applied Physics Letters*, 65(4): 415–417, 1994.
71. B. A. Morrow, I. A. Cody, and L. S. M. Lee, Infrared studies of reactions on oxide surfaces. 7. Mechanism of the adsorption of water and ammonia on dehydroxylated silica, *The Journal of Physical Chemistry*, 80(25): 2761–2767, 1976.
72. B. A. Morrow and A. J. McFarlan, Chemical reactions at silica surfaces, *Journal of Non-Crystalline Solids*, 120(1–3): 61–71, 1990.
73. K. Tiefenthaler and W. Lukosz, Grating couplers as integrated optical humidity and gas sensors, *Thin Solid Films*, 126(3–4): 205–211, 1985.
74. D. B. Asay and S. H. Kim, Evolution of the adsorbed water layer structure on silicon oxide at room temperature, *The Journal of Physical Chemistry B*, 109(35): 16760–16763, 2005.
75. J. Mathew, Y. Semenova, G. Rajan, and G. Farrell, Humidity sensor based on photonic crystal fibre interferometer, *Electronics Letters*, 46(19): 1341–1343, 2010.
76. F. C. Favero, J. Villatoro, and V. Pruneri, Microstructured optical fiber interferometric breathing sensor, *Journal of Biomedical Optics*, 17(3): 037006–037011, 2012.
77. X. Fu, G. Fu, and W. Bi, An experimental study about humidity sensors based on tapered optical fibers, *Proceedings of SPIE*, 8561: 856117, pp. 1–6, 2012.
78. Y.-C. Lin, Breath sensor based on reflective optical lensed fiber, *Microwave and Optical Technology Letters*, 55(2): 450–454, 2013.
79. M. Shao, X. Qiao, H. Fu, N. Zhao, Q. Liu, and H. Gao, An in-fiber Mach-Zehnder interferometer based on arc-induced tapers for high sensitivity humidity sensing, *IEEE Sensors Journal*, PP(99): 1–1, 2013.
80. K. P. Birch and M. J. Downs, An updated Edlén equation for the refractive index of air, *Metrologia*, 30(3): 155, 1993.
81. S. Mc Murtry, J. D. Wright, and D. A. Jackson, Sensing applications of a low-coherence fibre-optic interferometer measuring the refractive index of air, *Sensors and Actuators B: Chemical*, 72(1): 69–74, 2001.
82. A. Mehrabani and H. Golnabi, Investigation of humidity effect on the air refractive index using an optical fiber design, *Journal of Applied Sciences*, 11(16): 3022–3027, 2011.
83. H. Golnabi, Using three different optical fiber designs to study humidity effect on the air refractive index, *Optics and Lasers in Engineering*, 50(11): 1495–1500, 2012.
84. J. Shemshad, S. M. Aminossadati, and M. S. Kizil, A review of developments in near infrared methane detection based on tunable diode laser, *Sensors and Actuators B: Chemical*, 171–172: 77–92, 2012.
85. H. Riris, C. B. Carlisle, L. W. Carr, D. E. Cooper, R. U. Martinelli, and R. J. Menna, Design of an open path near-infrared diode laser sensor: Application to oxygen, water, and carbon dioxide vapor detection, *Applied Optics*, 33(30): 7059–7066, 1994.
86. R. A. Potyrailo, S. E. Hobbs, and G. M. Hieftje, Near-ultraviolet evanescent-wave absorption sensor based on a multimode optical fiber, *Analytical Chemistry*, 70(8): 1639–1645, 1998.
87. B. Schirmer, H. Venzke, A. Melling, C. S. Edwards, G. P. Barwood, P. Gill, M. Stevens, R. Benyon, and P. Mackrodt, High precision trace humidity measurements with a fibre-coupled diode laser absorption spectrometer at atmospheric pressure, *Measurement Science and Technology*, 11(4): 382, 2000.
88. M. G. Kozlov, M. A. Kustikova, and K. A. Tomskii, Optical hygrometer, *Measurement Techniques*, 48(7): 731–735, 2005.

89. C. Lauer, S. Szalay, G. Bohm, C. Lin, F. Kohler, and M. C. Amann, Laser hygrometer using a vertical-cavity surface-emitting laser (VCSEL) with an emission wavelength of 1.84 μm, *IEEE Transactions on Instrumentation and Measurement*, 54(3): 1214–1218, 2005.
90. S. Basu, D. E. Lambe, and R. Kumar, Water vapor and carbon dioxide species measurements in narrow channels, *International Journal of Heat and Mass Transfer*, 53(4): 703–714, 2010.
91. K. Sun, L. Xie, Y. Ju, X. Wu, J. Hou, W. Han, X. Wang et al., Compact fiber-optic diode-laser sensor system for wide-dynamic-range relative humidity measurement, *Chinese Science Bulletin*, 56(32): 3486–3492, 2011.
92. Q. Wang, J. Chang, C. G. Zhu, C. Li, F. J. Song, Y. N. Liu, and X. Z. Liu, Detection of water vapor concentration based on differential value of two adjacent absorption peaks, *Laser Physics Letters*, 9(6): 421, 2012.
93. S. Zhang, Q. Wang, Y. Zhang, F. Song, K. Chen, G. Chou, J. Chang et al., Water vapor detection system based on scanning spectra, *Photonic Sensors*, 2(1): 71–76, 2012.
94. K. Kali, D. J. Webb, C. Zhang, I. Johnson, X. F. G. Chen, D. S. Rodriguez, J. D. Barton et al., Applications of polymer fibre grating sensors, in *The 18th International Conference on Plastic Optical Fibers*, Sydney, New South Wales, Australia, 2009.
95. K. Kuang, S. Quek, C. Koh, W. Cantwell, and P. Scully, Plastic optical fibre sensors for structural health monitoring: A review of recent progress, *Journal of Sensors*, 2009: 1–13, 2009.
96. W. Zhang, D. J. Webb, and G.-D. Peng, Investigation into time response of polymer fiber Bragg grating based humidity sensors, *Journal of Lightwave Technology*, 30(8): 1090–1096, 2012.
97. E. Cussler, *Diffusion: Mass Transfer in Fluid Systems*, Vol. 1(2), p. 9, Cambridge University Press, Cambridge, U.K., 1984.
98. J. Crank, *The Mathematics of Diffusion*, Oxford University Press, Oxford, U.K., 1979.
99. D. Turner, Polymethyl methacrylate plus water: sorption kinetics and volumetric changes, *Polymer*, 23(2): 197–202, 1982.
100. C. Zhang, X. Chen, D. J. Webb, and G. D. Peng, Water detection in jet fuel using a polymer optical fibre Bragg grating, *Proceedings of SPIE*, 7503: 750380, 2009.
101. T. Li, C.-L. Zhao, X. Dong, W. Qian, Y. Jin, and S. Jin, Relative humidity sensor based on photonic crystal fiber with tapered and filled in polymer, *Proceedings of SPIE*, 7900: 79900G, 2010.
102. J. Witt, M. Steffen, M. Schukar, and K. Krebber, Investigation of sensing properties of long period gratings based on microstructured polymer optical fibres, *Proceedings of SPIE*, 7653: 76530I, 2010.
103. J. Witt, Humidity sensing based on microstructured POF long period gratings, *POF 2011 Bilbao (Spain) Conference Proceedings*, Bilbao, Spain, 2011.
104. C. Zhang, W. Zhang, D. J. Webb, and G. D. Peng, Optical fibre temperature and humidity sensor, *Electronics Letters*, 46(9): 643–644, 2010.
105. W. Zhang, D. J. Webb, and G. Peng, Investigation into time response of polymer fiber Bragg grating based humidity sensors, *Journal of Lightwave Technology*, 30(8): 1090–1096, 2012.
106. G. Rajan, Y. M. Noor, B. Liu, E. Ambikairaja, D. J. Webb, and G.-D. Peng, A fast response intrinsic humidity sensor based on an etched singlemode polymer fiber Bragg grating, *Sensors and Actuators A: Physical*, 203: 107–111, 2013.
107. F. Gu, L. Zhang, X. Yin, and L. Tong, Polymer single-nanowire optical sensors, *Nano Letters*, 8(9): 2757–2761, 2008.
108. M. M. Noor, N. Khalili, I. Skinner, and G. Peng, Optical relative humidity sensor based on a hollow core-photonic bandgap fiber, *Measurement Science and Technology*, 23(8): 085103, 2012.
109. M. Mohd Noor, N. Khalili, and G.-D. Peng, All-fiber optic humidity sensor based on photonic bandgap fiber and digital WMS detection, *IEEE Sensors Journal*, 13(5): 1817–1823, 2013.
110. M. M. Noor, N. Khalili, and G. Peng, SOA-based fiber ring laser with seed of DFB wavelength scanning for relative humidity measurement using an air-guided photonic crystal fiber, *Measurement Science and Technology*, 24(8): 085203, 2013.
111. M. M. Noor, N. Kassim, A. Supaat, M. Ibrahim, A. Azmi, A. Abdullah, and G. Peng, Temperature-insensitive photonic crystal fiber interferometer for relative humidity sensing without hygroscopic coating, *Measurement Science and Technology*, 24(10): 105205, 2013.
112. J. Mathew, Y. Semenova, and G. Farrell, Relative humidity sensor based on an agarose-infiltrated photonic crystal fiber interferometer, *IEEE Journal of Selected Topics in Quantum Electronics*, 18(5): 1553–1559, 2012.
113. M. Mohd Noor, G. Rajan, and G. Peng, Microstructured fiber sealed-void interferometric humidity sensor, *IEEE Sensors Journal*, 14(4): 1154–1159, 2013.
114. S. Akita, A. Seki, and K. Watanabe, A monitoring of breathing using a hetero-core optical fiber sensor, *Proceedings of SPIE*, 7981: 79812W, 2011.

115. Y. Kang, H. Ruan, Y. Wang, F. J. Arregui, I. R. Matias, and R. O. Claus, Nanostructured optical fibre sensors for breathing airflow monitoring, *Measurement Science and Technology*, 17(5): 1207, 2006.
116. F. J. Arregui, Y. Liu, I. R. Matias, and R. O. Claus, Optical fiber humidity sensor using a nano Fabry–Perot cavity formed by the ionic self-assembly method, *Sensors and Actuators B: Chemical*, 59(1): 54–59, 1999.
117. S. Muto, H. Sato, and T. Hosaka, Optical humidity sensor using fluorescent plastic fiber and its application to breathing-condition monitor, *Japanese Journal of Applied Physics Part 1 Regular Papers Short Notes And Review Papers*, 33: 6060, 1994.

16 Medical Applications of Fiber-Optic Sensors

Vandana Mishra

CONTENTS

16.1 INTRODUCTION

Light and optical technologies are essential for both diagnostic and therapeutic parts of healthcare. The necessity of illuminators for examining the internal organs was felt since ancient times, which later resulted in the concept of light guidance through optical fibers. Being made with dielectric materials, the fibers are intrinsically immune to electromagnetic interference (EMI) apart from being chemically inert, smaller in size, flexible, and lightweight. Hence, these light-guiding fibers have great application potential in the field of medicine. In fact, the first clinical application of fibers was as flexible endoscopes reported by Hirschowitz in 1958.[1,2] In the 1960s, a major breakthrough in the form of *laser* made them more relevant by opening up various new application areas in therapy and surgery while adding sophistication to viewing of internal organs. The spectral analysis of laser light sent and received from biological tissues through fibers is able to extract information about various aspects of human physiology. Though initial fiber endoscopes were developed for illuminating and imaging internal organs, subsequent technological developments resulted in endoscopes equipped with fiber-optic sensors (FOSs). Noninvasive *optical biopsy* systems based on fiber-optic diagnostics are slowly coming to reality.

During the 1970s and 1980s, the use of optical fibers to deliver laser power for imaging, therapy, and surgery became widespread; however, their applications as sensing elements were in initial experimental stages. For better imaging, state-of-the-art camera is replacing fibers, but where diagnosis is required, imaging fiber with sensing capabilities is a must. A number of companies started developing fiber endoscopes that integrate a variety of nonimaging sensors. One example is laser Doppler analysis (Section 16.5.6) of different cells in which scattered light is used to detect blood velocity and oxygen content to determine if sufficient blood is reaching vital organs.[3] Miniature sensors at the end of optical fiber were devised to measure pressures in the arteries, bladder, urethra, and rectum. Fiber-optic ultraminiature pressure sensors (Section 16.5.9) are capable of sensing intravascular blood pressure or intracranial pressure (ICP) during trauma.[4] Some biological and/or chemical analysis is also possible utilizing the phenomenon of luminescence. A later development is the use of a single multimode fiber as a thin miniaturized high-resolution endoscope without lens and actuator to visualize inside very small cavities.[5]

Today, novel specialty fibers are being manufactured, which are creating new roles for FOS as multifunctional tools. By modifying properties such as numerical aperture, structure, material, core-cladding diameters, and coating material, the fibers can be adopted for a variety of applications. For example, an optical fiber–based drug delivery device can be much safer option for the delivery of photosensitive chemicals to kill drug-resistant infections or tumors.[6]

The high rate at which new knowledge is being added to the theme of this chapter is a reflection of the interest that the researchers and the industry have on the subject. It will not be possible to cover each one of them in this chapter. This chapter is a contemporary review of classification, working, and emerging application areas of FOSs in the field of medicine.

16.2 ADVANTAGES OF FOS

Most of the optical fibers are fabricated using either silica SiO_2 or special polymers like polymethylmethacrylate (PMMA); both of these materials have intrinsic properties that make them valuable for health-care applications. Currently, fiber-optic industries routinely make customized biocompatible and nontoxic fibers for medical applications that can be used for in vivo measurements and can be left in their position for repeated or continuous monitoring. All those distinctive attributes that allow them to be safely used in a clinical setup are described subsequently.[7,8]

16.2.1 Immunity to Electromagnetic/Radiofrequency Interference and Chemically Inert Nature

Due to its dielectric construction, fibers can be used in high voltage; electrically noisy, high magnetic field; high temperature; or other harsh environments specially in the intensive care unit (ICU) or MRI rooms. Being chemically inert, it will not interfere with the clinical environment when used invasively.

16.2.2 Flexibility and Geometric Versatility

Fibers can withstand bend (*less than 10 mm radius in some cases*)[9] and are very flexible so that they can adhere to complex biological structures, for example, bones and tissues, and can make access to very small or difficult-to-reach parts of the human anatomy possible.

16.2.3 Small Size and Lightweight

Standard fibers such as single-mode optical fibers (SMFs) have a diameter of 125 μm comparable to normal surgical suture. Even with protective coatings, the diameter does not exceed 500 μm, making it easy to bundle multiple fibers together in one catheter or needle. The fiber materials have densities much lower than the metal wires. Thus, miniaturization is easy and especially useful for very minute sample size or for minimally invasive sensing.

16.2.4 Biocompatibility and Safety

For medical applications, optical fibers are made with biocompatible materials. Even when used in vivo, they are not immunogenic, thrombogenic, or carcinogenic.[10] Since the information communication is through visible or infrared (IR) light waves, there is no chance of electric shock to the patient.

16.2.5 Remote Sensing

Thanks to the revolution in fiber-optic communications, fibers are known to have lowest attenuation over long distances. It is possible to have all the electronic instrument parts far away from the sensor

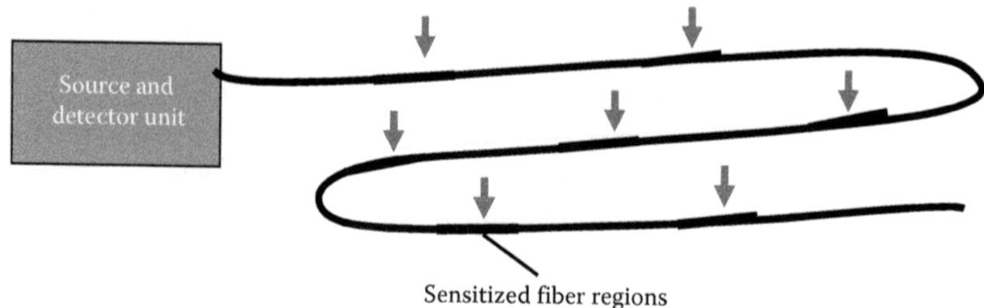

FIGURE 16.1 Schematic of distributed FOS.

fiber parts at the patient bedside. Single fiber serves both as sensor and as data transmitting channel and is compatible with communication fiber facilitating telemetry.

16.2.6 Multiparameter and Distributed Sensing

Each fiber of a fiber bundle can be used to sense various parameters with the help of suitable transducing and signal processing techniques. Moreover, several sensing elements can be multiplexed on a single fiber strand facilitating distributed sensing at various points. A common sensor platform can be created for either multiparameter sensing or distributed sensing of a single parameter to give a profile. Figure 16.1 shows a scheme of distributed sensor network.

16.2.7 No Cross Talk

There is *no cross talk* between close fibers so that multiple fibers having different functions can be housed in close vicinity, for example, a catheter. Some of these fibers can be used for sending and receiving light, while others can have sensing properties.

16.3 CLASSIFICATION

The simplest classification of FOSs with respect to their geometric construction consists of two basic categories referred to as *extrinsic*, or hybrid, and *intrinsic* or all-fiber FOSs. With the help of proper transducers, FOS architecture can be in the form of single sensor for single parameter, sensor array, distributed sensor, multiparameter sensor, and a combination of these.[7] Some of the sensor types mostly used in the field of medicine are briefly described in the subsequent section.

16.3.1 Extrinsic

Extrinsic FOSs are indirect sensors, making use of an additional sensing element, fixed at the fiber end, which causes optical modulation of the guided light. Many FOSs are of this type, since it is relatively easy to find materials that can change their own optical characteristics (i.e., refractive index [RI], reflectivity, absorbance, and fluorescence) by interacting with the parameter being tested.

The phenomenon of interference of light enables realization of many high-precision sensor systems for pressure and force. The use of optical fibers allows making such devices extremely compact and economic. Among various fiber-optic interferometers such as Mach–Zehnder, Sagnac, Michelson, and Fabry–Perot (FP), FP does not have two separate arms. Therefore, it can be made very small with capability of point measurement and is the most common example of extrinsic FOS.

In a typical fiber-optic *extrinsic Fabry–Perot interferometer* (*EFPI*, Figure 16.2), the interference occurs at the partially reflecting end-face surface of the fiber and an external mirror that can

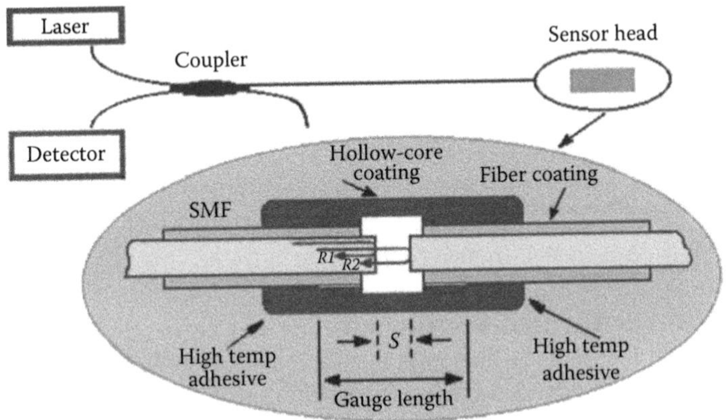

FIGURE 16.2 Schematic of extrinsic FP interferometer.

be another fiber, a mirror or a diaphragm. A single fiber acts as both the lead-in and lead-out fibers with a well-defined gauge length and width of the FP cavity.[11] This sensor system consists of a laser diode, which illuminates an FP cavity through a fused biconical tapered coupler.

The distance between the input/output and reflector attachment points in the silica capillary tube defines the gauge length L of the EFPI sensor. The first reference reflection, R1, is independent of applied perturbation. The second sensing reflection, R2, is dependent on the cavity length that, in turn, is modulated by the applied perturbation. The interference between the two signals is observed at the output as variations in optical intensity. Any strain over the gauge length of the EFPI will cause a change in cavity length s. The strain ε is determined as the ratio between the end-face displacement Δs and the gauge length L:

$$\varepsilon = \frac{\Delta s}{L}$$

where Δs is determined from the number of interference fringes shifted due to the applied strain. If m is the number of fringes shifted, then

$$\Delta s = \frac{m\lambda}{2}$$

Other parameters like RI can be obtained through the phase shift or the fringe contrast change of the interference signal in modified designs of the FP/hybrid structures.

16.3.2 Intrinsic

Intrinsic FOSs use the fiber itself as the sensing element and are generally indirect sensors in which modulation is induced through a modified portion of core or cladding. The natural cladding of the fiber can be chemically treated or stripped away and replaced with a layer so that the core–cladding interface acts as the sensing element. The core can be suitably doped, so that a modulation of absorption or fluorescence occurs.

The simplest and oldest example of intrinsic FOSs is microbending sensors (Figure 16.3). These sensors are light intensity–based sensors where intensity modulation is induced by bending losses in multimode plastic fibers. Several other parameters such as temperature, pressure, and force can also be measured by using suitable transducers. These sensors are low cost and robust, and when

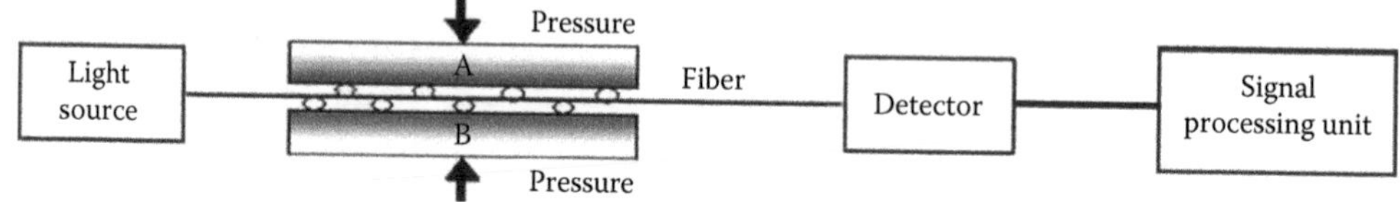

FIGURE 16.3 Microbending sensors.

used with optical time domain reflectometry techniques, they may be used for distributed sensing. Their overall accuracy is very low.

The FP interferometer described in previous section can be intrinsic as well by introducing in-fiber gratings and/or mirror at the fiber end.

Since the 1990s, the most common intrinsic sensors are in-fiber gratings where a periodic variation of RI in the fiber core is created. These gratings are categorized as *fiber Bragg gratings* (*FBGs*) and *long-period gratings* (*LPGs*) (Figure 16.4). In an FBG, a narrow band of wavelength is reflected, and there is a corresponding drop of intensity in the transmitted spectrum while LPGs work as wavelength-dependent loss elements exhibiting multiple loss resonance bands in the transmitted spectrum. A variation in parameters to be sensed can be detected easily by measuring corresponding wavelength shift. Apart from basic FBG and LPGs, there are some other distinct grating types that are formed when deliberate nonuniformity in the RI profile is introduced. *Blazed* or *tilted* fiber Bragg gratings (TFBGs) are resulted when grating planes are at an angle with the fiber axis and can be designed in a way that its core mode is coupled with some of the cladding modes. And, therefore, the Bragg wavelength becomes a function of ambient RI that can be used for *RI sensing* (Section 16.4.7). Another type is *chirped* FBG in which the grating period is aperiodic having a monotonic increase/decrease or nonuniformity longitudinally. In this type of gratings, each point has different Bragg wavelengths, and hence, its spectrum can be used to monitor a parameter profile by distributed sensing.[12]

FBG sensor works by monitoring Bragg wavelength shift due to external perturbation:

$$\Delta\lambda_B = 2\left[\Lambda\frac{\partial n}{\partial l} + n\frac{\partial \Lambda}{\partial l}\right]\Delta l + 2\left[\Lambda\frac{\partial n}{\partial T} + n\frac{\partial \Lambda}{\partial T}\right]\Delta T$$

where

Λ is the pitch or periodicity of the grating
n is the effective RI of the core
λ_B is the Bragg wavelength[12,13]
Δl is the change in grating length due to strain
ΔT is the change in ambient temperature

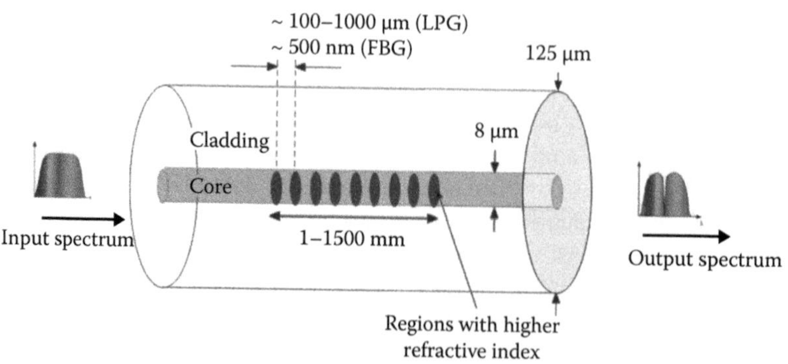

FIGURE 16.4 Schematic representation of FBG and LPG.

LPGs couple core mode to different cladding modes making it sensitive to the RI of the surrounding medium as well. This sensitivity is expressed in terms of either wavelength(s) shift

$$\frac{d\lambda_i}{dn_{sur}} = \frac{d\lambda_i}{dn_{ieff}^{cl}} \cdot \frac{dn_{ieff}^{cl}}{dn_{sur}}$$

where n_{sur} is the RI of the surrounding medium

or as variation in intensity of the loss resonance peak

$$I = \frac{\int_r \int_\phi \psi_{core} \psi_{clad}^* \, r dr d\phi}{\sqrt{\int_r \int_\phi \psi_{core} \psi_{core}^* \, r dr d\phi} \sqrt{\int_r \int_\phi \psi_{clad} \psi_{clad}^* \, r dr d\phi}}$$

where Ψs are the electromagnetic field components of the two coupling modes and r and Φ are radial and azimuthal coordinates, respectively.[14] This is the basic principle of an LPG-based RI and concentration sensor.[15,16]

In the category of intrinsic FOS, specialty fibers like photonic crystal fibers (PCFs)[17] with hollow core or microstructured fibers with structured holes in the cladding make a good sensing platform.

Another simpler classification relevant in biomedical applications can be either *invasive* or *noninvasive* sensors. As the name suggests, the noninvasive sensors do not involve puncture of the skin or other entry into the body.

16.4 WORKING

The basic principle of FOSs involves modulation of one or more of the optical properties (phase, intensity, wavelength, and polarization state) of the light guided through the optical fiber in response to the parameter under investigation. FOS system comprises four essential components, a light source, an optical fiber, a transducer element, and a light detector. The transducer, which can be the fiber itself in some cases, causes the light property to change with respect to certain parameter. The detector is a part of signal processing electronic unit that analyzes this change.

Spectral measurements involve different types of spectroscopic technique, for example, absorption, reflectance, fluorescence/phosphorescence, bioluminescence, Raman, and surface plasmon resonance (SPR); almost all the bio- or chemical sensors are dependent on one or more intensity-based spectral measurements.

The absorbance and fluorescence analyses need a specific measuring cell at the fiber end, so as to establish the optical path and provide back transmission. Reflectance analyses, on the contrary, are performed using bare fibers, the numerical aperture of which determines the view angle inspected. Fibers are thus customized in terms of material, structure, and tip geometries to fit into various measurement systems.

16.4.1 Absorbance Measurements

The simplest optical biosensors use absorbance measurements to determine any changes in the concentration of analytes that absorb a given wavelength of light. The system works by transmitting light through an optical fiber to the sample a sensitive layer of biological receptor is immobilized either close to or directly on the surface of the optical fiber.[7,18] On interaction with the analyte, variation of the absorbance properties of the sensitive layer occurs, which is detected through the same fiber or a second fiber.

Light absorption in a sample is governed by the *Beer–Lambert law* that relates absorbance A to an analyte dispersed in a solution of concentration C by the following relation:

$$A = \varepsilon LC$$

where
L is the optical path length
ε is the wavelength-dependent molar absorption coefficient

By measuring the solution absorbance, it is possible to estimate bio-analyte concentration in aqueous solutions.

This transduction mechanism is used to sense biological pathogen or gas (as oxygen or carbon dioxide) concentrations and to measure pH in solutions. Another important sensing platform uses near-infrared spectroscopy (NIRS). Most of the biological tissues have very good transparency in the near-IR range while oxygenated hemoglobin (HbO_2) and deoxyhemoglobin (Hb) have good absorbance between 700 and 1000 nm, and the transmission of light is relatively unaffected by water in the same region. Thus, the NIRS is favorable for the measurement of these parameters. Cerebral blood volume in human infants had been quantified by NIRS utilizing the differences in the visible and near-IR absorbances of oxygenated and deoxygenated hemoglobin.[19]

16.4.2 Reflectance Measurements

When light comes into contact of tissue, it undergoes absorption or scattering. The light scattered from the tissue in the backwards direction is reflectance that contains information regarding the scattering and absorption properties of the tissue. Unlike absorption, reflectance measurement is noninvasive as it uses surface probes. Normally, for reflectance measurement, one source fiber injects light onto the tissue surface, and reflected light is detected by several detection fibers at varying distances surrounding the source fiber. The tissue properties are extracted through measured reflectance data by fitting an analytic equation. Depending on the specific application, a reflectance sensor can be designed to be depth sensitive.[20]

Diffuse reflectance spectroscopy (DRS) spectra in the 1250–2500 nm regions were used to estimate the regional differences of water content in the skin. Measurements were performed in vivo for the skin of the forehead, cheek, jaw, elbow, volar forearm, palm, knee, and heel of seven healthy volunteers, using a Fourier transform near-infrared (FT-NIR) spectrophotometer with a fiber-optic probe.[21] DRS was developed to distinguish between healthy and malignant tissues directly utilizing multispectral imaging camera connected to preprogrammed computer to record and process the images sequentially. The role of optical fiber is limited to launching and receiving light for referencing and not a part of sensing scheme. One multispectral DR imaging method was investigated for estimating melanin concentration, blood concentration, and oxygen saturation in human skin tissue and validated in in vivo experiments on fingers.[22]

Reflectance measurement form sensitive biosensors in surface plasmon resonance or surface grating geometries where, by linking receptor molecules to the surface, complementary binding molecules can be detected without using any kind of fluorescent probe or label.[23,24]

16.4.3 Fluorescence Measurements

Fluorescence measurements can detect naturally fluorescent or artificially labeled analyte. The naturally fluorescent tissue components called fluorophores are aromatic amino acid components (tryptophan, tyrosine, and phenylalanine) of proteins, nucleic acids, and nucleotide coenzymes.[25] Changes in intrinsic fluorescence show changes in protein structure and can be used to distinguish between normal and cancerous tissues. The artificially labeled compounds indirectly monitor formation or consumption of a transducer and are more common in several biomedical applications ranging from glucose sensing to cancer detection.[26]

Since fluorescence intensity is proportional to the excitation intensity, even weak signals can be observed. Like absorption and reflectance measurements, conventional fluorescent sensor consists of an excitation fiber, a collection fiber, and a spacer.

Two possible configurations, that is, single fiber and multifiber, exist. Single-excitation-collection fiber configurations are useful for a more localized fluorescence measurement, while multifiber configuration with fiber sensors at different detection locations achieves a larger and broader investigation volume. For example, microspheres containing suitable fluorophores are deposited onto the wells of etched fiber bundles for the detection of sodium, potassium, calcium, and chloride in a sample. The fluorescence emission characteristics of individual microspheres were observed and were found to selectively and rapidly change with sample composition.[27]

16.4.4 Bioluminescence/Chemiluminescence Measurements

These techniques are similar to fluorescence with different photoemission principles; one uses live organisms (bioluminescence) and the other uses immune reagents (chemiluminescence). Other fluorometric methods require an excitation source to stimulate photoemission of the fluorescent-tagged species and filters to block high background. Bio-/chemiluminescence measurements do not require filters or sources and therefore are suitable for lab-on-chip analysis. Moreover, they couple the bioreceptor intimately to the surface of a transducer; thus, signal collection is much better, making them more sensitive than absorption spectroscopy and fluorometry.[28]

Progress in molecular biology and biotechnology has resulted in (a) more efficient substrates and (b) recombinant/mutated enzymes and genes as reporters or probes leading to the development of ultrasensitive and selective bioanalytical tools.[29]

In lieu of their higher sensitivity and stand-alone configuration, bio-/chemiluminescence-based FOSs are appropriate for miniaturized lab-on-chip or μ-total analysis systems (μ-TAS), for example, microarrays, microfluidic devices, and high-density-well microtiter plates, with very small sample volumes. One lab-on-chip bioluminescence detection system reported by Eltoukhy et al.[30] can be customized for various applications such as determining gene expression using reporter gene assays, determining intracellular ATP, immunoassays, and sequencing DNA. They created a lab-on-chip tool consisting of a fiber-optic faceplate* with immobilized luminescent reporters/probes directly coupled to detection and processing system on a chip.

16.4.5 Raman Spectroscopy

Raman spectroscopy offers certain advantages compared to other techniques for optical biopsies with its ability to provide fingerprint biochemical information that can predict the onset of cancer and other life-threatening illnesses. When an incident light interacts with molecules in a sample, a small proportion of it is inelastically scattered with the frequency shift function of the specific Raman active vibration modes of the molecules. Thus, peaks in the Raman spectrum correspond to known molecular signatures in the tissue, thereby providing more structured and detailed information without labels or other indicators. Because disease symptoms are always preceded by changes in the molecular composition of tissue or body fluids, Raman spectroscopy offers the ability to precisely diagnose a wide range of disease states.

There had been certain hindrance in the development of the Raman-based technique. Foremost among these was *fluorescence interference* due to highly fluorescent nature of tissue in the ultraviolet (UV) region of the spectrum. Second, the Raman signal is very weak; for every one million photons applied to

* A fiber-optic faceplate is an optical mosaic in which fibers less than an inch in length are fused together to form a vacuum-tight glass plate. It precisely transmits an image from its input surface to its output surface.

tissue, 1000 are fluorescent and 1 is Raman. To obtain spectra with good signal-to-noise ratios, researchers have had to use laser power densities that are unacceptably high—prompting biological changes or damage in the tissue—and collection times that are too long to be practical in a clinical setting. To resolve these problems, new fiber-optic probes were designed that used filters and collected scattered light in the fiber tip. Raman instruments with a mobile setup consisting of a NIR laser to avoid fluorescence, a suitable fiber-optic probe, high-quality charge-coupled device (CCD) detector, and spectrometer optimized for the NIR region were developed.[31] A series of in vitro and in vivo trials have demonstrated application of Raman instruments with suitable fiber probes in real-time diagnosis, for example, detection and characterization of skin diseases, (pre)malignant lesions, and pathogenic microorganisms.

The magnitude of the Raman scattering efficiency can be enhanced manifold when a compound is adsorbed on or near special metal surfaces. Thus, for higher sensitivity and faster response, fiber-optic surface-enhanced Raman scattering (SERS) sensor probe optimized with nanoscale roughness of silver or gold is being investigated.[32,33] For example, SERS technique has been demonstrated for the simultaneous detection of three bacterial meningitis pathogens.[34]

16.4.6 Fiber Evanescent Wave Spectroscopy

Since the advent of the Fourier transform infrared (FTIR) spectrometer, IR spectroscopy has become an indispensable and efficient tool of analytical studies. The fiber is employed not only to transmit the IR beams from the spectrometer to the sample but also as a sensor probe. For this, fiber evanescent wave spectroscopy (FEWS) is useful where interaction takes place at the interface between the sample and the fiber. The principle of measurement is based on an exponentially decaying evanescent wave around the fiber partially absorbed by the sample during light propagation. The penetration depth that describes the extent of evanescent wave is too small for sensing biological cells or proteins that are one or two orders larger. The penetration depth can be increased to facilitate mode coupling by bending, tapering, or changing the characteristic of light. Tapered FEWS biosensors have been widely used for the measurement of cells, proteins, and DNA. A review of biosensing application including the target detected, matrix, limit of detection (LOD), sensor geometry, fiber type, and principle of detection can be found in Ref. [35].

IR optical fibers transmit radiation with wavelengths >2 mm and are able to collect the highly specific vibrational spectrum of organic chemicals and biomolecules. Although their communication properties are very poor as compared to conventional fibers, they are excellent candidate as evanescent wave fiber sensors and image bundles in fiber lengths <2–3 m.[36] One such fiber fabricated with chalcogenide glass has been reported for the study of the brain metabolism modification by mid-infrared (MIR) spectroscopy, after a transient focal cerebral ischemia provoked artificially in the rat brain though not in humans.[37]

16.4.7 Refractive Index Measurements

The RI sensing is the backbone of various biological processes and can be indicative of crucial parameters in health care. There are various FOS configurations capable of measuring minute changes in surrounding RI including etched FBGs; TFBGs; LPGs; concatenated LPGs forming a Mach–Zehnder interferometer; interferometer using multimode/single-mode/multimode (MSM) fibers, or single-mode/multimode/single-mode (SMS) fibers spliced in sequence; micro- and/or nanofiber (MNF) refractometers; structured fiber tips; and PCFs.[11–16,38,39]

16.4.7.1 Surface Plasmon Resonance Technology

SPR occurs at metal–dielectric interface where evanescent component of light in dielectric causes excitation of electron density oscillations (known as surface plasmon wave [SPW]). For RI sensing, the angle of the incident beam is kept constant, and wavelength is varied so that resonance occurs at a particular wavelength. The resonance parameter is a function of the RI of the dielectric medium.

In optical fibers, the evanescent wave propagates along the core–cladding interface. Therefore, fiber-based SPR sensor involves removing a section of the fiber cladding and depositing an SPR-active thin metal layer symmetrically around the fiber core. The SPR effect can be exploited by deposition of metal thin film on different fiber structures like tapered fibers, unclad fibers, D-fibers, or polished fibers (where cladding is removed partially).[40,41]

16.4.7.2 Fiber Grating Technology

These are the most versatile FO sensing tools mostly because the measurand information is wavelength encoded. In normal FBGs, the optical signals are confined to the fiber core regions making them insensitive to external RI variation, but they can be made sensitive either by writing tilted grating or by etching the cladding part.[42,43]

LPGs on the other hand are intrinsically sensitive to external RI and can provide direct detection of large molecules, by using suitable coatings (Figure 16.5). The binding of the specific target results in RI change around the LPG region, and thus, a wavelength shift is observed. The magnitude of the observed shift can be correlated to target concentration. The wavelength shift is more if hydrogel-like coatings are used because there is strain effect due to swelling of the hydrogel in addition to RI change.[44] The sensitivity of LPG-based sensor can be increased by using concatenated LPGs that form a Mach–Zehnder interferometer. Reducing diameter of the fiber section between two gratings further increases sensitivity.[13–15,45]

16.4.7.3 Photonic Crystal Fiber

PCF or microstructured fibers are special fiber with air holes running throughout the fiber length. They act as refractometer by infusion of fluids through one or more air holes, tapering or writing LPG on its core. In some cases, hollow-core PCFs can be used as microfluidic channels.[46–49]

16.4.7.4 Microfibers and Nanofibers

MNFs are fiber tapers with a waist region size comparable to the wavelength of transmitting light. Their unique properties such as large evanescent fields, strong optical confinement, flexibility, and configurability can be utilized in various refractometer configurations. Straight taper waist, MNF loop, and knot- and coil-resonant sensors exploit strong evanescent field to interact with surrounding medium. A comprehensive description and comparison can be found in a review article by Chen et al.[50]:

- A lab-on-chip-type sensor configuration based on the FP interferometer, with the possibility of multiple sensor integration, was reported. In this refractometer, the FP cavity was fabricated by aligning two metal-deposited, SMF end faces inside a microchannel on a silicon chip, and the RI change within this cavity was determined by demodulating the transmission spectrum phase shift.[51]

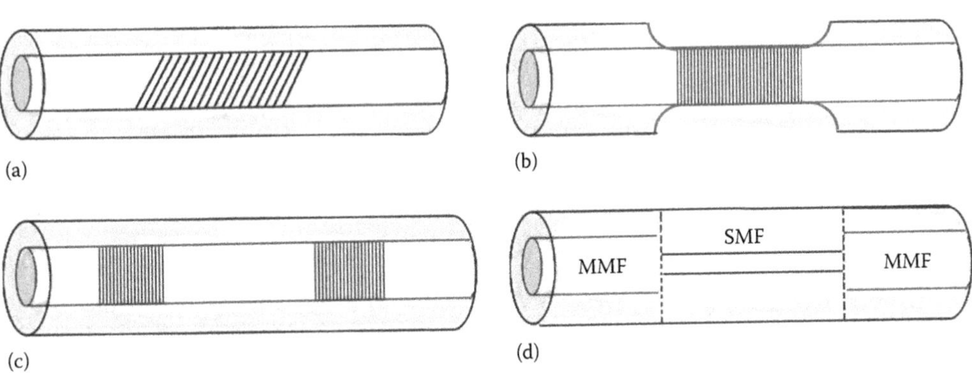

FIGURE 16.5 (a) TFBG, (b) etched FBG, (c) concatenated LPG, and (d) MSM structure for RI sensing.

- Tapered optical fiber tips can be simple and effective RI sensors for minute analyte volumes.[52] A simple in-line FP cavity produced near the tip of an SMF by laser micromachining can provide temperature-independent measurement of any RI larger than that of air with a good linearity and a high resolution. The FP sensor head was fabricated by micromachining a circular hole at the center of the cross section of an SMF.[53]
- A compact probe based on a micronotch cavity fabricated by focused ion beam micromachining was reported by Kou et al. They later developed a metal–dielectric–hybrid probe with milled gratings replacing the microcavity.[54,55]
- Another compact single-mode/multimode/single-mode fiber tip (SMST) is recently reported. It consists of an input/output SMF, an MMF half taper section followed by a gold film at the end of the tip as reflection mirror. This RIS probe used focused ion beam milling to have a clean fiber tip cut that was then coated by a layer of gold to increase reflection.[56]

16.5 APPLICATION AREAS

The application areas of FOS in biomedicine are diverse as evident by the number of articles and book chapters on the topic available in the literature. This section covers some of the existing and emerging application areas in the field.[57–60]

16.5.1 Arterial Gas Sensors

Monitoring of arterial blood gases is essential in the management of critically ill patients. Earlier technique required intermittent arterial blood sampling, was analyzed by electrode-based analyzers, had a significant turnaround time, exposed the health-care professionals to the patient's blood, and resulted in iatrogenic blood loss. These problems can be avoided with a miniaturized FOS that can be placed intravascularly to continuously measure the changes in pH, pCO_2, and pO_2, thus forming continuous intra-arterial blood gas monitoring systems (CIBGMs).

For some analysis, pulse oximetry (described in later section) can be used, but such noninvasive technologies cannot replace arterial blood gas analysis as they cannot detect a high pO_2 or define a safe lower saturation limit and cannot differentiate between various dyshemoglobinemias (*abnormal hemoglobin derivatives that are incapable of binding O_2*).[61]

CIBGMs have been developed by several manufacturers commercially and are routinely being used clinically. These devices are based on either fluorescent or absorbance measurements. FOS designs consist of fiber bundle having each fiber equipped with nontoxic dye specific for each analyte. The analyte should have good absorption and fluorescence emission characteristics. Finally, the detection unit is coupled with customized software for rapid data analysis. Various CIBGMs with their operating principles, applications, performance, limitations, costs, and overall impact on patient treatment have been reported.[62,63]

16.5.2 Biosensors/Immunosensors

A fiber-optic biosensor is an analytical system that can be used to convert the existence of a biomolecule or compound into a measurable optical signal. With some exceptions like fiber grating sensors, most of the fiber-optic biosensors require fluorescent or colorimetric labeling of biomolecules. When the labeled biomolecule interacts with its target, the fluor is excited by the light passing through the fiber, and the resulting fluorescence is measured.

Highly selective bioaffinity interactions of the antibody and antigen give rise to immunosensors. Wei et al. have designed the FOB-3, a portable fiber-optic biosensor based on the principle of evanescent wave and that detected *Yersinia pestis** antibodies against the F1 antigen of *Y. pestis*,

* Causes plague outbreak.

B. anthracis, Staphylococcal enterotoxin B, and *yellow fever virus* sensitively. The FOB-3 use sandwich immunoassay with *capture antigen* immobilized on the fiber probes and *detection antigen* labeled with fluorescent dye.[64]

An Israeli research group has developed chemiluminescent FOSs to antibodies from *hepatitis C*, *West Nile* and *ebola viruses*, *cholera toxin*, and *ovarian cancer* using either silane or electropolymerization on indium tin oxide (ITO)-coated fiber optics. This technique was more sensitive than conventional ELISA methods and could be used at point-of-care real-time detection.[65–67]

16.5.2.1 Bilirubin

Bilirubin (BR), one of the bile pigments in the blood, is a by-product of hemoglobin metabolism. As an index of liver function, serum BR levels indicate how well the liver is processing BR. Elevated levels of BR are indicative of necrotic liver diseases, various hemolytic disorders, and neonatal jaundice, while its low level indicates an iron deficiency and can be linked to coronary artery disease (CAD). The feasibility of FOS application has been established by using spectrophotometric procedure for BR determination. An FOS was developed by following the variation of its yellow absorption band. In the study, hemoglobin in the presence of glucose oxidase (GOD) was used to oxidize BR to biliverdin resulting in a decrease of BR yellow absorption band. In another method, both the enzyme and a fluorescence indicator were immobilized in a polymeric matrix that was attached covalently to an optical fiber tip to miniaturize FOS size. The response of the sensors was based on the dependence of the fluorescence intensity of ruthenium diimine complexes on the oxygen concentration in the analyte solution. The enzymatic oxidation of BR depleted the level of molecular oxygen in the analyte, thus decreasing the fluorescence quenching efficiency of the ruthenium complex by oxygen and, as a result, increasing the fluorescence intensity of the sensor.[68,69]

16.5.2.2 Celiac Disease

For the diagnosis of *celiac disease*, a tapered OFS was used to determine the presence of antigliadin antibodies (AGAs). A nanofilm of gliadin antigen was deposited on the taper waist surface using the ESA method. An evanescent wave interacts selectively with the antibodies present in the surrounding without any interference from other molecules. The interaction resulted in a change of optical power proportional to the change in the concentration of bound antibodies.[70]

16.5.2.3 Cholera

A chemiluminescent-based FOS was developed to detect *cholera* antitoxin IgA immunoglobulins. For this, fiber tips were immobilized with the cholera toxin B subunit. The cholera antitoxin analyte is marked by a secondary antibody labeled with horseradish peroxidase to generate chemiluminescent signal.[71] Later, the same group used electropolymerization on ITO-coated fiber. This fiber-optic electroconductive surface modification is done by the deposition of a thin layer of indium tin oxide upon which biotin-pyrrole monomers were electropolymerized and then exposed to avidin. Avidin–biotin interactions were used to modify the fiber optics with biotin-conjugated cholera toxin B subunit molecules.[72]

16.5.2.4 Cholesterol

Luminescence-based FOS was used for free cholesterol determination in serum samples. A luminescent material, sensitive to oxygen changes, was used as an optical transducer of the oxidation of cholesterol by cholesterol oxidase. The biocatalyst is entrapped in a graphite powder layer deposited onto the dyed silicone film. A bifurcated fiber bundle was used to carry the light from and to the flow cell. The measured signal is related to the cholesterol concentration.[73]

16.5.2.5 DNA Hybridization

DNA hybridization is a technique to measure the degree of genetic similarity between pools of DNA sequences. It is important for DNA sequencing, early diagnosis of infection and genetic diseases, and decoding of the human genome.

Most common FOS for DNA hybridization monitoring employs fluorometric detection where biotinylated single-stranded DNA (ssDNA) capture probes are immobilized on fiber by avidin–biotin affinity system. Hybridization with fluorescein-labeled complementary strands is monitored in real time by fluorescence detection. An evanescent wave fluorescent FOS based on quartz fiber[74] and silica fiber[75] had been reported. The fiber surface was regenerated either thermally or chemically for consecutive measurements.[76–78]

For simultaneous analysis of multiple DNA sequences, FOS array has been applied. A fiber bundle with each fiber carrying a different capture DNA probe immobilized on its distal end was assembled. Hybridization of fluorescently labeled complementary oligonucleotides to the array was monitored by observing the increase in fluorescence that accompanied binding. Chemical regeneration of the probes was achieved by dipping the fiber tip in a solution of formamide in TE buffer and retained their sensing capabilities for months if stored at 4°C.[79]

RI measurement–based FOS can monitor DNA hybridization in real time without any labeling. One such fiber grating sensor was reported employing etched FBGs to immobilize ssDNA. The sensor was kept in the hybridizing buffer solution for an hour. The hybridization of DNA strands was visualized by the wavelength shift. Stability and repeatability of etched FBG-based sensor are yet to be addressed.[80] LPG sensors on the other hand are more robust for such experiments. After the wavelength shift due to hybridization, stripping of the hybridized DNA or regeneration caused the wavelength to shift back. The hybridization process was repeated again to demonstrate reusability of the sensor.[81]

A RI-based biconical tapered fiber* probe was immobilized with capture ssDNA. Hybridization with the complementary target ssDNA changed the RI of the surrounding medium that leads to changes in the light propagation characteristics of the tapered fiber and hence in the output power of the sensor.[82]

Triple-helical DNAs are used as sequence-specific artificial nucleases, modulators of DNA-binding proteins/gene expression, materials for genomic mapping, and sensitive screening reagents to detect mutations within duplex DNA. FOS-based fluorometric detection of triplex formation in both parallel and antiparallel configuration has been reported.[83]

Another label-free FOS involves a novel *molecular beacon* (MB) DNA probe. MBs are ssDNA molecules that possess a stem-and-loop structure. The loop portion of the molecule can form a double-stranded DNA in the presence of a complementary nucleic acid. Biotinylated MB has been designed and immobilized on an optical fiber surface by means of biotin–(strept)avidin interactions. The MB-based DNA biosensor can be used for real-time detection of target DNA and RNA molecules without the use of competitive assays. The sensor is rapid, stable, highly selective, and reproducible.[84]

Wang et al. have reported fiber FP interferometry for a simple cost-effective direct in-cell application. The capture DNA is immobilized onto the surface of a fiber tip by a layer-by-layer electrostatic self-assembly technique. Hybridization of target DNA with complementary capture DNA increases the optical thickness of the fiber tip. This phenomenon can be detected by demodulation of the spectrum of an FP cavity.[85]

Microspheres-based technique described in Section 16.4.3 has also been employed for DNA hybridization detection. It is based on the attachment of different DNA probes to microspheres assembled in microwells created at the distal end of an optical fiber. A high-density array of DNA-functionalized microspheres randomly distributed in the microwells at the end faces of a fiber bundle comprise 6000–50,000 fused optical fibers. The positions of microspheres in the random array are determined by an optical encoding scheme. This platform provides a microarray with the smallest individual feature sizes, is able to detect nanomolar concentration of the analyte, and can be used for high-density genotyping and gene expression.[86,87]

* Fabricated by heating–pulling method using a CO_2 laser.

Other label-free schemes involve SPR, evanescent wave, or microstructured fiber–based techniques. The DNA probe technology proved to be a unique type of biorecognition element that is highly selective and stable and can be easily synthesized in the laboratory. Specific DNA probes can be tailored for unique biosensor development and genetics studies.[88]

16.5.2.6 Dengue

Dengue fever is caused by the dengue virus (DENV), transmitted mainly by the female *Aedes aegypti* mosquito. Chemiluminescent optical fiber immunosensor was developed for the detection of anti-DENV immunoglobulin M (IgM) in human serum samples, compared to other techniques, and found to be reliable, sensitive, fast, and cost effective.[89] Recently, a simple localized surface plasmon resonance (LSPR)–based FOS has been reported. The sensor system consists of a white-light source, a detector (spectrum analyzer), a 2 × 1 optical fiber coupler (OFC), and a standard multimode fiber with gold nanoparticles (AuNPs) deposited tip. Dengue anti-NS1 antibody was immobilized on AuNPs, and then the tip was incubated by dipping it in a solution of NS1 antigen for an hour. The sensor was shown to have a good correlation between wavelength shifts of the LSPR dip in the reflected signals and NS1 antigen concentration. It could potentially be used for dengue diagnosis in the acute phase of the infection.[90]

16.5.2.7 *E. coli* O157:H7

E. coli O157:H7 strain of *Escherichia coli* is known to cause severe life-threatening illness in humans. A fluorescent immunoassay system has been reported for rapid detection of *E. coli* O157:H7 using a flow cell that enables *in-reaction* measurement by analyzing the rate of fluorescence.[91]

Dead *Escherichia coli* (*E. coli*) O157:H7 has recently been identified without interference from live *E. coli* O157:H7 and six other *E. coli* serotypes using an antibody-activated combination tapered fiber as probe and propidium iodide (PI) as the fluorescent dye.[92]

16.5.2.8 Glucose

Most of the FOSs for glucose detection involve taking out blood sample and using various techniques to quantify glucose concentration. Though measurement can be fast and accurate, these methods are not practical or patient friendly to be used for self-monitoring. There is a need for developing techniques to monitor glucose levels in noninvasive and real-time continuous ways. Various researchers are investigating easier, more accurate, and instant self-monitoring methods.[93–95]

An FOS that measures glucose concentration by fluorescence resonance energy transfer (FRET) assay based on the selective binding of saccharides by the bioreceptor concanavalin A (Con A) was designed and built by Meadows and Schultz. The Con A was immobilized onto the inner side of a section of dialysis tubing. This tubing was sealed at one end, with the other end fixed to an optical fiber. The sensor was sensitive to temperature fluctuations and pH and had very slow response time. In a later improved version, the sensor probe indirectly measures glucose concentrations from the level of fluorescence quenching caused by the homogeneous competition assay between tetramethylrhodamine isothiocyanate (TRITC)-labeled Con A (receptor) and fluorescein isothiocyanate (FITC)-labeled dextran (ligand). The FITC signal is used as an indicator for glucose concentrations, and the TRITC signal is used for internal calibration.[96,97]

LPG RI sensors were reported for nonselective analysis of highly concentrated glucose solutions. For specific and more sensitive detection of glucose, a specific enzyme GOD was immobilized on suitably treated surface of an LPG. The immobilization of GOD caused the conversion of glucose to gluconic acid during the analysis (*a well-known reaction in earlier generation of glucose monitors*) changing the media's RI even with dilute glucose solutions. The utility of the sensor was demonstrated by determining glucose concentration in blood samples.[98,99]

A cladding stripped SMF with GOD immobilized at the core surface and heterodyne interferometry was utilized to achieve rapid glucose concentration measurement.[100] For a noninvasive approach, an FOS probe attached on the skin surface was reported. The probe consisted of one central illuminating fibers surrounded by several collecting fibers arranged in circle. Absorption spectra in the NIR wavelength range were measured at the forearm of volunteers, from which the blood glucose levels were calculated. Partial least square regression (PLSR) analysis had shown good correlation between these calculated values of the glucose level and directly measured values from blood samples. This method is not very efficient as the absorption coefficient of glucose in the NIR band is low.[101] Other noninvasive methods explored in laboratories were Raman spectroscopy and optical coherence tomography (OCT), but nothing has yet been commercialized.[102]

An *implantable* FOS that measures glucose concentration by FRET assay based on chemical reactions from the subcutaneous end of the fiber had been evaluated for continuous in vivo glucose monitoring in animals and found to be feasible. The tiny FOS was chronically implanted in a subject's skin. Attached to the end of the fiber inside the skin is a polyethylene-glycol polymer matrix interspersed with pairs of tightly bound chemicals, each tagged with a different fluorescent molecule. Under UV light from portable analyzer, the bound molecules shine at one wavelength. When the researchers place the matrix in a glucose solution, glucose molecules knock out and replace one of the chemicals. As a result, the chemical complex starts emitting at two different wavelengths. The ratio of the fluorescence intensities at the two wavelengths is in proportion to the glucose concentration.[103,104]

For minimally invasive high-precision in vivo glucose determination, small fiber-optic attenuated total reflection (ATR) sensor using MIR spectroscopy with a tunable CO_2 laser source was proposed, and the potential of this method was evaluated in vitro. This sensor is intended to be subcutaneously implanted in tissue to continuously monitor the body's glucose levels within the interstitial fluid (ISF), where the glucose concentration is considered to be strongly correlated with the concentration of glucose in the blood.[105]

The first preclinical demonstration of continuous glucose monitoring (CGM) by fluorophore acrylodan–labeled and genetically engineered glucose/galactose-binding protein (GGBP) immobilized at a fiber tip was reported in 2011. The GGBP immobilized tip fiber sensors contained in a stainless steel needle were inserted subcutaneously into Yucatan and Yorkshire swine, and the sensor response to changing glucose levels was monitored over a 7-day period. The performance of this sensor in swine models indicates its potential for CGM in humans.[106]

A CGM system based on a fluorescent biosensor placed on the tip of an optical fiber was investigated. This clinical trial was performed for 2 weeks at the Diabetes Institute in St. Joseph Hospital in Heidelberg, Germany. FiberSense was inserted into subcutaneous tissue (abdomen and upper arm) of diabetic patients. Data collected at several measurement sessions under glucose challenge conditions (oral carbohydrate uptake and insulin administration) were compared to the capillary blood glucose (evaluated by laboratory method) and to a commercially available CGM system placed on the contralateral abdominal body area and were found accurate.[107]

16.5.2.9 Hemoglobin

As described earlier, NIRS provides a noninvasive means to monitor oxyhemoglobin (HbO_2), deoxyhemoglobin (Hb), and total hemoglobin concentration (HbO_2 + Hb) that is considered as total blood volume (HbT) as well as oxygen saturation in the living tissue. Optical fiber probes were used in a novel, highly sensitive NIR continuous wave (CW-NIR) spectroscopy instrument prototype designed for noninvasive analysis of the two main forms of hemoglobin.[108]

16.5.2.10 HIV Protein p24

LPG-based biosensors are well suited for the detection of large molecules like proteins. LPG biosensors developed by Luna Analytics (Blacksburg, VA) had been used for detecting human immunodeficiency virus (HIV) proteins. Different concentrations of the p24 HIV protein were quantitated by using antibodies coated onto the LPG and by plotting the wavelength, which is related to analyte concentration or bound target mass. The sensitivities of these LPG sensors were found to be comparable to those of ELISA techniques (10–20 ng/mL).[44]

16.5.2.11 Protein C

It is one of the anticoagulants, the deficiency of which can lead to massive thrombotic complications. Monoclonal antibodies against protein C (PC) have been immobilized on the surface of a quartz fiber enclosed in a 300 μL chamber. After incubation with the sample containing PC, the fiber was probed with a fluorophore-tagged secondary antibody, the fluorescence intensity of which is correlated with the concentration of the protein. The device has a linear working range of 0.2–2.0 μg/mL, which includes the PC range of interest.[109]

The same sensing scheme has been employed by other authors for detection and quantification of other small biomolecules like *D-dimer* and other cross-linked fibrin degradation products, the most specific marker of *sepsis syndrome* and *thrombotic disorders*. Fiber-optic immunosensor analysis required significantly less time and skill than the standard ELISA method.[110]

16.5.2.12 Visceral Leishmaniasis (Kala Azar)

It is caused by a human blood protozoal parasite *Leishmania donovani* and is the most severe form of leishmania. It is a serious health hazard prevalent throughout tropical and temperate regions including Africa, China, India, Nepal, southern Europe, Russia, and South America.

An evanescent wave technique for the detection of antibodies specific for parasite *L. donovani* in human serum sample was developed. The method is based on a tapered decladded fiber immobilized with the antigen *L. donovani* by covalent bonding. This fiber was made to react first with the *L. donovani*–infected serum sample and then with the FITC-labeled antihuman IgG to generate the fluorescent signal. The signal intensity was proportional to *L. donovani*–specific antibodies present in the test sera. The system did not show any false-negative or false-positive results, indicating the specificity of the sensor.[111]

16.5.2.13 Fiber-Optic Nanobiosensor

Fiber-optic *nanobiosensor* is a term coined for biosensor that enables measurements in individual living cells or probe individual chemical within a cell. Vo-Dinh and Kasili have developed a fiber-optic immunosensor for the measurement of human health–related biomarkers associated with exposure to polycyclic aromatic hydrocarbon (PAH). The distal end of a tapered fiber having nanometer-sized diameters was the sensor probe. The probe was coated with antibodies for benzo[a]pyrene tetrol (BPT), a metabolite of the carcinogen benzo[a]pyrene to selectively measure its concentration within individual cells of two different cell lines, human mammary carcinoma cells and rat liver epithelial cells.[112]

16.5.3 Brain Research

In the human cognitive neuroscience research, fiber-based spectroscopic methods have been investigated not only to quantify the blood volume in the brain[18] but also to capture images showing in real time when and where in vivo neurons activate in response to various stimuli.[36,113–117]

After the first demonstration of NIR spectroscopy for noninvasive brain monitoring way back in 1977, it has been used to provide spatiotemporal maps of intact human brain and to monitor cerebral oxygenation in patients with traumatic brain injury or brain pathology. The main absorber for NIR light in brain tissue is hemoglobin with its light absorption properties depending upon oxygenation level associated with brain activity.[114]

An optical fiber delivers NIR signal from a source to a specific scalp location, and another fiber located at a different scalp location (typically 3–5 cm away from the illumination point) collects light that has probed the cerebral tissue. Several types of sources (incandescent lamps, light-emitting diodes [LEDs], laser diodes, and regular lasers) and detectors (light-sensitive diodes, photomultiplier tubes, and CCD cameras) can be used. By using a number of such source–detector pairs for various points on the scalp, brain mapping can be performed. Fast optical signals refer to changes in optical scattering of brain tissues concurrently with neural activity. Because these measurements are localized, they can measure brain activity with high (ms level) temporal resolution in specific brain areas. That is why this technique is called event-related optical signals (EROS). For validation of EROS technique, illumination and collection optical fibers were secured on the human volunteer's head by means of a motorcycle helmet. The results showed that noninvasive optical imaging can be used to map the brain activities with good spatial and temporal resolution. The test subjects were then exposed to various visual and auditory stimuli and/or memory tests. Measurement of the amount of light reaching the detector and the delay time between emission and detection provides information about when and where neurons become active. A comprehensive review of technology and various research activities for noninvasive brain studies like motor area activation, visual and cognitive stimulations and mental workload discrimination can be found in the literature.[115–117] NIRS has been used to monitor patients with traumatic brain injury and intracranial hemorrhage and patients undergoing carotid *endarterectomy*.[*,118,119]

Control of human mind with tiny implantable devices in the brain used to be a theme for science fiction,[†] but recently, implantable fiber-optic interface in mouse brain is being investigated. In one such study, a fiber-optic implant was permanently fixated onto the cranium of anesthetized mice with the assembly of the fiber-optic coupler connecting the implant to a solid-state laser. The implant is able to chronically photostimulate functional neuronal circuitry with less tissue damage. Recently, specifically designed implantable fiber-optic interface with parallel optical interrogation of distinctly separate, functionally different sites in the brain of freely moving mice allows the same groups of neurons to be accessed and interrogated over many weeks, providing a long-term dynamic detection of genome activity in response to a broad variety of pharmacological and physiological stimuli.[120,121] Such systems are valuable for in vivo optogenetics that allows stimulation of genetically targeted neurons in a safe manner. Arrest of spontaneous seizures using a real-time, closed-loop, response system and in vivo optogenetics in a mouse model with temporal lobe epilepsy has been investigated. The results demonstrate that spontaneous temporal lobe seizures can be detected and terminated by optically modulating specific cell populations in a spatially restricted manner.[122]

In a cost-effective approach, the University of Washington researchers designed and created a *fiber-optic pen* with computer-interfaced drawing pad to study the inside of the brains of children with learning disabilities.[123]

16.5.4 Cancer Detection

The use of fiber-optic diagnostic techniques for cancer detection is relatively new. In 2009, Brown et al. have reviewed various spectroscopic tools available for the study, diagnosis, prognosis, and treatment of cancer.[124]

The DRS and intrinsic fluorescence spectroscopy (IFS) are the most popular fiber-optic techniques for real-time cancer detection. Monte Carlo (MC)-model-based approaches for the analysis of fluorescence and DRS were used to extract the properties of malignant and nonmalignant tissues to diagnose breast cancer. The fluorescence and DRS were measured with a multiwavelength optical spectrometer, and the fiber-optic probe used consists of a central illumination core and a surrounding collection ring. More than 100 tissue samples of breast were obtained from patients undergoing either breast cancer surgery (lumpectomy or mastectomy) or breast reduction surgery and used ex vivo for analysis.[125] A similar methodology was also reported for cervical cancer screening in resource-limited settings in Leogane, Haiti. The group had employed a fiber-optic spectrometer and MC model for analysis of cervical tissues in 39 patients at Duke University Medical Center.[126] DRS were collected from 49 sites in the cervical transformation zone of 21 female patients aged 30–62 years from clinics for cervical cancer screening in Haiti. The highly portable and self-contained system consists of a self-calibration fiber-optic probe coupled to an LED, two miniature USB fiber-based spectrometers, and a laptop for control and power.[127]

---

* A surgical procedure used to prevent strokes in people who have carotid artery disease.

† Robin Cook, *Mindbend* (1985), William Gibson, *Neuromancer* (1984).

Reflectance and fluorescence spectra were collected using a fiber-optic-based clinical device known as fast excitation–emission matrix (FastEEM) from 43 patients at Boston Medical Center undergoing standard colposcopy* with directed biopsy for spectroscopic diagnosis of cervical dysplasia.[128]

The clinical need for rapid and reliable intraoperative margin assessment of excised surgical specimens was addressed using fiber-optic method. Existing margin assessment methods can be time consuming and inaccurate due to limited sampling. A portable, quantitative, optical fiber probe–based, spectroscopic tissue scanner is designed and tested in a proof-of-concept study in human tissue for breast cancer location. These studies showed that the tissue scanner can detect small foci of breast cancer in a background of normal breast tissue. This tissue scanner is simpler in design, images a larger field of view at higher resolution, and provides a more physically meaningful tissue diagnosis than other imaging systems.[129]

Fiber-optic probes have been developed for FTIR spectroscopy in living systems, allowing for cancer detection mostly for excised human tissues. Advances in silver halide fiber probes are enabling clinical applications of FTIR for blood analyses and detection of malignant tissues in the skin, oral cavity, colon, and bladder. Silver halide is preferred because of its transparency in the MIR and the fact that it is nontoxic, flexible, nonhydroscopic, stable, and has low optical attenuation and low energy density. Mackanos and Contag have reviewed this method for distinguishing premalignant and malignant tissues both after tissue sampling (biopsy) and in vivo.[130]

An integrated OCT, ultrasound (US), and photoacoustic imaging (PAI) prototype endoscopy system has been reported for ovarian tissue characterization. Combining OCT, US, and PAI provides complementary tissue optical absorption, scattering information, and deep tissue structures. The system consists of a ball-lensed OCT sample arm probe, a multimode fiber having the output end polished at 45° angle so as to deliver the light perpendicularly for PAI and a high-frequency US transducer. The performance of this system prototype was demonstrated for ex vivo ovarian tissue characterization of a healthy porcine ovary, an abnormal ovary from a 44-year-old premenopausal patient with endometriosis and a malignant ovary of a 61-year-old postmenopausal patient. The results have shown that this integrated device can be applied for ovarian cancer detection and characterization.[131]

Recently, a fiber-optic confocal Raman probe coupled with a ball lens has been developed by a team of researchers from National University of Singapore. The probe is capable of enhancing in vivo epithelial tissue Raman measurements at endoscopy. With the help of extensive epithelial Raman signatures acquired, this probe was used to objectively target dysplasia in Barrett's esophagus or BE (a metaplastic precursor of esophageal adenocarcinoma). The investigation illustrates its efficacy for in vivo precancer diagnosis and tissue characterization of BE at the molecular level during gastrointestinal endoscopy. In the system, the fiber-optic probe collects a biomolecular fingerprint of tissues while the customized online software enables biomolecular information extraction and analysis, with a diagnostic result presented in real time during endoscopic examination. The system has been used in more than 500 patients in Singapore with stomach, esophageal, colon, rectum, head and neck, and cervical cancers.[132,133]

16.5.5 Force Sensors

Paul et al. had suggested the use of five FBG force sensors to measure *handgrip strength* through a grip holder. Handgrip strength monitoring is rated as 1 of the top 10 fitness tests to evaluate different physical and functional disorders related to health care. The conventional methods (dynamometer) are rough and uncomfortable and do not provide individual finger strengths, thus not suitable especially for rehabilitation programs.[134]

An optical fiber force sensor was demonstrated for in vivo monitoring of the *subbandage pressure*. This sensor consisting of two arrays of FBG entwined in a double-helix form and packaged with contact-force sensitivity has inherent temperature immunity and is capable of real-time,

* A medical diagnostic procedure that involves visual inspection and biopsy of the cervix for signs of disease.

distributed sensing of subbandage pressure.[135] A graduated compression bandaging of the diseased limbs is believed to control edema and assist venous blood circulation. The success of the therapy depends on the amount and the distribution of the compressive pressure applied. Insufficient pressure would impair the efficacy of the healing process while excessive compression can result in aggravated tissue damage. Furthermore, the subbandage pressure is likely to change over time due to various reasons.

A smart sensing scissor blade for application in minimally invasive robotic surgery has been reported using FBG sensors by researchers at Dublin Institute of Technology. The system enables the measurement of *instrument–tissue interaction forces* for the purpose of force reflection and tissue property identification. Force sensing is required in robotic surgery as the surgeon has a restricted view of the actual operating site; there can be discrepancy in eye–hand coordination. Moreover, there are frictions of minimally invasive robotic surgery instruments during insertion into the body. Thus, real-time force measurement improves accuracy and minimizes errors. Conventional strain gauges are not suitable in the clinical atmosphere due to EMI. OFS on the other hand can be easily integrated with the miniaturized robotic instruments at crucial points.[136,137] Another research group has presented backbone shape allay force sensor using FBG to realize a bending force sensing in a compact stick shape implementation. Due to its structure, the bending motion is converted to the expansion and compression motion that is measured by FBG sensor. The sensor has the capability of sterilization, output amplification, and temperature compensation. The prototype and evaluation tests showed promising results to be applied to minimally invasive robotic surgery tools.[138]

In the case of retinal microsurgery, accurate and precise tool tip control is crucial. Nowadays, microforceps are used to peel epiretinal membrane (a micron scale layer of scar tissue formed on the retinal surface causing distorted and/or blurred vision) without damaging its neurons. For a better control and precision, built-in fiberscope and a force-sensing FBG to provide force feedback have been reported.[139,140]

16.5.6 Laser Doppler Velocimetry

Laser Doppler velocimetry (LDV) in medicine involves measuring the Doppler shift of light reflected by moving red blood cells (RBCs) and can detect abnormality in cardiac functions. The use of optical fiber helps to miniaturize the sensor head enabling it to be safely inserted into the blood vessels by using a catheter or a needle. In fact, there are a diverse range of applications of fiber optics in laser velocimetry. Fibers are used not only for beam delivery and scattered light collection but also for imaging the flow field in planar velocimetry system and in signal processing.[141,142]

LDV measures blood velocity in retinal arterioles and venules from which total blood flow in a single vessel can be calculated. The distributions of RBC velocity and volumetric blood flow rate in individual vessels of human retina were determined as a function of vessel diameter using bidirectional LDV that allows direct quantitative and noninvasive measurements. This approach is helpful in the understanding of retinal physiology, blood flow regulation, and the morphology of the vascular transport system.[143,144]

A miniaturized fiber-optic LDV sensor, which can be directly inserted into a blood vessel, is recently developed for clinical use. A convex lens-like surface was formed by a chemical etching on the fiber's tip. A laser beam emitted from the fiber's tip was focused and formed the measuring volume. The local velocity and flow profile of the pulsatile blood flow can be successfully measured by this FOS.[145]

16.5.7 Optical Coherence Tomography

An optical analogue of ultrasound, OCT is an imaging technology that performs high-resolution, cross-sectional tomographic imaging of internal structure in biological systems. OCT uses low-coherence Michelson interferometry to produce real-time in vivo cross-sectional images of biological tissues. With fiber-optic components, OCT can be integrated with a wide range of medical instruments. In fact, fiber-optic OCT probes are a key component in various deployment cases like endoscopic, intravascular, and needle based.

The first implementation of in vivo OCT was reported for human retina, and till date most dominating field of OCT biomedical application is *ophthalmology*, though new areas such as gastroenterology, dermatology, and dentistry are emerging. Functional images of biological tissue can also be obtained with the modification of OCT method. Color Doppler optical coherence tomography (CDOCT) is such a functional extension of OCT that can image flow in turbid media.[146–151]

Apart from being dominant tool in tissue imaging its potential application as a sensor for intraoperative control is being explored. With sensing ability, it can avoid hazardous excisional biopsy, improve accuracy, guide other microsurgical procedures, and prevent accidental nerve and vascular damage during surgery. Having high resolution, high penetration depth, and potential for functional imaging, it will be able to perform *optical biopsy* without extinction of tissue samples.

Recently, a handheld smart micromanipulation aided robotic-surgery tool (SMART) microforceps guided by a fiber-optic common-path optical coherence tomography (CP-OCT) sensor was reported in a proof-of-concept study. A fiber-optic CP-OCT distance and motion sensor was integrated into the shaft of a microforceps. The basic grasping and peeling functions of the microforceps were evaluated in dry phantoms and in a biological tissue model. As compared to freehand use, targeted grasping and peeling performance assisted by active tremor compensation significantly improves microforceps user performance. Strategic incorporation of SMART functions into microforceps can enhance surgical capabilities, safety, and efficiency during microsurgery.[152]

16.5.8 pH and Oxygen Sensor

Continuous monitoring of blood pH and gases (CO_2 and O_2) is essential during surgery and in the ICU. Fluorimetry-based FOSs in catheter capable of monitoring multiple parameters were commercially available since the 1980s.[153] Those probes consisted of three different fibers, one each for each measurand encapsulated in a polymer enclosure along with a thermocouple for temperature monitoring.

Most of the pH sensors are based on absorption and fluorescence of pH indicators deposited on the fiber surface. Various pH indicators that have been reported include acidochrome dye, fluorescein acrylamide, thymol blue, ethyl violet dye, neutral red, eosin, the mixture of dipicrylamine and victoria blue, or the mixture of cresol red, bromophenol blue, and chlorophenol red. Goicoechea et al. demonstrated an FP nanocavity pH sensor by coating the nanostructured film on the fiber-end surface, and white-light interferometry is employed for the measurement of pH-induced swelling of the film.[154] A nanocoating-deposited LPG pH sensor was reported with resonant wavelength shift indicating pH-induced RI changes. There are two types of coatings: the first one is based on polyallylamine hydrochloride (PAH) and polyacrylic acid (PAA) and the second one is incorporating the pigment Prussian blue (PB) in the PAH/PAA matrix. Linear response in the pH range 4–7 with repeatability and high sensitivity was obtained. Gu proposed a simple sensor based on the thin-core fiber modal interferometer.[155,156]

A TFBG deposited with electrostatic self-assembly multilayer of pH-sensitive polymeric film, polydiallyldimethylammonium chloride (PDDA), and PAA was reported. The PDDA/PAA film exhibits a reduction in RI by swelling in different pH solutions resulting in wavelength shift.[157]

Tissue pH mapping gives a regional measure of anaerobic metabolism and low blood flow state since malignant tumors induce a decrease in the pH. A nontoxic pH-dependent indicator was injected in the tissues to be analyzed, and the twin-fiber probe illuminated the tissue and measured the fluorescence intensities at 465 and 490 nm. The ratio of these fluorescence intensities is related to the pH of the tissue.[158]

Fiber-optic and electrochemical pH sensors were developed for tracking the brain tissue pH. The FOS design consists of the immobilization of a pH-sensitive dye, seminaphthorhodamine-1 carboxylate (SNARF-1C) within a silica sol–gel matrix, and a miniature optoelectronics package to acquire data. The device can be useful for stroke patients or people with traumatic brain injury.[159] Another approach utilized NIR spectroscopy where NIR light passes through the skin to probe muscle and tissue at depths of 5–10 mm. A fiber-optic probe was used to collect light both through skin and directly from internal organs during surgical procedures, to investigate the measurement of tissue pH in a number of settings. A successful measurement of tissue pH during complete blood flow cessation (vessel occlusion) and during reduced blood flow conditions such as hemorrhagic shock was demonstrated. It was shown, in a series of animal studies, that optically measured tissue pH is accurate under different conditions.[160]

A broad-range pH FOS based on evanescent wave absorption was presented. The sensor was prepared by immobilizing a mixture of three pH-sensitive indicators (dyes) onto the unclad fiber surface using sol–gel technology. The sensor had shown a linear, reversible, and repeatable response over a broad range of pH values between 4.5 and 13.0.[161]

16.5.9 Pressure/Strain Sensors

For pressure or strain measurements in medical setup, miniaturized piezoresistive or capacitative transducers used to be integrated into catheter. They were expensive, fragile, and not suitable in many applications due to EMI, making FOS an obvious alternative.

Starting with light intensity modulation schemes through microbending or diaphragm/reflector (Figure 16.6) placed at the fiber extremities,[162] other configurations such as FP interferometer cavity and FBG have been developed for the possibility of in vivo pressure measurements. A wide variety of FOSs are available for specific pressure measurement applications.

16.5.9.1 Blood Pressure

Initially, sensors with vibrating membranes located at the tip or at the side of a catheter were used in in vivo blood pressure measurement. This arrangement was very large and impractical. FP-based sensors are much better alternative as they are very small and cost effective. Several companies are providing FP-based sensors to monitor the arterial pressure during intra-aortic balloon pump therapy.[163] In a proof-of-concept demonstration, FBG Fabry–Perot interferometer (FBGI) sensor configuration was also investigated as optical blood pressure manometer by adhering the sensor section of fiber to the skin over the person's radial artery.[164] This manometer not only measured

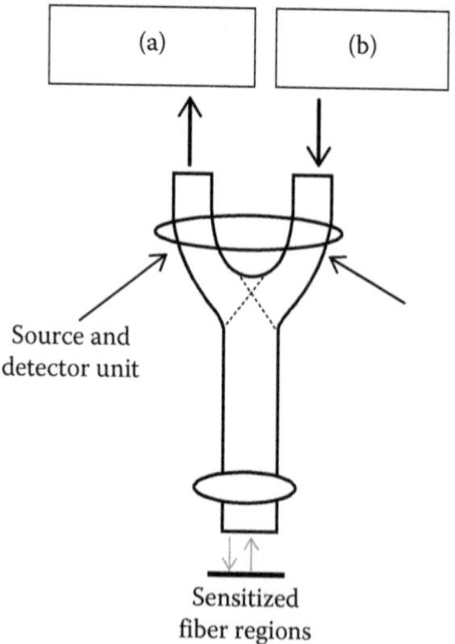

FIGURE 16.6 Basic diaphragm-based pressure FOS.

accurate systolic and diastolic blood pressures once it was calibrated but also provided a continuous pressure waveform quite comparable with conventional system readings.

16.5.9.2 Intramuscular Pressure

Kaufman et al. proposed an FP-based *intramuscular pressure* sensor to measure the force output of muscles.[165] An FBG pressure sensor was demonstrated in multielement catheter for the diagnosis of gastrointestinal motility disorders in the human esophagus.[166]

16.5.9.3 Intracranial Pressure

ICP is an important metric for patients with head trauma or tumors. When ICP is elevated, the cerebral perfusion pressure is decreased that can cause ischemia, intracranial hemorrhages, irreversible brain damage, and death. In the 1990s, commercially available FOS (Camino ICP Monitor) was evaluated along with conventional strain gauge and found to be accurate and reliable.[4,167] These pressure sensors were based on the formation of a low-finesse FP cavity between the polished end of a fiber and a reflective undersurface of a diaphragm that deflects under pressure. The device is fast and simple and has small dimensions. That is why there is reduced risk of inflammation or infection and its insertion through a catheter into small vessels and cavities is easier.

16.5.9.4 Tracheal Pressure

Respiratory monitoring requires minimally invasive sensors for direct measurements of tracheal pressure.[168] Commercial FPI-based pressure sensors were far too coarse to be employed for pediatric or neonatal intensive care applications. A smaller diameter fiber with a microtransducer at its tip was introduced by Samba Sensors, AB, Göteborg, Sweden.[169] The clinical application of the transducer tested in five pediatric patients proved easy and tolerant to nursing procedures. Respiratory motion sensing is a part of futuristic smart textile projects where wearable fabrics will monitor various health parameters. A prototype bandage fabrics embedded with FOSs for respiratory motion sensing is shown in Figure 16.7.

A temperature-independent FBG or LPG can be designed for noninvasive measurement of the torso movement during respiration to understand respiratory physiology and to monitor the lung function. In a preliminary work,[170] FBG sensor was employed using an elastic belt to hold it in place on the chest to detect thorax movements during artificial ventilation, even in the presence of electrical bursts caused by electrodes situated on the chest. Expansion of the thorax cage during respiration

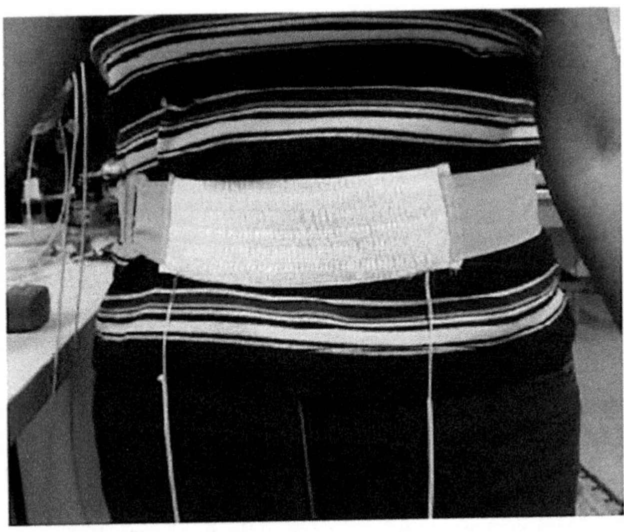

FIGURE 16.7 Prototypes of fiber-embedded textile for respiratory motion sensing manufactured by ELASTA Ind nv (Belgium) and Multitel, Research Centre (Belgium) as a part of OFSETH project (Section 16.6).

was accordingly transmitted to the sensor grating and caused it to deform under the strain. First application of a multiplexed LPG array on curvature sensing garment used to monitor the thoracic and abdominal movements of a human during respiration was reported by Allsop et al.[171,172] They have shown that it is possible to generate a geometric profile of the chest and abdomen in three dimensions with an array of 20 sensors. An FBG sensor is mounted on a springboard placed between the bed mattress and the back of a person for recording the ballistocardiographic signal during an MRI examination and determining the heart rate and respiratory rate on its basis.[173]

16.5.9.5 Intravaginal

The measurement of the pelvic muscle pressure was demonstrated using FBG-based intravaginal probe in Portugal.[174] This measurement is essential for understanding pelvic floor disorder pathophysiology. Existing methods for this purpose have drawback of either indirect measurement or inconvenience in usage. The reported probe had two FBG sensors in a single optical fiber: one directly attached to the mechanical apparatus that transduces radial pressure into axial load and the other to measure temperature. The system was tested in a sample of patients with known pelvic floor disorders, and the preliminary investigations indicated good sensitivity to radial pressure changes within the pelvic floor due to normal breathing cycle of the patient.

16.5.9.6 Intervertebral

Intervertebral disc pressure response to compressive load in five lumbar functional spine units was measured using FBG in a patented configuration. The pressure measurement with FBGs is less disruptive than other existing techniques due to smaller size of FOS. In an improved configuration, FBG sensor placed in silicone-filled needle was applied for intervertebral disc pressure measurements to a cadaveric porcine functional spinal unit, and the results were in agreement to those obtained with the standard strain gauge sensor.[175,176]

Many more applications of fiber-optic pressure/strain sensor in the tendon, ligament, bones, dental biomechanics, etc., are being investigated by several research groups globally.[177–180]

16.5.9.7 Pressure Mapping

With the distributed sensing capabilities, FOS can construct pressure-mapping devices to be placed in between the body parts and supporting surfaces. In a study carried out at Nanyang Technological University, Singapore researchers have reported an FBG-based sensor in instrumented tibial spacer (ITS) to correct misalignment during total knee replacement surgery. The sensor with sampled chirped gratings inscribed on each fiber to generate four to five subgratings that act as sensing points was embedded in a fiber-reinforced composite. During a total knee joint replacement procedure, the ITS sensor can slide in place of the prosthetic spacer. The femur can be rolled over the ITS sensor and the alignment checked from the pressure map. When the required alignment is achieved, the ITS sensor can be replaced by the actual tibial prosthetic spacer, and the knee joint can be sutured.[181]

FOS foot pressure mapping along with the development of multiple neural network can be used for *human gait analysis*, and a preliminary study for diabetic patients had been reported.[182] Existing systems of gait analysis use large numbers of sensors or complex imaging systems, whereas single fiber with multiple FBGs can be embedded within materials forming a surface without loss of material strength.

Distributed FBG sensors can be bonded at strategic locations on the patients' bed to continuously monitor their movements from a remote station. The concept of a *smart bed* was investigated in Singapore.[183] A series of 12 FBG sensors underneath the patient's mattress with suitable algorithm give pressure profile and respiratory rate of the patient while another set of gratings placed on top of the mattress detect heart rate. Existing methods require different techniques for each individual parameter while this single system shown schematically in Figure 16.8 can monitor respiratory rate, heart rate, pressure points, and occupancy of patient on bed in a continuous, nonintrusive, and robust manner.

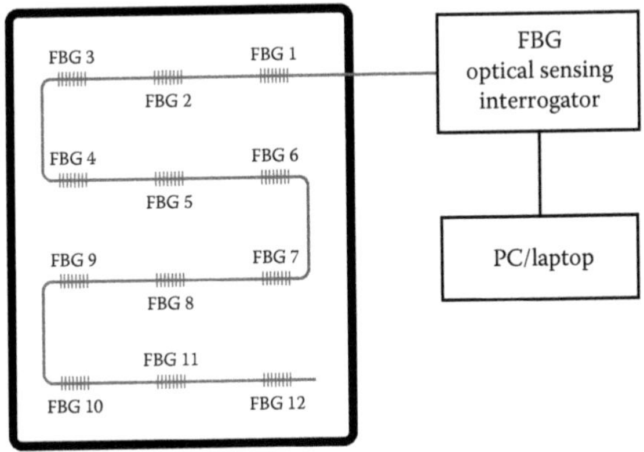

FIGURE 16.8 Smart bed/seat.

A project entitled "Intelligent Adaptable Surface with Optical Fiber Sensing for Pressure-Tension Relief" (IASIS) of the European Commission has utilized FBG-based 2D and 3D sensing systems for utilization in *smart* fiber-based human machine interfaces (HMIs) employed in clinical beds, amputee sockets, and wheelchair seating systems, targeting pressure ulcer and wound treatment.[184]

16.5.10 Pulse Oximeter

A pulse oximeter is a simple noninvasive tool for monitoring the percentage of oxygen in the patient's blood to detect hypoxemia that in extreme cases can cause multiple organ failure.

Pulse oximetry is based on two facts: (a) the presence of a pulsatile signal caused by changes in arterial blood volume associated with cardiac contraction is independent of nonpulsatile arterial blood, venous blood, and other tissues and (b) oxygenated hemoglobin absorbs more IR light than red light, whereas deoxygenated hemoglobin absorbs more red light. The oximeter exploits the difference in the absorption spectra of HbO_2 and Hb to measure oxygen saturation of blood. The measurement assumes that the attenuation of light by the body segment is caused by the three independent components: arterial blood, venous blood, and tissues. Thus, subtracting the DC component of the attenuation from the total attenuation leaves only the cardiac-synchronous pulsatile component.

In 1992, fiber-optic pulse oximeter from Nonin Medical Inc., Plymouth, MN, was implemented successfully to obtain and transmit physiological signals from patients having an MRI. The pulse oximeter probe was attached to either the fingers or the toes of patients (Figure 16.9) throughout the MR examination process.[185]

Pulse oximetry is a well-established tool routinely used to determine a patient's arterial oxygen saturation and heart rate. The decreasing size of pulse oximeters over the recent years has broadened their spectrum of use such as for the diagnosis and treatment of sleep-disordered breathing and cardiorespiratory stability.[186]

The pulse oximeters available today are not suitable for human abdominal organs, such as the intestine, bowel, liver, and kidney, as they are bulky and their clip-on sensor can compress and damage the tissues. In an attempt to achieve sensor that is suitable for use in the human abdomen both operatively and postoperatively, a design of a new FOS is suggested. Hickey et al. proposed to use a small electrically isolated and sterilizable FOS and stitch it in the organ of interest before the completion of the surgery. The probe is inserted in a commercial surgical drain tube that is passed outside the abdominal wall with its other end connected to the processing system. When the monitoring is complete, the probe can be pulled out with all other drainage tubes after operations. The fiber-optic probe can also be used for the monitoring of perfusion in the stomach wall and jejunum. For such application, the probe can be inserted orally/nasally with the guidance of an endoscope. The researchers have reported the design, technical development, and validation of such FOS systems.[187]

FIGURE 16.9 Nonin Medical's 8000 FC FOS with finger wrap (in the inset). (Reprinted with the permission of Nonin Medical, Inc, Plymouth, MN, © 2014.)

A feasibility study proposed and explored *textile-based pulse oximeters* made of plastic optical fibers (POF) fabrics that were used as light emitters and light detectors to human tissue. Two POF emitters and one POF detector textile were incorporated into a cotton glove at the forefinger tip position, and preliminary experiments were performed.[188]

Although pulse oximetry has come a long way in terms of its development, there is a scope for further improvement to overcome its limitations and to provide quantifying information on other blood chromophores. Pulse oximeters using up to 12 light wavelengths have recently been developed by Masimo Corp. (Irvine, CA, U.S.A.). These new *Rainbow Pulse CO-oximeter* instruments can estimate blood levels of carboxyhemoglobin, methemoglobin, and total hemoglobin (SpHb), as well as the conventional SpO_2 value. The accuracy of these new measurements has been studied in human volunteers and clinical trials.[189]

Researchers from City University London have demonstrated that a fiber-optic pulse oximeter can be used to estimate arterial oxygen saturation from the brain tissue of patients recovering from neurosurgery or head injury. The use of optical fibers results in a probe small enough to pass through a cranial bolt of the type already in use for ICP monitoring. A two-wavelength fiber-optic reflectance photoplethysmography (PPG) system was developed and evaluated. These custom-made probes are suitable for real-time monitoring in body cavities and solid organs.[190]

16.5.11 Temperature Sensors

Temperature measurements are necessary for patient monitoring during MRI, for thermal therapy, and for cardiac output monitoring by means of the thermodilution technique. Thermal therapy involves destruction of redundant tissue by heating or freezing without surgery. Examples of thermal therapies include treatment of benign prostatic diseases, ablation therapy of cardiac arrhythmia, and microwave-induced hyperthermia for radiotherapy in cancer treatment, during which temperature of both diseased and normal tissues should be monitored. Temperature monitoring of a patient during MRI procedure is crucial in general and in particular to ensure that the implantable devices do not heat up under strong fields of MRI. Fiber-optic thermometry is better suited for medical applications as a fiber probe can be placed inside an NMR machine with high magnetic field for accurate temperature measurement.[57]

Initially, photoluminescence-/fluorescence-based temperature sensors were developed. Temperature was measured by decay time of a fluorescent compound like semiconductor gallium arsenide with strong temperature-dependent near-IR fluorescence at the fiber tip.[191] The higher the temperature, the faster the decay. Though accurate and useful, those sensors measure temperature at one point only.

For temperature profiling applications, FBG sensors are better because of their multiplexing capability. An FBG-based temperature profile monitoring system was first proposed by researchers from the United Kingdom and Australia using four strain-free FBG arrays that meet the established medical requirements and tested the feasibility of FBG temperature probe inside an NMR machine with high magnetic field.[192,193]

The first in vivo trials of such temperature probes incorporating five FBGs along a single fiber were undertaken successfully at the Cancer Research Institute in Perth, Australia, by the same group[194] on rabbits undergoing hyperthermia treatment of the kidney and liver via inductive heating of metallic implants. A distributed FBG sensor system was tested effectively in temperature range of −195.8°C through 100°C for in vivo use during freezing of porcine liver, for their mechanical stability and MR compatibility.[195]

Recently, the concept of focal therapy for cancer is emerging to replace open surgeries to avoid nerve damages. Cryosurgery is one of the focal therapies currently available to focally ablate any tumor, including prostate cancer. Initially used as a salvage treatment for radiation-failed prostate cancer, cryotherapy has been used as a primary treatment for localized and locally advanced prostate cancer.[196,197]

In cryosurgery, it is very critical to monitor exact temperature profile throughout the surgical procedure to ensure adequate freezing (−40°C) in the cancerous region precisely but preventing low temperatures in the attached nerve fibers, blood vessels, or nearby organs for their safety. An FBG-based reusable multipoint temperature monitoring system was developed recently that can record temperatures in a continuum of either four points at 10 mm intervals or eight points at 5 mm intervals.[198]

16.6 DISCUSSION

Optical fibers have been used for a variety of sensing applications in the area of medicine. Various FOS instruments are already available commercially to monitor pressure, temperature, force, shape, and biological/chemical species.[199] Advancement in fiber optics has resulted in disposable sensors of small size that can be incorporated into catheters or endoscopes and made imaging, pulse oximetry, and OCT more functional.

More FOS techniques have been emerging for in vivo measurements of pressure/strain and force at different parts of human anatomy. For example, FBG-based sensor reported in Refs. [175,176] is the smallest disc pressure sensor to date and with just 0.4 mm in diameter could be used in regions of the human spine that other sensors cannot reach.[200]

With technological advances in related biomedical, electrical engineering, and computer science, the FOS ballistocardiographic measurements are possible in novel setups in bed, backrest, seat, scale, and headrest positions.[173,201,202]

In the last 20 years of time, the integration of nanotechnology, biology, and photonics (fiber optics + optical processing) has been a revolutionizing realm of sensing at the molecular and cellular level. This has catalyzed the emergence of innovative hybrid technologies. Consequently, novel concepts like optical biopsies (Section 16.5.7), shape and tactile sensing,[203–205] targeted drug delivery, human gait and balance analysis, robotic surgery (Sections 16.5.5 and 16.5.7), and lab-on-chip and nonsurgical therapies are being explored and are within reach. In a feasibility study for targeted drug delivery, nanofiber was shown to optically trap nanoparticles that were released by turning off the optical power at the target. Similarly, photodynamic therapy of tumors could benefit from a fiber optic that guides the photosensitizer to a specific location. For this, an optical fiber with a maneuverable miniprobe tip has been developed that sparges O_2 gas and photodetaches photosensitizer molecules.[206,207]

Collaborative projects, like smart textiles (part of collaborative Optical Fibre Sensors Embedded into technical Textile for Healthcare [OFSETH] project from five Europe countries)[188,208,209] smart bed with intelligent HMIs (part of IASIS project),[184] are currently being pursued globally. OFSETH is a Europe-specific research project targeted to develop wearable systems that can monitor vital physiological parameters such as respiration movements, cardiac activity, pulse oximetry, and temperature of the body. It depends upon pure FOS technologies like FBG, NIRS, and others in technical or smart textiles for wearable health monitoring. Another project in the United Kingdom will develop a fiber-optic device to detect potentially fatal lung conditions in intensive care patients and to continuously monitor the blood without the need for blood sampling.[210] A lot of research and technological development is going on for integration of a suitable FOS and its data processing and networking. Thus, it can be predicted that FOS system with integration of cross-disciplinary tools is able to provide efficient solutions to a variety of diagnostic and therapeutic problems in medicine.

REFERENCES

1. Hirshowitz, B. I., L. E. Curtiss, C. W. Peters, and H. P. Pollard. Demonstration of a new gastroscope the "fiberscope." *Gastroenterology* 35(1): 50–53, 1958.
2. Singh, N., V. Mishra, and P. Kapur Medical fiber endoscopy. In *Trends in Medical Physics and Biomedical Instrumentation*, D. V. Rai and R. Bahadur (eds.). New Era Book Agency, Chandigarh, India, 2009. ISBN 978-81-290-0039-2.
3. Katzir, A. Optical fibers in medicine. *Scientific American* 260: 120, 1989.
4. Totsu, K., Y. Haga, and M. Esashi. Ultra-miniature fiber-optic pressure sensor using white light interferometry. *Journal of Micromechanics and Microengineering* 15(1): 71, 2005.
5. Čižmár, T. and K. Dholakia. Exploiting multimode waveguides for pure fiber-based imaging. *Nature Communications* 3: 1027, 2012.
6. Ma, N., F. Gunn-Moore, and K. Dholakia. Optical transfection using an endoscope-like system. *Journal of Biomedical Optics* 16(2): 028002, 2011.
7. Udd, E. and W. B. Spillman Jr. *Fiber Optic Sensors: An Introduction for Engineers and Scientists*. John Wiley & Sons, Hoboken, NJ, 2011. ISBN: 978-1-118-01408-0.
8. Mishra, V. and N. Singh. Optical fiber gratings in perspective of their applications in biomedicine. In *Biomedicine*, Lin, C. (ed.). InTech Open, 2012. ISBN 979-953-307-565-4.
9. Bending the Truth—Get the straight story about Corning® ClearCurve® multimode fibers, www.corning.com/cablesystems. Posted on January 2011.
10. Yang, C., C. Zhao, L. Wold, and K. R. Kaufman. Biocompatibility of a physiological pressure sensor. *Biosensors and Bioelectronics* 19(1): 51–58, 2003.
11. Kashyap, R. *Fiber Bragg Gratings*, 2nd edn. Academic Press, London, U.K., 2009.
12. Murphy, K. A., M. F. Gunther, A. M. Vengsarkar, and R. O. Claus. Quadrature phase-shifted, extrinsic Fabry–Perot optical fiber sensors. *Optics Letters* 16(4): 273–275, 1991.
13. Rao, Y.-J. In-fibre Bragg grating sensors. *Measurement Science and Technology* 8(4): 355, 1997.
14. Mishra, V., V. K. Jindal, R. P. Bajpai, N. Singh, S. C. Jain, P. Kaur, R. Luthra, and H. Singla. Refractive index and concentration sensing of solutions using mechanically induced long period grating pair. *Optical Engineering* 44(9): 094402, 2005.
15. James, S. W. and R. P. Tatam. Optical fibre long-period grating sensors: Characteristics and application. *Measurement Science and Technology* 14(5): R49–R61, 2003.
16. Patrick, H. J., A. D. Kersey, and F. Bucholtz. Analysis of the response of long period fiber gratings to external index of refraction. *Journal of Lightwave Technology* 16(9): 1606, 1998.
17. Russell, P. Photonic crystal fibers. *Science* 299(5605): 358–362, 2003.
18. Yu, F. T. S. and S. Yin. *Fiber Optic Sensors*. CRC Press, Boca Raton, FL, 2002.
19. Wyatt, J. S., M. Cope, D. T. Delpy, C. E. Richardson, A. D. Edwards, S. Wray, and E. O. Reynolds. Quantitation of cerebral blood volume in human infants by near-infrared spectroscopy. *Journal of Applied Physiology* 68(3): 1086–1091, 1990.
20. Kim, A., M. Roy, F. Dadani, and B. C. Wilson. A fiberoptic reflectance probe with multiple source-collector separations to increase the dynamic range of derived tissue optical absorption and scattering coefficients. *Optics Express* 18(6): 5580–5594, 2010.
21. Egawa, M., H. Arimoto, T. Hirao, M. Takahashi, and Y. Ozaki. Regional difference of water content in human skin studied by diffuse-reflectance near-infrared spectroscopy: Consideration of measurement depth. *Applied Spectroscopy* 60(1): 24–28, 2006.
22. Nishidate, I., T. Maeda, K. Niizeki, and Y. Aizu. Estimation of melanin and hemoglobin using spectral reflectance images reconstructed from a digital RGB image by the wiener estimation method. *Sensors* 13(6): 7902–7915, 2013.

23. Chau, L.-K., Y.-F. Lin, S.-F. Cheng, and T.-J. Lin. Fiber-optic chemical and biochemical probes based on localized surface plasmon resonance. *Sensors and Actuators B: Chemical* 113(1): 100–105, 2006.
24. Cunningham, B., P. Li, B. Lin, and J. Pepper. Colorimetric resonant reflection as a direct biochemical assay technique. *Sensors and Actuators B: Chemical* 81(2): 316–328, 2002.
25. Wolfbeis, O. S., L. J. Weis, M. J. P. Leiner, and W. E. Ziegler. Fiber-optic fluorosensor for oxygen and carbon dioxide. *Analytical Chemistry* 60(19): 2028–2030, 1988.
26. Richards-Kortum, R. Fluorescence spectroscopy of turbid media. In *Optical Thermal Response of Laser-Irradiated Tissue*, A. J. Welch and M. J. C. van Gemert (eds.). Plenum, New York, 1995, pp. 667–706.
27. Wygladacz, K., and E. Bakker. Imaging fiber microarray fluorescent ion sensors based on bulk optode microspheres. *Analytica Chimica Acta* 532(1): 61–69, 2005.
28. Turner, G. K. Measurement of light from chemical or biochemical reactions. In *Bioluminescence and Chemiluminescence: Instruments and Applications*, Vol. 1, N. Van Dyke, C. Van Dyke, and K. Woodfork (eds.). 1985, pp. 43–78.
29. Roda, A., P. Pasini, M. Mirasoli, E. Michelini, and M. Guardigli. Biotechnological applications of bioluminescence and chemiluminescence. *Trends in Biotechnology* 22(6): 295–303, 2004.
30. Eltoukhy, H., K. Salama, and A. El Gamal. A 0.18-μm CMOS bioluminescence detection lab-on-chip. *IEEE Journal of Solid-State Circuits* 41(3): 651–662, 2006.
31. Choo-Smith, L.-P., H. G. M. Edwards, H. Ph Endtz, J. M. Kros, F. Heule, H. Barr, J. S. Robinson, H. A. Bruining, and G. J. Puppels. Medical applications of Raman spectroscopy: From proof of principle to clinical implementation. *Biopolymers* 67(1): 1–9, 2002.
32. Hartley, J. S., S. Juodkazis, and P. R. Stoddart. Optical fibers for miniaturized surface-enhanced Raman-scattering probes. *Applied Optics* 52(34): 8388–8393, 2013.
33. Lucotti, A., A. Pesapane, and G. Zerbi. Use of a geometry optimized fiber-optic surface-enhanced Raman scattering sensor in trace detection. *Applied Spectroscopy* 61(3): 260–268, 2007.
34. Gracie, K., E. Correa, S. Mabbott, J. A. Dougan, D. G., R. Goodacre, and K. Faulds. Simultaneous detection and quantification of three bacterial meningitis pathogens by SERS. *Chemical Science* 5: 1030–1040, 2014.
35. Leung, A., P. M. Shankar, and R. Mutharasan. A review of fiber-optic biosensors. *Sensors and Actuators B: Chemical* 125(2): 688–703, 2007.
36. Boussard-Pledel, C., S. Hocde, G. Fonteneau, H. L. Ma, X.-H. Zhang, K. Le Foulgoc, J. Lucas, B. Perio, and J. Hamelin. Infrared glass fibers for evanescent wave spectroscopy. In *Specialty Fiber Optics for Medical Applications*, A. Katzir and J. A. Harrington (eds.). SPIE, San Jose, CA, 1999, Vol. 3596, pp. 91–98.
37. Anne, M.-L., J. Keirsse, V. Nazabal, K. Hyodo, S. Inoue, C. Boussard-Pledel, H. Lhermite et al. Chalcogenide glass optical waveguides for infrared biosensing. *Sensors* 9(9): 7398–7411, 2009.
38. Villatoro, J. and D. Monzón-Hernández. Low-cost optical fiber refractive-index sensor based on core diameter mismatch. *Journal of Lightwave Technology* 24(3): 1409–1413, 2006.
39. Singh, N., V. Mishra, S. C. Jain, K. V. Sreekanth, G. C. Poddar, and P. Kapur. Enhanced sensitivity refractive index sensor based on a segmented fiber with bending. *Indian Journal of Pure and Applied Physics* 47: 655–657, Sept. 2009.
40. Homola, J. Present and future of surface plasmon resonance biosensors. *Analytical and Bioanalytical Chemistry* 377(3): 528–539, 2003.
41. Sharma, A. K., R. Jha, and B. D. Gupta. Fiber-optic sensors based on surface plasmon resonance: A comprehensive review. *IEEE Sensors Journal* 7(8): 1118–1129, 2007.
42. Maguis, S., G. Laffont, P. Ferdinand, B. Carbonnier, K. Kham, T. Mekhalif, and M.-C. Millot. Biofunctionalized tilted fiber Bragg gratings for label-free immunosensing. *Optics Express* 16(23): 19049–19062, 2008.
43. Dagenais, M., A. N. Chryssis, S. S. Saini, S. M. Lee, H. Yi, and W. E. Bentley. High sensitivity bio-sensor based on an etched fiber Bragg grating. In *Integrated Photonics Research and Applications*. Optical Society of America, San Diego, CA, 2005, paper IWD3.
44. Pennington, C., M. Jones, M. Evans, R. VanTassell, and J. Averett. Fiber optic based biosensors utilizing long period grating (LPG) technology. In *Proceedings of SPIE—The International Society for Optical Engineering*, Bellingham, WA, 2001, vol. 4255, pp. 53–62.
45. Ding, J.-F., A. P. Zhang, L.-Y. Shao, J.-H. Yan, and S. He. Fiber-taper seeded long-period grating pair as a highly sensitive refractive-index sensor. *IEEE Photonics Technology Letters* 17(6): 1247–1249, 2005.
46. Jensen, J., P. Hoiby, G. Emiliyanov, O. Bang, L. Pedersen, and A. Bjarklev. Selective detection of antibodies in microstructured polymer optical fibers. *Optics Express* 13(15): 5883–5889, 2005.

47. Li, C., S.-J. Qiu, Y. Chen, F. Xu, and Y.-Q. Lu. Ultra-sensitive refractive index sensor with slightly tapered photonic crystal fiber. *IEEE Photonics Technology Letters* 24(17): 1771, 2012.
48. Rindorf, L., J. B. Jensen, M. Dufva, L. H. Pedersen, and O. Bang. Photonic crystal fiber long-period gratings for biochemical sensing. *Optics Express* 14(18): 8224–8231, 2006.
49. Fan, X. and I. M. White. Optofluidic microsystems for chemical and biological analysis. *Nature Photonics* 5(10): 591–597, 2011.
50. Chen, G. Y., M. Ding, T. P. Newson, and G. Brambilla. A review of microfiber and nanofiber based optical sensors. *The Open Optics Journal* 7: 32–57, 2013.
51. Tian, Y., W. Wang, N. Wu, X. Zou, C. Guthy, and X. Wang. A miniature fiber optic refractive index sensor built in a MEMS-based microchannel. *Sensors* 11(1): 1078–1087, 2011.
52. Tai, Y.-H. and P.-K. Wei. Sensitive liquid refractive index sensors using tapered optical fiber tips. *Optics Letters* 35(7): 944–946, 2010.
53. Ran, Z. L., Y. J. Rao, W. J. Liu, X. Liao, and K. S. Chiang. Laser-micromachined Fabry-Perot optical fiber tip sensor for high-resolution temperature-independent measurement of refractive index. *Optics Express* 16(3): 2252–2263, 2008.
54. Kou, J.-L., J. Feng, Q.-J. Wang, F. Xu, and Y.-Q. Lu. Microfiber-probe-based ultrasmall interferometric sensor. *Optics Letters* 35(13): 2308–2310, 2010.
55. Kou, J.-L., S.-J. Qiu, F. Xu, Y.-Q. Lu, Y. Yuan, and G. Zhao. Miniaturized metal-dielectric-hybrid fiber tip grating for refractive index sensing. *IEEE Photonics Technology Letters* 23(22): 1712–1714, 2011.
56. Ding, M., P. Wang, J. Wang, and G. Brambilla. FIB-milled gold-coated singlemode-multimode-singlemode fiber tip refractometer. *IEEE Photonics Technology Letters*, 26(3): 239–241, 2014.
57. Mignani, A. G. and F. Baldini. Biomedical sensors using optical fibres. *Reports on Progress in Physics* 59(1): 1–28, 1996.
58. Baldini, F. Optical, chemical, and biochemical sensors in medicine. *The International Society for Optical Engineering Newsroom*, 2005, pp. 1–3.
59. Katzir, A. Novel optical fiber techniques for medical applications. *Proceedings of SPIE* 494: 89, 1984.
60. Peterson, J. I. and G. G. Vurek. Fiber-optic sensors for biomedical applications. *Science* 224(4645): 123–127, 1984.
61. Venkatesh, B. Continuous intra-arterial blood gas monitoring. *Critical Care and Resuscitation* 1(2): 140, 1999.
62. Myklejord, D. J., M. R. Pritzker, D. M. Nicoloff, A. M. Emery, and R. W. Emery. Clinical evaluation of the on-line Sensicath TM blood gas monitoring system. *Heart Surgery Forum* 1(1): 60–64, 1998.
63. Ganter, M. and A. Zollinger. Continuous intravascular blood gas monitoring: Development, current techniques, and clinical use of a commercial device. *British Journal of Anaesthesia* 91(3): 397–407, 2003.
64. Wei, H., Z. Guo, Z. Zhu, Y. Tan, Z. Du, and R. Yang. Sensitive detection of antibody against antigen F1 of *Yersinia pestis* by an antigen sandwich method using a portable fiber optic biosensor. *Sensors and Actuators B: Chemical* 127(2): 525–530, 2007.
65. Konry, T., A. Novoa, Y. Shemer-Avni, N. Hanuka, S. Cosnier, Arielle Lepellec, and R. S. Marks. Optical fiber immunosensor based on a poly(pyrrole-benzophenone) film for the detection of antibodies to viral antigen. *Analytical Chemistry* 77(6): 1771–1779, 2005.
66. Herrmann, S., B. Leshem, S. Landes, B. Rager-Zisman, and R. S. Marks. Chemiluminescent optical fiber immunosensor for the detection of anti-West Nile virus IgG. *Talanta* 66(1): 6–14, 2005.
67. Leshem, B., G. Sarfati, A. Novoa, I. Breslav, and R. S. Marks. Photochemical attachment of biomolecules onto fibre-optics for construction of a chemiluminescent immunosensor. *Luminescence* 19(2): 69–77, 2004.
68. Vidal, M. M., I. Delgadillo, M. H. Gil, and J. Alonso-Chamarro. Study of an enzyme coupled system for the development of fibre optical bilirubin sensors. *Biosensors and Bioelectronics* 11(4): 347–354, 1996.
69. Li, X. and Z. Rosenzweig. A fiber optic sensor for rapid analysis of bilirubin in serum. *Analytica Chimica Acta* 353(2): 263–273, 1997.
70. Corres, J. M., I. R. Matias, J. Bravo, and F. J. Arregui. Tapered optical fiber biosensor for the detection of anti-gliadin antibodies. *Sensors and Actuators B: Chemical* 135(1): 166–171, 2008.
71. Marks, R. S., M. M. Levine, E. Bassis, and A. Bychenko. Chemiluminescent optical fiber immunosensor for detecting cholera antitoxin. *Optical Engineering* 36(12): 3258–3264, 1997.
72. Konry, T., A. Novoa, S. Cosnier, and R. S. Marks. Development of an electroptode immunosensor: Indium tin oxide-coated optical fiber tips conjugated with an electropolymerized thin film with conjugated cholera toxin B subunit. *Analytical Chemistry* 75(11): 2633–2639, 2003.
73. Marazuela, M. D., B. Cuesta, M. C. Moreno-Bondi, and A. Quejido. Free cholesterol fiber-optic biosensor for serum samples with simplex optimization. *Biosensors and Bioelectronics* 12(3): 233–240, 1997.

74. Graham, C. R., D. Leslie, and D. J. Squirrell. Gene probe assays on a fibre-optic evanescent wave biosensor. *Biosensors and Bioelectronics* 7(7): 487–493, 1992.
75. Piunno, P. A., U. J. Krull, R. H. Hudson, M. J. Damha, and H. Cohen. Fiber-optic DNA sensor for fluorometric nucleic acid determination. *Analytical Chemistry* 67(15): 2635–2643, 1995.
76. Abel, A. P., M. G. Weller, G. L. Duveneck, M. Ehrat, and H. M. Widmer. Fiber-optic evanescent wave biosensor for the detection of oligonucleotides. *Analytical Chemistry* 68(17): 2905–2912, 1996.
77. Kleinjung, F., F. F. Bier, A. Warsinke, and F. W. Scheller. Fibre-optic genosensor for specific determination of femtomolar DNA oligomers. *Analytica Chimica Acta* 350(1): 51–58, 1997.
78. Niu, S.-Y., S.-J. Wang, C. Shi, and S.-S. Zhang. Studies on the fluorescence fiber-optic DNA biosensor using p-hydroxyphenylimidazo [f] 1,10-phenanthroline ferrum (III) as indicator. *Journal of Fluorescence* 18(1): 227–235, 2008.
79. Ferguson, J. A., T. C. Boles, C. P. Adams, and D. R. Walt. A fiber-optic DNA biosensor microarray for the analysis of gene expression. *Nature Biotechnology* 14(13): 1681–1684, 1996.
80. Chryssis, A. N., S. S. Saini, S. M. Lee, H. Yi, W. E. Bentley, and M. Dagenais. Detecting hybridization of DNA by highly sensitive evanescent field etched core fiber Bragg grating sensors. *IEEE Journal of Selected Topics in Quantum Electronics* 11(4): 864–872, July–August 2005.
81. Hine, A. V., X. Chen, M. D. Hughes, K. Zhou, E. Davies, K. Sugden, I. Bennion, and L. Zhang. Optical fibre-based detection of DNA hybridization. *Biochemical Society Transactions* 37(2): 445, 2009.
82. Zibaii, M. I., H. Latifi, E. Ghanati, M. Gholami, and S. M. Hosseini. Label free detection of DNA hybridization by refractive index tapered fiber biosensor. *Proceedings of the SPIE 7715, Biophotonics: Photonic Solutions for Better Health Care II*, Jürgen Popp; Wolfgang Drexler; Valery V. Tuchin; Dennis L. Matthews, Brussels, Belgium, 77151Z, May 20, 2010. DOI:10.1117/12.853974.
83. Uddin, A. H., P. A. Piunno, R. H. Hudson, M. J. Damha, and U. J. Krull. A fiber optic biosensor for fluorimetric detection of triple-helical DNA. *Nucleic Acids Research* 25(20): 4139–4146, 1997.
84. Liu, X. and W. Tan. A fiber-optic evanescent wave DNA biosensor based on novel molecular beacons. *Analytical Chemistry* 71(22): 5054–5059, 1999.
85. Wang, X., K. L. Cooper, A. Wang, J. Xu, Z. Wang, Y. Zhang, and Z. Tu. Label-free DNA sequence detection using oligonucleotide functionalized optical fiber. *Applied Physics Letters* 89(16): 163901, 2006.
86. Steemers, F. J., J. A. Ferguson, and D. R. Walt. Screening unlabeled DNA targets with randomly ordered fiber-optic gene arrays. *Nature Biotechnology* 18(1): 91–94, 2000.
87. Epstein, J. R., A. P. Leung, K.-H. Lee, and D. R. Walt. High-density, microsphere-based fiber optic DNA microarrays. *Biosensors and Bioelectronics* 18(5): 541–546, 2003.
88. Wang, Y.-M., X.-F. Pang, and Y.-Y. Zhang. Recent advances in fiber-optic DNA biosensors. *Journal of Biomedical Science & Engineering* 2(5):312–317, 2009.
89. Atias, D., Y. Liebes, V. Chalifa-Caspi, L. Bremand, L. Lobel, R. S. Marks, and P. Dussart. Chemiluminescent optical fiber immunosensor for the detection of IgM antibody to dengue virus in humans. *Sensors and Actuators B: Chemical* 140(1): 206–215, 2009.
90. Camara, A. R., P. M. Gouvêa, A. C. Dias, A. Braga, R. F. Dutra, R. E. de Araujo, and I. Carvalho. Dengue immunoassay with an LSPR fiber optic sensor. *Optics Express* 21(22): 27023–27031, 2013.
91. Miyajima, K., T. Koshida, T. Arakawa, H. Kudo, H. Saito, K. Yano, and K. Mitsubayashi. Fiber-optic fluoroimmunoassay system with a flow-through cell for rapid on-site determination of *Escherichia coli* O157: H7 by monitoring fluorescence dynamics. *Biosensors* 3(1): 120–131, 2013.
92. Liu, T., Y. Zhao, Z. Zhang, P. Zhang, J. Li, R. Yang, C. Yang, and L. Zhou. A fiber optic biosensor for specific identification of dead *Escherichia coli* O157:H7. *Sensors and Actuators B: Chemical* 196: 161–167, 2014.
93. Jiang, D., E. Liu, X. Chen, and J. Huang. Design and properties study of fiber optic glucose biosensor. *Chinese Optics Letters* 1(2): 108–110, 2003.
94. Ganesh, A. B. and T. K. Radhakrishnan. Employment of fluorescence quenching for the determination of oxygen and glucose. *Sensors & Transducers* 60(10): 439–445, 2005.
95. Rosenzweig, Z. and R. Kopelman. Analytical properties of miniaturized oxygen and glucose fiber optic sensors. *Sensors and Actuators B: Chemical* 36(1): 475–483, 1996.
96. Meadows, D. and J. S. Schultz. Fiber-optic biosensors based on fluorescence energy transfer. *Talanta* 35(2): 145–150, 1988.
97. Meadows, D. L. and J. S. Schultz. Design, manufacture and characterization of an optical fiber glucose affinity sensor based on an homogeneous fluorescence energy transfer assay system. *Analytica Chimica Acta* 280(1): 21–30, 1993.
98. Kim, D. W., Y. Zhang, K. L. Cooper, and A. Wang. In-fiber reflection mode interferometer based on a long-period grating for external refractive-index measurement. *Applied Optics* 44(26): 5368–5373, 2005.

99. Deep, A., U. Tiwari, P. Kumar, V. Mishra, S. C. Jain, N. Singh, P. Kapur, and L. M. Bharadwaj. Immobilization of enzyme on long period grating fibers for sensitive glucose detection. *Biosensors and Bioelectronics* 33(1): 190–195, 2012.
100. Lin, T.-Q., Y.-L. Lu, and C.-C. Hsu. Fabrication of glucose fiber sensor based on immobilized GOD technique for rapid measurement. *Optics Express* 18(26): 27560–27566, 2010.
101. Maruo, K., J. Chin, and M. Tamura. Noninvasive blood glucose monitoring by novel optical-fiber probe. In *International Symposium on Biomedical Optics*. International Society for Optics and Photonics, Bellingham, WA, 2002, pp. 20–27.
102. Tura, A., A. Maran, and G. Pacini. Non-invasive glucose monitoring: Assessment of technologies and devices according to quantitative criteria. *Diabetes Research and Clinical Practice* 77(1): 16–40, 2007.
103. Liao, K.-C., T. Hogen-Esch, F. J. Richmond, L. Marcu, W. Clifton, and G. E. Loeb. Percutaneous fiber-optic sensor for chronic glucose monitoring in vivo. *Biosensors and Bioelectronics* 23(10): 1458–1465, 2008.
104. Liao, K.-C., S.-C. Chang, C.-Y. Chiu, and Y.-H. Chou. Acute response in vivo of a fiber-optic sensor for continuous glucose monitoring from canine studies on point accuracy. *Sensors* 10(8): 7789–7802, 2010.
105. Yu, S., D. Li, H. Chong, C. Sun, H. Yu, and K. Xu. In vitro glucose measurement using tunable mid-infrared laser spectroscopy combined with fiber-optic sensor. *Biomedical Optics Express* 5(1): 275–286, 2014.
106. Weidemaier, K., A. Lastovich, S. Keith, J. B. Pitner, M. Sistare, R. Jacobson, and D. Kurisko. Multi-day pre-clinical demonstration of glucose/galactose binding protein-based fiber optic sensor. *Biosensors and Bioelectronics* 26(10): 4117–4123, 2011.
107. Müller, A. J., M. Knuth, K. S. Nikolaus, R. Krivánek, F. Küster, and C. Hasslacher. First clinical evaluation of a new percutaneous optical fiber glucose sensor for continuous glucose monitoring in diabetes. *Journal of Diabetes Science and Technology* 7(1): 13–23, 2013.
108. Timm, U., D. McGrath, E. Lewis, J. Kraitl, and H. Ewald. Sensor system for non-invasive optical hemoglobin determination. In *Sensors, 2009 IEEE*. IEEE, Christchurch, NZ, 2009, pp. 1975–1978.
109. Spiker, J. O. and K. A. Kang. Preliminary study of real-time fiber optic based protein C biosensor. *Biotechnology and Bioengineering* 66(3): 158–163, 1999.
110. Rowe, C. A., J. S. Bolitho, A. Jane, P. G. Bundesen, D. B. Rylatt, P. R. Eisenberg, and F. S. Ligler. Rapid detection of D-dimer using a fiber optic biosensor. *Thrombosis and Haemostasis* 79(1): 94–98, 1998.
111. Nath, N., S. R. Jain, and S. Anand. Evanescent wave fibre optic sensor for detection of *L. donovani* specific antibodies in sera of kala azar patients. *Biosensors and Bioelectronics* 12(6): 491–498, 1997.
112. Vo-Dinh, T. and P. Kasili. Fiber-optic nanosensors for single-cell monitoring. *Analytical and Bioanalytical Chemistry* 382(4): 918–925, 2005.
113. Jobsis, F. F. Noninvasive, infrared monitoring of cerebral and myocardial oxygen sufficiency and circulatory parameters. *Science* 198(4323): 1264–1267, 1977.
114. Son, I.-Y. and B. Yazici. Near infrared imaging and spectroscopy for brain activity monitoring. In *Advances in Sensing with Security Applications*, J. Byrnes and G. Ostheimer (eds.). Springer, Dordrecht, the Netherlands, 2006, pp. 341–372.
115. Gratton, G. and M. Fabiani. Fast optical signals: Principles, methods, and experimental results. In *In Vivo Optical Imaging of Brain Function*, R. D. Frostig (ed.), 2nd edn. CRC Press, Boca Raton, FL, 2009, Chapter 15.
116. Gratton, G. and M. Fabiani. Fast optical imaging of human brain function. *Frontiers in Human Neuroscience* 4: 52, 2010.
117. Sassaroli, A., F. Zheng, L. M. Hirshfield, A. Girouard, E. T. Solovey, R. J. K. Jacob, and S. Fantini. Discrimination of mental workload levels in human subjects with functional near-infrared spectroscopy. *Journal of Innovative Optical Health Sciences* 1(2): 227–237, 2008.
118. Kirkpatrick, P. J., P. Smielewski, M. Czosnyka, D. K. Menon, and J. D. Pickard. Near-infrared spectroscopy use in patients with head injury. *Journal of Neurosurgery* 83(6): 963–970, 1995.
119. Kirkpatrick, P. J., P. Smielewski, P. C. Whitfield, M. Czosnyka, D. Menon, and J. D. Pickard. An observational study of near-infrared spectroscopy during carotid endarterectomy. *Journal of Neurosurgery* 82(5): 756–763, 1995.
120. Ung, K. and B. R. Arenkiel. Fiber-optic implantation for chronic optogenetic stimulation of brain tissue. *Journal of Visualized Experiments* 68: e50004, 2011.
121. Doronina-Amitonova, L. V., I. V. Fedotov, O. I. Ivashkina, M. A. Zots, A. B. Fedotov, K. V. Anokhin, and A. M. Zheltikov. Implantable fiber-optic interface for parallel multisite long-term optical dynamic brain interrogation in freely moving mice. *Scientific Reports* 3: 3265, 2013.
122. Krook-Magnuson, E., C. Armstrong, M. Oijala, and I. Soltesz. On-demand optogenetic control of spontaneous seizures in temporal lobe epilepsy. *Nature Communications* 4: 1376, 2013.

123. Reitz, F., T. Richards, K. Wu, P. Boord, M. Askren, T. Lewis, and V. Berninger. A low-cost, computer-interfaced drawing pad for fMRI studies of dysgraphia and dyslexia. *Sensors* 13(4): 5099–5108, 2013.
124. Brown, J., K. Vishwanath, G. M. Palmer, and N. Ramanujam. Advances in quantitative UV–visible spectroscopy for clinical and pre-clinical application in cancer. *Current Opinion in Biotechnology* 20(1): 119–131, 2009.
125. Volynskaya, Z., M. Fitzmaurice, J. Nazemi, R. R. Dasari, R. Shenk, A. S. Haka, K. L. Bechtel, N. Wang, and M. S. Feld. Diagnosing breast cancer using diffuse reflectance spectroscopy and intrinsic fluorescence spectroscopy. *Journal of Biomedical Optics* 13(2): 024012, 2008.
126. Zhu, C., G. M. Palmer, T. M. Breslin, J. Harter, and N. Ramanujam. Diagnosis of breast cancer using fluorescence and diffuse reflectance spectroscopy: A Monte-Carlo-model-based approach. *Journal of Biomedical Optics* 13(3): 034015, 2008.
127. Chang, V. T.-C., D. Merisier, B. Yu, D. K. Walmer, and N. Ramanujam. Towards a field-compatible optical spectroscopic device for cervical cancer screening in resource-limited settings: Effects of calibration and pressure. *Optics Express* 19(19): 17908–17924, 2011.
128. Mirkovic, J., C. Lau, S. McGee, C. Crum, K. Badizadegan, M. Feld, and E. Stier. Detecting high-grade squamous intraepithelial lesions in the cervix with quantitative spectroscopy and per-patient normalization. *Biomedical Optics Express* 2(10): 2917–2925, 2011.
129. Lue, N., J. W. Kang, C.-C. Yu, I. Barman, N. C. Dingari, M. S. Feld, R. R. Dasari, and M. Fitzmaurice. Portable optical fiber probe-based spectroscopic scanner for rapid cancer diagnosis: A new tool for intraoperative margin assessment. *PLoS ONE* 7(1): e30887, 2012. DOI:10.1371/journal.pone.0030887.
130. Mackanos, M. A. and C. H. Contag. Fiber-optic probes enable cancer detection with FTIR spectroscopy. *Trends in Biotechnology*, 28 (6): 317–323, June 2010.
131. Yang, Y., X. Li, T. Wang, P. D. Kumavor, A. Aguirre, K. K. Shung, Q. Zhou, M. Sanders, M. Brewer, and Q. Zhu. Integrated optical coherence tomography, ultrasound and photoacoustic imaging for ovarian tissue characterization. *Biomedical Optics Express* 2(9): 2551–2561, 2011.
132. Bergholt, M. S., W. Zheng, K. Y. Ho, M. Teh, K. G. Yeoh, J. B. Yan So, A. Shabbir, and Z. Huang. Fiberoptic confocal Raman spectroscopy for real-time in vivo diagnosis of dysplasia in Barrett's esophagus. *Gastroenterology* 146(1): 27–32, 2014.
133. Dubay, L., http://www.biopticsworld.com/articles/2014/02/real-time-invivo-molecular-diagnostic-system-diagnoses-precancerous-tissues-during-endoscopy.html. Posted on February 12, 2014.
134. Paul, J., L. Zhao, and B. K. A. Ngoi. Fiber-optic sensor for handgrip-strength monitoring: Conception and design. *Applied Optics* 44(18): 3696–3704, 2005.
135. Wang, D. H.-C., N. Blenman, S. Maunder, V. Patton, and J. Arkwright. An optical fiber Bragg grating force sensor for monitoring sub-bandage pressure during compression therapy. *Optics Express* 21(17): 19799–19807, 2013.
136. Rajan, G., D. Callaghan, Y. Semenova, M. McGrath, E. Coyle, and G. Farrell. A fiber Bragg grating-based all-fiber sensing system for telerobotic cutting applications. *IEEE Sensors Journal* 10(12): 1913–1920, 2010.
137. Callaghan, D. Force sensing surgical scissor blades using fibre Bragg grating sensors. Doctoral thesis. Dublin Institute of Technology, Dublin, Ireland, 2013.
138. Arata, J., S. Terakawa, and H. Fujimoto. Fiber optic force sensor for medical applications within a backbone-shape structure. *Procedia CIRP* 5: 66–69, 2013.
139. He, X., M. A. Balicki, J. U. Kang, P. L. Gehlbach, J. T. Handa, R. H. Taylor, and I. I. Iordachita. Force sensing micro-forceps with integrated fiber Bragg grating for vitreoretinal surgery. In *SPIE BiOS. Proc. SPIE 8218, Optical Fibers and Sensors for Medical Diagnostics and Treatment Applications XII.* International Society for Optics and Photonics, San Francisco, CA, 2012, paper 82180W.
140. Kuru, I., B. Gonenc, M. Balicki, J. Handa, P. Gehlbach, R. H. Taylor, and I. Iordachita. Force sensing micro-forceps for robot assisted retinal surgery. In *Engineering in Medicine and Biology Society (EMBC), 2012 Annual International Conference of the IEEE.* IEEE, San Diego, CA, 2012, pp. 1401–1404.
141. Jones, J. D., R. K. Chan, M. Corke, A. D. Kersey, and D. A. Jackson. Fibre optic laser Doppler velocimetry. In *Fibre Optics'84.* International Society for Optics and Photonics, London, U.K., 1984, pp. 229–236.
142. Charrett, T. O. H., S. W. James, and R. P. Tatam. Optical fibre laser velocimetry: A review. *Measurement Science and Technology* 23(3): 032001, 2012.
143. Riva, C. E., J. E. Grunwald, S. H. Sinclair, and B. L. Petrig. Blood velocity and volumetric flow rate in human retinal vessels. *Investigative Ophthalmology & Visual Science* 26(8): 1124–1132, 1985.
144. Feke, G. T., D. G. Goger, H. Tagawa, and F. C. Delori. Laser Doppler technique for absolute measurement of blood speed in retinal vessels. *IEEE Transactions on Biomedical Engineering* 9: 673–800, 1987.
145. Kohri, S., T. Tajikawa, and K. Ohba. Development of a miniaturized fiber-optic LDV sensor for local blood velocity measurement. *Biomedical Engineering Research* 2(3): 131–138, 2013.

146. Huang, D., E. A. Swanson, C. P. Lin, J. S. Schuman, W. G. Stinson, W. Chang, M. R. Hee, T. Flotte, K. Gregory, and C. A. Puliafito. Optical coherence tomography. *Science* 254(5035): 1178–1181, 1991.
147. Swanson, E. A., J. A. Izatt, C. P. Lin, J. G. Fujimoto, J. S. Schuman, M. R. Hee, D. Huang, and C. A. Puliafito. In vivo retinal imaging by optical coherence tomography. *Optics Letters* 18(21): 1864–1866, 1993.
148. Fercher, A. F., W. Drexler, C. K. Hitzenberger, and T. Lasser. Optical coherence tomography-principles and applications. *Reports on Progress in Physics* 66(2): 239, 2003.
149. Xie, T., D. Mukai, S. Guo, M. Brenner, and Z. Chen. Fiber-optic-bundle-based optical coherence tomography. *Optics Letters* 30(14): 1803–1805, 2005.
150. McLaughlin, R. A. and D. D. Sampson. Clinical applications of fiber-optic probes in optical coherence tomography. *Optical Fiber Technology* 16(6): 467–475, 2010.
151. Rollins, A. M., S. Yazdanfar, J. K. Barton, and J. A. Izatt. Real-time in vivo color Doppler optical coherence tomography. *Journal of Biomedical Optics* 7(1): 123–129, 2002.
152. Song, C., D. Y. Park, P. L. Gehlbach, S. J. Park, and J. U. Kang. Fiber-optic OCT sensor guided "SMART" micro-forceps for microsurgery. *Biomedical Optics Express* 4(7): 1045–1050, 2013.
153. Wolfbeis, O. S. Fibre-optic sensors in biomedical sciences. *Pure and Applied Chemistry* 59(5): 663–672, 1987.
154. Goicoechea, J., C. R. Zamarreño, I. R. Matias, and F. J. Arregui. Utilization of white light interferometry in pH sensing applications by mean of the fabrication of nanostructured cavities. *Sensors and Actuators B: Chemical* 138(2): 613–618, 2009.
155. Corres, J. M., I. Del Villar, I. R. Matias, and F. J. Arregui. Fiber-optic pH-sensors in long-period fiber gratings using electrostatic self-assembly. *Optics Letters* 32(1): 29–31, 2007.
156. Gu, B., M.-J. Yin, A. P. Zhang, J.-W. Qian, and S. He. Low-cost high-performance fiber-optic pH sensor based on thin-core fiber modal interferometer. *Optics Express* 17(25): 22296–22302, 2009.
157. Shao, L.-Y., M.-J. Yin, H.-Y. Tam, and J. Albert. Fiber optic pH sensor with self-assembled polymer multilayer nanocoatings. *Sensors* 13(2): 1425–1434, 2013.
158. Devoisselle, J.-M., V. Maunoury, S. R. Mordon, and D. Coustaut. Measurement of in vivo tumorous/normal tissue pH by localized spectroscopy using a fluorescent marker. *Optical Engineering* 32(2): 239–243, 1993.
159. Grant, S. A., K. Bettencourt, P. Krulevitch, J. Hamilton, and R. Glass. In vitro and in vivo measurements of fiber optic and electrochemical sensors to monitor brain tissue pH. *Sensors and Actuators B: Chemical* 72(2): 174–179, 2001.
160. Soller, B. R., N. Cingo, and T. Khan. Fiber optic sensing of tissue pH to assess low blood flow states. In *Sensors, 2002. Proceedings of IEEE*. IEEE, Orlando, FL, 2002, vol. 1, pp. 266–269.
161. Dong, S., M. Luo, G. Peng, and W. Cheng. Broad range pH sensor based on sol–gel entrapped indicators on fibre optic. *Sensors and Actuators B: Chemical* 129(1): 94–98, 2008.
162. Tohyama, O., M. Kohashi, M. Sugihara, and H. Itoh. A fiber-optic pressure microsensor for biomedical applications. *Sensors and Actuators A: Physical* 66(1): 150–154, 1998.
163. Roriz, P., O. Frazão, A. B. Lobo-Ribeiro, J. L. Santos, and J. A. Simões. Review of fiber-optic pressure sensors for biomedical and biomechanical applications. *Journal of Biomedical Optics* 18(5): 050903, 2013.
164. van Brakel, A., P. L. Swart, A. A. Chtcherbakov, and M. G. Shlyagin. Blood pressure manometer using a twin Bragg grating Fabry-Perot interferometer. In *Photonics Asia 2004*. International Society for Optics and Photonics, Beijing, China, 2005, pp. 595–602.
165. Kaufman, K. R., T. Wavering, D. Morrow, J. Davis, and R. L. Lieber. Performance characteristics of a pressure microsensor. *Journal of Biomechanics* 36(2): 283–287, 2003.
166. Arkwright, J. W., N. G. Blenman, I. D. Underhill, S. A. Maunder, M. M. Szczesniak, P. G. Dinning, and I. J. Cook. *In vivo* demonstration of a high resolution optical fiber manometry catheter for diagnosis of gastrointestinal motility disorders. *Optics Express* 17(6): 4500–4508, 2009.
167. Morton, R., T. H. Lucas, A. Ko, S. R. Browd, R. G. Ellenbogen, and R. M. Chesnut. Intracerebral abscess associated with the Camino intracranial pressure monitor: Case report and review of the literature. *Neurosurgery* 71(1): E193–E198, 2012.
168. De Blasi, R. A., G. Conti, M. Antonelli, M. Bufi, and A. Gasparetto. A fibre optics system for the evaluation of airway pressure in mechanically ventilated patients. *Intensive Care Medicine* 18(7): 405–409, 1992.
169. Sondergaard, S., S. Karason, A. Hanson, K. Nilsson, S. Hojer, S. Lundin, and O. L. A. Stenqvist. Direct measurement of intratracheal pressure in pediatric respiratory monitoring. *Pediatric Research* 51(3): 339–345, 2002.

170. Wehrle, G., P. Nohama, H. J. Kalinowski, P. I. Torres, and L. C. G. Valente. A fibre optic Bragg grating strain sensor for monitoring ventilatory movements. *Measurement Science and Technology* 12(7): 805, 2001.
171. Allsop, T., T. Earthrowl-Gould, D. J. Webb, and I. Bennion. Embedded progressive-three-layered fiber long-period gratings for respiratory monitoring. *Journal of Biomedical Optics* 8(3): 552–558, 2003.
172. Allsop, T., M. Miller, I. Bennion, K. Carroll, G. Lloyd, and D. J. Webb. Application of long-period-grating sensors to respiratory plethysmography. *Journal of Biomedical Optics* 12(6): 064003, 2007.
173. Dziuda, Ł., F. W. Skibniewski, M. Krej, and P. M. Baran. Fiber Bragg grating-based sensor for monitoring respiration and heart activity during magnetic resonance imaging examinations. *Journal of Biomedical Optics* 18(5): 057006, 2013.
174. Ferreira, L. A., F. M. Araújo, T. Mascarenhas, R. M. N. Jorge, and A. A. Fernandes. Dynamic assessment of women pelvic floor function by using a fiber Bragg grating sensor system. In *Biomedical Optics 2006. Proc. SPIE 6083, Optical Fibers and Sensors for Medical Diagnostics and Treatment Applications VI.* International Society for Optics and Photonics, San Jose, CA, 2006, p. 60830H.
175. Dennison, C. R., P. M. Wild, D. R. Wilson, and P. A. Cripton. A minimally invasive in-fiber Bragg grating sensor for intervertebral disc pressure measurements. *Measurement Science and Technology* 19(8): 085201, 2008.
176. Dennison, C. R., P. M. Wild, D. R. Wilson, and M. K. Gilbart. An in-fiber Bragg grating sensor for contact force and stress measurements in articular joints. *Measurement Science and Technology* 21(11): 115803, 2010.
177. Ren, L., G. Song, M. Conditt, P. C. Noble, and H. Li. Fiber Bragg grating displacement sensor for movement measurement of tendons and ligaments. *Applied Optics* 46(28): 6867–6871, 2007.
178. Mishra, V., N. Singh, U. Tiwari, and P. Kapur. Fiber grating sensors in medicine: Current and emerging applications. *Sensors and Actuators A: Physical* 167(2): 279–290, 2011.
179. Mishra, V., N. Singh, D. V. Rai, U. Tiwari, G. C. Poddar, S. C. Jain, S. K. Mondal, and P. Kapur. Fiber Bragg grating sensor for monitoring bone decalcification. *Orthopaedics & Traumatology: Surgery & Research* 96(6): 646–651, 2010.
180. Tjin, S. C., Y. K. Tan, M. Yow, Y.-Z. Lam, and J. Hao. Recording compliance of dental splint use in obstructive sleep apnoea patients by force and temperature modelling. *Medical and Biological Engineering and Computing* 39(2): 182–184, 2001.
181. Mohanty, L., S. C. Tjin, D. T. T. Lie, S. E. C. Panganiban, and P. K. H. Chow. Fiber grating sensor for pressure mapping during total knee arthroplasty. *Sensors and Actuators A: Physical* 135(2): 323–328, 2007.
182. Hao, J. Z., K. M. Tan, S. C. Tjin, C. Y. Liaw, P. R. Chaudhuri, X. Guo, and C. Lu. Design of a foot-pressure monitoring transducer for diabetic patients based on FBG sensors. In *Lasers and Electro-Optics Society, 2003. LEOS 2003. The 16th Annual Meeting of the IEEE*, Tucson, AZ, October 27–28, 2003, vol. 1, pp. 23–24.
183. Hao, J., M. Jayachandran, P. L. Kng, S. F. Foo, P. W. Aung Aung, and Z. Cai. FBG-based smart bed system for healthcare applications. *Frontiers of Optoelectronics in China* 3(1): 78–83, 2010.
184. Pleros, N., G. T. Kanellos, and G. Papaioannou. Optical fiber sensors in orthopedic biomechanics and rehabilitation. In *Ninth International Conference on Information Technology and Applications in Biomedicine, 2009 (ITAB 2009)*. IEEE, Larnaca, Cyprus, 2009, pp. 1–4.
185. Shellock, F. G., S. M. Myers, and K. J. Kimble. Monitoring heart rate and oxygen saturation with a fiber-optic pulse oximeter during MR imaging. *AJR. American Journal of Roentgenology* 158(3): 663–664, 1992.
186. Netzer, N., A. H. Eliasson, C. Netzer, and D. A. Kristo. Overnight pulse oximetry for sleep-disordered breathing in adults: A review. *CHEST Journal* 120(2): 625–633, 2001.
187. Hickey, M., N. Samuels, N. Randive, R. M. Langford, and P. A. Kyriacou. Measurement of splanchnic photoplethysmographic signals using a new reflectance fiber optic sensor. *Journal of Biomedical Optics* 15(2): 027012, 2010.
188. Rothmaier, M., B. Selm, S. Spichtig, D. Haensse, and M. Wolf. Photonic textiles for pulse oximetry. *Optics Express* 16(17): 12973–12986, 2008.
189. Barker, S. J. and J. J. Badal. The measurement of dyshemoglobins and total hemoglobin by pulse oximetry. *Current Opinion in Anesthesiology* 21(6): 805–810, 2008.
190. Phillips, J. P., R. M. Langford, S. H. Chang, K. Maney, P. A. Kyriacou, and D. P. Jones. Cerebral arterial oxygen saturation measurements using a fiber-optic pulse oximeter. *Neurocritical Care* 13(2): 278–285, 2010.
191. Sun, M. H., K. A. Wickersheim, and J. Kim. Fiberoptic temperature sensors in the medical setting. In *OE/LASE'89*, Los Angeles, CA, January 15–20, 1989. Optical Fibers in Medicine IV, Abraham Katzir. International Society for Optics and Photonics, Los Angeles, CA 1989, pp. 15–21.

192. Rao, Y. J., B. Hurle, D. J. Webb, D. A. Jackson, L. Zhang, and I. Bennion. In-situ temperature monitoring in NMR machines with a prototype in-fibre Bragg grating sensor system. In *Optical Fiber Sensors*. Optical Society of America, Williamsburg, VA, 1997, p. OFB7.
193. Rao, Y.-J., D. J. Webb, D. A. Jackson, L. Zhang, and I. Bennion. In-fiber Bragg-grating temperature sensor system for medical applications. *Journal of Lightwave Technology* 15(5): 779–785, 1997.
194. Webb, D. J., S. Jones, L. Zhang, I. Bennion, M. W. Hathaway, and D. A. Jackson. First in vivo trials of a fiber Bragg grating based temperature profiling system. *Journal of Biomedical Optics* 5(1): 45–50, 2000.
195. Samset, T. Mala, R. Ellingsen, I. Gladhaug, O. Søreide, and E. Fosse. Temperature measurement in soft tissue using a distributed fibre Bragg-grating sensor system. *Minimally Invasive Therapy & Allied Technologies* 10(2): 89–93, 2001.
196. Lindner, U., J. Trachtenberg, and N. Lawrentschuk. Focal therapy in prostate cancer: Modalities, findings and future considerations. *Nature Reviews Urology* 7(10): 562–571, 2010.
197. Gowardhan, B. and D. Greene. Salvage cryotherapy: Is there a role for focal therapy? *Journal of Endourology* 24(5): 861–864, 2010.
198. Gowardhan, B. and D. Greene. Cryotherapy for the prostate: An in vitro and clinical study of two new developments; advanced cryoneedles and a temperature monitoring system. *BJU International* 100(2): 295–302, 2007.
199. Mendez, A. Medical applications of fiber-optics: Optical fiber sees growth as medical sensors, 2011. http://www.laserfocusworld.com/articles/2011/01/medical-applic…
200. Roriz, P., L. Carvalho, O. Frazão, J. L. Santos, and J. A. Simões. From conventional sensors to fibre optic sensors for strain and force measurements in biomechanics applications: A review. *Journal of Biomechanics* 47(6): 1251–1261, 2014.
201. Zhu, Y., H. Zhang, M. Jayachandran, A. K. Ng, J. Biswas, and Z. Chen. Ballistocardiography with fiber optic sensor in headrest position: A feasibility study and a new processing algorithm. In *Engineering in Medicine and Biology Society (EMBC), 2013 35th Annual International Conference of the IEEE*. IEEE, Osaka, Japan, 2013, pp. 5203–5206.
202. Dziuda, Ł. and F. W. Skibniewski. A new approach to ballistocardiographic measurements using fibre Bragg grating-based sensors. *Biocybernetics and Biomedical Engineering* 34(2): 101–116. 2014.
203. Cowie, B. M., D. J. Webb, B. Tam, P. Slack, and P. N. Brett. Fibre Bragg grating sensors for distributive tactile sensing. *Measurement Science and Technology* 18(1): 138, 2007.
204. Cutkosky, M. R., R. D. Howe, and W. R. Provancher. Force and tactile sensors. In *Springer Handbook of Robotics*. B. Siciliano and O. Khatib (eds.), 2008, Springer, pp. 455–476.
205. Xie, H., H. Liu, S. Luo, L. D. Seneviratne, and K. Althoefer. Fiber optics tactile array probe for tissue palpation during minimally invasive surgery. In *2013 IEEE/RSJ International Conference on Intelligent Robots and Systems (IROS)*. IEEE, Tokyo, Japan, 2013, pp. 2539–2544.
206. Xin, H. and B. Li. Targeted delivery and controllable release of nanoparticles using a defect-decorated optical nanofiber. *Optics Express* 19(14): 13285–13290, 2011.
207. Zamadar, M., G. Ghosh, A. Mahendran, M. Minnis, B. I. Kruft, A. Ghogare, D. Aebisher, and A. Greer. Photosensitizer drug delivery via an optical fiber. *Journal of the American Chemical Society* 133(20): 7882–7891, 2011.
208. Grillet, A., D. Kinet, J. Witt, M. Schukar, K. Krebber, F. Pirotte, and A. Depré. Optical fiber sensors embedded into medical textiles for healthcare monitoring. *IEEE Sensors Journal* 8(7): 1215–1222, 2008.
209. http://www.ofseth.org.
210. Phones, probes and sensors to transform healthcare, http://www.epsrc.ac.uk/newsevents/news/2013/Pages/transformhealthcare.aspx. Press release on May 9, 2013.

17 Optical Fiber Sensors for Smart Composite Materials and Structures

Manjusha Ramakrishnan, Yuliya Semenova, Gerald Farrell, and Ginu Rajan

CONTENTS

17.1 INTRODUCTION

Composite material structures [1] are widely used in the aerospace, marine, aviation, transport, sport/leisure, and civil engineering industries [1]. Such composite material structures are frequently subjected to external perturbations and varying environmental conditions, which may cause the structures to suffer from fatigue damage and/or failures, and thus require real-time structural health monitoring (SHM). Of necessity, the diagnosis process and condition monitoring of composite structures are usually carried out during their working life [2]. The goal of such diagnosis is to detect, identify, locate, and assess the defects that may affect the safety or performance of a structure. Sensors that are commonly employed for SHM are resistance strain gauges, optical fiber sensors

(OFSs), piezoelectric sensors, eddy current sensors, and microelectromechanical systems (MEMS) sensors [3]. Traditional nondestructive evaluation (NDE) techniques such as ultrasonic inspection, acoustography, low-frequency methods, radiographic inspection, shearography, acousto-ultrasonic, and thermography are effective in SHM of composite materials and structures, but it is difficult to use them in an operational structure due to the size and weight of the devices [4].

OFSs on the other hand are suitable candidates for SHM of composite materials during operation since they are capable of achieving the goals of diagnostics as well as condition monitoring and can also be embedded into such structures acting as a nervous system [5]. This powerful and potential rich technology is being currently implemented in a wide variety of applications. In the composite materials area, compared to traditional NDE techniques, OFSs offer unique capabilities: monitoring the manufacturing process of composite parts, performing nondestructive testing once fabrication is complete, and enabling health monitoring and structural control. Investigations of OFS embedded in composite structures indicate that OFS technology is capable of monitoring stress/strain, temperature, composite cure process, vibration, humidity, delamination, and cracks and thus has great potential for condition monitoring of a variety of composite materials applications [6,7]. Because of their minimal weight, small size, high bandwidth, and immunity to electromagnetic interference, OFSs have significant performance advantages over traditional sensors. Furthermore, optical fibers are steadily becoming more cost effective due to advances in the telecommunication and optoelectronic industries. Also, to ensure the integrity of composite structures, an important requirement for the sensor is that it should be possible to embed it into the host composite without modifying the host's properties and functions. This can be achieved by OFSs, which are compatible with reinforcement fabrics and can be embedded within the structure without affecting its strength and thereby act as the nerves of the system and provide vital information on the internal mechanical health of the structure.

Recognizing the increase in the use of composite structures, this chapter describes an overview of recent advances in the area of OFS technology for condition monitoring applications in composite materials, arranged as a number of sections. An overview of different types of most commonly used composite materials and their properties is given in Section 17.2. Then the demand for the SHM in composite material structures is discussed in Section 17.3, following which the common fabrication methods of composite materials with embedded fiber sensors are detailed in Section 17.4 while the issues of composite degradation associated with embedding of fiber sensors are discussed in Section 17.5. An overview of different types of fiber sensors for strain/temperature measurements in composite materials is presented in Section 17.6 and an introduction to lightweight composite material is presented in Section 17.7. The recent trends, issues, and future challenges of the OFS technology are discussed in Section 17.8. The aim of this chapter is to provide a state of the art in the area of composite materials embedded with OFSs and its potential application in SHM.

17.2 COMPOSITE MATERIALS AND THEIR CLASSIFICATION

In this section, we discuss important classifications of typical composite materials and their properties. In general, fiber composite materials have two constituents: reinforcement and matrix [8], which when combined together can produce a material with properties superior to those of the constituent materials. It is known that the mechanical properties of the composite materials differ depending on the matrix and the holding reinforcing materials used to fabricate the composite material [8].

The functions of the matrix in a composite material are to transfer the load to the reinforcement fibers, provide temperature resistance and chemical resistance, and maintain the reinforcement fibers in a fixed orientation [9]. Matrix materials are generally divided into two categories: thermosets and thermoplastics [10]. Thermoset matrices are the most common, being low cost and solvent resistant. The curing process in the thermosets is an irreversible chemical process. The most common type of thermoset is epoxy resin, which possesses superior performance and is relatively low cost.

Thermoplastic-based matrices can be softened by heating to an elevated temperature. Then to form a thermoplastic composite material, the softened matrix is mixed with the reinforcement fiber, a process that is reversible. The most common thermoplastic is polypropylene, which has a limited temperature range of up to 150°C. The typically used thermoset and thermoplastic matrices are classified according to the extent of their use in Figure 17.1a and b, respectively.

The second constituent of a composite material is the reinforcement fiber [8]. The composite material's tensile properties, stiffness, and impact resistance are influenced by the type of the fiber reinforcement [9]. Thus, for a manufacturer, it is important to select an appropriate fiber reinforcement taking into account the end user's application of the fabricated composite structure. The reinforcement fiber materials are mainly categorized into two types: glass fiber and advanced fiber [10]. There are different kinds of glass fibers available that are made of different kinds of glass, ranging from normal silica glass to high purity quartz glass, and each of these glass fibers offers its own set of properties. There are also advanced fiber materials that offer high strength and high stiffness but with low weight [10]. Boron, silicon, carbide, carbon, Aramid (Kevlar® or Twaron®), polyamide, nylon, self-reinforced polymer fibers (e.g., Curv®, Pure®, and Tegris®), basalt fiber, and natural fibers (e.g., flax, jute, coir, hemp, and cotton) are examples of such advanced fibers. These reinforcement fibers can be arranged in a woven, knitted, braided, or nonwoven style. In the case of a multilayered composite material fabricated using woven-style reinforcement fiber fabrics, it is found that the mechanical properties are also determined by the fiber reinforcement fabrics' woven style, density, and ply configuration [7,8]. Typically available categories of composite material are shown in Figure 17.2.

As regards applications, fiber-reinforced composite materials are commonly used for fabrication of various structural parts such as aircraft tails, wings, fuselages, propellers, helicopter rotor blades, wind turbine blades, and parts of racing cars, boats, etc. [7]. One of the prominent challenges for a composite structural part designer is a fail-safe design solution, which requires optimization of multiple parameters. Typically, given the performance specifications for the composite part [8], the areas that need optimization are the following: the selection of most appropriate reinforcement and

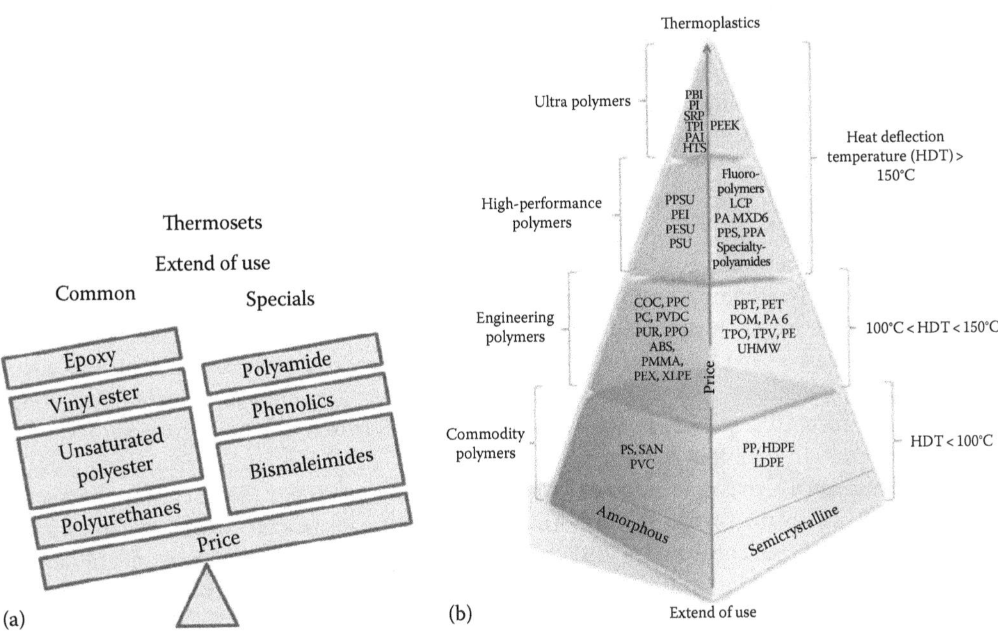

FIGURE 17.1 Classification of typical thermosets (a) and thermoplastic matrices (From http://nptel.ac.in/courses/Webcourse-contents/IISc-BANG/Composite%20Materials/pdf/Lecture_Notes/LNm1.pdf) (b) and the extent of their use and price. (Courtesy of SpecialChem, Paris, France.)

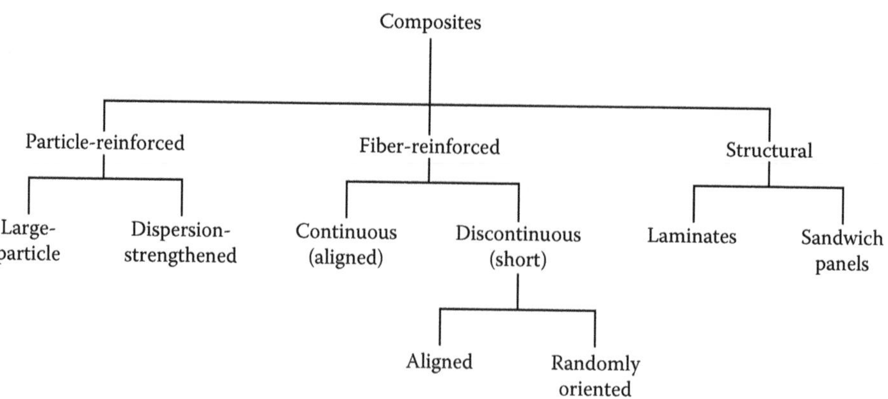

FIGURE 17.2 Typically available categories of composite materials.

matrix constituents to satisfy the requirements for strength and stiffness of the particular composite part with a minimum weight; the selection of the composite part's geometry; a careful analysis of stress distribution within the part; the minimization of moisture ingress; optimization of toughness; and an analysis of failure modes. Prior to opting for a particular design for mass production, the manufacturer also needs to consider the repeatability of the composite part's performance specification, ease of production, cost efficiency, and quality assurance mechanisms [9,10]. Finally, even though the designed structural part is optimized for fail-safe performance, there is a possibility of damage during operation in extreme environmental conditions and mechanical failure of the structure due to external perturbations [3]. All of this necessitates the requirement for nondestructive SHM techniques throughout the lifetime of the composite structural part.

17.3 DEMAND FOR THE SHM IN COMPOSITE MATERIAL STRUCTURES

The demand for SHM in composites is in the first instance driven by the increased use of composite materials. One of the main reasons that drive industry to use composite materials is the benefit of lower weight, resulting in improved energy efficiency and less CO_2 emissions. This increasing use of composite materials brings with it a need to establish inspection and/or monitoring regimes to ensure structural integrity and safe operation throughout service life. This can make the cost of ownership high, with downtime being an important issue [11]. Therefore, any SHM methodology that could increase the inspection intervals for a structure and indicate damage before costly failures occur would be very advantageous, financially and in many cases in terms of safety.

The global use of carbon fiber–reinforced plastic (CRP) and glass fiber–reinforced plastic (GRP) composite materials based on structural parts has significantly increased in application areas such as aircraft, sport vehicles, wind turbines, and infrastructure constructions, because of the inherent advantages of composite materials. For example, in aircraft, the advantages are well known: lighter weight for the aircraft, reduced requirements for maintenance, and increased passenger comfort. The use of composite materials in aircraft is growing as shown in Figure 17.3a. Composite-based aircrafts consume less fuel and result in a high cost benefit that yields energy savings up to 18%. Some existing commercial aircraft models such as the Boeing 787 and Airbus A350 comprise circa 50% composite materials by weight. A market research conducted in 2013, in terms of the global CRP market by Bernhard Jahn and the European GRP market by Dr. Elmar Witten, indicates that there will be a growing demand for such aircraft in the coming years [12]. Figure 17.3b and c shows the overall use of CRP composites by different industries and GRP production in different European industries, respectively.

The second driving force behind the need for SHM in composite material is the necessity to monitor composite parts during operation, for safety and early detection of failure. During a typical 20-year

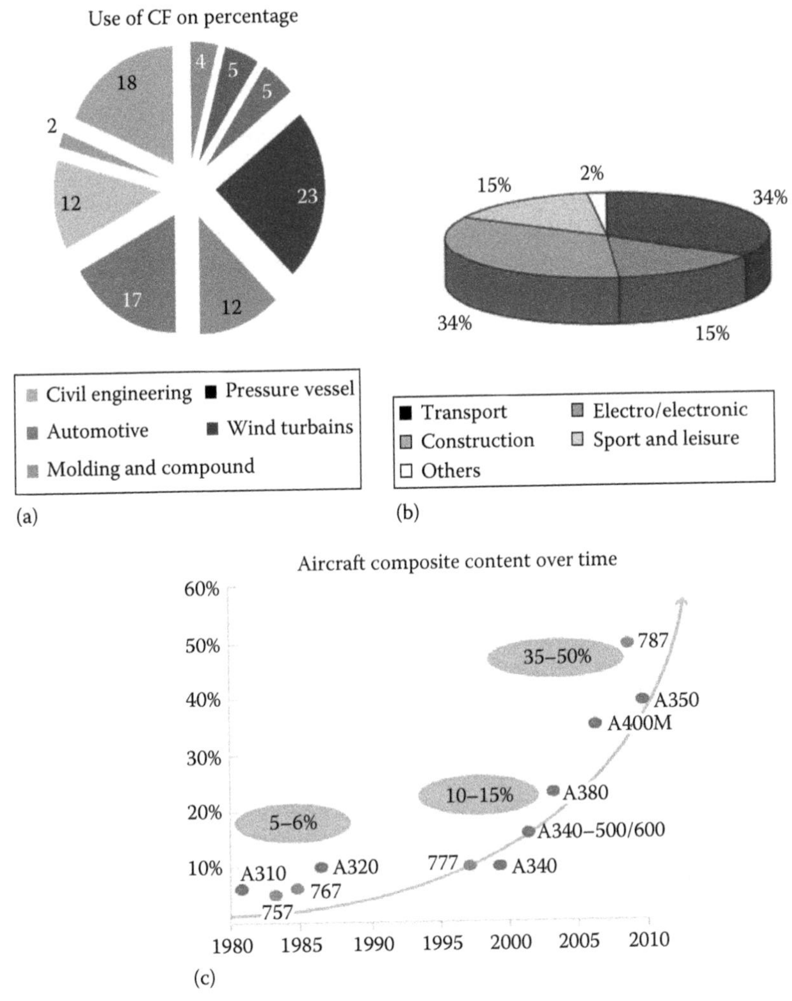

FIGURE 17.3 (a) Use of CF composites by industry (From The global CRP market by CCeV, http://www.carbon-composites.eu/sites/carbon-composites.eu/files/anhaenge/13/09/17/ccev-avk-marktbericht_2013-final-englisch-bj.pdf), (b) GRP production in Europe for different application industries (From European GRP market by AVK, http://www.carbon-composites.eu/sites/carbon-composites.eu/files/anhaenge/13/09/17/ccev-avk-marktbericht_2013-final-englisch-bj.pdf), and (c) development of composite aerospace applications in the last 40 years. (From The Hexcel Corp. Aerostrategy, http://www.pluscomposites.eu/sites/default/files/Technical%20series%20-%20Part%202%20-%20Market%20and%20market%20developments_0.pdf.)

service life, composite structures such as wind turbine blades, helicopter blades, and aircraft parts are subjected to static and dynamic lifts, and drag and inertial loads over a wide range of temperatures and often severe environmental conditions. As the growth of production of composite parts has increased, the composite industry has increasingly focused on damage/failure-free composite structures and nondestructive techniques for SHM over their lifetime. Compared to conventional nondestructive sensing techniques, OFSs have achieved wide acceptance due to their attractive properties such as small size, immunity to electromagnetic interference, and low cost [1]. OFSs embedded in various structures are very useful for strain/temperature monitoring [2] applications in extreme environmental conditions. For example, issues such as bend loading and icing can be monitored and avoided by implementing smart composite structures with embedded OFSs. Such smart composite materials with embedded OFSs can enhance the safety of advanced machines, structures, and devices.

17.4 COMPOSITE MATERIALS WITH EMBEDDED FIBER SENSORS: FABRICATION METHODS

In this section, we discuss the commonly employed sample fabrication methods for composite materials with embedded OFSs. The fabrication processes commonly adopted by laboratory-based and small-scale industrial manufacturers are the expertise-intensive hand layup and pre-preg layup [13] methods. The hand layup method is a process for arranging fiber-reinforced layers in a laminate and shaping the laminate to fabricate the desired part. For this, the reinforcement fibers or fabrics are stacked one over another by applying the matrix in between them. After stacking, curing (or polymerization) of the resin matrix of the multiple laminate layered composite, the sample can be shaped as per the manufacturer's specifications. Pre-pregs are single laminates of *pre-impregnated* composite fibers with a matrix material such as epoxy resin. In the pre-preg layup method, multiples of composite pre-preg laminates are stacked one over another and the curing can be done unaided or by applying heat and/or pressure. The molding process can be either vacuum-bag molding or an autoclave molding. Typical autoclave curing conditions involve a temperature range from 120°C to 200°C and with an applied pressure up to 100 psi.

For embedding optical fibers in samples prepared by both the hand layup and the pre-preg layup methods, the process is similar: before the curing process, optical fibers are placed on the corresponding composite layer and some prestrain is applied to make sure the optical fibers remain free of bends [14]. Positioning of the optical sensor is highly application specific and depends on the location of the areas where parameters need to be monitored. For example, for strain and temperature measurements of composite material structures, it is reported that OFSs were embedded within the farthest layer from the middle layer to achieve the highest measurement sensitivity [15]. The different steps for embedding OFSs by the hand layup and pre-preg methods are shown in Figure 17.4a and b, respectively. For larger composite parts, optical fibers can be embedded manually with expert assistance as shown in Figure 17.4c, but for large-scale production of smart composite structure materials embedded with OFSs, automated robotic systems are now common and an example of such is a robotic system by *SABCA Limburg and Sonaca/Ingersoll* shown in Figure 17.4d. One of the most important concerns while embedding OFS inside composite structures is their influence on the structural integrity of the composite part. In the next section, we have detailed some methods to reduce the risk of composite material degradation and to maintain the structural integrity of composite material structures.

17.5 COMPOSITE DEGRADATION ASSOCIATED WITH EMBEDDING OF FIBER SENSORS

One of the major concerns of OFS for composite materials is the potential for degradation of the composite material mechanical properties and possible increased failure rate due to the presence of an embedded optical fiber [16]. Various studies have been carried out to analyze the influence of the embedded OFS on composite material tensile/compressive strength, stiffness, interlaminar fracture toughness, and fatigue resistance. Lee et al. reported that having embedded optical fibers passing through or parallel to ply drops in a laminate does not have any significant effects on the static strength of the laminate [17,18]. This is valid even for optical fibers placed in the most critical locations. In principle, any potential degradation of strength and the modulus of a composite material will be a function of the orientation of the optical fiber relative to the nearest plies, the overall thickness of the laminate, the optical fiber diameter, and the type of protective coating on the optical fiber [17]. Degradation becomes increasingly severe with an increasing angle between the optical fiber and ply directions. Another issue is the larger diameter of the optical fiber (with its buffer coating) compared to the ply fibers of the composite. In general, commercially available optical fibers have diameters from 125 to 230 μm, which is about 10–15 times larger than the average E glass fiber or carbon fiber. The optical fiber orientation perpendicular to the fiber reinforcement

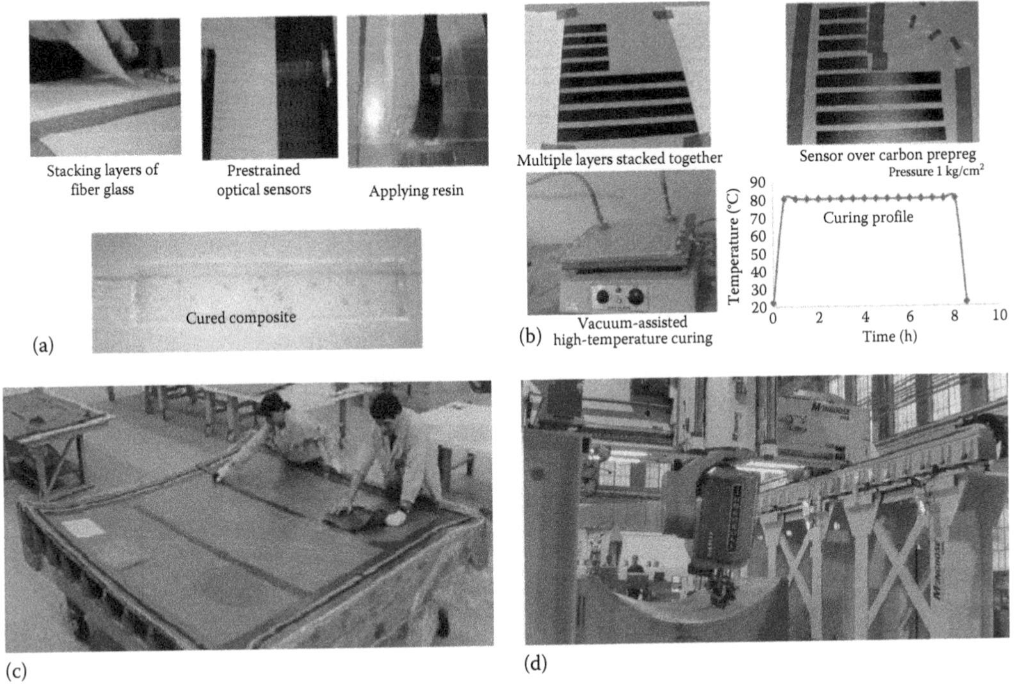

FIGURE 17.4 Embedding fiber sensors inside composite materials (a) hand layup, (b) pre-preg method, (c) expert-assisted manufacturing of composite part embedded with OFS, and (d) automated robotic-assisted manufacturing of composite structures embedded with OFS. (Photo (d): SABCA Limburg and Sonaca/Ingersoll.)

can result in a characteristic *eye* pattern or pocket within the resin, which acts as a defect center in the composite part that could lead to premature failure in the form of delamination [18]. However, testing conducted at a number of labs worldwide concluded that such delaminations were insignificant if the OFS density was low as shown in Figure 17.5. Wojciech et al. reported that if the optical fiber is laid along the direction of reinforcement fiber, there is a uniform consolidation around the optical fiber with minimum defects and thus the laminate's mechanical parameters are least affected [6,17,18]. Also, the optical fibers laid along the direction of fiber reinforcement have limited influence on the mechanical behavior of the composite structure because of the optical fiber's own inherent load-carrying ability [17].

An analysis [19] on the flexural strength of a composite material, embedded with OFSs, that bears tensile loading showed that the flexural strength did not suffer any noticeable degradation when the OFSs were embedded in the tensile region either in the longitudinal direction or in the transverse direction. For compressive loading, the situation is similar if the sensor is embedded in the longitudinal direction but it is found that if the OFS is embedded in the transverse direction with

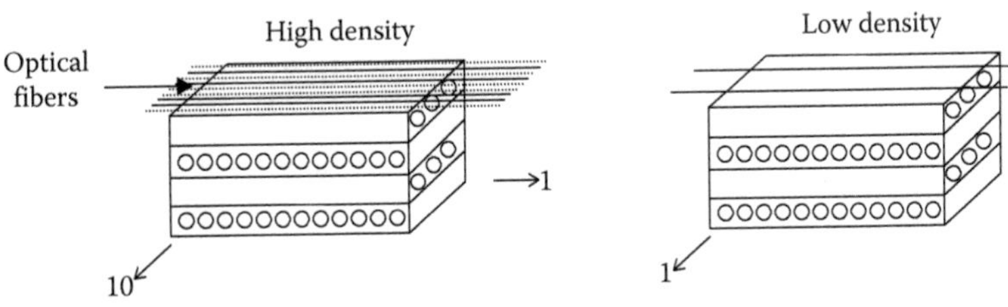

FIGURE 17.5 Optical fibers in composite materials embedded with high and low density.

respect to the compressive region, the flexural strength is degraded significantly [21]. In order to realize SHM for composite material components in many applications, it is essential to measure the strain and temperature of composite materials, and in the next section we discuss the different types of OFS strain and temperature sensors that can be embedded in composite materials.

17.6 TYPES OF OPTICAL FIBER SENSORS FOR COMPOSITE MATERIALS

There are a wide variety of condition monitoring applications in composite materials, including cure process monitoring, vibration measurements, and detection of delamination and cracking [20,21]. For all these applications, the key requirement is to measure either strain or temperature or both parameters. In this section, we discuss the different types of OFSs that can be used with composite materials to measure strain/temperature when embedded inside a composite material. The different types of OFSs reported for strain/temperature measurements for composite materials are fiber Bragg grating (FBG) sensors [22], interferometric OFSs [23], polarimetric sensors [24], distributed sensors (using techniques such as Rayleigh scattering, Raman scattering, and Brillouin scattering) [25,26], and hybrid sensors [20,21,27,28].

17.6.1 Fiber Bragg Grating Sensor for Composite Materials

FBGs are the most commonly employed OFSs in SHM applications for composite materials [22,29]. An FBG sensor comprises a grating region with a periodic change in refractive index in the core region of an optical fiber. Such a periodically modulated refractive index structure enables the light to be coupled from the forward propagating core mode into a backward propagating core mode generating a reflection response, as shown in the schematic illustrating the principle of an FBG provided in Figure 17.6 [30]. The light reflected due to periodic variations of the refractive index of the Bragg grating with a central wavelength is given by

$$\lambda_B = 2n_0\Lambda \tag{17.1}$$

where

n_0 is the effective refractive index of the fiber core

Λ is the grating pitch

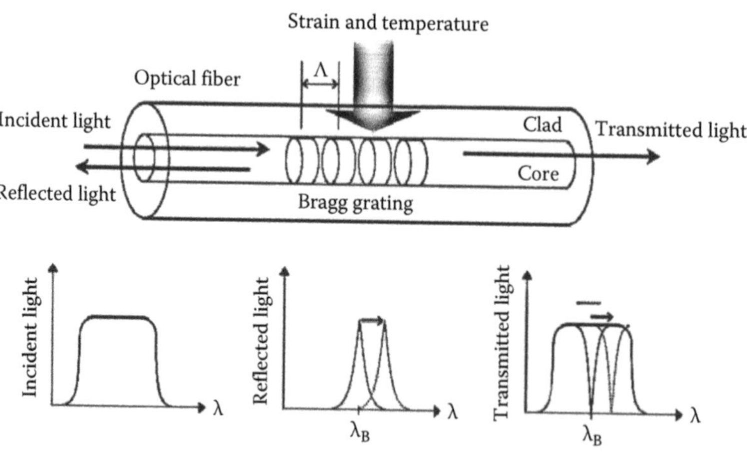

FIGURE 17.6 Schematic of a fiber Bragg grating and its operation.

The strain sensitivity of the Bragg wavelength arises from the change in the pitch of the fiber grating due to the strain and changes in the refractive index arising from the strain-optic effect. When a strain is applied to the grating, the Bragg reflected wavelength changes. The wavelength shift $\Delta\lambda_\varepsilon$ for a value of elongation ΔL is given by

$$\Delta\lambda_\varepsilon = \lambda_B \cdot \left(\frac{1}{\Lambda}\frac{\partial \Lambda}{\partial L} + \frac{1}{n_0}\frac{\partial n_0}{\partial L} \right) \cdot \Delta L \tag{17.2}$$

In practice, the applied strain value can be estimated from the measurement of reflected wavelength as it changes due to the applied strain. The typical strain sensitivity of an FBG at 1550 nm is ~1.2 pm/με.

Bragg wavelength shift can also occur due to changes in temperature. For a temperature variation of ΔT, the corresponding wavelength shift $\Delta\lambda_T$ is given by

$$\Delta\lambda_T = \lambda_B \cdot \left(\frac{1}{\Lambda}\frac{\partial \Lambda}{\partial T} + \frac{1}{n_0}\frac{\partial n_0}{\partial T} \right) \cdot \Delta T \tag{17.3}$$

The temperature sensitivity of the Bragg wavelength arises from the change in the grating pitch associated with the thermal expansion of the fiber and the change in the refractive index arising from the thermo-optic effect. Thus, Equation 17.3 can also be written as

$$\Delta\lambda_T = \left(\alpha_0 + \beta_0\right) \cdot \lambda_B \cdot \Delta T \tag{17.4}$$

where

α_0 is the coefficient of thermal expansion (CTE) of the fiber
β_0 is the fiber refractive index variation with temperature

The values of α_0 and β_0 are constants for silica optical fiber and are 0.55×10^{-6}/°C and 6.6×10^{-6}/°C, respectively. The typical temperature sensitivity of an FBG at 1550 nm is ~11.6 pm/°C.

17.6.1.1 Strain and Temperature Measurements Using Embedded FBGs

Commonly, FBGs are used to measure axial strain in the longitudinal direction as this provides higher sensitivity and temperature. For example, the measured wavelength shift of the embedded FBG sensors placed at different locations inside the composite for different deflection values during a three-point bending strain test on composite material is shown in Figure 17.7a. The measured wavelength shift of the embedded FBG sensors at different temperatures of composite materials is shown in Figure 17.7b. One can see that a red shift of the Bragg wavelength arises as a result of an increase in temperature. Other than standard Bragg gratings with uniform period, there are chirped gratings with a gradual period variation [31], and tilted fiber Bragg gratings (TFBGs) [31] with gratings written at an angle to the fiber axis are also employed for strain and temperature measurements in composite materials. The chirped FBG sensor has a gradual distribution of the grating period [32]. This variation in the grating period provides a one-to-one correspondence between the wavelength in the spectrum and the position of the gauge section within the composite material, which is the significant advantage of chirped FBG sensors over the conventional FBG sensors [32]. The main disadvantage of conventional FBG sensors is the cross-sensitivity between temperature and strain. In a different manner from conventional FBGs, the wave vector of a TFBG has a certain angle with respect to the fiber axis, making the resonance wavelengths of core mode and cladding mode to be sensitive to temperature, while their transmission power is temperature independent [33]. Thus, based on these unique characteristics of TFBG, it

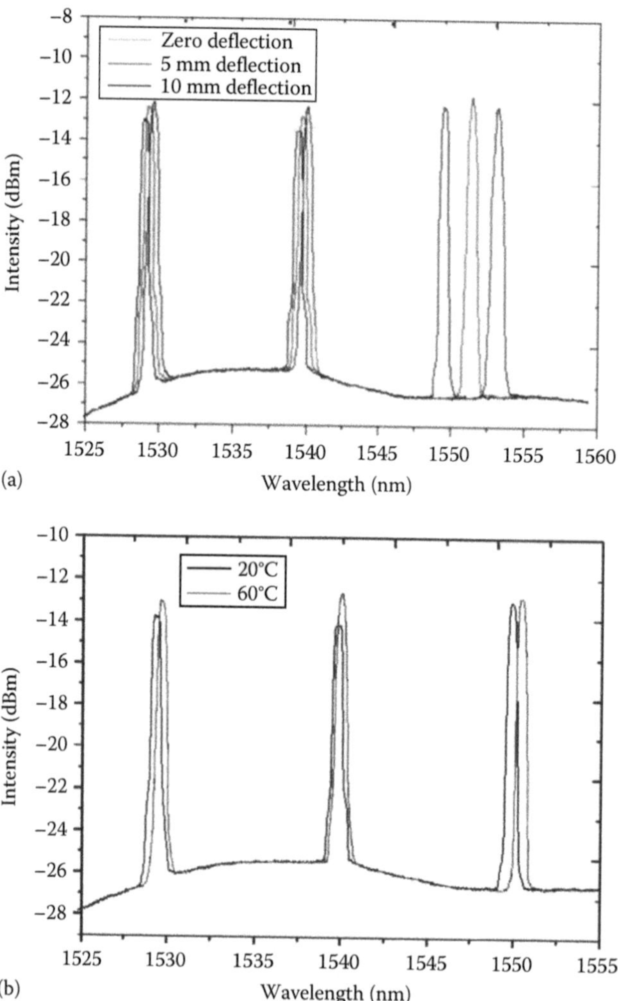

FIGURE 17.7 Wavelength shift for the FBG sensors embedded in composite materials at different locations (a) at different deflection values and (b) at different temperatures.

is possible to achieve simultaneous discrimination of mechanical perturbations and temperature [31,33]. Different methods have been reported by other authors to compensate for the cross-sensitivity effects between temperature and strain; these methods include introduction of a reference FBG [34], dual-wavelength superimposed FBGs [35], combined FBG and a long-period grating (LPG) [36], different cladding diameter FBGs [37], combination of an FBG and a Fabry–Perot interferometer [38], and superstructure FBG method [39].

It is also reported that for coating-stripped FBGs, the transverse CTE for a composite material as well as the uniform lateral strain can cause Bragg peak distortion and broadening [40]. Such a nonuniform lateral strain distribution in the composite material can be measured by using an FBG written in a microstructured high-birefringence fiber [41].

17.6.1.2 Vibration Measurements Using Embedded FBGs

Composite material structures are frequently subjected to external excitations over a variety of vibration frequency ranges. Such dynamic interference may cause the structures to suffer from fatigue damage and/or catastrophic failures if the excitation frequency approaches natural

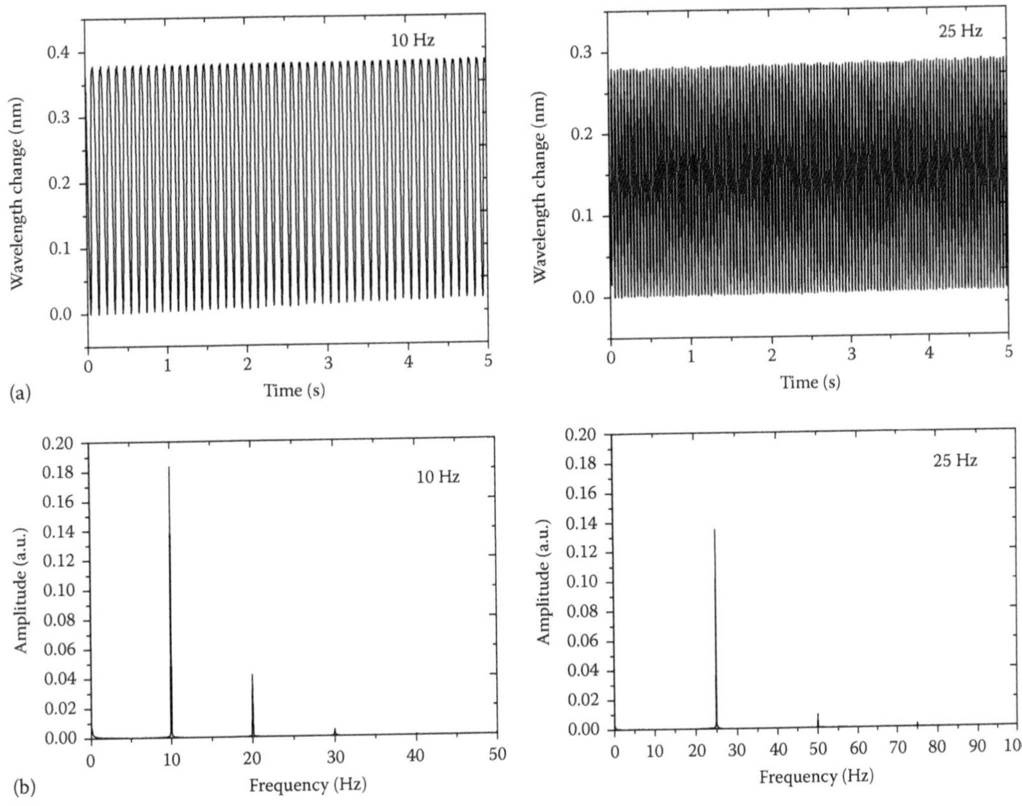

FIGURE 17.8 (a) Temporal and (b) FFT spectrum of the FBG sensor at different frequencies. (From Rajan, G. et al., *IEEE Sensors J.*, 12(5), 1365, 2012.)

frequency of the structures. The microstructural mechanisms of damage accumulation include fiber breakage and matrix cracking, debonding, transverse ply cracking, and delamination. These mechanisms may occur independently or sometimes interactively, and the predominance of one or the other may be strongly affected by both material variables and testing conditions [42]. FBGs can be effectively used for vibration measurements of composite materials. Rajan et al. demonstrated that FBGs embedded in composite materials can be used to measure vibrations up to a frequency of 25 Hz [43]. The time-domain responses of the embedded FBG sensor at various frequencies are shown in Figure 17.8a, and the corresponding frequency spectra are shown in Figure 17.8b.

17.6.1.3 Effect of Thermal Expansion of the Composite Material on FBG Sensitivity

For an FBG sensor embedded inside a composite material, three effects can take place due to change in temperature: (1) wavelength shift induced by the thermo-optic effect and thermal expansion of the polymer fiber; (2) wavelength shift induced by a longitudinal strain resulting from the thermal expansion of the composite sample; and (3) wavelength shift and spectral broadening corresponding to a transverse strain (or thermal stress) induced by thermal expansion of the composite sample. Due to the combined effect of (1) and (2), the temperature sensitivity for the embedded FBG sensor will be different from that for a free-space grating and if the transverse strain is effectively transferred to the grating, additionally a spectral broadening can be observed. Based on the type of reinforcement used, orientation of the reinforcement ply, and matrix materials, the magnitude of the wavelength shift due to thermal expansion may vary.

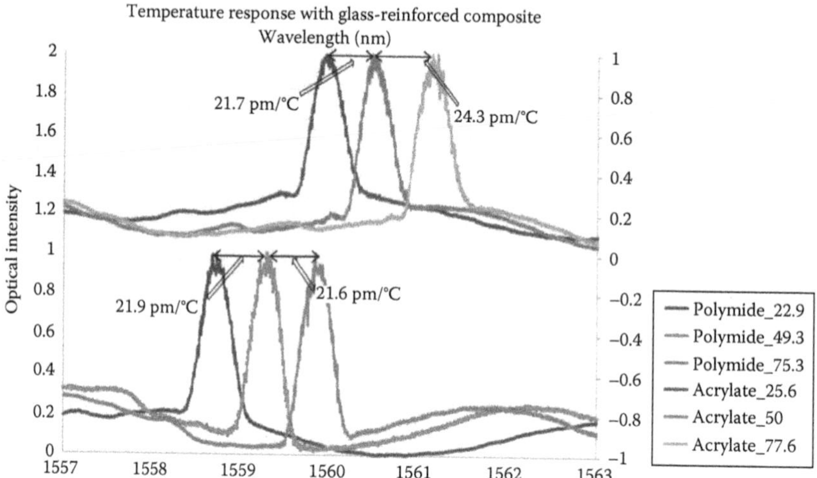

FIGURE 17.9 Comparison of wavelength shift of acrylate- and polyimide-coated FBGs embedded in glass fiber pre-preg composite material. (From Rajan, G. et al., *Proc. IEEE Sens.*, 351, 2011.)

Considering only the temperature effect on the FBG and the thermal expansion–induced mechanical strain, the net wavelength shift for an embedded FBG with temperature can be written as [14]

$$\Delta\lambda_T = \left(\alpha + \xi + k_c\right)\lambda_B \Delta T, \tag{17.5}$$

where

α is the CTE

ξ is the thermo-optic coefficient of the optical fiber

λ_B is the Bragg wavelength of the FBG

The coefficient k_c corresponds to the effective mechanical strain on the FBG due to the thermal expansion of the composite material

Figure 17.9 shows the temperature sensitivity of the embedded silica FBGs with polyamide and acrylate coating, which shows significantly high temperature sensitivity compared to the free-space temperature sensitivity that arises from the strain induced on the grating due to thermal expansion.

17.6.1.4 Acoustic Emission Measurements Using Phase-Shifted FBGs

Among the FBG-based sensing techniques for acoustic emission (AE) measurement, one of the proven and most reliable techniques is based on phase-shifted FBG (PS-FBG) sensors and its corresponding interrogation system. PS-FBGs are widely used in optical fiber communications and optical fiber sensing applications as a wavelength multiplexer and also as a strain sensor. The extremely sharp resonance in the transmission spectrum of the PS-FBG enables it to develop a high sensitivity interrogation system capable of measuring very small strain changes, even at high acoustic frequencies. A typical transmission spectrum of a PS-FBG interrogated using a tunable laser form is shown in Figure 17.10. Using this method, it is proved that microscale strain can be measured and is suitable for AE sensing in composite structures [45].

17.6.2 FBG Written in Microstructured Fibers as a Sensor for Composite Materials

Conventional FBG sensor's primary response is to an axial strain and thus transverse strain has very little influence on such an FBG sensor's response. For strain mapping in some SHM applications,

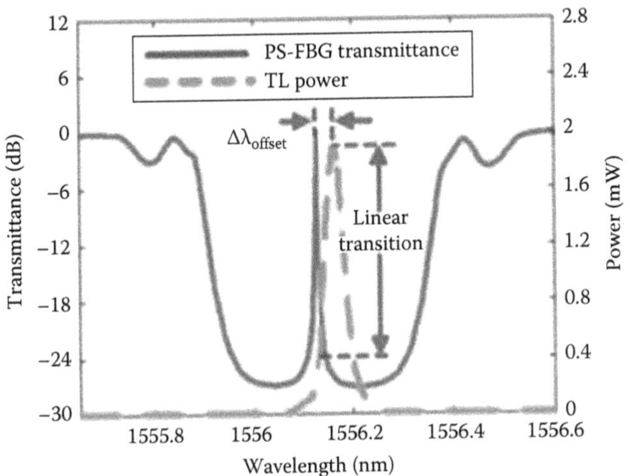

FIGURE 17.10 Spectral response of a PS-FBG and its interrogation technique based on a narrow band laser signal. (From Azmi, A.I. et al., *Photon. Sens.*, 3(2), 184, 2013.)

for example, for the detection of damage, cracks, and delamination, a multiaxial strain measurement is required. This in turn results in the importance of development of a sensing scheme that provides measurements of axial strain together with lateral stain. Luyckx et al. reported that FBGs written in highly birefringent (HB) fibers and highly birefringent microstructured optical fibers (HB-MOFs) have real potential to measure transverse strain and axial strain simultaneously [41,46]. As represented in Figure 17.11a, an FBG written in an HB fiber displays two Bragg peaks, corresponding to both orthogonally polarized modes. The Bragg peak separation depends on the phase modal birefringence variation induced by transverse load and temperature. The properties of an MOF and its sensitivity to different measurands are determined by the type of the fiber used. But it is reported that FBGs written in HB fibers such as bow-tie type and Panda type have the disadvantage of greater temperature and strain cross-sensitivity.

The cross-sensitivity issue can be resolved by writing FBGs in low temperature–sensitive MOFs, such as highly birefringent polarization-maintaining photonic crystal fibers (HB-PM-PCF). For example, a comparison response of the embedded FBG sensors written in conventional birefringent optical fibers (bow tie) and an HB-PM-PCF when the composite material is exposed to controlled mechanical and thermal loads are presented in Figure 17.11b and c, respectively. For the FBG in a bow-tie-type fiber, the Bragg peak separation varies in accordance with transverse stain and temperature as shown in the figures. However, in the case of an FBG in an MOF, the Bragg peak separation varies with transverse stain only. A temperature-independent axial strain measurement in a composite material can be carried out using an FBG written in an HB-PM-PCF [46]. It is also reported that the two Bragg peaks in the case of FBG in PM-PCF are farther separated than for an FBG in a bow-tie fiber, and therefore, FBGs written in MOF/PM-PCF allow for more accurate measurements of the peak wavelengths and thus are more suitable for composite material sensing applications [41,46].

17.6.3 Polymer FBG Sensor for Composite Materials

Polymer FBGs have attracted much interest among application engineers and scientists due to their unique advantages compared to the silica counterparts, such as the inherent fracture resistance, low Young's modulus, high flexibility, high temperature sensitivity, large strain measurement range, and low density [47–49]. In recent years, advancements in the fabrication of single-mode polymer optical fiber (POF) have also enhanced the research and development of Bragg grating sensors in

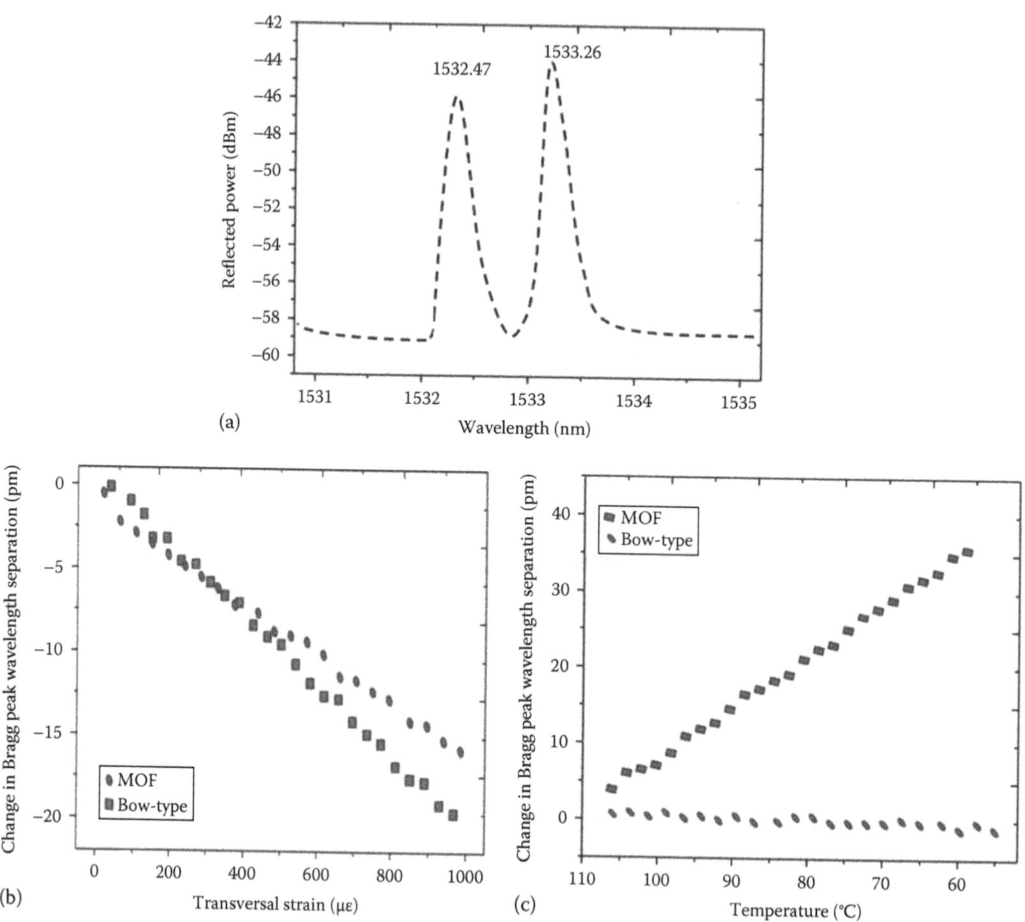

FIGURE 17.11 (a) Reflection spectra for an FBG written in an HB-PM-PCF with two peaks corresponding to slow axis and fast axis, (b) change in the peak separation with transverse strain for FBGs written in MOF and bow-tie-type fibers, and (c) change in peak separation with temperature. (From Luyckx, G. et al., *IEEE Photon. Technol. Lett.*, 21, 1290, 2009.)

POF [50]. Sensors for different applications based on single-mode POF gratings have been reported [51,52]. For composite materials, the attractive characteristics of polymer FBGs include their high temperature sensitivity, large strain range, and the absence of buffer coating. These distinct features of the polymer FBGs might give an edge over the standard silica FBGs in measuring some of the parameters of the composite materials.

Rajan et al. conducted studies with polymer FBGs embedded in glass-reinforced composite materials [53]. For the polymer FBG embedded in composite materials, due to temperature change, spectral broadening is observed together with the shift in the peak reflected wavelength. The observed wavelength shift of the polymer FBG embedded in a glass-reinforced composite and its comparison with silica FBG and free-space FBG sensors are shown in Figure 17.12a. A blue shift in the wavelength is observed for the polymer FBG due to the negative thermo-optic coefficient, while a red shift is observed for the silica FBG. The observed temperature sensitivity of the embedded polymer FBG is 92.28 ± 2 pm/°C, which is close to the free-space temperature sensitivity of 90 ± 6 pm/°C.

The change in bandwidth of the reflection spectra of the polymer FBG at different temperatures and its comparison with silica FBG is shown in Figure 17.12b. It can be seen that the bandwidth of the polymer FBG increases as the temperature increases. The measured bandwidth change for the

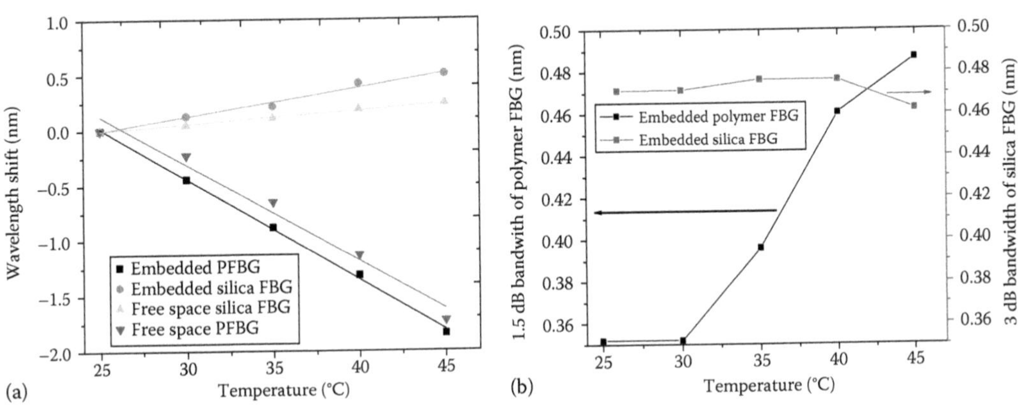

FIGURE 17.12 (a) Temperature-induced wavelength shift of the embedded polymer and silica FBGs and its comparison with free-space FBGs, (b) measured 1.5 dB bandwidth of polymer FBG and 3 dB bandwidth of silica FBG at different temperatures.

embedded polymer FBG within the temperature range of 30°C–45°C was 8.5 pm/°C. Therefore, from the observed spectral broadening and distortion, it can be concluded that the thermal expansion induced stress is effectively transferred to the polymer fiber and can be measured using a polymer FBG. It is assumed that the reason for this is the absence of a buffer coating for the polymer fiber, which results to a direct transfer of the surrounding physical phenomena to the core and cladding of the polymer fiber. Stress induced by localized microbends in the composite material could also contribute to the chirping effect.

The strain sensitivity of the embedded polymer FBG and its comparison with silica FBG is shown in Figure 17.13. From the figure, it can be seen that the strain sensitivity of the embedded polymer and silica FBGs are very close to each other. In free space, the polymer FBG had a slightly higher strain sensitivity (1.340 ± 0.015 pm/με) compared to that of silica FBG (1.2 ± 0.01 pm/με). The similarity in the measured strain sensivity of the embedded silica and polymer FBGs underlines the fact that longitudinal strain in the composite is not effectively transferred to the polymer fiber. This can be attributed to the differences in mechanical properties of the polymer fiber and

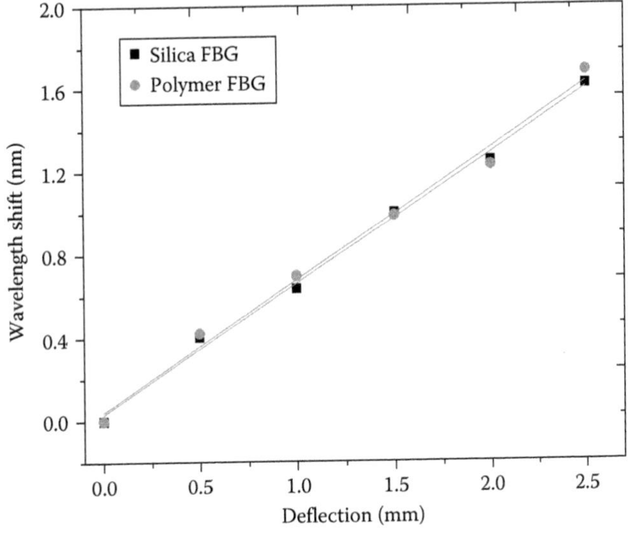

FIGURE 17.13 Wavelength shifts of the polymer and silica FBGs with deflection in the middle of the composite material.

the composite material, which resulted in the mechanical strain not being effectively transferred to the polymer FBGs as compared to the case of silica FBG. With further works in this area, polymer FBG has the potential in measuring temperature and thermal expansion of the composite material simultaneously.

17.6.4 Interferometric Fiber Sensors for Composite Materials

Interferometric fiber sensors [24] can also be employed for strain/temperature measurements in composite materials and this is discussed in detail in this section. There are different types of interferometric fiber sensors that differ in their operating principles and strain/temperature sensing characteristics. Among the interferometric fiber sensors most commonly employed for composite sensing applications are extrinsic Fabry–Perot interferometers (EFPIs), microhole collapsed modal interferometers, and Sagnac fiber loop mirror (FLM) sensors.

The schematic experimental arrangement of the EPFI sensor is shown in Figure 17.14b, which comprises a superluminescent light diode (SLD) source, a coupler, and a spectrometer. Any applied longitudinal strain to the FPI sensor alters the physical length of the cavity, which results in a phase difference between reflected and transmitted beams. By measuring the shift of the wavelength spectrum, the strain applied to the FPI can be measured. It is found that the shorter the optical path difference (OPD), the larger will be the free spectral range (FSR), resulting in a wider dynamic range for a sensor [24,54]. Therefore, the dynamic range of the sensor can be tuned by varying the cavity length, which in turn changes the OPD of the FPI sensor [54].

EFPI sensors based on PCFs are proved to be good candidates for measurements of axial strain in composite materials since their sensitivity to transverse strain and temperature is insignificant [55–57]. Figure 17.15a illustrates experimentally measured strain using embedded EFPI sensor during the three-point bending test of a composite material sample [57].

Recent advancements in the area of PCF sensors [58] have opened new possibilities for the development of temperature-insensitive microhole collapsed PCF modal interferometric sensors. Microhole collapsed PCF modal interferometers can be used for localized strain measurements in a composite material. It can be fabricated by fusion slicing of a PCF section between two standard single-mode fibers. During the splicing process, at both ends of the PCF section, the holes of the PCF are collapsed in a microscopic region. The first collapsed region allows for the excitation of multiple modes in the PCF and at the second collapsed region at the other end the modes recombine and thus the PCF section forms an interferometer [58]. Temperature-independent strain measurement in composites using such a microhole collapsed PCF interferometer fabricated from a microstructured PM-PCF fiber (LMA-10) was reported by Rajan et al. An example of the output of a microhole collapsed modal interferometer embedded in composite beam undergoing three-

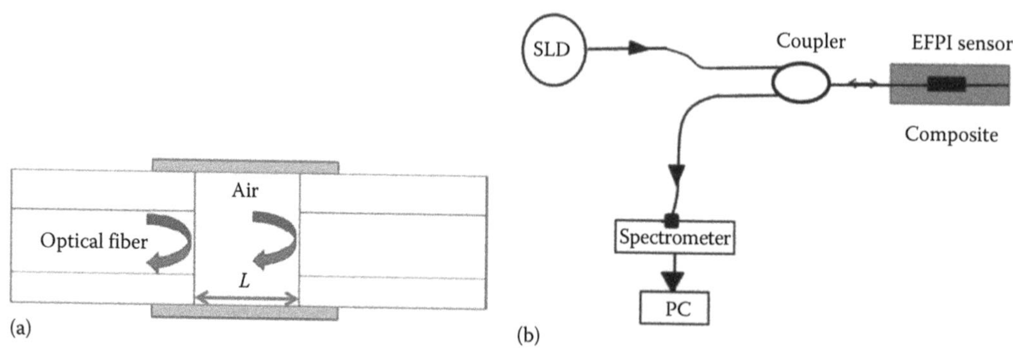

FIGURE 17.14 (a) One of the typical EFPI sensor designs and (b) schematic experimental arrangement for the EPFI sensor.

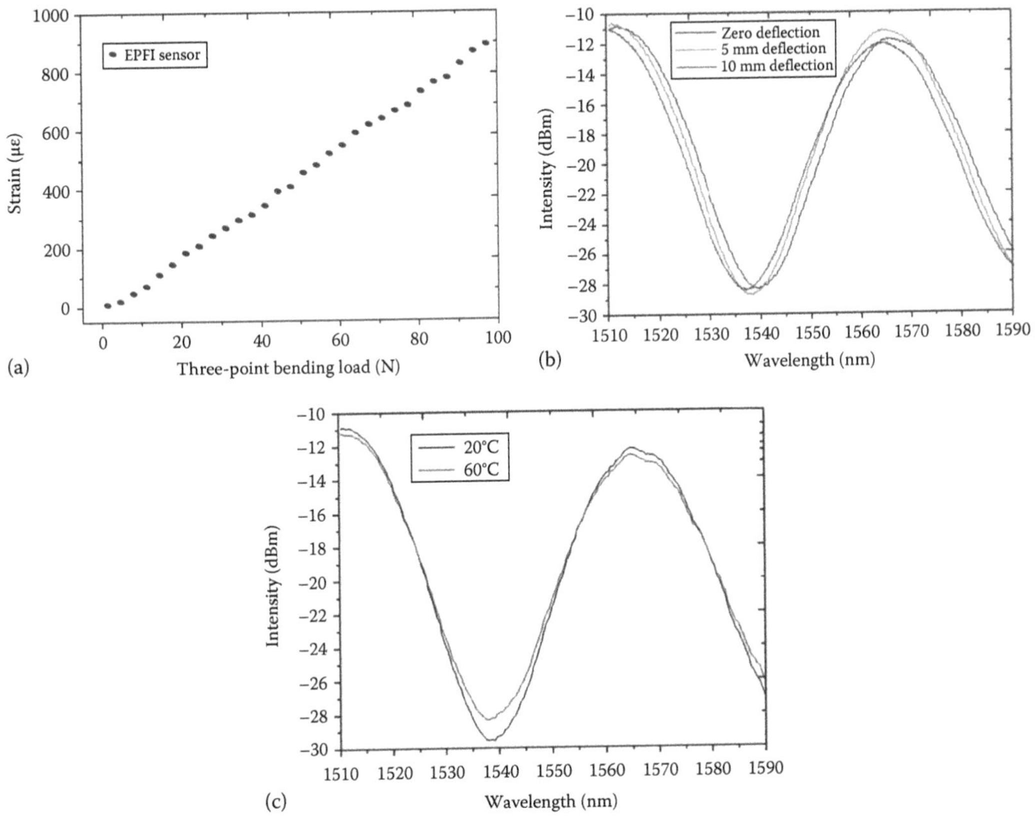

FIGURE 17.15 (a) Experimentally measured strain using embedded EFPI sensor during a three-point bending test in a composite material (From Leng, J.S. et al., *Sensor. Actuat. A-Phys.*, 103, 330, 2003), (b) responses of the PCF-based sensors embedded in the composite material sample during deflections based on a three-point bending test, and (c) at different temperatures of the composite sample.

point bending deflection is shown in Figures 17.14b and c illustrates that such a sensor embedded in composite material has very low temperature dependence [58].

Fiber-optic Sagnac interferometers (SIs) are another promising candidate for composite material sensing applications. In a fiber-optic SI, input light is split into two parts propagating in the opposite directions by a 3 dB fiber coupler, and these two counterpropagating beams are combined again at the same coupler as shown in Figure 17.16. The fabrication of such an interferometer can be

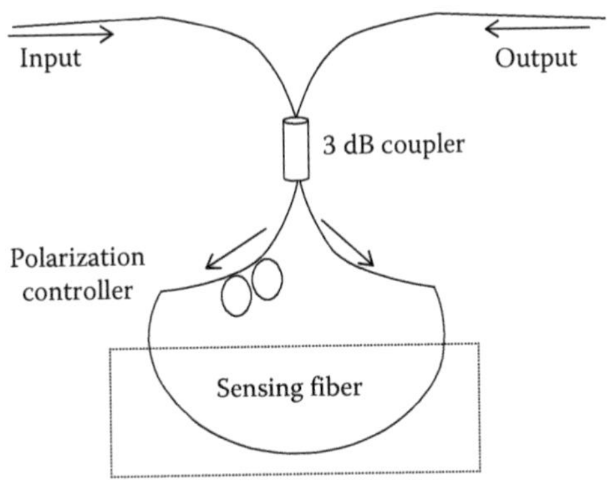

FIGURE 17.16 Schematic of the sensor based on an SI.

simply achieved by connecting the ends of a conventional 3 dB coupler. HB fibers or polarization-maintaining fibers (PMFs) are typically utilized as sensing fibers since HB fibers maximize the polarization dependence of the signal within the SIs. In order to control the input light polarization, a polarization controller (PC) is connected to the sensing fiber. The signal at the output port of the fiber coupler is a result of interference between the beams polarized along the slow axis and the fast axis. The phase of the interference is simply given as

$$\delta_{SL} = \frac{2\pi}{\lambda} BL \tag{17.6}$$

where

$B = | n_f - n_s |$ is the birefrigent coefficient of the sensing fiber
L is the length of the fiber
n_f and n_s are the effective indices of the fast and slow modes

Fiber-optic-based SIs can be used for measuring parameters such as strain, temperature, pressure, and twist. Such fiber-optic-based SIs provide the value of the sensed parameter averaged over the length of the sensor. However, a disadvantage of the SI is its significant temperature and strain cross-sensitivity. Rajan et al. reported that the temperature strain cross-sensitivity issue can be eliminated by employing PM-PCFs [58]. The output of a low temperature–sensitive PM-PCF SI-based sensor embedded in composite beam undergoing a three-point bending deflection is shown in Figure 17.17. Such low temperature–sensitive HB fiber–based SIs are an appropriate option for measuring strain acting on a composite material over the length of the sensing fiber.

17.6.5 Optical Fiber Polarimetric Sensors for Composite Materials

The polarization properties of light propagating through an optical fiber can be affected by stress, strain, pressure, and temperature acting on a measuring fiber, and in a fiber polarimetric sensor, the polarization change is detected to retrieve the sensing parameter [59,60]. A symmetric deformation effect or temperature variation in a single-mode fiber influences the propagation constant (β) for every mode because of the changes in the fiber length (L) and the refractive indices of the core and

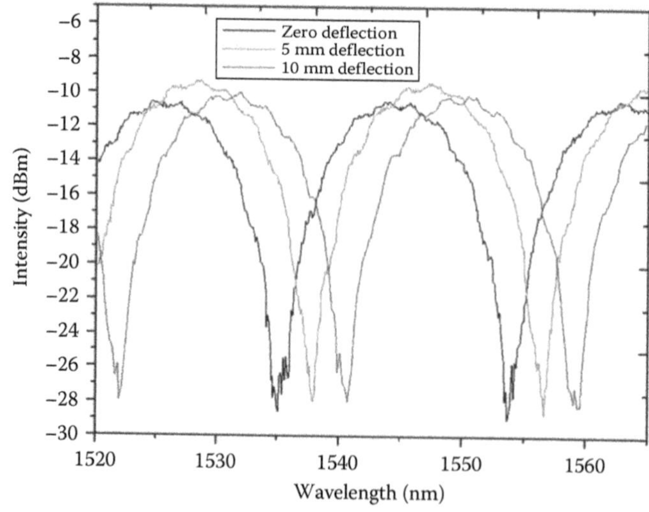

FIGURE 17.17 Responses of the PM-PCF SI-based sensor embedded in the composite material sample during deflections based on a three-point bending test.

the cladding [24,61]. Under the influence of longitudinal strain (ε) or temperature (*T*), for single-mode fiber polarimetric sensors, the change in the phase difference can be written as [61]

$$\frac{\delta(\Delta\Phi)}{\delta X} = \Delta\beta\frac{\partial L}{\partial X} + L\frac{\partial(\Delta\beta)}{\partial X} \tag{17.7}$$

where *X* stands for temperature, pressure, or strain.

Optical fiber polarimetric sensors can be realized by different types of PM fibers such as Panda fiber, bow-tie-type fiber, side-hole fibers, and PM-HB MOF fibers. Optical fiber polarimetric sensors can be embedded in composite materials to measure the average strain/temperature over the sensor length [24,61]. It is possible to vary the strain/temperature sensitivity of optical fiber polarimetric sensors by selecting a PM fiber type with appropriate birefringence and length [61,62].

For optical fiber polarimetric sensors, the phase difference between the two orthogonal polarizations can be extracted using an experimental setup consisting of a tunable laser source and a polarimeter/polarization control system [63]. Optical fiber polarimetric sensors can also be operated in intensity domain with the help of a polarizer–analyzer arrangement, and the experimental setup for operation in intensity domain is shown in Figure 17.18. For polarimetric sensors, the change in the output intensity at a wavelength λ due to externally applied longitudinal strain can be described by the following formula [24,61]:

$$I_s(\lambda) = \frac{I_0}{2}\left[1 + \cos(\Delta\Phi)\right] \tag{17.8}$$

For such polarimetric sensors operating in the intensity domain, periodic variations in the output intensity can be associated with applied strain or temperature [64]. Wolinski et al. reported that low strain–sensitive polarimetric sensors such as PM-PCF polarimetric sensor guarantee a linear response for a wide range of applied strain values [61] (Figure 17.19). Moreover, the insignificant temperature sensitivity of PM-PCFs makes them the most appropriate candidates for composite strain measurements [64]. However, for local strain/temperature measurements of large composite structures, polarimetric sensors cannot be employed as the polarimetric sensors measure average strain over the length of the sensing fiber.

When embedded in the composite material, OFSs are influenced by various physical parameters of the composite material such as thermal expansion composite materials [14]. It is found that PM-PCF polarimetric fiber sensor with an acrylate coating embedded in the composite material shows the same response as the one in free space while the coating-stripped fiber polarimetric sensor shows significant temperature sensitivity when embedded in the composite material as shown in Figure 17.20a and b. This is due to the stress-induced change in birefringence created by the thermal expansion of the composite material, while in the case of a buffer-coated fiber, the effect is considerably reduced as the thermal stress is largely eliminated by the buffer coating. But a coating-stripped fiber polarimetric sensor shows a significant increase in temperature sensitivity when embedded in the composite material as shown in Figure 17.20c and d, which is due to the stress-induced change in birefringence created by thermally induced strain. A comparison of temperature sensitivities of PM-PCF sensors in various configurations (with and without buffer coating) at different layers

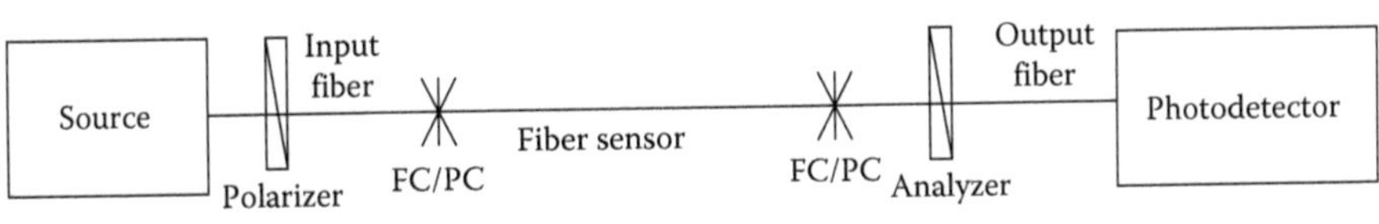

FIGURE 17.18 Experimental setup for measurements with fiber-optic polarimetric sensors in intensity domain.

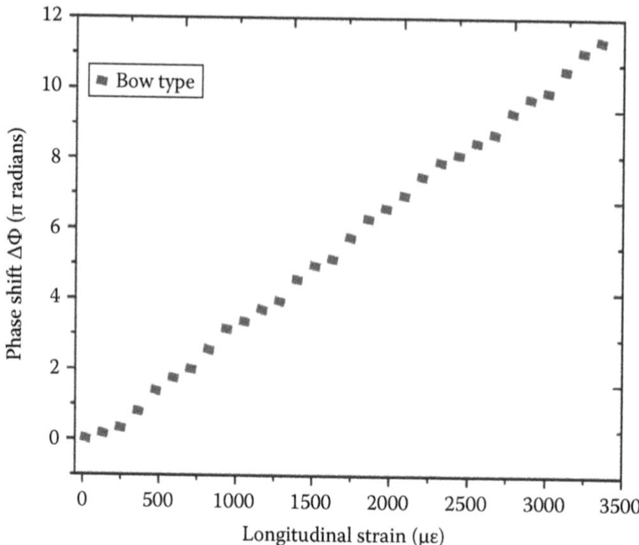

FIGURE 17.19 Change in the polarization of fiber-optic polarimetric sensors as a function of strain applied to a composite sample. (From Domański, A.W. et al., *Photon. Lett. Poland*, 1, 121, 2009.)

is shown in Figure 17.20e, and it is clear that the presence of a coating preserves the temperature independence of a PM-PCF sensor when it is embedded in a composite. The thermal expansion of the composite material is the main source of error in strain and temperature measurements using embedded polarimetric fiber sensors, and more accurate strain and temperature measurements can be obtained with buffer-coated polarimetric fiber sensors.

It is also found that the temperature-induced phase change in buffer stripped PM-PCF polarimetric sensors embedded in different type composite material samples varies with the amount of thermal elongation induced strain over a temperature for the particular material. Figure 17.21 shows the observed phase shift for stripped PM-PCF sensors embedded in different composite samples for temperature variation from 0°C to 65°C. The information from embedded sensors opens up the possibility of measuring the CTE value of a composite material itself, based on the observed phase change in the embedded buffer stripped PM-PCF polarimetric sensor. The CTE of carbon–epoxy, E glass–epoxy, and E glass–unsaturated polyester resin composite material samples is measured experimentally using the PM-PCF sensor and it shows good agreement with the theoretically calculated CTE values as shown in Figure 17.22. Thus, PM-PCF-based polarimetric sensor can provide a vital key information for the composite manufacturing industry by further developing nondestructive methods based on embedded OFSs for the characterization of composite material at different stages in their life cycle, from fabrication through operation to predicting the onset of failure.

17.6.6 Distributed Optical Fiber Sensors

Distributed optical fiber sensors (DOFSs) are capable of providing a continuous measurand profile over the length of the optical fiber and thus are very promising for strain/temperature measurements in large structures such as bridges, buildings, and pipelines [25,65,66]. However, given the nature of composite structures, the length is normally limited to 80 m or less with a strain or temperature requirement of at least 1°C or 20 με with a few centimeters resolution. DOFSs are categorized into several types based on the sensing technology and the physical effect underpinning the operating principle: (1) optical time-domain reflectometry (OTDR) and optical frequency-domain reflectometry (OFDR), both based on Rayleigh scattering; (2) Raman optical time-domain reflectometry (ROTDR) and Raman optical frequency-domain reflectometry (ROFDR), both based on Raman

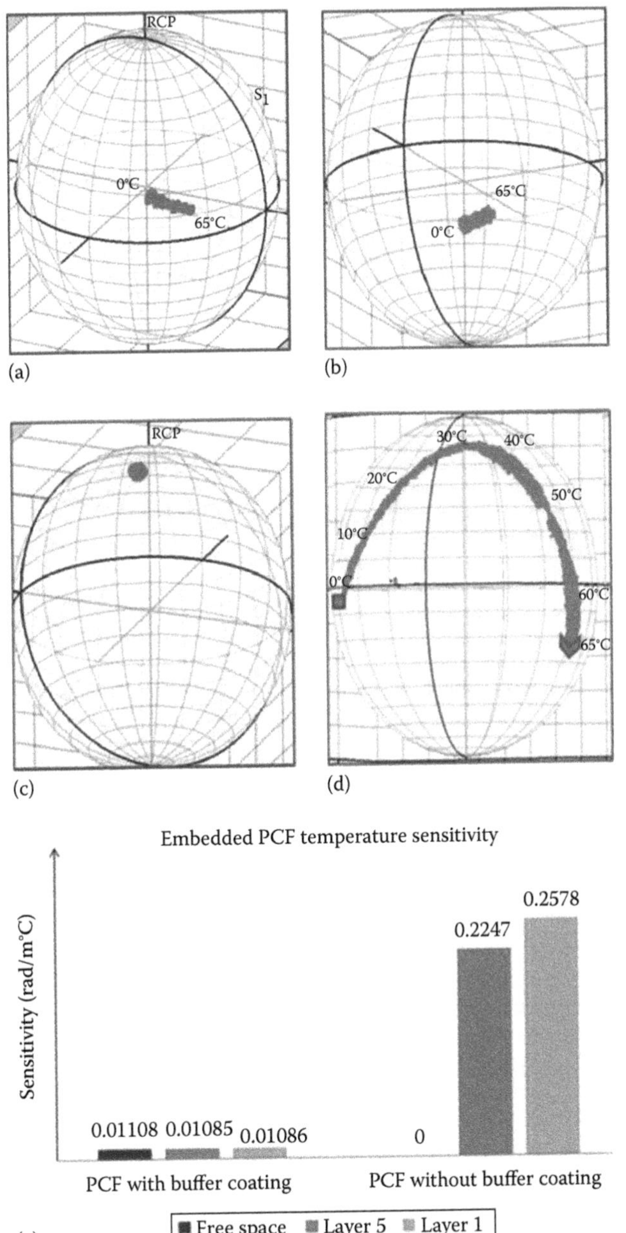

FIGURE 17.20 (a) Observed phase shift in coated PM-PCF fibers for a temperature variation from 0°C to 65°C (a) free space (b) after embedding. Observed phase shift in coating-stripped PM-PCF fibers for a temperature variation from 0°C to 65°C (c) in free space (d) after embedding and (e) comparison of the temperature sensitivity of PM-PCF polarimetric sensors at different configurations.

scattering; and (3) Brillouin optical time-domain reflectometry (BOTDR) and Brillouin optical frequency-domain reflectometry (BOFDR), both based on Brillouin scattering [67].

OTDR and OFDR are the first generation of optical fiber distributed sensors, based on the use of Rayleigh scattering to reflect the attenuation profiles of long-range optical fiber links [68]. An example of strain measurement by OFDR technique for various loading conditions is shown in Figure 17.23. An optical pulse is launched into an optical fiber and the power of the Rayleigh

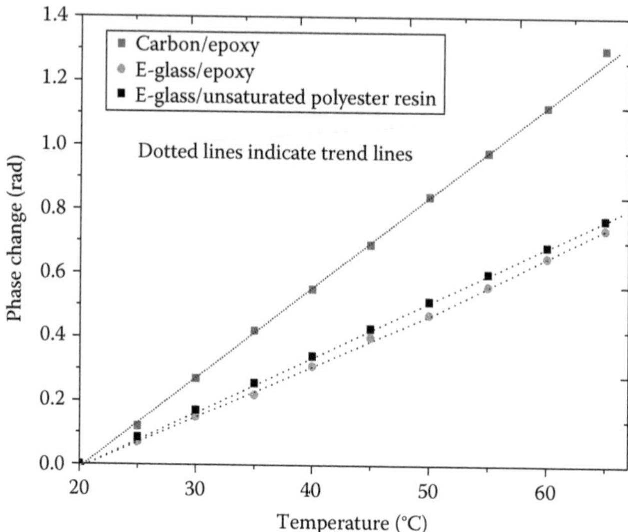

FIGURE 17.21 Observed phase shift for stripped PM-PCF sensors embedded in different composite samples for temperature variation from 0°C to 65°C.

backscattered light is detected using a photodetector as the light pulse propagates along the fiber link [67]. Frequency-based Brillouin method can provide rapid strain/temperature sensing [69]. BOFDR can be employed to measure strain/temperature variations as small as 1°C or 20 με with a few centimeters of resolution [69,70]. For BOFDR-based long-distance measurements, single-mode fibers are used [69,71]. ROTDR is an appropriate candidate for temperature measurements in composites since the intensity ratio between anti-Stokes components and Stokes components of the ROTDR response provides information about temperature [71]. ROTDR can be employed to measure temperature variations as small as 0.1°C with a few meters of resolution [71,72]. Multimode fibers are commonly used for ROTDR-based short-distance measurements [72]. In the case of composite strain/temperature measurements from the aforementioned categories of DOFS, an appropriate sensing technology can be selected based on the application and its requirements; specifically, BOFDR is more suitable for strain measurements and ROTDR is more suitable for temperature measurements in composite material [71–73].

17.6.7 Hybrid Sensing Approaches for Simultaneous Strain and Temperature Measurements in Composite Materials

In most of practical applications of composite materials, the key requirement is to measure more than one parameter. For example, for SHM applications in composite parts such as helicopter rotor blades, wind turbines, and aircraft structures, a simultaneous monitoring of temperature and strain is favorable. Several OFSs such as FBGs and PM polarimetric sensors are capable of measuring strain and temperature, but there is always an issue of accurate discrimination between these two influencing factors [23,25]. One of the simplest and promising methods for simultaneous measurement of strain and temperature in composite parts is the hybrid optical fiber sensing approach; in this, two or more OFSs operate in a combined manner to eliminate the disadvantages of individual OFSs providing accurate and independent strain/temperature information [74]. Many researchers have investigated different hybrid sensing approaches for simultaneous strain and temperature measurements such as using a combination of FBG sensors with various types of sensors such as LPGs [36,74], Fabry–Perot interferometer sensors [37], PCF modal interferometers [75], and FLMs using a small-core MOF [76]. Sensors based on gratings written in MOFs and standard optical fibers have

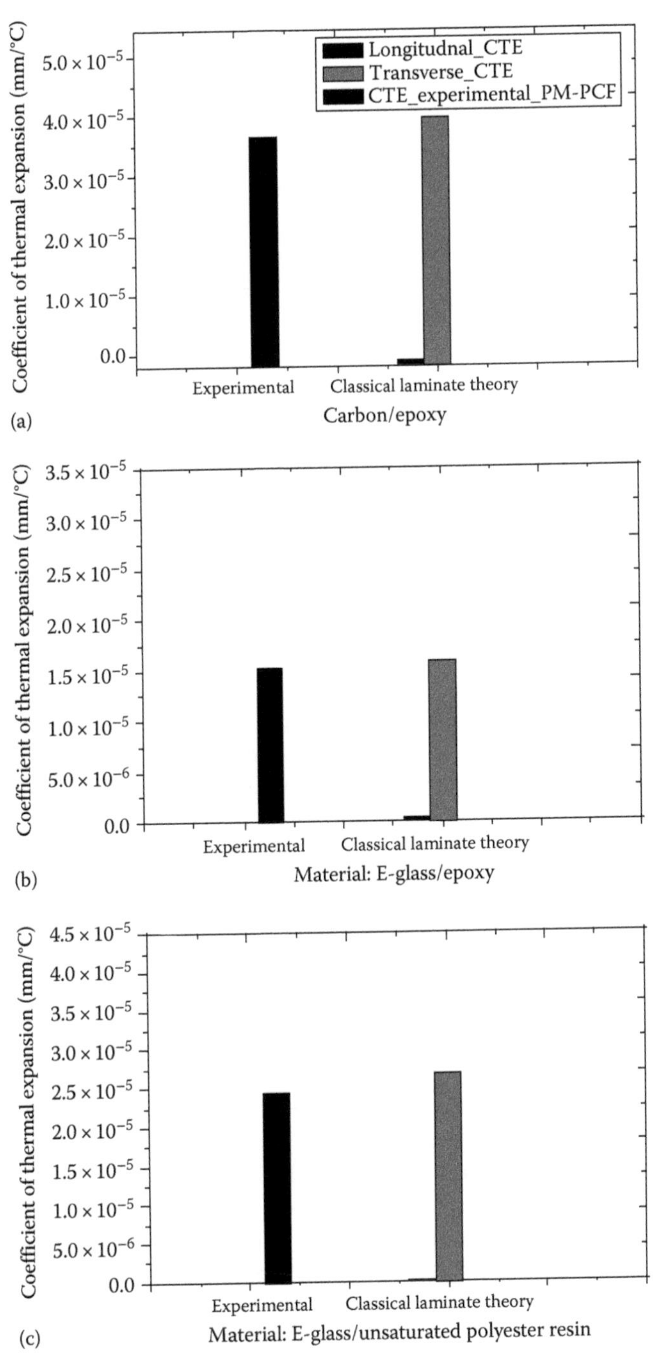

FIGURE 17.22 Comparison of theoretical predictions with experiment, (a) carbon–epoxy and (b, c) E glass–epoxy/E glass–resin composite samples.

also been reported for discrimination between strain and temperature [77]. Previously, we reported a hybrid approach that involves an FBG sensor and a PM-PCF polarimetric sensor for simultaneous measurement of strain and temperature [78] of composite materials. In the context of composite structures, the temperature sensitivity of an OFS is largely influenced by the thermal strain of the composite material [14]. Sensors based on stripped PM-PCF-based polarimetric sensors have been

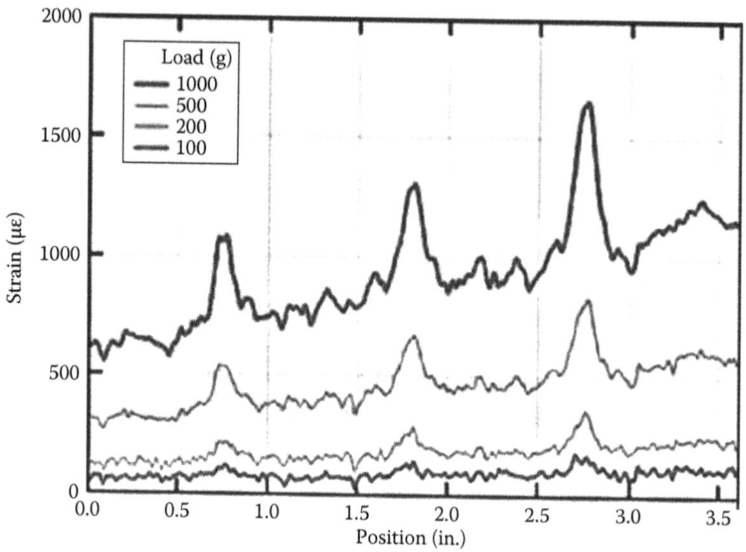

FIGURE 17.23 Strain measured by the optical fiber bonded to a composite sample for various loading conditions. (From Luna Engineering Note EN-FY1317, http://lunainc.com/wp-content/uploads/2013/09/LT_TD_EN-FY1317_DFS-Composite-Applications_Rev-3.pdf.)

reported for thermal strain and thermal expansion measurements in composite materials [13]. A hybrid OFS embedded in a composite material that is capable of discriminating between strain, temperature, and thermal strain has great potential for future OFS-based SHM applications.

17.7 LIGHTWEIGHT SANDWICH COMPOSITES WITH EMBEDDED FIBER SENSORS

Among the different types of composite material laminates, it is common in advanced manufacturing industries to construct composite laminates using a *sandwich* construction, consisting of an outer layer of reinforcing fiber in conjunction with a core filling, such as a solid foam core or a honeycomb core [79]. The combination of carbon fiber–reinforced composite materials with structurally efficient sandwich laminate design offers novel opportunities for ultra-lightweight structures. This sandwich combination allows the structural element with a very high bending stiffness to weight ratio as well as bending strength to weight ratio [80].

Many applications including marine and automobile industry [81] use sandwich composite construction to save weight and increase stiffness, because the low density foam is lightweight while ensuring that the reinforcing fibers sustain the majority of the stress at the extremes from the neutral bending axis. However, one of the main concerns related to sandwich composite structures is the fact that their load-carrying ability can be significantly reduced by the presence of delamination between the foam core and the reinforcement fiber. Such damages usually cannot be observed from outside the sandwich construction but can cause a serious reduction in load-bearing capacity of the structure. For sandwich structures, due to the presence of the core material, the performance of the embedded sensor could be different from those of the standard composite laminates.

FBG embedded in sandwich composites is used to measure strain, impact, and temperature of the composites. Rajan et al. demonstrated a carbon foam core sandwich composite (Figure 17.24) embedded with composite materials and the strain experienced within the material is measured [82]. Other works carried out by Hackney and Peters showed that FBGs can be used to impact measurement in foam core sandwich composites [83]. Rajan et al. also demonstrated the temperature measurements of sandwich composites using embedded FBGs. The measured wavelength shift for

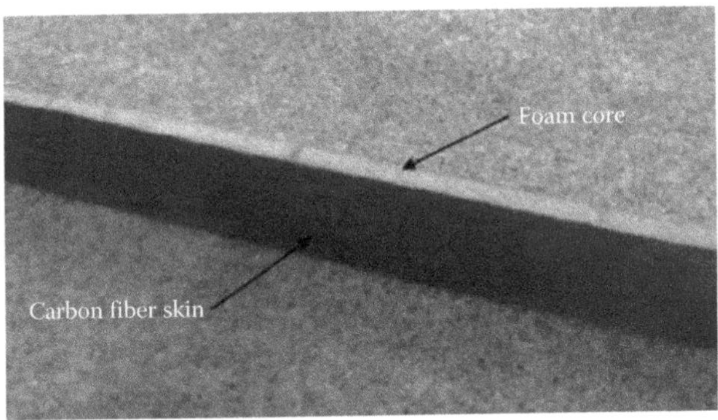

FIGURE 17.24 A carbon fiber foam core sandwich composite.

the embedded FBG with change in temperature is shown in Figure 17.25a. From the graph, it can be seen that the carbon fiber foam composite specimen shows a temperature sensitivity of approximately 9.5 pm/°C, which is close to the free-space sensitivity of the FBG (10 pm/°C). In comparison with embedded FBGs in standard glass and carbon fiber composites, which shows considerably high sensitivity (12–20 pm/°C) [14] due to the stress/strain induced by the thermal expansion of the composite, the results obtained for the carbon foam composite are highly promising, as there was no considerable effect of the material properties on the sensitivity of the embedded grating, and hence, the embedded sensor can accurately measure the true temperature inside the composite.

Another important factor to be considered in the case of a foam core sandwich composite is the relaxation of the foam during bending measurements and its effects on the performance of the embedded FBG. To evaluate this, the reflection spectra of the FBG are monitored after removing the physical pressure applied to the laminate. It was observed that the FBG peak reflected wavelength continues to shift slightly even after the complete removal of the load as shown in Figure 17.25b. This is assumed to be due to the relaxation of the foam core in the composite sandwich. The observed wavelength shift was approximately 0.02 nm for a time period of 60 s, which is equivalent to a strain of 16 με given a strain sensitivity of 12 pm/με. This adds an error of ±2.5% in the strain measurement using the embedded FBG in the case of the sandwich laminates. However, for the foam relaxation, the percentage of error also depends on the amount of load applied and the type of load applied. Further research is required to quantitatively predict the suitability of using embedded FBG for accurate impact/strain and temperature measurement for sandwich composites, and many groups are currently focusing on SHM of these types of lightweight composites.

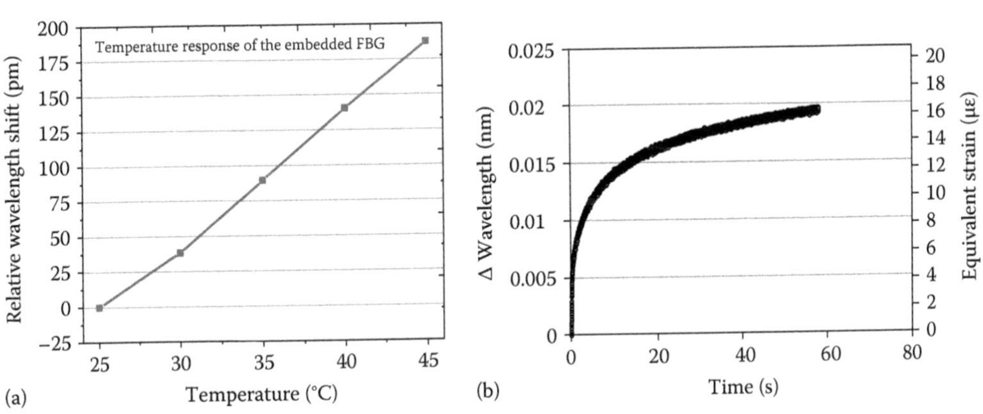

FIGURE 17.25 (a) Temperature response of the embedded FBG in a sandwich composite and (b) wavelength shift of the FBG due to the gradual relaxation of the foam core.

17.8 RECENT TRENDS, ISSUES, AND FUTURE CHALLENGES OF THE OFS TECHNOLOGY

Embedding an optical fiber inside composite materials is a minimally invasive technique; however, for industrial applications, there still exist a few issues that are under investigation and in this section we discuss those challenging issues. One of the challenges with composite structures embedded with a fiber sensor is the provision for a reliable method for accessing the sensor [84,85]. Custom-designed connectors with ingress/egress optical fiber ends are a promising technique, and one example of such a connector is shown in Figure 17.26. However, such egress optical fibers may become brittle at the edges of the smart composite structures [21,86]. Also, connecting to such a connector is very hard and edge trimming of the composite after the connector installation is impossible.

The second issue is the structural damage to the composite material due to the fact that optical fibers have larger diameter compared to reinforcement fibers. As a solution, researchers are considering new optical fibers with a reduced diameter and an optimized coating. One example of such fibers is a thinner optical fiber with 80 μm diameter known as Draw Tower Grating (DTG) fiber, which is proven to be less invasive when embedded within laminate composite structures or textiles [87].

Another issue related to embedding of fiber sensor in a composite part in weight- or space-sensitive applications is the large size of conventional OFS interrogators, which could present difficulties for sensing of composite parts that, for example, are in constant motion, such as helicopter rotor blades and wind turbine structures [88]. To an extent, this issue can be resolved by adopting miniaturized interrogation modules based on photonic integrated circuits (PICs) or flexible polymer waveguides [89,90]. Such surface attachable flexible interrogators allow for integration of photodetector arrays with wireless communication technology and thus have strong potential in smart sensing of composite parts in motion.

Finally, it must be recognized that the OFS embedding procedure is often a labor-intensive task. Therefore, ideally it requires a reliable automated optical fiber placement system that matches well with the existing industrial composite production processes. Several composite manufactures, such as Airborne, have introduced automated fiber placement system that provide control over embedding depth, prestrain, position, and alignment. Such an all-automated fiber placement system is shown in Figure 17.27. The biggest challenge in using most of the automated fiber sensor embedding systems is that it is difficult to ensure an adequate repeatability of the fiber sensor placement. In addition, during such an automated fabrication process, the safety of delicate areas of optical fiber (such as grating written area of FBG, spliced and buffer coating removed areas of sensing optical fiber) and control over the alignment of specialized fibers such as MOFs are not well assured. Thus, an automated fiber placement system together with an X-ray-based microcontrolled tomography could be a suitable solution for quality control of the embedded fiber sensor and a realistic option for appropriate alignment of optical fibers for future smart sensing applications.

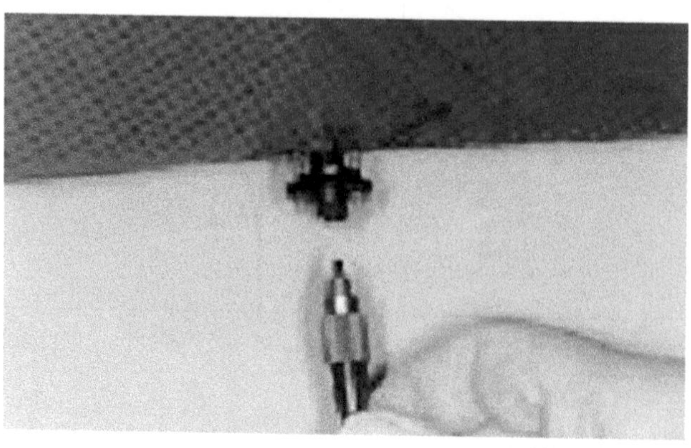

FIGURE 17.26 Egress optical fiber with connector.

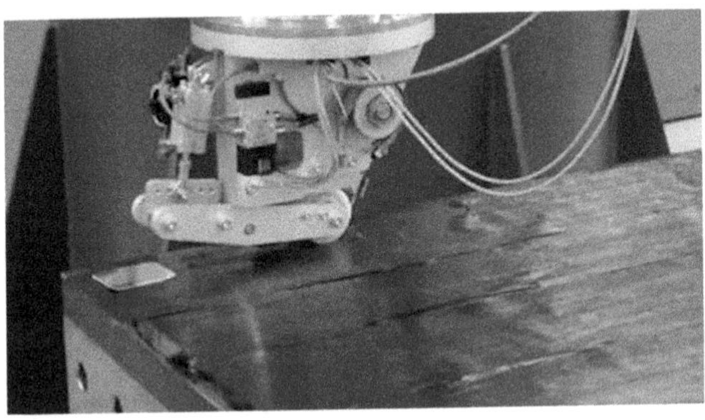

FIGURE 17.27 Automated fiber placement system. (From Airborne, http://www.airborne-international.com.)

17.9 CONCLUSION

This chapter presents a comprehensive overview of the OFSs used for sensing of composite materials and provides essential information regarding many aspects, including different types of composite materials, properties, and their performance, the relevance of OFS for composite material sensing applications, future challenges of embedded OFSs, and recent trends. Embedded OFSs are useful to the smart sensing applications of composite materials in aerospace, structural, civil, and sports industry and this technology will be commercially available and beneficial to the society after a few more years of research work.

REFERENCES

1. Garg, D. P., Zikry, M. A., Anderson, G. L., Gros, X. E., Current and potential future research activities in adaptive structures: An ARO perspective, *Smart Materials and Structures*, 10, 610–623, 2001.
2. Balageas, D., Introduction to structural health monitoring, ISTE, pp. 16–43, 2006.
3. Jian, C., Lei, Q., Shenfang, Y., Lihua, S., PeiPei, L., Dong, L., Structural health monitoring for composite materials, in Hu, N. (ed.), *Composites and Their Applications*, InTech, Rijeka, Croatia, 2012. ISBN: 978-953-51-0706-4.
4. Zhu, Y. K., Gui, Y. T., Rong, S. L., Hong, Z., A review of optical NDT technologies, *Sensors*, 11, 7773–7798, 2011.
5. Gros, X. E., Current and future trends in non-destructive testing of composite materials, *Annales de Chimie Science Matériaux*, 25, 539–544, 2000.
6. Błażejewski, W., Gąsior, P., Kaleta, J., Application of optical fibre sensors to measuring the mechanical properties of composite materials and structures, in *Advances in Composite Materials: Ecodesign and Analysis*, B. Attaf (Ed.), InTech, Rijeka, Croatia, 2011. ISBN: 978-953-307-150-3.
7. Méndez, A., Csipkes, A., Overview of fibre optic sensors for NDT applications, in *Nondestructive Testing of Materials and Structures*, Springer, the Netherlands, 2013, pp. 179–184.
8. Chung, D. D. L., *Composite Materials: Science and Applications*, 2nd edn., Springer Publications, London, U.K., 2010.
9. Kaw, A. K., *Mechanics of Composite Materials*, CRC Press, Boca Raton, FL, 2010.
10. Bunsell, A. R., Jacques, R., *Fundamentals of Fibre Reinforced Composite Materials*, CRC Press, Boca Raton, FL, 2010.
11. Eaton, M. J., Pullin, R., Holford, K. M., Acoustic emission source location in composite materials using delta T mapping, *Composites Part A*, 43(6), 856–863, 2012.
12. Elmar, W., Bernhard, J., Composites market report, The European GRP market (AVK), The global CRP market (CCeV), 2013. http://www.carbon-composites.eu/sites/carbon-composites.eu/files/anhaenge/13/09/17/ccev-avk-marktbericht_2013-final-englisch-bj.pdf.
13. Ramakrishnan, M., Rajan, G., Semenova, Y., Boczkowska, A., Domański, A., Wolinski, T., Farrell, G., Measurement of thermal elongation induced strain of a composite material using a polarization maintaining photonic crystal fibre sensor, *Sensors and Actuators A: Physical*, 190, 44–51, 2013.

14. Ramakrishnan, M., Rajan, G., Semenova, Y., Lesiak, P., Domanski, A., Wolinski, T., Boczkowska, A., Farrell, G., The influence of thermal expansion of a composite material on embedded polarimetric sensors, *Smart Materials and Structures*, 20, 125002, 2011.
15. Lesiak, P. et al., Influence of lamination process on optical fibre sensors embedded in composite material, *Measurement*, 45, 2275–2280, 2012.
16. Jensen, D. W., Pascual, J., Degradation of graphite/bismaleimide laminates with multiple embedded fibre-optic sensors, *Proceedings of SPIE*, 1370, 228–237, 1990.
17. Lee, D. C., Lee, J. J., Yun, S. J., The mechanical characteristics of smart composite structures with embedded optical fibre sensors, *Composite Structure*, 32, 39–50, 1995.
18. Güemes, J., Perez, J. S., *Fibre Optics Sensors in New Trends in Structural Health Monitoring*, Springer, Vienna, Austria, 2013, pp. 265–316.
19. Emmanwori, L., Shivakumar, K. N., Structural performance of composite laminates with embedded fibre optic sensor under tension and compression loads, *Proceedings of the 43rd Annual Conference of American Institute of Aeronautics and Astronautics*, Denver, CO, 2002.
20. Ciang, C. C., Jung, R. L., Hyung, J. B., Structural health monitoring for a wind turbine system: A review of damage detection methods, *Measurement Science and Technology*, 19, 122001, 2008.
21. Zhou, G., Sim, L. M., Damage detection and assessment in fibre-reinforced composite structures with embedded fibre optic sensors: Review, *Smart Materials and Structures*, 11, 925–939, 2002.
22. Du, W., Tao, X. M., Tam, H. Y., Choy, C. L., Fundamentals and applications of optical fibre Bragg grating sensors to textile structural composites, *Composite Structures*, 42, 217–229, 1998.
23. Rao, Y. J., Recent progress in fibre-optic extrinsic Fabry–Perot interferometric sensors, *Optical Fibre Technology*, 12, 227–237, 2006.
24. Murukeshan, V. M., Chan, P. Y., Ong, L. S., Asundi, A., On-line health monitoring of smart composite structures using fibre polarimetric sensor, *Smart Materials and Structures*, 8, 544–548, 1999.
25. Murayama, H., Kazuro, K., Hirosh, N., Akiyoshi, S., Kiyoshi, U., Application of fibre-optic distributed sensors to health monitoring for full-scale composite structures, *Journal of Intelligent Material Systems and Structures*, 14, 3–13, 2003.
26. Lopez-Higuera, J. M., Cobo, L. R., Incera, A. Q., Cobo, A., Fibre optic sensors in structural health monitoring, *Journal of Lightwave Technology*, 29(4), 587–608, 2011.
27. Gerard, F. F., Fibre optic sensor systems for monitoring composite structures, *RP Asia 2005 Conference*, Bangkok, Thailand, August 25–26, 2005.
28. Montanini, R., D'Acquisto, L., Simultaneous measurement of temperature and strain in glass fibre/epoxy composites by embedded fibre optic sensors: I. Cure monitoring, *Smart Materials and Structures*, 16, 1718, 2007.
29. Glisic, B., Inaudi, D., *Fibre Optic Methods for Structural Health Monitoring*, 1st edn., John Wiley & Sons, Ltd., Chichester, U.K., 2007.
30. Hill, K. O., Meltz, G., Fibre Bragg grating technology fundamentals and overview, *Journal of Lightwave Technology*, 15, 1263–1276, 1997.
31. Raman, K., *Fibre Bragg Gratings*, Academic Press, San Diego, CA, 1999.
32. Yashiro, S., Okabe, T., Toyama, N., Takeda, N., Monitoring damage in holed CFRP laminates using embedded chirped FBG sensors, *International Journal of Solids and Structures*, 44(2), 603–613, 2007.
33. Dong, X., Zhang, H., Liu, B., Miao, Y., Tilted fibre Bragg gratings: Principle and sensing applications, *Photonic Sensors*, 1(1), 6–30, 2011.
34. Zhao, Y., Yanbiao, L., Discrimination methods and demodulation techniques for fibre Bragg grating sensors, *Optics and Lasers in Engineering*, 41, 1–18, 2004.
35. Xu, M. G., Archambault, J. L., Reekie, L., Dakin, J. P., Discrimination between strain and temperature effects using dual-wavelength fibre grating sensors *Electronics Letters*, 30, 1085–1087, 1994.
36. James, S. W., Dockney, M. L., Tatam, R. P., Simultaneous independent temperature and strain measurement using in-fibre Bragg grating sensors, *Electronics Letters*, 32, 1133–1134, 1996.
37. Samer, K. B., Sun, T., Grattan, K. T. V., Simultaneous measurement of temperature and strain with long period grating pairs using low resolution detection, *Sensors and Actuators*, 144, 83–89, 2008.
38. Oliveira, R. D., Ramos, C. A., Marques, A. T., Health monitoring of composite structures by embedded FBG and interferometric Fabry-Perot sensors, *Computers and Structures*, 86, 340–346, 2008.
39. Guan, B. O., Tam, H. Y., Tao, X. M., and Dong, X. Y., Simultaneous strain and temperature measurement using a superstructure fibre Bragg grating, *IEEE Photonics Technology Letters*, 12, 675–677, 2000.
40. Luyckx, G., Eli, V., Nicolas, L., Joris, D., Strain measurements of composite laminates with embedded fibre Bragg gratings: Criticism and opportunities for research, *Sensors*, 11, 384–408, 2011.

41. Sonnenfeld, C., Sulejmani, S., Geernaert, T., Eve, S., Lammens, N., Luyckx, G., Eli, V., Thienpont, H., Microstructured optical fibre sensors embedded in a laminate composite for smart material applications, *Sensors*, 11, 2566–2579, 2011.
42. Reid, S. R., *Impact Behaviour of Fibre-Reinforced Composite Materials and Structures*, CRC Press, Boca Raton, FL, 2003.
43. Rajan, G., Ramakrishnan, M., Semenova, Y., Boczkowska, A., Domanski, A., Woliński, T., Farrell, G., Analysis of vibration measurement in composite materials using embedded PM-PCF polarimetric sensor and an FBG sensor, *IEEE Sensors Journal*, 12(5), 1365–1371, 2012.
44. Rajan, G., Ramakrishnan, M., Semenova, Y., Farrell, G., Domanski, A., Boczkowska. A., and Wolinski, T., Performance analysis of composite materials embedded with optical fibre sensors, *Proceedings of the IEEE Sensors*, 351–354, 2011.
45. Azmi, A. I., Peng, G.-D., Raju, Failure monitoring of E-glass/vinylester composites using fibre grating acoustic sensor, *Photonic Sensors*, 3(2), 184–192, 2013.
46. Luyckx, G. et al., Response of FBGs in microstructured and bow tie fibres embedded in laminated composite, *IEEE Photonics Technology Letters*, 21, 1290–1292, 2009.
47. Peng, G.-D., Xiong, Z., Chu, P. L., Photosensitivity and gratings in dye-doped polymer optical fibres, *Optical Fibre Technology*, 5, 242–251, 1999.
48. Xiong, Z., Peng, G.-D., Wu, B., Chu, P. L., Highly tunable Bragg gratings in single-mode polymer optical fibres, *IEEE Photonics Technology Letters*, 11, 352–354, 1999.
49. Kuang, K. S. C., Quek, S. T., Koh, C. G., Cantwell, W. J., Scully, P. J., Plastic optical fibre sensors for structural health monitoring: A review of recent progress, *Journal of Sensors*, 2009, 312053, 2009.
50. Webb, D. J., Polymer optical fibre Bragg gratings, in Bragg gratings, photosensitivity and poling in glass waveguides, OSA Technical Digest, Paper Btu3E.4.4, 2012.
51. Kalli, K., Dobb, H. L., Webb, D. J., Carroll, K., Themistos, C., Komodromos, M., Peng, G.-D., Fang, Q., Boyd, I. W., Development of an electrically tuneable Bragg grating filter in polymer optical fibre operating at 1.55 μm, *Measurement Science and Technology*, 17, 3155–3164, 2007.
52. Zhang, W., Webb, D. J., Peng, G.-D., Investigation in to time response of polymer fibre Bragg grating based humidity sensors, *Journal of Lightwave Technology*, 30, 1090–1096, 2012.
53. Rajan, G., Ramakrishnan, M., Semeonva, Y., Ambikairajah, E., Farrell, G., Peng, G.-D., Experimental study and analysis of a polymer fibre Bragg grating embedded in a composite material, *Journal of Lightwave Technology*, 32(9), 1726–1733, 2014.
54. Lee, B. H., Kim, Y. H., Park, K. S., Eom, J. B., Kim, M. J., Rho, B. S., Choi, H. Y., Interferometric fibre optic sensors, *Sensors*, 12, 2467–2486, 2012.
55. Liu, T., Wu, M., Rao, Y., Jackson, D. A., Fernando G. F., A multiplexed optical fibre-based extrinsic Fabry-Perot sensor system for in-situ strain monitoring in composites, *Smart Materials and Structures*, 7, 550, 1998.
56. Leng, J. S., Asundi, A., Structural health monitoring of smart composite materials by using EFPI and FBG sensors, *Sensors and Actuators A: Physical*, 103, 330–340, 2003.
57. Fernando, G. F., Liu, T., Crosby, P., Doyle, C. A., Martin, D. B., Ralph, B., Badcock, R., A multi-purpose optical fibre sensor design for fibre reinforced composite materials *Measurement Science and Technology*, 8, 1065, 1997.
58. Rajan, G., Ramakrishnan, M., Lesiak, P., Semenova, Y., Wolinski, T., Boczkowska, A., Farrell, G., Composite materials with embedded photonic crystal fibre interferometric sensors, *Sensors and Actuators A: Physical*, 182, 57–67, 2012.
59. Wolinski, T. R., Polarimetric optical fibres and sensors, *Progress in Optics*, 40, 1–75, 2000.
60. Domanski, A. W., Tomasz, R. W., Wojtek, J. B., Polarimetric fibre optic sensors: State of the art and future, *International Conference on Interferometry-94, International Society for Optics and Photonics*, pp. 21–28, Poland, 1994.
61. Wolinski, T. R., Lesiak, P., Domanski, A. W., Polarimetric optical fibre sensors of a new generation for industrial applications, *Bulletin of the Polish Academy of Sciences, Technical Sciences*, 56, 125–132, 2008.
62. Ma, Ji., Asundi, A., Structural health monitoring using a fibre optic polarimetric sensor and a fibre optic curvature sensor-static and dynamic test, *Smart Materials and Structures*, 10, 181–188, 2001.
63. Hogg, W. D., Roderick, D. T., Polarimetric fibre optic structural strain sensor characterization, *Proceedings of SPIE*, 1170, 542–550, 1990.
64. Domański, A. W., Lesiak, P., Karolina, M., Boczkowska, A., Budaszewski, D., Ertman, S., Woliński, T. R., Temperature-insensitive fibre optic deformation sensor embedded in composite material, *Photonics Letters of Poland*, 1, 121–123, 2009.
65. Adachi, S., Distributed optical fibre sensors and their applications, *IEEE SICE Annual Conference*, 2008.

66. Niklès, M., Fibre optic distributed scattering sensing system: Perspectives and challenges for high performance applications, *Third European Workshop on Optical Fibre Sensor*, International Society for Optics and Photonics, Napoli, Italy, 2007.
67. Grattan, K. T. V., Sun, T., Fibre optic sensor technology: An overview, *Sensors and Actuators A: Physical*, 82, 40–61, 2000.
68. Kersey, A. D., A review of recent developments in fibre optic sensor technology, *Optical Fibre Technology*, 2, 291–317, 1996.
69. Bao, X., Liang, C., Recent progress in Brillouin scattering based fibre sensors, *Sensors*, 11, 4152–4187, 2011.
70. Hartog, A. H., Progress in distributed fibre optic temperature sensing, *Proceedings of the Environmental and Industrial Sensing*, International Society for Optics and Photonics, Chicago, IL, 2002.
71. Culshaw B., Optical fibre sensor technologies: Opportunities and-perhaps-pitfalls, *Journal of Lightwave Technology*, 22, 39–50, 2004.
72. Ravet, F., Distributed Brillouin sensor application to structural failure detection, in *New Developments in Sensing Technology for Structural Health Monitoring*, Springer, Berlin, Germany, 2011, pp. 93–136.
73. Zeng, X., Xiaoyi, B., Chia, Y. C., Theodore, W. B., Anthony, W. B., Michael, D. D., Graham, F., Alexander, L. K., Anastasis, V. G., Strain measurement in a concrete beam by use of the Brillouin-scattering-based distributed fibre sensor with single-mode fibres embedded in glass fibre reinforced polymer rods and bonded to steel reinforcing bars, *Applied Optics*, 41, 5105–5114, 2002.
74. Patrick, H. J., Williams, G. M., Kersey, A. D., Pedrazzani, J. R., Hybrid fibre Bragg grating/long period grating sensor for strain/temperature discrimination, *IEEE Photonics Technology Letter*, 8, 1223–1225, 1996.
75. Dong, B., Hao, J., Liaw, C., Lin, B., Tjin, S. C., Simultaneous strain and temperature measurement using a compact photonic crystal fibre inter-modal interferometer and a fibre Bragg grating, *Applied Optics*, 49, 6232–6235, 2010.
76. Andre, R. M., Marques, M. B., Roy, P., Frazao, O., Fibre loop mirror using a small core micro-structured fibre for strain and temperature discrimination, *IEEE Photonics Technology Letter*, 22, 1120–1122, 2010.
77. Frazão, O., Carvalho, J. P., Ferreira, L. A., Araújo, F. M., Santos, J. L., Discrimination of strain and temperature using Bragg gratings in microstructured and standard optical fibres, *Measurement Science and Technology*, 16, 2109, 2005.
78. Rajan, G., Ramakrishnan, M., Semenova, Y., Karolina, M., Lesiak, P., Domanski, A. W., Wolinski, T. R., Farrell, G., A photonic crystal fibre and fibre Bragg grating-based hybrid fibre-optic sensor system, *IEEE Sensors Journal*, 12(5), 39–43, 2012.
79. Sokolinsky, V. S., Shen, H., Vaikhanski, L., Nutt, S. R., Experimental and analytical study of nonlinear bending response of sandwich beams, *Composite Structures*, 60, 219–2295, 2003.
80. Donaldson, S. L., Miracle, D. B., *ASM Handbook Composites*, Vol. 21, ASM International, Materials Park, OH, 2001.
81. Bakis, C. E. et al., Fibre-reinforced polymer composites for construction-state-of-the-art review, *Journal of Composites for Construction*, 6(2), 73–87, 2002.
82. Rajan, G., Darmanin, E., Peng, G.-D., Prusty, G., Carbon fibre-foam sandwich composite laminate embedded with fibre Bragg grating sensors, *23rd OFS Conference*, Santander, Spain, 2014.
83. Hackney, D., Peters, K., Damage identification after impact in sandwich composites through embedded fibre Bragg sensors, *Journal of Intelligent Material Systems and Structures*, 22, 1305–1316, 2011.
84. Kang, H. K., Park, J. W., Ryu, C. Y., Hong, C. S., Kim, C. G., Development of fibre optic ingress/egress methods for smart composite structures, *Smart Materials and Structures*, 9, 149–156, 2000.
85. Green, A. K., Zaidman, M., Shafir, E., Tur, M., Gali, S., Infrastructure development for incorporating fibre-optic sensors in composite materials, *Smart Materials and Structures*, 9, 316–321, 2000.
86. Friebele E. J. et al., Optical fibre sensors for spacecraft applications, *Smart Materials and Structures*, 8, 813–838, 1999.
87. Chojetzki, C., Rothhardt, M., Ommer, J., Unger, S., Schuster, K., Mueller, H. R., High-reflectivity draw-tower fibre Bragg gratings—Arrays and single gratings of type II, *Optical Engineering*, 44, 060503, 2005.
88. Palmer, D., Engelbart, R., Vaccaro, C., Future directions relative to NDE of composite structures, SAE Technical Paper 2004-01-2817, 2004.
89. Van, H., Bram, G. L., Erwin, B., Jeroen, M., Sandeep, K., Oliver, M., David, J. W., Kate, S., Peter, V. D., Geert, V. S., Ultra small integrated optical fibre sensing system, *Sensors*, 12, 12052, 2012.
90. Missinne, J., Geert, V. S., Bram, V. H., Kristof, V. C., Tim, V. G., Peter, D., Jan, V., Peter, V. D., An array waveguide sensor for artificial optical skins, In *SPIE OPTO: Integrated Optoelectronic Devices*, International Society for Optics and Photonics, San Francisco, CA, pp. 722105–722105, 2009.

18 Future Perspectives for Fiber-Optic Sensing

Brian Culshaw

CONTENTS

18.1 INTRODUCTION: FIBER SENSING AND PHOTONICS

Half a century ago in 1967, the first fiber sensor paper appeared.[1] This described the Fotonic sensor, which remains available as a surface finish diagnostic tool capable of subnanometer resolution. The basic idea is shown in Figure 18.1—it is simply an illuminated reflector, which may or may not be specular or diffuse together with a cunningly arranged ray of collection fibers organized to provide self-referencing capability. This basic device still features in the occasional study of more exotic variations of the general theme.[2,3]

This same half century has seen optics as a generic discipline travel from essentially the domain of the spectroscopist and photographer into an increasingly important aspect of engineering. Indeed, many argue that optics is most definitely *the electronics of the twenty-first century*. The most frequent manifestation of this progress sits, of course, in the everyday Internet, which requires the extremely high bandwidth capabilities of optical fiber transmission. We also have the ubiquitous digital photograph and a host of everyday artifacts from thermally activated lamps to your CD and DVD players, all of which rely intrinsically on the principles of optic, or, as it increasingly becomes defined, photonics.[4]

A key parallel enabler in all this has been the relentless increase in the capacity of digital processing equipment in its many manifestations. Not only has the simple number crunching contributed but also the ever-increasing capability of precision lithographic processing. When the first fiber sensor papers were published, lithographic definition for integrated circuits relied on Rubylith,

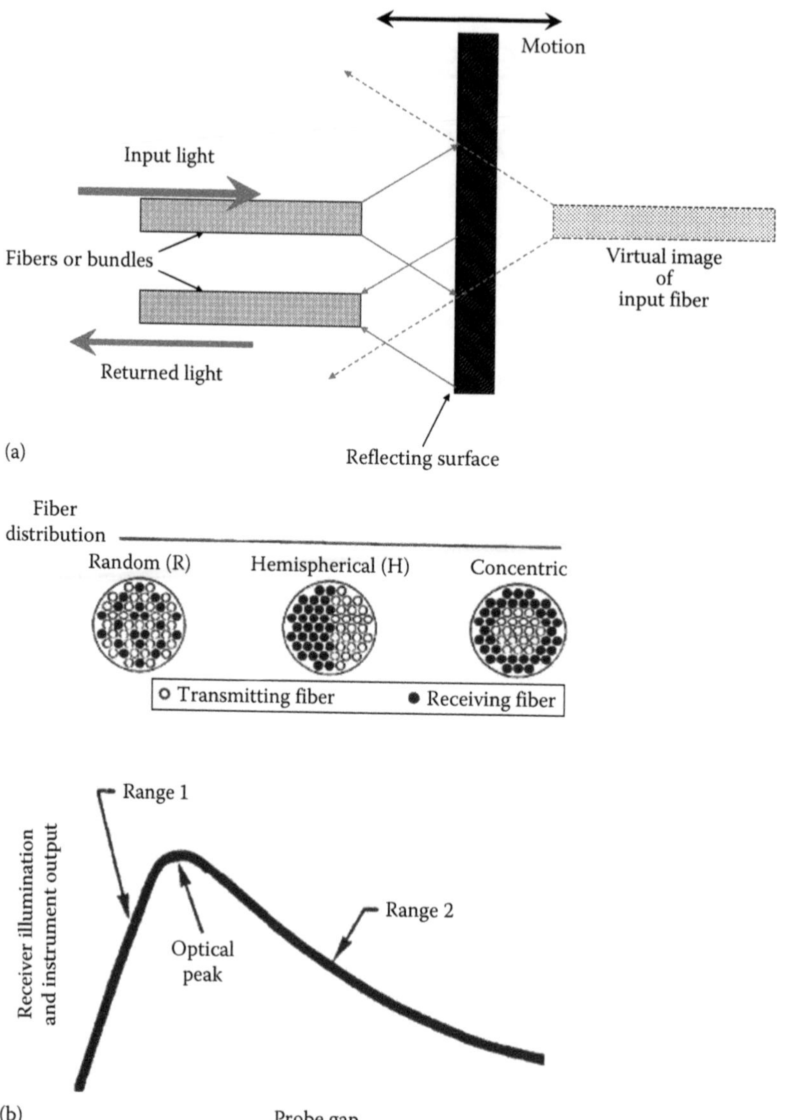

FIGURE 18.1 The essential elements of the first fiber sensor. (a) The basic principle illustrating how the collected light may vary with distance and (b) some embodiments of the fiber transmit and receive geometries as outlined in Ref. [1].

the precision of which was ultimately determined by the sharpness of a knife. The same operation now relies on an electron beam milling and similar tools. This in turn has facilitated what is, in effect, the ever-increasingly precise printed circuit board. Half a century ago, *microstrip* began to push the printed circuit into the gigahertz region. Now *nanophotonics* pushes printed circuits into the optical domain.

The result is an ever-increasing toolbox from which the photonic engineer can evolve novel concepts and innovative devices. In parallel, the economic and cultural environment in which these devices may become useful becomes more accepting toward such innovation and gradually recognizes that these exotic processes are becoming everyday accessories.

In the tiny corner of photonics that is optical fiber sensing, this points toward enormous opportunities for the future evolution of the discipline and more importantly the application of the discipline as an effective tool contributing to society. This process will embrace three generic themes. The first is principally innovative research to utilize the new tools in photonics and explore their prospects as enablers for measuring the world around us. The second may in many ways be regarded as a consolidation through which currently available and currently evolving fiber-optic sensing tools adapt themselves into present society. The third lies in examining the future challenges that our burgeoning population is facing and ascertaining how the new photonics, the new fiber-optic sensors, and indeed the currently maturing technologies may most effectively contribute to meeting the needs imposed by our anticipated 10 billion population just a few short decades from now. These themes form the framework for the remainder of this chapter, but beforehand it is useful to briefly reflect on the successes and frustrations in the evolution of fiber sensing technologies to the present time.

18.2 BRIEFEST HISTORY

The early days of fiber sensing bristled with almost unbridled (as it happens totally misguided) optimism since fiber-optic communications offered to totally revolutionize communication networks—which they did—then the same should be true for sensor technologies. The researcher, who is also the inevitable sources of such speculation, is quite appropriately somewhat disconnected from the real world and is almost without exception unfamiliar with the actualities of sensing and measurement.

Those who undertake measurements do so almost invariably because they have to. Measuring is a means to an end whether it involves guiding a space probe to Mars or monitoring the fuel level in an automobile. Measurements must also be reliable and correct to the extent that in most cases the user is totally unaware that measurements are even being made. The outcome is that those who install measuring equipment much prefer to install something that someone else has already characterized and thoroughly explored. Furthermore, they are totally oblivious to the technology involved to making the measurement. These sensor technologies—fiber optic or otherwise—must operate within this culture of cautious conservative reluctance to adopt anything new.

The early demonstrations of fiber-optic sensing and measurement simply duplicated measurements available already using other techniques and often at a cost that far exceeded those achievable using current technologies.

It took some time before new measuring modalities became apparent, most notably distributing sensing. Consequently, it took time for the sectors in which fiber optics could become disruptive technology to consolidate. Currently, the anticipated world market for fiber sensor techniques sits around the $1 billion per annum level addressed predominantly through a large number of small and even very small specialist companies.[5] The research continues with a backdrop of optimism toward significant expansion in the coming decade, and the word is slowly propagating among the user community that the technology does work, can be installed, can be relied upon, and can, indeed, be a very attractive proposition.

18.3 *NEW* PHOTONICS: WHAT IT INVOLVES

18.3.1 The Basics

The scientific concepts underlying photonics[4] embrace, at a very simple level, the understanding of the behavior of light in materials and structures and the application of this understanding. This has numerous facets to it of which Figure 18.2 endeavors to present the most important features. This figure also highlights the *new* functions that have become available, thanks to the *new* optics facilitated through a combined evolution of our understanding of the principles and

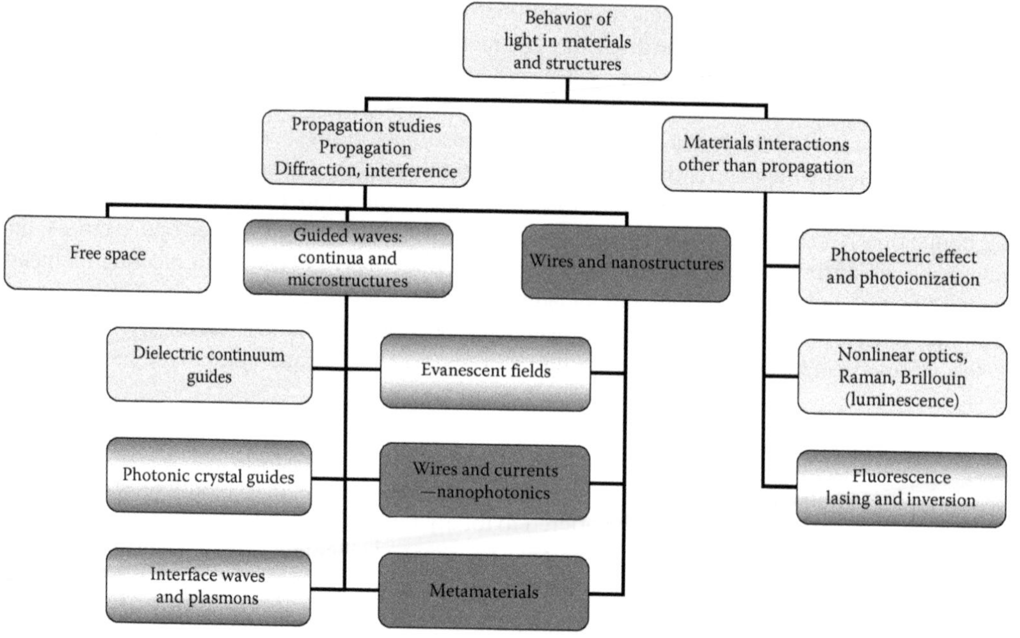

FIGURE 18.2 Optoelectronics—the study light in materials and structures and the exploitation thereof. The dark shaded areas are *new* to nanophotonics, while the combined graded areas represent concepts evolved for guided wave optics but which are applicable to nanophotonic structures, sometimes with appropriate modifications. The light shaded concepts are well established *classical* optics.

the availability of the tools, principally precision lithography, through which these concepts may be addressed.

The concepts that have consolidated from this combination of enhanced knowledge of the physics and the processes and the enhanced capability of fabrication tools include the following:

- Slow (and fast) light, which is in effect the subtle and elegant understanding and exploitation of dispersion curves both within materials themselves and within structurally controlled artifacts superimposed upon the base material.
- Squeezed light, a technique that in principle can significantly suppress the inherent noise levels within an optical system. Initially, this phenomenon was implemented through effects in bulk materials, but more recently, structural artifacts have also been brought to bear offering the tantalizing prospect of integrated squeezed light sources.
- Nanophotonics is a far-reaching enabling concept, which, in effect, translates the ideas of inductors, capacitors, and resistors up to optical frequencies. It is, of course, well known that even at low frequencies, inductors and capacitors exhibit far from ideal behavior and that these deviations become more apparent as the frequency increases. While the LCR simplification is certainly a useful aid, it is far from the whole story. Metamaterials are, in effect, based upon this generic process and are one very important subset of nanophotonics. The quantum dot also sits within this domain. Just as electrical circuits and metallic cavities significantly modify the radio-frequency properties of the materials from which they are made similarly, the quantum dot has profound influence on the optical properties of the constituent materials. Nanophotonics also embraces the capacity to fabricate tiny, tiny structures as chemical reactors, as sample manipulation tools—enabling the *lab in a fiber*.[6]

- Combining structures and materials with the facility for the electrical tuning thereof opens yet another dimension of flexibility for new optical systems. While computer-generated holograms and spatial light modulators emerged half a century or so ago, their evolution into high-performance and surprisingly low-cost tools like projection TV promises potential in the domain of, for example, precision sample handling compatible with biological cells and adaptive electronically controllable switched sources and detection modalities. Few, if any, have yet found their way into the optical fiber sensor portfolio.

In the coming sections, we shall speculate on how just a few of these techniques might introduce interesting modalities to fiber sensing.

18.3.2 Slow Light in Fiber Sensing

The basic concept that drives interest in slow light as a potential tool for enhancing the performance of fiber-optic sensors is simply expressed as if light goes slow, then it will stay in the medium of interest for a longer period of time and therefore interact more strongly.

Slow light exploits the characteristics of a dispersion curve in a material or structure around resonance (Figure 18.3).[7] In very simple terms, a resonator like a Fabry–Perot or a ring resonator may be construed as a slow light structure since a given photon at the resonant frequency stays in the structure for something of the order of the Q factor or the finesse times as long as it would if the resonator were absent (Figure 18.4). In the sensing context, the measurand reacts with the light via the geometry of the structure, so the longer the light stays within the structure, the greater the interaction. Hence, slow light produces sensitivity gains in resonators. Indeed, it could be argued that in this regard, slow light is a slightly more exotic term for the well-known resonant enhancement.

The effects of resonances within the transmission material, typically induced through, for example, Brillouin-based scatter processes (Figure 18.5), itself in the slow light context result in exactly the same effective behavior—a given photon stays within the transmission material for a longer period of time than it does off resonance. However, in this case, it is intuitively obvious that if the measurement interaction reaction is through processes outside the material, then trapping the light within the material will have no net effect on the measurement process. In contrast, if, for example, the measurement process involves the transmission through the sample, as in the case of gas or liquid spectroscopy, perhaps then the resonance is being used regardless so that once more there is no perceptible difference.

This train of thought appears to conclude that in the sensing context, slow light is something of a distraction[8] though of course it may be useful in packaging research proposals. In other

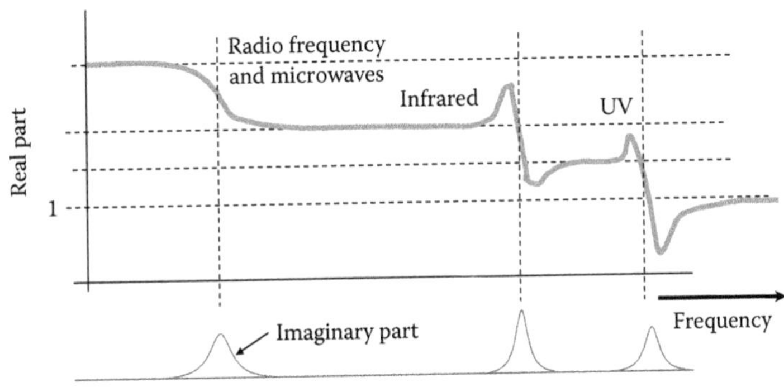

FIGURE 18.3 A very schematic representation of the behavior of dielectric constant versus frequency with polar, rotational, molecular stretch, electronic, and in the X-ray region, nuclear contributions. The number of terms in the Clausius–Mossotti equation contributes falls with frequency until eventually we observe dielectric constant <1 at very high frequencies (and sometimes just close to resonances as observed in some forms of *slow* light).

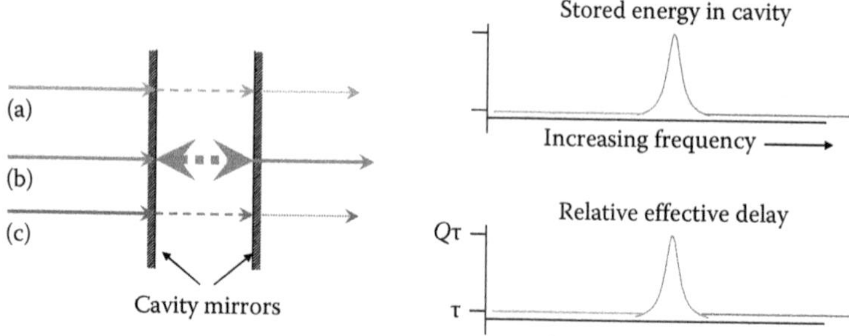

FIGURE 18.4 Slow light in a passive optical cavity. This simplified representation indicates the slowing of light at resonance (b) due to energy storage in the cavity. At the points above (a) and below (c) this resonance, the light passes through the cavity in essentially a single pass. At resonance, the light passes through the cavity of the order of *Q* times, where *Q* is the quality factor, or finesse. From a sensing perspective, the light distribution around the cavity is unaffected by the resonance, so that there is on the order *Q* times more interaction with the outside world at resonance.

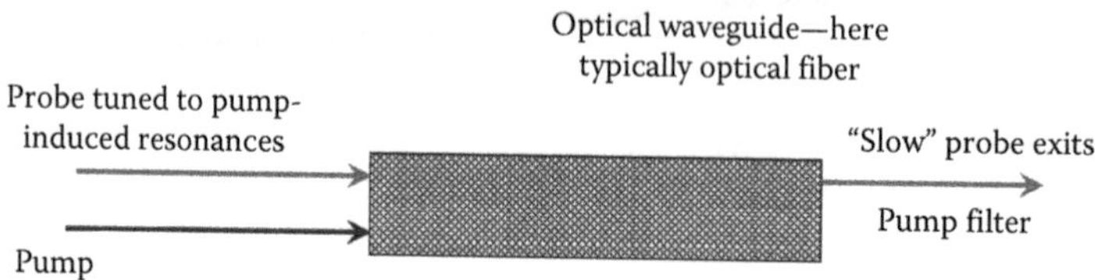

FIGURE 18.5 Slow light induced by pump-induced nonlinear material processes typically Brillouin scatter rather than structural interactions. Here, the light is absorbed and delayed within the material itself and so cannot interact with an external measurand—so material-induced slow light interactions produce negligible changes in sensor sensitivity.

contexts—data storage is an example—controllable delays within the transmission medium may be extremely useful, and it is in this direction that researching slow light in guided systems appears to offer the most promise.[9] Furthermore, it is within the very nature of research that interpretations like the one made here in the sensing context can be questioned and often reversed.

18.3.3 Nanophotonics

Nanophotonics as a generic topic has gained considerable momentum over the last couple of decades.[10,11] While it is surrounded by mystique, it is really simply the study of the properties of materials and the way in which these materials interact with light when the material dimensions slip below the wavelength. Indeed, the analogy of electrical circuits at optical frequencies, while having its conceptual limitations, appears to be a viable starting point for the nonspecialist. With appropriate care, many of the concepts of circuits and circuit components can be extrapolated into the optical domain. Figure 18.6 is an example of such an extrapolation. Here, we consider the hypothetical case of a gold wire 1 wavelength in length and 1/10th of a wavelength in diameter. The calculation shows the resistance of the wire accounting for skin depths as a function of propagating wavelength.

The reason for including this very simple, rough, and ready calculation is to emphasize that any thought of using metallic conductors of any significant length at optical frequencies is clearly entirely impractical due to the ohmic losses in even very small (wavelengths) lengths of conductor. Additionally, the graph also indicates the inductive impedance of such a length of wire, again estimated using conventional circuit-based approaches. The inductive component clearly dominates (as indeed is often the case in the UHF and microwave domains). So while the circuit analogy is perhaps useful as a conceptual tool, the normal ideas of component geometries need to be carefully reconsidered.[12]

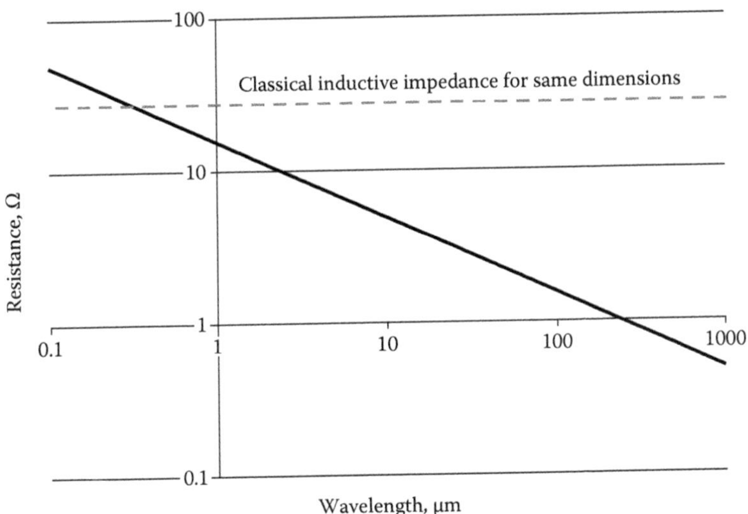

FIGURE 18.6 An illustrative indication of the behavior of *conventional* conductors at photonic frequencies. This shows the resistance of an imaginary wire λ/10 in diameter and λ in length indicating that very large ohmic losses are inevitable in all but the smallest of nanophotonic systems.

There are, however, some major features in common across the frequency spectrum that can also be extrapolated with care into the nanophotonic domain. Nanophotonics is really about considering systems that manipulate and detect an optical signal as a voltage or current rather than as an integrated wave front—an operation that has been assigned the awe-inspiring term of superresolution. However, electronic engineers use superresolution measurements on a daily basis every time an oscilloscope or voltmeter is interfaced to a circuit. This will give a superresolution image of a component, for example, at the frequency at which it is operated. In the nanophotonic context, one obvious example of such a tool is shown in Figure 18.7. This almost exactly replicates the microwave

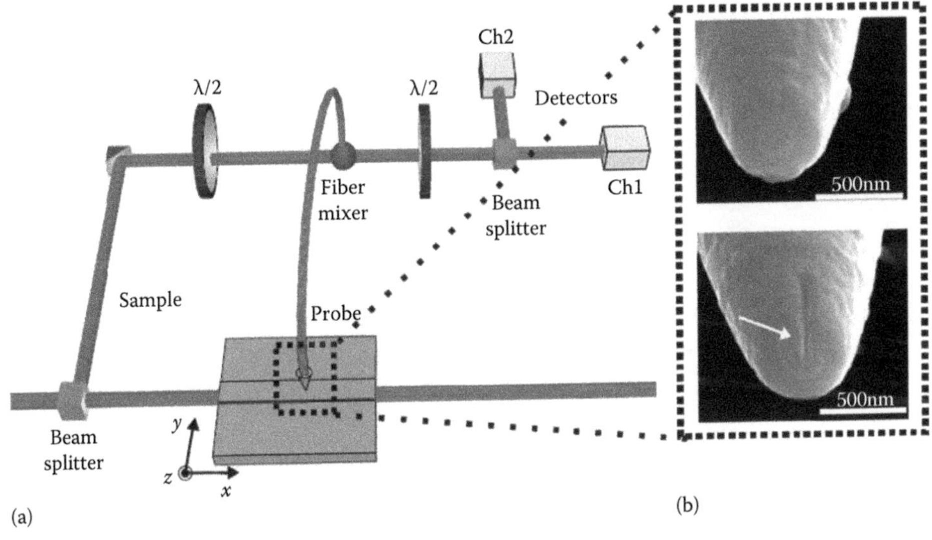

FIGURE 18.7 The nanophotonic *slotted line* using a homodyne-based probe to indicate the relative phase and amplitude of the optical field seen by the nanoprobe (inset): (a) is the schematic of the necessary equipment and (b) is a scanning electron microscope picture of the nanoprobe used. (From Buressi, M. et al., *Science*, 326, 550, 2009.)

engineer's slotted line[13] through which voltage waveforms as a function of position can be accurately determined. In the sensor context, there are probably three subsets of nanophotonics, namely, quantum dots, metamaterials, and superresolution imaging, which may find a place and these we shall very briefly explore in turn.

The tantalizing prospect of the metamaterial was first proposed in the mid-1990s promising the *invisibility cloak* that if placed around a solid object would, in effect, render it transparent.[14] The first demonstrations were targeted at proof of principle at microwave frequencies where the necessary resonance circuit elements can be readily fabricated (Figure 18.8).

It is perhaps the concept of the synthetic material with a refractive index of −1, which offers the most intrigue of the many features of so-called metamaterials.[15] This facilitates an imaging system, which no longer suffers from the diffraction limit—assuming of course that the pixel-type detector in the image plane has the necessary resolution. This might then enable *visual* images of nanoscale particles and in turn the somewhat fanciful nanoscale chemical analysis system shown in Figure 18.9 and based upon optical fibers might—who knows—be viable. The system as idealistically drawn does, of course, highlight some issues. While the illumination from a single-mode fiber could, in principle, excite a superresolution image (indeed it always does), the collection optics needs critical thought since identifiable paths are needed for each pixel and these cannot be wires simply because the losses are too high, as we have discussed, and dielectric guides that will have the necessary loss characteristics require a second metamaterial image collection system. Point-by-point scanning would, of course, achieve the

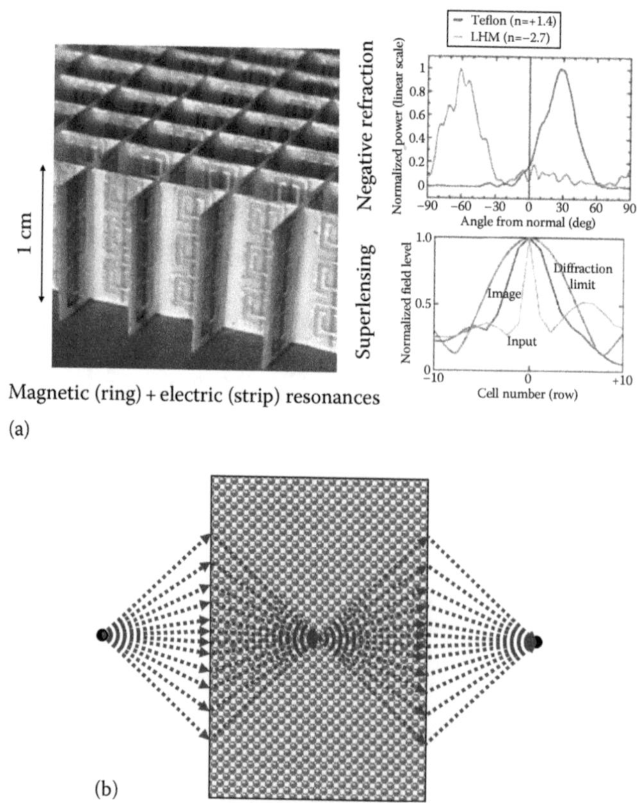

FIGURE 18.8 (a) The original metamaterial *superlensing* demonstration at microwave frequencies illustrating the use of macroscale *circuit elements* to approach the negative index conditions required. More recently, nanoscale circuit elements have approached the similar function at optical frequencies (From Smith, D.R. et al., *Science*, 305, 788, 2004.) and (b) a conceptual diagram of the operation of the perfect superlens with $n = -1$.

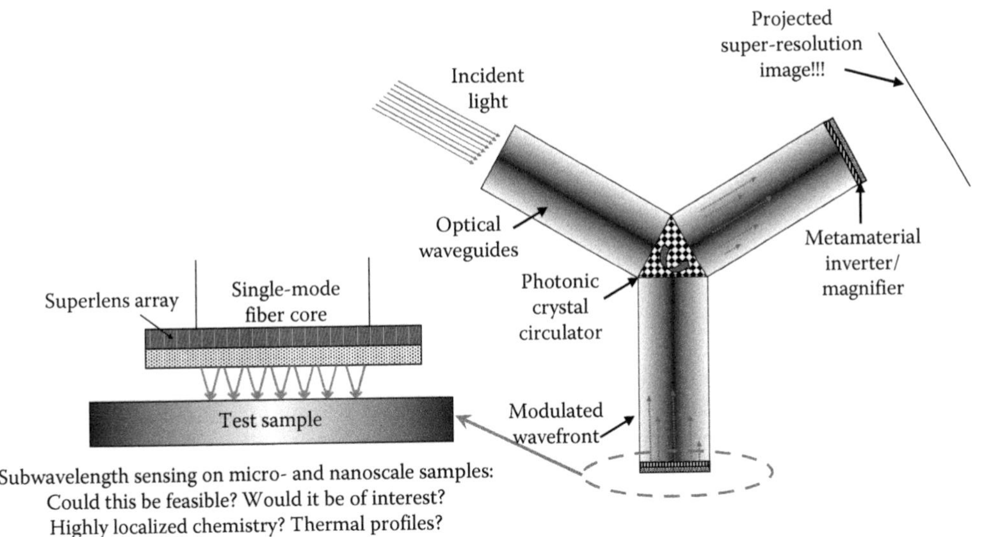

FIGURE 18.9 A possibly fanciful excursion into the possible implications of metamaterial-based super-resolution imaging for fiber sensors.

functionality and could, in principle, be realized. Indeed, the scanning in the near-field optical microscope is, in effect, almost there already. Whole-field imaging clearly has enormous benefits.

Quantum dots—in the form of gold and nanoparticles—have been used in colloidal solutions since ancient times as pigments especially in stained glass.[16] Depending on the size of the colloidal gold, the particles appear in different colors, ranging from the reds to the blues. The particles, typically of the order of 100 nm in diameter, are now small enough for their structural dimensions to modify the energy levels within the material and thereby alter its absorption spectrum and change its color with the larger ones absorbing more of the red and therefore looking blue. In more recent times, their application has ventured from the glass shop and into biomedicine where the particles have been used as markers and enhancers for processes such as surface-enhanced Raman spectroscopy (SERS). Indeed, some have found such application in the context of fiber optic–based sensor systems.[17,18] There is also yet another fuzzy boundary between the quantum dot, metallic particle, and the area, which has evolved under the title of plasmonics.

Quantum dots appear in other guises too, most frequently based upon semiconductors though again the modification of internal energy levels under the influence of dimensional resonances is the basic operating mode. These nanodevices can appear in the guise of modified fluorophores, optical sources, and detectors particularly in the context of quantum cryptography.[19,20]

Plasmonics forms another subset of nanophotonics. Perhaps one could argue that the original surface plasmon devices (Figure 18.10) were the first examples of nanophotonics albeit only in one dimension.[21,22] Certainly, the impact of the thickness of that metallic layer was recognized at a very early stage. Surface plasmon–based devices, effectively waveguide versions of the traditional device architecture in Figure 18.10, have been investigated for well over 20 years configured both as polarizers[23] and as chemical (strictly refractive index)[24] sensors.

Plasmonics reduces to a 2D, sometimes 3D, version of this original configuration.

So what does plasmonics offer to the future of fiber sensing? There could be a spatially multiplexed multiple-analyte detection array, all of which could sit on the end of a single-mode optical fiber. This would return a spectrum dependent upon the analytical composition of the picoliter or less sample localized on the core. Furthermore, the adaptability of the basic platform to address straightforward absorption spectroscopy, fluorescence, Raman, and more so SERS offers potential

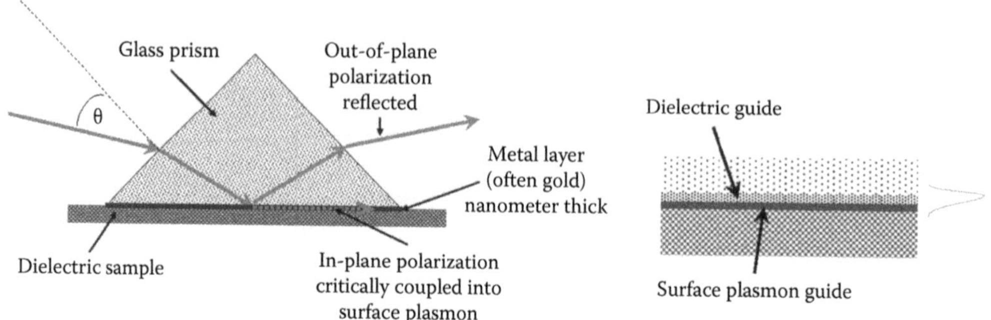

FIGURE 18.10 Surface plasmons in very thin (subskin depth) metal films. The Kretschmann configuration (a) for chemical (i.e., refractive index) sensing by detecting the critical angle as illustrated and (b) a waveguide equivalent that couples from the upper all-dielectric guide into the surface plasmon guide comprising a thin metal field and a dielectric substrate when the propagation constants in the two coincide. A phase-matching structure can be placed between the two to enhance the overall tunability.

flexibility albeit at the price of, to current standards, processing and alignment complexities. Figure 18.11 illustrates some conceptual thought outlines. The *lab in a fiber* is beginning to emerge as a tangible, workable, and achievable concept—a means through which innovative biochemical measurements may be realized and exploited.

The whole area of nanophotonics and its many side shoots will coalesce into an array of engineering tools, which will become significantly utilized over the coming decades. We are beginning to take the first tentative steps in addressing its potential in a practical domain. There are many authoritative reviews and texts available on the subject at the basic level,[23,24] but there remains immense scope for the imagination to link the potential that the tool offers in manipulating light into a huge diversity of real-world applications, of which fiber-based sensing is just one.[25,26]

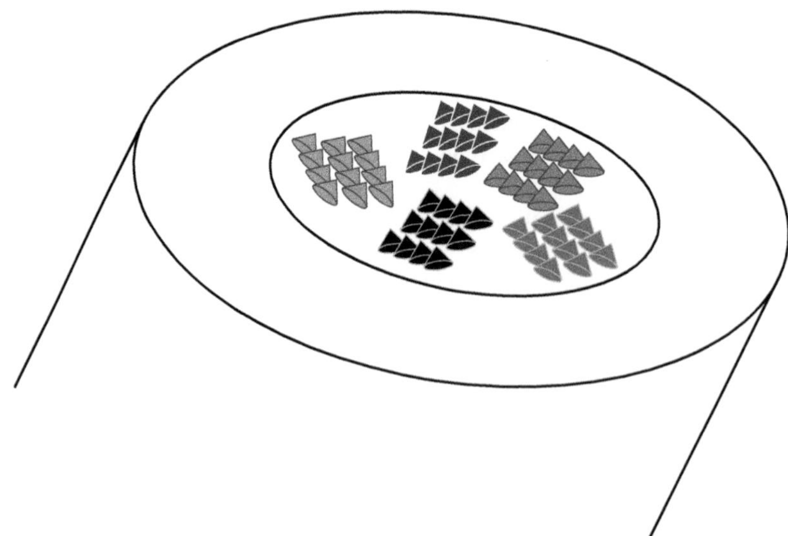

FIGURE 18.11 A *concept diagram* of a possible surface nanophotonics–based multiparameter sensor system fabricated on the end of a fiber. This hypothetical array of nanophotonic *optical wire* cones could be fabricated to be sensitive to either physical and/or (bio)chemical phenomena. There will be interrogation issues, signal-to-noise ratio considerations, and host of other practicalities to consider, but when nanophotonics becomes *optical wiring*, much will be achieved.

18.3.4 Squeezing Light

Optical detection processes are ultimately limited through the shot noise introduced by the random rate of arrival of photons in the optical detection system. The noise manifests itself within the bandwidth of some form or other signal, which is modulated onto the optical carrier, and this modulation process—as has been mentioned many times in this book—can affect either the phase of the optical carrier or its amplitude. The shot noise is down converted onto the modulated signal via the detection process in the receiver. To a first approximation, the receiver is usually configured to endeavor to ignore intensity noise for phase-modulated carrier systems and vice versa for intensity-modulated systems. There are certainly many examples of fiber-optic sensors that operate very close to the shot noise limit including fiber gyroscope and many of the interferometric and polarimetric sensor architectures.

Classically, the shot noise of an optical source manifests itself after detection equally in phase and amplitude components (Figure 18.12). The squeezing process can implement, in principle, significant reduction in either the phase[27,28] or the amplitude noise and the expense of an equivalent increase in the complementary noise component, also as indicated in Figure 18.12.

The processing of squeezing light is inherently quite complex and described in far more detail elsewhere.[29] It is, however, interesting to make some observations on the potential for the process in the context of fiber-optic sensing. Inevitably, if the intensity noise is squeezed, then the phase noise is correspondingly magnified since any optical detection system—at least of those observed and used to date—responds to intensity. It appears to follow that any possible gains going through the somewhat elaborate process of squeezing the light can only accrue in systems operated such that the impact of phase noise on the optical carrier can be neglected. Phase-modulated systems like the fiber gyro that operate on the zero fringe do, in fact, come into this category. Furthermore, since the demonstrated processes for successfully squeezing light have predominantly relied upon large bulk optic configurations, then possibly only bulk optic sensing systems will be candidates for such treatment.

There is indeed an excellent example of such a system. The LIGO gravitational telescope operates using a squeezed light source, which in turn gives a few (up to possibly 10 dB) improvement in signal-to-noise ratio.[30] The system (Figure 18.13) is clearly very complex and apparently well beyond the scope of those interested in fiber-optic sensing. Little has, as yet, appeared in this context, and

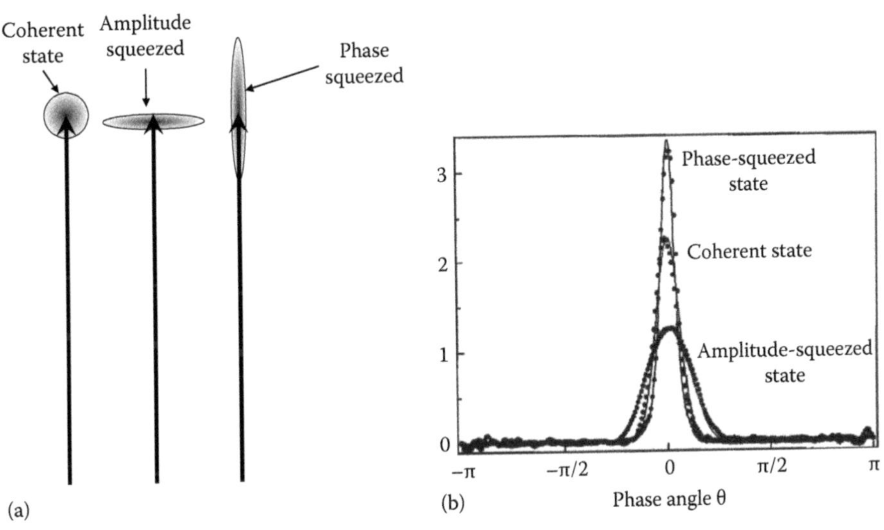

FIGURE 18.12 The basic concepts of squeezed light—source noise is the basic limit, so trading phase noise versus amplitude noise may reduce the detected noise level and therefore enhance sensor system sensitivity. (a) A phasor representation with the noise probability distribution function represented at the tip of the phasor and (b) a spectral representation of the process of squeezing. (From Breitenbach, G. et al., *Nature*, 387, 471, 1997.)

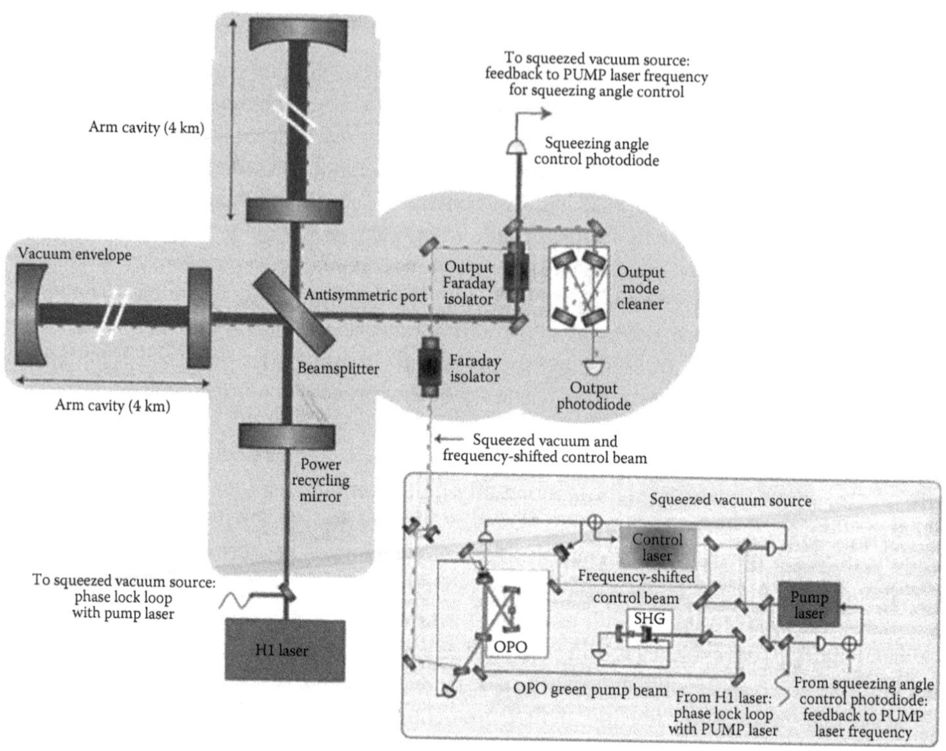

FIGURE 18.13 Squeezed light in action—the LIGO H1 interferometer with the squeezed light source at bottom right. (From The LIGO Scientific Collaboration, *Nat. Photon.*, 7(8), 613, 2013.)

the one example of which this author is aware is shown in Figure 18.14.[31] The squeezed source and detection system illustrated in the diagram is clearly extremely complex, but squeezing of a little over 4 dB was observed, and while it is still far from a practical, realizable gyroscope, it at least provides encouragement that perhaps for highly sensitive systems this is worthy of investigation.

One of the clear issues with the squeezed light source is its complexity and its apparent incompatibility with waveguide optics and fiber component technologies. However, one of the key essential components of the squeezed light system is a carefully configured optical parametric oscillator (OPO). While the OPO is traditionally thought of in the context of large laser systems, there have been numerous successful guided wave implementations in both fiber[32,33] and integrated optic waveguide formats. In recent times, there have been a few ventures into the guided wave squeezed source[34,35] with some modest success (Figure 18.15). The OPO system can also be effectively implemented using optically activated mechanics—now facilitated through the nanophotonic technologies discussed earlier. These approaches too have shown some success,[36,37] and it appears to be simply a matter of time before useful improvements in fiber-optic sensor sensitivity can be realized using these approaches. The practicalities of such a system remain up for question, and indeed, there are relatively few occasions where the shot noise limit really inhibits the actual application of a fiber-optic sensor. However, the prospect is undoubtedly intriguing, and there are likely to be interesting results evolving in this domain in the future.

18.3.5 Computer-Controlled Optics

The spatial light modulator dates back conceptually over at least 60 years.[38] The basic idea is that the pixels in the spatial light modulator can be computer controlled to produce near- or far-field idealizations of an optical wave front. The radical change is that over the past couple of decades, spatial light modulator systems have become commonplace whether in digital projectors or in

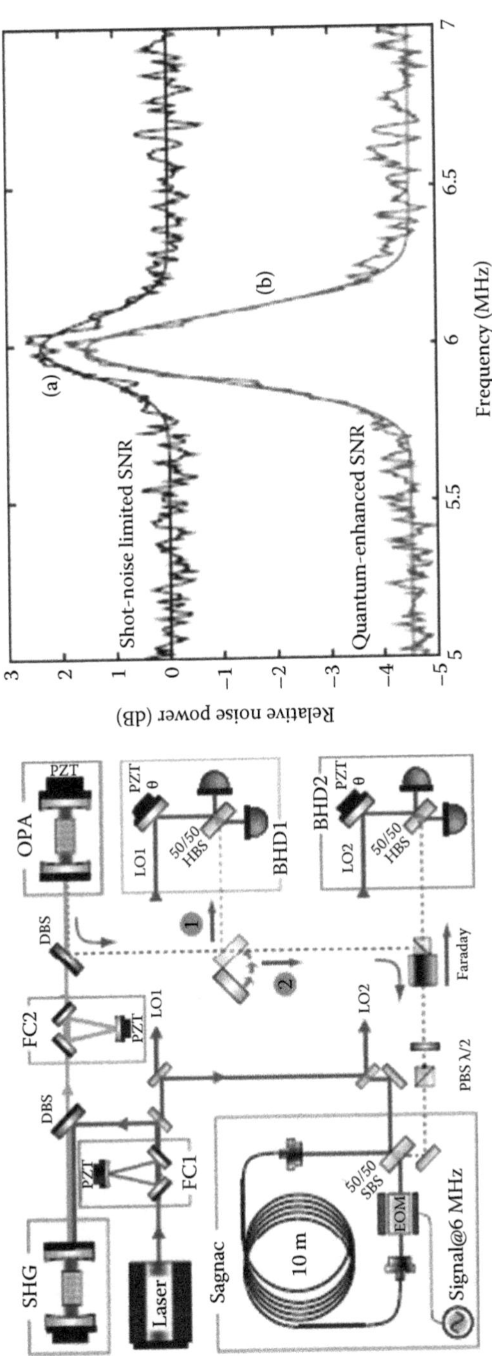

FIGURE 18.14 A demonstration of squeezed light in a FOG/Sagnac interferometer: (a) the squeezed source with the interferometer in the bottom left and (b) the spectral noise reductions observed—here about 4.5 dB around 6 MHz. (From Mehmet, M. et al., *Opt. Lett.*, 35(10), 1665, 2010.)

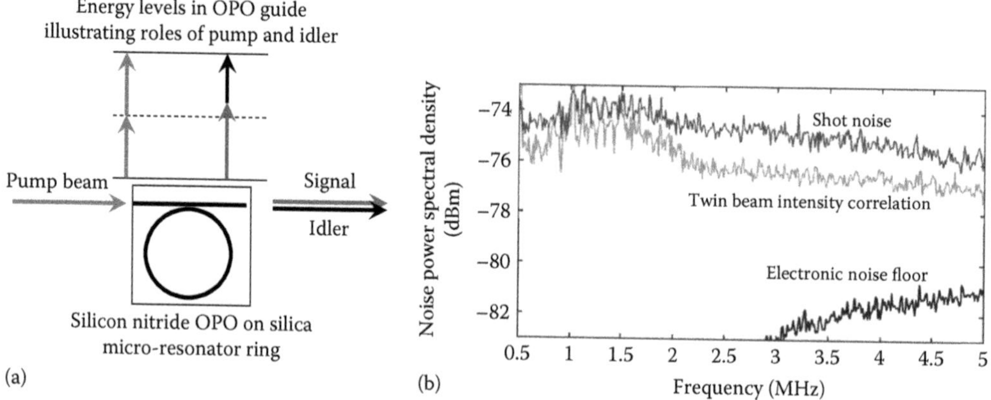

FIGURE 18.15 (a) A much simplified representation of the *squeezing on a chip* described in Ref. [35]. The resonator dimensions are compatible with operation on a 500 mm square chip area, and the results (b) utilize the correlation between noise distributions in idler and signal output beams to achieve effective squeezing (shown at right—centre line).

projection television sets. The spatial light modulator has also found some scientific application, for example, in adaptive transform plane microscope imaging, in beam steering, and in adaptive focusing systems.[39,40]

There are several apparently obvious possibilities in fiber sensing though as yet there have been little, if any, report of their implementation. Such possibilities include adaptive optical source interfacing to a fiber sensor or sensor network, adaptive interfacing of a single source into an array of possible sensing locations, and the possibilities in adaptive optical tweezers for handling micro- and nanoparticles or biological cell structures (Figure 18.16).

The relatively slow take-up on the application of such a potentially attractive tool could stem from a number of factors. The need is probably not currently apparent, and additionally, the viability of the concept to open up new sensing modalities remains to be demonstrated and conveyed to an appropriate user community. There is also a lingering impression that such systems are inherently expensive and cumbersome clearly contradicting their application in mass market

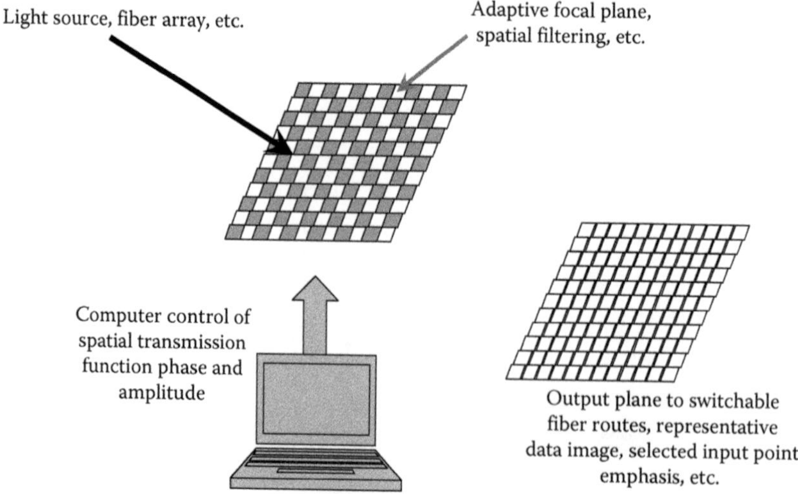

FIGURE 18.16 Computer-controlled optics—its use has been little explored in fiber sensing, if at all. Current spatial light modulator technology can provide flexible routing, optical domain sensor array processing, and numerous other functions.

visual display systems. This appears to be a tool awaiting exploitation, and no doubt in time, as increasingly complex analysis becomes necessary, particularly in the biochemical domain, the need will begin to knock on the door of the possibility, and interesting prospects will undoubtedly emerge.

18.3.6 Some Concluding Comments

This brief discussion has endeavored to match a few of the intriguing tools emerging in photonics overall to the basic architecture of optical fiber sensing, as yet most of these tools have found but modest application beyond curiosity in the research laboratory. However, the existence of these tools and no doubt many others that have either been overlooked in the present discussion or have not yet emerged will significantly enhance the capability of basic fiber sensor technology. Furthermore, as the demands from society for sensing capability increase, there is little doubt that the flexibility offered by integrating these concepts will find a useful place in an emerging world.

18.4 APPLICATIONS ARENA: SOME GENERAL POINTS

Sensors and sensor technology are remarkably diverse. They are also quite specialized dependent upon application and performance requirements. For example, look around an average domestic environment for things that measure temperature. There are probably 20 or so different devices looking entirely different but all based on one of the few technologies—thermistors, bimetallic strips, and very occasionally thermocouples. This simple observation emphasizes two very important basic points concerning sensors and their implementation. The first is that sensors are very specific to a specific application need, and the second is that the same technology in an entirely different package can fulfill a range of sometimes radically different requirements. There is a third important observation concerning sensing, namely, that sensing is always there to facilitate something else. In other words, it has to meet a need.

This leads to the question as to how, in an environment that is on the face of it appears to be adequately measured, new technologies can begin to contribute. Sensor industry gurus have evolved three tests, of which at least one must be demonstrably applicable to our new technology. The tests are, expressed very crudely,

- To perform the same job to the same specification for 1/10th the cost
- To perform the same job but 10 times as well for the same cost
- To realize a measurement that has been hitherto impossible and that can be persuasively demonstrated to meet a need

There are examples of optical fiber sensors in each of these categories, and Figure 18.17 endeavors to highlight the way these categories work for the fiber sensor.

The benefits of fiber sensing have been well rehearsed, but these benefits are only an advantage when they present a gain to the potential user compared to what is currently available. This in turn depends upon market sector and the current and projected needs within that market sector. Linked into this are the necessary issues of stability, repeatability, resolution, accuracy, and reliability across a range of environmental conditions and over time. Add to the mix, cultural momentum, which is particularly evident in conservative industries like aerospace and link with that the need for established codes of practice prior to adopting a particular technology, and the somewhat contradictory parallel need to adopt the particular technology in order to verify the established code of practice and the whole complex conundrum of actually exploiting new sensor technologies in the real world becomes apparent.

Figure 18.18 has attempted to bring together the benefits of fiber sensors, the functionality of fiber sensors, and the inherent requirements of particular sector into a representation of the

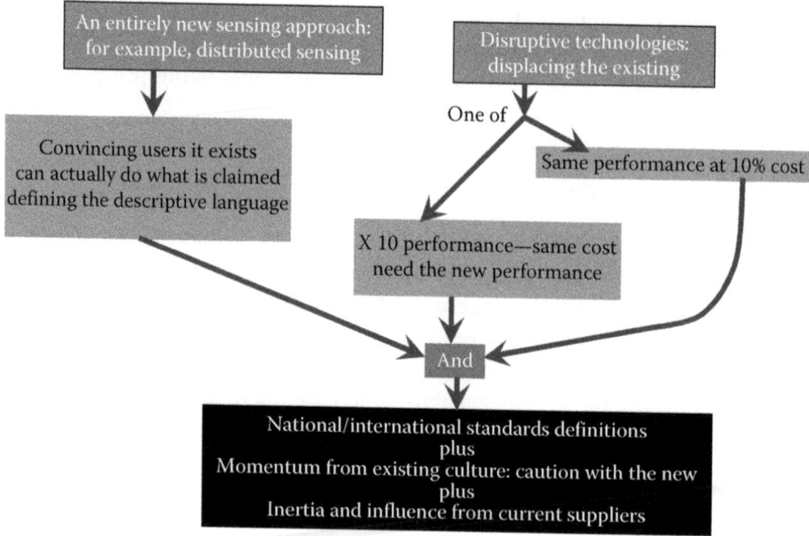

FIGURE 18.17 A summary of the elements involved in gaining acceptance for a new sensor technology. Fiber sensing offers opportunities in both entirely new approaches (especially distributed sensing) and in displacing the existing, for example, in power distribution network current and voltage monitoring. Getting the technology to work convincingly is only the start!

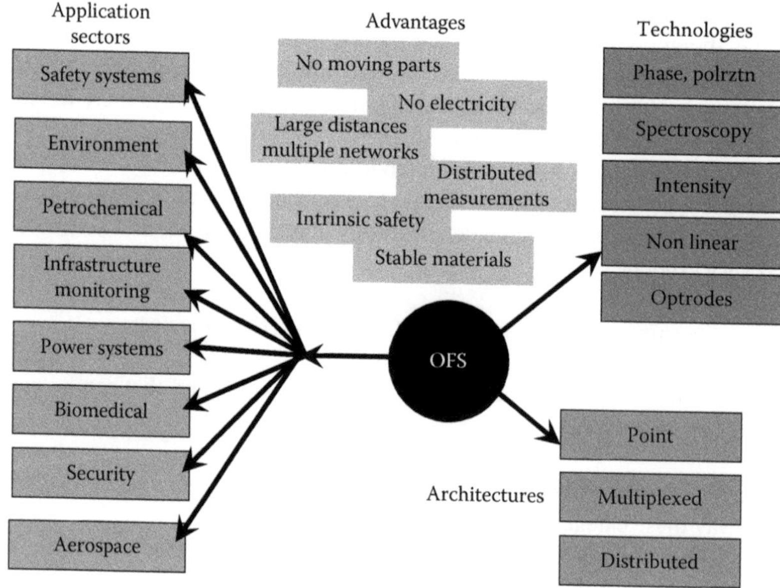

FIGURE 18.18 An attempt to highlight application sectors, principal features, and principal benefits of fiber-optic sensors. Most applications can benefit from one or more of the advantages, using one or other of the sensing technologies and operating with one or other architecture. Each application is separate and distinctive: the potential for considerable fragmentation in the modest US$1 billion or so world market is very obvious!

interactions required for successful implementation. This encapsulates the simple observations that one of the principal requirements for the successful implementation is a combination of patience and persuasion. It is tempting to project that since optical fiber communications has dominated long-haul transmission, so fiber sensing should do the same in the sensor arena. This naïve assertion does ignore three factors, namely, the benefits of fiber communications are enormous compared to

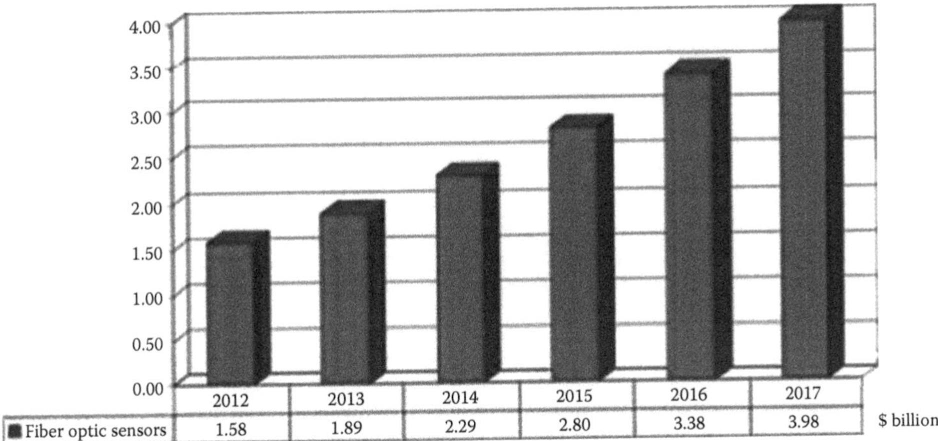

FIGURE 18.19 Projected market growth to 2017 for fiber-optic sensors. This is from a report by Electronicast LLC. Similar conclusions have also been reached in other recent (2013) reports.

anything else, the communications market place is technically uniform in that everyone requires the same functionality, and the costs of transmission per unit bandwidth are hugely persuasive compared to any alternative. Sensing is an entirely different world.

18.4.1 Current Applications Arena for Fiber Sensing

There are, of course, those analysts who present market forecasts.[41,42] These market forecasts, of which one is shown in Figure 18.19, differ somewhat from one to the other but agree in some general trends. These are the following:

- The current overall fiber sensor market is approximately $1 billion per year and is set to expand by the order of a factor of four within the next 5 or 6 years. While this might appear outrageously optimistic, there are several factors that support this general conclusion. The technology of fiber sensing is becoming accepted in the user community and the more widely known, the benefits of distributed sensing and passive multiplexed networks can be very persuasive, and the systems can be installed and maintained by technically skilled staff.
- This $1 billion per annum is less than 1% of the total sensor market, and this very small section is in turn highly fragmented into perhaps a dozen different application areas. Each of these areas has its own cultural requirements even for the same technology. Consequently, fiber-optic sensing tends to be served by, with a few exceptions, relatively small companies specializing by sector and technology.
- The overall market is dominated perhaps to as much as 80% by the distributed sensor architecture and by large passive multiplexed networks.

18.5 CONSOLIDATING THE PRESENCE FOR FIBER-OPTIC SENSING

18.5.1 Analysis of Current Success Stories

First and foremost, fiber sensors have probably never made a presence on the basis of simple cost advantage—but with one possible exception. The Faraday rotation current probe and associated fiber link voltage probes are now the technology of choice for monitoring high-voltage networks (Figure 18.20). They replace a cumbersome and expensive current transformer with stringent demands on the insulation between the windings with a fiber optic all-dielectric linkage into the

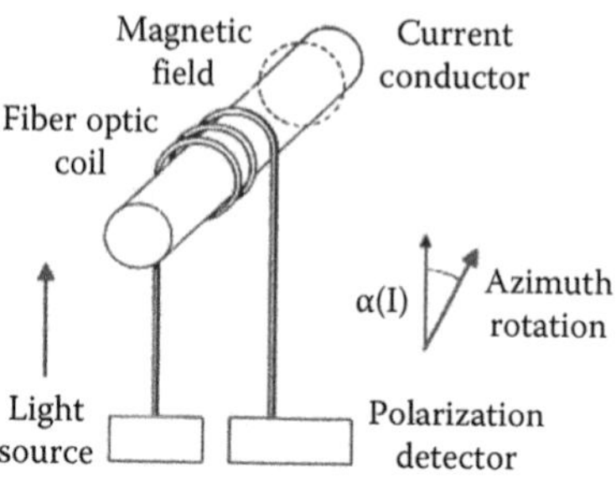

FIGURE 18.20 The fiber-optic current monitor system, based on Faraday rotation in an optical fiber, which has displaced previous technology and offers both significant cost savings and operational benefits. (This example Courtesy of T Bosselmann–Siemens, Erlangen, Germany.)

high field area.[43,44] The whole process is consequently much less expensive using fiber optics. Not only is it much less expensive, it also performs far better with much higher bandwidth and sensing fidelity. Furthermore, it can be corrected relatively easily for influences such as temperature and, when suitably engineered, will withstand the rigors of out-of-doors operation. The whole situation is also made much more straightforward from a certification and qualification perspective by the fact that the manufacturer is frequently also the user of the sensor and therefore takes responsibility for its operation. Alternatively, in the cases where relatively small companies are involved, the company has a very close relationship with its customer possibly even to the extent of substantial investment. Finally, since the sensor performs a monitoring function rather than a safety critical assessment function, the certification is to user specifications and is largely independent of national and international standards and codes of practice.

There are also a few examples where fiber technology offers significant performance enhancement for similar operating cost to old technologies and the significant performance benefits are persuasive enough for the user community to adopt the new technique. Perhaps the best example of this is the fiber-optic gyroscope (FOG) (Figure 18.21). The FOG appeared in the lab in 1977[44] and in the Boeing 777 in 1997, initially to a relatively modest technical specification. The benefits of the FOG compared to alternative gyroscope technologies, including around the laser gyro, are very substantial.[45] There are no physical moving parts (even the laser gyro has the dither), and therefore, there are no bearings to wear out. With suitable care in the packaging—itself a nontrivial task—the FOG can be rendered as virtually temperature insensitive, vibration insensitive, and stray magnetic field insensitive. The last of these requires suitable but practical screening. There are more subtle benefits too. The FOG only responds to rotation signals and, unlike moving mass gyroscope, is not affected by linear accelerations. Additionally, in terms of physical size, weight, volume, and cost, it is similar or far better than the rotating mass equivalents.

The fiber gyroscope is now very competitive in the inertial navigation market (the systems that can guide aircraft across continents without reference to GPS). It has also found presence in space exploration where extremely high accuracy and total immunity to linear acceleration are both necessities and in some subsea vehicles. At the more modest performance level, the FOG has found its place in antenna stabilization systems where small weight and immunity to electromagnetic radiation are important factors. This is then an example of similar costs and extremely persuasive

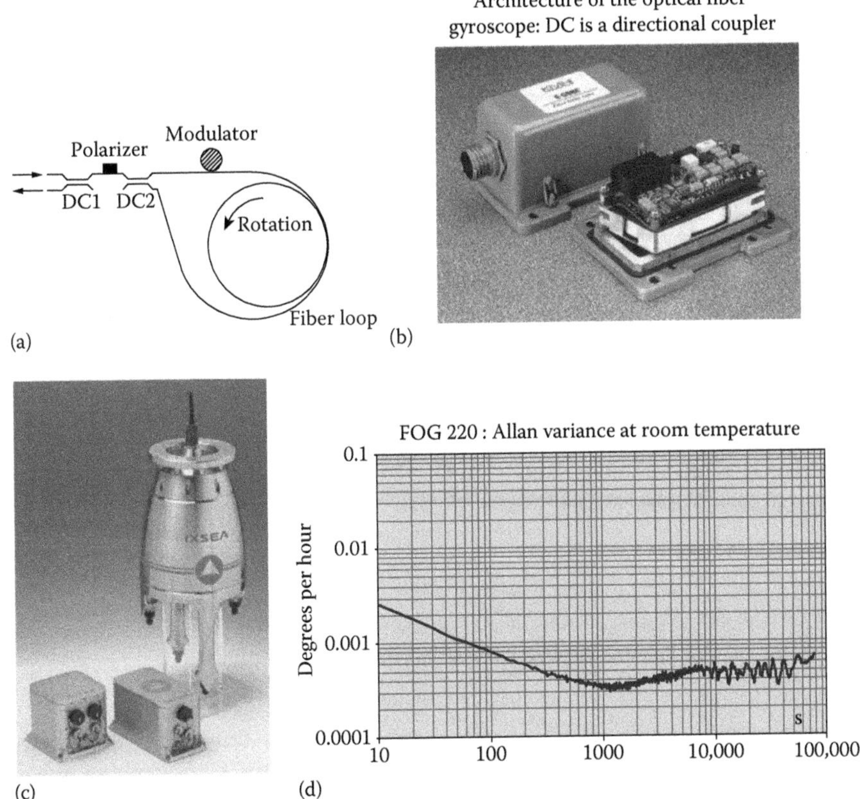

FIGURE 18.21 The FOG—(a) the so-called basic configuration, (b) a modest performance production version. (Courtesy of KVH.) (c) A space-qualified high-performance gyroscope (Courtesy of IxBlue) with (d) noise levels of a few × 10^{-3} degrees per hour (one revolution per century).

performance winning the day for fiber sensors. The actual detailed requirements for each particular gyroscope span an enormous range as will be evident from the preceding discussion. Detection limits range from one rotation per century down to one rotation per hour. The fragmentation is, therefore, apparent.

There is also an interesting observation on the evolution of standards for the fiber gyroscope. The community recognized this instrument as a very strong contributor and so, in the early 1990s, convened a group to determine an international standard. The final draft was issued in 1999 for consideration and correction by the community. The draft remains a draft up to this day. There are probably three contributory factors here that obviate the need for standard in the conventional sense of the word. The first is that many major manufacturers of gyroscopes, especially in the aerospace industry, are themselves system providers for the aerospace industry and take full responsibility for the product at the system level. They will test to their own satisfaction, but there is no need to refer back to international standards since the final customer's procurement specification determines what is what. The second is that for the very highly demanding applications, for example, space probes, the relationship between the final user, typically NASA or ESA, and the equipment provider is very direct, and again, reference to outside standards is unnecessary. The use of internationally approved components will typically be adequate. The final reason, which in many ways is a consequence of the previous comments, is that the fiber gyroscope is not a commodity catalogue item, obviating the necessity for an obvious *plug and play* interchangeability from one manufacturing source to the other. It is typically sold as a subsystem within a more complex engineering artifact, and it is this artifact that must plug in rather than the gyroscope as a component.

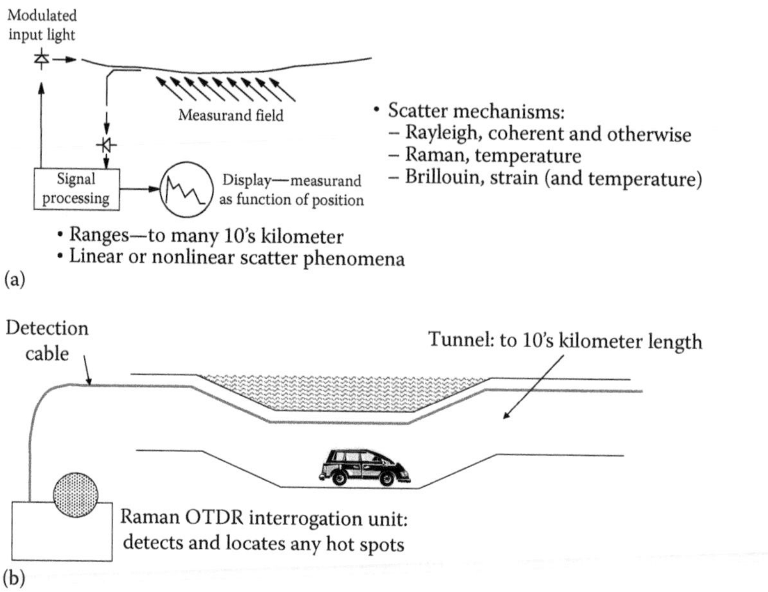

FIGURE 18.22 Distributed sensing—a technique unique to fiber-optic sensors. (a) A schematic of the basic concepts and (b) a sketch of Raman distributed fire sensing in a highway tunnel.

Probably the highest profile application for fiber sensors has centered on distributed optical fiber sensor technologies (Figure 18.22).[46,47] These emerged as a curiosity—nothing like it had ever appeared before. It was all too good to be true—simply lay out an optical fiber along the artifact you wish to monitor typically a pipeline or an electric cable, and the fiber sensor will produce a map of temperature versus position with spatial resolutions of the order of meters or better and thermal resolutions of the order of 1°. Potential users needed significant persuasion, a process that took many years, many trial reference sites, concerted attempts for mutual understanding between manufacturer and customer, and inevitably a few errors along the way. However, a few successes notably in the oil exploration industry especially downhole thermal monitoring proved to be persuasive and credibility emerged. Distributed sensing is now the major contributor to sales of fiber-optic sensors. Again, this is provided typically by organizations like the oilfield instrumentation contractors who will manufacture the sensor in a small division, often originally a small start-up company. Whilst this is initially for their own demands and their own markets, this does establish credibility for other applications. Technological benefits are so persuasive that even though one manufacturer's concept for spatial, temporal, and thermal resolution may differ from another, the final users are content with mutually agreed performance specifications and incorporating professional standard components within the system manufacture.

The fiber Bragg grating (FBG) is the most promising of the passive multiplexed network architectures demonstrated thus far with fiber optics (Figure 18.23). The FBG makes the same measurement as the strain gauge only without enormous bundles of wires and with the physical long range and electric interference immunity that characterizes fiber sensors overall.[48,49] There have been countless demonstrations of complex arrays on bridges, airframes and turbine blades, and a host of others. There are some real products that typically involve close relationships between system supplier and user. However, the aspiration to replace the strain gauge as the preferred means for measuring strain remains elusive despite the enormous technical and operational benefits. The international strain gauge industry is, however, interested in exploring this option, and for this to happen, three factors need to be put in place. The first is that the strain gauge should be reliable when attached to the structure under test and this reliability should be repeatedly demonstrable regardless of FBG

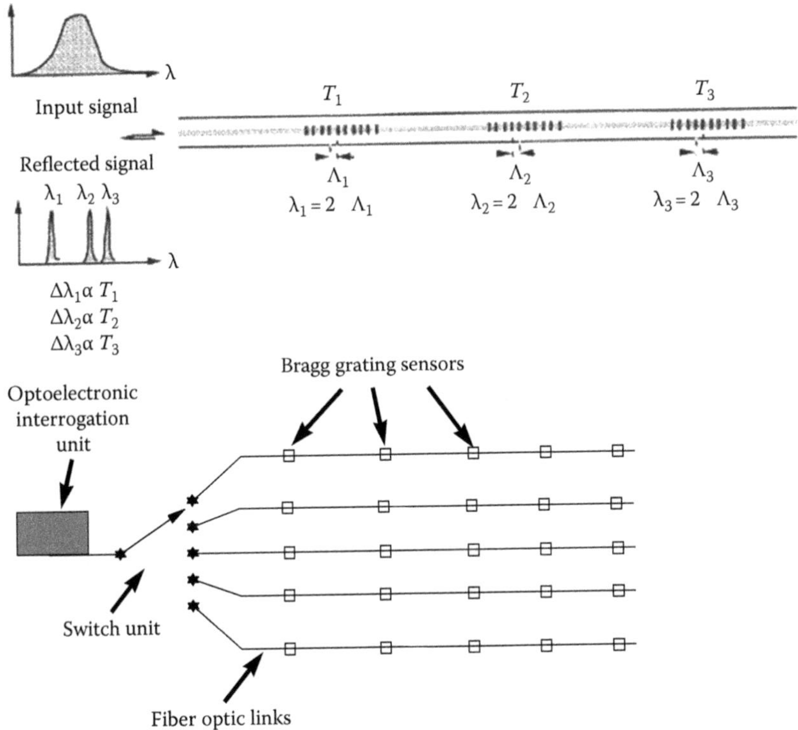

FIGURE 18.23 Fiber Bragg grating arrays for very large (to many hundreds in the lower configuration) passive sensor systems. There can be many tens of gratings on a single fiber.

strain gauge manufacturer. The second is that FBG *stick on* strain gauges should be interchangeable among the manufacturers and suppliers. The third is that the interrogation systems should also be interchangeable. If these conditions are met, then we have a commodity FBG strain gauge exactly analogous to the commodity electrical strain gauge.

This requires considerable iteration and negotiation to be put in place. There are many ways of measuring the sensor wavelength of reflection of an FBG, and these can, and often do, give different results with the variation in the shape of the reflected spectrum. There are variations in the wavelength reference standards implemented in FBG interrogators. These all need agreement in convincing technical definition. Additionally, the strain gauge itself must have demonstrated reliability; test procedures agreed through which this reliability can be demonstrated and defined interface processes for sample attachment. There is a considerable effort required to make this happen with a corresponding significant investment. The question remains—is it worth it especially given current activity involving close relationships between supplier and user, most notably in the wind turbine sector? There most certainly is some interest in evolving an FBG standard, and the International Electrotechnical Commission[50] has published a number of drafts of the document. Unlike the fiber gyro equivalent, this particular document is likely to emerge since here we are dealing with a specialized component, which may integrate into a variety of systems in contrast to the FOG, which is made available as a system that in turn meets the system definition in its own right.

There is also considerable promise in other sectors for the fiber sensor industry, and a much vaunted benefit of fiber sensor technology lies in its intrinsic safety. Indeed, the whole area of safety and security is one that appears to be potentially attractive.[51] Safety critical systems—the key to, for example, protecting workers in the petrochemical industry—have, however, yet to emerge as an important part of the commercially attractive fiber-optic sensor portfolio. As confidence in the technology consolidates, and this has most certainly emerged within the past decade, then the very skeptical (correctly so) procurement processes within this sector will become more accustomed to working with this *new* approach, and national licensing authorities are gradually becoming more adept at interpreting their specifications for intrinsically safe fiber systems.

Returning to that optimistic market forecast for significant expansion within the coming years, the preceding few paragraphs, which are really all about the reasons for considerable consolidation in user confidence and about interpreting fiber sensors for specialized and demanding applications, certainly give considerable confidence. Maybe these optimistic projections are based on a sound interpretation of the current situation after all.

18.6 SENSING IN FUTURE SOCIETIES: HOW MIGHT FIBER-OPTIC SYSTEMS CONTRIBUTE

Numerous organizations have endeavored to apply clairvoyant skills to future technological need. Some concentrate on technologies, for example, recent contributions in both the European Union's Photonics 21 Group and the National Academies in the United States.[52,53] These both present a justifiably optimistic perspective on the prospects for those skilled in optics and photonics and their applications.

The US National Academies of Engineering took a different approach and endeavored to identify engineering's *grand challenges* for the next half century.[54] These were initially published in 2008 but at the time of writing still stand and are still stimulating debate—the engineering community in the United Kingdom examined these within recent months. The commentary behind the headlines in Figure 18.24 is certainly thought provoking having made assumptions about our social scientific and technological evolution, which are well justified.

Looking introspectively at the technology, which is the subject of this book, it soon becomes apparent that indeed there are sectors in here to which fiber sensors are very likely to make a contribution (dark shaded in the diagram) and certainly others where good arguments to be made for the technology to play a part (light shaded).

Meanwhile, the United Nations has declared that 2015 will be *The International Year of Light*[55] with the professional societies already making plans to extol the virtues of the science, the technologies, and their prospects to the broadest possible audience ranging from senior politicians to students in schools. It is interesting to reflect that over the past century, the term *electronics* has slipped into the everyday vocabulary and is recognized as a mysterious technique, which facilitates a whole range of everyday artifacts. Light is most certainly beginning to play a similar role at the beginning of the twenty-first century to that played by electricity at the beginning of the twentieth century.

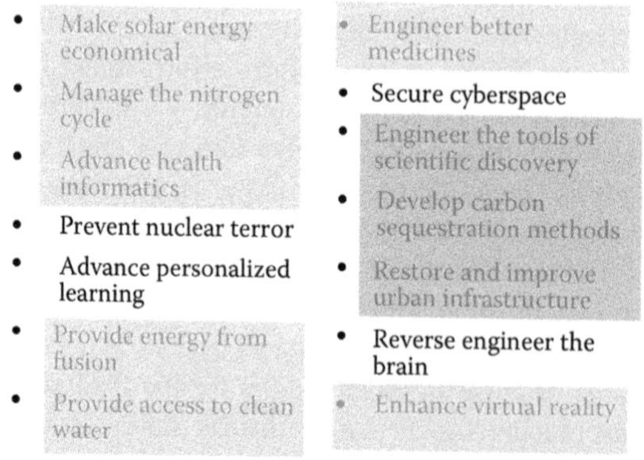

- Make solar energy economical
- Manage the nitrogen cycle
- Advance health informatics
- Prevent nuclear terror
- Advance personalized learning
- Provide energy from fusion
- Provide access to clean water

- Engineer better medicines
- Secure cyberspace
- Engineer the tools of scientific discovery
- Develop carbon sequestration methods
- Restore and improve urban infrastructure
- Reverse engineer the brain
- Enhance virtual reality

FIGURE 18.24 Society and engineering's grand challenges—National Academy of Engineering, U.S.A.—where may OFS contribute? The dark shaded sections are very strong prospects, with good opportunities also in the light shaded sections.

Our current world is beginning to appreciate photonics—fiber communications is everywhere—and as the coming decades progress, photonics appears likely to make its mark as the twenty-first centuries reply to electronics. Its contributions will be substantial.

REFERENCES

1. Menadier C, Kissinger C, and Adkins H, The fotonic sensor, *Instrum. Control Syst.*, 40, 114, 1967.
2. Patil SS and Shaligram AD, Modelling and experimental studies on retro-reflective fiber optic micro-displacement sensor with variable geometrical properties *Sens. Actuat. A-Phys.*, 172(2), 428–433, 2011.
3. Cao H, Chen Y, Zhou Z, and Zhang G, Theoretical and experimental study on the optical fiber bundle displacement sensors, *Sens. Actuat. A-Phys.*, 136(2), 580–587, 2007.
4. Rogers AJ, *Essentials of Photonics*, CRC Press, Boca Raton, FL, 2009.
5. Culshaw B and Kersey A, Fiber optic sensing: A historical perspective, *IEEE J. Lightwave Technol.*, 26(9), 1064–1078, 2008.
6. Cusano A, Consales M, Crescitelli A, Ricciardi A. (Eds.), *Lab-on-Fiber Technology*, Springer Series in Surface Sciences, Vol. 56, pp. XXII, 184, 213, 359. Springer Verlag, 2014.
7. Thevenaz L, Slow and fast light in optical fibres, *Nat. Photon.*, 2(8), 474–481, 2008; Slow light for the rest of us, *J. Phys. Rev. Focus*, 3, 37, June 29, 1999.
8. Dicaire I, De Rossi A, Combrié S, and Thévenaz L, Probing molecular absorption under slow light propagation using a phonic crystal waveguide, *Proceeding of the SPIE 8421, OFS2012, 22nd International Conference on Optical Fiber Sensors*, Beijing, China, 2012.
9. Thévenaz L, Chin, S-H, Berger P, Bourderionnet J, Sales S, and Sancho-Dura J, Slow light fiber systems in microwave photonics, *Proc. SPIE*, 7949, 79490B, Advances in Slow and Fast Light IV, San Francisco, CA2011.
10. Gaponenko SV, *Introduction to Nanophotonics*, Cambridge University Press, New York, 2010. ISBN: 9780521763752.
11. Baxter J, Nanophotonics circuit design, *Nat. Photon.*, 6, 213, 2012.
12. Yablonovitch E, Photonic crystals: Semiconductors of light, *Sci. Am.*, 285(6), 47–55, December 2001.
13. Gardiol F, *Introduction to Microwaves*, Artech House, Dedham, MA, 1984, p. 325. ISBN 0-89006-3-134-3.
14. Billings L, Metamaterials world, *Nature*, 500, 138–140, August 8, 2013.
15. Hendry JB and Smith DR, Reversing light with negative refraction, *Phys. Today*, 57, 37–43, 2004.
16. Sharma V, Park K, and Srinivasarao M, Colloidal dispersion of gold nanorods: Historical background optical properties, seed-mediated synthesis, shape separation and self-assembly, *Mater. Sci. Eng. Rep.*, 65(1–3), 1–38, 2009.
17. Vasudevanpillai B, Tamitake I, Abdulaziz A, Athiyanathil S, and Mitsuru I, Semiconductor quantum dots and metal nanoparticles: Syntheses, optical properties and biological applications, *Anal. Bioanal. Chem.*, 391(7), 2469–2495, August 2008.
18. Monticone F and Alu A, Metamaterial enhanced nanophotonics, *Opt. Photon. News*, 24(12), 35, December 2013.
19. Matsumoto Y, and Takagahara T (Eds.), *Semiconductor Quantum Dots: Physics, Spectroscopy and Applications*, Springer Verlag, Berlin, Germany, 2002.
20. Michler P (Ed.), *Single Semiconductor Quantum Dots*, Springer Verlag, Berlin, Germany, 2009.
21. Kretschmann E, The determination of the optical constants of metals by excitation of surface Plasmon's, *Z. Phys. A*, 241, 313, 1971.
22. Tudos, AJ and Schasfoort RBM (Eds.), *Handbook of Surface Plasmon Resonance*, RSC Publishing, London, U.K., 2008.
23. Johnstone W, Stewart G, Culshaw B, and Hart T, Surface plasmon polaritons in thin metal films and their role in fiber optic polarising devices, *J. Lightwave Technol.*, 8(4), 538–544, April 1990.
24. Sharma AK, Jha R, and Gupta BD, Fiber optic sensors based on surface plasmon resonance: A comprehensive review, *IEEE Sens. J.*, 7, 1118, 2007.
25. Maier S, *Plasmonics: Fundamentals and Applications*, Springer Verlag, New York, 2007. ISBN 978-0-387-33150-8.
26. Sadowski JW, Korhonen I, and Peltonen J, Characterization of thin films and their structures in surface plasmon resonance measurements, *Opt. Eng.*, 34(9), 2581–2586, 1995.
27. Grangier P, Slusher RE, Yurke B, and Laporta A, Squeezed light enhanced polarisation interferometer, *Phys. Rev. Lett.*, 59, 2153–2156, November 1987.

28. Vahlbruch H et al., Observation of squeezed light with 10 dB quantum noise reduction, *Phys. Rev. Lett.*, 100(3), 033602, 2008.
29. Garrison J and Chiao R, *Quantum Optics*, Oxford University Press, New York, 2014.
30. The LIGO Scientific Collaboration, Enhanced sensitivity of the LIGO gravitational wave detector by using squeezed states of light, *Nat. Photon.*, 7(8), 613–619, 2013.
31. Mehmet M, Eberle T, Steinlechner S, Vahlbruch H, and Schnabel R, Demonstration of a quantum-enhanced fiber Signac interferometer, *Opt. Lett.*, 35(10), 1665–1667, 2010.
32. Hansryd J et al., Fiber-based optical parametric amplifiers and their applications, *Sel. Top. Quant. Electron. IEEE J.*, 8(3), 506–520, 2002.
33. Sohler W et al., Integrated optical devices in lithium niobate, *Opt. Photon. News*, 19(1), 24–31, 2008.
34. Shelby RM et al., Generation of squeezed states of light with a fiber optic ring interferometer, *Phys. Rev. A*, 33(6), 4008, 1986.
35. Dutt A et al., On-chip optical squeezing, Cornell University, Ithaca, NY, 2013.
36. Safavi-Naeini AH, Gröblacher S, Hill JT, Chan J, Aspelmeyer M, and Painter O, Squeezed light from a silicon micromechanical resonator, *Nature*, 500, 185–187, August 8, 2013.
37. Purdy TP et al., Strong optomechanical squeezing of light, *Phys. Rev. X*, 3(3), 031012, 2013.
38. Efron U (Ed.), *Spatial Light Modulator Technology Materials, Devices and Applications*, CRC Press, New York, 1994, p. 47.
39. Psaltis D and Mok F, Holographic memories, *Sci. Am.*, 273, 70–76, 1995.
40. Hardy JW, *Adaptive Optics for Astronomical Telescopes*, Oxford University Press, New York, 1998.
41. Photonic sensor consortium market survey report fiber optic sensors, Information Gatekeepers Inc., Brighton, MA, June 2013.
42. Global fiber optic sensor market 2014–2018, December 2013, TechNavio inc via reportlinker.com.
43. Bosselmann T, Innovative applications of fibre optic sensors in energy and transportation, *Proc. SPIE Int. Soc. Opt. Eng.*, 5855, 188–193, 2005.
44. Bohnert K et al., Highly accurate fiber optic DC current sensor for the electrowinning industry, *IEEE Trans. Ind. Appl.*, 43(1), 180–187, 2007.
45. LeFevre HC, *The Fibre Optic Gyroscope*, 2nd edn., Artech House, Boston, MA, 2014.
46. Hartog A, *Distributed Fibre Optic Sensors*, CRC Press, Boca Raton, FL, 2014.
47. Kersey AD, Fibre optic sensors shine bright: Industrial applications where FOS bring differentiated performance/value, *Proc. SPIE 8421, OFS2012 22nd International Conference on Optical Fiber Sensors*, 84210F, 17 October 2012, doi:10.1117/12.2008245.
48. Schröder K et al., An approach to continuous on-site monitoring of contact forces in current collectors by a fibre optic sensing system, *Opt. Lasers Eng.*, 51(2), 172–179, 2013.
49. Udd E, Fire and ice: 25 years of fiber grating sensor technology, *Speciality Optical Fibers*, OSA Technical Digest, 2012.
50. Habel WR, Schukar VG, and Kusche N, Fiber optic strain sensors are making the leap from lab to industrial use—Reliability and validation as a precondition for standards, *Meas. Sci. Technol.*, 24(9), 094006, 2013.
51. The EU COST TD 1001 specifically addresses optical fiber sensors for safety and security, ww3.ul.ie/td1001/.
52. Photonics our vision for a key enabling technology of Europe, May 2011. Report available for free download from http://www.photonics21.org.
53. Optics and photonics: Essential technologies for our nation, 2013. Report available for free download from http://www.nap.edu/catalog/php?record id=13491. See also US National Photonics Initiative https://spie.org/x8893.xml.
54. The NAE Grand Challenges discourse, www.engineeringchallenges.org/.
55. The International Year of Light for 2015, spie.org/x93905.xml.
56. G. Breitenbach, S. Schiller, and J. Mlynek, Measurement of the quantum states of squeezed light, *Nature*, 387, 471–475, 29 May 1997.

Index

A

B

C

E

F

J

K

L

M

N

O

N

O

T

U

V

W

Milton Keynes UK
Ingram Content Group UK Ltd.
UKHW051904071024
449327UK00025B/2083

9 780367 656058